Condensed Matter in a Nutshell

Condensed Matter in a Nutshell

Gerald D. Mahan

PRINCETON UNIVERSITY PRESS · PRINCETON AND OXFORD

Published by Princeton University Press, 41 William Street,
Princeton, New Jersey 08540

In the United Kingdom: Princeton University Press,
6 Oxford Street, Woodstock, Oxfordshire OX20 1TW

press.princeton.edu

Library of Congress Cataloging-in-Publication Data
Mahan, Gerald D.
Condensed matter in a nutshell / Gerald D. Mahan.
p. cm. — (In a nutshell)
Includes bibliographical references and index.
ISBN 978-0-691-14016-2 (casebound : alk. paper) 1. Condensed matter—Textbooks. I. Title.
QC173.454.M34 2011
530.4'1—dc22 2010005193

British Library Cataloging-in-Publication Data is available

This book has been composed in Scala

Printed on acid-free paper. ∞

Printed in the United States of America

10 9 8 7 6 5 4 3 2

Contents

Preface

This book is a text for a graduate course in material science, condensed matter physics, or solid state physics. The student is expected to have taken previous courses in quantum mechanics and electromagnetic theory. No prior knowledge or course in condensed matter physics is required.

The earlier chapters introduce basic concepts, such as crystal structures, energy-band theory, phonons, and types of crystal binding. Intermediate chapters discuss the basic features of transport and optical properties of solids. Later chapters discuss current research topics, such as magnetism, superconductivity, and nanoscience. There is an extensive treatment of metals, from the viewpoints of free electrons, tight binding, and strong correlations. There is an extensive discussion of semiconductors, from the viewpoints of both intrinsic, and then extrinsic properties. All chapters except the first have homework problems. These problems have been worked by a generation of students.

I have taught this course many times. The course syllabus in the first semester is fairly standard, and covers the basic material. This course material for the second semester varies from year to year, depending on what topic is "hot" in condensed matter physics. In writing his book, I have included many of the hot topics of the past.

I wish to thank my wife, Sally, for her patience as I wrote still another textbook. I also thank Princeton University Press for encouraging me to finish this project, which was half-done for years. I also thank the Physics Department at Penn State for allowing me to teach this course for several years as I finished the manuscript.

Condensed Matter in a Nutshell

1 Introduction

The history of material science is closely tied to the availablility of materials. Experiments must be done on samples. In the early days of the twentieth century, most of the available materials were found in nature. They were minerals or compounds.

1.1 1900–1910

Scanning the table of contents of the *Physical Review* for the decade 1900–1910, one finds that experiments were done on the following elements and compounds:

- Alkali metals: Na, K, Rb
- Noble metals: Cu, Ag, Au
- Divalent metals: Zn, Cd
- Multivalent metals: Al, Sn, Hg, Bi, Pb
- Transition metals: Ti, Fe, Ni, Mo, Rh, Ta, W, Ir, Pt
- Rare earth metals: Er
- Semiconductors: C, Si, Se, P
- Binary compounds: CaO, MgO, ZnS, HgS, CdS, H_2O, AgCl, AgBr, NaF, NaBr, NaCl, LiCl, KCl, TlCl, TlBr, $PbCl_2$, $PbCl_2$, PbI_2
- Oxides: KNO_3, $LiNO_3$, $NaNO_3$, $AgNO_3$, $K_2Cr_2O_2$, $NaClO_3$

The binary compounds were identified by their chemical name, such as cadmium sulfide, calcium oxide, or ice.

Table 1.1 A partial list of minerals that were used in experiments reported in the *Physical Review* during the period 1900–1910

Name	Formula
anatase	TiO_x
aragonite	$CaCO_3$
brookite	TiO_2
calcite	$CaCO_3$
cinnabar	HgS
eosin	$C_{20}H_8Br_4O_5$
fluorite	CaF_2
glass	a-SiO_2
pyrites	FeS_2
magnetite	Fe_3O_4
molybdenite	MoS_2
mica	silicates
quartz	SiO_2
sidot blende	ZnS

Among the most interesting materials were minerals. They were usually, and often only, identified by their mineral name. A partial list is given in table 1.1. Several minerals we were unable to identify from their names. The point of this list is that all of these compounds are found in nature as crystals. The samples were not grown in the laboratory, they were found in caves or mines. ZnS was then called sidot blende, but today is called zincblende.

A few materials were actually grown in a laboratory. One was silicon, which was grown in the research laboratory of the General Electric Company. Other artificial materials used in experiments were rubber, brass, asphalt, steel, constantan, and carborundum.

1.2 Crystal Growth

Today nearly all materials used in experiments are either grown in a laboratory or purchased from a company that grew them in a laboratory. The techniques were discovered one by one during the twentieth century. Some notable landmarks:

1. Jan Czochralski [3] invented a method of pulling crystals from their melt in 1917. His apparatus is shown in fig. 1.1. The crystals are pulled vertically, slowly, starting with a small seed crystal. Today the crystal is rotated to ensure that inhomogenieties in the liquid do not make the crystals inhomogeneous. Two-thirds of crystals are grown using the Czochralski method. Large single crystals are prepared this way. For example, silicon crystals used in the manufacturing of integrated circuits are pulled.

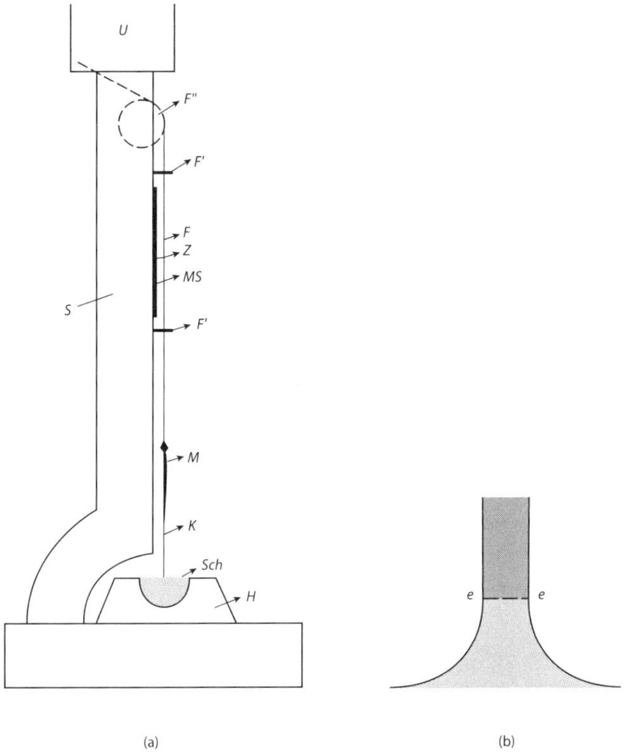

FIGURE I.1. (a) Czochralski apparatus for pulling a crystal from the melt. Melt is at *Sch*, and a tiny seed particle (not shown) is at the end of the silk thread *K*, *F*. (b) Details of region where the crystal grows.

2. Percy Bridgman [2] reported the Bridgman method in 1925. A hollow cylinder is packed with powder or small crystals. It is pulled slowly through a hot region, where the material is melted and recrystalized. Large single crystals can be made this way. The cylinder can be moved vertically or horizontally.

3. William Pfann [4] invented the method of zone refining in 1952, whereby a crystal is pulled through a hot area that locally melts and recrystallizes it. Zone refining generally purifies a crystal, by pushing impurities to the end of the crystal. A crystal may be zone refined several times to obtain a low density of impurities.

4. Large single crystals may be grown from a melt. A supersturated solution of the compound will precipitate the excess material. At the right temperature, it precipitates by growing single crystals. This process happens daily in the author's pantry, as large sugar crystals are grown in the container of maple syrup. This rock candy is a family favorite.

5. Small crystals can be grown in a vapor. The material is inserted into a container, often a glass tube. Then it is heated, so the vapor is supersaturated. At the right temperature, it will grow crystals. This process is slow, but is used for laboratory samples.

The above methods are all traditional, and make three-dimensional, homogeneous samples. Many crystals today are grown using *epitaxy*. Epitaxy is the technique of growing a crystal, layer by layer, on the atomically flat surface of the same, or another, crystal. The atoms are brought to the surface by a variety of methods.

- *Molecular beam epitaxy* (MBE) uses a beam of atoms, or molecules, that are directed toward the surface. John Arthur [1] reported this method in 1968 for growing layers of GaAs. The particle beams originate in a small furnace that creates a vapor of the material, and a hole in the furnace lets atoms out. This process is very slow, but is widely practiced.
- *Chemical vapor deposition* (CVD) uses a vapor of the material in contact with the surface. This method is also called vapor-phase epitaxy (VPE).
- *Liquid phase epitaxy* (LPE) has a liquid of the material in contact with the surface. It is a variation of the solution method mentioned earlier.

1.3 Materials by Design

There are about 92 ∼ 100 stable elements in the periodic table. Around 10^4 binary compounds can be formed from pairs of different atoms. Not all pairs form a compound, but many pairs form several different crystals. Putting three elements together has about 10^6 possible compounds, and putting four elements together has about 10^8 possible compounds. The number of new materials that are grown for the first time is thousands each year. Most of these new compounds have rather ordinary properties. However, occasionally one is found that is a high-temperature superconductor, a high-field magnet, or an excellent thermoelectric. Condensed matter physics continues to be an exciting area of research, because new crystals are constantly being discovered. There seems to be no end to this process, since the number of possible new compounds is endless.

An interesting challenge is to try to make this process more efficient. At the moment the scientific community grows thousands of new materials, and a few turn out to be interesting. This process is obviously inefficient. I challenge you, the reader, to find the answer to the following questions:

- What material is the best superconductor? It would have the highest transition temperature T_c to the superconducting phase. Do not tell me the electronic properties or the best density of states. Tell me which atoms are in the crystal, and in what arrangement.
- What material is the best ferromagnet?
- What material has the best magnetoresistance? It is used in computer memories.
- What semiconductor has the highest mobility?
- What material is the best thermoelectric? Typical thermoelectrics have a high figure of merit over a limited ($\Delta T \sim 100^\circ$C) temperature range. So there are several answers to this question for different temperature regions.

- What material is the best conductor of heat? Actually, it is probably impossible to beat diamond.
- What material has the lowest value of thermal conductivity? New low values are still being reported.
- What material is the best ferroelectric?
- What material has the best nonlinear optical properties?

All of these questions need answers. How do we invent new crystal structures that will have these desired properties?

Today the reverse process is easy. Say a new crystal is grown and its atomic coordinates are measured by x-ray scattering. Modern computer codes can

- Calculate all of the electronic energy bands.
- Calculate all of the vibrational modes (phonons).
- Calculate the transport coefficients, such as electrical resistance, Seebeck coefficient, and thermal conductivity. These coefficients determine a crystal's thermoelectric properties.
- The calculation of magnetic properties is still difficult, but is improving.
- The calculation of a crystal's superconducting properties is not yet possible.
- The calculation of its ferroelectric properties is routine by experts such as Karin Rabe or David Vanderbilt.

Given the atomic coordinates, theorists know how to calculate many of properties of the material. However, the reverse process is still not possible. We can not predict new arrangements of atoms that give a material with designed properties.

A less ambitious objective would be to predict structures of new crystals from selected elements. The prediction of crystal structure is still an art rather than a science. Given a proposed structure, the computer codes can tell whether the structure is stable, and give the dimensions of the unit cell. However, there are only a few cases where theory has dreamt up new, previously unknown, structures.

1.4 Artificial Structures

The prior section discussed designing three-dimensional, homogeneous crystals for a desired application. Another option is to design an artificially structured material.

Epitaxy can also be used to grow artificial structures such as superlattices. One band of material has n atomic layers of material A. Another band has m atomic layers of material B. Usually the material is periodic, with $n-A$, $m-B$, $n-A$, $m-B$, etc. The final structure has alternate bands of these two materials. If electrons, or holes, prefer to reside in band A instead of band B, they can be confined to two-dimensional motion in band A.

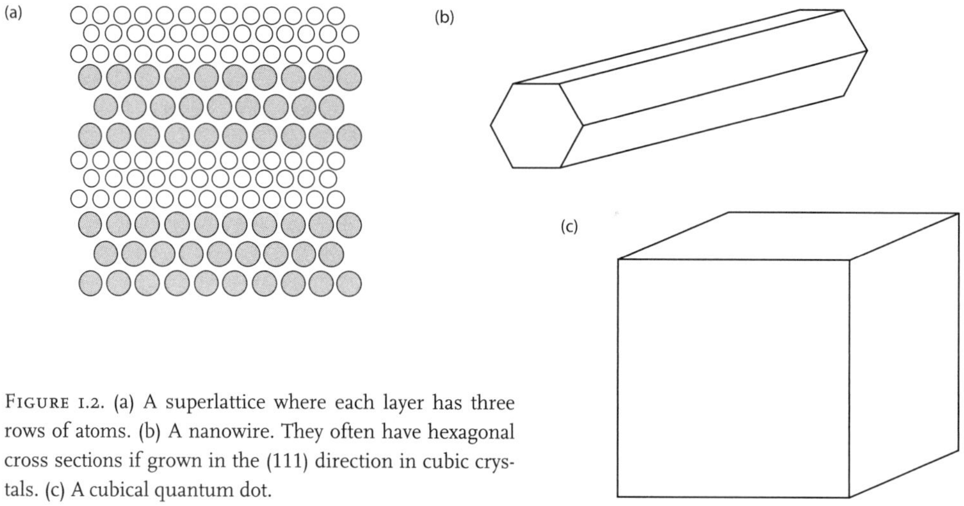

FIGURE 1.2. (a) A superlattice where each layer has three rows of atoms. (b) A nanowire. They often have hexagonal cross sections if grown in the (111) direction in cubic crystals. (c) A cubical quantum dot.

Superlattices have the property that we reduce the effective dimensionality of the motion from three to two. These interesting systems are discussed in chapter 15.

Experimentalists routinely grow nanowires of many materials using a variety of methods. The nanowires have a radius of 10–50 nm, and can have a length of many micrometers. Often they are single crystals, or contain only a few twin boundaries. In these systems the motion of the electrons, or holes, is largely one dimensional. Carbon or BN nanotubes are another type of one-dimensional conductor. These one-dimensional systems have many interesting properties, and may have useful applications in electronic devices.

Quantum dots are small nanocrystals of material. An isolated Qdot has interesting electrical properties if one can pass current through it, either by tunneling or by wire contacts. The optical properties may also be interesting. However, large crystals can be grown that are periodic arrays of Qdots: they can be two- or even three-dimensional crystals. These new systems are just starting to be investigated, and have many interesting properties. One application is to engineer photon energy gaps that trap electromagnetic radiation at selected frequencies.

Figure 1.2 shows a superlattice, a nanowire, and a Qdot. Another two-dimensional system is a single layer of carbon with the structure found in graphite. This single layer is called *graphene*. The graphene can be cut into strips, using an electron beam, and it is then a one-dimensional object. If it is cut into a finite area, it becomes a Qdot. Graphene is discussed in chapter 15.

If your goal is to design new materials with specific properties, you are not limited to three dimensions. You should consider one- and two-dimensional configurations of materials. These increase the number of options.

Nanofabrication facilities exist in many countries. They can engineer new structures. Figure 1.3 shows some typical structures produced in nanofab facilities. Figure 1.3a shows a Qdot transistor in the center. Pointed objects are gates to control electron density. Figure 1.3b shows a toadstool structure. The electrons in the circular Qdots are very isolated.

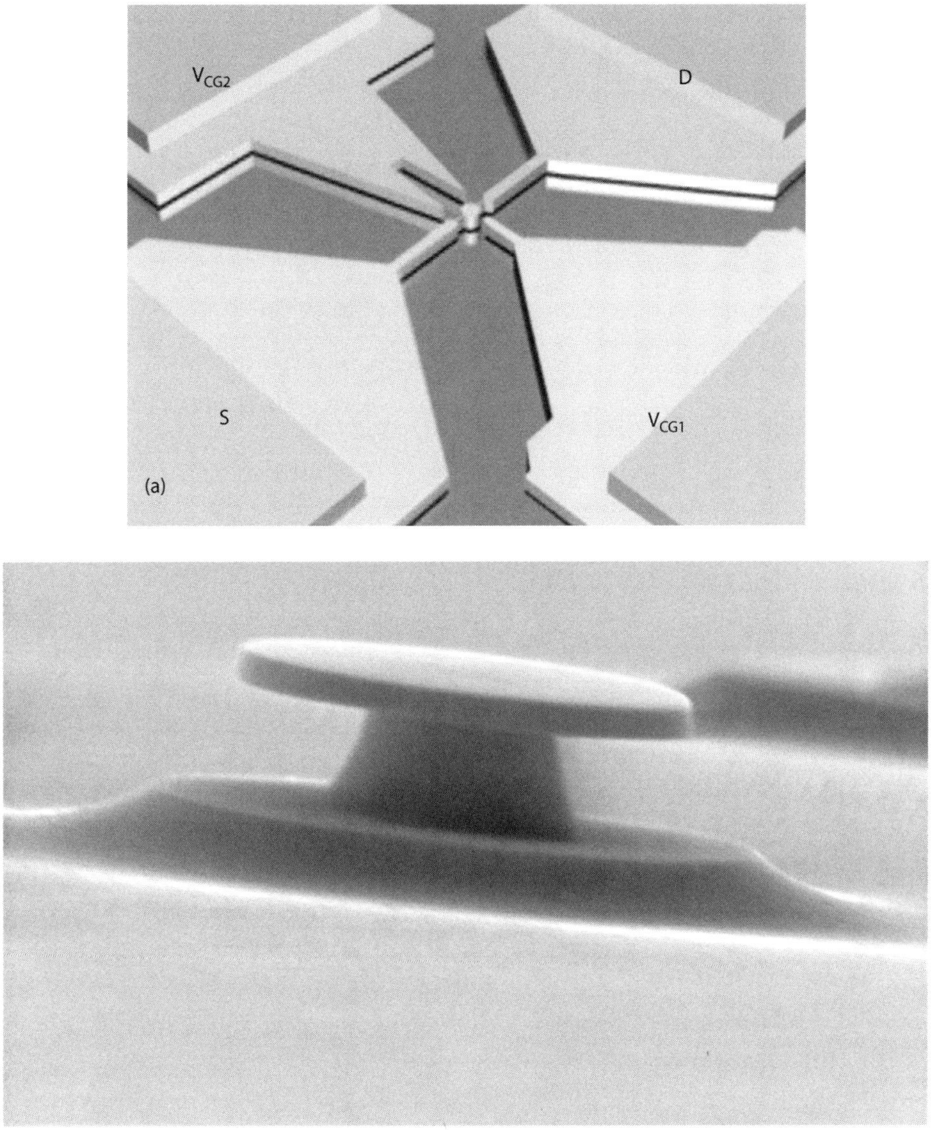

FIGURE 1.3. (a) Qdot transistor on a surface. From S. H. Son et al., *Physica E* **32**, 532 (2006). Used with permission of Elsevier. (b) Toadstool structure. Figure provided by Nitin Samarth (unpublished).

The devices shown in fig. 1.3 are only a few of many possible structures grown in nanofab facilities. Many new types of patterns are possible, which increases the opportunity to create new types of phenomena. You, the reader, should begin to think about what new kinds of physics you can do with all of this exciting technology. Many new tools are available for the next generation of condensed matter physics.

References

1. J. R. Arthur, Interaction of Ga and As_2 molecular beams with GaAs surfaces, *J. Appl. Phys.* **39**, 4032–4034 (1968)
2. P. W. Bridgman, Certain physical properties of single crystals of tungsten, antimony, bismuth, tellurium, cadmium, zinc, and tin, *Proc. Amer. Acad. Arts and Sci.* **60**, 305–383 (1925)
3. J. Czochralski, Ein neues verfahren zur messung der kristallisationsgeschwindigkeit der metalle. *Z. phys. chem. stöchiom. verwandtshafts.* **92**, 219–221 (1917) [A new method of measuring the velocity of crystallization of metals]
4. W. G. Pfann, Principles of zone-melting, *Trans. Am. Inst. Mining Metal. Eng.* **194**, 747–753 (1952)
5. M. A. Reed, Quantum dots. *Sci. Am.*, 118 (January, 1963)
6. S. H. Son, Y. S. Choi, S. W. Hwang J. I. Lee, Y. J. Park, Y. S. Yu, and D. Ahn, Gate bias controlled NDR in an IPGQDT, *Physica E* **32**, 532–535 (2006)

2 Crystal Structures

2.1 Lattice Vectors

The definition of a solid is quite subtle. One might define a solid as a hard material that retains its shape when you push on it. However, the alkali metals are regarded as solids but are very soft. They generally do not keep their shape when deformed. Another possible definition of a solid is that it keeps its shape indefinitely. However, many amorphous solids will change their shape over very long time periods, like decades. They are a type of glass, and glasses are really more like frozen liquids.

The definition of a crystal is much clearer: the atoms are in a regular array. So if one knows the positions of the atoms in one corner of the material, one can predict accurately their positions everywhere. In this chapter we will discuss the crystal arrangments found in different solids.

Crystals have a building block called a *unit cell.* Crystals are usually named after the atoms in the unit cell. For example, barium titanate has the chemical formula $BaTiO_3$. Each unit cell has one barium (Ba), one titanium (Ti), and three oxygen atoms (see fig. 2.1). The barium atoms have a *simple cubic* arrangement, abreviated as sc. If one were to stack identical cubes together, the Ba atoms would be at the corners. The titanium atoms are at the centers of each of the cubes. Of course, one could say with equal accuracy that the Ti atoms form the sc structure, and the Ba atoms are in the center of the cube. The oxygen atoms are at the center of the six faces of each of the cubes. The cube edge has a length a and volume $\Omega_0 = a^3$.

Lattice vectors ($\mathbf{a}_1$, $\mathbf{a}_2$, $\mathbf{a}_3$) define the *Bravais lattice.* One focuses on one of the atoms in the unit cell, such as Ba in barium titanate. Then one determines three noncoplanar vectors that go to three neighboring Ba atoms that are at identical sites in the neighboring unit cells. Any other Ba atom in the crystal can be reached by the lattice vector

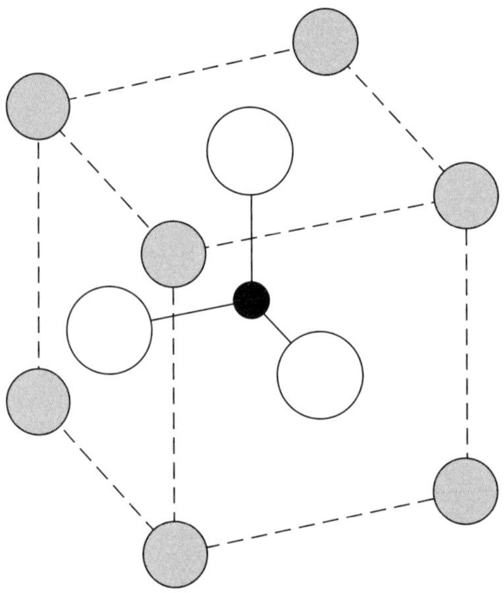

FIGURE 2.1. Unit cell of $BaTiO_3$. Shaded circles are Ba, the small solid circle is Ti, and the large hollow circles are O.

$$\mathbf{R}_{\ell,m,n} = \ell\mathbf{a}_1 + m\mathbf{a}_2 + n\mathbf{a}_3 \tag{2.1}$$

where (ℓ, m, n) are three integers. The Ba atom is being used as the basis for the structure since it comes first in the list of atoms. One of the other atoms, Ti or O, could also be chosen.

The number of first neighbors is called the *coordination number* and is denoted z. Any three of the z vectors can be chosen as lattice vectors as long as they are not coplanar. For barium titanate, with the sc structure, the three lattice vectors could be

$$\mathbf{a}_1 = a(100) \quad \text{or} \quad \mathbf{a}_1 = a(\bar{1}00) \tag{2.2}$$

$$\mathbf{a}_2 = a(010) \quad \text{or} \quad \mathbf{a}_2 = a(0\bar{1}0) \tag{2.3}$$

$$\mathbf{a}_3 = a(001) \quad \text{or} \quad \mathbf{a}_3 = a(00\bar{1}) \tag{2.4}$$

where the overbar denotes a minus number$(\bar{1} \equiv -1)$. It matters not whether one chooses the plus or minus sign for the lattice vectors. There are six vectors to nearest neighbors, of equal length. The vector from a Ba site to a Ti site, in the unit cell, is $\tau_{\text{Ti}} = a(111)/2$. The lattice vector from any Ba site to any Ti site is

$$\mathbf{R}_{\ell,m,n,\text{Ti}} = \ell\mathbf{a}_1 + m\mathbf{a}_2 + n\mathbf{a}_3 + \tau_{\text{Ti}} \tag{2.5}$$

The O sites can be located in a similar manner, using their three locations vectors from the Ba site:

$$\tau_{\text{O1}} = \frac{a}{2}(110), \quad \tau_{\text{O2}} = \frac{a}{2}(101), \quad \tau_{\text{O3}} = \frac{a}{2}(011) \tag{2.6}$$

Note that only plus signs were chosen. This choice is arbitrary, and one can also choose minus signs, such as $\tau_{01} = a(1\bar{1}0)/2$,

The lattice vectors also determine the volume of the unit cell, which is found from the absolute magnitude of the scalar product of the three lattice vectors:

$$\Omega_0 = \mathbf{a}_1 \cdot (\mathbf{a}_2 \times \mathbf{a}_3) \tag{2.7}$$

which is $\Omega_0 = a^3$ for barium titanate.

2.2 Reciprocal Lattice Vectors

The lattice vectors are expressed as in eqn. (2.1). The vector $\mathbf{R}_{\ell,m,n}$, with an appropriate choice of (ℓ, m, n) takes one to the same atom in any unit cell in the crystal. The reciprocal lattice vectors $\mathbf{G}$ obey the relationship

$$e^{i\mathbf{G}\cdot\mathbf{R}_{\ell,m,n}} = 1 \tag{2.8}$$

which is achieved when the scalar product $\mathbf{G} \cdot \mathbf{R}$ is equal to an integer multiple of 2π.

Three reciprocal lattice vectors are

$$\mathbf{G}_1 = \frac{2\pi}{\Omega_0}\mathbf{a}_2 \times \mathbf{a}_3 \tag{2.9}$$

$$\mathbf{G}_2 = \frac{2\pi}{\Omega_0}\mathbf{a}_3 \times \mathbf{a}_1 \tag{2.10}$$

$$\mathbf{G}_3 = \frac{2\pi}{\Omega_0}\mathbf{a}_1 \times \mathbf{a}_2 \tag{2.11}$$

These three vectors will not be coplanar as long as $(\mathbf{a}_1, \mathbf{a}_2, \mathbf{a}_3)$ are not coplanar. If (α, β, γ) are any three integers, then the most general reciprocal lattice vector is

$$\mathbf{G}_{\alpha,\beta,\gamma} = \alpha\mathbf{G}_1 + \beta\mathbf{G}_2 + \gamma\mathbf{G}_3 \tag{2.12}$$

Note the identities

$$\mathbf{G}_\varepsilon \cdot \mathbf{a}_j = 2\pi\delta_{\varepsilon j} \tag{2.13}$$

$$\mathbf{G}_{\alpha,\beta,\gamma} \cdot \mathbf{R}_{\ell,m,n} = 2\pi(\alpha\ell + \beta m + \gamma n) \tag{2.14}$$

$$\exp[i\mathbf{G}_{\alpha,\beta,\gamma} \cdot R_{\ell,m,n}] = 1 \tag{2.15}$$

The first equation states that $\mathbf{G}_1 \cdot \mathbf{a}_1 = 2\pi$ but that $\mathbf{G}_1 \cdot \mathbf{a}_{2,3} = 0$. The second equation states that the vector product of a lattice vector and a reciprocal lattice vector is always an integer multiple of 2π. The third equation states that the exponential of this phase factor is always unity, since the phase factor is a multiple of 2π.

Excitations in crystals are eigenstates of crystal wave vector $\mathbf{k}$. The allowed values of wave vector depend on the reciprocal lattice vectors. The reciprocal lattice vectors play a fundamental role in describing excitations in crystals.

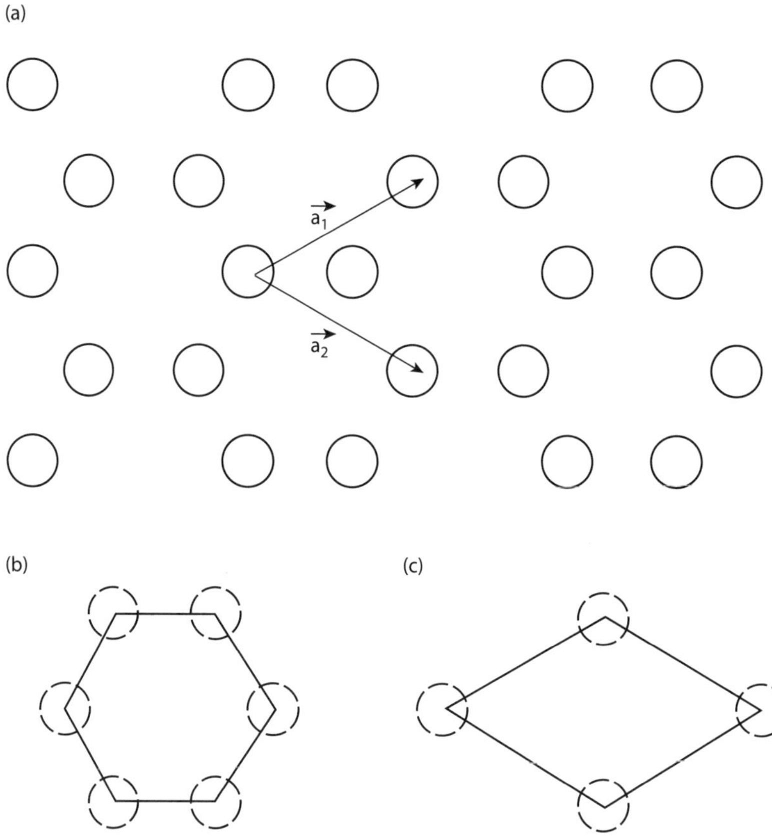

FIGURE 2.2. (a) Structure of graphene. A carbon atom is at each vertex. (b) Wigner-Seitz unit cell. (c) Another choice of unit cell.

One way to determine the shape of the unit cell is to start with the z vectors $\mathbf{a}_j$ to the neighboring cells. Bisect each of these vectors with a plane, and the miminal enclosed volume is the fundamental unit cell of volume Ω_0. This construction defines the *Wigner-Seitz unit cell*. It is not the only unit cell that can be chosen. Any unit cell is valid as long as it (i) fills all space, and (ii) has the volume Ω_0. Figure 2.2a shows an example for the two-dimensional honeycomb lattice, which is the structure of a single sheet of graphite (called *graphene*). There are two carbon atoms in each unit cell $(N_c = 2)$, which are usually called the A and the B atoms. Each A atom has three nearest-neighbor B atoms. Each A atom has $z = 6$ vectors $\mathbf{a}_j$ to second nearest-neighbor A atoms. The Wigner-Seitz unit cell is shown in fig. 2.2b. Figure 2.2c shows another choice of unit cell that has the same area. In two dimensions, only two lattice vectors $(\mathbf{a}_1, \mathbf{a}_2)$ are needed to define the lattice and reciprocal lattice vectors. The third unit vector is perpendicular to the plane.

The *Brillouin zone*, abbreviated BZ, is the fundamental unit cell in the space defined by the reciprocal lattice vectors. One finds all of the z_G reciprocal lattice vectors with the same nonzero length $\mathbf{G}_0$. Draw them in three dimensions: this is called the *star*. Then bisect them all with a plane, and the enclosed volume is the Brillouin zone. The polygons of the unit cell in real space and in reciprocal space are often very different. The volume of the BZ is

$$\Omega_{BZ} = \mathbf{G}_1 \cdot (\mathbf{G}_2 \times \mathbf{G}_3) \tag{2.16}$$

It is easy to prove in d-dimensions that

$$\Omega_0 \Omega_{BZ} = (2\pi)^d \tag{2.17}$$

The above procedure defines the *first Brillouin zone*. The second, third, and other BZs are found by going further out in wave vector space. They each must have exactly the same volume (3D) or area (2D) as the first BZ. Examples are given below.

2.3 Two Dimensions

Two dimensions is the easiest way to visualize crystal structure. Many two-dimensional systems are found in nature:

1. The *square lattice* is denoted sq. The distance between atoms is a. It has

$$\mathbf{a}_1 = a(10) \quad \mathbf{G}_1 = \frac{2\pi}{a}(10) \tag{2.18}$$

$$\mathbf{a}_2 = a(01) \quad \mathbf{G}_2 = \frac{2\pi}{a}(01) \tag{2.19}$$

$$\Omega_0 = a^2 \quad \Omega_{BZ} = \frac{(2\pi)^2}{a^2} \tag{2.20}$$

In constructing the reciprocal lattice vectors in two dimensions, take $\mathbf{a}_3 = \hat{z}$. The lattice structure is shown in fig. 2.3a. The Wigner-Scitz cell is a square, as is the first BZ. Figure 2.3b shows the first four BZs of the sq lattice. They are created by taking many reciprocal lattice vectors and bisecting the line that joins them to the origin. Each BZ has the same area when adding together the areas of the separate pieces. Each of these pieces can be moved to the first BZ by a reciprocal lattice vector. When all of the pieces are moved to the first BZ, they exactly fill up the area.

2. The *plane triangular lattice* is denoted pt. It is based on the hexagon. Each atom has six neighbors ($z = 6$). Any two nonparallel vectors can be chosen for the lattice vectors:

$$\mathbf{a}_1 = a(10) \quad \mathbf{G}_1 = \frac{4\pi}{a\sqrt{3}}\left(\frac{\sqrt{3}}{2}, \frac{\bar{1}}{2}\right) \tag{2.21}$$

$$\mathbf{a}_2 = a\left(\frac{1}{2}, \frac{\sqrt{3}}{2}\right) \quad \mathbf{G}_2 = \frac{4\pi}{a\sqrt{3}}(01) \tag{2.22}$$

$$\Omega_0 = \frac{\sqrt{3}}{2}a^2 \quad \Omega_{BZ} = \frac{8\pi^2}{a^2\sqrt{3}} \tag{2.23}$$

The Wigner-Seitz cell and the BZ are both hexagons. They are not oriented identically, but one is rotated 30° with respect to the other. The lattice is shown in fig. 2.4.

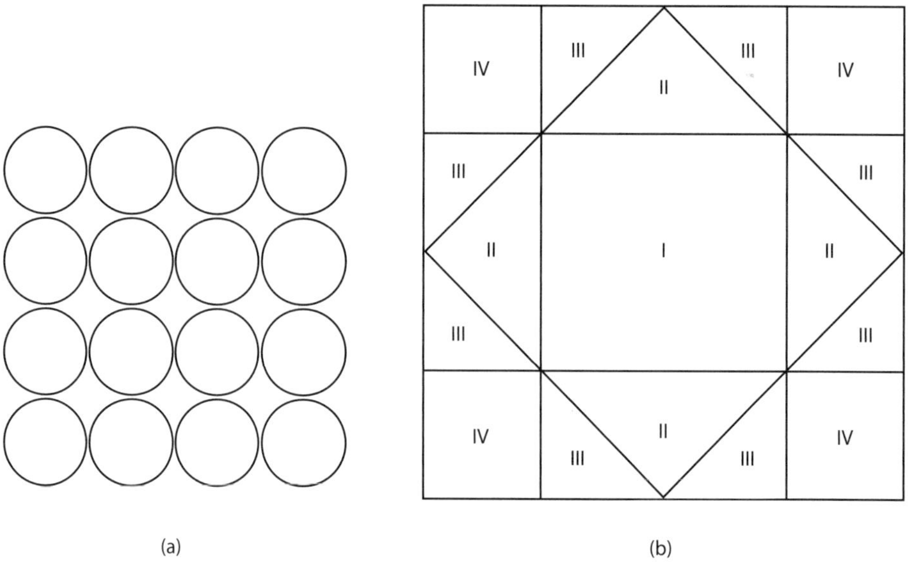

FIGURE 2.3. (a) Two-dimensional square lattice. (b) Labels I, II, III, and IV show parts of first, second, third, and fourth Brillouin zones.

3. One can fill all two-dimensional space with squares or hexagons. One cannot fill all space with pentagons. It was long thought that fivefold symmetry is not allowed in solids. It came as a big surprise when alloys of Al and Mn had x-ray diffraction patterns that showed fivefold symmetry. These alloys are made by rapid cooling from the liquid state, and were thought to be amorphous.

It is now realized that the alloy has strong short-range order, which causes the diffraction pattern. They do not have long-range order, and are truly amorphous. Because they have short-range order, they are called *quasicrystals*. Roger Penrose [3] introduced a way of producing fivefold short-range order. His method is called *Penrose tiling* and is discussed by Martin Gardner [1].

Penrose tiling is produced by creating two tiles, as shown in fig. 2.5. The fatter tile has an angle $\phi_1 = 2\pi/5$, while the skinny tile has an angle $\phi_2 = 2\pi/10$. Both tiles have the same edge length. Part (b) of the figure shows a typical tiling pattern. Note the many areas

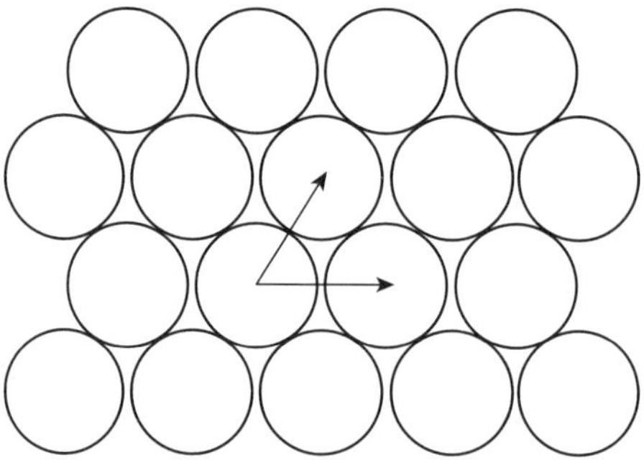

FIGURE 2.4. Two-dimensional plane triangular lattice.

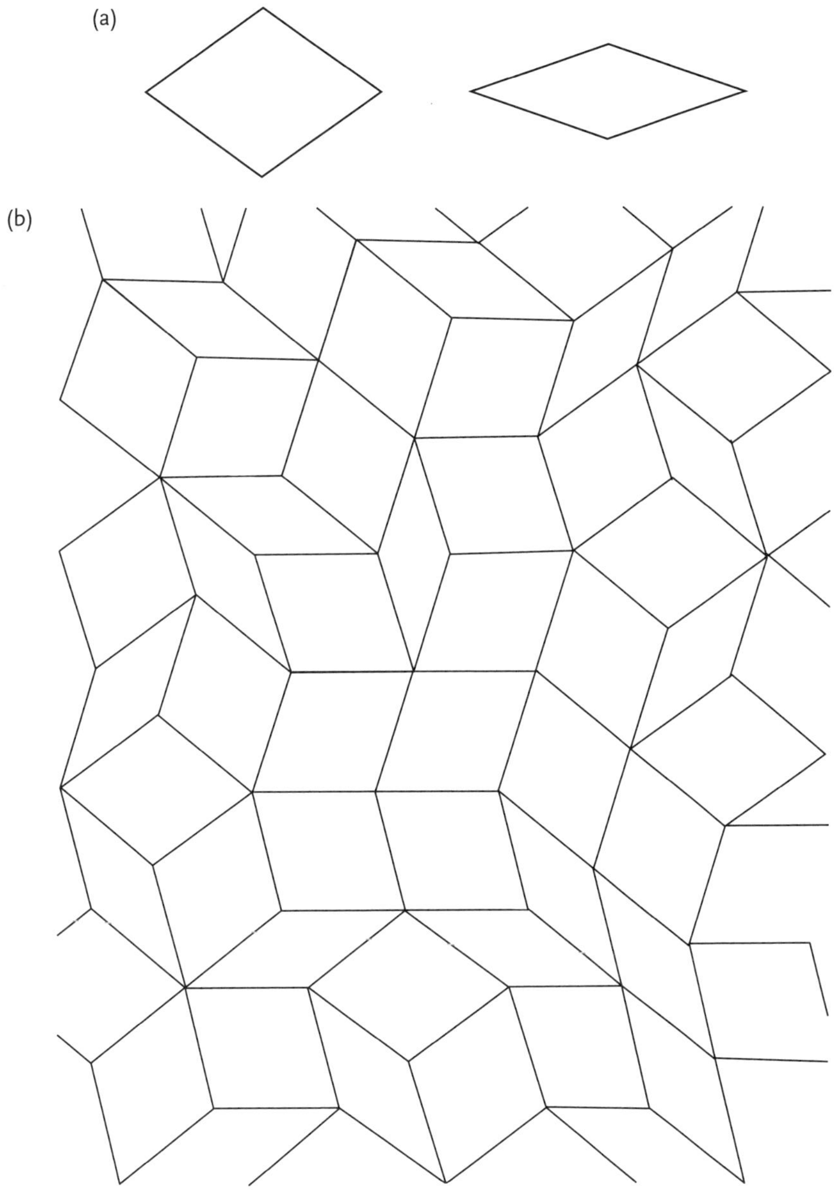

FIGURE 2.5. (a) The two tiles in Penrose tiling. The fatter one has an angle $\phi_1 = 2\pi/5$, while the skinny tile has an angle $\phi_2 = 2\pi/10$. (b) A typical tiling pattern.

that have local fivefold symmetry. The microscopic crystal structure of Al–Mn alloys are believed to have this kind of arrangement. The material is layered and tiling is found in the layers.

2.4 Three Dimensions

The following are some simple unit cells and their Brillouin zones.

- The simple cubic (sc) lattice has unit vectors: $\mathbf{a}_1 = a\hat{x}$, $\mathbf{a}_2 = a\hat{y}$, $\mathbf{a}_3 = a\hat{z}$. Then $z = 6$, $\Omega_0 = a^3$. The unit cell is a cube. Three reciprocal lattice vectors are

$$\mathbf{G}_1 = \frac{2\pi}{a}\hat{x},\ \mathbf{G}_2 = \frac{2\pi}{a}\hat{y},\ \mathbf{G}_3 = \frac{2\pi}{a}\hat{z} \tag{2.24}$$

Then $z_G = 6$, $\Omega_{BZ} = (2\pi/a)^3$, and the BZ is also a cube. The only element with this crystal structure is polonium.

- The body-centered cubic (bcc) structure has two interpenetrating sc lattices. Take an sc structure and put another atom in the center of each cube. The lattice structure is shown in fig. 2.6a. Each atom has eight first neighbors. Any three of these eight lattice vectors can be used to define the unit cell, as long as they are not coplanar. An acceptable set of unit vectors are

$$\mathbf{a}_1 = \frac{a}{2}(111),\ \mathbf{a}_2 = \frac{a}{2}(11\bar{1}),\ \mathbf{a}_3 = \frac{a}{2}(1\bar{1}1) \tag{2.25}$$

The notation $(1\bar{1}1)$ means a vector in the $(\hat{x}, -\hat{y}, \hat{z})$ direction. The volume of the unit cell is $\Omega_0 = a^3/2$, since each cube of volume a^3 has two atoms. Three reciprocal lattice vectors are

$$\mathbf{G}_1 = \frac{2\pi}{a}(110),\ \mathbf{G}_2 = \frac{2\pi}{a}(101),\ \mathbf{G}_3 = \frac{2\pi}{a}(011) \tag{2.26}$$

The bcc structure has $z = 8$ and $z_G = 12$. Some elements with the bcc structure are the alkali metals and other metals as shown in table 2.1.

- The face-centered cubic (fcc) is an sc structure with additional atoms at the center of each face of the cube. The lattice and reciprocal lattice vectors are

$$\mathbf{a}_1 = \frac{a}{2}(110),\ \mathbf{a}_2 = \frac{a}{2}(101),\ \mathbf{a}_3 = \frac{a}{2}(011) \tag{2.27}$$

$$\mathbf{G}_1 = \frac{2\pi}{a}(111),\ \mathbf{G}_2 = \frac{2\pi}{a}(11\bar{1}),\ \mathbf{G}_3 = \frac{2\pi}{a}(1\bar{1}1) \tag{2.28}$$

and $\Omega_0 = a^3/4$. Here $z = 12$, $z_G = 8$.

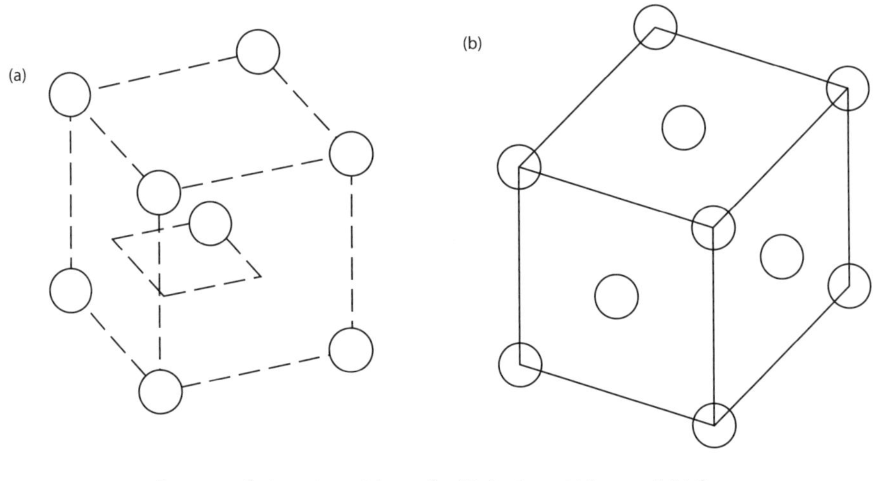

FIGURE 2.6. Atomic positions of cubic lattices: (a) bcc, and (b) fcc.

Table 2.1 Lattice constants *a* (in Å) of some bcc metals

Metal	a	Metal	a
Li	3.5093	Cr	2.8839
Na	4.2906	α-Fe	2.8665
K	5.247	Nb	3.3004
Rb	5.605	Mo	3.1473
Cs	6.067	Tl	3.882

Note. Measurements of a at different temperatures.

Table 2.2 Lattice constants *a* (in Å) of some fcc metals and rare-gas solids

Metal	a	Solid	a	Metal	a
Cu	3.61496	Ne	4.429	Ca	5.576
Ag	4.0862	Ar	5.256	Sr	6.0847
Au	4.07825	Kr	5.721	La	5.296
Al	4.04958	Xe	6.197	Ce	5.1612
Pb	4.9505	Co	3.548	Ni	3.5239

Note. Measurements of a at different temperatures.

Note that the bcc and fcc lattice have a reciprocal relationship. One has lattice vectors (111) and reciprocal lattice vectors (110), while the other lattice has them reversed. The Wigner-Seitz cell of one has the same shape as the Brillouin zone of the other. Some metals with the fcc structure are shown in table 2.2. Figure 2.7a shows the Wigner-Seitz cell for bcc, and (b) shows the BZ for bcc. Alternately, fig. 2.7a shows the BZ for fcc and (b) shows the Wigner-Seitz cell for fcc.

Figure 2.7 shows capital letters at symmetry points. This notation is taken from the theory of finite groups. At these symmetry points, group theory can prove theorems regarding the degeneracy of the energy bands, as well as permissible functions for basis states. The center of the BZ, at $\mathbf{k} = 0$, is always called the Γ point. These symmetry labels will show up in later chapters on energy bands.

- The *diamond lattice* has two carbon atoms per unit cell. The lattice structure is fcc. One carbon is found at the corners and faces of the cube. The second carbon is found at a tetrahedral position equidistant from four carbons: one is found at $\tau = a(111)/4$. Inside the cube of volume a^3 there are eight such tetrahedral sites, and four are occupied. The volume of the unit cell is $\Omega_0 = a^3/4$. The diamond lattice is found for group IV elements: carbon, silicon, germanium, and one form of tin.

- The *hexagonal close-packed* lattice is denoted hcp. It is one of the two structures (the other is fcc) that provide the closest packing of spheres. Start by packing the spheres onto a plane, as

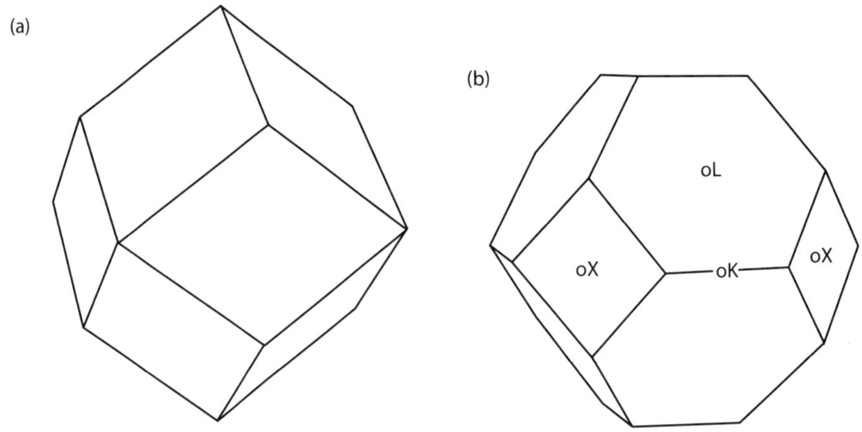

FIGURE 2.7. (a) The Wigner-Seitz cell for bcc, and (b) the BZ for bcc.

shown in fig. 2.8. The closest packing of spheres in a plane is the pt structure. After filling the plane with spheres, start packing them on the next layer. The second layer also has the pt structure. The spheres sit in the hole created by the vertex of three spheres in the first level. Call the first layer A, and the second layer B. There are two choices for B, since there are two possible vertices. The third layer also has two choices:

1. One is to return to the same arrangement as the A layer. This sequence of stacking is denoted ABABAB. . . . It is the hcp structure.
2. The second option is to stack the third layer at the point C, and then the fourth layer is again A. This sequence is ABCABCABC. . . . It is actually the fcc structure if one views the crystal along the body diagonal (111).

The hcp lattice has lattice vectors

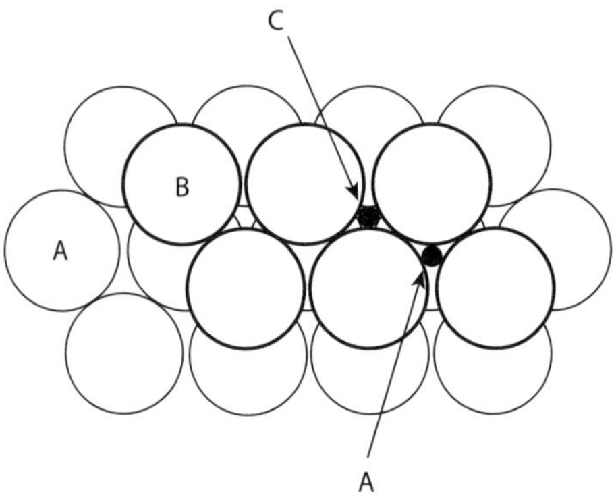

FIGURE 2.8. Close packing of spheres. Solid circles are first layer. Darker circles are second layer. Third layer can be stacked on top of first layer (A), or in C position.

Table 2.3 Room temperature lattice constants (in Å) of some divalent hcp metals

Metal	a	c	c/a
Be	2.2866	3.5833	1.567
Cd	2.97887	5.61765	1.886
Mg	3.20927	5.21033	1.624
Zn	2.6648	4.9467	1.856

$$\mathbf{a}_1 = a(100) \quad \mathbf{G}_1 = \frac{4\pi}{a\sqrt{3}}\left(\frac{\sqrt{3}}{2},\frac{1}{2},0\right) \tag{2.29}$$

$$\mathbf{a}_2 = a\left(\frac{\bar{1}}{2},\frac{\sqrt{3}}{2}\right) \quad \mathbf{G}_2 = \frac{4\pi}{a\sqrt{3}}(010) \tag{2.30}$$

$$\mathbf{a}_3 = c(001) \quad \mathbf{G}_3 = \frac{2\pi}{c}(001) \tag{2.31}$$

The distance c is between two A layers. There are two layers per unit cell along the axis perpendicular to the plane. For ideal close-packing of spheres, the length c obeys

$$\frac{c}{a} = \sqrt{\frac{8}{3}} \approx 1.633\ldots \tag{2.32}$$

Actual hcp metals have a ratio of c/a that is close to this value. Most divalent metals have the hcp structure, as shown in table 2.3.

2.5 Compounds

The above structures are for crystals composed of single elements. Other important structures are found for compounds. Some of these are now described.

- Rocksalt is the structure of NaCl and other alkali halide crystals. Sodium (Na) is the cation and donates an electron, while chlorine (Cl) is the anion that receives it. Start with the sc structure and make alternate atoms Na and Cl. The crystal structure of the Na atoms, or of the Cl atoms, is fcc. The constant a is for the fcc lattice constant. Table 2.4 shows some lattice constants of a few of the many crystals with the rocksalt structure. Most tables list them alphabetically. Here they are listed in the order they occur in the periodic table. Reading down a column, or across a row, the atoms with larger atomic number have a larger lattice constant. The ion size increases with atomic number. It is interesting that AgF, AgCl, and AgBr have the rocksalt structure. The other halides of Cu or Ag have the zincblende structure.

- The zincblende structure is named for ZnS. It is a binary crystal that is obtained from the diamond lattice by making alternate atoms Zn or S. The crystal symmetry is fcc. It is

Table 2.4 Room temperature lattice constants (in Å) of some binary solids with the rocksalt structure

Solid	a	Solid	a	Solid	a	Solid	a
LiF	4.0173	NaF	4.620	KF	5.347	AgF	4.92
LiCl	5.1295	NaCl	5.6406	KCl	6.2929	AgCl	5.547
LiBr	5.5013	NaBr	5.9732	KBr	6.600	AgBr	5.7745
LiI	6.000	NaI	6.4728	KI	7.0656		
MgO	4.216	CaO	4.811	SrO	5.160	BaO	5.536
MgS	5.19	CaS	5.67	SrS	6.020	BaS	6.37

an important structure since most III–V semiconductors (such as GaAs) and many II–VI semiconductors (such as ZnS) have this structure. Some examples are in table 2.5.

- The *wurtzite* structure is based on the hcp lattice. ZnO and CdS have this structure. Just as zincblende is obtained from fcc by putting atoms in tetrahedral locations, wurtzite is obtained from hcp by putting atoms in tetrahedral positions. The easiest way to visualize this structure is to put another atom beneath every atom in the hcp structure. For example, put Zn atoms in the hcp structure, and O atoms beneath them. Then alternate layers are Zn atoms and O atoms. Each is in a tetrahedral position. The lattice symmetry is hexagonal. Some examples are in table 2.6. Note that the nitrides of Al, Ga, and In have this structure.

- The *fluorite* structure is named after the mineral CaF_2. The Ca atoms are in the fcc lattice. The flourine atoms occupy all of the eight tetrahedral sites of the unit cube. Many crystals have this structure.

The three lattice vectors ($\mathbf{a}_1$, $\mathbf{a}_2$, $\mathbf{a}_3$) define a three-dimensional parallelpiped. It has sides (a_1, a_2, a_3) and the corners have three angles (α_1, α_2, α_3). The sides can all be the same value or all be different; similarly with the three angles. There are seven distinct possibilities:

Table 2.5 Room temperature lattice constants (in Å) of some binary solids with the zincblende structure

Solid	a	Solid	a	Solid	a	Solid	a
AlP	5.464	GaP	5.451	InP	5.869	CuF	4.255
AlAs	5.661	GaAs	5.653	InAs	6.058	CuCl	5.4057
AlSb	6.135	GaSb	6.096	InSb	6.479	CuBr	5.6905
ZnS	5.409	CdS	5.825	HgS	5.851	CuI	6.0427
ZnSe	5.668	CdSe	6.052	HgSe	6.084	AgI	6.473
ZnTe	6.089	CdTe	6.480	HgTe	6.461		

Table 2.6 Room temperature lattice constants (in Å) of some binary solids with the wurtzite structure

Solid	a	c	Solid	a	c
AlN	3.111	4.978	BeO	2.698	4.380
GaN	3.180	5.166	ZnO	3.2495	5.2069
InN	3.533	5.693	CdS	4.1348	6.7490

- *Triclinic crystals* have all three sides and all three angles different: ($a_1 \neq a_2 \neq a_3$, $\alpha_1 \neq \alpha_2 \neq \alpha_3$).
- *Monoclinic crystals* have two angles equal to 90°, but the sides are all different: ($a_1 \neq a_2 \neq a_3$, $\alpha_1 = \alpha_2 = 90° \neq \alpha_3$).
- *Orthorhombic crystals* have all sides different, but all angles equal to 90°: ($a_1 \neq a_2 \neq a_3$, $\alpha_1 = \alpha_2 = \alpha_3 = 90°$).
- *Tetragonal crystals* have a unit cell with all right angles. Two sides of the unit cell have the same lattice constant a, while the third distance c is different ($a_1 = a_2 \neq a_3$, $\alpha_1 = \alpha_2 = \alpha_3 = 90°$).
- *Trigonal crystals* have all three sides equal and all three angles equal, but the angles are not 90°: ($a_1 = a_2 = a_3$, $\alpha_1 = \alpha_2 = \alpha_3 \neq 90°$). Trigonal crystals are also called *rhombohedral*.
- *Cubic crystals* have all sides equal, and all angles equal to 90°: ($a_1 = a_2 = a_3$, $\alpha_1 = \alpha_2 = \alpha_3 = 90°$).
- *Hexagonal crystals* have two sides equal, and two angles equal to 90°: ($a_1 = a_2 \neq a_3$, $\alpha_1 = \alpha_2 = 90°$, $\alpha_3 = 120°$).

2.6 Measuring Crystal Structures

Crystal structures are usually measured by one of three different scattering methods:

- x-ray scattering
- electron scattering
- neutron scattering

For the determination of structure, only elastic scattering is important. The first two measurements determine the density of electronic charge in the solid. Neutrons scatter from nuclei, so measure the arrangements of the nuclei in the solid. Since the electrons bound to the nucleus have the same arrangement as the nuclei themselves, these various methods give the same crystal arrangement.

2.6.1 X-ray Scattering

X-ray scattering is done with photon energies in the kilovolt regime. X-ray sources are available in most laboratories. Such high energies are needed to get the photon wavelength to be as short as the interatomic distances.

X-ray scattering is discussed in detail. First, prove a theorem:

Any physical function $f(\mathbf{r})$ can be expanded in a Fourier series. If the potential is periodic, $f(\mathbf{r} + \mathbf{R}) = f(\mathbf{r})$, its only nonzero Fourier components are reciprocal lattice vectors.

This theorem is quite useful in discussing crystal physics. For any periodic function $n(\mathbf{r})$ in three dimensions,

$$n(\mathbf{r} + \mathbf{R}_{\ell,m,n}) = n(\mathbf{r}) \tag{2.33}$$

$$n(\mathbf{r}) = \sum_{\alpha,\beta,\gamma} n(\alpha,\beta,\gamma)\, e^{i\mathbf{G}_{\alpha,\beta,\gamma}\cdot\mathbf{r}} \tag{2.34}$$

$$n(\alpha,\beta,\gamma) = \int \frac{d^3r}{\Omega_0} n(\mathbf{r})\, e^{i\mathbf{G}_{\alpha,\beta,\gamma}\cdot\mathbf{r}} \equiv n(\mathbf{G}) \tag{2.35}$$

where the integral is over the unit cell. The phase factor in eqn. (2.34) has the feature that for any lattice vector $\mathbf{R}_j$

$$e^{i\mathbf{G}_{\alpha,\beta,\gamma}\cdot(\mathbf{r}+\mathbf{R}_j)} = e^{i\mathbf{G}_{\alpha,\beta,\gamma}\cdot\mathbf{r}} \tag{2.36}$$

so that $n(\mathbf{r} + \mathbf{R}_j) = n(\mathbf{r})$.

The symbol $n(\mathbf{r})$ is used to apply the theorem to the electron density. The Hamiltonian of electrons interacting with an electromagnetic field is

$$H = \frac{1}{2m}\sum_j [\mathbf{p}_j - e\mathbf{A}(\mathbf{r}_j)]^2 \tag{2.37}$$

where $\mathbf{p}_j$ is the momentum of the electrons, $\mathbf{A}(\mathbf{r}_j)$ is the vector potential of the x-rays at the electron positions $\mathbf{r}_j$, and (e, m) are the charge and mass of the electron. The nonlinear interaction of the x-rays with the electron is the last term:

$$V = \frac{e^2}{2m}\sum_j \mathbf{A}^2(\mathbf{r}_j) = \frac{e^2}{2m}\int d^3r\, n(\mathbf{r})\mathbf{A}^2(\mathbf{r}) \tag{2.38}$$

The last expression introduces the electron density.

If the initial wave vector of the photon is $\mathbf{k}_i$ and the final wave vector is $\mathbf{k}_f$, then write the vector potential as

$$\mathbf{A}(\mathbf{r}) = A_i e^{i\mathbf{k}_i\cdot\mathbf{r}_j} + \mathbf{A}_f e^{i\mathbf{k}_f\cdot\mathbf{r}_j} \tag{2.39}$$

$$|\mathbf{A}(\mathbf{r})|^2 = |A_i|^2 + |A_j|^2 + 2\mathbf{A}_i\cdot\mathbf{A}_f e^{i(\mathbf{k}_i-\mathbf{k}_f)\cdot\mathbf{r}_j} \tag{2.40}$$

The last term provides the scattering. The effective matrix element is

$$V = \frac{e^2}{m}\mathbf{A}_i\cdot\mathbf{A}_f\sum_{\mathbf{G}} n(\mathbf{G})\int d^3r\, e^{i\mathbf{r}\cdot(\mathbf{k}_i+\mathbf{G}-\mathbf{k}_f)} \tag{2.41}$$

The integral over $\int d^3r$ is nonzero only when $\mathbf{k}_f = \mathbf{k}_i + \mathbf{G}$. The elastic scattering changes the photon wave vector by a reciprocal lattice vector. The scattering matrix element is proportional to $n(\mathbf{G})$.

If the incoming beam of x-rays is tightly focused, then scattering by **G** will produce a small spot of photons at some scattering angle. A solid has many reciprocal lattice vectors, so there will be many that scatter, and the scattering pattern will be many such spots. From the array of spots, the values of **G** can be deduced, and hence the structure.

2.6.2 *Electron Scattering*

Electron scattering is done with electron kinetic energies in the kilovolt regime. These electrons have very short wavelengths, much less than an angstrom. Low-energy electrons scatter strongly, and multiple scattering effects are hard to analyze. High energies reduce the cross section, so usually one scattering event occurs. Also, electron exchange effects in scattering can be neglected at high energies.

2.6.3 *Neutron Scattering*

For neutrons to have a wavelength around $\lambda \sim 3$ Å, the kinetic energy must be about 0.03 eV. This value is a thermal energy (k_BT) at room temperature. The traditional source of such neutrons has been nuclear reactors: One drills a hole in the reactor shielding, and thermal neutrons are emitted. The spallation neutron source provides more energetic neutrons. One accelerates a beam of protons and then shoots them at a target of liquid mercury. Mercury converts some of the protons to neutrons, which are used to generate a beam of energetic neutrons.

Neutrons have a large cross section for scattering from atomic hydrogen, so neutron scattering is used to determine the location of this atom in many organic solids and in biomaterials. In contrast, x-rays and electron scattering measure the locations of the electrons, and they are usually unable to locate hydrogen bonds.

Neutrons scatter from the nuclei. Both particles are small on the scale of atomic dimensions. The scattering can be considered to occur at a point, so the interaction between the neutron at **r** and the nuclei at $\mathbf{R}_j$ is

$$V(\mathbf{r}) = \frac{2\pi\hbar^2}{m_n}\sum_j [b_j + c_j\mathbf{s}_n \cdot \mathbf{S}_j]\delta^3(\mathbf{r} - \mathbf{R}_j) \tag{2.42}$$

where m_n is the mass of the neutron. The first term b_j gives the isotropic scattering, which is found for all nuclei. This parameter has the units of length and has a typical size of $b \sim 10$ fm. The second term is due to the spin $(\mathbf{S}_j)$ of the nuclei, which interacts with the spin $(\mathbf{s}_n)$ of the neutron. This scattering is for nuclei with a net nuclear moment. That term is not treated here, since its analysis is quite complex. It is discussed in Shirane et al [5].

Since the scattering is weak, it is accurate to use the Born approximation to scattering. The differential cross section for elastic scattering has a matrix element that is the Fourier transform of the potential:

$$V(\mathbf{q}) = \frac{2\pi\hbar^2}{m_n}\sum_j [b_j + c_j \mathbf{s}_n \cdot \mathbf{S}_j] e^{i\mathbf{q}\cdot\mathbf{R}_j} \tag{2.43}$$

$$\frac{d\sigma}{d\Omega} = \left(\frac{m_n}{2\pi\hbar^2}\right)^2 |V(\mathbf{q})|^2 \tag{2.44}$$

Crystal structure is determined by Bragg scattering, where the scattering wave vector $\mathbf{q}$ equals one of the reciprocal lattice vectors. In that case, the phase factor is unity. When the magnetic scattering is ignored, Bragg scattering gives

$$V(\mathbf{G}) = \frac{2\pi\hbar^2}{m_n}\sum_j b_j = \frac{2\pi\hbar^2}{m_n} N\bar{b} \tag{2.45}$$

where $\bar{b}$ is the average value of b among all of the N atoms in the crystal. Then the differential cross section is

$$\frac{d\sigma}{d\Omega} = N^2 (\bar{b})^2 |_{\mathbf{q}=\mathbf{G}} \tag{2.46}$$

These formulas describe the *coherent neutron scattering*. It is coherent since one is scattering from all of the nuclei equally, in a coherent fashion. Often the factor of N^2 is replaced by a delta function:

$$\frac{d\sigma}{d\Omega} = \frac{(2\pi)^3}{\Omega_0} N (\bar{b})^2 \sum_{\mathbf{G}} \delta^3(\mathbf{q} - \mathbf{G}) \tag{2.47}$$

where Ω_0 is the volume of the crystal unit cell.

Most crystals also have an *incoherent cross section*. The incoherent cross section occurs when there is some random variation in b_j among the nuclei. This most often occurs due to isotopes of the same element. Most elements have several different stable isotopes, and each has its own interaction b_j with the neutron. For example, germanium has five abundant isotopes. Since the average interaction is $\bar{b}$, each isotope has a fluctuation $\delta b_j = b_j - \bar{b}$ from the average. The scattering cross section depends on the square of the matrix element:

$$|V(q)|^2 = \left(\frac{2\pi\hbar^2}{m_n}\right)^2 \left|\sum_J \left(\bar{b} + \delta b_j\right) e^{i\mathbf{q}\cdot\mathbf{R}_j}\right|^2 \tag{2.48}$$

$$= \left(\frac{2\pi\hbar^2}{m_n}\right)^2 \left[N^2 (\bar{b})^2 |_{\mathbf{q}=\mathbf{G}} + N\langle (\delta b)^2 \rangle\right] \tag{2.49}$$

The cross-term $\langle \delta b_j \rangle$ averages to zero. The last term is the incoherent scattering. It comes from the double summation

$$\sum_{j\ell} \delta b_j \delta b_\ell e^{i\mathbf{q}\cdot(\mathbf{R}_j - \mathbf{R}_\ell)} = \sum_{j=\ell} (\delta b_j)^2 + \sum_{j\neq\ell} \delta b_j \delta b_\ell e^{i\mathbf{q}\cdot(\mathbf{R}_j - \mathbf{R}_\ell)} \tag{2.50}$$

If the fluctuations are uncorrelated, the second term averages to zero. The first term gives $N\langle (\delta b)^2 \rangle$. The incoherent scattering is the same for all $\mathbf{q}$ and is isotropic.

The present discussion is confined to the measurement of crystal structure by elastic scattering. Inelastic neutron scattering is also done to measure the dispersion relation of excitations such as phonons and spin waves.

The above analysis assumed that all nuclei are in their ideal crystalline positions. That is true at very low temperature. At higher temperatures, the atoms vibrate due to thermal motion. Neutron scattering has a temperature-dependent intensity called the *Debye-Waller factor*. It is derived in a later chapter on phonons. It is a function of $F(\mathbf{q}^2)$, so the scattering is no longer isotropic.

2.7 Structure Factor

Equation (2.34) states that any periodic function of position in a crystal can be expanded using Fourier transforms, and the only nonzero wave vector components are the reciprocal lattice vectors $\mathbf{G}_\alpha$. This theorem applies to the density of electron charge:

$$n(\mathbf{r}) = \sum_{\mathbf{G}} n(\mathbf{G}) e^{i\mathbf{G}\cdot\mathbf{r}} \tag{2.51}$$

where $n(\mathbf{G})$ is the Fourier transform of the charge density in the unit cell

$$n(\mathbf{G}) = \frac{1}{\Omega_0}\int_{\Omega_0} d^3r e^{-i\mathbf{G}\cdot\mathbf{r}} n(\mathbf{r}) \tag{2.52}$$

Suppose that the unit cell has N_c atoms at positions $\boldsymbol{\tau}_j$, and $n_j(\mathbf{G})$ is the Fourier transform of the atomic charge at site $\mathbf{R}_j$. Then

$$n(\mathbf{G}) = \sum_{j=1}^{N_c} e^{i\mathbf{G}\cdot\boldsymbol{\tau}_j} n_j(\mathbf{G}) \tag{2.53}$$

If the unit cell has N_ℓ identical atoms, they will all have the same value for $n_j(\mathbf{G})$. For this subset of atoms, the above summation is only over the phase factors. This latter summation is called the *structure factor*:

$$S(\mathbf{G}) = \sum_{j=1}^{N_\ell} e^{i\mathbf{G}\cdot\boldsymbol{\tau}_j} \tag{2.54}$$

As an example, consider the diamond lattice, which is found for elements such as carbon or silicon. It has two $(N_c = 2 = N_\ell)$ identical atoms in the unit cell, which are at positions

$$\boldsymbol{\tau}_1 = (0,0,0),\ \boldsymbol{\tau}_2 = \frac{a}{4}(1,1,1) \tag{2.55}$$

The lattice symmetry is fcc. Table 2.7 shows the value of $S(\mathbf{G})$ for the first fifty-nine reciprocal lattice vectors. The third column shows different phase factors that are found for different choices of $\mathbf{G}$ that have the same $|\mathbf{G}|$. They all have the same value of $\exp(i\mathbf{G}\cdot\boldsymbol{\tau}_2)$, within a sign. Some of the $S(\mathbf{G})$ are two, which has the phase of the two atoms being multiples of

Table 2.7 Structure factor of the diamond structure

$a\mathbf{G}/(2\pi)$	#	$\mathbf{G}\cdot\boldsymbol{\tau}_2$	$S(\mathbf{G})$
(000)	1	0	2
(111)	8	$\pm\frac{3\pi}{2}, \pm\frac{\pi}{2}$	$1 \pm i$
(200)	6	$\pm\pi$	0
(220)	12	$0, \pm 2\pi$	2
(311)	24	$\pm\frac{\pi}{2}, \pm\frac{3\pi}{2}, \pm\frac{5\pi}{2}$	$1 \pm i$
(222)	8	$\pm\pi, \pm 3\pi$	0

2π, while some others are zero, which means the two terms cancel. If $S(\mathbf{G}) = 0$, there is no Bragg scattering of x-rays at that value of reciprocal lattice vector.

2.8 EXAFS

EXAFS is an acronym for *extended x-ray absorption fine structure*. X-rays are high-frequency photons. They can be absorbed in a solid by the excitation of an electron from an atomic core state to a conduction band state. Such measurements can also be done for a free atom, which provides data on the cross section for x-ray absorption. An interesting question is the degree that the cross section is changed or affected by the crystalline environment. The experimental answer is that, in x-ray absorption in a crystal, the curve of absorption coefficient $\alpha(\omega)$ shows small oscillations as a function of ω. These oscillations are shown in fig. 2.9a, where the smooth part of the x-ray absorption has been subtracted.

EXAFS is due to a double-scattering event. The electron, after absorbing the photon, becomes a free particle and leaves the atom as an outgoing spherical wave. This electron wave encounters the neighboring atoms in the solid and scatters from them. Some of this scattered wave comes back to the original atom, and its wave function interferes with the original one. This interference gives rise to the oscillatory absorption spectra. Each outgoing wave has the form

$$G \propto \frac{e^{ikr}}{r} \tag{2.56}$$

The double scattering produces an eigenfunction:

$$\psi_f = \frac{e^{ikr}}{r}\left[1 + A_0 \frac{e^{2ikr_j + 2i\delta(k)}}{r_j^2}\right] \tag{2.57}$$

$$|\psi_f|^2 = \frac{1}{r^2}\left\{1 + \frac{2A_0}{r_j^2}\cos[2kr_j + 2\delta(k)] + O(1/r_j^4)\right\} \tag{2.58}$$

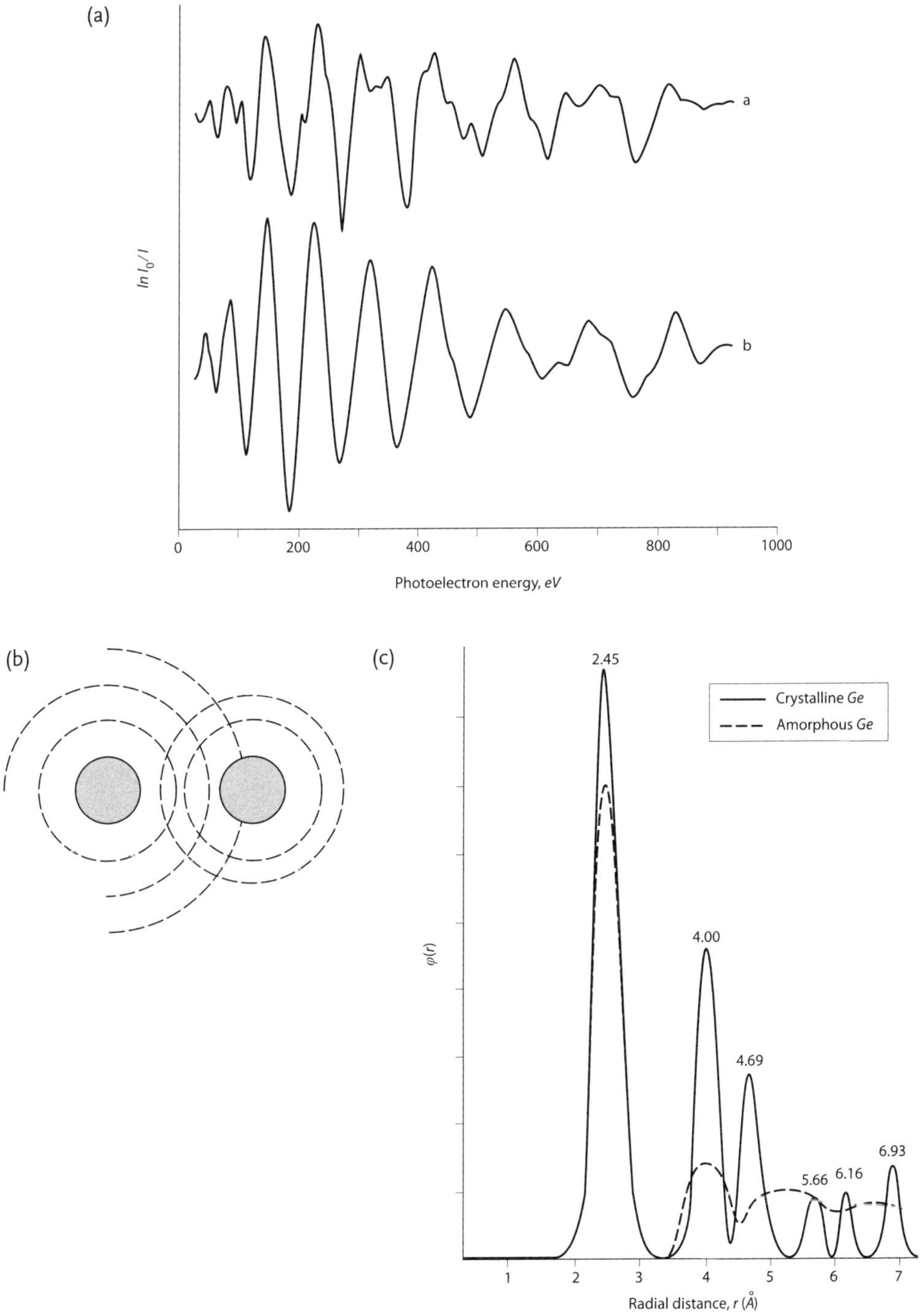

FIGURE 2.9. (a) EXAFS spectra of crystalline (spectrum a) and amorphous (spectrum b) germanium. From Sayers, Stern, and Lytle, *Phys. Rev. Lett.* **27**, 1204 (1971). Used with permission of the American Physical Society. (b) Real space drawing of an outgoing electron wave scattering from a neighboring atom, and scattering back to the original atom. (c) Nearest neighbors in real space.

where r_j is the distance to the first neighbor, and $\delta(k)$ is the phase shift of the electron scattering from that neighbor. The oscillatory term $\cos[2kr_j + 2\delta(k)]$ gives the EXAFS oscillations. This term can be Fourier transformed to determine the nearest-neighbor distance r_j. It is a very useful measurement method in glasses or disordered solids, in which neighboring distances may not be known. In fig. 2.9c, it is found that the first two neighbors in amorphous Ge are at the same bonding distance as in the crystal.

2.9 Optical Lattices

Atomic physics has used lasers to create a new kind of artificial lattice of atoms. They use a potential generated by lasers to make periodic potential minima in space. Atoms can be put in these potentials, which creates a dilute solid. The lattice constants are related to the wavelength of light.

A neutral atom has a polarizability α. The application of an external electric field $\mathbf{E}(t)$ induces a dipole on the atom of $\mathbf{p} = 4\pi\varepsilon_0\alpha\mathbf{E}$. The energy of the atom is lowered by the amount

$$E_G = -2\pi\varepsilon_0\alpha\mathbf{E}^2 \tag{2.59}$$

Let the electric field be a standing wave from a laser, with a Gaussian cross section $[\mathbf{r} = (\boldsymbol{\rho}, z)]$:

$$\mathbf{E}(\mathbf{r}, t) = E_0\hat{x}\cos(k_z z)\cos(\omega t)e^{-\alpha^2\rho^2/2} \tag{2.60}$$

$$E_G = -2\pi\varepsilon_0\alpha E_0^2\cos^2(k_z z)e^{-\alpha^2\rho^2}\cos^2(\omega t) \tag{2.61}$$

The factor $\cos^2(k_z z)$ makes periodic minima, which serve as potential mimima for atoms. The laser intensity is proportional to E_0^2. A single beam makes only a one-dimensional lattice. Two- and three-dimensional lattices can be made by crossing two or three laser beams. The Gaussian factor serves to confine the atoms in the optical lattice to a small region of space.

The polarizability α is a function of frequency:

$$\alpha(\omega) = \frac{e^2}{4\pi\varepsilon_0 m}\sum_n \frac{f_n}{\omega_n^2 - \omega^2} \tag{2.62}$$

where f_n is the oscillator strength, and ω_n is the frequency of an electronic transition in the atom. By tuning the laser frequency $\omega \approx \omega_n$, then $\alpha(\omega)$ is near resonance and becomes very large. One can make the polarizability negative if $\omega > \omega_n$.

References

1. M. Gardner, Extraordinary nonperiodic tiling that enriches the theory of tiles. *Sci. Am.* **236**, 110–21 (1977)

2. H. Metcalf and P. van der Stratten, *Laser Cooling and Trapping* (Springer, New York, 1999)
3. R. Penrose, Set of tiles covering a surface. U.S. Patent 4,133,152, issued on January 9, 1979
4. D. E. Sayers, E. A. Stern, and F. W. Lytle, New technique for investigating noncrystalline structures: Fourier analysis of the extended x-ray absorption fine structure. *Phys. Rev. Lett.* **27**, 1204–1207 (1971)
5. G. Shirane, S. M. Shapiro, and J. M. Tranquada, *Neutron Scattering with a Triple-axis Spectrometer* (Cambridge University Press, Cambridge, UK, 2002)
6. B. K. Teo, *EXAFS: Basic Principles and Data Analysis* (Springer-Verlag, Berlin, 1986)
7. R.W.G. Wyckoff, *Crystal Structures*, 2nd Ed. (Krieger, Malabar, FL, 1982)

Homework

1. What is the angle, in degrees, between two first neighbors in a tetrahedral bond?

2. Packing fraction is the fraction of area that pennies would cover in two-dimensional arrays. For the sq lattice, it is $f = [\pi a^2]/(2a)^2 = \pi/4$.

(a) What is the packing fraction of the pt lattice in two dimensions?

(b) In three dimensions, the packing fraction of the sc lattice is $f = (4\pi a^3/3)/(2a)^3 = \pi/6$. What are the packing fractions of the bcc and fcc lattices?

3. In the simple cubic lattice, each atom has six first neighbors, which are a away. There are twelve second neighbors at a distance of $a_2 = a\sqrt{2}$ away. Table 2.8 can be constructed. Construct the same table for the fcc and bcc lattices, up to fourth neighbors. The parameter a is always the cube edge.

Table 2.8 Neighbors of the sc lattice

neighbor	#	$\mathbf{a}/a$	$\lvert\mathbf{a}/a\rvert$
1	6	(100)	1
2	12	(110)	$\sqrt{2}$
3	8	(111)	$\sqrt{3}$
4	6	(200)	2

4. Consider in two dimensions a sq lattice of constant a. The atoms at each site have a magnetic moment. At one temperature the moments are all parallel, in a ferromagnetic array. At another temperature, alternate sites have up and down spin, in an antiferromagnetic array.

(a) What are the areas A_0 and A_{BZ} in the ferromagnetic array?

(b) What are the areas A_0 and A_{BZ} in the antiferromagetic array? Here the lattice vectors must go between sites with the same spin orientation.

5. Consider a pt lattice in two dimensions, where a is the distance between atoms.

(a) What are the lattice vectors? What is the area of a unit cell?

(b) What are the reciprocal lattice vectors? Draw the Brillouin zone.

(c) Draw the first three BZs for the two-dimensional pt lattice. Use a sheet of graph paper to make a nice drawing.

6. The two tiles that make up Penrose tiling can separately be made into a simple lattice. One is shown in fig. 2.10. What are the reciprocal lattice vectors and Brillouin zone of this lattice?

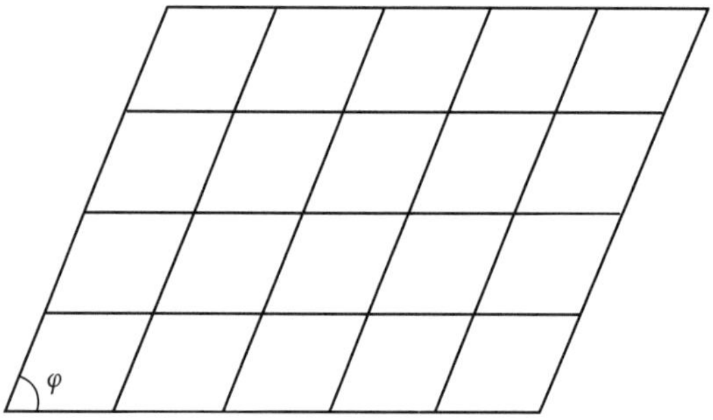

FIGURE 2.10. Two-dimensional lattice using one of the two Penrose tiles.

7. For elastic scattering of x-rays by reciprocal lattice vectors, show that the maximum allowed value of $|\mathbf{G}|$ depends on the photon wave vector k.

8. Graphene has two carbon atoms per unit cell, arranged in the honeycomb lattice. What is the structure factor for (i) $\mathbf{G} = 0$, (ii) the lowest set of nonzero reciprocal lattice vectors, and (iii) the second lowest set of reciprocal lattice vectors? All vectors with the same value of G^2 are in the same set.

9. In some high-temperature superconductors, the CuO_2 planes are arranged with lattice vectors between the Cu atoms of $\mathbf{a}_1 = \pm a\hat{x}$, $\mathbf{a}_2 = \pm a\hat{y}$, $\mathbf{a}_3 = (a/2)(\pm\hat{x} \pm \hat{y}) \pm c\hat{z}$. Find the volume V_0 of the unit cell and the G's. Draw the BZ for $c = 2a$.

3 Energy Bands

The Hamiltonian of a system of electrons in a crystal is

$$H = H_0 + V_{ei} + V_{ee} + V_{ep} \tag{3.1}$$

$$H_0 = \sum_j \frac{p_j^2}{2m} \tag{3.2}$$

The kinetic energy term is H_0. The second term V_{ei} is the electron–ion interaction, and is discussed in this chapter. The electron–electron interaction V_{ee} and the electron–phonon interaction V_{ep} are discussed in later chapters. If electron–electron interactions are ignored, then each electron proceeds through the crystal independently of the other electrons. The total Hamiltonian can be expressed as the summation of the Hamiltonians of the individual electrons:

$$H = \sum_j^N H_j \tag{3.3}$$

$$H_j = \frac{p_j^2}{2m} + U(r_j) \tag{3.4}$$

$$U(\mathbf{r}) = \sum_\ell v(\mathbf{r} - \mathbf{R}_\ell), \quad U(\mathbf{r} + \mathbf{R}_i) = U(\mathbf{r}) \tag{3.5}$$

The electron–ion interaction $U(\mathbf{r})$ is a summation over $v(\mathbf{r} - \mathbf{R}_\ell)$, which is the potential energy between an electron at $\mathbf{r}$ and an ion at $\mathbf{R}_\ell$.

3.1 Bloch's Theorem

Felix Bloch proved an important theorem in solid state physics. If an electron moves in a periodic potential, where $U(\mathbf{r} + \mathbf{R}_\ell) = U(\mathbf{r})$, then the eigenfunctions have the form

$$\psi(\mathbf{k}, \mathbf{r}) = \frac{e^{i\mathbf{k}\cdot\mathbf{r}}}{\sqrt{\Omega}} u(\mathbf{k}, \mathbf{r}) \tag{3.6}$$

$$u(\mathbf{k}, \mathbf{r} + \mathbf{R}_\ell) = u(\mathbf{k}, \mathbf{r}) \tag{3.7}$$

Here $\mathbf{R}_\ell$ is any lattice vector. The eigenfunction $\psi(\mathbf{k}, \mathbf{r})$ has a prefactor that resembles a plane wave, and then another factor $u(\mathbf{k}, \mathbf{r})$, which is called the *cell-periodic term*. The cell-periodic term is a periodic function of position in the lattice. This form of the eigenfunction applies to all crystals and all energy bands, whether nearly free electron bands in metals or valence bands in diamond.

3.3.1 Floquet's Theorem

Bloch's theorem is an application of a result known in mathematics as Floquet's theorem (1883). It states, in dimensionless form, that the equation

$$0 = \left[\frac{d^2}{dx^2} + v(x) + \varepsilon\right]\psi(x) \tag{3.8}$$

with a periodic potential $v(x + a) = v(x)$ has a general solution that obeys

$$\psi(x + a) = e^{ika}\psi(x) \tag{3.9}$$

where k is related to the eigenvalue ε. The general method of solving this equation can be applied to energy-band structures in one dimension. Define two solutions $\phi_1(x)$, $\phi_2(x)$, which have the following features:

- $\phi_1(x)$ is a solution to eqn. (3.8) with initial conditions $\phi_1(0) = 1, \phi_1(0)' = 0$, where the prime denotes a derivative.
- $\phi_2(x)$ is a solution to eqn. (3.8) with initial conditions $\phi_2(0) = 0, \phi_2(0)' = 1$.

The most general solution to eqn. (3.8) is

$$\psi(x) = C_1\phi_1(x) + C_2\phi_2(x) \tag{3.10}$$

where (C_1, C_2) are constants. Apply this solution to eqn. (3.9):

$$C_1\phi(x + a) + C_2\phi_2(x + a) = e^{ika}[C_1\phi_1(x) + C_2\phi_2(x)] \tag{3.11}$$

- At $x = 0$ eqn. (3.11) is

$$C_1\phi(a) + C_2\phi_2(a) = C_1e^{ika} \tag{3.12}$$

- Take the derivative of eqn. (3.11) and then set $x = 0$

$$C_1\phi(a)' + C_2\phi_2(a) = C_2e^{ika} \tag{3.13}$$

Equations (3.12) and (3.13) can be written in matrix form, with $\lambda = \exp(ika)$:

$$0 = \begin{pmatrix} \phi_1(a) - \lambda & \phi_2(a) \\ \phi_1'(a) & \phi_2'(a) - \lambda \end{pmatrix}\begin{pmatrix} C_1 \\ C_2 \end{pmatrix} \tag{3.14}$$

The solution is found by setting to zero the determinant of the matrix. The quantity λ is recognized as the eigenvalue. If it is a complex number of absolute magnitude of one, then k is real. If λ is a real number, but not one, then k is imaginary. Imaginary values of k, in energy band theory, occur at energy gaps.

As an example, consider in one dimension the band structure of a particle moving in a periodic array of repulsive delta function potentials at half-integer values of a:

$$v(x) = A\sum_n \delta(x - na - a/2) \tag{3.15}$$

In between the delta functions, the particles have plane wave properties with a wave vector given by $p = \sqrt{\varepsilon}$, assuming that ε is positive. To solve for the two functions, set $\phi_j(x)$ at $x = 0$ to have the correct initial conditions. Set up a general solution in the region to the right of the first delta function, and then match them at $x = a/2$.

- At $x = 0$ set $\phi_1(x) = \cos(px)$, which has the correct initial conditions. In the next unit cell, the eigenfunction must be a linear combination of a wave going to the right, and one going to the left:

$$\phi_1(x) = \begin{cases} \cos(px) & -a/2 < x < a/2 \\ Ie^{ipx} + Re^{-ipx} & a/2 < x < 3a/2 \end{cases} \tag{3.16}$$

Match $\phi_1(x)$ and its derivative at $x = a/2$, which gives two equations for the coefficients (I,R). The matching of derivatives involves the delta function:

$$\cos(\theta) = Iz + R/z, \ \theta = pa/2, \ z = e^{i\theta} \tag{3.17}$$

$$A\cos(\theta) = -p\sin(\theta) - ip[Iz - R/z] \tag{3.18}$$

Solving gives

$$I = \frac{1}{2}e^{-i\theta}\left[e^{i\theta} + i\frac{A}{p}\cos(\theta)\right] \tag{3.19}$$

$$R = \frac{1}{2}e^{i\theta}\left[e^{-i\theta} - i\frac{A}{p}\cos(\theta)\right] \tag{3.20}$$

$$\phi_1(a) = \cos(2\theta) - \frac{A}{2p}\sin(2\theta) \tag{3.21}$$

$$\phi_1'(a) = -p\sin(2\theta) - A\cos^2(\theta) \tag{3.22}$$

- The second function $\phi_2(x)$ is chosen to give a derivative of one at the origin:

$$\phi_2(x) = \begin{cases} \sin(px)/p & -a/2 < x < a/2 \\ [I'e^{ipx} + R'e^{-ipx}]/p & a/2 < x < 3a/2 \end{cases} \tag{3.23}$$

The matching of wave functions and derivatives at $x = a/2$ gives

$$\sin(\theta) = I'z + R'/z, \tag{3.24}$$

$$A\sin(\theta) = p\cos(\theta) - ip[I'z - R'/z] \tag{3.25}$$

Solving gives

$$I' = \frac{-i}{2}e^{-i\theta}\left[e^{i\theta} - \frac{A}{p}\sin(\theta)\right] \tag{3.26}$$

$$R' = \frac{i}{2}e^{i\theta}\left[e^{-i\theta} - \frac{A}{p}\sin(\theta)\right] \tag{3.27}$$

$$\phi_2(a) = \frac{1}{p}\left[\sin(2\theta) - \frac{A}{p}\sin^2(\theta)\right] \tag{3.28}$$

$$\phi_2'(a) = \cos(2\theta) - \frac{A}{2p}\sin(2\theta) = \phi_1(a) \tag{3.29}$$

Note that $\phi_2'(a) = \phi_1(a)$. A theorem states this identity is always obeyed, for any v(x). The eigenvalue of the matrix is

$$\lambda = \phi_1(a) \pm \sqrt{\phi_1'(a)\phi_2(a)} \tag{3.30}$$

The function in the square root can have either sign:

- If $\phi_1'(a)\phi_2(a) > 0$, the eigenvalues are real, k is imaginary, and the energy band has a gap.
- If $\phi_1'(a)\phi_2(a) < 0$, the eigenvalues are complex and k is usually real:

$$\lambda_{1,2} = \phi_1(a) \pm i\sqrt{-\phi_1'(a)\phi_2(a)} \tag{3.31}$$

In the present case,

$$\phi_1'(a)\phi_2(a) = -\left[\sin^2(pa)\left(1 - \frac{A^2}{4p^2}\right) + \frac{A}{2p}\sin(2pa)\right] \tag{3.32}$$

In the region where there are energy bands, then for j=(1,2)

$$|\lambda_j|^2 = \phi_1(a)^2 - \phi_1'(a)\phi_2(a) = 1 \tag{3.33}$$

The eigenvalues λ_j are pure phase factors. In eqn. (3.31) the two terms on the right must be $\cos(ka) \pm i\sin(ka)$, which gives the identity

$$\cos(ka) = \phi_1(a) = \cos(pa) - \frac{A}{2p}\sin(pa) \tag{3.34}$$

This equation determines the energy bands. For values of p where the right-hand side has a magnitude less than one, then k is real and energy bands are allowed. Band gaps occur where the right-hand side has a magnitude larger than one.

Turn eqn. (3.8) into a Schrödinger equation:

$$0 = \left[-\frac{\hbar^2}{2m}\frac{d^2}{dx^2} - E + V(x)\right]\phi(x) \tag{3.35}$$

$$V(x) = V_0\sum_n \delta(x - an - a/2) \tag{3.36}$$

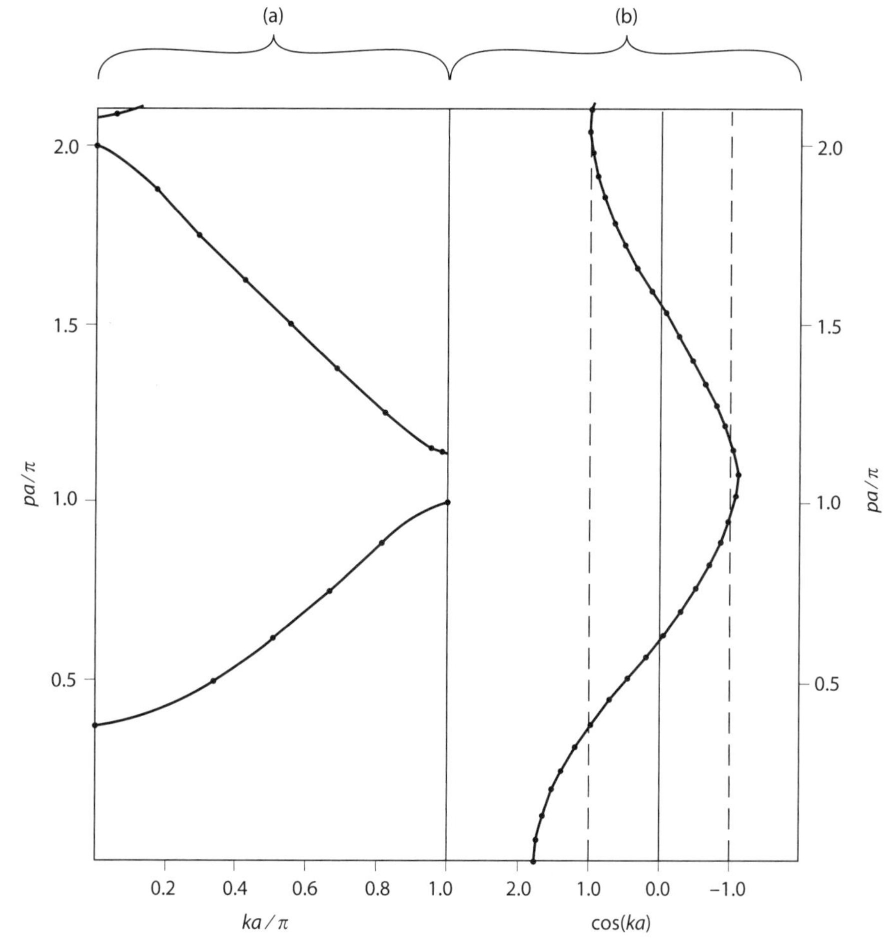

FIGURE 3.1. (a) Graph of the energy bands: pa/π as a function of wave vector ka/π. (b) Plot of the right-hand side of eqn. (3.38) vs. pa.

$$\varepsilon = \frac{2mE}{\hbar^2} = p^2, \; A = -\frac{2mV_0}{\hbar^2} \tag{3.37}$$

Since A has the units of wave vector, let $A = -2p_0$:

$$\cos(ka) = \cos(pa) + \frac{p_0}{p}\sin(pa) \tag{3.38}$$

Calculate the right-hand side as a function of pa.

A numerical example is shown in fig. 3.1. The curve was drawn using $p_0 a = \pi/4$. Part (b) shows the right-hand side of eqn. (3.38) as a function of pa. Vertical dotted lines denote $\pm$ 1. When this function is in the range $-1 < \cos(ka) < 1$ then one can determine ka and graph energy bands, When this function has a magnitude greater than one, there are band gaps. Figure 3.1a shows the result, graphed as a band structure, so the vertical axis is

pa/π and the horizontal axis is ka/π. There are obvious energy gaps where no propagating modes are allowed. The gaps get smaller with increasing values of pa.

3.2 Nearly Free Electron Bands

Free electron energy bands are usually called *nearly free electron bands*. The bands are free, with kinetic energy $\propto k^2$, except near symmetry points. The deviations from free motion are due to the potential energy of the ions.

3.2.1 *Periodic Potentials*

The potential energy $U(\mathbf{r})$ of the electron can be expanded in a Fourier series. This potential is periodic,

$$U(\mathbf{r} + \mathbf{R}_{\ell,m,n}) = U(\mathbf{r}) \tag{3.39}$$

The only wave vectors in the Fourier series are the reciprocal lattice vectors. One can always write

$$U(\mathbf{r}) = \sum_{\alpha,\beta,\gamma} u(\alpha,\beta,\gamma)\, e^{i\mathbf{G}_{\alpha,\beta,\gamma}\cdot\mathbf{r}} \tag{3.40}$$

$$u(\alpha,\beta,\gamma) = \int \frac{d^3r}{\Omega_0} U(\mathbf{r})\, e^{i\mathbf{G}_{\alpha,\beta,\gamma}\cdot\mathbf{r}} \equiv u(\mathbf{G}) \tag{3.41}$$

where the integral is over the unit cell. This theorem is quite useful in discussing crystal physics.

The physics is easily understood in one dimension. The Hamiltonian for a single electron in a periodic potential is

$$H = -\frac{\hbar^2}{2m}\frac{d^2}{dx^2} + \sum_{\alpha} u_{\alpha} e^{ixG_{\alpha}}, \quad G_{\alpha} = \frac{2\pi\alpha}{a} \tag{3.42}$$

The potential energy has been expanded in reciprocal lattice vectors. Expand the eigenfunction using the same vectors:

$$\psi(k, x) = \sum_{\alpha} C_{\alpha} e^{ix(k+G_{\alpha})} \tag{3.43}$$

The stationary solutions to Schrödinger's equation are

$$\begin{aligned} E(k)\,\psi(k,x) &= H\psi(k,x) \\ &= \sum_{\alpha} C_{\alpha} e^{ix(k+G_{\alpha})}\left[\varepsilon(k+G_{\alpha}) + \sum_{\alpha'} u_{\alpha'} e^{ixG_{\alpha'}}\right] \end{aligned} \tag{3.44}$$

$$\varepsilon(k+G) = \frac{\hbar^2(k+G)^2}{2m} \tag{3.45}$$

Multiply eqn. (3.44) by $\exp[-ix(k + G_\beta)]$ and integrate dx from 0 to a. Since the reciprocal lattice vectors are orthogonal over this unit cell,

$$E(k)\,C_\beta = \varepsilon(k + G_\beta)C_\beta + \sum_{\alpha'} u_{\alpha'} C_{\beta-\alpha'} \tag{3.46}$$

The above equation is solved to find the eigenvalues $E_n(k)$ in terms of the free electron energies $\varepsilon(k + G_n)$. It is a matrix equation, and can be written as

$$0 = \begin{bmatrix} E - \varepsilon(k) - u_0 & u_{\bar{1}} & u_1 & \cdots \\ u_1 & E - \varepsilon(k + G_1) - u_0 & u_2 & \cdots \\ u_{\bar{1}} & u_{\bar{2}} & E - \varepsilon(k + G_{\bar{1}}) - u_0 & \cdots \\ \vdots & \vdots & \vdots & \ddots \end{bmatrix} \begin{bmatrix} C_0 \\ C_1 \\ C_{\bar{1}} \\ \vdots \end{bmatrix} \tag{3.47}$$

For the potential energy to be real, $u_{\bar{n}} = u_n$, and the above matrix is symmetric. It is of infinite dimension, so it cannot be diagonalized exactly. However, the eigenvalues lowest in energy can be found by diagonalizing a finite matrix of large dimension on the computer. The eigenvectors determine the C_α and hence the eigenfunction.

The energy-band structure is shown in fig. 3.2. The horizontal axis is the wave vector k. It is marked at regular intervals by the points $G_\alpha = \alpha G_1$, where α has all integer values. The BZ is between $-\pi/a < k < \pi/a$. At each value of reciprocal lattice vector, draw the free electron energy bands $\varepsilon(k - G_\alpha)$ as the dashed line. These bands cross at symmetry

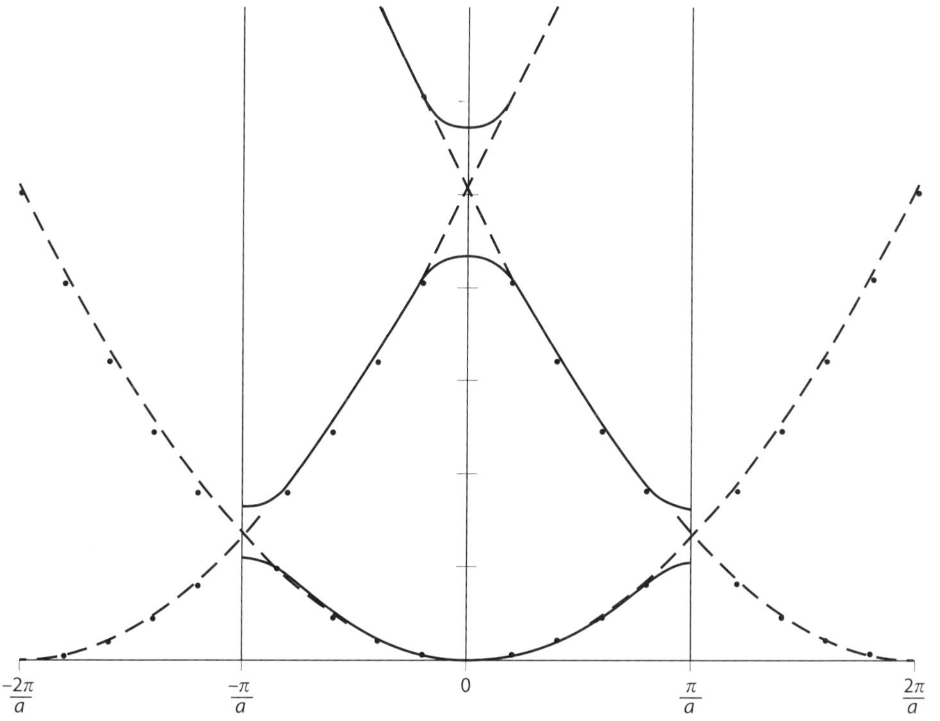

FIGURE 3.2. Nearly free electron energy bands in one dimension.

points, such as the edge of the BZ, and at $k = 0$. Where bands cross there are energy gaps due to nonzero values of u_α. The resulting energy bands are a periodic structure. It is sufficient to show these energy bands in the first BZ. Each different band is labeled $E_n(k)$, where k is in the BZ.

Where two free electron bands cross, it is often accurate to approximate the energy bands in this local region by the dispersion of the two bands that cross. For example, in fig. 3.2, the first crossing at the zone edge $k = \pi/a$, where $\varepsilon(k) \approx \varepsilon(k - G_1)$, $G_1 = 2\pi/a$, is approximated by

$$0 = \begin{bmatrix} E - \varepsilon(k) - u_0 & u_1 \\ u_1 & E - \varepsilon(k - G_1) - u_0 \end{bmatrix} \begin{bmatrix} C_0 \\ C_{\bar{1}} \end{bmatrix} \tag{3.48}$$

with the eigenvalues

$$E_\pm(k) = u_0 + \frac{1}{2}\left[\varepsilon(k) + \varepsilon(k - G_1) \pm \sqrt{[\varepsilon(k) - \varepsilon(k - G_1)]^2 + 4u_1^2}\right] \tag{3.49}$$

At precisely the zone edge, where $k = \pi/a$ and $\varepsilon(k) = \varepsilon(k - G_1)$, the eigenvalues are

$$E_\pm(\pi/a) = u_0 + \varepsilon(\pi/a) \pm u_1 \tag{3.50}$$

The energy gap at that point is $E_G = E_+ - E_- = 2|u_1|$. The Fourier coefficients u_α of the potential energy give the energy gaps at symmetry points. Nearly free electron bands describe the energy bands in many metals. They will be used in later chapters in discussing electron–electron and electron–phonon interactions.

Figure 3.3 shows the electron energy bands of metallic sodium in the bcc lattice. The bands are shown along principle symmetry directions, using the labels in fig. 2.7. Sodium has bands that show nearly free electron behavior.

3.3 Tight-binding Bands

There are other sets of basis states besides plane waves. One obvious set is the atomic orbitals of the atoms that comprise the crystal. Although each atom has an infinite number of possible orbitals, usually only the few orbitals are used that comprise the conduction and valence bands. For example, a good description of the valence bands of silicon is achieved using four atomic orbitals: the $3s$ and $3p_j$ states, where j is (x, y, z).

3.3.1 s-*State Bands*

As a simple example, consider a solid with one atom per unit cell. The tight-binding theory is developed with each atom having one s-state orbital $\phi(\mathbf{r} - \mathbf{R}_j)$ at site $\mathbf{R}_j$. Then a function with the symmetry of the unit cell is

$$\psi(\mathbf{k}, \mathbf{r}) = \sum_j e^{i\mathbf{k}\cdot\mathbf{R}_j}\phi(\mathbf{r} - \mathbf{R}_j) = e^{i\mathbf{k}\cdot\mathbf{r}}u(\mathbf{k}, \mathbf{r}) \tag{3.51}$$

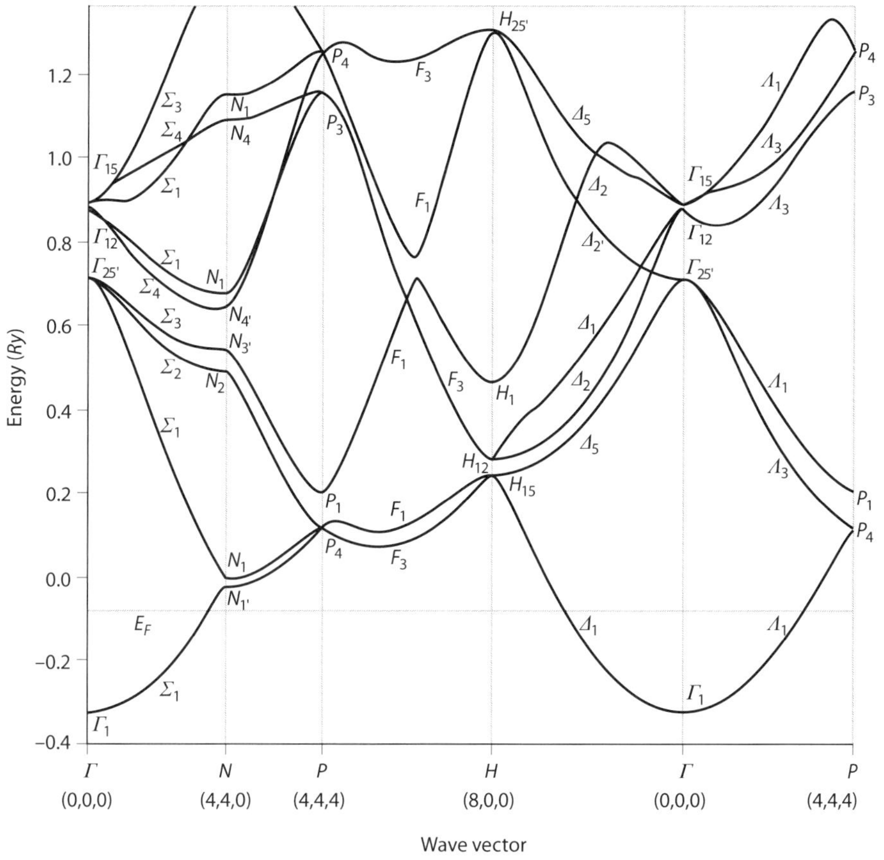

FIGURE 3.3. Electron energy bands of metallic sodium. From Ching and Callaway, *Phys. Rev. B* **11**, 1324 (1975). Used with permission of the American Physical Society.

$$u(\mathbf{k}, \mathbf{r}) = \sum_j e^{-i\mathbf{k}\cdot(\mathbf{r}-\mathbf{R}_j)} \phi(\mathbf{r} - \mathbf{R}_j) \tag{3.52}$$

the cell-periodic part if $u(\mathbf{k}, \mathbf{r})$. This function is a tight-binding eigenfunction, as shown below.

Start with a general function

$$R(\mathbf{r}) = \sum_j C_j \phi(\mathbf{r} - \mathbf{R}_j) \tag{3.53}$$

where C_j are unknown coefficients. Put this guess into Schrödinger's equation. The periodic potential $V(\mathbf{r})$ is assumed to be a summation of atomic potentials:

$$H = \frac{p^2}{2m} + \sum_j v(\mathbf{r} - \mathbf{R}_j) \tag{3.54}$$

$$\begin{aligned} ER = HR = \sum_\ell C_\ell \Bigg\{ \left[\frac{p^2}{2m} + v(\mathbf{r} - \mathbf{R}_\ell) \right] \\ + \sum_{j \neq \ell} v(r - R_j) \Bigg\} \phi(\mathbf{r} - \mathbf{R}_\ell) \end{aligned} \tag{3.55}$$

The potential energy $v(\mathbf{r} - \mathbf{R}_j)$ is between an electron at $\mathbf{r}$ and an atom located at $\mathbf{R}_j$. The atomic problem gives an eigenvalue:

$$\left[\frac{p^2}{2m} + v(\mathbf{r} - \mathbf{R}_\ell)\right]\phi(\mathbf{r} - \mathbf{R}_\ell) = E_s\phi(\mathbf{r} - \mathbf{R}_\ell) \tag{3.56}$$

Multiply eqn. (3.55) by $\phi(\mathbf{r} - \mathbf{R}_n)^*$ and integrate over all space. Assume the orbitals on different sites are orthogonal:

$$\int d^3r\phi^*(\mathbf{r})\phi(\mathbf{r} - \mathbf{R}_j) = \delta_{\mathbf{R}_j=0} \tag{3.57}$$

Atomic orbitals, when centered at different sites, are not naturally orthogonal.

They can be made orthogonal by a simple procedure. For example, if the overlap between first neighbors is large, but the overlap between distant neighbors is negligible, define

$$\tilde{\phi}(\mathbf{r} - \mathbf{R}_j) = \phi(\mathbf{r} - \mathbf{R}_j) - B\sum_{\delta=1}^{z}\phi(\mathbf{r} - \mathbf{R}_j - \boldsymbol{\delta}) \tag{3.58}$$

where $\boldsymbol{\delta}$ is the distance to a first neighbor. Taking the overlap integral between neighbors,

$$\begin{aligned} 0 &= \int d^3r\tilde{\phi}^*(\mathbf{r})\tilde{\phi}(\mathbf{r} - \boldsymbol{\delta}) \approx \int d^3r[\phi(\mathbf{r}) - B\phi(\mathbf{r} - \boldsymbol{\delta})][\phi(\mathbf{r} - \boldsymbol{\delta}) - B\phi(\mathbf{r})] \\ &= S(1 + B^2) - 2B \end{aligned} \tag{3.59}$$

$$S = \int d^3r\phi^*(\mathbf{r})\phi(\mathbf{r} - \boldsymbol{\delta}) \tag{3.60}$$

The middle equation can be solved for B, which makes $\tilde{\phi}(\mathbf{r})$ orthogonal to its neighbors.

Assuming orthogonality, we get from eqn. (3.55) after integrating $\int d^3r\phi(\mathbf{r} - \mathbf{R}_n)^*$

$$EC_n = E_sC_n + \sum_{\ell,j\neq\ell} C_\ell\langle n|v_j|\ell\rangle \tag{3.61}$$

$$\langle n|v_j|\ell\rangle = \int d^3r\phi(\mathbf{r} - \mathbf{R}_n)^*v(\mathbf{r} - \mathbf{R}_j)\phi(\mathbf{r} - \mathbf{R}_\ell) \tag{3.62}$$

The matrix element $\langle n|v_j|\ell\rangle$ has many possible values for the three indices (n, j, ℓ). The largest matrix elements are when pairs of indices are alike.

- One choice is $\ell = n \neq j$, which gives a term

  $$\delta V = \sum_{j\neq n}\langle n|v_j|n\rangle \tag{3.63}$$

  It is a Hartree term, where the electron on one orbital interacts with the potential energy from all of the other atomic sites.

- Another choice is to have $j = n$ and ℓ be a nearest neighbor of n. Remember that $\ell \neq j$, since the term $\ell = j$ is in E_s. Define $w = \langle n|v_n|n+\delta\rangle$.

Collecting all of these terms gives

$$C_n(E - E_s - \delta V) = w\sum_{\delta=1}^{z} C_{n+\delta} \tag{3.64}$$

This equation can be solved using Fourier transforms. Define a collective state

$$C(k) = \frac{1}{\sqrt{N}} \sum_n C_n e^{ik \cdot R_n} \tag{3.65}$$

Multiply eqn. (3.64) by $\exp[i\mathbf{k} \cdot \mathbf{R}_n]$ and sum over all $\mathbf{R}_n$:

$$0 = C(\mathbf{k})[E - \varepsilon(\mathbf{k})] \tag{3.66}$$

$$\varepsilon(\mathbf{k}) = E_s + \delta V + w\gamma(\mathbf{k}), \ \gamma(\mathbf{k}) = \sum_{\delta=1}^{z} e^{i\mathbf{k} \cdot \delta} \tag{3.67}$$

where z is the number of nearest neighbors. The eigenvalue for this collective state is $\varepsilon(\mathbf{k})$. The eigenfunction is the result presented earlier in eqn. (3.51). The function $\gamma(\mathbf{k})$ always appears as the dispersion in s-state tight-binding models where the interaction is limited to nearest neighbors.

3.3.2 p-*State Bands*

Many atomic orbitals are not *s*-states. An example using *p*-states is a two-dimensional square lattice of an imaginary material with the elements DO_2. Here O is oxygen and D is an cation of valence four to get charge neutrality (e.g, Si, Ge, Pb). As shown in fig. 3.4, the D atoms are arranged into a square lattice of lattice constant a, while the oxygen atoms are located midway between them in the bond positions. This example is inspired by the cuprate materials that are high-temperature superconductors, where the D atoms are copper. Copper has bonding by *d*-orbitals. The present example is simplified by having the D atoms contain a single *s*-orbital. Each O atom has three orbitals: p_x, p_y, p_z. There are a total of seven orbitals in each unit cell. The seven energy bands are derived from these seven orbitals.

The *s*-state orbital of the D atom is called $\phi_d(\mathbf{r} - \mathbf{R}_j)$ and has an eigenvalue E_s. The oxygen atom along the *x*-axis has the *p*-orbitals $\phi_\mu(\mathbf{r} - \mathbf{R}_j - a\hat{x}/2)$, where $\mu = (x, y, z)$, and the one along the *y*-axis is $\phi_\mu(\mathbf{r} - \mathbf{R}_j - a\hat{y}/2)$. Their eigenvalue is E_p.

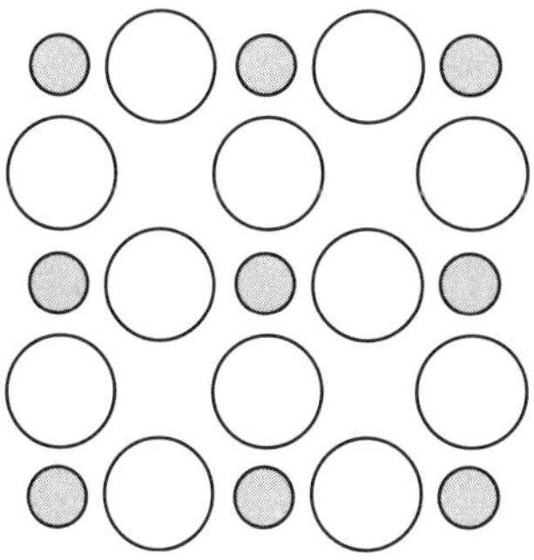

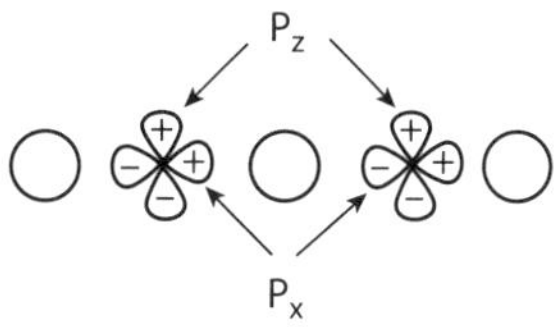

FIGURE 3.4. (a) Two-dimensional lattice structure of DO_2. (b) p_x orbital points toward the D atom, with plus orbital in one direction and minus in the other. p_z orbitals are perpendicular to the plane.

The two states $\phi_z(\mathbf{r} - \mathbf{R}_j - a\hat{x}/2)$, $\phi_z(\mathbf{r} - \mathbf{R}_j - a\hat{y}/2)$, which are perpendicular to the plane of the crystal, have nonzero matrix elements only with themselves and with each other. The p-orbitals have a form such as $\phi_\mu(\mathbf{r}) \propto zf(r)$. They are odd with respect to a reflection about the (xy)-plane. Since the other orbitals are even with respect to this reflection, parity dictates that $\langle n|j|\ell\rangle = 0$ if only one of (n, ℓ) is z. If one sits at a p-site, the tight-binding summation to the four nearest p-state neighbors is

$$\sum_{\delta=1}^{4} e^{i\mathbf{k}\cdot\delta} = (e^{i\theta_x} + e^{-i\theta_x})(e^{i\theta_y} + e^{-i\theta_y}) = 4\cos(\theta_x)\cos(\theta_y) \tag{3.68}$$

$$\theta_\mu = k_\mu a/2 \tag{3.69}$$

These two eigenstates give an effective Hamiltonian of

$$H = \begin{bmatrix} E_p & 4w_2\cos(\theta_x)\cos(\theta_y) \\ 4w_2\cos(\theta_x)\cos(\theta_y) & E_p \end{bmatrix} \tag{3.70}$$

$$E = E_p \pm 4w_2\cos(\theta_x)\cos(\theta_y) \tag{3.71}$$

The eigenvalue is the last line. The overlap constant w_2 is usually small since it is a second-neighbor interaction. The first neighbor of each oxygen is a D atom.

Consider the two orbitals on the oxygen along the x-axis. One (ϕ_x) points directly at the D atom, while the other (ϕ_y) is perpendicular to the D atom. The one that is perpendicular to the D atom also has a zero matrix element with ϕ_s, which again vanishes by parity arguments. Here the parity is along the y-axis. First-order matrix elements exist only between three orbitals: $\phi_d(\mathbf{r} - \mathbf{R}_j)$, $\phi_x(\mathbf{r} - \mathbf{R}_j - a\hat{x}/2)$, and $\phi_y(\mathbf{r} - \mathbf{R}_j - a\hat{y}/2)$. Figure 3.4b shows the p_x-orbital pointing along the x-axis. It has a plus lobe in one direction and a minus lobe in the other. The matrix element is plus in one direction, and minus in the other. The tight-binding result is

$$w_1\left[e^{i\theta_x} - e^{-i\theta_x}\right] = 2iw_1\sin(\theta_x) \tag{3.72}$$

The matrix element changes sign, due to the sign change in the p-orbitals. The effective Hamiltonian for these three atoms, considering only nearest-neighbor tight-binding terms, is

$$H = \begin{bmatrix} E_s & 2iw_1\sin(\theta_x) & 2iw_1\sin(\theta_y) \\ -2iw_1\sin(\theta_x) & E_p & 0 \\ -2iw_1\sin(\theta_y) & 0 & E_p \end{bmatrix} \tag{3.73}$$

$$E = E_p, \frac{1}{2}\left[E_s + E_p \pm \sqrt{(E_s - E_p)^2 + 16w_1^2[\sin^2(\theta_x) + \sin^2(\theta_y)]}\right] \tag{3.74}$$

One eigenvalue has no dispersion, while the other two have dispersion due to the overlap of the s- and p-orbitals on neighboring atoms.

The final two eigenvalues are obtained from the overlap of $\phi_y(\mathbf{r} - \mathbf{R}_j - a\hat{x}/2)$, $\phi_x(\mathbf{r} - \mathbf{R}_j - a\hat{y}/2)$. They have zero overlaps with the D atoms, but nonzero second-order overlaps

w'_2 with each other. Again we must keep track of the plus and minus overlaps of the orbitals, which gives

$$E = E_p \pm 4w'_2 \sin(\theta_x)\sin(\theta_y) \tag{3.75}$$

This completes the discussion of the seven energy bands due to the overlap of the seven orbitals that are most important in a tight-binding model of this imaginary material.

3.3.3 Wannier Functions

In covalent crystals, the orbitals are shared between atoms. *Wannier functions* are a set of orbitals in a crystal that are usually localized. They are obtained by first calculating all of the energy bands $E_n(\mathbf{k})$ and eigenfunctions $\psi_n(\mathbf{k}, \mathbf{r})$ in the crystal. The latter have the usual Bloch form:

$$\psi_n(\mathbf{k}, \mathbf{r}) = \frac{e^{\mathbf{k}\cdot\mathbf{r}}}{\sqrt{\Omega}} u_n(\mathbf{k}, \mathbf{r}) \tag{3.76}$$

where the wave vectors $\mathbf{k}$ are confined to the first Brillouin zone (BZ). Wannier functions are constructed for a single energy band n and centered around a site $\mathbf{R}_j$:

$$\begin{aligned} W_n(\mathbf{r} - \mathbf{R}_j) &= \sqrt{\Omega_0}\int_{BZ} \frac{d^3k}{(2\pi)^3} \psi_n(\mathbf{k}, \mathbf{r} - \mathbf{R}_j) \\ &= \sqrt{\Omega_0}\int_{BZ} \frac{d^3k}{(2\pi)^3} e^{-i\mathbf{k}\cdot\mathbf{R}_j} \psi_n(\mathbf{k}, \mathbf{r}) \end{aligned} \tag{3.77}$$

where Ω_0 is the volume of the crystal unit cell. The eigenfunction is obtained by the inverse transform:

$$\psi_n(\mathbf{k}, \mathbf{r}) = \frac{1}{\sqrt{\Omega_0}} \sum_j e^{i\mathbf{k}\cdot\mathbf{R}_j} W_n(\mathbf{r} - \mathbf{R}_j) \tag{3.78}$$

One advantage of Wannier functions is that they are orthogonal:

$$\begin{aligned} \int d^3r W_n^*(\mathbf{r} - \mathbf{R}_j) W_m(\mathbf{r} - R_\ell) = \Omega_0 \int_{BZ} \frac{d^3k}{(2\pi)^3} \int_{BZ} \frac{d^3k'}{(2\pi)^3} e^{i(\mathbf{k}\cdot\mathbf{R}_j - \mathbf{k}'\cdot\mathbf{R}_\ell)} \\ \times \int d^3r \psi_n^*(\mathbf{k}, \mathbf{r}) \psi_m(\mathbf{k}', \mathbf{r}) \end{aligned} \tag{3.79}$$

The integral d^3r is zero unless $n = m$ and $\mathbf{k}' = \mathbf{k}$. Then the d^3k integral forces $\mathbf{R}_j = \mathbf{R}_\ell$, so we prove

$$\int d^3r W_n^*(\mathbf{r} - \mathbf{R}_j) W_m(\mathbf{r} - R_\ell) = \delta_{nm}\delta_{j\ell} \tag{3.80}$$

A disadvantage of Wannier functions is that they are not eigenstates of any Hamiltonian. Their main virtue, besides being an ortho-normal set of functions, is that they are often localized. A simple tight-binding eigenfunction

$$\psi(\mathbf{k},\mathbf{r}) = \frac{1}{\sqrt{N}}\sum_j e^{i\mathbf{k}\cdot\mathbf{R}_j}\phi(\mathbf{r}-\mathbf{R}_j) \tag{3.81}$$

gives the Wannier function for a simple cubic lattice:

$$W(\mathbf{r}) = \frac{1}{\sqrt{\Omega_0}}\sum_j j_0(\theta_{xj})j_0(\theta_{yj})j_0(\theta_{zj})\,\phi(\mathbf{r}-\mathbf{R}_j) \tag{3.82}$$

$$j_0(z) = \frac{\sin(z)}{z},\quad \theta_{\mu j} = \frac{\pi}{a}R_{\mu j} \tag{3.83}$$

The amplitude falls off in an oscillatory manner for distances away from $r = 0$.

Wannier functions work well for energy bands $E_n(\mathbf{k})$ that do not overlap in energy with other bands. When energy bands overlap, one finds that the Wannier functions are not localized. Another problem is that all eigenfunctions, including Bloch functions, may have an arbitrary phase factor $\exp[i\phi(\mathbf{k})]$. Such a phase factor does not depend on position and does not influence the eigenvalues. However, it will change the Wannier function. This feature can be used to alter the Wannier functions. The phase $\phi(\mathbf{k})$ can be adjusted to improve the localization by reducing the spatial extent.

3.4 Semiconductor Energy Bands

The simplest view of semiconductor energy bands starts with diamond. It is composed of carbon, whose atomic electron configuration is $(1s)^2(2s)^2(2p)^2$. There are four electrons in the 2*s*- and 2*p*-orbitals. Each carbon atom in the crystal has four neighbors, arranged in tetrahedral coordination. There is a bond with each first neighbor. Each carbon contributes one electron to the bond. Each bond has a total of two electrons and is filled. The bonds are the valence bands. The conduction bands are composed of other orbitals. In semiconductors, the valence electrons are in bonds with the neighboring atoms. This behavior is in contrast with insulators, where the valence electrons are attached to either the anions or the cations. In intrinsic semiconductors and insulators, the valence bands are usually full of electrons, and the conduction bands are usually empty of electrons.

3.4.1 *What Is a Semiconductor?*

Before discussing the properties of semiconductors, one should first define the class of materials that are semiconductors. Where is the boundary between semiconductors and insulators? Some elements in the fourth column of the periodic table, such as silicon and germanium, are semiconductors. Other elements in the fourth column, tin and lead, are metals. Carbon is found in many different forms: diamond, graphite, C_{60}, nanotubes, etc. Diamond is generally regarded as a semiconductor, although that definition could be challenged because of its wide band gap. Binary solids made up of one atom from the III column of the periodic table and one from the V column, such as GaAs or InP, are usually semiconductors. There are exceptions to this statement, as discussed below. There

Table 3.1 Electronegativity scale of some elements, according to Pauling

Li	Be			B	C	N	O	F
1.0	1.5			2.0	2.5	3.0	3.5	4.0
Na	Mg			Al	Si	P	S	Cl
0.9	1.2			1.5	1.8	2.1	2.5	3.0
K	Ca	Cu	Zn	Ga	Ge	As	Se	Br
0.8	1.0	1.9	1.6	1.6	1.8	2.0	2.4	2.8
Rb	Sr	Ag	Cd	In	Sn	Sb	Te	I
0.8	1.0	1.9	1.7	1.7	1.8	1.9	2.1	2.5

are also semiconductors composed of binary compounds made from elements in rows IV and VI of the periodic table, such as PbTe or GeTe.

Many new materials have been made recently, and many old materials restudied. A general classification is needed. Some define a semiconductor as any crystal with a small (say, less than 2–3 eV) energy gap between conduction and valence band. An alternate definition is any material with a high mobility of electrons and holes.

Carver Mead [11] and his associates came up with an interesting way of predicting whether a new compound is a semiconductor. They used the electronegativity scale described by Linus Pauling [13]. Pauling defines electronegativity as "the power of an atom . . . to attract electrons to itself." The electronegativity (X) is a dimensionless number. A complete table for all elements is given on page 93 of [13], and some of these values are shown in table 3.1. In a binary material, the important parameter is the amount of charge transfer between the two different atoms in the unit cell. This charge transfer is related to the difference in the electronegativity of the two elements:

$$\Delta X = X_A - X_B \tag{3.84}$$

If $|\Delta X|$ is greater than a critical value ΔX_c, the binary compound is an insulator, and if $|\Delta X| < \Delta X_c$, the binary compound is a semiconductor. The critical value is about $\Delta X_c \sim 0.8$. The alkali halides, such as NaCl or KCl, all have $|\Delta X| > 1$ and are indeed insulators. Most III–V binary compounds, such as GaAs and InP, have $|\Delta X| < 1$, and are semiconductors. An exception is AlN, which has a very wide energy gap and should be treated as an insulator. Most oxides are insulators. In general, if ΔX is large, the atoms exchange electrons and become ions, and the material becomes an insulator.

Mead and associates found that the transition between semiconductor and insulator is very abrupt when several physical properties are plotted as a function of $|\Delta X|$. Figure 3.5 is taken from their paper [11], and shows three physical properties. The first, which they call S, is the slope of a line. Whenever a metal is deposited on the clean surface of a semiconductor, usually by evaporation, a surface state is formed. In semiconductors such as silicon or GaAs, the energy of this surface state is independent of the metal.

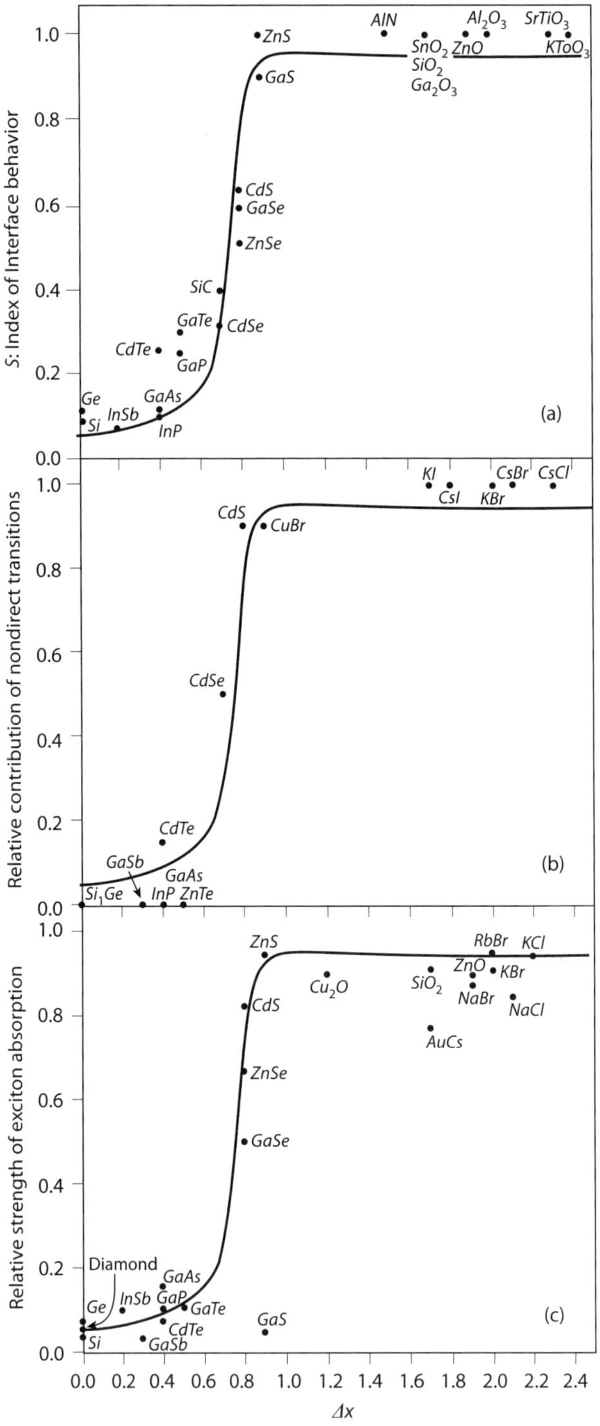

FIGURE 3.5. (a) Slope of surface-state energy vs. ΔX. (b) Relative contribution of indirect transitions. (c) Graph of exciton absorption intensity vs. ΔX. Adapted from Kurtin, McGill and Mead, *Phys. Rev. Lett.* **22**, 1433 (1969) [Used with permission of the American Physical Society]}

So a graph of surface-state energy vs. the electronegativity of the metal has a zero (or small) slope S. This is indicated by the points on the left of this figure. However, when the surface-state energy in NaCl is graphed as a function of the electronegativity of the metal, the slope is almost one. The slope is shown for other materials. It changes value abruptly at $X_c \sim 0.8$.

The third graph shows the intensity of the exciton absorption band. An exciton is a bound state between an electron in the conduction band and a hole from the valence band. Excitons are discussed in chapter 12 on optical properties. Insulators have *Frenkel excitons*. They have the electron and hole in the same unit cell, so the state has an atomic character. The oscillator strength for the absorption band is quite strong. Semiconductors have a *Wannier exciton*. The electron and hole are bound weakly, and their bound orbit extends over many unit cells of the crystal. The absorption peak for Wannier excitons is quite small. The graph shows that the transition between Wannier and Frenkel exciton behavior changes abruptly over small changes in ΔX.

The most interesting feature of these curves is that the II–VI binary semiconductors are mostly on the boundary. CdTe is universally regarded as a semiconductdor. ZnO is usually regarded as a semiconductor, but the Mead plot clearly puts it into the insulator category. Many others (CdSe, CdS, ZnSe, etc.) are on the boundary. We follow custom and treat them as semiconductors, but one should never overlook their ionic character.

The title of this section (What Is a Semiconductor?) is now answered. We follow Mead et al. and define a semiconductor as having $|\Delta X| < 0.8$.

3.4.2 *Si, Ge, GaAs*

Three important semiconductors are silicon (Si), germanium (Ge), and gallium arsenide (GaAs). Si and Ge have the diamond structure. GaAs has a similar structure, which is called zincblende. It is the diamond structure, with every other atom Ga or As. The unit cell of all three crystals is fcc. Their energy-band structure is shown in fig. 3.6. These three energy bands have several interesting features:

- The valence bands look alike for the three semiconductors. The valence bands can be derived using a simple tight-binding model with the p-orbitals of the atoms.
- All three valence bands have their maximum energy at the Γ-point, which is the center of the BZ at $\mathbf{k} = 0$. The bands would be sixfold degenerate at this point except for a small energy splitting due to the spin–orbit interaction. The atomic spin–orbit splitting of the p-orbitals is also found in the crystal, with nearly the same value.
- The conduction bands of the three atoms are all different. They cannot be derived with a simple nearest-neighbor, tight-binding model. However, these energy bands are easily calculated with modern band structure methods described in a later section.

The most important point in the conduction band of each semiconductor is the point of minimum energy. It is different for the three crystals.

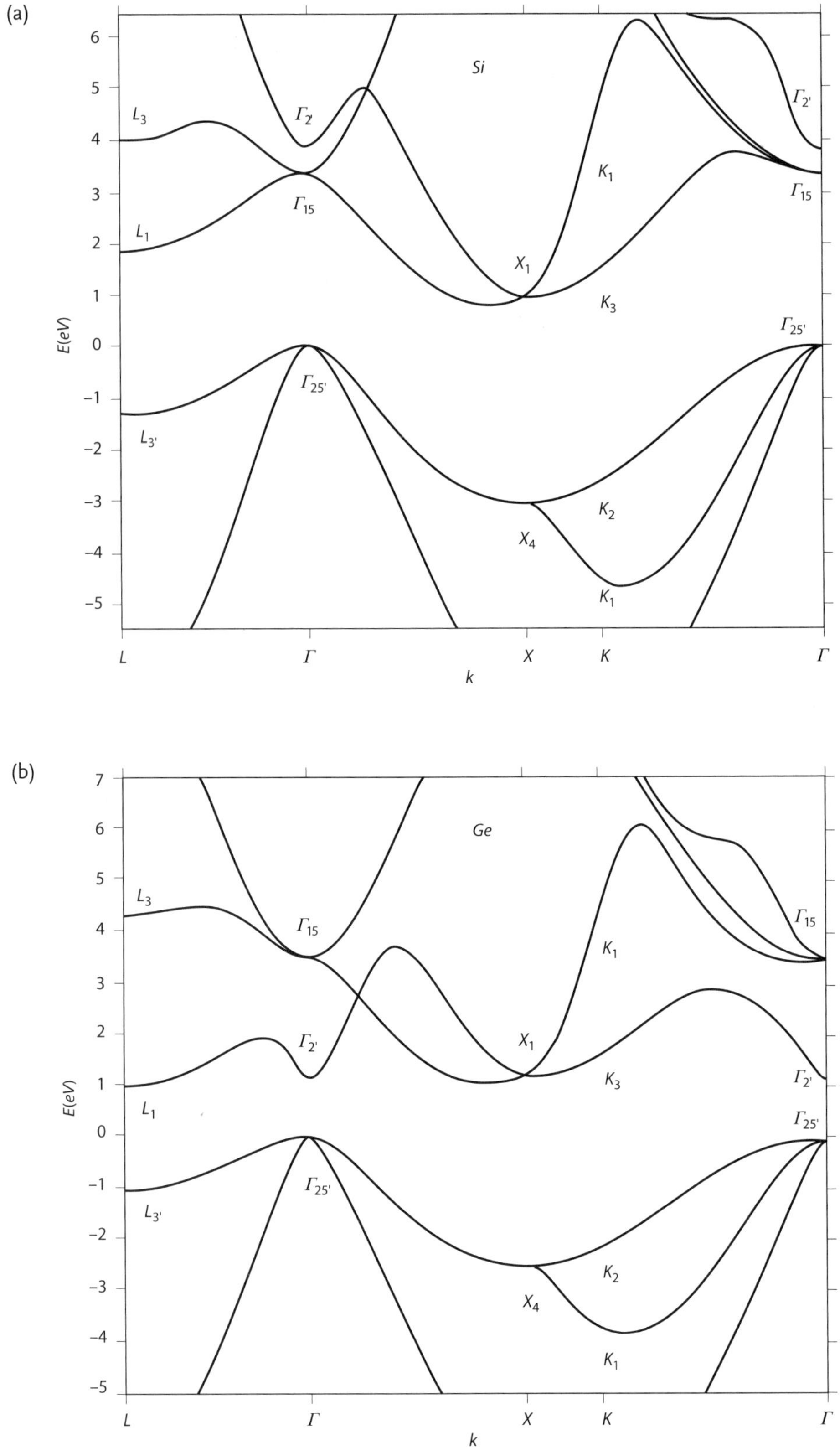
(a)
Si
E(eV)
6
5
4
3
2
1
0
−1
−2
−3
−4
−5
L
Γ
X
K
Γ
k
L3
L1
L3'
Γ2'
Γ15
Γ25'
X1
X4
K1
K3
K2
K1
Γ2'
Γ15
Γ25'
(b)
Ge
E(eV)
7
6
5
4
3
2
1
0
−1
−2
−3
−4
−5
L
Γ
X
K
Γ
k
L3
L1
L3'
Γ15
Γ2'
Γ25'
X1
X4
K1
K3
K2
K1
Γ15
Γ2'
Γ25'

(c)

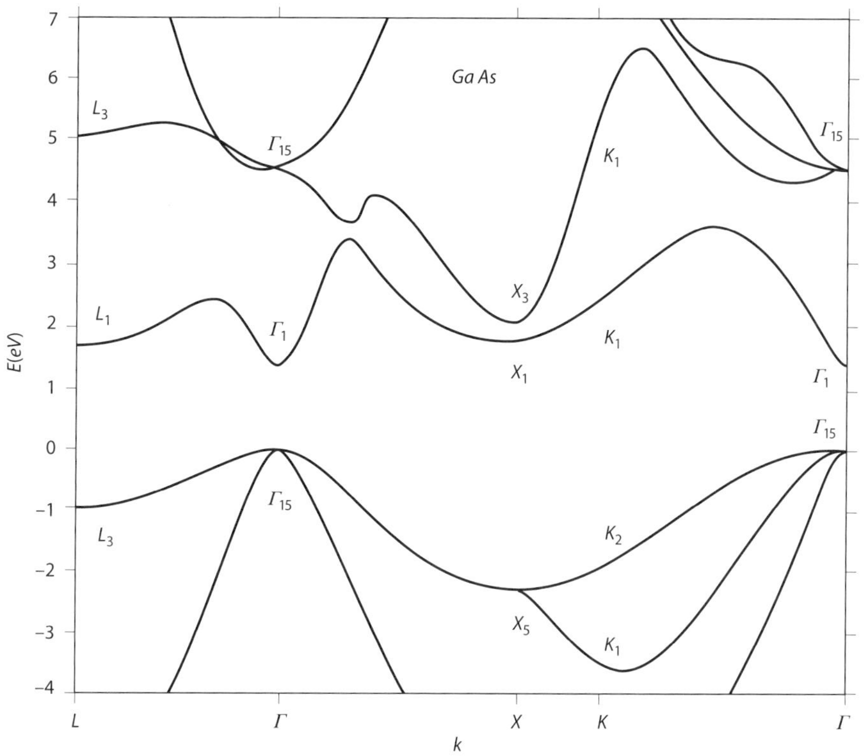

FIGURE 3.6. Electron energy bands of Si, Ge, and GaAs. From Cohen and Bergstresser, *Phys. Rev.* **141**, 789 (1966). Used with permission of the American Physical Society.

- Si has the minimum of the conduction band along the direction from Γ to *X*, which is the (100) direction in the BZ. There are six identical minima along the $\pm\hat{x}, \pm\hat{y}, \pm\hat{z}$ directions. Along each direction the surface of constant energy, near the bottom of the band, is an ellipsoid. For example, along the (100) direction it is

$$\varepsilon_{100}(\mathbf{k}) = E_{C0} + \frac{\hbar^2}{2m_\perp}(k_y^2 + k_z^2) + \frac{\hbar^2}{2m_\parallel}(k_x - k_0)^2 \tag{3.85}$$

$$m_\parallel = m_\ell = 0.916 m_e \tag{3.86}$$

$$m_\perp = m_t = 0.19 m_e \tag{3.87}$$

where the minimum is approximately at $k_0 \approx 1.4\pi/a$. There are two values of the effective mass, for motion perpendicular or parallel to the *x*-axis.

- Ge has its conduction-band minimum at the L-point, which is the edge of the BZ. There are eight equivalent L-points, at $\mathbf{k}_L = \pi(\pm 1, \pm 1, \pm 1)/a$. Each of the eight L-points contains half of an ellipsoid of constant energy, so Ge has four equivalent ellipsoids. Each one also has a separate mass for motion perpendicular and parallel to the motion along the direction $\mathbf{k}_L$:

$$m_\parallel = m_\ell = 1.57 m_e \tag{3.88}$$

$$m_\perp = m_t = 0.0807 m_e \tag{3.89}$$

- GaAs has its conduction-band minimum at the Γ-point $k = 0$. There is only one minimum, and the motion is isotropic:

$$E_c(\mathbf{k}) = E_{c0} + \frac{\hbar^2 k^2}{2m_c} \tag{3.90}$$

where in GaAs $m_c \approx m_e/16$, and m_e is the mass of an electron. At very low temperatures $m_c = 0.0665 m_e$, while at room temperatures, where the gap is smaller, $m_c = 0.0636 m_e$. The small value of the conduction-band mass in GaAs is the reason for the large mobility of its conduction electrons. GaAs is a *direct band gap* material, since the conduction-band minimum and valence-band maximum occur at the same point in the BZ. Si and Ge have *indirect band gaps*.

Besides Si and Ge, the other group IV semiconductors are diamond and tin. Diamond has the same structure, and also has an indirect energy gap. Tin (Sn) has several common crystalline structures.

The III–V semiconductors are those with a cation from the third column of the periodic table and an anion from the fifth column: GaAs is an example. Most have the zincblende structure. Some are direct gap materials, and some are indirect gap.

3.4.3 HgTe and CdTe

Both HgTe and CdTe are II–VI semiconductors with the zincblende structure, but their energy bands are very different. CdTe is a conventional semiconductor, with a conduction that has s-symmetry, with the band minimum at the Γ-point ($\mathbf{k} = 0$). The three valence bands have *p*-symmetry, with their maximum at the Γ-point. The band gap is direct, with a value at low temperature of $E_G = 1.5$ eV.

HgTe has a band inversion. The *s*-symmetry band Γ_6 is lower in energy than the *p*-symmetry bands Γ_7 and Γ_8. Its energy bands are shown in fig. 3.7, which was adapted from Cade and Lee [1]. On the left of this figure are the energy bands of CdTe, and on the right are those of HgTe. Now count electrons. In CdTe, the three valence bands are full of electrons. In HgTe, the same number of electrons fill up the Γ_6- and Γ_7-bands, and one of the Γ_8 bands. So two of the *p*-symmetry bands are filled and one is empty. Since two bands meet at the Γ-point, HgTe has no band gap. It is a semimetal. That feature gives it interesting electronic properties.

CdTe and HgTe have identical crystal structures, and nearly the same lattice constant:

$$a_{\mathrm{CdTe}} = 6.480\ \text{Å} \qquad a_{\mathrm{HgTe}} = 6.429\ \text{Å} \tag{3.91}$$

They can be alloyed continuously: $Hg_{1-x}Cd_xTe$. The various energy bands, at the Γ-point, change linearly with alloy concentration x. There are values of x for which the energy gap is very small. This small gap is useful for detecting infrared radiation. This alloy is widely used for this purpose; e.g., in night vision goggles.

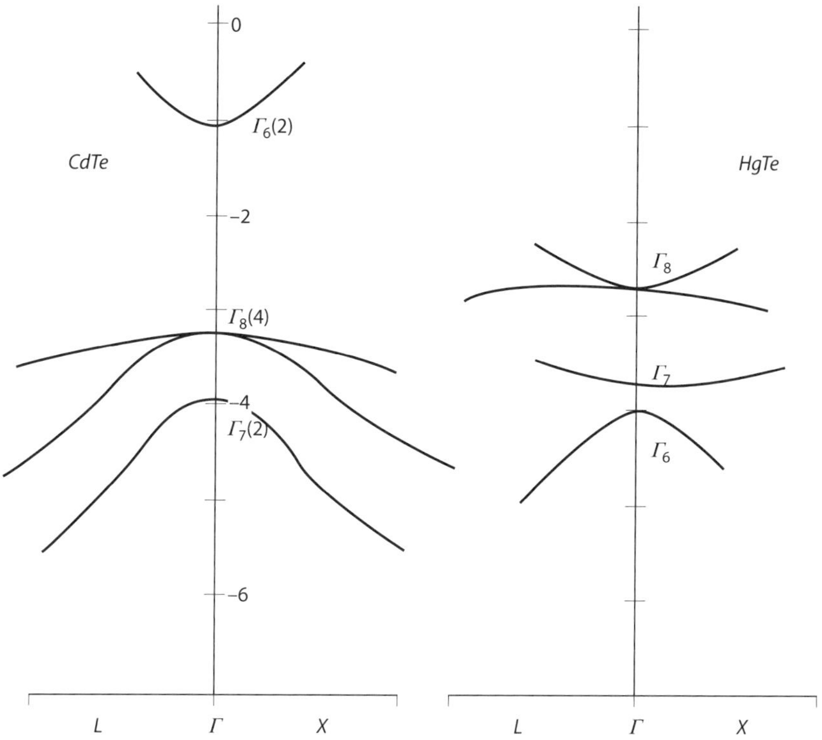

FIGURE 3.7. Energy bands of HgTe and CdTe near to $k = 0$.

3.4.4 k · p *Theory*

Most holes spend their time near the top of the valence band, and most conduction band electrons spend their time near the bottom of the conduction band. The properties of the bands near these extrema is quite important. In direct gap materials, the properties of the conduction and valence bands are related. This relationship is determined by $k \cdot p$ theory.

Consider the case of GaAs. The conduction-band minimum is composed of mostly an s-state of Ga. The valence band is composed of the three p-state orbitals of As. Assume that the states at $\mathbf{k} = 0$ are in fact the atomic orbitals, as in the tight-binding model. At small but nonzero values of wave vector we can use these states as the basis of a perturbation expansion. The form for the eigenfunction is

$$\psi(\mathbf{k}, \mathbf{r}) = e^{i\mathbf{k}\cdot\mathbf{r}} u(\mathbf{k}, \mathbf{r}) \tag{3.92}$$

$$u(\mathbf{k}, \mathbf{r}) = a_s \phi_s(\mathbf{r}) + a_x \phi_x(\mathbf{r} - \boldsymbol{\tau}) + a_y \phi_y(\mathbf{r} - \boldsymbol{\tau}) + a_z \phi_z(\mathbf{r} - \boldsymbol{\tau}) \tag{3.93}$$

where ϕ_s is the Ga s-orbital, and $\phi_\mu(\mathbf{r} - \boldsymbol{\tau})$ is the As p_μ-orbital. The As atom is a distance τ away from the Ga. The unknown coefficients (a_s, a_x, a_y, a_z) are functions of wave vector $\mathbf{k}$. At $\mathbf{k} = 0$ the eigenvalue is

$$H = \frac{p^2}{2m_e} + U(\mathbf{r}) \tag{3.94}$$

$$H\psi(0, \mathbf{r}) = E_s a_s \phi_s(\mathbf{r}) + E_p[a_x \phi_x(\mathbf{r} - \boldsymbol{\tau}) + a_y \phi_y(\mathbf{r} - \boldsymbol{\tau}) + a_z \phi_z(\mathbf{r} - \boldsymbol{\tau})] \tag{3.95}$$

Now evaluate the eigenvalue for nonzero $\mathbf{k}$. The kinetic energy operator gives

$$p^2 \psi(\mathbf{k}, \mathbf{r}) = e^{i\mathbf{k}\cdot\mathbf{r}} \left(\frac{\hbar}{i}\right)^2 \left[(i\mathbf{k})^2 u(\mathbf{k}, \mathbf{r}) + \nabla^2 u(\mathbf{k}, \mathbf{r}) + 2i(\mathbf{k}\cdot\nabla) u(\mathbf{k}, \mathbf{r})\right] \tag{3.96}$$

The term with ∇^2 just gives the atomic eigenvalue. The other terms give

$$\begin{aligned} E\psi &= H\psi = \varepsilon(\mathbf{k})\psi + V \\ &\quad + e^{i\mathbf{k}\cdot\mathbf{r}}\{E_s a_s \phi_s(\mathbf{r}) + E_p[a_x \phi_x(\mathbf{r} - \boldsymbol{\tau}) + a_y \phi_y(\mathbf{r} - \boldsymbol{\tau}) + a_z \phi_z(\mathbf{r} - \boldsymbol{\tau})]\} \end{aligned} \tag{3.97}$$

$$V = -i\frac{\hbar^2}{m_e} e^{i\mathbf{k}\cdot\mathbf{r}} (\mathbf{k}\cdot\nabla) u(\mathbf{k}, \mathbf{r}) = \frac{\hbar}{m_e} e^{i\mathbf{k}\cdot\mathbf{r}} (\mathbf{k}\cdot\mathbf{p}) u(\mathbf{k}, \mathbf{r}) \tag{3.98}$$

$$\varepsilon(\mathbf{k}) = \frac{\hbar^2 k^2}{2m_e} \tag{3.99}$$

The term V is called the $k \cdot p$ interaction. The term $\varepsilon(\mathbf{k})$ is the free electron kinetic energy. Take eqn. (3.97), multiply by $\exp(-i\mathbf{k}\cdot\mathbf{r})\phi_s^*(\mathbf{r})$ from the left, and integrate over all volume. The three orbitals are orthogonal, by parity:

$$Ea_s = [\varepsilon(\mathbf{k}) + E_s]a_s + \frac{\hbar}{m_e}\int d^3 r \phi_s^*(\mathbf{r})(\mathbf{k}\cdot\mathbf{p}) u(\mathbf{k}, \mathbf{r}) \tag{3.100}$$

The interaction V has odd parity and ϕ_s^* has even parity. The integral is zero unless a term in $u(\mathbf{k}, \mathbf{r})$ has odd parity. Write $\mathbf{k}\cdot\mathbf{p} = k_x p_x + k_y p_y + k_z p_z$. The integral is nonzero in the term $k_x p_x$ only for the term in $u(\mathbf{k}, \mathbf{r})$ that has odd parity with respect to x. Similarly, the term $k_y p_y$ is nonzero only for the term in $u(\mathbf{k}, \mathbf{r})$ that has odd parity with respect to y. The result is

$$Ea_s = [\varepsilon(k) + E_s]a_s + \frac{\hbar}{m_e}[a_x k_x p_{cv,x} + a_y k_y p_{cv,y} + a_z k_z p_{cv,z}] \tag{3.101}$$

$$p_{cv,\mu} = \frac{\hbar}{i}\int d^3 r \phi_s^* \frac{\partial}{\partial r_\mu}\phi_\mu \tag{3.102}$$

The symbol $p_{cv,\mu}$, $(\mu = x, y, z)$, is the matrix element of momentum between the conduction and valence bands. It has the same value for all three choices of μ, so call it p_{cv}. Next take eqn. (3.97), multiply by $\phi_x^*(\mathbf{r} - \boldsymbol{\tau})$ from the left, and integrate over all volume:

$$Ea_x = [\varepsilon(k) + E_p]a_x + \frac{\hbar}{m_e}k_x p_{cv} a_s \tag{3.103}$$

Repeat the last operation using ϕ_y and ϕ_z. The results can be expressed as the Hamiltonian matrix

$$0 = \begin{bmatrix} E_s + \varepsilon - E & w_x & w_y & w_z \\ w_x & E_p + \varepsilon - E & 0 & 0 \\ w_y & 0 & E_p + \varepsilon - E & 0 \\ w_z & 0 & 0 & E_p + \varepsilon - E \end{bmatrix} \begin{bmatrix} a_s \\ a_x \\ a_y \\ a_z \end{bmatrix} \tag{3.104}$$

$$w_\mu = \frac{\hbar}{m_e} k_\mu p_{cv} \tag{3.105}$$

The four eigenvalues are

$$E_{1,2} = E_p + \varepsilon(\mathbf{k}), E_p + \varepsilon(\mathbf{k}), \text{ and} \tag{3.106}$$

$$E_\pm = \varepsilon(\mathbf{k}) + \frac{1}{2}\left[E_s + E_p \pm \sqrt{(E_s - E_p)^2 + 4w^2}\right] \tag{3.107}$$

$$w^2 = w_x^2 + w_y^2 + w_z^2 = \frac{\hbar^2 k^2}{m_e^2} p_{cv}^2 \tag{3.108}$$

Two of the valence bands are unaffected by the $k \cdot p$ interaction V, and have an eigenvalue $E = E_p + \varepsilon(\mathbf{k})$. Actually, they interact, using the $k \cdot p$ interaction, with other bands above the conduction band. One valence band interacts with the conduction band, and this interaction determines the effective mass. At small values of k^2, $2w < (E_s - E_p)$ and expand the argument of the square root. The energy gap $E_G = E_s - E_p$, since $E_s = E_{c0}$, $E_p = E_{v0}$:

$$E_c(\mathbf{k}) = E_+ = E_{c0} + \varepsilon(\mathbf{k}) + \frac{w^2}{E_G} + O(k^4) = E_{c0} + \frac{\hbar^2 k^2}{2m_c} + O(k^4) \tag{3.109}$$

$$E_v(\mathbf{k}) = E_- = E_{v0} + \varepsilon(\mathbf{k}) - \frac{w^2}{E_G} + O(k^4) = E_{v0} - \frac{\hbar^2 k^2}{2m_v} + O(k^4) \tag{3.110}$$

where the effective masses of the conduction and valence bands are

$$\frac{m_e}{m_c} = 1 + \frac{C}{E_G}, \quad C = \frac{2p_{cv}^2}{m_e} \tag{3.111}$$

$$\frac{m_e}{m_v} = \frac{C}{E_G} - 1 \tag{3.112}$$

The valence band has three bands: a *light-hole band*, a *heavy-hole band*, and a *split-off band*. The split-off band is split off by the spin–orbit interaction. Each of these three bands has a degeneracy of two due to spin. The effective mass m_v is for the light-hole band. Its effective mass is related to the conduction-band effective mass m_c by the constant C, which has the units of energy. An example is GaAs. At room temperature, the energy gap and two masses are

$$E_G = 1.43 \text{ eV} \tag{3.113}$$

$$\frac{m_e}{m_c} = \frac{1}{0.0636} = 15.7 \tag{3.114}$$

$$\frac{m_e}{m_v} = \frac{1}{0.083} = 12.4 \tag{3.115}$$

Averaging the two numbers on the right gives

$$\frac{C}{E_G} = 14, \; C = 14E_G = 20 \text{ eV} \tag{3.116}$$

This value of C is typical for the group of III–V semiconductors. The small effective mass of the conduction band and the light-hole band are due to the large value of C and the small value of the energy gap. The interband matrix element p_{cv} is useful for evaluating other quantities, such as the interband optical absorption in chapter 12.

A better $k \cdot p$ theory includes the spin–orbit interaction. For the conduction-band mass, including the effect of the spin–orbit splitting Δ, gives

$$\frac{m_e}{m_c} = 1 + C\left[\frac{2}{3E_G} + \frac{1}{3(E_G + \Delta)}\right] \tag{3.117}$$

since one-third of the oscillator strength is in the split-off band. Here Δ is the spin–orbit splitting. This formula is useful for semiconductors where the energy gap is small. Some values for other direct gap semiconductors are shown in table 3.2. In InSb the spin–orbit energy is much larger than the energy gap. Note that $2P^2/m$ has about the same value in the three different semiconductors, although they have very different energy gaps and band masses.

A better theory of the valence band includes perturbations quadratic in wave vector, and gives [8]

$$E_v(\mathbf{k}) = E_{v0} - \frac{\hbar^2}{2m_e}\left[Ak^2 + \sqrt{B^2k^4 + D^2(k_x^2k_y^2 + k_y^2k_z^2 + k_z^2k_x^2)}\right] \tag{3.118}$$

In Si, $A = -4.0$, $|B| = 1.1$, and $|D| = 4.1$. The valence band is not isotropic, but is warped.

Another formula from $k \cdot p$ theory is the effective g value of the magnetic moment of an electron in the conduction band:

$$g^* = 2 - \frac{\Delta C}{3E_G(E_G + \Delta)}, \quad C = \frac{2p_{cv}^2}{m} \tag{3.119}$$

A free electron has $g \approx 2$, while the second term is due to $k \cdot p$ theory. This formula was first derived by Roth et al. [15], and a derivation is in Kittel [8].

Table 3.2 ***k* · *p* parameters for several semiconductors**

	$E_G(eV)$	m_c/m_e	Δ(eV)	$2P_{cv}^2/m_e$(eV)
GaAs	1.43	0.067	0.34	21.3
InP	1.34	0.073	0.11	17.5
InSb	0.18	0.013	0.98	19.0

3.4.5 Electron Velocity

Theorem: The group velocity of the electron is the gradient of the energy-band dispersion:

$$\hbar \mathbf{v}_{\mathbf{k}} = \nabla_k E_n(\mathbf{k}) = \int d^3 r \psi_n^*(\mathbf{k}, \mathbf{r}) \frac{\mathbf{p}}{m} \psi_n(\mathbf{k}, \mathbf{r}) \tag{3.120}$$

For quadratic band dispersion, the effective mass also determines the electron group velocity. For the conduction band $\mathbf{v}_k = \hbar \mathbf{k}/m_c$.

The theorem is proved by taking the derivative of the band dispersion. It is an eigenfunction of the Hamiltonian:

$$E_n(\mathbf{k}) = \int d^3 r \psi_n^*(\mathbf{k}, \mathbf{r}) H \psi_n(\mathbf{k}, \mathbf{r}) \tag{3.121}$$

Take the gradient with respect to wave vector

$$\begin{aligned} \nabla_k E_n(\mathbf{k}) &= \int d^3 r \nabla_k \left[e^{-i\mathbf{k}\cdot\mathbf{r}} u_n^*(\mathbf{k}, \mathbf{r}) H e^{i\mathbf{k}\cdot\mathbf{r}} u_n(\mathbf{k}, \mathbf{r}) \right] \\ &= \int d^3 r \left[u_n^*(\mathbf{k}, \mathbf{r}) \nabla_k \left(e^{-i\mathbf{k}\cdot\mathbf{r}} H e^{i\mathbf{k}\cdot\mathbf{r}} \right) u_n(\mathbf{k}, \mathbf{r}) \right. \\ &\quad \left. + E_n(\mathbf{k}) \nabla_k |u_n(k, r)|^2 \right] \end{aligned} \tag{3.122}$$

The last term is zero:

$$E_n(\mathbf{k}) \nabla_k \int d^3 r |u_n^*(\mathbf{k}, \mathbf{r})|^2 = E_n(\mathbf{k}) \nabla_k (1) = 0 \tag{3.123}$$

The first term contains

$$\nabla_k \left(e^{-i\mathbf{k}\cdot\mathbf{r}} H e^{i\mathbf{k}\cdot\mathbf{r}} \right) = -i \left(e^{-i\mathbf{k}\cdot\mathbf{r}} [\mathbf{r}, H] e^{i\mathbf{k}\cdot\mathbf{r}} \right) = \left(e^{-i\mathbf{k}\cdot\mathbf{r}} \frac{\mathbf{p}}{m} e^{i\mathbf{k}\cdot\mathbf{r}} \right) \tag{3.124}$$

Collecting these results proves the theorem in eqn. (3.120). The group velocity of an electron in an energy band is merely the slope of the band at that point.

3.5 Density of States

The density of states $N_d(E)$ of an energy band is an important function. In d-dimensions, where $d = (1, 2, 3)$, it is defined as

$$N_d(E) = \sum_n \int \frac{d^d k}{(2\pi)^d} \delta[E - \varepsilon_n(\mathbf{k})] \tag{3.125}$$

The integral $d^d k$ is over the Brillouin zone. The summation over n is over all of the energy bands in the BZ. This definition is the density of states for each spin configuration. In paramagnetic systems, the result is usually multiplied by two to include both spin states.

An interesting feature of the density of states is values of energy E where it has singular points or abrupt changes of slope. These are called *Van Hove singularities*, even when they are not singular. The nature of the singularity very much depends on the dimension of the energy band. The minimum energy of an electron in the conduction band and the maximum electron energy in the valence band are always at Van Hove singularities. They also show up as sharp features in the optical absorption spectra of solids.

We give three examples of these singularities: one dimension, two dimensions, and three dimensions. Each case assumes a very simple model: a simple cubic tight-binding energy band.

1. In one dimension, the energy dispersion for a single band is $\varepsilon(k) = A + B\cos(ka)$, and the density of states is

$$\begin{aligned} N_1(E) &= \int_{-\pi/a}^{\pi/a} \frac{dk}{2\pi} \delta[E - A - B\cos(ka)] \\ &= \frac{1}{\pi a\sqrt{B^2 - (E-A)^2}} \end{aligned} \tag{3.126}$$

The band dispersion ranges between the two points $A - |B| \le E \le A + |B|$. There is an inverse square root singularity at both end points, as shown in fig. 3.8a. These are the Van Hove singularities, and they are indeed singular. Such singularities are found in all energy bands in one dimension.

2. In two dimensions the band dispersion is

$$\varepsilon(k_x, k_y) = A + B[\cos(k_x a) + \cos(k_y a)] \tag{3.127}$$

$$N_2(E) = \int_{-\pi/a}^{\pi/a} \frac{dk_x}{2\pi} \int_{-\pi/a}^{\pi/a} \frac{dk_x}{2\pi} \delta\{E - A - B[\cos(k_x a) + \cos(k_y a)]\} \tag{3.128}$$

To evaluate this integral, define the dimensionless constants ξ, θ_μ, and change integration variables to positive angles, since the integrand is symmetric:

$$\xi = \frac{E-A}{B}, \quad \theta_\mu = k_\mu a \tag{3.129}$$

$$\begin{aligned} N_2(E) &= \frac{1}{\pi^2 a^2 B} \int_0^\pi d\theta_x \int_0^\pi d\theta_y \delta[\xi - \cos(\theta_x) - \cos(\theta_y)] \\ &= \frac{1}{\pi^2 a^2 B} K(q), \quad q = \frac{1}{2}\sqrt{4 - \xi^2} \end{aligned} \tag{3.130}$$

Note that $-2 \le \xi \le 2$. The density of states is given as an elliptic integral. It is shown in fig. 3.8b. There is a Van Hove singularity at $\xi = 0$. Here the divergence is $O[\ln(\xi)]$, which is a weak singularity. Such logarithmic singularities are found in all energy bands in two dimensions. There are also Van Hove singularities at the end points $\xi = \pm 2$.

3. The energy-band dispersion in three dimensions is

$$\varepsilon(k_x, k_y, k_z) = A + B[\cos(k_x a) + \cos(k_y a) + \cos(k_z a)] \tag{3.131}$$

$$N_3(E) = \int \frac{d^3k}{(2\pi)^3} \delta\{E - A - B[\cos(k_x a) + \cos(k_y a) + \cos(k_z a)]\} \tag{3.132}$$

The integral cannot be evaluated with simple functions. Again define ξ as in eqn. (3.129). Now the range of values is $-3 \le \xi \le 3$. A numerical evaluation of the integral is shown in fig. 3.8c. There are Van Hove singular points at $\xi = \pm 3, \pm 1$. In three dimensions, the

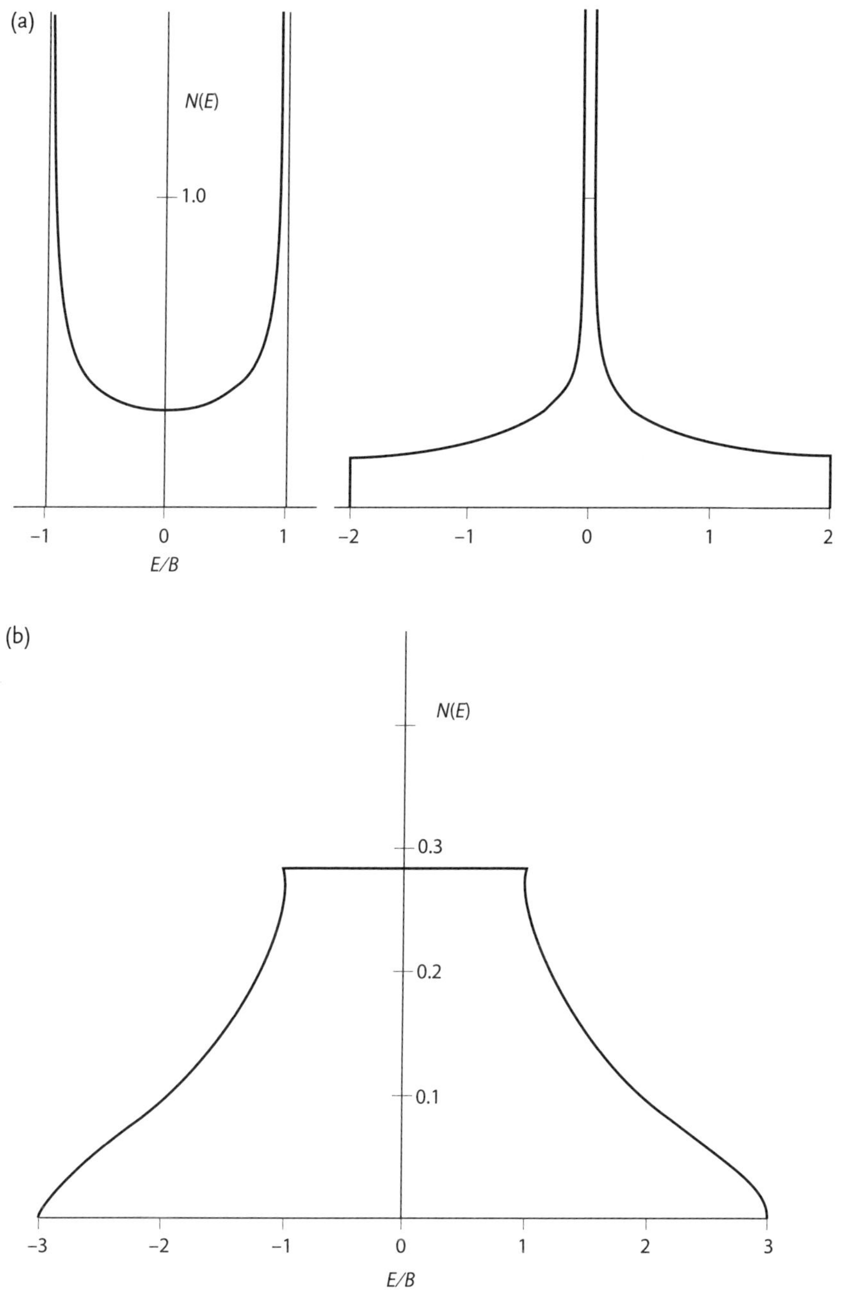

FIGURE 3.8. Tight-binding density of states for (a) one, (b) two, and (c) three dimensions.

Van Hove points do not show actual singularities in the density of states. Instead, there are abrupt changes in slope.

4. The above results can be extended to n-dimensions. Set $A = 0$ and get

$$N_n(E) = \int \frac{d^n\theta}{(2\pi)^n}\delta[E - B\gamma(\theta_i)] \tag{3.133}$$

$$\gamma(\theta_i) = \sum_{j=1}^{n} \cos(\theta_j) \tag{3.134}$$

The integral for general values of n can be evaluated by replacing the delta function by its integral definition, and then inverting the order of the integrals:

$$\delta[E - B\gamma(\theta_i)] = \int_{-\infty}^{\infty} \frac{ds}{2\pi} \exp\{is[E - B\gamma(\theta_i)]\} \tag{3.135}$$

$$N_n(E) = \int_{-\infty}^{\infty} \frac{ds}{2\pi} e^{isE} I(s)^n \tag{3.136}$$

$$I(s) = \int_{-\pi}^{\pi} \frac{d\theta}{2\pi} e^{-isB\cos(\theta)} = J_0(sB) \tag{3.137}$$

$$N_n(E) = \int_{-\infty}^{\infty} \frac{ds}{2\pi} e^{isE} J_0(sB)^n \tag{3.138}$$

where $J_0(z)$ is a Bessel function.

The above analysis calculates the density of states of simple model systems. It is interesting to show the density of electronic states for some actual solids. Figure 3.9 shows the calculated density of states of three simple solids: (a) silicon (semiconductor), (b) sodium (metal), and (c) lithium chloride (insulator). Silicon has a small energy gap, while LiCl has a large one. Various peaks correspond to van Hove singularities.

3.5.1 Dynamical Mean Field Theory

Metzner and Vollhardt [12] introduced *dynamical mean field theory* (DMFT) as the name for the limit that $n \to \infty$. This model has a number of advantages for the theory of highly correlated systems, such as the Hubbard model. The limit of large n has the disadvantage that the bandwidth gets infinitely large. Metzner and Vollhardt noted that this feature could be corrected by assuming that the hopping term B scaled with n according to $B = t/\sqrt{n}$, where t is just a constant. Then the limit of $n \to \infty$ is well behaved mathematically.

Take the limit of $n \to \infty$ in eqn. (3.138). $J_0(z)$ is always less than one, and taking a small number to the nth power makes it smaller still. The important contributions to the integral over s are found at small values. So use the series expansion for the Bessel function of small argument:

$$J_0(z) = 1 - \frac{z^2}{4} + O(z^4) \tag{3.139}$$

$$\lim_{n\to\infty} J_0(st/\sqrt{n})^n = \exp[n\ln(J_0)] = \exp\{n\ln[1 - t^2s^2/4n + O(1/n^2)]\}$$

Expand the argument of the log function:

$$\lim_{n\to\infty} J_0(st/\sqrt{n})^n \approx \exp\left[-\frac{t^2s^2}{4} + O(1/n)\right] \tag{3.140}$$

$$N_\infty(E) = \frac{1}{\sqrt{\pi}t} \exp\left[-\frac{E^2}{t^2}\right] \tag{3.141}$$

The factor of $I(s)^n$ becomes a Gaussian, and the density of states is a Gaussian.

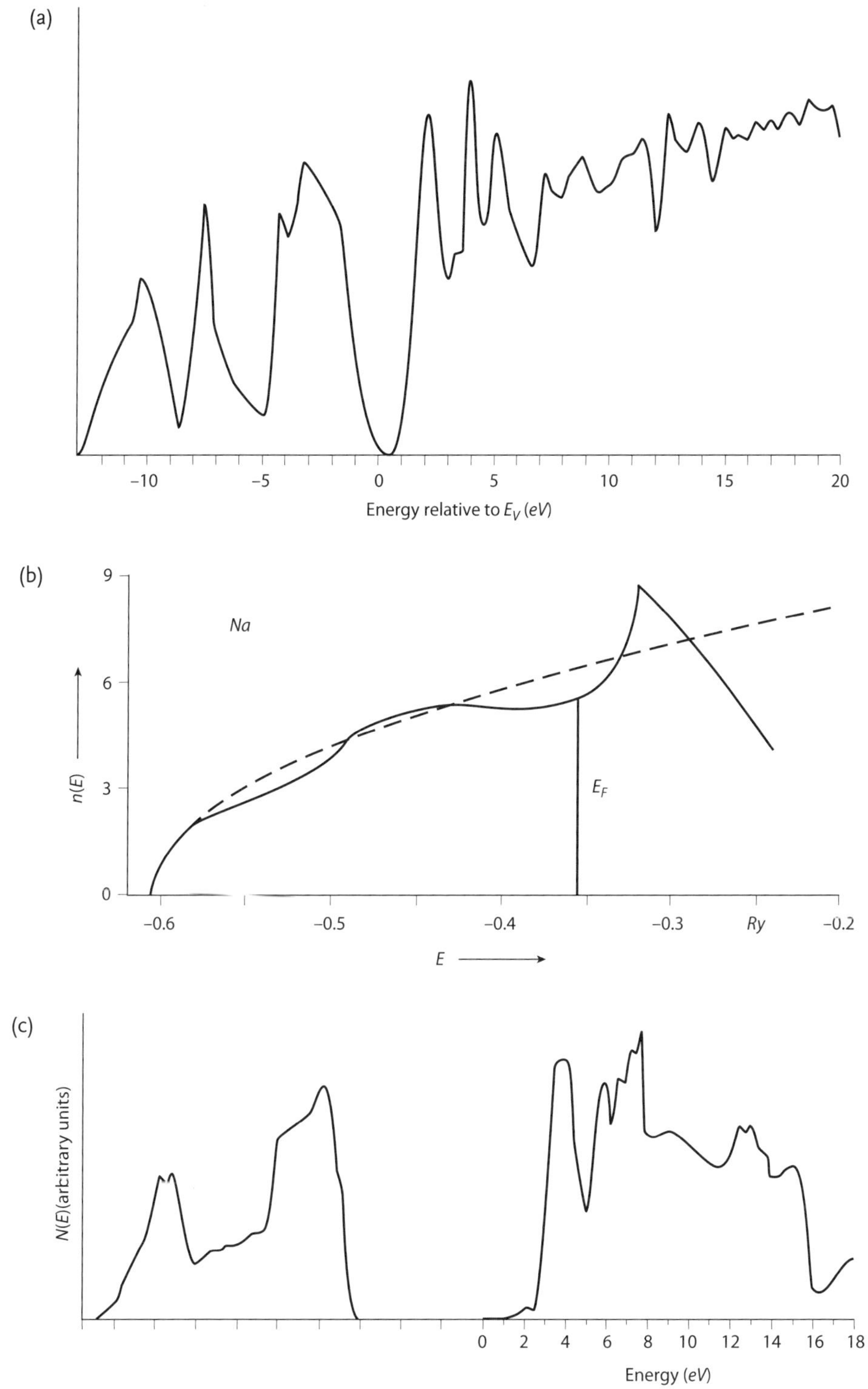

FIGURE 3.9. Density of states of (a) silicon [Chelikowsky et al., *Phys. Rev. B* **40**, 9644 (1989), used with permission of the American Physical Society)], (b) sodium [Cornwell, *Proc. R. Soc. (London) Ser., A* 261 (1961), used with permission of the Royal Society)], and (c) lithium chloride [Kunz, *Phys. Rev. B* **2**, 505 (1970), used with permission of the American Physical Society].

Crystals lattices of finite dimension have the feature that an electron can leave a site by hopping to a neighbor site. It can return to the original site by hopping backward, or by hopping around a loop to get to the original site. It is these latter processes, paths that are loops, that make the theory difficult. The reason the Bethe lattice is easy to solve is that there are no paths that are loops: if an electron hops off of a site in a certain direction, it can return only by reversing its path. DMFT has the same feature as the Bethe lattice. When the number of neighbors n becomes large, it is much less probable to hop back by an alternate path. When $n = \infty$ it becomes impossible. Then mean field theory becomes an exact theory.

3.6 Pseudopotentials

An important interaction term for an electron is the potential energy from the ion cores. We will assume here that the ions have a closed shell of electrons, but that is not always the case in many solids. The electron–ion potential $v_{ei}(r)$ is the potential energy between a conduction electron and one ion. The ion has a nuclear charge of $-Z_n e$, and Z_v electrons bound in the closed shells. The net charge on the ion is $-eZ = -e(Z_n - Z_v)$. At long distances from the ion, the potential is purely Coulombic:

$$\lim_{r \gg a_0} v_{ei}(r) = -\frac{Ze^2}{4\pi\varepsilon_0 r} \tag{3.142}$$

As the electron gets closer to the ion, the potential deviates from Coulombic. This change is partly due to the finite size of the bound electron states. Another important feature is that the conduction-band electron has a wave function that must be orthogonal to that of the bound electrons, since they are eigenstates of the same Hamiltonian. This required orthogonality has a big influence on the wave function of the conduction electrons.

Rather than deal with this complex behavior, it is customary to use a simple approximate potential for $v_{ei}(r)$. It is called a *pseudopotential*. The point of the pseudopotential is to eliminate the electrons in the filled atomic core of the ion, and to replace them by an effective potential function. One popular choice is

$$v_{ei}(r) = \begin{cases} A_\ell & r < a \\ -\frac{Ze^2}{4\pi\varepsilon_0 r} & r > a \end{cases} \tag{3.143}$$

The radius a is chosen to be the radius of the ion—we prefer to use the radii suggested by Pauling. An example is shown in fig. 3.10.

For distances $r < a$, the potential is a constant $A_\ell(E)$ that depends on angular momentum ℓ and energy E. This constant can be chosen using information about the bound-state energies of atoms. For example, take the ion Mg^{2+}, which is a filled shell. Add one more electron and calculate its binding energy. For *s*-waves, adjust the constant A_0 to give the correct binding energy for the lowest atomic state 4*s*. This process gives $A_0(E_{4s})$. Then solve again for the next higher binding energy, which gives a value of $A_0(E_{5s})$. Repeat this

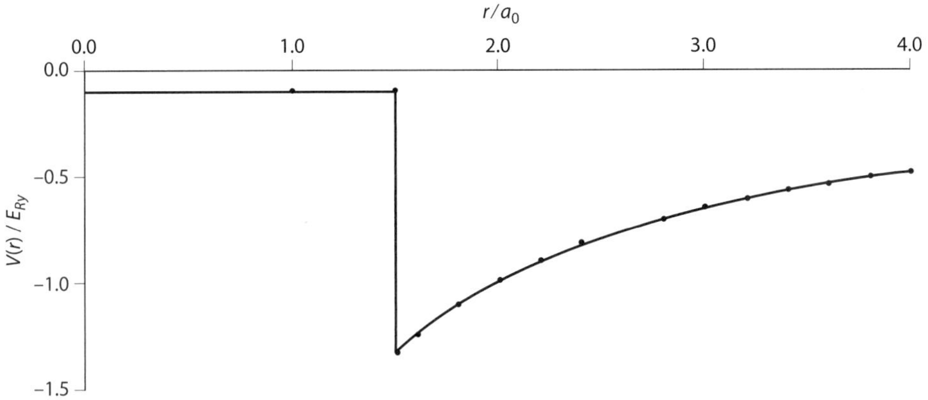

FIGURE 3.10. An electron–ion pseudopotential. The potential is a constant for $r < a$ and is Coulombic for $R > a$. Here $a = 1.5a_0$ and $Z = 2$.

process for several atomic s states. Then make a graph of the values of $A_0(E_{ns})$ vs. E_{ns}. One finds a straight line, as shown in fig. 3.11. This linear function $A_0(E)$ can be used to evaluate the electron–ion interaction for electrons of any energy in metallic magnesium.

The actual radial wave function $R_{n\ell}(r)$ of an unbound electron inside an atomic closed shell has many oscillations, in order to be orthogonal to the bound-state radial wave functions. This correct radial wave function is being replaced with a pseudo-wave function

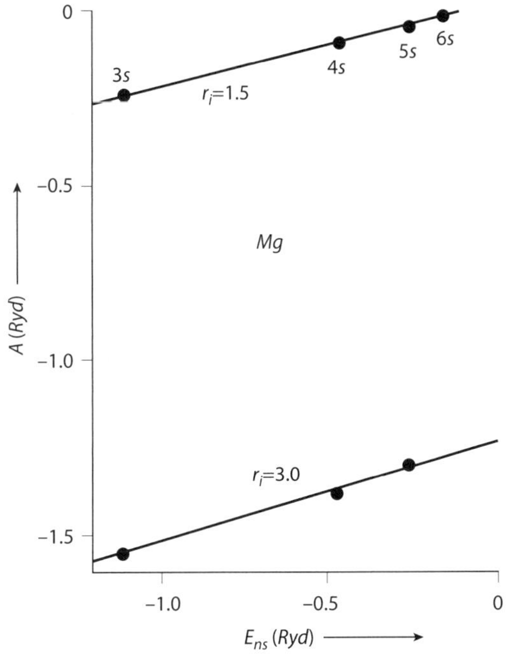

FIGURE 3.11. $A_0(E_{ns})$ for bound states in atomic Mg^{2+}. Lines are two different core radii $a = 4_i$.

$\chi(r)$, which has few or no oscillations. This approximation is the cause of having a potential that is energy dependent. In the Dirac notation, let $|j\rangle$ be an atomic orbital in the closed shell of electrons. They are orthogonal:

$$\langle j|j'\rangle = \delta_{jj'} \tag{3.144}$$

The correct wave function $|R\rangle$ is the pseudo-wave function $|\chi\rangle$ with the core states projected out:

$$|R\rangle = |\chi\rangle - \sum_{j=1}^{Z_v} |j\rangle\langle j|\chi\rangle \tag{3.145}$$

This choice has the correct feature that $|R\rangle$ is orthogonal to all of the bound orbitals. If $|v\rangle$ is one of the core orbitals, then

$$\langle v|R\rangle = \langle v|\chi\rangle[1 - 1] = 0 \tag{3.146}$$

If $H|R\rangle = E|R\rangle$, then the Hamiltonian equation for $|\chi\rangle$ is

$$E\left[|\chi\rangle - \sum_{j=1}^{Z_v} |j\rangle\langle j|\chi\rangle\right] = H\left[|\chi\rangle - \sum_{j=1}^{Z_v} |j\rangle\langle j|\chi\rangle\right] = H|\chi\rangle - \sum_{j=1}^{Z_v} E_j|j\rangle\langle j|\chi\rangle \tag{3.147}$$

$$E|\chi\rangle = [H + V_P(r)]|\chi\rangle \tag{3.148}$$

$$V_P(r) = \sum_{j=1}^{Z_v} (E - E_j)|j\rangle\langle j| \tag{3.149}$$

Using the pseudopotential introduces an auxillary potential $V_P(r)$ that depends linearly on energy. That is the reason that $A_\ell(E)$ depends linearly on energy as shown in fig. 3.11.

3.7 Measurement of Energy Bands

The energy bands in a solid can be measured by several techniques. They are discussed in historical order.

3.7.1 *Cyclotron Resonance*

Cyclotron resonance measures the motion of electrons in a magnetic field B. A free electron rotates in a circular orbit, perpendicular to the magnetic field, with the cyclotron frequency

$$\omega_c = \frac{eB}{m_e} \tag{3.150}$$

The sample is put into a microwave cavity, where the microwave frequency is ω_0. The magnetic field B is varied until $\omega_0 = \omega_c$, at which point the microwaves are being absorbed.

In quantum mechanics, the microwave absorbs by exciting an electron between Landau levels. Then one has a measurement of the electron mass:

$$m_e = \frac{eB}{\omega_0} \tag{3.151}$$

Cyclotron resonance was first done on semiconductors. There the cyclotron response is given by the effective mass: for the conduction band it is m_c. In GaAs, where the effective mass is isotropic, the above formula is altered to

$$m_c = \frac{eB}{\omega_0} \tag{3.152}$$

In Si or Ge, with their six or four ellipsoids, one measures a function of the parallel and perpendicular effective masses.

Obtaining a sharp resonance at $\omega_0 = \omega_c$, with a large Q, requires that $\omega_c\tau \gg 1$, where τ is the lifetime of the electron. The experiments are done at low temperatures and in very pure semiconductors, with as few impurities as possible. Of course, there is a negligible density of conduction-band electrons under those conditions, so one needs a light source to excite electrons to the conduction band during the experiment. The same experiment can be done on metals. It is discussed in the next chapter, in the section on the Fermi surface.

3.7.2 *Synchrotron Band Mapping*

Unoccupied energy bands of a solid can be measured by a photoemission experiment:

- A light source generates a monochromatic beam of photons of frequency ω and wave vector $\mathbf{q}$.
- This light is absorbed by exciting an electron from an initial wave vector $\mathbf{k}_i$ to a final wave vectors $\mathbf{k}_f = \mathbf{k}_i + \mathbf{q}$. For photons in the ultraviolet, then $\mathbf{q} \ll \mathbf{k}_i$ and $\mathbf{k}_f \approx \mathbf{k}_i$.
- If the velocity $\mathbf{v}(\mathbf{k}_f)$ is pointed at the surface, the electron can go to the surface in this conduction-band state, and exit the surface.
- The electron is captured by a counter that registers its angle of emission and its kinetic energy.

The electron wave refracts at the surface. A clean surface conserves the wave vector parallel to the surface plane. If the wave vector is $\mathbf{k}_o$ outside of the surface, conservation of parallel wave vector means that

$$k_{xf} = k_{xo}, \quad k_{yf} = k_{yo}, \quad k_{zo}^2 = k_{zf}^2 - \frac{2mV_0}{\hbar^2} \tag{3.153}$$

where $-V_0$ is the energy at the bottom of the band, with respect to vacuum. By measuring the energy and direction of the emitted particle, then one knows all three values (k_{xo},

k_{yo}, k_{zo}), all three values of $\mathbf{k}_f$, and all three values of $\mathbf{k}_i$, as well as all of the energies. Then one knows that point on the occupied and unoccupied energy bands. By measuring emitted electrons at different angles, one gets a band map. Usually a synchrotron is used as the light source, in order to have an intense beam of polarized, monochromatic light, whose frequency can be varied continuously.

References

1. N. A. Cade and P.M. Lee, Self-consistent energy band structures for HgTe and CdTe. *Solid State Commun.* **56**, 637–641 (1985)
2. J. R. Chelikowsky, T. J. Wagener, J. H. Weaver, and A. Jin, Valence-and conduction-band densities of states for tetrahedral semiconductors. *Phys. Rev. B* **40**, 9644 (1989)
3. W. Y. Ching and J. Callaway, Energy bands, optical conductivity, and Compton profile of sodium. *Phys. Rev. B* **11**, 1324–1328 (1975)
4. M. L. Cohen and T. K. Bergstresser, Band structures and pseudopotential form factors for fourteen semiconductors of the diamond and zinc-blende structures. *Phys.Rev.* **141**, 789–796 (1966)
5. J. F. Cornwell, Electronic energy bands of alkali metals. *Proc. R. Soc. London Ser. A* **261**, 551 (1961)
6. G. Dresselhaus, A. F. Kip, and C. Kittel, Cyclotron resonance of electrons and holes in silicon and germanium crystals. *Phys. Rev.* **98**, 368–384 (1955)
7. S.D. Kevan, *Angle-resolved Photoemission* (Elsevier, Amsterdam, 1992)
8. C. Kittel, *Quantum Theory of Solids* (Wiley, New York, 1963)
9. R. de L. Kronig and W. G. Penney, Quantum mechanics of electrons in crystal lattices. *Proc. R. Soc. London, Ser. A* **130**, 499–513 (1931)
10. A. B. Kunz, Energy bands and the optical properties of LiCl. *Phys. Rev. B* **2**, 5015 (1970)
11. S. Kurtin, T. C. McGill, and C. A. Mead, Fundamental transition in the electronic nature of solids. *Phys. Rev. Lett.* **22**, 1433 (1969)
12. W. Metzner and D. Vollhardt, Correlated lattice fermions in $d = \infty$ dimensions. *Phys. Rev. Lett.* **62**, 324–327 (1989)
13. L. Pauling, *The Nature of the Chemical Bond*, 3rd ed. (Cornell University Press, Ithaca, NY, 1960)
14. J. B. Pendry, *Low Energy Electron Diffraction* (Academic Press, New York, 1974)
15. L. M. Roth, B. Lax, and S. Zwerdling, Theory of optical magneto-absorption effects in semiconductors. *Phys. Rev.* **114**, 90 (1959)

Homework

1. Kronig and Penney [9] did the first band structure calculation in one dimension. Their potential $V(x)$ was a periodic array of square well potentials of width b and height V_0, as shown in fig. 3.12. Calculate $\phi_1(a)$ for this periodic potential, assuming $E > V_0$.

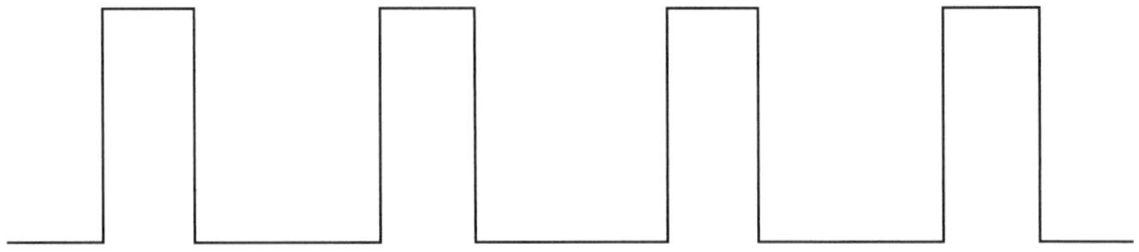

FIGURE 3.12. Kronig-Penney model.

2. Graphene is the name for a single sheet of graphite. It is a pure carbon material with the carbon atoms arranged into a two-dimensional honeycomb lattice.

   (a) What is the unit cell in real space?

   (b) What is the Brillouin zone?

   (c) Three electrons per carbon are in the sigma bonds, which are fully occupied. The other electron per atom is in the π-bands generated by the p_z orbitals that are orthogonal to the xy-plane. Since the p_z-orbitals are the same on each atom, one can calculate the band structure of the π bands using an isotropic tight-binding model. Derive these bands, and draw them along the two major axes in the BZ. At which point is the chemical potential? *Note*: Corner at zone edge is K-point, while center of line at zone edge is M-point. What are the band energies at these two points?

3. Graphene has two carbon atoms per unit cell, arranged in the honeycomb lattice. What is the structure factor for (i) $\mathbf{G} = 0$, (ii) the lowest set of nonzero reciprocal lattice vectors, and (ii) the second lowest set of reciprocal lattice vectors? All vectors with the same value of G^2 are in the same set.

4. Consider the pt lattice in 2D. The energy dispersion in the tight-binding model is

$$\gamma(\mathbf{q}) = \sum_{\delta=1}^{6} \exp[i\mathbf{q}\cdot\boldsymbol{\delta}] \tag{3.154}$$

   Evaluate this expression at three points in the Brillouin zone: (i) at $\mathbf{q} = 0$, (Γ-point), (ii) at any corner of the BZ (K-point), and (iii) at the center of any edge of the BZ (M-point: see fig. 15.5). Recall that the BZ is a hexagon.

5. For the square lattice of DO_2 atoms, what is the contribution to the Hamiltonian matrix from second-neighbor overlaps? Consider $s-s$ overlaps w_{2s} and $p-p$ overlaps w_{2p}.

6. Solve for the tight-binding energy bands of a sq lattice in two dimensions. There is one atom per unit cell, which has three orbitals: $\phi_s \sim g(r)$, $\phi_x \sim xf(r)$, and $\phi_y \sim yg(r)$. The two atomic eigenvalues are E_s and E_p. The overlap matrix elements between neighbors are w_{ss}, w_{sp}, w_{pp}. Derive the eigenvalue matrix, but do not try to solve the eigenvalue equation.

7. The dice lattice has the chemical formula DO_2, but is based on the pt lattice. Each D atom has six neighbors of O. Each O atom has three neighbors of O, and three of D.

   (a) Draw the lattice.

   (b) Calculate the energy bands for p_z-orbitals, assuming those on D have a site energy of E_D, and those on O have a site energy of E_O. The D atoms bond to all six of their neighbors, but the O atoms only bond to the D atoms.

8. In a magnetic field, write the Hamiltonian with a vector potential $\mathbf{A}$:

$$H = \frac{1}{2m}(\mathbf{p} - e\mathbf{A})^2 + U(\mathbf{r}) \tag{3.155}$$

where $U(\mathbf{r})$ is a periodic potential. Use $k \cdot p$ theory to prove that in the conduction band of a semiconductor, one can write the orbital part of the magnetic field as

$$H_{\text{eff}} = E_{c0} + \frac{1}{2m_c}(\mathbf{p} - e\mathbf{A})^2 \tag{3.156}$$

where m_c is the conduction band mass. This theorem is the basis for the quantum Hall effect!

9. Make a Mead plot using the high-frequency (ε_∞) and low-frequency (ε_0) dielectric constants. They can be found in Landolt-Börnstein, New Series, Group III, Volume 17 (many volumes), on the reference shelf with call number QC61.L332. Plot something like

$$Y = \frac{\varepsilon_0 - \varepsilon_\infty}{\varepsilon_0 + \varepsilon_\infty} \tag{3.157}$$

Your graph should have at least 20 data points.

10. A semiconductor has a simple cubic structure, and the energy-band dispersion for the conduction and valence bands are given by the tight-binding model:

$$E_c(\mathbf{k}) = E_{c0} + A[\cos(k_x a) + \cos(k_y a) + \cos(k_z a)] \tag{3.158}$$

$$E_v(\mathbf{k}) = E_{v0} - B[\cos(k_x a) + \cos(k_y a) + \cos(k_z a)] \tag{3.159}$$

(a) Show that the band gap is at the corner of the BZ, and find the expression for the gap.

(b) What are the effective masses of conduction and valence bands?

11. In two dimensions, the square lattice has a Brillouin zone in the shape of a square, where $(-\pi < \theta_x < \pi, -\pi < \theta_y < \pi)$, where $\theta_\mu = k_\mu a$. The dispersion of the electron is found to be

$$\varepsilon = E_0 \cos(\theta_x) \cos(\theta_y) \tag{3.160}$$

Find:

- The line of points in the BZ that have $\varepsilon = 0$
- The places in the BZ where ε has its maximum value.
- The places in the BZ where ε has its minimum value.
- Shade in the Fermi surface when the band is half-full $(E_0 > 0)$.
- The effective mass at the band minimum.

12. Derive the density of states in three dimensions when

$$\varepsilon(k) = E_0 + \frac{\hbar^2}{2}\left[\frac{k_x^2}{m_x} + \frac{k_y^2}{m_y} + \frac{k_y^2}{m_z}\right] \tag{3.161}$$

Apply this result to the conduction bands of Si and Ge. What is their density of states?

13. Derive the density of states $N_2(E)$ for the two-dimensional pt lattice. At what points in the BZ does the density of states diverge?

14. Derive the form of the density of states $N_n(E)$ for a lattice of n-dimensions when the bands have quadratic dispersion

$$N_n(E) = \int \frac{d^n p}{(2\pi)^n} \delta\left[E - \frac{\hbar^2}{2m}\sum_{j=1}^{n} p_j^2\right] \tag{3.162}$$

The density of states has the general form

$$N_n(E) = C_n E^{\alpha(n)} \tag{3.163}$$

Find the exponent $\alpha(n)$. The prefactor C_n is harder to find.

16. Use $k \cdot p$ theory to evaluate the integral

$$M_{cv}(\mathbf{q}) = \int d^3 r u_c^*(\mathbf{k} + \mathbf{q}, \mathbf{r}) u_v(\mathbf{k}, \mathbf{r}) \tag{3.164}$$

Assume that $\mathbf{q}$ is a photon wave vector, and is small. *Hint*: The $k \cdot p$ theory can be evaluated around any k-point.

4 Insulators

4.1 Rare Gas Solids

The simplest insulators are composed of single, neutral, elements from the last column of the periodic table: Ne, Ar, Kr, Xe. Solid He is not on the list, since it has significant quantum fluctuations that make it a complicated solid. The rare gas solids are thought to be held together by van der Waals interactions. The potential energy between any pair of neutral atoms is written as

$$V(R) = 4\varepsilon\left[\left(\frac{\sigma}{R}\right)^{12} - \left(\frac{\sigma}{R}\right)^{6}\right] \tag{4.1}$$

The two terms in this expression have a different origin. The term with R^{-6} is from the van der Waals interaction:

$$V_{\mathrm{vdW}} = -\frac{C_6}{R^6}, \quad C_6 = 4\varepsilon\sigma^6 \tag{4.2}$$

The constant C_6 is known for many pairs of atoms and ions. The term in eqn. (4.1) with R^{-12} is phenomenological. It is added as a repulsive interaction to keep the atoms apart. The exponent of -12 is a guess, and perhaps not a good one. But it is now well entrenched. The distance σ is where the potential function vanishes, and is sort of a hard-sphere distance. The energy ε is the depth of the potential well.

The ground-state energy of the rare gas solid is given by the summation of this potential over all pairs of atoms:

$$E_G = \frac{1}{2}\sum_{i\neq j} V(R_{ij}) = \frac{N}{2}\sum_{j\neq 0} V(R_{j0}) \tag{4.3}$$

where the factor of $\frac{1}{2}$ comes from counting each pair twice. The two power laws are summed over the crystal structure:

Table 4.1 Van der Waals summations for three lattices

Crystal	A_6	A_{12}	$A_6^2/2A_{12}$
fcc	14.4519	12.1319	8.6078
hcp	14.4548	12.1353	8.6088
bcc	12.2519	9.1142	8.2349

$$A_\ell = \sum_{j\neq 0}\left(\frac{d}{R_{j0}}\right)^\ell \tag{4.4}$$

where $d = a/\sqrt{2}$ is the nearest-neighbor distance for an fcc or hcp crystal. The hcp lattice is assumed to be ideal, with twelve nearest neighbors. Both the fcc and hcp have the same number (12) of nearest neighbors, and their values in table 4.1 are similar. The value for A_{12} comes almost entirely from first neighbors. The bcc lattice has eight first neighbors, so its values are different.

The ground-state energy per atom is

$$E_G/N = 2\varepsilon\left[A_{12}\left(\frac{\sigma}{d}\right)^{12} - A_6\left(\frac{\sigma}{d}\right)^6\right] \tag{4.5}$$

The ratio (d/σ) is varied to find the nearest-neighbor spacing d with the lowest ground-state energy. Instead of varying d, choose another parameter Y to vary:

$$Y = \left(\frac{\sigma}{d}\right)^6 \tag{4.6}$$

$$E_G(Y)/N = 2\varepsilon[A_{12}Y^2 - A_6 Y] \tag{4.7}$$

$$0 = \frac{1}{N}\frac{dE_G}{dY} = 2\varepsilon[2A_{12}Y_0 - A_6] \tag{4.8}$$

$$Y_0 = \frac{A_6}{2A_{12}} \tag{4.9}$$

$$E_G(Y_0)/N = -\varepsilon\frac{A_6^2}{2A_{12}} = -8.608\varepsilon \tag{4.10}$$

The crystal with the lowest ground-state energy has the largest value of $A_6^2/2A_{12}$. According to table 4.1, this is hcp. The rare gas solids (except He) all have fcc lattices, so this prediction is incorrect. Table 4.2 shows the binding energy per atom, in units of electron volts, for the rare gas solid. The predictions of eqn. (4.10) are accurate.

4.2 Ionic Crystals

Ionic crystals are solids composed of ions. In NaCl, for example, the Na atom has a filled 2p-shell plus an extra 3s-valence electron. The latter leaves the sodium atom and is

Table 4.2 Ground-state energy per atom, in electron volts

Solid	a(Å)	ε(eV)	σ(Å)	E_G (theory)	E_G(exp)
Ne	4.4644	0.0031	2.74	−0.027	−0.02
Ar	5.4676	0.0104	3.40	−0.090	−0.08
Kr	5.744	0.0140	3.65	−0.120	−0.11
Xe	6.350	0.0200	3.98	−0.172	−0.17

transferred to the chlorine atom. Chlorine has five-valence electrons, and the extra electron fills its p-shell. We view the solid as composed of Na^+ and Cl^- ions. The two types of ions are packed together in the crystal. The electron orbitals of nearest neighbors overlap a bit where they are pressed together. If one tries to count how many electrons are on the sodium ion and how many are on the chlorine ion, it gets confusing in the regions where the charges overlap. Some workers interpret this confusion as showing that the charge transfer is not complete and that the separate ions are partially ionized. We disagree. We think they are indeed ions with integer valence, but that the electron orbitals overlap when nearest neighbors are pressed together. The charge densities are sums of overlapping spherical distributions. This model of complete ionization gives a satisfactory account of many properties of these crystals.

Bragg scattering by x-rays measures the charge density $n(\mathbf{r})$ of electrons in the crystal. Precisely, it measures the Fourier transform $n(\mathbf{G})$ of the charge density at wave vectors equal to the reciprocal lattice vectors $\mathbf{G}$. These experimental values can be used to construct a real-space density plot of the electron density:

$$n(\mathbf{r}) = \sum_{\mathbf{G}} n(\mathbf{G})\, e^{i\mathbf{G}\cdot\mathbf{r}} \tag{4.11}$$

Figure 4.1 shows the density of electrons in CaF_2 as measured by Witte and Wölfel. Numbers denote lines of constant electron density. These lines are very circular. If one adds the charge density in the regions of the unit cell occupied by the two ions, one finds complete charge transfer. Ca and F have a large value of electronegativity difference ΔX, so the valence electrons of Ca are completely transferred to the F ion.

The ground-state energy of an ionic crystal can be calculated relatively accurately without great use of computers. Several different energy terms are discussed, starting with the largest.

The International System of units is employed. It uses two constants.

1. The magnetic permeability is $\mu_0 = 4\pi \times 10^{-7}$ newtons/ampere^2.

2. The dielectric permitivity is defined using μ_0 and the speed of light c:

$$\varepsilon_0 = \frac{1}{\mu_0 c^2} = 8.854188 \times 10^{-12} \text{ farad/meter} \tag{4.12}$$

The units of ε_0 are also coulombs^2/newton $\cdot$ m^2.

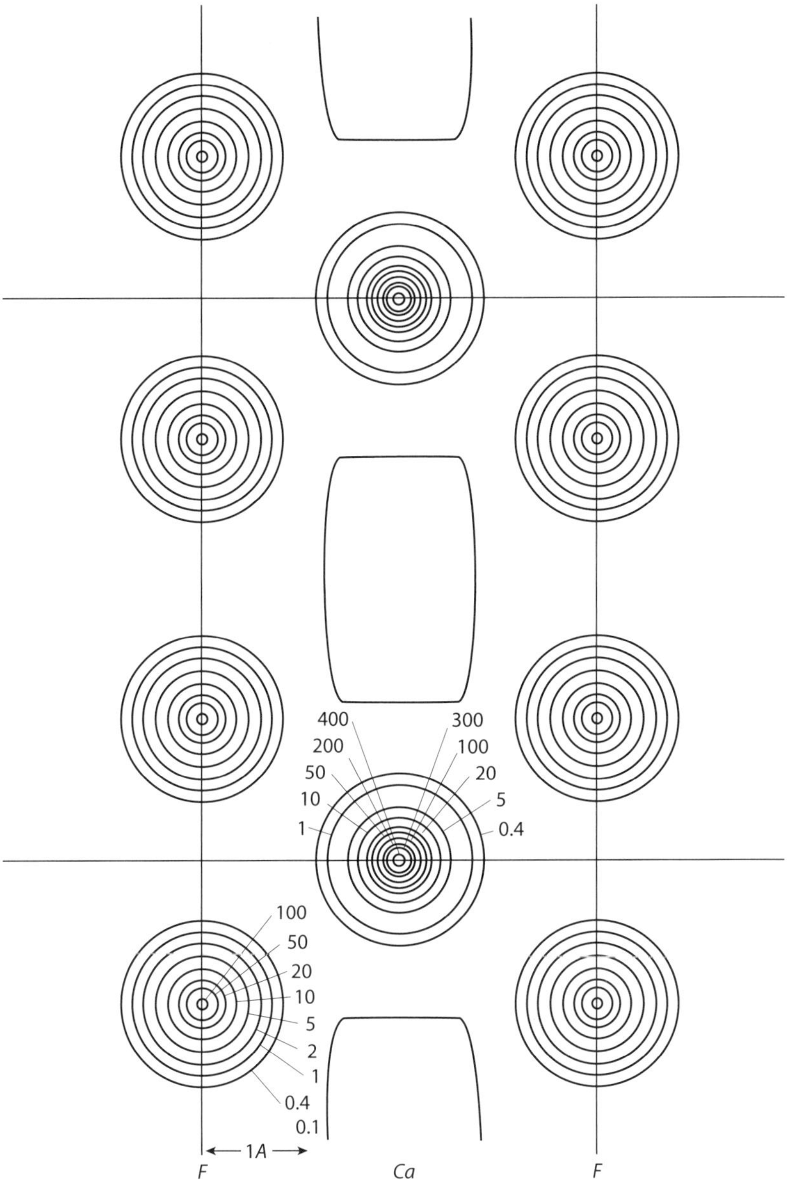

FIGURE 4.1. Charge density map in the *xxz* plane of CaF_2. By Witte and Wölfel, *Rev. Mod. Phys.* **30**, 51 (1958). Used with permission of the American Physical Society.

Electrical charges, such as e, are in coulombs. Potentials are in volts, and electric fields are in volts per meter.

4.2.1 Madelung Energy

For integer valence, the ion charges are $q_j = Z_j e$, where e is the electron charge, and Z_j is the valence. The Coulomb energy of a collection of charges q_j located at sites R_j is

Table 4.3 Madelung constants of several binary lattices

Structure	α
NaCl	1.74756
ZnS	1.63805
CsCl	1.76267

Table 4.4 Ionic binding energies, in eV per unit cell, of several crystals with the rocksalt lattice

Crystal	a	E_M	$(11/12)E_M$	U_0
NaF	4.620	−10.894	−9.99	
NaCl	5.6406	−8.923	−8.18	−8.0
NaBr	5.7932	−8.688	−7.96	−7.7
KF	5.347	−9.412	−8.63	
KCl	6.2929	−7.998	−7.33	−7.3
KBr	6.600	−7.626	−6.99	−7.1

Note. E_M is the madelung energy per unit cell, and $11E_M/12$ is corrected for repulsion. U_0 is the experimental cohesive energy per unit cell.

$$E_M = \frac{1}{4\pi\varepsilon_0} \sum_{i>j=1}^{N} \frac{q_i q_j}{|\mathbf{R}_i - \mathbf{R}_j|} \tag{4.13}$$

Let N denote the number of unit cells, which is the number of molecules. The subscript M denotes *Madelung energy*. The notation $i > j$ means to count each pair once. This calculation is relatively easy in binary crystals, which have two ions in the unit cell, with equal and opposite charges. One assumes the ions are point charges ($q_i = \pm Ze$), and finds

$$E_M = -\frac{e^2 Z^2}{4\pi\varepsilon_0 d} \alpha_M N \tag{4.14}$$

where $d = a/2$ is the distance between nearest neighbors: in NaCl it is the distance between a Na^+ and Cl^- ion. The other constant, α_M, is the Madelung constant, which depends on crystal structure. The ions are not point charges, and neighboring ions have overlapping charge distributions. However, if all distributions are spherical, this value is still accurate. Table 4.3 gives the value of this constant for several lattices favored by binary materials.

The Madelung constant α_M cannot be calculated by summing locally over a few neighbors. Ions far away contribute to the value. One must use *Ewald methods,* which are described in an appendix.

Table 4.4 shows the Madelung energy for several crystals with the rocksalt lattice. The second column is the lattice constant. The third column is the Madelung energy. It is the largest energy term, and is similar in value to the experimental value U_0. Many other lattice constants are given in chapter 2.

4.2.2 *Polarization Interactions*

The Coulomb interaction between ions is the largest energy term. Another energy term is due to the polarizability of the ions. If one ion has a charge q, it creates an electric field $\mathbf{E} = q\mathbf{r}/4\pi\varepsilon_0 r^3$. When this field acts on another atom or ion, it induces a dipole moment $\mathbf{p} = 4\pi\varepsilon_0 \alpha_d \mathbf{E}$, where α_d is the polarizability in units of volume, and $\mathbf{p}$ has the units of coulomb · meter. This dipole moment creates an electric field that acts back on the original charge. Since this is a self-energy effect, the energy term is

$$E_\alpha = -\frac{q^2\alpha_d}{8\pi\varepsilon_0 r^4} \tag{4.15}$$

This interaction is found between an ion and an atom in gases. It is *not* a term that contributes to the ground-state energy of most ionic crystals. In NaCl, each Cl^- has six Na^+ neighbors in an octahedral arrangement. This symmetrical arrangement does not produce an electric field, or a dipole moment, on the Cl^- ion. There is no induced dipole moment. The moral is that, when dealing with polarization forces, one cannot just add up pairwise interactions.

If a crystal has a static dielectric constant $\varepsilon = K\varepsilon_0$, the Coulomb interaction is screened by this dielectric constant:

$$V_{12} = \frac{q_1 q_2}{4\pi\varepsilon_0 K R_{12}} \tag{4.16}$$

Should the Madelung energy have a factor of inverse dielectric constant K? *No!* Again, dielectric screening is due to induced dipole moments on the ions. Since each ion has no net electric field due to the symmetrical arrangement of its neighbors, it has no net dipole and there is no factor of dielectric constant in the Madelung energy.

Are their any polarization corrections to the ground-state energy? The answer is yes. The energy term in eqn. (4.15) is the first term of a multipole expansion:

$$V = -\frac{q^2}{8\pi\varepsilon_0 r}\left[\frac{\alpha_d}{r^3} + \frac{\alpha_q}{r^5} + \frac{\alpha_o}{r^7} + \cdots\right] \tag{4.17}$$

where α_q, α_o are the quadrupole and octupole polarizabilities of the ion. A quadrupole has the units of qr^2 and an octupole has the units of qr^3. They are induced in the atom by the first and second derivatives of the electric field. Because of the symmetrical arrangement of the six neighbors in rocksalt, there is no quadrupole or octupole field on any ion when at its equilibrium site.

Zincblende structures have the four first-neighbor atoms in a tetrahedral arrangement. This symmetry does induce an octupole field at each ion site, which induces an octupole energy for each ion. Let $\mathbf{r} = (x, y, z)$ be the coordinates of an electron on an ion, as measured from its nucleus. Let $\boldsymbol{\delta}$ be the vector to a first neighbor. Assume $|\boldsymbol{\delta}| \gg |\mathbf{r}|$. The Coulomb energy between an electron and a neighbor ion of charge q is

$$V = \frac{-eq}{4\pi\varepsilon_0|\boldsymbol{\delta} - \mathbf{r}|} = -\frac{eq}{4\pi\varepsilon_0\delta}\left[1 + \frac{\boldsymbol{\delta}\cdot\mathbf{r}}{\delta^2} + \frac{3(\boldsymbol{\delta}\cdot\mathbf{r})^2 - r^2\delta^2}{2\delta^4} + \frac{5(\boldsymbol{\delta}\cdot\mathbf{r})^3 - 3(\boldsymbol{\delta}\cdot\mathbf{r})r^2\delta^2}{2\delta^6} + \cdots\right] \tag{4.18}$$

Evaluate this expression by summing over the four nearest neighbors in the tetrahedral bond:

$$\boldsymbol{\delta}_j = \frac{a}{4}(1, 1, 1),\ \frac{a}{4}(1, -1, -1),\ \frac{a}{4}(-1, -1, 1),\ \frac{a}{4}(-1, 1, -1) \tag{4.19}$$

The dipole and quadrupole terms in the expansion average to zero. In the fourth term, a nonzero contribution is

$$\sum_{j=1}^{4}(\boldsymbol{\delta}\cdot\mathbf{r})^3 = \left(\frac{a}{4}\right)^3[(x+y+z)^3+(x-y-z)^3+(-x-y+z)^3 + (-x+y-z)^3] = 24xyz\left(\frac{a}{4}\right)^3 \tag{4.20}$$

The octupole field at an ion site in zincblende from its four neighbors is

$$V_o = -\frac{eq}{4\pi\varepsilon_0}\frac{20}{\sqrt{3}}\frac{xyz}{d^4} \tag{4.21}$$

where the first-neighbor distance is $d = |\boldsymbol{\delta}| = a\sqrt{3}/4$. The factor of xyz is proportional to r^3 times an angular function $P_3^2(\theta)\sin(2\phi)$. The above interaction is evaluated using second-order perturbation theory for the central ion:

$$\Delta E = -\sum_m \frac{|\langle m|V_0|g\rangle|^2}{E_m - E_g} = -\left(\frac{40}{9}\right)\frac{q^2\alpha_o}{4\pi\varepsilon_0 d^8} \tag{4.22}$$

The above result can be derived using the definitions

$$\alpha_o = 2\frac{e^2}{4\pi\varepsilon_0}\sum_I \frac{|\langle g|r^3 P_3(\theta)|I\rangle|^2}{E_I - E_g} \tag{4.23}$$

$$\alpha_{xyz} = 2\frac{e^2}{4\pi\varepsilon_0}\sum_I \frac{|\langle g|xyz|I\rangle|^2}{E_I - E_g} = \frac{\alpha_o}{15} \tag{4.24}$$

The octupole polarizability is large for both the anions and cations. In silver and copper halides, which have the zincblende arrangement, this energy term is about one eV per unit cell. Considering the contribution from both ions in the unit cell, the ground-state energy is

$$\Delta E = -\left(\frac{40}{9}\right)\frac{q^2}{4\pi\varepsilon_0 d^8}[\alpha_{A,o} + \alpha_{C,o}] \tag{4.25}$$

For CuF, the two octupole polarizabilities are

$$\alpha_{Cu,o} = 1.6\ \text{Å}^7, \quad \alpha_{F,o} = 1.0\ \text{Å}^7, \quad d = 1.842\ \text{Å} \tag{4.26}$$

$$\Delta E = -\ 1.25\ \text{eV} \tag{4.27}$$

The octupole polarizability makes an important contribution to binding. Perhaps this explains why highly polarizable ions prefer the zincblende arrangement, compared to the rocksalt arrangement. The choice is partly due to ion radius, but partly due to polarization. Zincblende, with only four first neighbors, has smaller values for (i) the Madelung energy, (ii) the van der Waals energy, and (iii) the repulsive energy. It has the advantage over rocksalt of having a polarization energy.

A chemist describes an ion according to its propensity to form a covalent bond. An octupole moment is a covalent bond. Here we have treated covalency in the language of ionic polarization.

4.2.3 Van der Waals Interaction

The van der Waals interaction is a pairwise polarization interaction between two atoms. It has the general form, for interaction between atoms or ions (ij), of

$$V_{\rm vdW}(r) = -\frac{C_{6,ij}}{r^6} \tag{4.28}$$

In solids, the interaction energy between each pair of atoms or ions is calculated and added up for the ground-state energy. The difference from the previous case is that the force is due to fluctuations in polarization on each atom. Although the neighbors are arranged in a symmetrical configuration, they are all fluctuating independently. Modern computer codes enable accurate calculation of $C_{6,ij}$ between any pair of ions or atoms. In NaCl the van der Waals interaction between pairs of chlorine atoms is larger than between Na and Cl, so these pairs must be considered also. The Na^+ ions are in an fcc arrangement, so summing over all pairs of these ions gives the coefficient A_6 in table 4.1, and similarly for the Cl–Cl summation. For the Na–Cl summation, a new coefficient is needed, which is $B_6 \approx 6.6$, where the first six neighbors make the largest contribution. Divide by two to count each pair once, and the energy per unit cell from van der Waals attraction is

$$E_{\rm vdW} = -\frac{C}{d^6} \tag{4.29}$$

$$C = \left[B_6 C_{\rm NaCl} + A_6 \frac{C_{\rm ClCl} + C_{\rm NaNa}}{2(\sqrt{2})^6}\right] \tag{4.30}$$

In rare gas solids, and also in molecular crystals, the binding of the crystal is due entirely to van der Waals interactions. In ionic crystals, the main binding is due to the Madelung energy. The van der Waals energy makes a smaller contribution to the binding energy. The only calculation available for the constant C was by Mayer in 1933 [13]. Including it improves the estimates of the cohesive energy.

4.2.4 Ionic Radii

The next largest energy term is the repulsive interaction between neighboring ions. The long-range Coulomb interaction gives the Madelung energy, which draws the ions together. The short-range repulsive interaction keeps neighbors from getting too close together.

The range of the repulsive energy is determined by the ion radii. Charge densities of atoms and ions vanish at large distance. The radius of an ion or atom is not a precisely defined concept. Yet having a value is very useful when contemplating atomic structure. An extensive table of ionic radii was given by Pauling. Some of his values are presented in table 4.5. These radii are for the ions. The radius of a neutral atom has a different value. Values for double-negative ions, such as Se^{2-}, are omitted since they depend on the nearest-neighbor distance.

Table 4.5 Pauling radii r_i in Å

Ion	r_i	Ion	r_i	Ion	r_i
Li^+	0.60	F^-	1.36	Mg^{2+}	0.65
Na^+	0.95	Cl^-	1.81	Ca^{2+}	0.99
K^+	1.33	Br^-	1.95	Zn^{2+}	0.74
Rb^+	1.48	I^-	2.16	Sr^{2+}	1.13
Cs^+	1.69	Cd^{2+}	0.97	Ba^{2+}	1.35
Cu^+	0.96	Ag^+	1.26	Au^+	1.37

It is easy to construct a table of ionic radii. In a binary crystal such as KCl, the separation between ions is the sum of the two radii:

$$d = r_K + r_{Cl} \tag{4.31}$$

If one knows one of these values and also knows d, then the other radius is found by subtraction. The problem in this procedure is getting started. One has to know the radius of one ion, and the others are found from different ion pairs in different crystals. The easiest values to find are those of the anions. In KCl the two ions have similar radius, and the repulsive interaction, which determines the lattice constant, is between these two ions. However, the Li^+ ion has a very small radius. In LiCl, LiBr, and LiI, the repulsive interaction that determines the lattice constant is between the anions, which are second neighbors. Figure 4.2 shows a drawing of ion radii in KCl and LiCl. It is obvious that in KCl the ions that press together are first neighbors K^+ and Cl^-. In LiCl, the ions that press together are the second-neighbors Cl^-s. Since the second-neighbor distance in LiCl is $d_2 = a/\sqrt{2} = 3.63$ Å, the ionic radius of $Cl^- = d_2/2 = 1.81$ Å. From this value one can construct the full Pauling table of ionic radii.

4.2.5 *Repulsive Energy*

Another term in the ground-state energy is the repulsive interaction. It keeps the ions from getting too close together. The first two energy terms, Madelung and van der Waals,

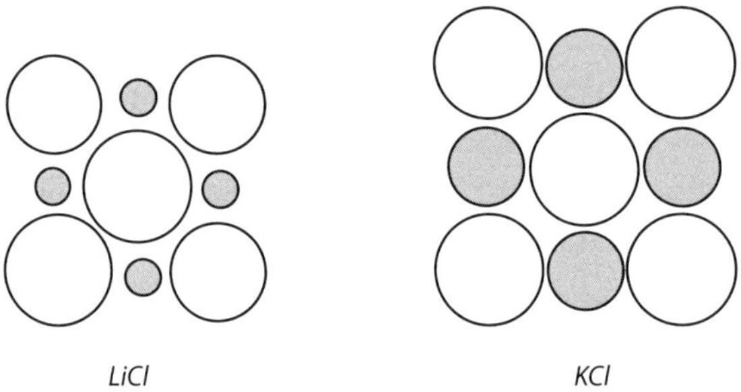

FIGURE 4.2. (a) Ion radii in LiCl. (b) Ion radii in KCl.

are very well characterized. The repulsive energy is not well known. Whether it is from cation–anion repulsion or from anion–anion repulsion is not known. So take it to be an unknown coefficient with a simple power law:

$$E_R = \frac{B}{d^n} \tag{4.32}$$

where (B, n) are adjustable parameters. Write the ground-state energy per unit cell as the sum of the three energy terms:

$$E(d) = \frac{B}{d^n} - \frac{C}{d^6} - \frac{A}{d} \tag{4.33}$$

The optimal value of the nearest-neighbor distance d is found from

$$0 = \frac{\delta E}{\delta d} = -\frac{nB}{d^{n+1}} + \frac{6C}{d^7} + \frac{A}{d^2} \tag{4.34}$$

Since d is known from experimental values, use the above expression to determine B:

$$B = \frac{d^n}{n}\left(\frac{6C}{d^6} + \frac{A}{d}\right) \tag{4.35}$$

$$E(d) = -\frac{C}{d^6}\left(1 - \frac{6}{n}\right) - \frac{A}{d}\left(1 - \frac{1}{n}\right) \tag{4.36}$$

Using these formulas, the repulsive interaction and the final ground-state energy is easily calculated. Table 4.4 shows the correction to the Madelung energy when $n = 12$. The term $(11/12)\, E_M$ is almost identical to the experimental binding energy.

4.2.6 Phonons

Phonons are the quantized vibrations of the ions. They are discussed in detail in chapter 7. The are described by harmonic oscillator statistics. At zero temperature, there is a zero-point motion due to all of the phonon frequencies $\omega_\lambda(\mathbf{q})$. Here $\mathbf{q}$ is a wave vector in the first BZ, while λ is a mode index. Since each ion can vibrate in three directions, the number of modes is three times the number of atoms in the unit cell of the crystal. In rocksalt there are six values of λ. Denote the highest phonon frequency as ω_x. Then $\omega_\lambda(\mathbf{q}) \le \omega_x$. An upper bound on the zero-point energy for N atoms is

$$E_P = \frac{1}{2N}\sum_{\lambda\mathbf{q}} \hbar\omega_\lambda(\mathbf{q}) < 3\hbar\omega_x \tag{4.37}$$

The maximum phonon energy $\hbar\omega_x$ in most crystals is less than 0.1 eV. It is a small contribution to the binding in ionic crystals.

The interesting feature of this term is that it contains the kinetic energy of the ions. In simple bound-state systems, such as the hydrogen atom, the electron kinetic energy is a Rydgerg energy E_{Ry}, the potential energy is $-2E_{Ry}$. The kinetic energy has a large magnitude. Here the kinetic energy is roughly half of the phonon zero-point energy, but that is a very small fraction of the binding energy of the ions. The kinetic energy in these systems makes a negligible contribution to crystal binding.

Excitons are another type of boson wave in the crystal. When an electron is excited to the conduction band of an insulator, it leaves a hole in the valence band. The hole is regarded as an excitation with a positive mass and positive charge. Hydrogenic-like bound states can form between the positive hole and the negative electron. In the crystal, it is a type of polarization wave. An individual ion has this excitation, but the atomic excitation is localized on that ion. In the crystal, the exciton can move from ion to ion and has dispersion. A contribution to the ground-state energy of the crystal is from the zero-point energy difference between the isolated ions $\hbar\omega_A/2$ and the crystal:

$$\delta E = \frac{\hbar}{2N}\left[\sum_{\mathbf{k}} \omega_x(\mathbf{k}) - N\omega_A\right] \tag{4.38}$$

This term makes an important contribution to the ground-state energy. However, *we have already included it!* It is the van der Waals energy. Optical transitions to create exciton states involve the interband dipole matric element $\langle c|\hat{\varepsilon}\cdot\mathbf{p}|v\rangle$, which is the same matrix element that enters into the definition of the polarizability, and of the van der Waals coefficient. See Mahan [11] for a discussion of evaluating the van der Waals energy as exciton zero-point energy.

4.3 Dielectric Screening

Many insulators have ionic bonding. One can view the solid as a collection of ions, where each has integer charge. An example is table salt, which can be written accurately as Na^+Cl^-. The optical and dielectric properties of these materials can be described by a simple model. Each ion, whether cation or anion, is assumed to be a spherical small entity with a polarizability α_j, which is assumed to be isotropic. An electric field $\mathbf{E}_j$ acting on the ion induces a dipole moment:

$$\mathbf{p}_j = 4\pi\varepsilon_0\alpha_j\mathbf{E}_j \tag{4.39}$$

The polarizability has the units of volume, the dipole moment has the units of charged times distance, and the electric field has the units of volts per meter. The dielectric constant ε_0 of vacuum has the units of coulomb^2 per volt-meter. A similar model is applied to molecular solids. Molecules are generally not spherically symmetric, in which case the polarizability is a tensor: $\alpha_{\mu\nu}$. For ions with filled electronic shells, the polarizabillity is isotropic: it is a diagonal tensor.

4.3.1 *Dielectric Function*

Consider that an electromagnetic wave is propagating through the crystal with a wave vector $\mathbf{k}$. There are two electric fields: (i) the *applied electric field*

$$\mathbf{E}_a = \mathbf{E}_0 e^{i(\mathbf{k}\cdot\mathbf{r} - \omega t)} \tag{4.40}$$

and (ii) the *local electric field* at each site $\mathbf{R}_j$

$$\mathbf{E}_\ell(\mathbf{r}) = \mathbf{E}_{\ell 0} e^{i(\mathbf{k}\cdot\mathbf{r} - \omega t)} \tag{4.41}$$

The local field differs from the applied field because of dipole–dipole interactions. The electric field at each ion site induces a dipole moment $\mathbf{p}_j = 4\pi\varepsilon_0\alpha_j\mathbf{E}_\ell$ proportional to the local electric field. The induced dipole provides another part of the local electric field due to dipole–dipole interactions:

$$\begin{aligned} \mathbf{E}_{\ell 0} e^{i(\mathbf{k}\cdot\mathbf{R}_j - \omega t)} &= \mathbf{E}_0 e^{i(\mathbf{k}\cdot\mathbf{R}_j - \omega t)} - \frac{1}{4\pi\varepsilon_0}\sum_{n\neq j}\phi(\mathbf{R}_{jn})\cdot\mathbf{p}_n \\ &= \mathbf{E}_0 e^{i(\mathbf{k}\cdot\mathbf{R}_j - \omega t)} - \sum_{n\neq j}\phi(\mathbf{R}_{jn})\cdot\alpha_n\mathbf{E}_{\ell 0} e^{i(\mathbf{k}\cdot\mathbf{R}_n - \omega t)} \end{aligned} \tag{4.42}$$

$$\phi_{\mu\nu}(\mathbf{R}) = \left(\frac{\delta_{\mu\nu}}{R^3} - \frac{3R_\mu R_\nu}{R^5}\right) \tag{4.43}$$

The induced dipole is created by the local field, not the applied field.

The simple case is when all atoms are identical, as in diamond, solid argon, or neon. Each atom has the same local field amplitude $\mathbf{E}_\ell$. Write the equations as

$$\mathbf{E}_l = \mathbf{E}_0 - \alpha\mathbf{T}(\mathbf{k})\cdot\mathbf{E}_\ell \tag{4.44}$$

$$T_{\mu\nu}(\mathbf{k}) = \sum_{n\neq j}\phi_{\mu\nu}(\mathbf{R}_{jn})e^{i\mathbf{k}\cdot(\mathbf{R}_n - \mathbf{R}_j)} \tag{4.45}$$

The matrix $T_{\mu\nu}(\mathbf{k})$ can be evaluated using Ewald methods. For optical waves, where the wave vector k is small, the result is quite simple. If a is the lattice constant, small wave vector means that $ka \ll 1$. The other length scale is the dimensions of the sample L. If $kL \leq 1$, the matrix $T_{\mu\nu}(\mathbf{k})$ depends on the shape of the sample: whether it is a needle or a platelet. This limit is important in nanostructures, where L is about the same size as the wavelength of light. The case of macroscopic samples have $kL \gg 1$. Then the matrix does not depend on the shape of the sample, but does depend on crystal symmetry:

$$\lim_{1/a \gg k \gg 1/L} T_{\mu\nu}(\mathbf{k}) = \frac{4\pi}{\Omega_0}\left[\frac{k_\mu k_\nu}{k^2} - t_{\mu\nu}\right] \tag{4.46}$$

$$\sum_{\mu=1}^{3} t_{\mu\mu} = 1 \tag{4.47}$$

The constant tensor $t_{\mu\nu}$ depends on the crystal structure. For cubic symmetry, the second-order tensor can only have diagonal elements, which must be identical. Then $t_{\mu\nu} = \delta_{\mu\nu}/3$. The volume of the unit cell is Ω_0, and denote $\tilde{\alpha} = \alpha/\Omega_0$ as the dimensionless polarizability per unit volume. The matrix equation for the electric field is now

$$\sum_\nu\left[\delta_{\mu\nu}\left(1 - \frac{4\pi\tilde{\alpha}}{3}\right) + 4\pi\tilde{\alpha}\hat{k}_\mu\hat{k}_\nu\right]E_{\ell,\nu} = E_{0,\mu} \tag{4.48}$$

Electromagnetic waves in vacuum have $\mathbf{k} \perp \mathbf{E}$. This is not true in crystals, unless the value of wave vector is small, or else the wave vector is going along directions of high symmetry. That case is assumed here, which gives

$$E_{\ell,\mu} = \frac{E_{0,\mu}}{1 - 4\pi\tilde{\alpha}/3} \tag{4.49}$$

The longitudinal components of the electric field obey

$$\hat{k} \cdot \mathbf{E}_\ell = \frac{\hat{k} \cdot \mathbf{E}_0}{1 + 8\pi\tilde{\alpha}/3} \tag{4.50}$$

The local electric field has a local field correction indicated in eqn. (4.49). In elementary textbooks, this factor is obtained by summing locally over atoms near to $\mathbf{R}_j$. Such summations give zero, and that derivation is incorrect. Other derivations use the integral over a spherical hole. That derivation does not indicate the proper dependence on photon wave vector. The above derivation is the correct way to get this local field. Using this local field, the full formula for the *Lorenz-Lorentz* dielectric function is $\varepsilon = K\varepsilon_0$, where

$$K = 1 + \frac{4\pi\tilde{\alpha}}{1 - 4\pi\tilde{\alpha}/3} \tag{4.51}$$

$$E_{\ell,\mu} = \frac{K+2}{3} E_{0,\mu} \tag{4.52}$$

The local transverse electric field is proportional to the applied field, with the ratio given by $(K+2)/3$. These important formulas are the basis for a wide variety of phenomena in cubic solids. An alternate way to present the above formula is as the *Clausius-Mossotti relation*:

$$\frac{K-1}{K+2} = \frac{4\pi\alpha}{3\Omega_0} \tag{4.53}$$

4.3.2 *Polarizabilities*

The above formula can be used to determine the polarizability of the atoms. In the rare gas solids, one finds the results of table 4.6. Tessman, Kahn, and Shockley [18] used the dielectric constants of the 20 crystals with the alkali–halide structure to deduce the polarizability of the alkali cations and halide anions. They assumed that the ions have the same polarizability in all crystals. This approximation is not very accurate. A better method was proposed much earlier by Joe Mayer [13]: he assumed that the cation polarizability was transferable, since the alkalis have tightly bound closed shells of electrons. In contrast, the halide anions are loosely bound, and the polarizatility changes from crystal to crystal. However, the changes are quite smooth when graphed as a function of the lattice constant. In all cases, the total polarizability of the solid is obtained from the Clausius-Mossotti relation. The halide polarizability α_H is found as

Table 4.6 Experimental dipole polarizabilities of the rare gas atoms, and alkali ions, in units of cubic angstroms (Å^3)

Atom	α	Ion	α
He	0.205	Li^+	0.030
Ne	0.396	Na^+	0.147
Ar	1.64	K^+	0.81
Kr	2.48	Rb^+	1.35
Xe	4.04	Cs^+	2.34

$$\alpha_H = \frac{3}{4\pi\Omega_0}\frac{K-1}{K+2} - \alpha_A \tag{4.54}$$

Figure 4.3 shows a graph of anion polarizability of halides (in units of Å^3) versus the lattice constant of the crystal.The four points on each line are for Li, Na, K, Rb. These graphs show that the anion polarizabilities are not the same in all crystals. A similar graph can be made for chalcogenide polarizability. In making these graphs, the alkali ion polarizability in table 4.6 was used for α_A. The quadrupole (α_q) and octupole (α_0) polarizabilities of the anions also depend significantly on the crystalline environment.

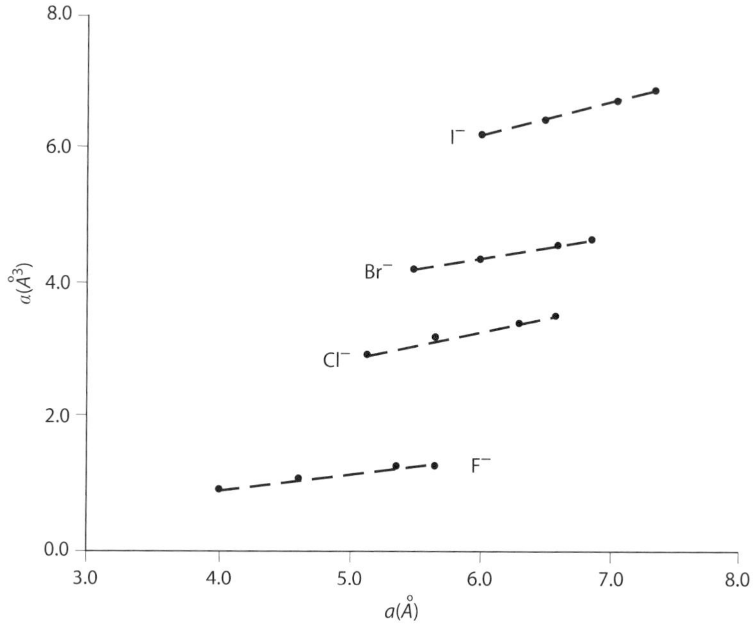

FIGURE 4.3. Anion polarizabilities in rock-salt structures as a function of lattice constant. Data from Mahan and Subbaswamy [12].

4.4 Ferroelectrics

Ferroelectrics are crystals that have a spontaneous electric polarization. Given that

$$\mathbf{D} = \varepsilon_0 \mathbf{E} + \mathbf{P} \tag{4.55}$$

Ferroelectrics have a fixed value of polarization **P**, which has the units of Coulombs per square meter. The macroscopic polarization is due to microscopic dipoles in the unit cell of the crystal. The microscopic dipoles can arise from either electronic polarization or from ionic polarization. Often both electrons and ions contribute to the polarization.

At high temperatures, the crystal is not ferroelectric, but still has interesting dielectric properties. The high-temperature phase is called a *paraelectric*. The transition temperature T_c is where the material changes from the high-temperature paraelectric phase to the low-temperature ferroelectric phase. The low-temperature ferroelectric phase usually has a different crystal structure than the paraelectric phase. The change in structure is usually due to small displacements of the atoms in the unit cell.

Some crystals, such as barium titanate ($BaTiO_3$), have several different low-temperature ferroelectric phases, and each has a different transition temperature. The values of T_c are different for each material. Examples for some perovskites are shown in table 4.7. Above the highest value listed the material is in its cubic phase. At the first (highest) value of T_c the crystal distorts into a lower symmetry phase.

Many ferroelectric crystals have the perovskite strucure. An example of a nonperovskite ferroelectric crystal is GeTe, with $T_c = 670$ K. It has the rocksalt structure. It has a small value of polarization, and is not commercially interesting. However, it is one of the few ferroelectrics that is also a semiconductor.

A dramatic feature of the ferroelectric transition is the divergence of the static dielectric constant as the temperature T approaches T_c. The behavior is described by the formula

$$\varepsilon(T) = \frac{C}{|T - T_c|} \tag{4.56}$$

This behavior is usually found only for $T > T_c$. The usual way to graph the data is as the inverse dielectric constant vs. temperature. Figure 4.4 shows this plot for $BaTiO_3$.

Table 4.7 Ferroelectric transition temperatures in kelvins in some perovskites

Crystal	T_c (K)
$SrTiO_3$	105
$BaTiO_3$	403, 278, 183
$PbTiO_3$	763
$NaTaO_3$	903, 823, 753
$NaNbO_3$	916, 845, 793, 753, 638, 73
$KNbO_3$	691, 498, 263

Note. Multiple values indicate multiple ferroelectric phases.

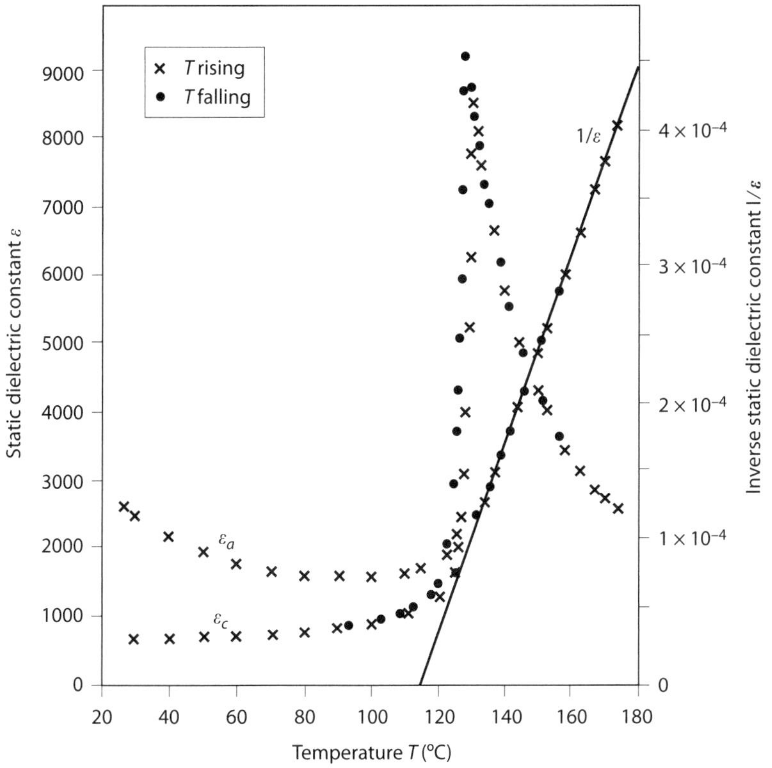

FIGURE 4.4. $\varepsilon_0/\varepsilon(T)$ vs. T for $BaTiO_3$. Data from Johnson, *Appl. Phys. Lett.* **7**, 221 (1965). Used with permission from the American Institute of Physics.

4.4.1 *Microscopic Theory*

A microscopic theory of ferroelectric ordering is found by assuming each unit cell has one or several dipoles $p_{j\mu}$ that are created during the transition. These dipoles could arise from small displacements of ions or from electronic charge polarization. The dipoles are assumed to mutually interact by the dipole–dipole interaction. The potential energy is

$$V = -\frac{1}{8\pi\varepsilon_0}\sum_{ij,\mu\nu} p_{i\mu}\phi_{\mu\nu}(\mathbf{R}_{ij})p_{j\nu} - \sum_i p_{i\mu}E_\mu \tag{4.57}$$

where E_μ is an applied field. The dipole–dipole tensor $\phi_{\mu\nu}$ is defined in eqn. (4.43). In actual ferroelectrics, elastic interactions between dipoles are also important. They are discussed in a later chapter.

The force on one dipole is determined by its local field:

$$E_{\ell\mu}(\mathbf{R}_i) = -\frac{\delta V}{\delta p_{i\mu}} = \frac{1}{4\pi\varepsilon_0}\sum_{j\nu}\phi_{\mu\nu}(\mathbf{R}_{ij})p_{j\nu} + E_\mu \tag{4.58}$$

The summation on the right extends over many lattice sites, since the dipolar interaction declines slowly with distance. It is a good approximation to replace $p_{j\nu}$ by its average in the crystal, which is $\langle p_\nu\rangle$. The summation over j becomes a constant tensor called $T_{\mu\nu}$, which has the units of inverse volume. Then the average local electric field on each dipole is the same:

$$\langle E_{\ell\mu} \rangle = \frac{1}{4\pi\varepsilon_0} \sum_\nu T_{\mu\nu} \langle p_\nu \rangle + E_\mu \tag{4.59}$$

$$T_{\mu\nu} = \sum_j \phi_{\mu\nu}(\mathbf{R}_{ij}) \tag{4.60}$$

The approximation of replacing $p_{j\nu}$ by its average is called by a variety of names, such as the *molecular field approximation* or the *mean field approximation*.

The tensor $T_{\mu\nu}$ is discussed earlier in this chapter, where it is a function of wave vector **k**. Here we need the result when $\mathbf{k} = 0$. If L is the dimension of the crystal, the value of $T_{\mu\nu}(\mathbf{k})$ becomes independent of wave vector whenever $kL > 1$. However, when $kL \ll 1$ the shape of the crystal determines the value of $T_{\mu\nu}$. Different values are found for a platelet than for a sphere or a needle. These differences arise from surface charges when the polarization fields are terminated at the boundaries.

The next step is to calculate the average value of the local dipole $\langle p_\nu \rangle$. It has an energy at each site of

$$\mathcal{E}_j = -\mathbf{p}_j \cdot \langle \mathbf{E}_{\ell,j} \rangle = -p_0 \langle E_\ell \rangle (\hat{n}_E \cdot \hat{n}_j) \tag{4.61}$$

where $\hat{n}_E$ is the direction of the field, $\hat{n}_j$ is the direction of the dipole, and p_0 is the magnitude of the dipole moment. In free space, the dipole can point in any direction. In a crystal, it is constrained by the atomic arrangements to point only in some directions. In cubic crystal, it may be constrained to point in the four directions of a tetrahedron or the six directions of an octahedron. This set of directions is denoted by $\hat{n}_j$. The thermal average of the dipole is

$$\langle p_\mu \rangle = p_0 \frac{\sum_j n_{j\mu} \exp(\phi_j)}{\sum_j \exp(\phi_j)} \tag{4.62}$$

$$\phi_j = \frac{p_0 \langle E_\ell \rangle}{k_B T} (\hat{n}_E \cdot \hat{n}_j) \tag{4.63}$$

To solve for the spontaneous polarization, use eqn. (4.59) for the average local field, and set the applied electric field $E_\mu = 0$. Then

$$\phi_j = \frac{1}{4\pi\varepsilon_0} \frac{p_0}{k_B T} \sum_{\mu\nu} n_{j\mu} T_{\mu\nu} \langle p_\nu \rangle \tag{4.64}$$

The above equation for ϕ_j, along with eqn. (4.62), gives a nonlinear equation for $\langle p_\nu \rangle$ as a function of temperature T. For $T > T_c$ the only solution is $\langle p_\nu \rangle = 0$. For $T < T_c$ it has solutions with nonzero values of $\langle p_\nu \rangle$. The transition temperature can be easily found since it is the temperature at which $\langle p_\nu \rangle \approx \eta$, where η is a very small value. Expand the right-hand side of eqn. (4.62) assuming η is very small:

$$\sum_{j=1}^{N_p} e^{-\phi_j} \approx N_p + O(\eta) \tag{4.65}$$

$$\sum_{j=1}^{N_p} n_{j\mu} e^{-\phi_j} \approx \sum_j \left[n_{j\mu} + \frac{p_0}{4\pi\varepsilon_0 k_B T} \sum_{\mu'\nu'} n_{j\mu} n_{j\mu'} T_{\mu'\nu'} \langle p_{\nu'} \rangle \right] \tag{4.66}$$

$$\sum_j n_{j\mu} n_{j\mu'} = \delta_{\mu\mu'} \frac{N_p}{3}, \quad \sum_j n_{j\mu} = 0 \tag{4.67}$$

$$\langle p_\mu \rangle = \frac{p_0^2}{12\pi\varepsilon_0 k_B T_c} \sum_\nu T_{\mu\nu} \langle p_\nu \rangle \tag{4.68}$$

In cubic crystals all second-rank tensors are diagonal, so $T_{\mu\nu} = \delta_{\mu\nu}/\Omega_a$, where Ω_a is a characteristic volume. The transition temperature is

$$T_c = \frac{p_0^2}{12\pi\varepsilon_0 k_B \Omega_a} \tag{4.69}$$

The same theory can be used to calculate the dielectric constant. Retain the applied electric field, so replace eqns. (4.64) and (4.68) by

$$\phi_j = \frac{p_0}{k_B T} \sum_\mu n_{j\mu} \left[\frac{1}{4\pi\varepsilon_0} \sum_\nu T_{\mu\nu} \langle p_\nu \rangle + E_\mu \right] \tag{4.70}$$

Calculate the average value of the dipole moment the same way:

$$\langle p_\mu \rangle = \frac{p_0^2}{3k_B T} \left[\frac{1}{4\pi\varepsilon_0} \sum_\nu T_{\mu\nu} \langle p_\nu \rangle + E_\mu \right] \tag{4.71}$$

Using $T_{\mu\nu} = \delta_{\mu\nu}/\Omega_a$ gives

$$\langle p_\mu \rangle = \chi E_\mu \tag{4.72}$$

$$\chi = \frac{p_0^2}{3k_B T} \left(\frac{1}{1 - T_c/T} \right) = \frac{p_0^2}{3k_B} \frac{1}{T - T_c} \tag{4.73}$$

$$\varepsilon = 1 + \frac{\chi}{\varepsilon_0 \Omega_0} \propto \frac{C}{T - T_c} \tag{4.74}$$

where Ω_0 is the volume per dipole. This dielectric function diverges as $T \to T_c$. This theory applies to the paraelectric phase when $T > T_c$.

Some insight into the ferroelectric transition is obtained from a one-dimensional model of an anharmonic potential for classical vibrations:

$$V(x) = -\frac{K_2}{2} x^2 + \frac{K_4}{4} x^4 \tag{4.75}$$

This potential energy is graphed in fig. 4.5. It is a symmetrical double well. At high energies, the oscillation is symmetrical. This behavior is found at high temperatures. At low temperatures, low-energy oscillations are confined to one well or another—we ignore tunneling between wells for a classical ion. As the temperature is lowered, there is a point (T_c) where the oscillator must choose one well or another and move into an off-center position. This behavior is the displacive motion in a ferroelectrics.

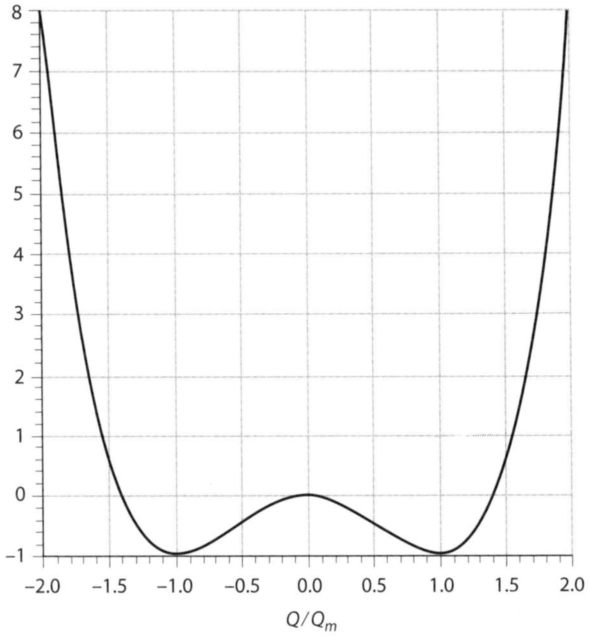

FIGURE 4.5. Particle oscillating in an anharmonic double well. At high energy, for large-amplitude oscillations, the motion is symmetrical. At low temperatures, for low-energy oscillation, the particle is confined to one of the two minima.

A feature of this model is that the vibrational frequency changes with energy. The horizontal line on this graph represents the total energy of the vibration, from $-x_0$ to $+x_0$. At the points $\pm x_0$ the kinetic energy is zero, and the total energy of the motion equals the potential energy:

$$E = -\frac{K_2}{2}x_0^2 + \frac{K_4}{4}x_0^4 \tag{4.76}$$

which can be solved to give

$$K_4 x_0^2 = K_2 + \sqrt{K_2^2 + 4K_4 E} \tag{4.77}$$

The frequency (ω) is related to the period (T) of the vibration, which changes with energy in a nonlinear way:

$$\omega^2 = \frac{1}{m}[K_4 x_0^2 - K_2] = \frac{1}{m}\sqrt{K_2^2 + 4K_4 E} \tag{4.78}$$

Since the energy is related to temperature, then the frequency of the vibration is temperature dependent. This quartic model predicts that $\omega^4 \propto E \propto k_B T$, whereas the experiments in ferroelectrics show that $\omega^2 \propto k_B(T - T_c)$. Other models [2] emphasize the competing forces from long-range dipolar interactions (a constant) and short-range near-neighbor fluctuations (T-dependent).

The Lyddane-Sachs-Teller relation between the longitudinal and transverse phonon frequencies and the high- and low-frequency dielectric functions is

$$\frac{\omega_{\rm TO}^2}{\omega_{\rm LO}^2} = \frac{\varepsilon_\infty}{\varepsilon_0} \propto \frac{\varepsilon_\infty}{C}|T_c - T| \tag{4.79}$$

In the last equality we have inserted the temperature dependence of the low-frequency dielectric function. The LST relation asserts that something is happening to the phonon frequencies at the ferroelectric transition. Experiments show that it is the transverse frequency that is tending toward zero:

$$\omega_{\rm TO}^2 \propto |T_c - T| \tag{4.80}$$

Since $\omega_{\rm LO} > \omega_{\rm TO}$, it is reasonable that $\omega_{\rm LO}$ is not vanishing. Neutron scattering experiments show that in some ferroelectrics, the transverse phonon modes are indeed tending toward zero. These are called *soft mode transitions.* The example of $PbTiO_3$ is shown in fig. 4.6. In this material T_c = 490°C, but the graphs for the square of the phonon frequency and the inverse dielectric constant are tending to zero at $T_0 \approx$ 450°C.

An important quantity in ferroelectrics is the degree of spontaneous polarization **P**. It has contributions from ion displacements and electronic charge distributions. The electronic term was initially confusing. After a theorist calculates the electron states in the ferroelectric and prints out all of the charge densities, how does one use this information to find the electronic part $\mathbf{P}_e$ of the polarization? The answer seems to depend on what shape is selected for the atomic unit cell, but there are several choices. This interesting question was answered by David Vanderbilt and his students [8]. They proved that the polarization was actually given by the Wannier functions $W_{n\sigma}(\mathbf{r})$:

$$\mathbf{P}_e = \frac{e}{\Omega}\sum_{\sigma,n=1}^{M}\int d^3r\, \mathbf{r}|W_{n\sigma}(\mathbf{r})|^2 \tag{4.81}$$

where *e* is the electron charge, and σ is electron spin. The summation is over all of the Wannier functions generated from occupied band orbitals in the insulator. This prescription is unique, and allows the polarization to be found from an electronic structure calculation.

4.4.2. Thermodynamics

An important phenomenological theory of phase transitions was introduced by Landau [9]. Here it is applied to ferroelectrics. A similar theory is applied to magnetism in chapter 13. It can be applied to other systems: liquid crystals, superconductors, etc. The spontaneous polarization $P(T)$ of a ferroelectric in zero electric field is a function of temperature. It is zero for $T > T_c$ and is nonzero for $T < T_c$. Landau proposed that the free energy was an analytic function of this polarization. It is treated as a scalar quantity:

$$\mathcal{F}(P) = A_0 + A_2P^2 + A_4P^4 - PE + \cdots \tag{4.82}$$

where E is the electric field, and the last term is usually written as $-\mathbf{E}\cdot\mathbf{P}$. Only even powers of P^n are contained in the series: the energy should not change if P is changed

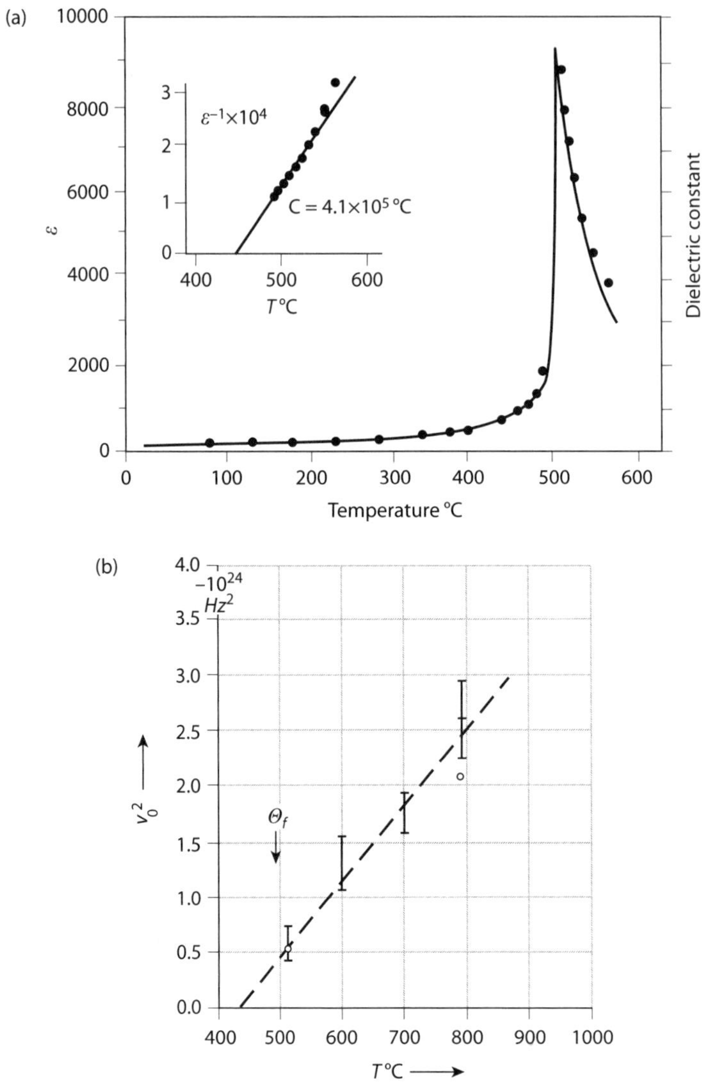

FIGURE 4.6. A graph of $1/\varepsilon(T)$ and ω_{TO}^2 in lead titanate as a function of temperature. Dielectric data from Remeika and Glass, *Mater. Res. Bull.* **5**, 37 (1970). Used with permission from Elsevier. Phonon data from Shirane et al., *Phys. Rev. B* **2**, 155 (1970). Used with permission of the American Physical Society.

to $-P$, while E is changed to $-E$. The constant A_0 is from other sources of energy, such as the Madelung energy. The key constants are A_2 and A_4. A_4 must be positive for the system to be stable at a nonzero value of P: If $A_4 < 0$ we would need a term $A_6 P^6$ with $A_6 > 0$.

The system will choose the value of P that minimizes this free energy:

$$0 = \frac{\delta \mathcal{F}}{\delta P} = 2A_2 P + 4A_4 P^3 - E \tag{4.83}$$

1. First consider the case of no external electric field ($E = 0$).

- If $A_2 > 0$, $A_4 > 0$, the only solution is $P = 0$.
- If $A_2 < 0$, $A_4 > 0$, there is a solution:

$$P = \pm\sqrt{\frac{-A_2}{2A_4}} \tag{4.84}$$

- Landau reasoned that the coefficient A_2 changed sign at T_c:

$$A_2(T) = a_2\left(\frac{T - T_c}{T_c}\right) \tag{4.85}$$

$$P(T) = \pm\sqrt{\frac{a_2}{2A_4}\left(\frac{T_c - T}{T_c}\right)}\Theta(T_c - T) \tag{4.86}$$

Landau theory predicts that the polarization is zero for $T > T_c$, while it increases according to $\sqrt{T_c - T}$ for $T < T_c$. These features agree with experimental observations.

2. When there is an electric field, several predictions can be made:
   At $T = T_c$, where $A_2 = 0$, then

$$P(T_c) = \left(\frac{E}{4A_4}\right)^{1/3} \tag{4.87}$$

- For $T \neq T_c$, the first term in eqn. (4.82) dominates, and

$$P = \frac{E}{a_2}\left(\frac{T_c}{T - T_c}\right) \tag{4.88}$$

The electric susceptibility is

$$\alpha = \left(\frac{\delta P}{\varepsilon_0 \delta E}\right)_{E=0} = \frac{1}{a_2 \varepsilon_0}\left(\frac{T_c}{T - T_c}\right) \tag{4.89}$$

- The dielectric constant is

$$\varepsilon = 1 + \frac{4\pi\alpha}{\Omega_0} \propto \frac{C}{T_c - T} \tag{4.90}$$

3. In the absence of an external field, the internal energy is

$$U(T) \propto -P^2(T) = -\Lambda(T_c - T), \quad T_c > T \tag{4.91}$$

The heat capacity is

$$C \propto \frac{dU}{dT} \propto \Lambda, \quad T_c > T \tag{4.92}$$

The heat capacity is discontinuous at the transition temperature.

These predictions agree with experiments, as discussed below.

4.4.3 $SrTiO_3$

Strontium titanate is a well-studied perovskite. It has two different features related to ferroelectrics.

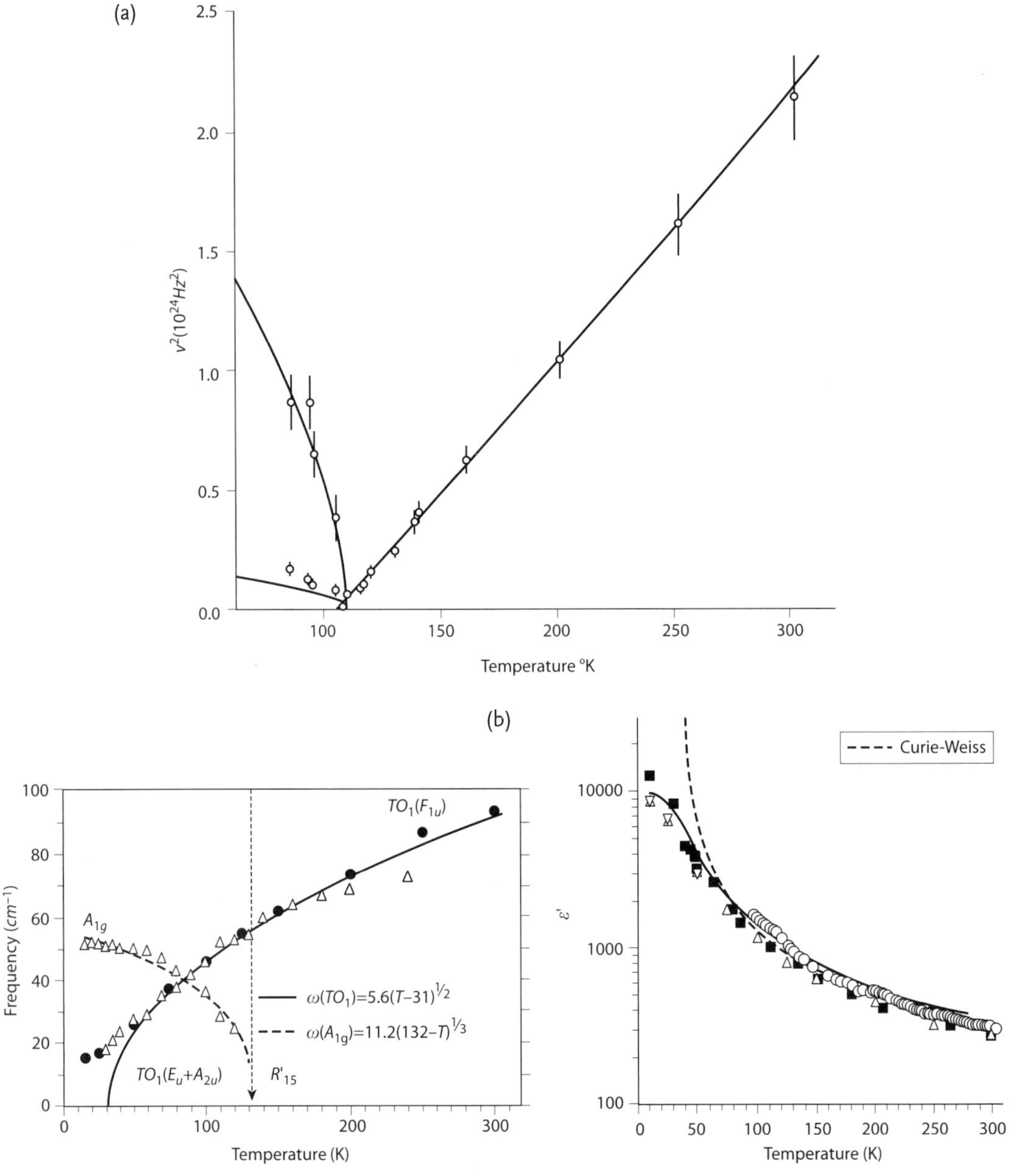

FIGURE 4.7. Data for $SrTiO_3$. (a) Phonon frequency squared near the $T_c = 103$ K structural phase transition. From Cowley et al., *Solid State Commun.* **7**, 181 (1969). Used with permission from Elsevier. (b) The TO phonon frequency, and dielectric constant at low temperatures. From Petzelt et al., *Phys. Rev. B* **64**, 184111 (2001). Used with permission of the American Physical Society.

1. There is a displacive phase transition at $T_c = 105$ K. It is a soft mode transition, where the transverse phonon is at the zone boundary: at wave vectors $\mathbf{q} = \pi(\pm 1, \pm 1, \pm 1)/a$. Figure 4.7a shows the square of this phonon frequency as a function of temperature. $\omega_t^2(T)$ has a linear plot, as revealed by neutron and x-ray scattering of phonons, which vanishes at T_c.

This phonon is not optically active and does not create long-range electric fields. Some writers call it an *antiferroelectric phase,* since there is no anomaly in the dielectric constant. Also, the change in heat capacity is modest, since there are no long-range electric fields to generate lots of energy. With five atoms per unit cell and fifteen phonon modes, a change in one mode affects the heat capacity by only $\Delta C/C \propto 1/15$. The same displacive transition is found in other perovskites, such as $KMnF_3$, $LaAlO_3$, and $RbCaF_3$.

2. Figure 4.7b shows a low-temperature graph of (i) the static dielectric function, and (ii) the TO zone center phonon mode. Both are going to zero at low temperature and appear to generate a low-temperature ferroelectric transition. However, it does not actually have a transition. Low-temperature quantum fluctuations stabilize the structure, so the dielectric constant saturates at a high value and the TO phonon stabilizes at a small value of frequency. Also note in the graph of the dielectric constant that nothing happens to it at $T = 105$ K, where the structural transition occurs.

4.4.4 $BaTiO_3$

Barium titanate is another well-studied ferroelectric. It has numerous ferroelectric phases. The transition at $T_c = 130°C = 403$ K is from the high-temperature cubic phase to the first ferroelectric phase, which has tetragonal structure. Figure 4.8 shows the heat capacity as

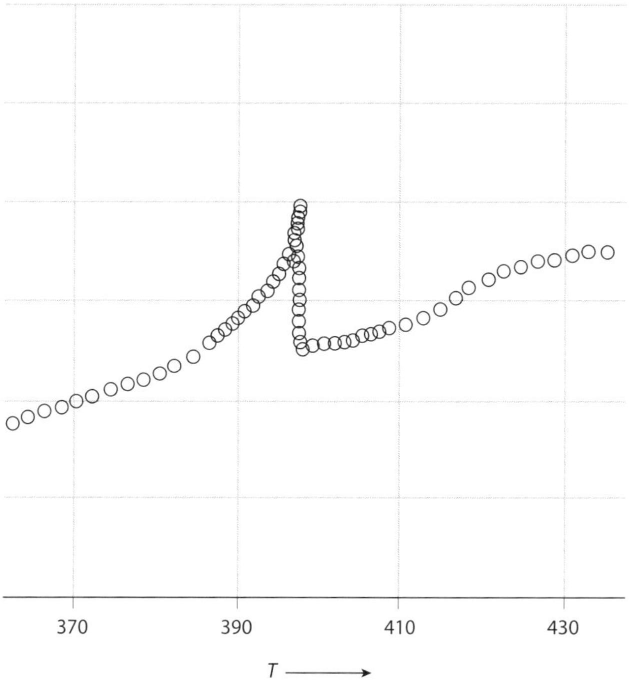

FIGURE 4.8. (a) Heat capacity of $BaTiO_3$. From Hatta and Ikushima, *J. Phys. Soc. Jpn.* **41**, 558 (1976). Used with permission from the Institute of Pure and Applied Physics.

a function of temperature [5]. There is a discontinuity at T_c, in agreement with Landau theory. Mean field theory, or Landau theory, describes very well the paraelectric region of temperature for $T > T_c$. However, Landau theory rarely describes the ferroelectric properties for $T < T_c$. The heat capacity is an example: it is not a constant for $T < T_c$, but changes rapidly with the lowering of temperature. It should not be surprising that Landau theory fails in this region, since the crystal has a structural change, a change in the Brillouin zone, and the emergence of new phonon modes. Figure 4.4 shows the dielectric function $\varepsilon(T)$, and also $1/\varepsilon(T)$ in the paraelectric phase [7]. There are two dielectric constants for $T < T_c$ due to the lower crystal symmetry in the ferroelectric phase. In the paraelectric phase, $\varepsilon(T)$ has the Landau form with $1/\varepsilon \propto (T - T_c)$. This ferroelectric phase is due to a softening of a zone-center TO phonons.

References

1. A. D. Bruce and R. A. Cowley, *Structural Phase Transitions*, (Taylor & Francis, London, 1981)
2. W. Cochran, Crystal stability and the theory of ferroelectricity. *Adv. Phys.* **9**, 387–423 (1960)
3. R. a. Cowley, Lattice dynamics and phase transitions of $SrTiO_3$. *Phys. Rev.* **134**, A981–A997 (1964)
4. R. A. Cowley, W.J.L. Buyers, and G. Dolling, Relationship of normal modes of vibration of $SrTiO_3$ and its antiferroelectric phase transition. *Solid State Commun.* **7**, 181 (1969)
5. I. Hatta and A. Ikushima, Temperature dependence of heat-capacity in $BaTiO_3$. *J. Phys. Soc. Jpn.* **41**, 558–564 (1976)
6. M. Holt, M. Sutton, P. Zschack, H. Hong, and T. C. Chiang, Critical x-ray scattering from $SrTiO_3$. *Phys. Rev. Lett.* **98**, 065501 (2007)
7. C. J. Johnson, Some dielectric and electro-optic properties of $BaTiO_3$ single crystals. *Appl. Phys. Lett.* **7**, 221 (1965)
8. R. D. King-Smith and D. Vanderbilt, Theory of polarization of crystalline solids. *Phys. Rev. B* **47**, 1651–1654 (1993)
9. L. D. Landau and E. M. Lifshitz, *Statistical Physics*, 2nd. ed. (Addison Wesley, Reading, MA, 1969), Chapter XIV
10. *Landolt-Börnstein*, Vol. 16, Group III, ed. T. Mitsui and S. Nomura, Ferroelectrics and Related Substances. (Springer-Verlag, Berlin, 1981)
11. G. D. Mahan, Van der Waals forces in solids. *J. Chem. Phys.* **43**, 1569–1574 (1965)
12. G. D. Mahan and K. R. Subbawamy, *Local Density Theory of Polarizability* (Plenum, New York, 1990)
13. J. E. Mayer, Dispersion and Polarizability and the van der Waals potential in the alkali halides. *J. Chem. Phys.* **1**, 270 (1933)
14. L. Pauling, *The Nature of the Chemical Bond*, 3rd ed. (Cornell, Ithaca, 1960)
15. J. Petzelt et al., Dielectric, infrared, and Raman response of undoped $SrTiO_3$ ceramics. *Phys. Rev. B* **64**, 184111 (2001)
16. J. P. Remeika and A. M. Glass, The growth and ferroelectric properties of lead titanate. *Mat. Sci. Bull.* **5**, 37–46 (1970)
17. G. Shirane, J. D. Axe, J. Harada, and J. P. Remeika, Soft ferroelectric modes in lead titanate. *Phys. Rev. B* **2**, 155–159 (1970)
18. J. R. Tessman, A. H. Kahn, and W. Shockley, Electronic polarizabilities of ions in crystals. *Phys. Rev.* **92**, 890–895 (1953)
19. H. Witte and E. Wölfel, Electron distributions in NaCl, LiF, CaF_2 and Al. *Rev. Mod. Phys.* **30**, 51–54 (1958)

Homework

1. Calculate the binding energy of a rare gas solid with the fcc structure using a 6-10 potential:

$$V(R) = A\left[\left(\frac{\tilde{\sigma}}{R}\right)^{10} - \left(\frac{\tilde{\sigma}}{R}\right)^{10}\right] \tag{4.93}$$

   (a) Relate $(A, \tilde{\sigma})$ to (ε, σ) of the 6-12 potential by assuming (i) the same value of C_6, and (ii) the same minimum value of the potential energy.

   (b) Calculate A_{10} for the fcc lattice. Sum over several sets of neighbors.

   (c) Then calculate the ground-state energy in the form

$$E_G/N = -C\varepsilon \tag{4.94}$$

   and find the value of C for the 6-10 potential.

2. The bulk modulus is defined as the second derivative, with respect to volume Ω, of the total ground-state energy:

$$B = \Omega\frac{d^2 E_G}{d\Omega^2} \tag{4.95}$$

   Evaluate this for the rare gas solids bound by a 6-12 potential. Show that one can write the result as

$$B = -C\frac{E_G}{\Omega} \tag{4.96}$$

   and find the dimensionless constant C.

3. Use the Ewald method to calculate the Madelung constant for the CsCl lattice. Do enough terms to get the result to 1%.

4. Write the dielectric constant in Lorenz-Lorentz form:

$$K = 1 + \frac{4\pi\bar{\alpha}}{1 - 4\pi\bar{\alpha}/3} \tag{4.97}$$

   The polarizability per unit volume has an electronic part $\bar{\alpha}_e$ and a paraelectric part:

$$\bar{\alpha} = \bar{\alpha}_e + \frac{C}{T - T_c} \tag{4.98}$$

   As one approaches the ferroelectric transition, one can write

$$\lim_{T \to T_c^*} K(T) \to \frac{C^*}{T - T_c^*} \tag{4.99}$$

   Derive expressions for T_c^* and C^* that give the dielectric screening of the dipolar interactions.

5 Free Electron Metals

5.1 Introduction

Paul Drude [3] first described the motion of electrons in metals by classical equations of motion. Later Arnold Sommerfeld [10] proposed that the conduction electrons in simple metals, such as lead or aluminum, could be treated quantum mechanically as noninteracting. This was a radical idea, since it assumed that one could ignore two important features: (i) the periodic potential of the ions, and (ii) the strong electron–electron interactions. Only much later were it realized that he was right, when the reasons for ignoring these two features were finally understood.

- The reason that one can ignore the periodic potential of the ions is due to Bloch's theorem. In such a periodic potential, the eigenfunction of an electron can be written as

$$H\psi_n(\mathbf{k}, \mathbf{r}) = E_n(\mathbf{k})\,\psi_n(\mathbf{k}, \mathbf{r}), \quad \psi_n(\mathbf{k}, \mathbf{r}) = e^{i\mathbf{k}\cdot\mathbf{r}}\frac{u_{k,n}(\mathbf{r})}{\sqrt{\Omega}} \tag{5.1}$$

$$H = \frac{p^2}{2m} + V(\mathbf{r}), \quad V(\mathbf{r}) = V(\mathbf{r} + \mathbf{R}_j) \tag{5.2}$$

where $\mathbf{R}_j$ is any lattice vector, and Ω is the volume of the crystal. Part of the wave function is periodic: $u_{k,n}(\mathbf{r} + \mathbf{R}_j) = u_{k,n}(\mathbf{r})$, where n is a band index. The cell-periodic part of the wave function is orthogonal for different bands:

$$\int_{\Omega_0} d^3r u^*_{k,n}(\mathbf{r})\, u_{k,m} = \delta_{nm} \tag{5.3}$$

where the integral is over the volume of one unit cell $\Omega_0 = \Omega/N$, and N is the number of unit cells in the crystal.

The rest of the wave function $[\exp(i\mathbf{k}\cdot\mathbf{r})/\sqrt{\Omega}]$ is the same as a free, noninteracting particle. Bloch's theorem tells us that if the ion potential is not too strong, then the dominent part of the wave function is the part that resembles a plane wave. The electron–ion potential is usually weak, since it is screened by the electron–electron interactions. The band structure of Bloch electrons is treated in chapter 3.

- Electron–electron interactions cannot actually be ignored. The prior paragraph mentioned the important role that electron–electron interactions play in screening the Coulomb interaction between the electrons and the ions. For this feature, the electron–electron interactions are responsible for making Sommerfeld's model accurate. Chapter 6 will explain the role that electron–electron interactions play in modifying the energy dispersion of the electrons. They give rise to an important collective mode, the plasmon, whose frequency ω_p is

$$\omega_p^2 = \frac{n_e e^2}{m\varepsilon_0} \tag{5.4}$$

where n_e is the density of conduction electrons. This frequency is quite large and the energy quantum $\hbar\omega_p$ is typically several electron volts. This energy is so large that the collective mode plays no role in electron motion. A thermalized electron does not have enough excitation energy to excite this high-energy mode, so it becomes irrelevant. The reason that electron–electron interactions play little role in the motion of thermal electrons, is that these interactions make a collective mode of very high energy, which can never be excited! Of course, if one injects an electron into a metal at very high energy, emitting these plasmons is the primary mechanism by which the electron loses its excess kinetic energy.

- Electron–electron interactions make a major contribution to the lifetime of electrons in metals. If $\varepsilon(\mathbf{k})$ is the band energy of the electrons and μ is the chemical potential, then the lifetime for electrons near to the Fermi surface has an energy dependence $\tau \sim 1/[\varepsilon(\mathbf{k}) - \mu]^2$. The lifetime is very long for thermally excited electrons, which have $\varepsilon(\mathbf{k}) - \mu \sim k_B T$. The longer lifetimes enable these thermally excited electrons to behave as "free," or noninteracting, quasiparticles. That is another reason that the Sommerfeld model works.

Metals are generally divided into two catagories: (i) free electron metals, and (ii) highly correlated metals. The Sommerfeld model applies only to the first category. The second category includes metals that have partially filled *d*- or *f*-shells in their atomic cores. These atoms usually have a magnetic moment, and the metals are often magnets or have a tendency for magnetic fluctuations. They are an important class of materials, which are treated in a later chapter. Here we concentrate on the Sommerfeld model.

Equation (5.4) for the plasma frequency is valid for three dimensions. In two dimensions the formula is

$$\omega_p^2(q) = \frac{n_e e^2}{2m\varepsilon_0} q \tag{5.5}$$

Here n_e is the density of electrons per unit area, and q is the wave vector of the plasmon. At small values of wave vector, the plasmon energy is not very large, so that such plasmons do affect the motion of electrons in two dimensions. The electron gas has very different features in two dimensions than in three dimensions. One dimension is different from either two or three.

5.2 Free Electrons

The Sommerfeld model assumes the electrons are perfectly free. Their eigenfunction is a simple plane wave:

$$H_0 = \frac{p^2}{2m} \tag{5.6}$$

$$\psi(\mathbf{k}, \mathbf{r}) = \frac{e^{i\mathbf{k}\cdot\mathbf{r}}}{\sqrt{\Omega}} \tag{5.7}$$

$$H_0\psi(\mathbf{k}, \mathbf{r}) = \varepsilon(\mathbf{k})\psi(\mathbf{k}, \mathbf{r}) \tag{5.8}$$

$$\varepsilon(\mathbf{k}) = \frac{\hbar^2 k^2}{2m} \tag{5.9}$$

In a metal the chemical potential μ is a positive number, if the energy zero is taken as the bottom of the conduction band. Then it is useful to relate kinetic energy with respect to the chemical potential:

$$\xi(k) = \varepsilon(\mathbf{k}) - \mu \tag{5.10}$$

If the metal crystal has a volume $\Omega = L_x L_y L_z$, the wave vector $k = (k_x, k_y, k_z)$ is a discrete variable:

$$k_x = \frac{2\pi\alpha}{L_x}, \quad k_y = \frac{2\pi\beta}{L_y}, \quad k_z = \frac{2\pi\gamma}{L_z} \tag{5.11}$$

where (α, β, γ) are integers that can be positive or negative. The allowed energy states are given by sets of points in wave vector space. Each point $\mathbf{k}$ permits two electrons states: one for spin-up, and the other for spin-down. Here the directions "up" or "down" can point along any axis of spin quantization.

5.2.1 *Electron Density*

At zero temperature, the electron system will be in its lowest energy state. If there are N electrons, the density is $n_0 = N/\Omega$. Each $\mathbf{k}$-point is occupied by two electrons, or by none. The set of occupied $\mathbf{k}$-points is those $N/2$ states with the lowest energy. They occupy a sphere in $\mathbf{k}$-space. The sphere has a radius k_F called the Fermi wave vector. It is found by summing up all wave vector states:

$$N = 2\sum_{\alpha,\beta,\gamma} \Theta(k_F - |\mathbf{k}|) \tag{5.12}$$

The theta function $\Theta(k)$ is one if the argument is positive, and zero if it is negative. Since N is a very large number, the summation is converted to a continuous integral:

$$N = 2\Omega \int \frac{d^3k}{(2\pi)^3} \Theta(k_F - |\mathbf{k}|) \tag{5.13}$$

$$\frac{N}{\Omega} = n_0 = \frac{k_F^3}{3\pi^2}, \quad \Omega_F = \frac{4\pi}{3} k_F^3 \tag{5.14}$$

The integral is easy, and yields an important relationship between the electron density n_0 and the Fermi wave vector k_F. Ω_F is the volume in phase space occupied by the electrons at zero temperature.

Luttinger's theorem is that this volume is unchanged by interactions. If one has a strongly interacting electronic system, which could be in a magnetic phase or in a superconducting state, the Fermi surface may no longer be a sphere. However, the volume enclosed by the Fermi surface is unchanged. This volume merely counts the number of occupied **k**-points, and that volume is invariant as long as the number of conduction electrons is unchanged.

Electron densities for the simple metals vary by a factor of forty. The density n_0 is the number of electrons per unit volume. A common way to represent the density uses the dimensionless parameter called r_s. It is the distance, in atomic units (the Bohr radius is a_0), that encloses one unit of charge:

$$1 = \frac{4\pi n_0}{3}(a_0 r_s)^3 \tag{5.15}$$

Many properties of the conduction electron can be expressed in terms of this parameter. For example, the Fermi wave vector is written as

$$k_F a_0 = \left(3\pi^2 n_e a_0^3\right)^{1/3} = \left(\frac{9\pi}{4}\right)^{1/3}\left(\frac{4\pi n_e a_0^3}{3}\right)^{1/3} = \frac{1.91916}{r_s} \tag{5.16}$$

The Fermi velocity and Fermi energy are then

$$v_F = \frac{\hbar k_F}{m} = \frac{\hbar}{m a_0}(k_F a_0) = \frac{4.198}{r_s}\ \text{Mm/s} \tag{5.17}$$

$$E_F = \frac{\hbar^2 k_F^2}{2m} = E_{\text{Ry}}(k_F a_0)^2 = \frac{50.11}{r_s^2}\ \text{eV} \tag{5.18}$$

where $E_{\text{Ry}} = 13.606$ eV is the Rydberg energy. The plasma frequency is

$$\hbar\omega_p = \hbar\sqrt{\frac{e^2 n_e}{m\varepsilon_0}} = \sqrt{12}\frac{E_{\text{Ry}}}{r_s^{3/2}} = \frac{47.13}{r_s^{3/2}}\ \text{eV} \tag{5.19}$$

Table 5.1 shows some values of r_s for simple metals, and the values of Fermi energy, plasma frequency, and Fermi velocity predicted by these formulas. Actual values in these metals vary somewhat, due to band structure effects. Also shown is the compressibility, which is discussed in the next chapter. It is defined as the change in volume with pressure:

$$K = -\frac{1}{\Omega}\left(\frac{\partial\Omega}{\partial P}\right) \tag{5.20}$$

5.2.2 *Density of States*

The density of states is the number of electron states per unit energy per unit volume. Sometimes the definition includes both spin states, while other times it is per spin state. If the energy bands are $E_n(\mathbf{k})$, then the density of states per spin is

Table 5.1 Free electron properties of simple metals.

Metal	r_s	E_F (eV)	$\hbar\omega_p$ (eV)	v_F (Mm/s)	K
Na	3.96	3.20	5.98	1.06	162
K	4.95	2.04	4.28	0.84	297
Rb	5.30	1.78	3.86	0.79	370
Cs	5.74	1.52	3.43	0.73	500
Mg	2.65	7.13	10.92	1.58	28.4
Ca	3.26	4.71	8.01	1.28	54.8
Sr	3.54	4.00	7.08	1.18	84.8
Ba	3.72	3.62	6.57	1.12	106
Al	2.07	11.69	15.83	2.03	12.7
Ga	2.19	10.44	14.54	1.91	20
In	2.41	8.62	12.60	1.73	25.7
Tl	2.49	8.08	12.00	1.68	26.7

Note. The compressibility K has units of $(10^{-12}/\text{Pa})$.

$$N(E) = \int \frac{d^3k}{(2\pi)^3} \sum_n \delta[E - E_n(\mathbf{k})] \tag{5.21}$$

This definition is useful for magnetic systems, where the different spin states may have different energy bands.

For a free electron system, the definition usually includes both spin states:

$$N(E) = 2\int \frac{d^3k}{(2\pi)^3} \delta[E - \varepsilon(\mathbf{k})] = 2\int \frac{d^3k}{(2\pi)^3} \delta\left[E - \frac{\hbar^2 k^2}{2m}\right] \tag{5.22}$$

$$= \frac{\sqrt{E}}{2\pi^2}\left(\frac{2m}{\hbar^2}\right)^{3/2} \tag{5.23}$$

The density of states of a free electron gas increases as the square root of the energy. The density of electrons at zero temperature is given by the integral over energy:

$$n_0 = \int_0^{\mu} dEN(E) = \frac{1}{3\pi^2}\left(\frac{2m\mu}{\hbar^2}\right)^{3/2} = \frac{k_F^3}{3\pi^2} \tag{5.24}$$

Similarly, the ground-state kinetic energy per unit volume is

$$E_G = \int_0^{\mu} EdEN(E) = \frac{\mu}{5\pi^2}\left(\frac{2m\mu}{\hbar^2}\right)^{3/2} = \frac{3}{5}n_0\mu \tag{5.25}$$

The average kinetic energy per electron is $3\mu/5$.

5.2.3 *Nonzero Temperatures*

At nonzero temperatures, one integrates over all values of energy, while including a Fermi-Dirac occupation factor:

$$n_0 = \int_0^\infty dEN(E)\,n_F(E) \tag{5.26}$$

$$n_F(E) = \frac{1}{e^{\beta(E-\mu)}+1}, \quad \beta = \frac{1}{k_BT} \tag{5.27}$$

If the density of electrons n_0 is a constant, then this integral defines the temperature dependence of the chemical potential $\mu(T)$. Most materials expand in size with increasing temperature, so that $n_0(T)$ slightly decreases. We can use this expression to determine the temperature dependence of the chemical potential at small temperatures. Let C denote the constant prefactor, and change the integration variable to $\xi = E - \mu$:

$$n_0 = C\int_{-\mu}^\infty \frac{d\xi\sqrt{\mu+\xi}}{e^{\beta\xi}+1}, \quad C = \frac{1}{2\pi^2}\left(\frac{2m}{\hbar^2}\right)^{3/2} \tag{5.28}$$

The first step is to integrate by parts:

$$\int udv = uv - \int vdu \tag{5.29}$$

$$dv = d\xi\sqrt{\mu+\xi}, \; v = \frac{2}{3}[\mu+\xi]^{3/2} \tag{5.30}$$

$$u = \frac{1}{e^{\beta\xi}+1}, \quad du = -\left(\frac{\beta}{2}\right)\frac{d\xi}{1+\cosh(\beta\xi)} \tag{5.31}$$

$$n_0 = \left(\frac{\beta C}{3}\right)\int_{-\mu}^\infty d\xi \frac{[\mu+\xi]^{3/2}}{1+\cosh(\beta\xi)} \tag{5.32}$$

where the constant term uv gives zero at its limits. At low temperatures β is a large number and $\cosh(\beta\xi)$ is very large unless ξ is near to zero. The denominator of the integrand gives a contribution only near to $\xi \sim 0$. In that case the numerator is expanded in a Taylor series:

$$[\mu+\xi]^{3/2} = \mu^{3/2} + \frac{3}{2}\xi\sqrt{\mu} + \frac{3}{8}\frac{\xi^2}{\sqrt{\mu}} + \cdots \tag{5.33}$$

Let the integration variable be $x = \beta\xi$. The lower limit of integration can be replaced by $-\infty$ since $\beta\mu \gg 1$. The three integrals are

$$\int_{-\infty}^\infty \frac{dx}{1+\cosh(x)} = 2 \tag{5.34}$$

$$\int_{-\infty}^\infty \frac{xdx}{1+\cosh(x)} = 0 \tag{5.35}$$

$$\int_{-\infty}^\infty \frac{x^2dx}{1+\cosh(x)} = \frac{2\pi^2}{3} \tag{5.36}$$

The integral for the density is now

$$n_0 = \frac{2C}{3}\left[\mu^{3/2} + \frac{\pi^2(k_BT)^2}{8\sqrt{\mu}} + \cdots\right] \tag{5.37}$$

Rearrange this expression to give

$$\mu(T)^{3/2} = \frac{3n_0}{2C} - \frac{\pi^2 (k_B T)^2}{8\sqrt{\mu}} + \cdots \tag{5.38}$$

The second term on the right is the correction term of $O(T^2)$. The first term on the right is $\mu(0)^{3/2}$. Since we are doing perturbation theory, replace $\sqrt{\mu}$ in the second term by $\sqrt{\mu(0)}$. The above expression becomes

$$\mu(T) = \mu(0)\left[1 - \frac{\pi^2 (k_B T)^2}{8\mu(0)^2} + \cdots\right]^{2/3} \tag{5.39}$$

$$= \mu(0) - \frac{\pi^2 (k_B T)^2}{12\mu(0)} + \cdots$$

The last formula gives the change in chemical potential as a function of temperature for simple metals like Mg and Al. $\mu(T)$ decreases quadratically with temperature. The change is small at ordinary temperatures since $k_B T/\mu(0) \ll 1$.

A similar Sommerfeld expansion can be done for the internal energy per unit volume:

$$U(T) = \int_0^\infty \frac{E dE N(E)}{e^{\beta(E-\mu)} + 1} = C\int_{-\mu}^\infty \frac{d\xi\,(\mu + \xi)^{3/2}}{e^{\beta\xi} + 1} \tag{5.40}$$

An integration by parts gives

$$U(T) = \left(\frac{\beta C}{5}\right)\int_{-\mu}^\infty d\xi \frac{[\mu + \xi]^{5/2}}{1 + \cosh(\beta\xi)} \tag{5.41}$$

$$[\mu + \xi]^{5/2} \approx \mu^{5/2} + \frac{5}{2}\xi\mu^{3/2} + \frac{15}{8}\xi^2\sqrt{\mu} + \cdots \tag{5.42}$$

Again use the three x-integrals in eqn. (5.35):

$$U(T) = \frac{2C}{5}\left[\mu^{5/2} + \frac{5\pi^2}{8}(k_B T)^2\sqrt{\mu} + O(T^4)\right] \tag{5.43}$$

Now make the correction for the temperature dependence of the chemical potential. To $O(T^2)$ write it as

$$U(T) = \frac{2C}{5}\mu(0)^{5/2}\left[\left(\frac{\mu(T)}{\mu(0)}\right)^{5/2} + \frac{5\pi^2}{8}\frac{(k_B T)^2}{\mu(0)^2}\right] \tag{5.44}$$

The prefactor is $U(T=0) = E_G = (\frac{3}{5})\mu(0)n_0$. Using the above result for $\mu(T)$ gives

$$U(T) = U(0)\left[1 + \frac{5\pi^2}{12}\frac{(k_B T)^2}{\mu(0)^2} + O(T^4)\right] \tag{5.45}$$

The heat capacity per unit volume is

$$C = \frac{dU}{dT} = U(0)\frac{5\pi^2}{6}\frac{k_B^2 T}{\mu(0)^2} + O(T^3) \tag{5.46}$$

$$= \frac{\pi^2}{2} n_0 \frac{k_B^2 T}{\mu(0)} \equiv \gamma T$$

$$\gamma = \frac{\pi^2}{2} n_0 \frac{k_B^2}{\mu(0)} = \frac{m k_F k_B^2}{3\hbar^2} \tag{5.47}$$

The low-temperature heat capacity of a Fermi gas is linear in temperature. The coefficient γ has the units of joules per volume per square of the temperature. The last equation used the definitions $n_0 = k_F^3/3\pi^2$ and $\mu(0) = \hbar^2 k_F^2/2m$ to get a simple formula. It contains fundamental constants ($\hbar$, k_B) plus the electron mass and Fermi wave vector. The Fermi wave vector k_F relates to the volume enclosed by the Fermi surface. By Luttinger's theorem, this cannot vary much. So variations in γ from the expected value are interpreted as changes in the mass of the electron. The effective mass can change due to (i) band structure, and (ii) electron–phonon effects.

A measurement of the heat capacity of some rare earth compounds found a value of γ that was over one hundred times larger than expected. This was attributed to a large effective mass, and these compounds were called *heavy fermion* materials. In some of these materials, the *f*-electrons form a band with a very large effective mass. In other heavy-fermion materials, the *f*-electrons are localized and have magnetic fluctuations. Then the large heat capacity is due to a Kondo resonance near to the Fermi surface. Our derivation assumed that the density of states $N(E)$ was a smooth function of E near to the Fermi energy. The existence of a resonance in the density of states makes our analysis incorrect. A correct calculation shows that the Kondo resonance predicts a large value of γ.

5.2.4 *Two Dimensions*

Two-dimensional electron gases are found in many places: in quantum wells, in inversion layers, and on the surfaces of solids and liquids. It is important to understand its properties. The density n_0 is the number of electrons per unit area. It is given by summing over all of the wave vector points within the Fermi circle:

$$n_0 = 2\int \frac{d^2k}{(2\pi)^2} \Theta(k_F - k) = \frac{k_F^2}{2\pi} \tag{5.48}$$

The Fermi energy has a simple relationship to the density:

$$E_F = \frac{\hbar k_F^2}{2m} = \pi \frac{\hbar^2 n_0}{m} = \mu(T=0) \tag{5.49}$$

which is also the chemical potential at zero temperature. The density of states in two dimensions is a constant when the dispersion is quadratic:

$$N_2(E) = 2\int \frac{d^2k}{(2\pi)^2} \delta\left[E - \frac{\hbar^2 k^2}{2m}\right] = \frac{m}{\pi\hbar^2} = \frac{n_0}{\mu(0)} \tag{5.50}$$

Find the temperature dependence of the chemical potential starting from

$$n_0 = 2\int \frac{d^2k}{(2\pi)^2} \frac{1}{e^{\beta(\varepsilon(k)-\mu)}+1} = N_2 \int_{-\mu}^{\infty} \frac{d\xi}{e^{\beta\xi}+1} \tag{5.51}$$

$$= N_2 k_B T \ln[1+e^{\beta\mu}]$$

The integral over $d\xi$ is evaluated exactly. There is no need for a Sommerfeld expansion. The chemical potential is

$$\mu(T) = k_B T \ln[e^{\beta n_0/N_2} - 1] \tag{5.52}$$

$$\frac{\beta n_0}{N_2} = \frac{\mu(0)}{k_B T} \tag{5.53}$$

The latter identity follows from eqns. (5.49) and (5.50). Take a factor of $\exp[\mu(0)\beta]$ from inside the logarithm and get

$$\mu(T) = \mu(0) + k_B T \ln[1 - e^{-\beta\mu(0)}] \tag{5.54}$$

The chemical potential does *not* decrease as $O(T^2)$ at low temperatures.

5.2.5 *Fermi Surfaces*

The energy bands of free electron metals are $\varepsilon(k) = \hbar^2 k^2/2m_e$ except near to symmetry points in the BZ. At thosc points the energy bands developed band gaps. This behavior is found in one, two, and three dimensions. The energy bands are parabolic except near to the edges and other symmetry points.

The Fermi surface is the line of points at zero temperature that divides the occupied from the empty electron states.

- In one dimension, the free electron Fermi surface is two points in the BZ, at $\pm k_F$, where $n_e = 2k_F/\pi$.
- In two dimensions, the free electron Fermi surface is a circle of radius k_F, where $n_e = k_F^2/2\pi$.
- In three dimensions, the free electron Fermi surface is a sphere of radius k_F, where $n_e = k_F^3/3\pi^2$.

These shapes, circles and spheres, get altered near symmetry points.

Figure 5.1 shows an example in two dimensions. It shows the BZ of a square lattice. The circle is Fermi surface with two electrons per unit cell ($n_e = 2/a^2$). Write the density of electrons on the sq lattice as $n_0 = n/a^2$, where n_0 is the density, and n is the number of valence electrons per atom. The Fermi wave vector is

$$k_F = \sqrt{2\pi n_0} = \frac{\sqrt{2\pi n}}{a} \tag{5.55}$$

Some values are shown in table 5.2. The edge of the BZ is at $G_0/2 = \pi/a$. If $n = 1$, the circle is entirely within the first BZ since $2.507 < \pi$, and the material is a conductor.

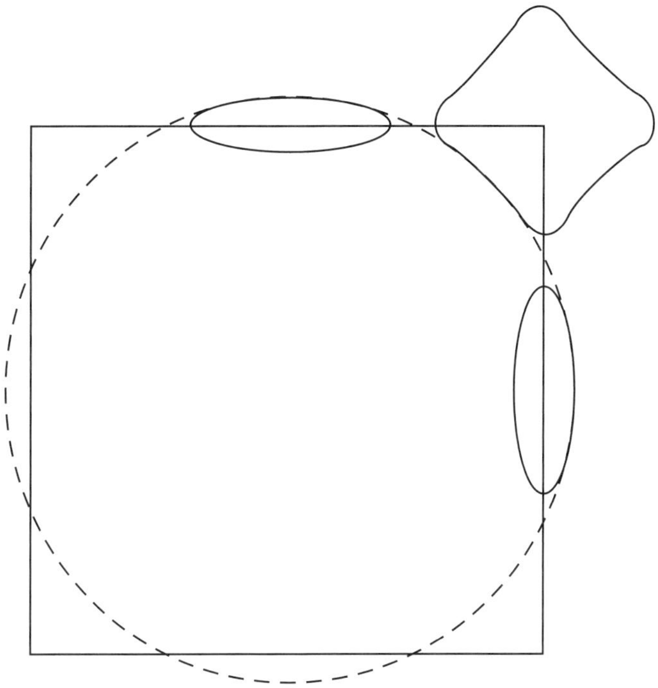

FIGURE 5.1. The square is the BZ of a two-dimensional square lattice. The dashed line is the free electron Fermi surface when there are two electrons per unit cell. The solid line is the actual Fermi surface. Various pockets of the Fermi surface are moved by reciprocal lattice vectors to make closed surfaces.

Table 5.2 Fermi wave vectors on the sq lattice

n	$k_F a$
1	2.507
2	3.545
3	4.342

Two electrons per atom could exactly fill the first BZ and make the material an insulator. However, in a nearly free electron metal, the Fermi surface is roughly a circle of radius $k_F a = 2\sqrt{\pi}$. This circle is the dashed line in fig. 5.1. The actual Fermi surface must have bands that are perpendicular to the edge of the BZ, sketched as the solid line. There are ellipsoids along the center of edges of the BZ that act as pockets of electrons, and stars at the corners that act as hole surfaces. The surfaces can be joined with the similar figures, separated by a reciprocal lattice vector, to make a complete ellipsoid or star. Even for this simple example, the Fermi surface has an interesting shape. In three dimensions, the shapes are quite remarkable.

Cyclotron resonance can be done on metals. The experiment measures the properties of electrons on the Fermi surface. They are in pockets, as shown in fig. 5.1. In a magnetic field perpendicular to the plane of fig. 5.1, the electrons at the Fermi energy have periodic orbits around the boundary of the pocket. The period of the orbit is determined by how long it takes the electron to go around the pocket.

An electron in a crystal changes its wave vector in response to an external force. In a magnetic field, the force is

$$\hbar\dot{\mathbf{k}} = e\mathbf{v}\times\mathbf{B} \tag{5.56}$$

where the overdot denotes time derivative. This motion does not change the energy of the particle. Denote the band energy as $\varepsilon(\mathbf{k})$. Then

$$\frac{d}{dt}\varepsilon(\mathbf{k}) = \nabla_k\varepsilon(\mathbf{k})\cdot\dot{\mathbf{k}} = e\mathbf{v}\cdot(\mathbf{v}\times\mathbf{B}) = 0 \tag{5.57}$$

$$\hbar\mathbf{v} = \nabla_k\varepsilon(\mathbf{k}) \tag{5.58}$$

Since the Fermi surface is the locus of points with constant energy, the magnetic field will move the particle in a periodic orbit while keeping it on the Fermi surface. This topic is discussed more in a later section.

5.2.6 *Thermionic Emission*

Early in the last century it was observed that a solid, when heated, would thermally emit electrons. Thermionic emission became the basis of glass tube rectifiers that made radio possible before the invention of the transistor. Careful experimental work established *Richardson's equation* for the emitted current from metal surfaces:

$$J = AT^2\exp\left[-\frac{W}{k_BT}\right] \tag{5.59}$$

where T is the absolute temperature, W is the *work function*, and the prefactor $A \approx 120$ amps/cm^2 K^2. Figure 5.2 shows the energy level diagram for a metal. The zero of energy is the vacuum. The chemical potential μ is a negative energy below the vacuum. The work function is the energy difference between the vacuum and the chemical potential. Since

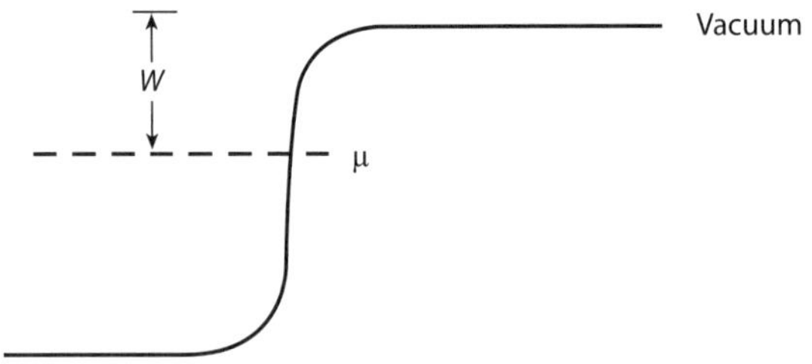

FIGURE 5.2. Surface work function W is the difference between the vacuum energy and the chemical potential μ.

the vacuum energy is a constant zero everywhere and the chemical potential in a solid in equilibrium is constant everywhere, one might think that the work function W should be independent of which crystal face is the surface. In fact, it does depend on crystal face! For each different crystal face, the outer layer of surface atoms/ions has a different arrangement. The distribution of electron charge is also different. So the work function depends slightly on crystal face.

The flow of current from the surface is determined by (i) the electron charge (e), (ii) times the density of electrons $n_F[\varepsilon(k)]d^3k$, and (iii) times the velocity of electrons going toward the surface, which is placed in the $+z$ direction. There is also a constraint that the kinetic energy in the $+z$ direction is larger than $E_c = W + \mu$ as measured from the bottom of the conduction band:

$$J = 2e\int_{k_z>0}\frac{d^3k}{(2\pi)^3}n_F[\varepsilon(k)]v_{kz}\Theta(E_z - E_c) \tag{5.60}$$

The factor of two is for spin degeneracy. Use polar coordinates where $d^3k = 2\pi dk_z k_\perp dk_\perp$. Since $v_{kz} = \hbar k_z/m$, we can change integration variables to energy:

$$v_{kz}d^3k = \frac{2\pi m}{\hbar^3}dE_z dE_\perp, \quad E_z = \frac{\hbar^2 k_z^2}{2m}, \quad E_\perp = \frac{\hbar^2 k_\perp^2}{2m} \tag{5.61}$$

$$J = \frac{em}{2\pi^2\hbar^3}\int_0^\infty dE_\perp \int_{E_c}^\infty dE_z n_F[\varepsilon(k)] \tag{5.62}$$

The cutoff energy E_c is very high, and only electrons contribute that have a large kinetic energy. The occupation function is approximated by

$$n_F[\varepsilon(k)] = \frac{1}{e^{\beta(\varepsilon-\mu)}+1} \approx e^{-\beta(\varepsilon-\mu)} = e^{-\beta(E_z+E_\perp-\mu)} \tag{5.63}$$

$$J = \frac{em}{2\pi^2\hbar^3}e^{\beta\mu}\int_0^\infty dE_\perp e^{-\beta E_\perp}\int_{E_c}^\infty dE_z e^{-\beta E_z} \tag{5.64}$$

$$= \frac{em}{2\pi^2\hbar^3}e^{\beta\mu}(k_BT)^2 e^{-\beta E_c} = AT^2 e^{-\beta W}$$

$$A = \frac{emk_B^2}{2\pi^2\hbar^3} = 120.17\frac{A^2}{cm^2K^2} \tag{5.65}$$

Thermionic emission is an important surface phenomenon. It is also an important interface phenomenon at the boundary between two materials. Electrons can always be thermally excited over a potential barrier, although this is less likely at low temperatures, where $\beta = 1/k_BT$ becomes very large.

5.3 Magnetic Fields

Numerous phenomena are found when subjecting a nearly free electron metal to a magnetic field. For a single electron, the nonrelativistic Hamiltonian is

$$H = \frac{1}{2m}(\mathbf{p} - e\mathbf{A})^2 \tag{5.66}$$

The vector potential $\mathbf{A}(\mathbf{r})$ is chosen to give a constant magnetic field $\mathbf{B} = \nabla \times \mathbf{A} = B\hat{z}$. Use the gauge that $\nabla \cdot \mathbf{A} = 0$. There are many ways of satisfying these relations. One choice is

$$\mathbf{A} = B(0, x, 0) \tag{5.67}$$

$$H = \frac{1}{2m}\left[p_x^2 + (p_y - eBx)^2 + p_z^2\right] \tag{5.68}$$

so that $A(\mathbf{r})$ has only a y-component. The Hamiltonian commutes with p_y and p_z; both are constants of motion. Write the eigenfunction as

$$\psi(\mathbf{r}) = \frac{e^{i(k_y y + k_z z)}}{\sqrt{L_y L_z}} f(x) \tag{5.69}$$

The x-direction has the Hamiltonian for a harmonic oscillator:

$$H\psi = \frac{1}{2m}\frac{e^{i(k_y y + k_z z)}}{\sqrt{L_y L_z}}\left[p_x^2 + (\hbar k_y - eBx)^2 + (\hbar k_z)^2\right] \tag{5.70}$$

$$= \frac{1}{2m}\frac{e^{i(k_y y + k_z z)}}{\sqrt{L_y L_z}}\left[p_x^2 + (eB)^2(x - x_0)^2 + (\hbar k_z)^2\right]$$

$$x_0 = \frac{\hbar k_y}{eB} \tag{5.71}$$

The first two terms in the square brackets are the Hamiltonian for a simple harmonic oscillator. The eigenfunctions and eigenvalues are

$$E_n(k_z) = \hbar\omega_c\left(n + \frac{1}{2}\right) + \varepsilon_z, \quad \omega_c = \frac{eB}{m}, \quad \varepsilon_z = \frac{\hbar k_z^2}{2m} \tag{5.72}$$

$$\psi_n(k_y, k_z; \mathbf{r}) = \frac{e^{i(k_y y + k_z z)}}{\sqrt{L_y L_z}}\phi_n(x - x_0) \tag{5.73}$$

where ϕ_n is the eigenfunction of the harmonic oscillator in one dimension. The cyclotron frequency is ω_c. The first term in the energy has equally spaced states separated by the the magnetic energy $\hbar\omega_c$, which are called *Landau levels.* It is useful to introduce a unit of magnetic length ℓ that will enter many equations:

$$\ell^2 = \frac{\hbar}{eB}, \quad x_0 = k_y \ell^2 \tag{5.74}$$

For a free particle of mass m_e the magnetic length can also be written as

$$\ell^2 = \frac{\hbar}{m\omega_c} \tag{5.75}$$

However, the first definition is more fundamental, since the length is defined only by the magnetic field and fundamental constants.

5.3.1 *Integer Quantum Hall Effect*

Consider electrons that are confined to the (x, y)-plane and are subjected to a magnetic field in the $\hat{z}$ direction. This situation is found when electrons are confined to the surface of liquid helium or confined to a single layer in a quantum well. The derivation is almost the same as above, but toss out the z-dependence:

$$H = \frac{1}{2m}\left[p_x^2 + \left(p_y - eBx\right)^2\right] - \mu B \sigma_z \tag{5.76}$$

$$E_{nm_s} = \hbar\omega_c\left(n + \frac{1}{2}\right) - \Delta m_s, \quad \Delta = \mu B, \quad m_s = \pm 1 \tag{5.77}$$

$$\psi_n(k_y; \mathbf{r}) = \frac{e^{ik_y y}}{\sqrt{L_y}}\phi_n(x - x_0) \tag{5.78}$$

The last term is from the spin. A free electron has a magnetic moment μ given by the Bohr magneton $\mu_0 = e\hbar/2m$. In a semiconductor, the moment is changed by $k \cdot p$ theory.

In two dimensions, the density of states is

$$N_2(E) = \frac{1}{L_x L_y}\sum\nolimits_{nk_y m_s} \delta(E - E_{nm_s}) \tag{5.79}$$

The right-hand side does not seem to depend on k_y. Convert the summation over k_y to an integral:

$$\frac{1}{L_y}\sum\nolimits_{k_y} = \int_0^{?} \frac{dk_y}{2\pi} \tag{5.80}$$

What are the limits of integration? The argument of the harmonic oscillator function is $x - x_0 = x - \ell^2 k_y$. The orbit must stay inside of the sample, so $0 < x < L_x$. Since the orbits are small, change this constraint to $0 < x_0 < L_x$, which is $k_y < L_x/\ell^2$. The above integral is

$$\frac{1}{L_y}\sum_{k_y} = \frac{L_x}{2\pi\ell^2} \tag{5.81}$$

$$N_2(E) = \frac{1}{2\pi\ell^2}\sum_{nm_s}\delta(E - E_{nm_s}) \tag{5.82}$$

The density of states is a series of delta functions. The quantity $2\pi\ell^2 = A_0$ is the area assigned to each electron. The total number of electrons per unit area in the layer is

$$n_e = \int dE N_2(E) n_F(E - \mu) = \frac{1}{2\pi\ell^2}\sum\nolimits_{nm_s} \frac{1}{e^{\beta(E_{nm_s} - \mu)} + 1} \tag{5.83}$$

The above expression is rewritten as

$$n_e = \frac{\nu}{2\pi\ell^2} \tag{5.84}$$

$$\nu = \sum_{nm_s} \frac{1}{e^{\beta(E_{nm_s} - \mu)} + 1} \tag{5.85}$$

where ν is the number of filled or partially filled Landau levels. Each spin state is counted separately.

Some typical experimental numbers are presented. The factor in the denominator is

$$2\pi\ell^2 = \frac{h}{eB} = \frac{\phi_0}{B} \tag{5.86}$$

The flux quantum is $\phi_0 = hc/e$ in cgs units, and is $\phi_0 = h/e$ in SI units. It has a numerical value of $\phi_0 = 4.13567 \times 10^{-15}$ in units of joules per ampere, which is also tesla square meters. A typical electron density in a quantum well is $n_0 \sim 10^{15}$ per square meters, which requires a magnetic field of $B \sim 4T$.

The quantum Hall effect is the observation of plateaus in the Hall voltage at certain values of magnetic field. The integer quantum Hall effect (IQHE) is when $\nu = 2\pi\ell^2 n_e$ has integer values, which occurs when Landau levels are exactly filled. An experimental example is shown in fig. 5.3. The IQHE was discovered by von Klitzing, Dorda, and Pepper in 1980 [12].

Consider the classical motion of an electron in two dimensions, with a magnetic field perpendicular to the plane. Newton's second law is

$$m\frac{d}{dt}\mathbf{v} = e[\mathbf{E} + \mathbf{v} \times \mathbf{B}] - \frac{m\mathbf{v}}{\tau} \tag{5.87}$$

where τ is the lifetime of the electron. The experiments are dc, so set $\dot{\mathbf{v}} = 0$ and find

$$v_x = \frac{e\tau}{m}[E_x + v_y B] \tag{5.88}$$

$$v_y = \frac{e\tau}{m}[E_y - v_x B] \tag{5.89}$$

The current is along the bar in the x-direction, as shown in fig. 5.4a. There is no current in the y-direction, so set $v_y = 0$:

$$v_x = \frac{e\tau}{m}E_x \tag{5.90}$$

$$E_y = v_x B \tag{5.91}$$

A longitudinal current (I_x, in amperes) is

$$I_x = en_e v_x L_y = \sigma E_x \tag{5.92}$$

$$\sigma = \frac{e^2\tau n_e}{m} \tag{5.93}$$

where L_y is the width of the sample, and n_e is the number of electrons per unit area. In two dimensions the electrical conductivity σ has the units of siemens $= 1/\text{ohms}$. The Hall voltage is $V_y = L_y E_y$. So the above formula gives

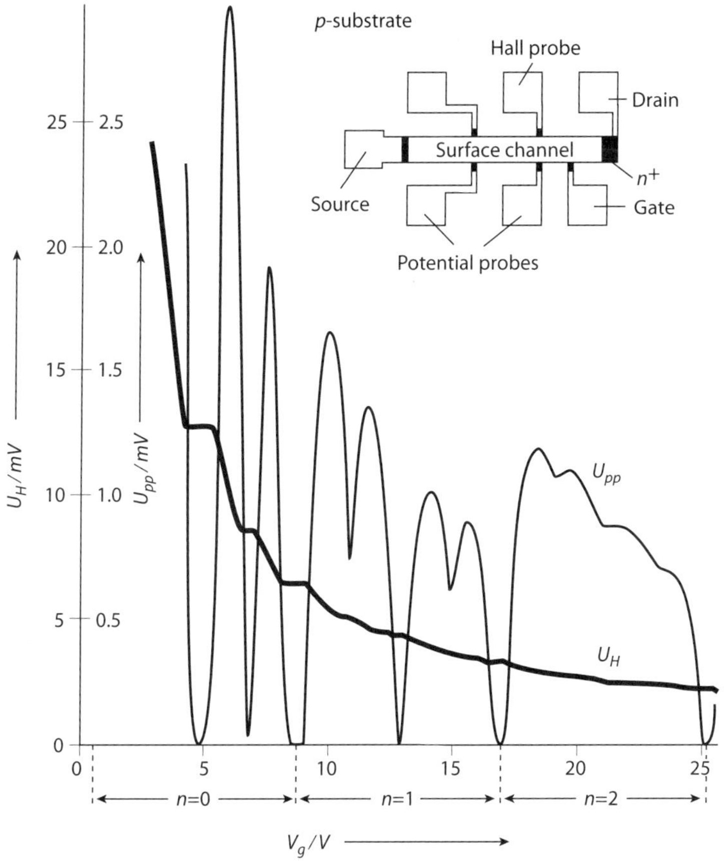

FIGURE 5.3. Transverse Hall voltage U_H and longitudinal voltage U_{PP} as a function of electron density (gate voltage), showing plateaus at integer values of ν. Magnetic field fixed at 18 T. From von Klitzing, Gorda, and Pepper, *Phys. Rev. Lett.* **45**, 494 (1980). Used with permission of the American Physical Society.

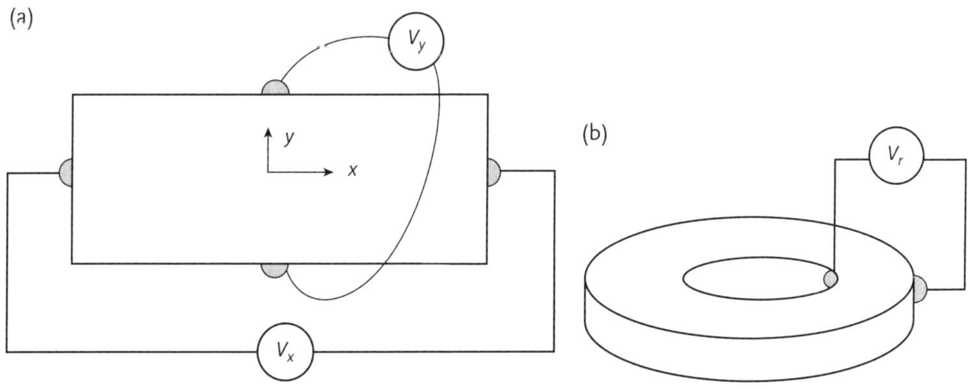

FIGURE 5.4. (a) Bar geometry. (b) Corbino disk.

$$V_y = E_y L_y = v_x B L_y = B\frac{I_x}{n_e e} = R_{xy} I_x \tag{5.94}$$

$$R_{xy} = \frac{B}{e n_e} \tag{5.95}$$

In classical physics, the transverse resistance R_{xy} is a linear function of the magnetic field B. The longitudinal resistance $R_{xx} = 1/\sigma$ is unaffected by the magnetic field.

In the IQHE, use eqn. (5.89) for n_e:

$$R_{xy} = \frac{2\pi \ell^2 B}{e\nu} = \frac{\phi_0}{e\nu} = \frac{R_0}{\nu} \tag{5.96}$$

$$R_0 = \frac{h}{e^2} = 25{,}812.8 \text{ ohms} \tag{5.97}$$

The fundamental unit of resistance is R_0. It is the inverse of the fundamental unit of conductance $\sigma_0 = 1/R_0 = 3.8740 \times 10^{-5}$ siemens. The transverse resistance R_{xy} is R_0 divided by the number of filled or partially filled Landau levels ν. In high-quality quantum wells of GaAs at low temperature and in a high magnetic field, it is found that the Hall resistance R_{xy} has plateaus when ν is an integer in eqn. (5.96). This phenomenon is the integer quantum Hall effect.

The simple interpretation of the IQHE is that ν is always an integer, so the Landau levels prefer to be exactly filled. The actual phenomenon is far more complicated. Semiconductors have impurities, such as donors and acceptors, that create local potential regions. Nearly all of the electrons in Landau orbits are bound to these local potentials. They are localized and unable to move. We started with a free electron in two dimensions. The magnetic field reduces the number of free dimensions to one, symbolized by the quantum number k_y. But any weak potential in one dimension can bind a particle.

The electrons contributing to the current are in orbits along the edge of the sample. The Corbino disk is shaped like a washer used to secure bolts: it is a squat, hollow cylinder, as shown in fig. 5.4b. In classical physics, one can put a voltage difference ΔV between the inner and outer edges and run a current radially outward between the two surfaces. When this geometry is used in the quantum Hall regime, no current is found to flow radially. There is current around the edge if two contacts are attached there. This is proof that conduction in the IQHE is related to movement along edges and not to conduction in the bulk of the material. Read Jain's book [5] for possible explanations of the plateaus in this case.

5.3.2 *Fractional Quantum Hall Effect*

Tsui, Störmer, and Gossard [11] discovered that there are also plateaus when ν is composed of simple fractions, such as $\frac{1}{3}$ or $\frac{2}{5}$. This latter phenomenon is called the *fractional quantum Hall effect*. The FQHE is due to strong correlations between the positions of the electrons. If we introduce electron–electron interactions into the Hamiltonian of a two-dimensional system in a strong magnetic field, the equations become very complicated.

Nevertheless, the resulting phenomenon can be explained in a simple fashion. The theory uses wave functions that were just written down by inspiration.

Discuss the FQHE using the Hamiltonian of an electron in two dimensions in the symmetric gauge:

$$\mathbf{A}(\mathbf{r}) = \frac{B}{2}(-y, x, 0) \tag{5.98}$$

$$\begin{aligned} H &= \frac{1}{2m}[(p_x + eA_x)^2 + (p_y + eA_y)^2] \\ &= \frac{1}{2m}[(p_x - \frac{eB}{2}y)^2 + (p_y + \frac{eB}{2}x)^2] \\ &= \frac{1}{2m}[p_x^2 + p_y^2 + \left(\frac{eB}{2}\right)^2 (x^2 + y^2)] + \frac{eB}{2m}(xp_y - yp_x) \end{aligned} \tag{5.99}$$

The electron charge is $-e$, $(e > 0)$. The last term is the z-component of angular momentum. The solutions have an angular momentum, which is assigned a quantum number n. The following eigenfunction will be shown to be an eigenstate of the Hamiltonian:

$$\begin{aligned} \phi_n(x, y) &= \rho^n e^{-in\theta} \exp[-\rho^2/(4\ell^2)] \\ &= (x - iy)^n \exp[-\rho^2/(4\ell^2)] \end{aligned} \tag{5.100}$$

$$\rho e^{-i\theta} = x - iy \tag{5.101}$$

where $x = \rho\cos(\theta)$, $y = \rho\sin(\theta)$. By direct differentiation $(p_\mu = -i\hbar\partial/\partial r_\mu)$

$$\frac{eB}{2m}(xp_y - yp_x)\phi_n = \frac{\hbar\omega_c}{2}\frac{\partial}{i\partial\theta}\phi_n = -\frac{\hbar\omega_c n}{2}\phi_n \tag{5.102}$$

$$\frac{1}{2m}(p_x^2 + p_y^2)\phi_n = \frac{\hbar^2}{2m\ell^2}\left[n + 1 - \frac{\rho^2}{4\ell^2}\right]\phi_n \tag{5.103}$$

$$\frac{\hbar^2}{2m\ell^2} = \frac{\hbar\omega_c}{2} \tag{5.104}$$

The term in ρ^2 cancels the other ρ^2 term in the Hamiltonian. The eigenvalue ε_n of the Hamiltonian is

$$\varepsilon_n = \frac{\hbar\omega_c}{2}[n + 1 - n] = \frac{\hbar\omega_c}{2} \tag{5.105}$$

The eigenvalue is one-half of the energy spacing between Landau levels. It is the lowest eigenvalue. It is interesting that it is not dependent on the angular momentum n. When solving using the other gauge, the eigenvalues did not depend on the extra quantum number k_y. In the symmetric gauge, they do not depend on the angular momentum n. So when adding electrons to the lowest Landau level, they are added with different angular momentum, which is done by increasing the value of n. For a system of N electrons, the Slater determinent for this state, when all have the same spin configuration, is

$$\Psi_1(\boldsymbol{\rho}_1, \boldsymbol{\rho}_2, \ldots, \boldsymbol{\rho}_N) = \frac{1}{\sqrt{N!}} \begin{vmatrix} \phi_0(\boldsymbol{\rho}_1) & \phi_1(\boldsymbol{\rho}_1) & \phi_2(\boldsymbol{\rho}_1) & \cdots & \phi_{N-1}(\boldsymbol{\rho}_1) \\ \phi_0(\boldsymbol{\rho}_2) & \phi_1(\boldsymbol{\rho}_2) & \phi_2(\boldsymbol{\rho}_2) & \cdots & \phi_{N-1}(\boldsymbol{\rho}_2) \\ \phi_0(\boldsymbol{\rho}_1) & \phi_1(\boldsymbol{\rho}_3) & \phi_2(\boldsymbol{\rho}_3) & \cdots & \phi_{N-1}(\boldsymbol{\rho}_3) \\ \vdots & \vdots & \vdots & \ddots & \vdots \\ \phi_0(\boldsymbol{\rho}_N) & \phi_1(\boldsymbol{\rho}_N) & \phi_2(\boldsymbol{\rho}_N) & \cdots & \phi_{N-1}(\boldsymbol{\rho}_N) \end{vmatrix} \tag{5.106}$$

Define $z_j = x_j - iy_j$, and the eigenfunctions are all

$$\phi_n(\boldsymbol{\rho}_j) = z_j^n \exp\left[-\rho_j^2/4\ell^2\right] \tag{5.107}$$

The exponential factor is the same for each term in a row and can be removed from the determinant:

$$\Psi_1(\boldsymbol{\rho}_1, \boldsymbol{\rho}_2, \ldots, \boldsymbol{\rho}_N) = \frac{\Psi_0}{\sqrt{N!}} \begin{vmatrix} 1 & z_1 & z_1^2 & \cdots & z_1^{N-1} \\ 1 & z_2 & z_2^2 & \cdots & z_2^{N-1} \\ 1 & z_3 & z_3^2 & \cdots & z_3^{N-1} \\ \vdots & \vdots & \vdots & \ddots & \vdots \\ 1 & z_N & z_N^2 & \cdots & z_N^{N-1} \end{vmatrix} \tag{5.108}$$

$$\Psi_0 = \exp\left[-\frac{1}{4\ell^2}\sum_{j=1}^{N}|z_j|^2\right] \tag{5.109}$$

The determinant was given by Vandermonde [7].

$$\Psi_1(\boldsymbol{\rho}_1, \boldsymbol{\rho}_2, \ldots, \boldsymbol{\rho}_N) = \frac{\Psi_0}{\sqrt{N!}} \Pi_{i>j}(z_i - z_j) \tag{5.110}$$

It is exactly given by the difference of all pairs of z-values. According to the exclusion principle, if two electrons are at the same point ($\boldsymbol{\rho}_i = \boldsymbol{\rho}_j$), the many-particle wave function must vanish. The Slater determinent has this feature, since a determinant vanishes if any two rows are identical. For the Vandermonde determinent, the wave function vanishes due to the factor of $z_i - z_j = 0$.

Another feature of the many-particle wave function is that it must change sign when interchanging the positions of any two electrons of identical spin. The Slater determinant has this property, since exchanging two rows changes the sign. In the Vandermonde determinant, the sign change comes from $(z_i - z_j) \to -(z_j - z_i)$.

Bob Laughlin [6] made the very important observation that the wave function maintains these symmetry properties if it is written

$$\Psi(\boldsymbol{\rho}_1, \boldsymbol{\rho}_2, \ldots, \boldsymbol{\rho}_N) = \frac{\Psi_0}{\sqrt{N!}} \Pi_{i>j}(z_i - z_j)^m \tag{5.111}$$

if m is an odd integer such as three or five. The Laughlin wave function cannot be derived from a single Slater determinant. It is a wave function of a many-electron state with a high degree of correlation. Numerous calculations have shown that it is a nearly a perfect description of the quantum Hall state with $\nu = 1/m$. The case that $m = 3$ describes the electron correlation in the electron plateaus when $\nu = 1/3$, and the choice $m = 5, 7$ describes $\nu = \frac{1}{5}, \frac{1}{7}$. The starting wave function Ψ_1 with $m = 1$ describes the plateau when $\nu = 1$.

5.3.3 Composite Fermions

As more experiments were performed on better samples and at higher magnetic fields, more fractions were observed. Only a few of these can be explained by the Laughlin wave function. All of the observed fractions can be explained by another wave function written down by Jainendra Jain [5]. If Ψ_n is the wave function for the ground state that has all Landau levels filled up to n, then

$$\Psi_\nu(\rho_1, \rho_2, \ldots, \rho_N) = P_{\text{LLL}} \Pi_{i>j}(z_i - z_j)^{2p} \Psi_n \tag{5.112}$$

where p is a positive integer. The prefactor P_{LLL} is a projection operator that projects all eigenfunctions onto the "lowest Landau level." Jain [5] showed that this wave function gives the fractional states

$$\nu = \frac{n}{2pn + 1}, \quad \nu = \frac{n}{2pn - 1} \tag{5.113}$$

The factor of $(z_i - z_j)^{2p}$ attaches two vortices to each electron. A vortex is a current loop created by the magnetic field. The Laughlin state has $m = 2p + 1$.

Jain defined the composite fermion as follows: *A composite fermion is the bound state of an electron and an even number of quantized vortices.* Composite fermions become the new quasiparticles for electrons in a magnetic field. Most of the magnetic field is tied up in the vortices, which are attached to the electrons. The remaining magnetic field B^* is

$$B^* = B - 2pn_e\phi_0 \tag{5.114}$$

where n_e is the two-dimensional density of electrons, and $\phi_0 = h/e$ is the quantum of magnetic flux. The remaining field B^* can be either positive or negative, which corresponds to the $\pm$ sign choice in the denominator of eqn. (5.113).

This composite fermion model explains all of the experimental data on the many different fractional states. The theory has no adjustable parameters. It predicts that the fractions have only odd denominators, which agrees with experiments.

5.3.4 deHaas–van Alphen Effect

There are many interesting phenomena in pure metals in large magnetic fields at low temperature. The criteria for their observation is that $\omega_c\tau > 1$, where ω_c is the cyclotron frequency, and τ is the lifetime of the electrons from scattering by defects or phonons. Defect scattering is important at low temperature, and phonon scattering is important at high temperature.

All of these phenomena show an oscillatory change in an experimental parameter as the magnetic field is varied. The oscillations are periodic when graphed as a function of the inverse magnetic field. Some of these phenomena are the following:

- The *deHaas-van Alphen effect*, which is a change in the induced magnetic moment M
- The *Shubnikov-deHaas effect*, which is a change in the electrical resistivity
- The acoustic attenuation
- The temperature, which is a magnetothermal effect
- The physical size of the sample, which is a magnetostriction effect
- The velocity of sound

As the magnetic field is increased, the quantum of energy $\hbar\omega_c$ increases and the Landau energy levels get further apart. Since the chemical potential μ is relatively fixed, then fewer Landau levels are occupied. Below we show how this process causes an oscillatory behavior.

Although this technique was developed fifty years ago, it is still being used. Indeed, many countries have established national laboratories for high magnetic fields, where measurements can be done at fields up to 100 teslas. These high magnetic fields achieve $\omega_c\tau > 1$ even for samples with large electron lifetimes.

In three dimensions the density of orbital states is

$$N_3(E) = \frac{1}{L_x L_y L_z}\sum_{nk_yk_z}\delta(E - E_n - \varepsilon_z), \quad E_n = \hbar\omega_c\left(n + \frac{1}{2}\right) \tag{5.115}$$

The summation over k_y is evaluated exactly as in eqn. (5.81). The summation over k_z is

$$\frac{1}{L_z}\sum_{k_z}\delta(E - E_n - \varepsilon_z) = \int\frac{dk_z}{2\pi}\delta(E - E_n - \varepsilon_z) = \frac{\sqrt{2m}}{2\pi\hbar}\frac{1}{\sqrt{E - E_n}} \tag{5.116}$$

$$N_3(E) = \frac{\sqrt{2m}}{(2\pi\ell)^2\hbar}\sum_{n=0}^{n_x}\frac{1}{\sqrt{E - \hbar\omega_c(n + 1/2)}} \tag{5.117}$$

The density of electrons in the conduction band is

$$n_e = 2\int dE n_F(E - \mu) N_3(E) \tag{5.118}$$

where the factor of two is for spin degeneracy. The integral can be evaluated at $T = 0$, where the Fermi-Dirac occupation factor is $n_F(E - \mu) = \Theta(\mu - E)$:

$$n_e = \frac{\sqrt{2m}}{(\pi\ell)^2\hbar}\sum_{n=0}^{n_x}\sqrt{\mu - \hbar\omega_c(n + 1/2)} \tag{5.119}$$

where n_x is the largest value of n with a positive argument for the square root. This expression gives the usual density of electrons in the limit of very small magnetic field. Change the summation to an integration, with variable $dx = \hbar\omega_c dn$. Recall that $\ell^2 = \hbar/m\omega_c$ and get

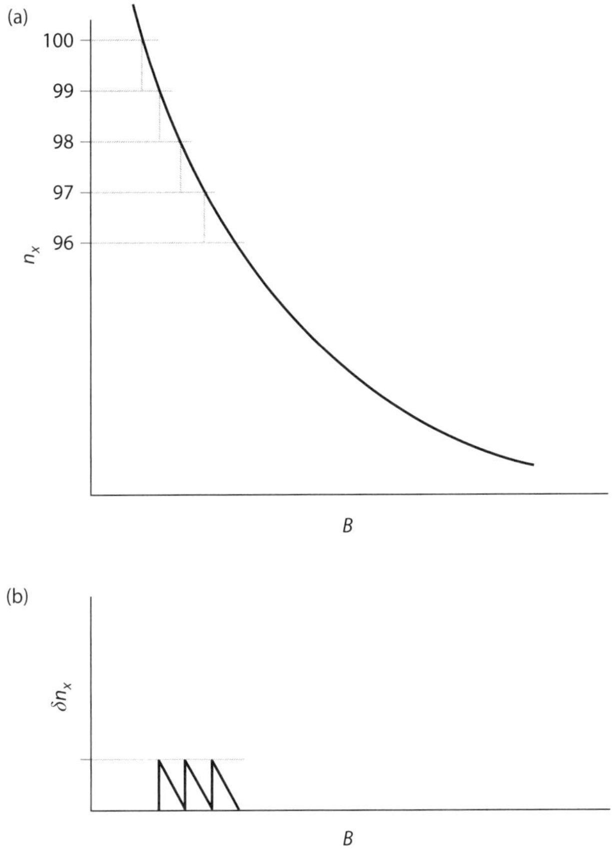

FIGURE 5.5. (a) Graph of n_x as a function of magnetic field. (b) Graph of $\delta n_x(B)$.

$$n_e = \frac{1}{2\pi^2}\left(\frac{2m}{\hbar^2}\right)^{3/2}\int_0^{\mu} dx\sqrt{\mu - x} = \frac{1}{3\pi^2}\left(\frac{2m\mu}{\hbar^2}\right)^{3/2} = \frac{k_F^3}{3\pi^2} \tag{5.120}$$

which is the correct answer for a spherical Fermi surface.

Equation (5.119) is an interesting expression. The density of electrons, on the left side, is a constant, independent of the magnetic field. Nearly every quantity on the right side, such as ℓ, n_x, μ, ω_c, depends on magnetic field. The largest occupied Landau level is n_x, which is the largest integer that satisfies

$$n_x \leq \frac{\mu}{\hbar\omega_c} - \frac{1}{2} \tag{5.121}$$

Keep in mind that for typical metals in typical magnetic fields, the values of n_x are integers of three or four digits.

Figure 5.5 shows how n_x depends on magnetic field. The right-hand side of the above equation mades a smooth curve that decreases with increasing magnetic field. The integer n_x makes a staircase of values. Define an oscillating function $\delta n_x(B)$ according to

$$n_x = \frac{\mu}{\hbar\omega_c} - \frac{1}{2} - \delta n_x(B) \tag{5.122}$$

Figure 5.5b shows that this function is a series of ramps. The height of the ramps is always one, but the width of the ramps increases with increasing magnetic field. For a free electron system, $\omega_c = eB/m$, $\mu = \hbar^2 k_F^2/2m$, and the cross-sectional area of the Fermi surface is $\mathcal{A}_F = \pi k_F^2$. Then

$$X = \frac{\mu}{\hbar\omega_c} = \frac{\hbar^2 k_F^2/2m}{e\hbar B/m} = \frac{(\pi k_F^2)\hbar}{2\pi eB} = \frac{\ell^2 \mathcal{A}_F}{2\pi} \tag{5.123}$$

Onsager [8] showed that the period of the oscillation depends upon the area $\mathcal{A}_F$ of the Fermi surface, regardless of its shape:

$$\delta n_x = \delta\left(\frac{\ell^2 \mathcal{A}_F}{2\pi}\right) \tag{5.124}$$

The area $\mathcal{A}_F$ is the maximum cross section of the BZ perpendicular to the direction of the magnetic field.

The relative spacing between ramps increases with magnetic field. If one graphs $\delta n_x(B)$ as a function of inverse magnetic field, one finds the ramps are equally spaced. So δn_x is a periodic function of inverse magnetic field. Other quantities also vary periodically with $1/B$, such as the chemical potential μ and the internal energy. Almost every measureable quantity of the metal has a small dependence on a periodic function of $1/B$.

As shown in fig. 5.5b, the quantity δn has a sawtooth shape when graphed as a function of X. Since the ramp is periodic, it can be expanded in a Fourier series of the form

$$\delta n(X, T = 0) = \sum_{p=1} \frac{1}{\pi p} \sin[2\pi p X] \tag{5.125}$$

Experimentally it is found that the sawtooth does not have a sharp point. It is smeared out for several possible reasons. One is temperature. At nonzero temperature, the Fermi surface is not sharp, but has a thermal smearing on the order of $O(k_B T)$. The above result is thermally smeared by recalling that $X = \mu/\hbar\omega_c$ and writing eqn. (5.125) as ($\xi = E - \mu$)

$$\delta n(X, T) = \sum_{p=1} \frac{1}{\pi p} \int_{-\infty}^{\infty} d\xi \sin\left[2\pi p \frac{(\mu + \xi)}{\hbar\omega_c}\right]\left(-\frac{dn_F(\xi)}{d\xi}\right) \tag{5.126}$$

$$= \frac{2\pi k_B T}{\hbar\omega_c} \sum_{p=1} \frac{\sin(2\pi p X)}{\sinh(2\pi p Y)}, \quad Y = \frac{\pi k_B T}{\hbar\omega_c}$$

The denominator has the factor of $\sinh(2\pi p Y)$, which grows with increasing integers p and eventually cuts off the summation over p.

Another possible factor for rounding the sawtooth shape is the broadening of the Landau levels due to the finite mean-free-path of the electron. Introduce the electron lifetime τ, which is caused by the electron scattering from impurities or other defects. It gives a Lorentzian broadening:

$$\delta n(X,T) = \sum_{p=1} \frac{1}{\pi p} \int_{-\infty}^{\infty} \frac{d\xi}{2\pi} \sin\left[2\pi p \frac{(\mu+\xi)}{\hbar\omega_c}\right] \frac{\hbar/\tau}{\xi^2 + (\hbar/2\tau)^2} \quad (5.27)$$

$$= \sum_{p=1} \frac{1}{\pi p} \sin[2\pi p X] \exp\left[-\frac{\pi p}{\omega_c \tau}\right]$$

These two types of broadening can be combined in the case that $2\pi p Y > 1$, so that

$$\frac{1}{\sinh(2\pi p Y)} \approx 2\exp[-2\pi p Y] \quad (5.128)$$

$$\delta n(X,T) = \frac{4\pi k_B T}{\hbar\omega_c} \sum_{p=1} \sin(2\pi p X) \exp\left[-\frac{2\pi^2 p k_B}{\hbar\omega_c}(T+T_D)\right] \quad (5.129)$$

$$T_D = \frac{\hbar}{2\pi k_B \tau} \quad (5.130)$$

where T_D is called the *Dingle temperature.* It is a broadening due to the lifetime of the electron, but enters the theory as an effective temperature. The result of these broadening terms is that the summation over p has only a few terms before the amplitude starts to get small. The sawtooth shape is rounded and agrees well the the observed oscillations. For more details see Shoenberg's book [9].

5.4 Quantization of Orbits

The above formulas apply for free particles. In a crystal, the electron orbits get altered by the crystalline potential. However, they are still quantized. The magnetic field can be imposed in any direction and is usually rotated during the experiment (usually, the magnetic field is fixed and the crystal is rotated). In each case, the magnetic field energy E_n represents the quantization of the extremal orbit perpendicular to the magnetic field. The extremal orbit has the largest value of n, and $k_z \sim 0$. If one could look inside the Brillouin zone along the direction of magnetic field, the Fermi surface would be observed as an area. The outer circumference of this area is the extremal orbit. As the magnetic field is rotated, this extremal area will change. The changes provide an opportunity to map the shape of the Fermi surface.

The derivation is simplified by making it two dimensional. Only simple periodic orbits are considered. The motion is periodic in wave vector space. It is also periodic in real space: the particle is going around a closed orbit in the crystal. This motion is quantized in real space: the magnetic flux enclosed in the periodic orbit is quantized. If A_r is the enclosed area in real space, then quantum mechanics shows that

$$BA_r = \phi_0\left(n + \tfrac{1}{2}\right) \quad (5.131)$$

where the flux quantum ϕ_0 is related to the magnetic length $2\pi\ell^2 = \phi_0/B$. For small values of quantum number n the orbits enclose a small area. For orbits around the Fermi surface the value of n in metals is typically quite large. As an example of this formula, consider the case of an extremal orbit in the shape of a circle, for a free particle. The particle velocities and positions are

$$v_x(t) = v_0 \cos(\omega_c t), \quad v_y(t) = v_0 \sin(\omega_c t) \tag{5.132}$$

$$x(t) = \frac{v_0}{\omega_c} \sin(\omega_c t), \quad y(t) = -\frac{v_0}{\omega_c} \cos(\omega_c t) \tag{5.133}$$

$$r = \frac{v_0}{\omega_c} \tag{5.134}$$

The motion is circular in wave vector space, in velocity space, and in real space. In real space, the radius is r and the area is $A_r = \pi r^2$. The quantization in eqn. (5.131) is

$$B\pi r^2 = \pi B\left(\frac{v_0}{\omega_c}\right)^2 = \phi_0\left(n + \frac{1}{2}\right) \tag{5.135}$$

The above formula can have its symbols rearranged into ($\phi_0 = h/e$):

$$v_0^2 = 2\frac{\omega_c}{eB}\frac{h}{2\pi}\omega_c\left(n + \frac{1}{2}\right) = \frac{2}{m}\hbar\omega_c\left(n + \frac{1}{2}\right) \tag{5.136}$$

$$K = \frac{m}{2}v_0^2 = \hbar\omega_c\left(n + \frac{1}{2}\right) = E_n \tag{5.137}$$

The kinetic energy K is replaced by E_n.

This result can be generalized to any shape of orbit. This generalization depends on a theorem: *The area of the periodic orbit in real space A_r is proportional to the area $\mathcal{A}_F$ enclosed in wave vector space according to*

$$A_r = \ell^4 \mathcal{A}_F \tag{5.138}$$

The proof is simple. The dynamics of Bloch electrons in a magnetic field are given by

$$\hbar\dot{\mathbf{k}} = \mathbf{F} = e\mathbf{v} \times \mathbf{B} \tag{5.139}$$

The force $\mathbf{F}$ in a magnetic field is the Lorentz force. In two dimensions, with $\mathbf{B} = B\hat{z}$,

$$\dot{k}_x = \left(\frac{eB}{\hbar}\right)v_y \tag{5.140}$$

$$\dot{k}_y = -\left(\frac{eB}{\hbar}\right)v_x \tag{5.141}$$

The part in brackets is the constant ℓ^{-2}. Integrate the equations in time and find

$$k_x(t) = \left(\frac{eB}{\hbar}\right)[y(t) - y_0] \tag{5.142}$$

$$k_y(t) = -\left(\frac{eB}{\hbar}\right)[x(t) - x_0] \tag{5.143}$$

The integration constants (x_0, y_0) locate the center of the orbit in real space, but do not influence the area of the periodic orbit. The integral for the area of the Fermi surface is now directly proportional to the integral for the area in real space:

$$\mathcal{A}_F = \int k_y dk_x = \left(\frac{eB}{\hbar}\right)^2 \int dy[x - x_0] = \frac{A_r}{\ell^4} \tag{5.144}$$

which proves the theorem. Rearrange eqn. (5.131) into

$$B\ell^4 \mathcal{A}_F = \phi_0\left(n + \frac{1}{2}\right) \tag{5.145}$$

Since $2\pi\ell^2 = \phi_0/B$, rearrange the above formula into

$$\mathcal{A}_F = (2\pi)^2 \frac{B}{\phi_0}\left(n + \frac{1}{2}\right) \tag{5.146}$$

The final formula is the generalization of eqn. (5.131) to an arbitrary shaped Fermi surface. The extremal area of the Fermi surface is quantized in a magnetic field. Since the area of the Fermi surface is generally fixed according to Luttinger's theorem, the above formula gives the value $n = n_F$ of the magnetic quantum number of those electrons right on the Fermi surface. As the magnetic field is increased, this value of magnetic quantum number must get smaller:

$$\delta n = \delta\left(\frac{\phi_0 \mathcal{A}_F}{(2\pi)^2 B}\right) \tag{5.147}$$

This change goes as the inverse power of the magnetic field. A measurement of any experimental quantity, such as magnetization or resistivity, as a function of inverse magnetic field gives a periodic oscillation. The period of this oscillation is a direct measure of the area $\mathcal{A}_F$ of the Fermi surface.

Wannier first realized that the area of the orbit in real space A_r must be a multiple of the area of the crystal unit cell, in the plane perpendicular to the magnetic field. Otherwise, the magnetic field breaks the spatial periodicity of the crystal, and Bloch's theorem is invalid. Wannier's student Hofstadter solved this problem [4] and derived the beautiful "butterfly pattern" for the allowed orbits for the two dimensional square lattice (see fig. 5.6). The horizontal axis is a dimensionless quantity similar to wave vector, and the vertical axis is a dimensionless quantity related to the eigenvalue.

5.4.1 *Cyclotron Resonance*

The cyclotron frequency for a free particle is $\omega_c = eB/m$. The symbol T is the period of the orbit: it is the time it takes the electron to go around the Fermi surface during its periodic orbit. For free particles

$$\omega_c T = 2\pi \tag{5.148}$$

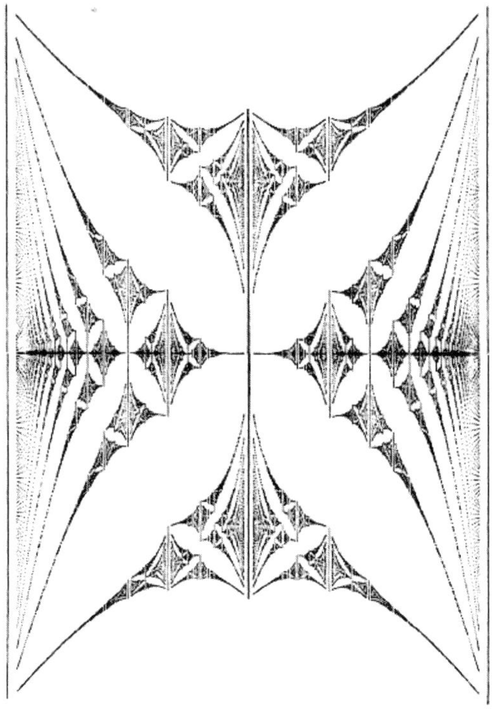

FIGURE 5.6. Allowed energy bands of a square lattice in a magnetic field. From Hofstadter, *Phys. Rev. B* **14**, 2239 (1976). Used with permission of the American Physical Society.

The period is not given simply by the area of the Fermi surface. Generally there is some loose scaling, since it takes longer to get around a big Fermi surface than a smaller one. However, the period T is a different measurement than the period of the deHaas-van Alphen effect discussed previously.

We give a nontrivial example. Consider the square lattice in two dimensions. If the energy bands are given by the tight-binding model, the dispersion is

$$\varepsilon(k) = -w[\cos(\theta_x) + \cos(\theta_y)], \quad \theta_\mu = k_\mu a \tag{5.149}$$

where a is the lattice constant, and w is a constant with the units of energy. The minus sign is chosen so that the band minimum is at $k = 0$. Let the band be less than half-full. A constant energy surface has

$$\cos(\theta_x) + \cos(\theta_y) \equiv \lambda = -\frac{\varepsilon(k)}{w} > 0 \tag{5.150}$$

Define θ_0 as the maximum value of the θ_μ. Then $\cos(\theta_0) = \lambda - 1$. A constant energy surface is shown in fig. 5.7 for $\lambda = 0.1$.

The particle velocities are

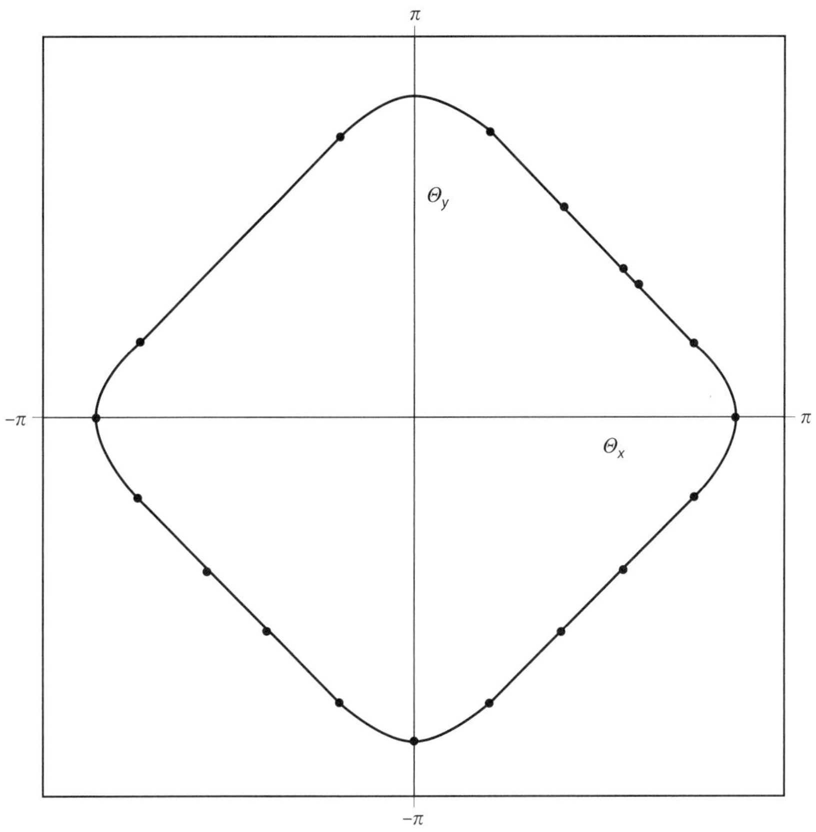

FIGURE 5.7. Tight-binding constant energy surface $\lambda = 0.1$ for the two-dimensional square lattice.

$$v_x = \frac{1}{\hbar}\frac{\partial \varepsilon(\mathbf{k})}{\partial k_x} = v_0 \sin(\theta_x), \quad v_0 = \frac{wa}{\hbar} \tag{5.151}$$

$$v_y = v_0 \sin(\theta_y) \tag{5.152}$$

In a magnetic field, the time evolution of the wave vector is

$$\frac{dk_x}{dt} = \frac{v_y}{\ell^2} = \frac{v_0}{\ell^2}\sin(\theta_y) \tag{5.153}$$

$$= \frac{v_0}{\ell^2}\sqrt{1 - [\cos(\theta_x) - \lambda]^2}$$

The dispersion relation (5.150) is used to relate $\sin(\theta_y)$ to $\cos(\theta_x)$. The above formula is rearranged into

$$dt = \frac{\ell^2}{av_0}\frac{d\theta_x}{\sqrt{1 - [\cos(\theta_x) - \lambda]^2}} \tag{5.154}$$

For electrons on the Fermi surface set $\lambda = -\mu/w$. It is sufficient to integrate both sides over one quadrant of the Fermi surface, and then find the period by multiplying by four:

$$T = \frac{4\ell^2}{av_0}\int_0^{\theta_0} \frac{d\theta_x}{\sqrt{1-[\cos(\theta_x)-\lambda]^2}} \tag{5.155}$$

The integral is recognized as the type that can be expressed as an elliptic function:

$$T = \frac{16}{2+\lambda}\frac{\ell^2}{a^2}\frac{\hbar}{w}K(m), \quad m = \left(\frac{2-\lambda}{2+\lambda}\right)^2 \tag{5.156}$$

If the Fermi surface is exactly half-full, then $\lambda = 0$, $m = 1$ and $K(m = 1)$ is infinity. The particle takes forever to go around the Fermi surface of the half-filled energy band! However, the area of this orbit is $\mathcal{A}_F = 2\pi^2/(a^2)$. This example shows that the period is not related to the area of the orbit.

The area of the Fermi surface can be evaluated as four times an integral over one quadrant:

$$\begin{aligned}\mathcal{A}_F &= \frac{4}{a^2}\int_0^{\theta_0} d\theta_x \theta_y \\ &= \frac{4}{a^2}\int_0^{\theta_0} d\theta_x \cos^{-1}[\lambda - \cos(\theta_x)]\end{aligned} \tag{5.157}$$

Use eqn. (5.150) to give θ_y in terms of θ_x. A more transparent expression is obtained after an integration by parts:

$$\mathcal{A}_F = \frac{4}{a^2}\int_0^{\theta_0} \frac{\theta_x d\theta_x}{\sqrt{1-[\cos(\theta_x)-\lambda]^2}} \tag{5.158}$$

A similar integral was evaluated for the period of oscillation. However, the integral for $\mathcal{A}_F$ has an additional factor of θ_x in the integrand, so the two integrals are different.

The period of the orbit is directly proportional to the derivative of the Fermi surface area with respect to the chemical potential. Take the derivative of eqn. (5.157), where $\lambda = -\mu/w$. λ occurs in the integrand, and in the upper limit $\Theta_0(\lambda)$. The integrand is zero at $\theta_x = \Theta_0$ so this second contribution is zero. Taking the derivative of the integrand gives

$$\frac{d\mathcal{A}_F}{d\mu} = \frac{4}{wa^2}\int_0^{\theta_0} \frac{d\theta_x}{\sqrt{1-[\lambda-\cos(\theta_x)]^2}} \tag{5.159}$$

This integral is identical to eqn. (5.155) that defines the period T. After sorting out the prefactors,

$$T = \hbar\ell^2 \frac{d\mathcal{A}_F}{d\mu} \tag{5.160}$$

The period in cyclotron resonance measures the energy derivative of the area of the Fermi surface. Although this formula was derived for the square lattice, it is valid for all Fermi surfaces.

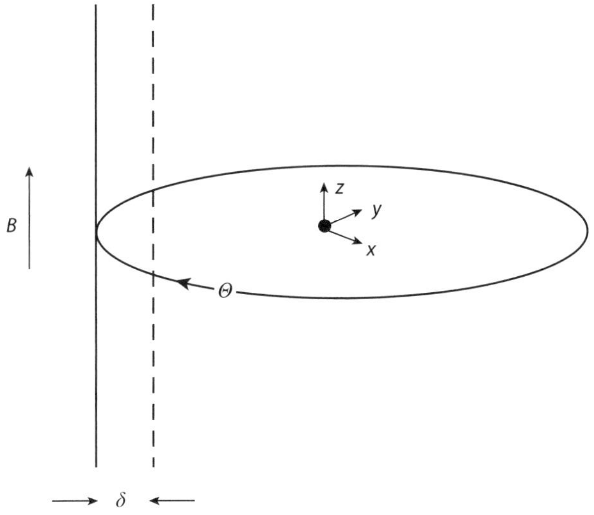

FIGURE 5.8. Azbel-Kaner geometry. Magnetic field is parallel to metal surface. Skin depth is shown. Electron orbits extend far into the metal.

Cyclotron resonance experiments are done by applying a microwave field of frequency ω_0 to the sample, and varying the magnetic field until reaching resonance $T\omega_0 = 2\pi$. This works well in pure semiconductors. In metals, the microwaves penetrate into the metal only a skin depth:

$$\delta = \frac{c}{\sqrt{2\pi\sigma\omega_0}} \tag{5.161}$$

Here σ is the dc electrical conductivity in units of inverse seconds. The Azbel-Kaner effect [1] utilizes the geometry shown in fig. 5.8. The magnetic field is parallel to the surface of the metal. The electron orbits penetrate far into the interior of the metal, well beyond the skin depth. The electrons are accelerated by the microwaves only while they are in the surface area defined by the skin depth. One can still observe cyclotron resonance in this geometry.

There are many other interesting phenomena related to Bloch electrons in magnetic fields. If the Fermi surface is connected in an extended zone scheme, there are open orbits where the electron does not make periodic motion in real space. Cu, Ag, and Au have Fermi surfaces with open orbits.

References

1. M. I. Azbel and E. A. Kaner, The theory of cyclotron resonance in metals. *Sov. Phys. JETP* **3**, 772–774 (1956)
2. W. J. De Haas and P. M. van Alphen, *Leiden Comm.* 208d, 212a (1930), and 220d (1932)
3. P. Drude, On the electron theory of metals, *Ann. der Phys.* **1**, 566–613 (1900), in German

4. D. R. Hofstadter, Energy levels and wave functions of Bloch electrons in rational and irrational magnetic fields. *Phys. Rev. B* **14**, 2239–2249 (1976)
5. J. K. Jain, *Composite Fermions*, (Cambridge, 2007)
6. R. B. Laughlin, Quantized Hall conductivity in two dimensions. *Phys. Rev.* **23**, 5632–5633 (1981)
7. G. D. Mahan, *Applied Mathematics* (Kluwer Academic, New York, NY, 2002), Chapter 1
8. L. Onsager, Interpretation of the de Haas-van Alphen effect. *Philos. Mag.* **43**, 1006–1008 (1952)
9. D. Shoenberg, *Magnetic Oscillations in Metals* (Cambridge University Press, Cambridge, 1984)
10. A. Sommerfeld, On the electron theory of metals based on Fermi statistics. *Z. Phys.* **47**, 1–32 (1928)
11. D. C. Tsui, H. L. Störmer, and A. C. Gossard, Two dimensional magnetotransport in the extreme quantum limit. *Phys. Rev. Lett.* **48**, 1559–1562 (1982)
12. K. von Klitzing, G. Dorda, and M. Pepper, New method for high accuracy determination of the fine-structure constant based on quantized Hall resistance. *Phys. Rev. Lett.* **45**, 494–497 (1980)

Homework

1. Consider an electron gas of density n_0 in three dimensions that is completely ferromagnetic: all electron spins point in the same direction. Derive:

(a) The Fermi wave vector in terms of n_0.

(b) The parameter r_s as the radius in atomic units that encloses one unit of charge.

(c) The average kinetic energy per electron.

2. A nearly free electron metal has a three-dimensional density of states $N(E)$, which is a smooth function of energy near to $E \approx \mu$.

(a) Derive an expression for the change in chemical potential with temperature $\mu(T)$.

(b) Derive an expression for the heat capacity.

Both results involve $N(\mu)$ and its derivatives and integrals.

3. The pressure P and compressibility K of an electron gas are defined as

$$P = -\left(\frac{dE_G}{d\Omega}\right)_N \tag{5.162}$$

$$\frac{1}{K} = -\Omega\left(\frac{dP}{d\Omega}\right)_N \tag{5.163}$$

where E_G is the total ground-state energy for a system of N-electrons, and Ω is the volume. Evaluate these two expressions for the noninteracting Fermi gas.

4. Derive the the internal energy $U(T)$ for the two-dimensional electron gas. Then derive a formula for the heat capacity at low temperature.

5. In a two-dimensional metal, at what temperature T_0 is $\mu(T_0) = 0$?

6. Derive the form of thermionic emission in two dimensions.

7. The lowest value of work function reported so far is $e\phi = 0.9$ eV when CsO is deposited on the surface of silicon. Use this value to calculate the thermionic current at $T =$ 1000 K.

8. The chemical potential in graphene is near the K-point. Show that near that point, we can write $J_0\gamma(k) \approx \hbar v(k_x + ik_y)$ and give the value for v, where here (k_x, k_y) are the values using the K-point as an origin. In a magnetic field B perpendicular to the plane, we can write the Hamiltonian as

$$H = \begin{bmatrix} 0 & \Lambda \\ \Lambda^\dagger & 0 \end{bmatrix} \tag{5.164}$$

$$\Lambda = v[p_x - eA_x + i(p_y - eA_y)] \tag{5.165}$$

Derive the eigenvalues of this Hamiltonian.

9. A Corbino disk has the shape of a hollow disk. Calculate the dc electrical resistance for current flow from the inner radius $(r = a)$ of the disk to the outer radius $(r = b)$ when there is a magnetic field B along the cylinder axis (i.e., the disk is in the (x, y)-plane, and the magnetic field is along the $\hat{z}$ direction). The inner and outer surfaces are each at constant voltage (V_i, V_o).

10. The Fock-Darwin problem is the energy levels of an electron in two dimensions, in two simultaneous potentials: (i) a harmonic oscillator, and (ii) a magnetic field. The Hamiltonian is

$$H = \frac{1}{2m}(\mathbf{p} - e\mathbf{A})^2 + \frac{m\omega_0^2}{2}(x^2 + y^2) \tag{5.166}$$

$$\mathbf{A} = -\frac{B}{2}[\hat{x}y - \hat{y}x] \tag{5.167}$$

Find the eigenvalues of this Hamiltonian.

11. In the fractional quantum Hall effect, the composite fermion picture predicts fractions

$$\nu = \frac{n}{2pn \pm 1} \tag{5.168}$$

What fractions does this predict for denominators ≤ 11?

12. An electron is constrained to move around a circular ring of radius R. Its quantum mechanical wave function is therefore a function of the azimuthal angle around the ring: $\psi(\phi)$.

    (a) Find the energy and eigenfunctions when the magnetic field $B = 0$.

    (b) Repeat when a magnetic field B is perpendicular to the plane of the ring.

(c) Show that if the flux through the ring is a multiple of the flux quantum, the result is identical to (a).

13. Suppose silicon is doped with donors so there is a chemical potential $\mu > E_{c0} + E_F$ where the Fermi energy E_F is positive in the conduction band. A magnetic field B is imposed along the (100) direction. What values of cyclotron period T are observed for the various ellipsoids? The effective masses of the ellipsoids are $m_\perp, m_\parallel$.

14. Consider a square lattice in two dimensions that has two electrons per site in a paramagnetic ground state. The Fermi surface consists of electron pockets at the zone faces, and hole pockets at the zone corner.

(a) What is the area of one electron pocket?

(b) Express (a) in terms of the chemical potential, take its derivative, and derive the period of cyclotron resonance.

6 Electron–Electron Interactions

Actual metals have significant electron–electron interactions between all of the electrons in the crystal. The interactions between the conduction electrons provide dielectric screening and cause collective excitations such as plasmons. The Hamiltonian for the conduction electrons is

$$H = \sum_{j=1}^{N_e} \frac{p_j^2}{2m} + V_C \tag{6.1}$$

$$V_C = \frac{e^2}{8\pi\varepsilon_0} \int d^3 r [\rho(\mathbf{r}) - \rho_i(\mathbf{r})] \int d^3 r' \frac{[\rho(\mathbf{r}') - \rho_i(\mathbf{r}')]}{|\mathbf{r} - \mathbf{r}'|} \tag{6.2}$$

$$\rho(\mathbf{r}) = \sum_{j=1}^{N_e} \delta^3(\mathbf{r} - \mathbf{r}_j) \tag{6.3}$$

where $\rho(\mathbf{r})$ is the charge density of the electrons, and $\rho_i(\mathbf{r})$ is the charge density of the ion cores, which are usually represented by a pseudopotential. The ion positions $\mathbf{R}_l$ are arranged on a crystal lattice, and $\rho_i(\mathbf{r} + \mathbf{R}_l) = \rho_i(\mathbf{r})$. For electron–electron interactions, the term of the ion charge interacting with itself is omitted, since it is just a constant. It is necessary to have this term to maintain charge neutrality

$$\int d^3 r \rho(\mathbf{r}) = \int d^3 r \rho_i(\mathbf{r}) \tag{6.4}$$

The constant term from ion–ion interactions does not influence electron dynamics. Write the Hamiltonian with potential energy terms due to electron–electron interactions, electron–ion interactions, and ion–ion interactions:

$$H = \sum_{j=1}^{N_e} \frac{p_j^2}{2m} + V_{ee} + V_{ei} + V_{ii} \tag{6.5}$$

$$V_{ee} = \frac{e^2}{8\pi\varepsilon_0}\int d^3r\rho(\mathbf{r})\int d^3r'\frac{\rho(\mathbf{r}')}{|\mathbf{r}-\mathbf{r}'|} \tag{6.6}$$

$$V_{ei} = -\frac{e^2}{4\pi\varepsilon_0}\int d^3r\rho(\mathbf{r})\int d^3r\frac{\rho_i(\mathbf{r}')}{|\mathbf{r}-\mathbf{r}'|} \tag{6.7}$$

$$V_{ii} = \frac{e^2}{8\pi\varepsilon_0}\int d^3r\rho_i(\mathbf{r})\int d^3r\frac{\rho_i(\mathbf{r}')}{|\mathbf{r}-\mathbf{r}'|} \tag{6.8}$$

The term V_{ii} is a constant. The term V_{ei} describes the electron interaction with the periodic potential $U(\mathbf{r})$ from the ions. It can also be written in terms of electron coordinates $\mathbf{r}_j$ as

$$V_{ei} = \sum_j U(\mathbf{r}_j) = \int d^3r\rho(\mathbf{r})U(\mathbf{r}) \tag{6.9}$$

$$U(\mathbf{r}) = -\frac{e^2}{4\pi\varepsilon_0}\int d^3r'\frac{\rho_i(\mathbf{r}')}{|\mathbf{r}-\mathbf{r}'|} \tag{6.10}$$

It provides the energy-band structure, which was described in chapter 3. The main interest here is in the first term: electron–electron interactions. V_{ee} can also be written as

$$V_{ee} = \frac{e^2}{8\pi\varepsilon_0}\sum_{ij}\frac{1}{|\mathbf{r}_i-\mathbf{r}_j|} \tag{6.11}$$

by using the definition (6.3) of electron density. This equation has an undesirable feature. It allows an electron to interact with itself, whenever $i = j$. Of course, electrons do interact with themselves, but that is a problem in field theory and not in condensed matter physics. These terms are eliminated. Also, a factor of $\frac{1}{2}$ is eliminated by counting each pair of electrons only once

$$V_{ee} = \frac{e^2}{8\pi\varepsilon_0}\sum_{i\neq j}\frac{1}{|\mathbf{r}_i-\mathbf{r}_j|} = \frac{e^2}{4\pi\varepsilon_0}\sum_{i>j}\frac{1}{|\mathbf{r}_i-\mathbf{r}_j|} \tag{6.12}$$

This form of the electron–electron interactions is the basis for the remaining chapter. The interactions are discussed in free-electron metals, such as Al, Sn, and Pb. A later chapter discusses strongly correlated metals, such as Fe and Ce.

6.1 Second Quantization

The eigenfunctions $\psi_n(\mathbf{k}, \mathbf{r})$ are a complete set of orthonormal eigenstates for band n and wave vector k:

$$\sum_{\mathbf{k},n}\psi_n^*(\mathbf{k},\mathbf{r})\psi_n(\mathbf{k},\mathbf{r}') = \delta^3(\mathbf{r}-\mathbf{r}') \tag{6.13}$$

$$\int d^3r\psi_{n'}^*(\mathbf{k}',\mathbf{r})\psi_n(\mathbf{k},\mathbf{r}) = \delta_{nn'}\delta_{\mathbf{k},\mathbf{k}'} \tag{6.14}$$

where the right side is a three-dimensional delta function. They can be used as a basis to expand any function of $\mathbf{r}$, including the most general wave function $\psi(\mathbf{r})$:

$$\psi(\mathbf{r}) = \sum_{n\mathbf{k}} C_{n,\mathbf{k}} \psi_n(\mathbf{k}, \mathbf{r}) \tag{6.15}$$

The normalization integral gives

$$1 = \int d^3 r |\psi(\mathbf{r})|^2 = \sum_{n\mathbf{k};n'\mathbf{k}'} C^\dagger_{n',\mathbf{k}'} C_{n,\mathbf{k}} \int d^3 r \psi^*_{n'}(\mathbf{k}', \mathbf{r}) \psi_n(\mathbf{k}, \mathbf{r}) \tag{6.16}$$

$$= \sum_{n\mathbf{k}} C^\dagger_{n,\mathbf{k}} C_{n,\mathbf{k}}$$

The above expansion resembles a standard Fourier series. In physics, there is an important difference. Fundamental particles such as electrons cannot be divided into half or thirds. An electron has to be in one state or another, but not in two at once. In the hydrogen atom, there is no stationary solution that has half an electron in the $1S$ state and the other half in the $2P_z$ state. This combination is possible as a virtual excited state, but not as a stationary state.

When writing the expansion (6.15), the symbols $C_{n,\mathbf{k}}$ are treated as operators, whose role is to ensure that there is only one electron in every eigenstate. This algebra is called *second quantization*. For fermions, such as electrons and holes, the operators obey anti-commutation relations. Let the label λ denote all of the quantum numbers that denote an eigenstate, such as $\lambda = (n, \mathbf{k}, \sigma)$, where $\mathbf{k}$ is wave vector, n is a band index, and σ is the spin quantum number. For electrons, the two possibilities are spin-up α or spin-down β, where "up" and "down" refer to any spatial direction chosen for spin quantization.

There are two kinds of operators:

- *Raising operators* $C^\dagger_\lambda$ create a fermion in the state λ:

  $$C^\dagger_\lambda |0_\lambda\rangle = |1_\lambda\rangle \tag{6.17}$$

  $$C^\dagger_\lambda |1_\lambda\rangle = 0 \tag{6.18}$$

  where $|0_\lambda\rangle, |1_\lambda\rangle$ denote the state with zero or one electron in the state λ. The first equation shows that one electron can be put into this state if it is initially empty. The second shows that one cannot put two electrons in any state, which is the exclusion principle. Raising operators are also called creation operators.

- *Lowering operators* C_λ remove a particle from the state. For fermions,

  $$C_\lambda |1_\lambda\rangle = |0_\lambda\rangle \tag{6.19}$$

  $$C_\lambda |0_\lambda\rangle = 0 \tag{6.20}$$

  The latter equation shows that one cannot remove an electron from a state if there are none there. Lowering operators are also called destruction operators.

Fermions allow only zero or one particle in each eigenstate. The symmetric combination of operators gives unity when operating on either of these two states:

$$[C_\lambda C^\dagger_\lambda + C^\dagger_\lambda C_\lambda]|0_\lambda\rangle = C_\lambda |1_\lambda\rangle = |0_\lambda\rangle \tag{6.21}$$

$$[C_\lambda C^\dagger_\lambda + C^\dagger_\lambda C_\lambda]|1_\lambda\rangle = C^\dagger_\lambda |0_\lambda\rangle = |1_\lambda\rangle \tag{6.22}$$

This combination of operators is represented by curly brackets. The above relations are summarized by writing them as

$$C_\lambda C_\lambda^\dagger + C_\lambda^\dagger C_\lambda = \{C_\lambda, C_\lambda^\dagger\} = 1 \tag{6.23}$$

For different eigenstates, one has

$$\{C_\lambda, C_{\lambda'}^\dagger\} = \delta_{\lambda\lambda'} \tag{6.24}$$

$$\{C_\lambda, C_{\lambda'}\} = \{C_\lambda^\dagger, C_{\lambda'}^\dagger\} = 0 \tag{6.25}$$

The latter two identities reflect the symmetry principle that exchanging two fermions must change the sign of the wave function. In our case, changing the order of the operators must also change the sign.

Any function of position $f(\mathbf{r})$ can be represented by these operators. First start with the particle density:

$$\rho(\mathbf{r}) = \psi(\mathbf{r})^\dagger \psi(\mathbf{r}) = \sum_{\lambda,\lambda'} C_\lambda^\dagger C_{\lambda'} \psi_\lambda^*(\mathbf{r}) \psi_{\lambda'}(\mathbf{r}) \tag{6.26}$$

Expressions such as the kinetic or potential energy are evaluated by an integration over space:

$$V(\mathbf{r}) \rightarrow V = \int d^3 r \rho(\mathbf{r}) V(r) = \sum_{\lambda,\lambda'} C_\lambda^\dagger C_{\lambda'} V_{\lambda\lambda'} \tag{6.27}$$

$$V_{\lambda\lambda'} = \int d^3 r V(r) \psi_\lambda^*(\mathbf{r}) \psi_{\lambda'}(\mathbf{r}) \tag{6.28}$$

$$\frac{p^2}{2m} \rightarrow \sum_{\lambda,\lambda'} C_\lambda^\dagger C_{\lambda'} \left(\frac{p^2}{2m}\right)_{\lambda\lambda'} \tag{6.29}$$

$$\left(\frac{p^2}{2m}\right)_{\lambda\lambda'} = -\frac{\hbar^2}{2m} \int d^3 r \psi_\lambda^*(\mathbf{r}) \nabla^2 \psi_{\lambda'}(\mathbf{r}) \tag{6.30}$$

The Hamiltonian for a single electron interacting with a simple potential is then

$$H = \sum_{\lambda,\lambda'} H_{\lambda\lambda'} C_\lambda^\dagger C_{\lambda'} \tag{6.31}$$

$$H_{\lambda\lambda'} = \left(\frac{p^2}{2m}\right)_{\lambda\lambda'} + V_{\lambda\lambda'} \tag{6.32}$$

Usually one tries to determine the exact eigenfunctions of the Hamiltonian H. If these exact eigenfunctions are used as the basis set, then

$$H_{\lambda\lambda'} = E_\lambda \delta_{\lambda,\lambda'} \tag{6.33}$$

$$H = \sum_\lambda E_\lambda C_\lambda^\dagger C_\lambda \tag{6.34}$$

For a system of particles in thermal equilibrium, the average value of the occupation numbers is

$$n_\lambda = C_\lambda^\dagger C_\lambda \tag{6.35}$$

$$n_F(E_\lambda) = \langle C_\lambda^\dagger C_\lambda \rangle = \frac{1}{e^{\beta(E_\lambda - \mu)} + 1} \tag{6.36}$$

where $\beta = 1/k_BT$ and μ is the chemical potential of the particles. This average is between zero and one. The instantaneous value of n_λ is either zero or one, depending whether the fermion state is $|0_\lambda\rangle$ or $|1_\lambda\rangle$.

6.1.1 Tight-binding Models

Any basis set can be used to expand the wave function $\psi(\mathbf{r})$. In many materials it is convenient to use atomic orbitals as the basis set, as is commonly done by chemists in describing electron states in molecules. For example, in high-temperature superconductors, the conduction electrons are confined to two-dimensional planes of copper and oxygen. The tight-binding basis uses the three p-orbitals of each oxygen and the five d-orbitals for each copper. This restricted basis set gives a good description of the electronic properties for conduction electrons whose energy is near to the chemical potential.

Another important example is the conduction bands of carbon when it has trigonal bonding, such as in graphite, graphene, fullerene, or nanotube. Graphene is a single sheet of graphite, where the carbons are arranged in a honeycomb lattice. The sigma bonds are in the plane between the atoms and are completely filled with electrons, and their energy is away from the chemical potential. The valence and conduction bands are provided by the carbon $2p_z$-orbital that is perpendicular to the graphene plane. A good description of these electronic states is provided by the basis function

$$\psi(\mathbf{r}) = \sum_{j\sigma} C_{j\sigma} \phi_{2p_z}(\mathbf{r} - \mathbf{R}_j) \tag{6.37}$$

where $C_{j\sigma}$ is the lowering operator for an electron of spin σ at the carbon atom site $\mathbf{R}_j$. Note that local orbitals are used as the basis set, rather than plane wave states. The expansion is not over a complete set of states $\psi_\lambda(\mathbf{r})$, but is over a very restricted basis set of one orbital per atom. Yet this procedure works quite well for many solids and is widely used.

For band motion, the local operator gets changed to a wave vector operator:

$$C_{j\sigma} = \frac{1}{\sqrt{N}} \sum_{\mathbf{k}} C_{\mathbf{k}\sigma} e^{i\mathbf{k}\cdot\mathbf{R}_j} \tag{6.38}$$

$$\psi(\mathbf{r}) = \sum_{\mathbf{k}\sigma} C_{\mathbf{k}\sigma} \phi(\mathbf{k}, \mathbf{r}) \tag{6.39}$$

$$\phi(\mathbf{k}, \mathbf{r}) = \frac{1}{\sqrt{N}} \sum_{j} e^{i\mathbf{k}\cdot\mathbf{R}_j} \phi_{2p_z}(\mathbf{r} - \mathbf{R}_j) \tag{6.40}$$

The latter eigenfunction is used to calculate band dispersion in tight-binding models.

6.1.2 Nearly Free Electrons

The complete Hamiltonian is derived for the Sommerfeld model. The electrons states are initially approximated as simple plane waves, and the various interaction terms are expressed in terms of raising and lowering operators:

$$H = H_0 + U(\mathbf{r}) + V_{ee} \tag{6.41}$$

$$H_0 = \sum_j \left[\frac{p_j^2}{2m} - \mu\right] \tag{6.42}$$

The term H_0 is for a set of free electrons with mass m. The energy has been normalized with respect to the chemical potential μ. The term $U(\mathbf{r})$ is from the periodic potential of the ions. The term V_{ee} is from electron–electron interactions. Other interactions, such as electron–phonon, are discussed in later chapters.

- The noninteracting Hamiltonian H_0 has eigenfunctions that are plane waves:

$$\phi_{\mathbf{p}\sigma}(\mathbf{r}) = \frac{e^{i\mathbf{p}\cdot\mathbf{r}}}{\sqrt{\Omega}}\chi_\sigma \tag{6.43}$$

$$\psi(\mathbf{r}) = \sum_{\mathbf{p}\sigma} C_{\mathbf{p}\sigma}\phi_{\mathbf{p}\sigma}(\mathbf{r}) \tag{6.44}$$

$$H_0 = \int d^3r\psi^\dagger(\mathbf{r})\left[\frac{p^2}{2m} - \mu\right]\psi(\mathbf{r}) = \sum_{\mathbf{p}\sigma}\xi(p)\,C^\dagger_{\mathbf{p}\sigma}C_{\mathbf{p}\sigma} \tag{6.45}$$

$$\xi(p) = \frac{p^2}{2m} - \mu \tag{6.46}$$

where χ_σ is the spin state: usually up (α) or down (β). The above derivation is for a single electron. If we have a set of N conduction electrons in the metal, a many-electron wave function with the correct symmetry is given by a Slater determinant:

$$\Psi(\mathbf{p}_1, \sigma_1, \mathbf{p}_2, \sigma_2, \cdots : \mathbf{r}_1, \mathbf{r}_2 \cdots \mathbf{r}_N) = \tag{6.47}$$

$$\frac{1}{\sqrt{N!}}\begin{vmatrix} \phi_{\mathbf{p}_1\sigma_1}(\mathbf{r}_1) & \phi_{\mathbf{p}_1\sigma_1}(\mathbf{r}_2) & \cdots & \phi_{\mathbf{p}_1\sigma_1}(\mathbf{r}_N) \\ \phi_{\mathbf{p}_2\sigma_2}(\mathbf{r}_1) & \phi_{\mathbf{p}_2\sigma_2}(\mathbf{r}_2) & \cdots & \phi_{\mathbf{p}_2\sigma_2}(\mathbf{r}_N) \\ \vdots & \vdots & \ddots & \vdots \\ \phi_{\mathbf{p}_N\sigma_N}(\mathbf{r}_1) & \phi_{\mathbf{p}_N\sigma_N}(\mathbf{r}_2) & \cdots & \phi_{\mathbf{p}_N\sigma_N}(\mathbf{r}_N) \end{vmatrix}$$

The proper evaluation of the Hamiltonian is

$$H_0 = \int d^3r_1 \cdots d^3r_N \Psi^* \sum_{j=1}^{N}\left[\frac{p_j^2}{2m} - \mu\right]\Psi \tag{6.48}$$

$$= \sum_{\mathbf{p}_i\sigma_i;i=1}^{N} \xi(\mathbf{p}_i)$$

The last summation is over all of the N momentum states $\mathbf{p}_i$ and spin states σ_i that are in the Slater determinant. The above result could be regarded as a theorem in matrix algebra. It is derived by expanding the determinant in minors. In taking the matrix element $\Psi^\dagger\Psi$ one also evaluates the spin states as $\langle\chi_i\chi_j\rangle = \delta_{ij}$.

A way to generalize this result is to write it as

$$H_0 = \sum_{\mathbf{p}\sigma}\xi(\mathbf{p})\,C^\dagger_{\mathbf{p}\sigma}C_{\mathbf{p}\sigma} \tag{6.49}$$

where the operator combination $C^{\dagger}_{p\sigma}C_{p\sigma}$ gives the number of electrons in the state $(p\sigma)$, which is zero or one. Of course, this is the same expression (6.45) derived above for the one-electron Hamiltonian. It is also exact for a many-electron Hamiltonian.

- The electron–ion interaction $U(\mathbf{r})$ is periodic in lattice vectors $\mathbf{R}_j$: $U(\mathbf{r}+\mathbf{R}_j) = U(\mathbf{r})$. Often this symmetry is achieved simply when the ion cores are smaller than the size of the unit cell and do not overlap. The periodic potential is then

$$U(\mathbf{r}) = \sum_j v_{ei}(\mathbf{r}-\mathbf{R}_j) \tag{6.50}$$

where $v_{ei}(\mathbf{r}-\mathbf{R}_j)$ is the potential energy of an electron with a single ion centered at the site $\mathbf{R}_j$. Usually a pseudopotential is used, as described in an earlier chapter.

Associated with the set of lattice vectors $\mathbf{R}_j$ are a set of reciprocal lattice vectors $\mathbf{G}_n$, such that $\mathbf{R}_j \cdot \mathbf{G}_n = 2\pi m_{jn}$, where m_{jn} is an integer that depends on j and n. They have the property

$$1 = \exp[i\mathbf{R}_j \cdot \mathbf{G}_n] \tag{6.51}$$

The Fourier transform of a periodic function such as $U(\mathbf{r})$ contains only reciprocal lattice vectors:

$$U(\mathbf{r}) = \sum_{\mathbf{G}} e^{i\mathbf{G}\cdot\mathbf{r}} u(\mathbf{G}) \tag{6.52}$$

$$U(\mathbf{r}+\mathbf{R}_j) = \sum_{\mathbf{G}} e^{i\mathbf{G}\cdot(\mathbf{r}+\mathbf{R}_j)} u(\mathbf{G}) = U(\mathbf{r}) \tag{6.53}$$

$$u(\mathbf{G}) = \int_{\Omega_0} d^3r e^{i\mathbf{G}\cdot\mathbf{r}} U(\mathbf{r}) = \int_{\Omega} d^3r e^{i\mathbf{G}\cdot\mathbf{r}} v_{ei}(\mathbf{r}) \tag{6.54}$$

The Fourier transform can be viewed as the integral of $U(\mathbf{r})$ over one unit cell of volume Ω_0, or else the Fourier transform of the single atom potential $v_{ei}(r)$ over all space Ω. This interaction is written using operators as

$$U = \int d^3r U(\mathbf{r}) \rho(\mathbf{r}) = \sum_{\mathbf{G}} u(\mathbf{G}) \rho_{\mathbf{G}} \tag{6.55}$$

$$\rho(\mathbf{r}) = |\psi(r)|^2 = \frac{1}{\Omega}\sum_{\mathbf{q}\mathbf{p}\sigma} e^{i\mathbf{q}\cdot\mathbf{r}} C^{\dagger}_{\mathbf{p}+\mathbf{q},\sigma} C_{\mathbf{p}\sigma} = \sum_{\mathbf{q}} e^{i\mathbf{q}\cdot\mathbf{r}} \rho_{\mathbf{q}} \tag{6.56}$$

$$\rho_{\mathbf{G}} = \frac{1}{\Omega}\sum_{\mathbf{p}\sigma} C^{\dagger}_{\mathbf{p}+\mathbf{G},\sigma} C_{\mathbf{p}\sigma} \tag{6.57}$$

The periodic potential $U(\mathbf{r})$ has a Fourier series that contains only wave vectors $\mathbf{G}$ of the reciprocal lattice. The potential scatters the electron by only these wave vectors. Using this interaction, chapter 3 discusses how to calculate the energy bands of the electron in the crystal.

- Electron–electron interactions are calculated from the classical formula between two charge distributions of density $\rho(\mathbf{r}) = |\psi(\mathbf{r})|^2$:

$$V_{ee} = \frac{e^2}{8\pi\varepsilon_0}\int d^3r_1 \int d^3r_2 \frac{|\psi(\mathbf{r}_1)|^2 |\psi(\mathbf{r}_2)|^2}{|\mathbf{r}_1-\mathbf{r}_2|} \tag{6.58}$$

A factor of 1/2 occurs because each pair of electrons is given twice in the double integral. Using the wave function in eqn. (6.44) gives

$$V_{ee} = \frac{1}{2\Omega}\sum_{\mathbf{kpq}ss'} v_q C^{\dagger}_{\mathbf{p}+\mathbf{q},s} C^{\dagger}_{\mathbf{k}-\mathbf{q},s'} C_{\mathbf{k},s'} C_{\mathbf{p},s} \tag{6.59}$$

$$v_q = \int d^3r \frac{e^2}{4\pi\varepsilon_0 r} e^{i\mathbf{q}\cdot\mathbf{r}} = \frac{e^2}{\varepsilon_0 q^2} \tag{6.60}$$

The electron–electron interaction is given in terms of v_q, which is the Fourier transform in three dimensions of Coulomb's law. In eqn. (6.59) the operators are written in a particular order: two raising operators on the left, and two lowering operators on the right. For there to be electron–electron interactions, there must be at least two electrons in the system, so that $C_\lambda C_{\lambda'}$ operating on the state gives nonzero. In understanding this formula keep in mind that electrons are not actually destroyed. They are simply scattered: one possible event is that the two electrons are initially in the states $(\mathbf{p}, s)$ and $(\mathbf{k}, s')$ and are scattered to $(\mathbf{p}+\mathbf{q}, s)$ and $(\mathbf{k}-\mathbf{q}, s')$. Wave vector is conserved. Since the electrons are identical, it is possible to have "exchange scattering," which can happen only if the two electrons have the same spin $(s = s')$. Then a possible process is $\mathbf{p} \to \mathbf{k}-\mathbf{q}$, $\mathbf{k} \to \mathbf{p}+\mathbf{q}$. The change in wave vector is $\mathbf{q} = \mathbf{k} - \mathbf{p}$. Exchange scattering is very important in metals, and is discussed below.

6.1.3 *Hartree Energy: Wigner-Seitz*

The electron–electron interaction (6.59) has an obvious divergence as $q \to 0$. The divergence occurs because a plasma of only electrons has a very large repulsive Coulomb energy, which diverges as the number of electrons N becomes large. In a metal, this divergence is absent since the ions have a positive charge, and the overall system is charge neutral. Let the ions have a charge density $-e\rho_i(\mathbf{r})$. The *Hartree energy* of an individual electron is the average Coulomb interaction from the other electrons and from the ions:

$$V_H(\mathbf{r}) = \frac{e^2}{4\pi\varepsilon_0}\int \frac{d^3r'}{|r-r'|}[\rho(\mathbf{r}') - \rho_i(\mathbf{r}')] \tag{6.61}$$

This potential energy is averaged over all of the electrons to give the Hartree energy of the system. Then one evaluates eqn. (6.2).

The metal must be charge neutral, and each unit cell is neutral. The contribution to the Coulomb interaction from distant unit cells is small. Rewrite the above integral as a summation over unit cells and an integral over the Wigner-Seitz cell:

$$V_H(\mathbf{r}) = \frac{e^2}{4\pi\varepsilon_0}\sum_j \int_{\Omega_0} \frac{d^3r'}{|\mathbf{r}-\mathbf{r}'-\mathbf{R}_j|}[\rho(\mathbf{r}') - \rho_i(\mathbf{r}')] \tag{6.62}$$

If the cells were spherical, then the interaction from distant cells would be identically zero. They are not spherical, but are symmetric polygons, and the contribution is small. By far the largest contribution is due to the interaction within the same cell where $\mathbf{R}_j = 0$. Since this term is so large, a reasonable approximation is to include only it.

Wigner and Seitz found a simple and clever way to evaluate this expression. They assumed

- The unit cell is spherical with a radius r_0.
- The ion is a point charge of valence Z, so that $\rho_i(\mathbf{r}) = Z\delta^3(\mathbf{r})$.
- The electrons have a uniform density $|\psi(r)|^2 = n_0$.

The calculation proceeds in several steps:

- The radius of the unit sphere is $r_0 = \eta a_0 r_s$, and is chosen so that the system is charge neutral. The dimensionless parameter r_s was introduced in chapter 5 and is defined as $4\pi n_0 (r_s a_0)^3/3 = 1$:

$$Z = n_0 \int_{r \le r_0} d^3 r = \frac{4\pi n_0}{3}(r_0)^3 = \frac{4\pi n_0}{3}(\eta a_0 r_s)^3 \tag{6.63}$$

$$Z = \eta^3 \tag{6.64}$$

- The interaction energy between the electrons and the ion is

$$E_{ei} = n_0 \int_{r \le r_0} d^3 r \left(-\frac{Ze^2}{4\pi\varepsilon_0 r} \right) = -\frac{e^2}{2\varepsilon_0} n_0 Z (r_0)^2 \tag{6.65}$$

$$= -\frac{3Z^{5/3}}{r_s} E_{\text{Ry}}$$

- The next energy is the interaction of the electrons with themselves. This term is easy to evaluate for a spherical charge distribution. Use Gauss's law that the electric field $E(r)$ at a distance r from the center is $E = en(r)/[4\pi\varepsilon_0 r^2]$, where $n(r)$ is the number of electrons enclosed in the sphere of radius r:

$$n(r) = \frac{4\pi r^3}{3} n_0 \tag{6.66}$$

$$eE = -\frac{dV_{ee}}{dr} = \frac{e^2}{3\varepsilon_0} n_0 r = \frac{Ze^2}{4\pi\varepsilon_0 r_0^3} r \tag{6.67}$$

The last equation can be integrated:

$$V_{ee} = \frac{Ze^2}{8\pi\varepsilon_0 r_0}\left[C - \left(\frac{r}{r_0} \right)^2 \right] \tag{6.68}$$

where C is a constant of integration. It is chosen to make the sphere appear to be charge neutral to an outside observer: $C = 3$, so that $V_{ee}(r_0) = Ze^2/4\pi\varepsilon_0 r_0$ at the edge of the sphere. This is equal and opposite to the potential energy from the ion charge. The energy from electron–electron interactions is then

$$E_{ee} = \frac{n_0}{2} \int_{r \le \eta a_0 r_s} d^3 r V_{ee}(r) = \frac{n_0 e^2 Z}{4\varepsilon_0 \eta r_s a_0} \int_0^{\eta r_s a_0} r^2 dr \left[3 - \left(\frac{r}{\eta r_s a_0} \right)^2 \right] \tag{6.69}$$

The integral is easy. Use eqn. (6.64) to convert n_0 to Z, and find

$$E_{ee} = \frac{6Z^{5/3}}{5r_s} E_{\text{Ry}} \tag{6.70}$$

The total energy combines the electron–ion and electron–electron energies:

$$E_T = \frac{Z^{5/3}}{r_s} E_{\text{Ry}} \left[-3 + \frac{6}{5} \right] = -\frac{1.8}{r_s} Z^{5/3} E_{\text{Ry}} \tag{6.71}$$

This famous result is a good approximation to the more rigorous derivation.

There are two kinds of energy calculations in systems of interacting particles. One is to calculate the energy of one particle, due to its interactions with the other particles. The second is to calculate the ground-state energy of the whole system of particles. Here we have done the latter. The Wigner-Seitz energy (6.71) is the Coulomb energy per electron of the system of electrons.

6.1.4 Exchange Energy

The exchange energy of the gas of conduction electrons was first calculated by Fock, and is often called the Fock energy. It is a very important term, since it is the large negative energy that results in metallic binding. Without it, there would be no metals! A theory that retains only the exchange energy and the Hartree energy is called the *Hartree-Fock approximation*. It is derived for the electron gas.

There are two different ways to calculate this energy, which give the same answer. One method uses operators, and the other uses Slater determinants. In the operator method, start with the term for electron–electron interactions:

$$V_{ee} = \frac{1}{2\Omega} \sum_{\mathbf{kpq}ss'} v_q C^{\dagger}_{\mathbf{p+q},s} C^{\dagger}_{\mathbf{k-q},s'} C_{\mathbf{k},s'} C_{\mathbf{p},s} \tag{6.72}$$

Take its expectation value in the ground state, which is denoted by angle brackets:

$$E_{ee} = \langle V_{ee} \rangle = \frac{1}{2\Omega} \sum_{\mathbf{kpq}ss'} v_q \langle C^{\dagger}_{\mathbf{p+q},s} C^{\dagger}_{\mathbf{k-q},s'} C_{\mathbf{k},s'} C_{\mathbf{p},s} \rangle \tag{6.73}$$

For a paramagnetic metal at zero temperature, the bracket denotes the ground state, which has all wave vector states occupied for $|\mathbf{k}| < k_F$, where k_F is the Fermi energy:

$$|\rangle = \Pi_{|\mathbf{k}|<k_F} C^{\dagger}_{\mathbf{k}\uparrow} C^{\dagger}_{\mathbf{k}\downarrow} |0\rangle \tag{6.74}$$

and $|0\rangle$ is the vacuum of no particles. In the bracket in eqn. (6.73), the first two operators remove two electrons from the ground state:

$$C_{\mathbf{k},s'} C_{\mathbf{p},s} |\rangle \tag{6.75}$$

which requires that the two states $(\mathbf{k}, \mathbf{p})$ are both less than k_F. The two raising operators in eqn. (6.73) must return the two electrons to the ground state: otherwise the operation by $\langle|$ would give zero. So the operators must be paired. The two possible pairings are as follows:

- Set $\mathbf{q} = 0$ so that the pairings are

$$\langle C^{\dagger}_{\mathbf{p},s} C_{\mathbf{p},s} \rangle \langle C^{\dagger}_{\mathbf{k},s'} C_{\mathbf{k},s'} \rangle = n_{\mathbf{k}s} n_{\mathbf{p}s'} \tag{6.76}$$

where $n_{\mathbf{k}s}$ and $n_{\mathbf{p}s}$ are occupation numbers that are zero or one, depending on whether the wave vectors are greater or less than k_F. This term is part of the Hartree energy, as discussed in the previous section.

- The other possible pairing has

$$-\langle C^{\dagger}_{\mathbf{p}+\mathbf{q},s} C_{\mathbf{k},s'} \rangle \langle C^{\dagger}_{\mathbf{k}-\mathbf{q},s'} C_{\mathbf{p},s} \rangle = -\delta_{ss'} \delta_{\mathbf{q}=\mathbf{k}-\mathbf{p}} n_{\mathbf{p}s} n_{\mathbf{k}s} \tag{6.77}$$

There is a sign change that comes from interchanging the order of the middle two operators. This sign change is a general property of fermion operators. Note that the two spins are alike ($s = s'$) and $\mathbf{q} = \mathbf{k} - \mathbf{p}$. This term is not part of the Hartree energy and is the exchange term. The contribution of exchange to the ground-state energy of the system is

$$E_X = -\frac{e^2}{2\varepsilon_0 \Omega} \sum_{\mathbf{k}\mathbf{p}s} \frac{n_{\mathbf{k}s} n_{\mathbf{p}s}}{|\mathbf{k}-\mathbf{p}|^2} \tag{6.78}$$

It is also useful to calculate the exchange energy of a particular electron, say of wave vector $\mathbf{p}$ and spin s. That can be done in the above expression by taking a functional derivative with respect to $n_{\mathbf{p}s}$:

$$\Sigma_X(\mathbf{p}) = \frac{\delta E_X}{\delta n_{\mathbf{p}s}} = -\frac{e^2}{\varepsilon_0 \Omega} \sum_{\mathbf{k}} \frac{n_{\mathbf{k}s}}{|\mathbf{k}-\mathbf{p}|^2} \tag{6.79}$$

The derivative has produced a factor of two: since $\mathbf{k}$ and $\mathbf{p}$ are variables of summation, they each are going to equal the wave vector $\mathbf{p}$ at some point, so there are two terms in the derivative.

The above integral is evaluated in three dimensions at zero temperature. Use spherical coordinates $d^3k = 2\pi d\nu k^2 dk$, where $\nu = \cos(\theta) = \hat{k} \cdot \hat{p}$ is the cosine of the angle between $\mathbf{k}$ and $\mathbf{p}$:

$$\begin{aligned}
\Sigma_X(\mathbf{p}) &= -\frac{e^2}{\varepsilon_0 (2\pi)^2} \int_0^{k_F} k^2 dk \int_{-1}^{1} \frac{d\nu}{k^2 + p^2 - 2pk\nu} \\
&= -\frac{e^2}{4\pi^2 \varepsilon_0 p} \int_0^{k_F} k dk \ln\left|\frac{p+k}{p-k}\right| \\
&= -\frac{e^2}{8\pi^2 \varepsilon_0 p} \left[2pk_F + (k_F^2 - p^2) \ln\left|\frac{p+k_F}{p-k_F}\right| \right] \\
&= -\frac{e^2 k_F}{4\pi^2 \varepsilon_0} S(p/k_F)
\end{aligned} \tag{6.80}$$

$$S(y) = 1 + \frac{1}{2y}(1 - y^2) \ln\left|\frac{1+y}{1-y}\right| \tag{6.81}$$

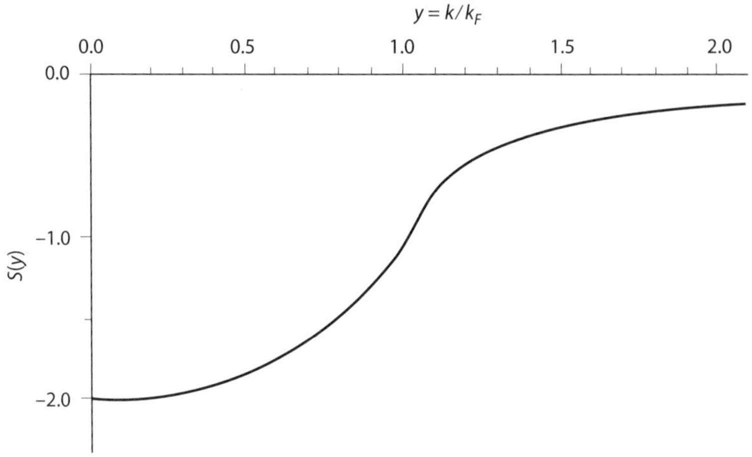

FIGURE 6.1. The function $S(y)$ for the exchange energy.

The function $S(y)$ is graphed in fig. 6.1. Some values are $S(0) = 2$, $S(1) = 1$, and $S(\infty) = 0$. The exchange energy of the system of electrons, per electron, is

$$E_X = \frac{1}{n}\int \frac{d^3p}{(2\pi)^3} n_p \Sigma_X(p) = -\frac{3e^2 k_F}{(4\pi)^2 \varepsilon_0} \tag{6.82}$$

The right-hand side should have a factor of $(\frac{2}{2})$: It is multiplied by two to account for the two spin states, and divided by two to compensate for counting each interaction twice.

The Hartree-Fock energy of the electron gas is composed of kinetic, Hartree, and exchange energies:

$$E_{\text{HF}} = E_K + E_H + E_X \tag{6.83}$$

$$E_K = \frac{1}{\Omega}\sum_{p\sigma} n_{p\sigma}\left(\frac{\hbar^2 p^2}{2m} - \mu\right) \tag{6.84}$$

The Hartree (E_H) and exchange (E_X) were given above. Similarly, the energy of an individual electron is

$$E_s(\mathbf{p}) = \frac{\hbar^2 p^2}{2m} - \mu + \Sigma_H + \Sigma_X(p) \tag{6.85}$$

The Hartree self-energy Σ_H is assigned as a problem. It is the energy of the electron from interacting with the ions [e.g., $U(r)$], and the other electrons in the ground state.

6.1.5 Compressibility

Fred Seitz proved an important theorem that relates the ground-state energy of a system $E(N)$ to the chemical potential. The chemical potential is the energy difference between two ground states: one with $N + 1$ electrons, and one with N:

$$\mu = E_G(N+1) - E_G(N) = \frac{dE_G(N)}{dN} + \frac{1}{2!}\frac{d^2E_G(N)}{dN^2} + \cdots \tag{6.86}$$

this difference is evaluated by a Taylor series. Since $E_G(N) \sim O(N)$, the first derivative is $O(N^0)$ and gives μ. The second and higher derivatives are $O(1/N)$ and are neglected for large N:

$$\mu = \frac{dE_G(N)}{dN} \tag{6.87}$$

This expression is valid in metals, but not in insulators. Insulators have an energy gap between adding an electron and removing one.

In metals, this identity is valid for each term in the energy expansion. Keep in mind that $k_F = (3\pi^2 N/\Omega)^{1/3}$.

- For the kinetic energy

$$E_K(N) = \frac{3}{5}N\frac{\hbar^2 k_F^2}{2m} = \frac{3\hbar^2(3\pi^2)^{2/3}}{10m\Omega^{2/3}}N^{5/3} \tag{6.88}$$

$$\mu_K = \frac{dE_K(N)}{dN} = \frac{\hbar^2}{2m}\left(3\pi^2\frac{N}{\Omega}\right)^{2/3} = \frac{\hbar^2 k_F^2}{2m} = E_F \tag{6.89}$$

- For the exchange energy

$$E_X = -N\frac{3e^2 k_F}{(4\pi)^2\varepsilon_0} = -\frac{3e^2(3\pi^2)^{1/3}}{(4\pi)^2\varepsilon_0\Omega^{1/3}}N^{4/3} \tag{6.90}$$

$$\mu_X = \frac{dE_X(N)}{dN} = --\frac{e^2 k_F}{4\pi^2\varepsilon_0} = \Sigma_X(k_F) \tag{6.91}$$

Seitz's theorem works for these two energy terms. It also works for the correlation energy.

The compressibility K of a system is its bulk spring constant. (Here K is not kinetic energy.) If you squeeze it, what is its restoring force? It is defined as a double derivative of the ground-state energy with respect to volume:

$$\frac{1}{K} = \Omega\left(\frac{d^2E_G(N)}{d\Omega^2}\right)_N = n_0^2\frac{d\mu}{dn_0} \tag{6.92}$$

The last identity is proved. Let $\mathcal{E}_G(n_0)$ be the ground-state energy per electron, where $n_0 = N/\Omega$ is the density. Then $E_G(N) = N\mathcal{E}_G(N/\Omega)$. Some derivatives are

$$\mu = \frac{d}{dN}[N\mathcal{E}_G(n_0)] = \mathcal{E}_G(n_0) + n_0\frac{d}{dn_0}\mathcal{E}_G(n_0) \tag{6.93}$$

$$\frac{dE_G(N)}{d\Omega} = -n_0^2\frac{d}{dn_0}\mathcal{E}_G(n_0) \tag{6.94}$$

$$\Omega\frac{d^2E_G(N)}{d\Omega^2} = n_0^2\left[2\frac{d\mathcal{E}_G(n_0)}{dn_0} + n_0\frac{d^2\mathcal{E}_G(n_0)}{dn_0^2}\right] = n_0^2\frac{d\mu}{dn_0} \tag{6.95}$$

The inverse compressibility is a derivative of the chemical potential with respect to density.

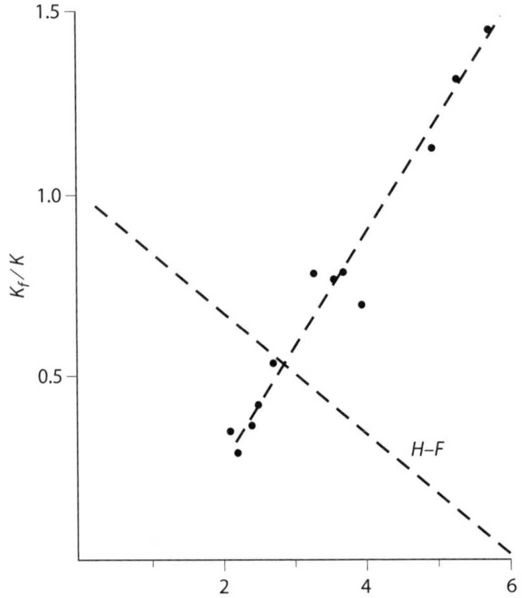

FIGURE 6.2. Inverse compressibility K_f/K as a function of r_s. Dashed line labeled H-F is Hartree-Fock prediction. Points are experimental values for free electron metals.

The compressibility is usually compared to that of a free-particle Fermi gas at zero temperature. There the only energy term is the kinetic energy. When taking the derivative with repect to the density, remember that $k_F \sim n_0^{1/3}$:

$$n_0 \frac{d}{dn_0}\left(\frac{\hbar^2 k_F^2}{2m}\right) = \frac{2}{3} E_F \tag{6.96}$$

The free-particle compressibility is $K_f = 3/(2E_F n_0)$. Most of the numerical results are presented as the ratio K_f/K, where K_f is the above result. An effective mass of unity is assumed.

As an example of using this formula, the compressibility is calculated in the Hartree-Fock approximation, The latter is writen as

$$\mu = \frac{\hbar^2 k_F^2}{2m} - \frac{e^2 k_F}{4\pi^2 \varepsilon_0} = E_F + \Sigma_X \tag{6.97}$$

$$n_0 \frac{d\mu}{dn_0} = \frac{2}{3} E_F + \frac{1}{3}\Sigma_X = \frac{2}{3} E_F\left(1 - \frac{1}{\pi k_F a_0}\right) \tag{6.98}$$

$$\frac{K_f}{K} = 1 - \frac{1}{\pi k_F a_0} = 1 - \frac{r_s}{6.03} \tag{6.99}$$

The Hartree-Fock prediction is that the inverse compressibility is a linear function of r_s, and becomes negative around $r_s = 6.03$. Figure 6.2 shows the Hartree-Fock prediction as the dashed line labeled H-F. The points are experimental values for K for free-electron metals. Obviously, the experiments and the theory do not agree! The experimental values are influenced by the potential energy from the ion cores.

6.2 Density Operator

The general expression for the electron density is expanded in terms of a set of quantum numbers (λ, λ'):

$$\rho(\mathbf{r}) = |\psi(\mathbf{r})|^2 = \sum_{\lambda\lambda'} C^{\dagger}_{\lambda} C_{\lambda'} \phi^{*}_{\lambda}(\mathbf{r})\phi_{\lambda'}(\mathbf{r}) \tag{6.100}$$

A special case, often used in metals, is when the eigenfunctions are plane waves:

$$\phi_{\lambda'}(\mathbf{r}) = \frac{e^{i\mathbf{p}\cdot\mathbf{r}}}{\sqrt{\Omega}}\chi_{\sigma'}, \quad \phi_{\lambda}(\mathbf{r}) = \frac{e^{i(\mathbf{p}+\mathbf{q})\cdot\mathbf{r}}}{\sqrt{\Omega}}\chi_{\sigma} \tag{6.101}$$

$$\rho(\mathbf{r}) = \frac{1}{\Omega}\sum_{\mathbf{pq}\sigma} C^{\dagger}_{\mathbf{p}+\mathbf{q},\sigma} C_{\mathbf{p}\sigma} e^{-i\mathbf{q}\cdot\mathbf{r}} = \sum_{\mathbf{q}} e^{-i\mathbf{q}\cdot\mathbf{r}}\rho_{\mathbf{q}} \tag{6.102}$$

$$\rho_{\mathbf{q}} = \frac{1}{\Omega}\sum_{\mathbf{p}\sigma} C^{\dagger}_{\mathbf{p}+\mathbf{q},\sigma} C_{\mathbf{p}\sigma} \tag{6.103}$$

Both operators have the same spin configuration since the spin matrix element $\langle \chi_\sigma | \chi_{\sigma'} \rangle = \delta_{\sigma\sigma'}$. The operator $\rho_{\mathbf{q}}$ describes charge fluctuations. At $\mathbf{q} = 0$ there are no fluctuations, and the operator is merely the average electron density n_e:

$$\rho_{\mathbf{q}=0} = \frac{1}{\Omega}\sum_{\mathbf{p}\sigma} C^{\dagger}_{\mathbf{p}\sigma} C_{\mathbf{p}\sigma} = \frac{1}{\Omega}\sum_{\mathbf{p}\sigma} n_{\mathbf{p}\sigma} = n_e \tag{6.104}$$

The electron–electron interaction term in a plane wave basis is

$$V_{ee} = \frac{1}{2\Omega}\sum_{\mathbf{q}\mathbf{p}\sigma\sigma'} v_q C^{\dagger}_{\mathbf{p}+\mathbf{q},\sigma} C^{\dagger}_{\mathbf{p}'-\mathbf{q},\sigma'} C_{\mathbf{p}'\sigma'} C_{\mathbf{p}\sigma} \tag{6.105}$$

The middle two operators have the form of a density operator:

$$\rho_{-\mathbf{q}} = \frac{1}{\Omega}\sum_{\mathbf{p}'\sigma'} C^{\dagger}_{\mathbf{p}'-\mathbf{q},\sigma'} C_{\mathbf{p}'\sigma'} = \rho^{\dagger}_{\mathbf{q}} \tag{6.106}$$

The outer two operators in V_{ee} also have the form of a density operator, but must be brought together. Commute the right-hand operator to the left in two hops:

$$\begin{aligned} C^{\dagger}_{\mathbf{p}'-\mathbf{q},\sigma'} C_{\mathbf{p}'\sigma'} C_{\mathbf{p}\sigma} &= -\, C^{\dagger}_{\mathbf{p}'-\mathbf{q},\sigma'} C_{\mathbf{p}\sigma} C_{\mathbf{p}'\sigma'} \\ &= -[-C_{\mathbf{p}\sigma} C^{\dagger}_{\mathbf{p}'-\mathbf{q},\sigma'} + \delta_{\sigma\sigma'}\delta_{\mathbf{p}=\mathbf{p}'-\mathbf{q}}] C_{\mathbf{p}'\sigma'} \end{aligned} \tag{6.107}$$

and the interaction is

$$V_{ee} = \frac{1}{2\Omega}\sum_{\mathbf{q}\mathbf{p}\sigma\sigma'} v_q [C^{\dagger}_{p+q,\sigma} C_{p\sigma} C^{\dagger}_{p'-q,\sigma'} C_{p'\sigma'} - \delta_{\sigma\sigma'}\delta_{p=p'-q} n_{p+q,\sigma}] \tag{6.108}$$

The last summation $\sum_{p\sigma} n_{p+q,\sigma} = N_e$ gives the total number of electrons in the conduction band. Write the electron–electron interaction as

$$V_{ee} = \frac{1}{2\Omega}\sum_{q} v_q[\Omega^2 \rho_q \rho_q^\dagger - N_e] \tag{6.109}$$

The interaction is often written in this fashion.

The density operator ρ_q is the product of two fermion operators. It has boson properties. So $\rho_q^\dagger$ makes a boson excitation of wave vector **q**, while the operator ρ_q destroys this boson excitation. For small values of **q**, the boson excitation is the plasma oscillation of the metal.

6.2.1 *Two Theorems*

Now prove two theorems.

1. $\rho^\dagger(\mathbf{q}) = \rho(-\mathbf{q})$, where the dagger signifies Hermitian conjugate. It is proved rather easily:

$$\rho_q^\dagger = \frac{1}{\Omega}\sum_{p\sigma}\left[C_{p+q,\sigma}^\dagger C_{p'\sigma'}\right]^\dagger \tag{6.110}$$
$$= \frac{1}{\Omega}\sum_{p\sigma} C_{p\sigma}^\dagger C_{p+q,\sigma}$$

Change the summation variable $\mathbf{p} \to \mathbf{p} - q$ and find

$$\rho_q^\dagger = \frac{1}{\Omega}\sum_{p\sigma} C_{p-q,\sigma}^\dagger C_{p\sigma} = \rho_{-q} \tag{6.111}$$

which completes the proof.

2. The second theorem is that two density operators commute for any values of wave vectors:

$$0 = [\rho_{q}, \rho_{q'}^\dagger] \tag{6.112}$$

In the bracket, the second factor could be written as $\rho_{-q'}$. Write out the definition of the operators and the commutator:

$$[\rho_q, \rho_{q'}^\dagger] = \frac{1}{\Omega}\sum_{p\sigma p'\sigma'}\left[C_{p+q,\sigma}^\dagger C_{p\sigma} C_{p'-q,\sigma'}^\dagger C_{p'\sigma'}\right.$$
$$\left. - C_{p'-q,\sigma'}^\dagger C_{p'\sigma'} C_{p+q,\sigma}^\dagger C_{p\sigma}\right] \tag{6.113}$$

In each term, commute the middle two operators, so that all of the raising operators are on the left and the lowering operators are on the right:

$$C_{p\sigma} C_{p'-q,\sigma'}^\dagger = -C_{p'-q,\sigma'}^\dagger C_{p\sigma} + \delta_{\sigma\sigma'}\delta_{p=p'-q} \tag{6.114}$$

$$C_{p'\sigma'} C_{p+q,\sigma}^\dagger = -C_{p+q,\sigma}^\dagger C_{p'\sigma'} + \delta_{\sigma\sigma'}\delta_{p=p'-q} \tag{6.115}$$

The terms with four operators are identical and cancel:

$$C_{p+q,\sigma}^\dagger C_{p'-q,\sigma'}^\dagger C_{p\sigma} C_{p'\sigma'} = C_{p'-q,\sigma'}^\dagger C_{p+q,\sigma}^\dagger C_{p'\sigma'} C_{p\sigma} \tag{6.116}$$

since one can interchange the order of two raising operators, or two lowering operators, with just a sign change. The terms generated by the delta function also cancel:

$$0 = \frac{1}{\Omega}\sum_{p\sigma p'\sigma'} \delta_{\sigma\sigma'}\delta_{p=p'-q}\left[C^{\dagger}_{p+q,\sigma}C_{p'\sigma'} - C^{\dagger}_{p'-q,\sigma'}C_{p\sigma}\right] \tag{6.117}$$

$$= \frac{1}{\Omega}\sum_{p\sigma}\left[C^{\dagger}_{p+q,\sigma}C_{p+q,\sigma} - C^{\dagger}_{p,\sigma}C_{p\sigma}\right] = \frac{1}{\Omega}\sum_{p\sigma}[n_{p+q\sigma} - n_{p\sigma}]$$

$$= \frac{N_e}{\Omega}[1-1]$$

The last term equals zero, since $\mathbf{p}$ is a variable of summation, and can be changed in the first term to $\mathbf{p}+\mathbf{q}\rightarrow\mathbf{p}$. The theorem is proved.

6.2.2 Equations of Motion

The density operator $\rho(\mathbf{q}, t)$ has a time dependence, which can be found from its equation of motion. One evaluates the right-hand side of the expression:

$$\frac{\partial}{\partial t}\rho_q(t) = \frac{i}{\hbar}[H, \rho_q(t)] \tag{6.118}$$

$$H = H_0 + V_{ee} \tag{6.119}$$

$$H_0 = \sum_{p\sigma}\xi(\mathbf{p})\,C^{\dagger}_{p\sigma}C_{p\sigma};\quad \xi(\mathbf{p}) = \varepsilon(\mathbf{p}) - \mu \tag{6.120}$$

$$V_{ee} = \frac{1}{2\Omega}\sum_{q} v_q[\Omega^2\rho_q\rho^{\dagger}_q - N_e] \tag{6.121}$$

Energy-band structure is ignored in this discussion, and only electron–electron interactions are included. In considering the commutator in eqn. (6.118), recall that all density operators commute and $[V_{ee}, \rho] = 0$. The answer comes from the kinetic energy term:

$$\frac{\partial}{\partial t}\rho_q(t) = \frac{i}{\hbar}[H_0, \rho_q(t)] \tag{6.122}$$

$$= \frac{i}{\hbar\Omega}\sum_{p\sigma,p'\sigma'}\xi(\mathbf{p})\left[C^{\dagger}_{p\sigma}C_{p\sigma}C^{\dagger}_{p'+q\sigma'}C_{p'\sigma'} - C^{\dagger}_{p'+q,\sigma'}C_{p'\sigma'}C^{\dagger}_{p\sigma}C_{p\sigma}\right]$$

Commute the middle two operators to bring all raising operators to the left:

$$C_{p\sigma}C^{\dagger}_{p'+q\sigma'} = -C^{\dagger}_{p'+q\sigma'}C_{p\sigma} + \delta_{\sigma\sigma'}\delta_{p=p'+q} \tag{6.123}$$

$$C_{p'\sigma'}C^{\dagger}_{p\sigma} = -C^{\dagger}_{p\sigma}C_{p'\sigma'} + \delta_{\sigma\sigma'}\delta_{p=p'} \tag{6.124}$$

The two terms with four operators now cancel and the terms produced by the delta functions give

$$\frac{\partial}{\partial t}\rho_q(t) = \frac{i}{\hbar\Omega}\sum_{p\sigma}C^{\dagger}_{p+q\sigma}C_{p\sigma}[\xi(\mathbf{p}+\mathbf{q}) - \xi(\mathbf{p})] \tag{6.125}$$

A simple expression is found when using a plane-wave dispersion for the electron:

$$\frac{\partial}{\partial t}\rho_{\mathbf{q}}(t) = \frac{i\hbar}{2m\Omega}\sum_{\mathbf{p}\sigma} C^{\dagger}_{\mathbf{p}+\mathbf{q}\sigma}C_{\mathbf{p}\sigma}\left[\mathbf{q}\cdot(\mathbf{q}+2\mathbf{p})\right] \tag{6.126}$$

$$= i\mathbf{q}\cdot\mathbf{j}_{\mathbf{q}}(t)$$

$$\mathbf{j}_{\mathbf{q}}(t) = \frac{\hbar}{m\Omega}\sum_{\mathbf{p}\sigma}\left[\mathbf{p}+\frac{\mathbf{q}}{2}\right]C^{\dagger}_{\mathbf{p}+\mathbf{q}\sigma}C_{\mathbf{p}\sigma} \tag{6.127}$$

The symbol $\mathbf{j}_{\mathbf{q}}(t)$ denotes the current operator in quantum mechanics:

$$\mathbf{j}_{\mathbf{q}}(\mathbf{r},t) = \frac{\hbar}{2mi}\left[\psi^{\dagger}\nabla\psi - \psi\nabla\psi^{\dagger}\right] \tag{6.128}$$

$$= \sum_{\mathbf{q}}\mathbf{j}_{\mathbf{q}}(t)\,e^{-i\mathbf{q}\cdot\mathbf{r}}$$

Equation (6.126) is equivalent to the *equation of continuity*:

$$0 = \frac{\partial\rho(\mathbf{r},t)}{\partial t} + \nabla\cdot\mathbf{j}_{\mathbf{q}}(\mathbf{r},t) \tag{6.129}$$

All fluids obey this equation, including the fluid of conduction electrons.

6.2.3 Plasma Oscillations

Another interesting equation of motion is obtained by taking a second commutator:

$$\frac{\partial^2}{\partial t^2}\rho_{\mathbf{q}}(t) = \frac{i^2}{\hbar^2}\left[H,\left[H_0,\rho(\mathbf{q},t)\right]\right] \tag{6.130}$$

$$= \frac{i^2}{\hbar^2}\left[H_0+V_{ee},\left[H_0,\rho_{\mathbf{q}}(t)\right]\right]$$

$$\left[H_0,\left[H_0,\rho_{\mathbf{q}}(t)\right]\right] = \frac{1}{\Omega}\sum_{\mathbf{p}\sigma}\left[\xi(\mathbf{p}+\mathbf{q})-\xi(\mathbf{p})\right]^2 C^{\dagger}_{\mathbf{p}+\mathbf{q}\sigma}C_{\mathbf{p}\sigma} \tag{6.131}$$

The second time derivative has two terms: the one from H_0 can be deduced from the above result, since each commutator gives a factor of $[\xi(\mathbf{p}+\mathbf{q}) - \xi(\mathbf{p})]$. The important term is from electron–electron interactions and takes more steps to evaluate:

$$\left[V_{ee},\left[H_0,\rho(\mathbf{q},t)\right]\right] = \frac{1}{2\Omega^2}\sum_{\mathbf{p}\sigma}\sum_{\mathbf{p}'\mathbf{k}'\mathbf{q}'ss'} v_{q'}\left[\xi(\mathbf{p}+\mathbf{q}) - \xi(\mathbf{p})\right] \tag{6.132}$$

$$\times\left[C^{\dagger}_{\mathbf{p}'+\mathbf{q}'s}C^{\dagger}_{\mathbf{k}-\mathbf{q}'s'}C_{\mathbf{k}s'}C_{\mathbf{p}'s},\ C^{\dagger}_{\mathbf{p}+\mathbf{q}\sigma}C_{\mathbf{p}\sigma}\right]$$

Evaluate the commutator by commuting all raising operators to the left and all lowering operators to the right. Each commute of a raising and lowering operator produces a delta function. The two terms cancel with six operators. The answer is found from the four delta functions that result from the four commutes:

$$\left[C^{\dagger}_{p'+q's'}C^{\dagger}_{k-q's'}C_{ks'}C_{p's'},\ C^{\dagger}_{p+q\sigma}C_{p\sigma}\right] = \delta_{s\sigma}\delta_{p'=p+q}C^{\dagger}_{p'+q's}C^{\dagger}_{k-q's'}C_{ks'}C_{p\sigma}$$
$$-\ \delta_{\sigma s'}\delta_{p+q=k}C^{\dagger}_{p'+q's}C^{\dagger}_{k-q's'}C_{p's}C_{p\sigma} - \delta_{\sigma s}\delta_{p=p'+q'}C^{\dagger}_{p+q\sigma}C^{\dagger}_{k-q's'}C_{ks'}C_{p's}$$
$$+\ \delta_{\sigma s'}\delta_{p=k-q'}C^{\dagger}_{p+q\sigma}C^{\dagger}_{p'+q's}C_{ks'}C_{p's} \tag{6.133}$$

which gives for the commutator

$$\left[V_{ee},[H_0,\rho_q(t)]\right] = \frac{1}{2\Omega^2}\sum_{pq'ss'} v_{q'}[\xi(p+q)-\xi(p)] \tag{6.134}$$
$$\times\left\{\sum_k\left[C^{\dagger}_{p+q+q's}C^{\dagger}_{k-q's'}C_{ks'}C_{ps} - C^{\dagger}_{p+qs}C^{\dagger}_{k-q's'}C_{ks'}C_{p-q's}\right]\right.$$
$$\left.+\sum_{p'}\left[C^{\dagger}_{p+qs'}C^{\dagger}_{p'+q's}C^{\dagger}_{p'+q's}C_{p's} - C^{\dagger}_{p+q-q's'}C^{\dagger}_{p'+q's}C_{ps'}C_{p's}\right]\right\}$$

In the second term, change summation variables to $\mathbf{p}-\mathbf{q}'\to\mathbf{p}$. In the third term, change $\mathbf{p}+\mathbf{q}'\to\mathbf{p}$, then $\mathbf{q}'\to-\mathbf{q}'$, and interchange the order of the last two operators, and find

$$\left[V_{ee},[H_0,\rho_q(t)]\right] = \frac{1}{2\Omega^2}\sum_{pq'ss'} v_{q'}[\xi(p+q)-\xi(p)-\xi(p+q+q')+\xi(p+q')] \tag{6.135}$$
$$\times\left[\sum_k C^{\dagger}_{p+q+q's}C^{\dagger}_{k-q's'}C_{ks'}C_{ps} + \sum_{p'} C^{\dagger}_{p+q+q's}C^{\dagger}_{p'-q's}C_{p's}C_{ps'}\right]$$

The two sets of operators are made identical by changing $\mathbf{p}'\to\mathbf{k}$ in the last term and interchanging (s,s'). Keep only one term and multiply by two. The combination of kinetic energy terms is

$$\xi(p+q)-\xi(p)-\xi(p+q+q)+\xi(p+q') = -\frac{\hbar^2}{m}\mathbf{q}\cdot\mathbf{q}' \tag{6.136}$$

and the double commutator is

$$\left[V_{ee},[H_0,\rho_q(t)]\right] = -\frac{\hbar^2}{m\Omega^2}\sum_{pq'k'ss'} v_{q'}(\mathbf{q}\cdot\mathbf{q}')\,C^{\dagger}_{p+q+q's}C^{\dagger}_{k-q's'}C_{ks'}C_{ps} \tag{6.137}$$

One tries to evaluate this type of expression by pairing one raising with one lowering operator into a number operator n_{ps}. In the above expression, there are four ways of doing this.

- The choice $\mathbf{q}'=0$ gives

  $$C^{\dagger}_{p+qs}C_{ps}\langle C^{\dagger}_{ks'}C_{ks'}\rangle \tag{6.138}$$

  It has the right form, but the term $\mathbf{q}'=0$ is absent from V_{ee}, since that term is part of the Hartree energy.

- The choice $\mathbf{q}'=-\mathbf{q}$ gives

  $$\langle C^{\dagger}_{ps}C_{ps}\rangle C^{\dagger}_{k+qs'}C_{ks'} \tag{6.139}$$

  This term gives plasmons. For $\mathbf{q}=-\mathbf{q}'$ the factor of $v_{q'}(\mathbf{q}\cdot\mathbf{q}')=-e^2/\varepsilon_0$.

- One exchange term has $s = s'$, $\mathbf{q}' = \mathbf{k} - \mathbf{p}$:

$$-\langle C^\dagger_{\mathrm{p}s} C_{\mathrm{p}s} \rangle C^\dagger_{\mathbf{k}+\mathbf{q}s} C_{\mathbf{k}s} \tag{6.140}$$

- Another exchange term has $s = s'$, $\mathbf{q}' = \mathbf{k} - \mathbf{p} - \mathbf{q}$:

$$-\langle C^\dagger_{\mathbf{k}s} C_{\mathbf{k}s} \rangle C^\dagger_{\mathbf{p}+\mathbf{q}s} C_{\mathbf{p}s} \tag{6.141}$$

The two exchange terms do not give plasmons, and are complicated to evaluate. Ignoring them gives

$$\left[V_{ee}, [H_0, \rho_{\mathbf{q}}(t)] \right] = \frac{e^2 \hbar^2}{m \varepsilon_0 \Omega^2} \left[\sum_{\mathbf{p}s} n_{\mathbf{p}s} \right] \left[\sum_{\mathbf{k}s'} C^\dagger_{\mathbf{k}+\mathbf{q}s'} C_{\mathbf{k}s'} \right] \tag{6.142}$$

The first set of brackets is the number of electrons N_e, while the second set is $\Omega \rho_{\mathbf{q}}(t)$:

$$\left[V_{ee}, [H_0, \rho_{\mathbf{q}}(t)] \right] = \hbar^2 \omega_p^2 \rho_{\mathbf{q}}(t), \quad \omega_p^2 = \frac{e^2 N_e}{m \varepsilon_0 \Omega} \tag{6.143}$$

$$\frac{\partial^2}{\partial t^2} \rho_{\mathbf{q}}(t) = -\omega_p^2 \rho_{\mathbf{q}}(t) - \frac{1}{\hbar^2} \frac{1}{\Omega} \sum_{\mathbf{p}\sigma} [\xi(\mathbf{p}+\mathbf{q}) - \xi(\mathbf{p})]^2 C^\dagger_{\mathbf{p}+\mathbf{q}\sigma} C_{\mathbf{p}\sigma} \tag{6.144}$$

The first term on the right gives plasma oscillations, which are an important feature of the gas of conduction electrons. The second term on the right is from the double commutator with H_0. This term, like the exchange terms, contribute to the dispersion of the plasmon $\omega_p(q)$.

6.2.4 Exchange Hole

The conduction electrons form a gas of particles that can be treated as a fluid. It is useful to introduce some standard fluid concepts such as the *pair distribution function* $g_{\sigma\sigma'}(r)$. It is defined as the probability that an electron of spin σ' is located at a distance r from another electron of spin σ. The probability is normalized to the average density n_e.

- In a ferromagnetic system, where all electrons have the same spin ($\sigma' = \sigma$), then $g_{\sigma\sigma}(r)$ goes to one at large values of r.
- In a paramagnetic system, with equal densities of up and down spin, then $g_{\sigma\sigma}(r)$ and $g_{\sigma-\sigma}(r)$ each go to one-half at large values of r.

Define

$$g_\sigma(r) = \sum_{\sigma'} g_{\sigma\sigma'}(r) \tag{6.145}$$

as the probability that an electron, of any spin, is at a distance r from an electron of spin σ located at $\mathbf{r} = 0$. It goes to one at large distance.

The quantity $g_\sigma(r) - 1$ is nonzero only over small distances, on the scale of the Bohr radius a_0, and possesses a Fourier transform:

$$S_\sigma(\mathbf{q}) = 1 + n_e \int d^3 r e^{i\mathbf{q}\cdot\mathbf{r}} [g_\sigma(r) - 1] \tag{6.146}$$

$S_\sigma(\mathbf{q})$ is called the *static structure factor*, or *liquid structure factor*. In paramagnetic systems, the function is the same for either value of spin, and the spin subscript is usually omitted.

The pair distribution function can be calculated in the Hartree-Fock approximation. Express the many-electron wave function as a Slater determinant and then integrate over all variables except two, say $(\mathbf{r}_1, \mathbf{r}_2)$:

$$\Psi(\mathbf{p}_1, \sigma_1, \mathbf{p}_2, \sigma_2, \cdots : \mathbf{r}_1, \mathbf{r}_2 \cdots \mathbf{r}_N) = \tag{6.147}$$

$$\frac{1}{\sqrt{N!}} \begin{vmatrix} \phi_{\mathbf{p}_1\sigma_1}(\mathbf{r}_1) & \phi_{\mathbf{p}_1\sigma_1}(\mathbf{r}_2) & \cdots & \phi_{\mathbf{p}_1\sigma_1}(\mathbf{r}_N) \\ \phi_{\mathbf{p}_2\sigma_2}(\mathbf{r}_1) & \phi_{\mathbf{p}_2\sigma_2}(\mathbf{r}_2) & \cdots & \phi_{\mathbf{p}_2\sigma_2}(\mathbf{r}_N) \\ \vdots & \vdots & \ddots & \vdots \\ \phi_{\mathbf{p}_N\sigma_N}(\mathbf{r}_1) & \phi_{\mathbf{p}_N\sigma_N}(\mathbf{r}_2) & \cdots & \phi_{\mathbf{p}_N\sigma_N}(\mathbf{r}_N) \end{vmatrix}$$

$$\rho_N(\mathbf{r}_1, \mathbf{r}_2) = \int d^3r_3 d^3r_4 \ldots d^3r_N |\Psi|^2 \tag{6.148}$$

$$= \frac{1}{2N_P} \sum_{\mathbf{p}_i\mathbf{p}_j\sigma_i\sigma_j} \begin{vmatrix} \phi_{\mathbf{p}_i\sigma_i}(\mathbf{r}_1) & \phi_{\mathbf{p}_i\sigma_i}(\mathbf{r}_2) \\ \phi_{\mathbf{p}_j\sigma_j}(\mathbf{r}_1) & \phi_{\mathbf{p}_j\sigma_j}(\mathbf{r}_2) \end{vmatrix}^2$$

where the double summation is over all pairs of electron states, and $N_P = N(N-1)/2$ is the number of such pairs. The notation $\int d^3r_j$ means to do the space integral, but also average over spin variables:

$$\int d^3r_j \phi_{\mathbf{p}_i\sigma_i}(\mathbf{r}_j)^* \phi_{\mathbf{p}_l\sigma_l}(\mathbf{r}_j) = \delta_{il} \tag{6.149}$$

The quantity $\rho_N(\mathbf{r}_1, \mathbf{r}_2)$ is the probability of having one electron at $\mathbf{r}_1$ and another at $\mathbf{r}_2$ in an N-particle system. The result depends on the spin configurations of the two electrons. Adding that feature to the notation, this density matrix is related to the pair distribution function:

$$\rho_N(\mathbf{r}_1, \sigma_1; \mathbf{r}_2, \sigma_2) = \frac{g_{\sigma_1\sigma_2}(\mathbf{r}_1, \mathbf{r}_2)}{\Omega^2} \tag{6.150}$$

One case has both electrons with the same spin: either both up or both down. In that case, the 2×2 determinant has a cross-term. Another case is a paramagnetic electron gas with equal numbers of up and down spin electrons, which is assumed here:

$$\rho_N(\mathbf{r}_1, \sigma; \mathbf{r}_2, \sigma) = \frac{1}{2N_P\Omega^2} \sum_{\mathbf{p}_i\mathbf{p}_j} |e^{i(\mathbf{p}_i\cdot\mathbf{r}_1 + \mathbf{p}_j\cdot\mathbf{r}_2)} - e^{i(\mathbf{p}_j\cdot\mathbf{r}_1 + \mathbf{p}_i\cdot\mathbf{r}_2)}|^2 \tag{6.151}$$

$$= \frac{1}{N_P\Omega^2} \sum_{\mathbf{p}_i\mathbf{p}_j} \{1 - \cos[(\mathbf{p}_i - \mathbf{p}_j)\cdot(\mathbf{r}_1 - \mathbf{r}_2)]\}$$

$$g_{\sigma\sigma}(\mathbf{r}_1 - \mathbf{r}_2) = \frac{1}{N_P} \sum_{\mathbf{p}_i\mathbf{p}_j} \{1 - \cos[(\mathbf{p}_i - \mathbf{p}_j)\cdot(\mathbf{r}_1 - \mathbf{r}_2)]\} \tag{6.152}$$

The above formula shows that, in the electron gas, in the Hartree-Fock approximation, $g_{\sigma\sigma}(r=0) = 0$. Two electrons with the same spin cannot be at the same point at the same time. This is the exclusion principle for fermions. Each electron carries with it a "hole," which is a region in which other electrons of the same spin are excluded. It also explains the origin of the exchange energy, since an electron with one spin, say up, does not have any other electrons with the same spin in its immediate vicinity. The hole reduces the repulsive interaction between electrons. Another important result is that

$$\lim_{r\to\infty} g_{\sigma\sigma}(r) = \frac{2}{N_p\Omega^2}\sum_{\mathbf{p}_i\mathbf{p}_j} 1 = \frac{1}{2} \tag{6.153}$$

The double summation over the two wave vectors of occupied electron states just gives $N_p/2$.

The other important case has the two electrons with opposite spins. In that case there is no cross term in the determinant, since $\langle\sigma|-\sigma\rangle = 0$, and

$$\begin{aligned}\rho_N(\mathbf{r}_1, \sigma; \mathbf{r}_2, -\sigma) &= \frac{1}{N_p\Omega^2}\sum_{\mathbf{p}_i\mathbf{p}_j}\left\{|e^{i(\mathbf{p}_i\cdot\mathbf{r}_1+\mathbf{p}_j\cdot\mathbf{r}_2)}|^2 + |e^{i(\mathbf{p}_j\cdot\mathbf{r}_1+\mathbf{p}_i\cdot\mathbf{r}_2)}|^2\right\}\\ &= \frac{2}{N_p\Omega^2}\sum_{\mathbf{p}_i\mathbf{p}_j} 1 = 1, \quad g_{\sigma-\sigma}(r) = \frac{1}{2}\end{aligned} \tag{6.154}$$

In the Hartree-Fock approximation, there is no correlation between the motion of up and down spin electrons, so that $g_{\sigma-\sigma}(r)$ is one-half everywhere. In the Hartree-Fock approximation, there is no "hole" for electrons of the opposite spin.

Hartree-Fock theory is an approximation. When including more terms in the perturbation series for electron–electron interactions, one finds there is a small amount of correlation between electrons of opposite spin. Even for an electron of opposite spin there is a small "hole" around each electron.

6.3 Density Functional Theory

Density functional theory (DTF) was created by Walter Kohn and collaborators. It is an approximate method of treating electron–electron interactions in real systems such as atoms, molecules, and solids. Although invented for use in condensed matter physics, it eventually became widely used in chemistry. Kohn's Nobel Prize for the development of DFT was in chemistry.

Large computers can be used to solve for the motion of a single electron in almost any kind of potential energy $V(\mathbf{r})$. For one electron, computer codes are available to solve a Hamiltonian of the type

$$H = -\frac{\hbar^2\nabla^2}{2m} + V(\mathbf{r}) \tag{6.155}$$

The many-electron problem is not in the above form, because of electron–electron interactions. It is in the form

$$H = \sum_{i=1}^{N_e} H_i + V_{ee} \tag{6.156}$$

$$H_i = -\frac{\hbar^2\nabla_i^2}{2m} + U(\mathbf{r}_i) \tag{6.157}$$

$$V_{ee} = \frac{1}{2}\sum_{i\neq j}\frac{e^2}{|\mathbf{r}_i - \mathbf{r}_j|} \tag{6.158}$$

The objective of DFT is to take the last term V_{ee} and include it in $V(\mathbf{r})$ so that one has an effective Hamiltonian for a single electron.

The physics of DFT is based on a theorem published by Hohenberg and Kohn [6]. It states that the ground-state energy E_G of a system of N electrons is a function only of the density $n(\mathbf{r})$ of the electrons.

The theorem is initially surprising. The eigenfunctions of individual electrons $\psi_{\mathbf{k}}(\mathbf{r})$ are usually complex numbers, with amplitude and phase. Yet the density is a real number. Somehow, all of the phase information is irrelevant! But the theorem is true, and is the basis of DFT. The effective potential energy function for an electron is obtained by taking a functional derivative:

$$V_{eff}(\mathbf{r}) = \frac{\delta E_G}{\delta n(\mathbf{r})} \tag{6.159}$$

The Hohenberg-Kohn theorem reduces the computational problem to finding out how the ground-state energy depends on density.

6.3.1 Functional Derivatives

Equation (6.159) is the basis of DFT. What does this equation mean? The concept of functional derivative needs to be understood. First, consider the integral

$$\phi(x) = \int^x dx' f(x') \tag{6.160}$$

Take the x-derivative of this equation:

$$\frac{d\phi(x)}{dx} = f(x) \tag{6.161}$$

The two equations (6.160) and (6.161) are equivalent. However, this is not a functional derivative. Instead, consider the expression

$$E_G = \int d^3r F[n(\mathbf{r})] \tag{6.162}$$

Let the density vary $n(\mathbf{r}) \rightarrow n(\mathbf{r}) + \delta n(\mathbf{r})$ and expand the function in a Taylor series:

$$F[n(\mathbf{r})] \rightarrow F[n(\mathbf{r}) + \delta n(\mathbf{r})] \approx F[n(\mathbf{r})] + \delta n(\mathbf{r})\left(\frac{dF}{dn}\right)_{n(\mathbf{r})} + \cdots \tag{6.163}$$

$$E_G + \delta E_G = \int d^3r\left\{F[n(\mathbf{r})] + \delta n(\mathbf{r})\left(\frac{dF}{dn}\right)_{n(\mathbf{r})} + \cdots\right\} \tag{6.164}$$

$$\delta E_G = \int d^3r \delta n(\mathbf{r})\left(\frac{dF}{dn}\right)_{n(\mathbf{r})} \tag{6.165}$$

The functional derivative of the ground-state energy, with respect to density, is interpreted as

$$\frac{\delta E_G}{\delta n(\mathbf{r})} = \left(\frac{dF}{dn}\right)_{n(\mathbf{r})} \tag{6.166}$$

Some examples are

$$E_G = \int d^3 r n(\mathbf{r}) f(\mathbf{r}), \quad \frac{\delta E_G}{\delta n(\mathbf{r})} = f(\mathbf{r}) \tag{6.167}$$

$$E_B = \int \frac{d^3 k}{(2\pi)^3} \varepsilon(\mathbf{k})\, n_F[\varepsilon(\mathbf{k})], \quad \frac{\delta E_B}{\delta n_F(\mathbf{k})} = \varepsilon(\mathbf{k}) \tag{6.168}$$

$$\frac{\delta E_B}{\delta \varepsilon(\mathbf{k})} = n_F[(\varepsilon(\mathbf{k})] + \varepsilon(\mathbf{k})\left(\frac{dn_F(x)}{dx}\right)_{x=\varepsilon(\mathbf{k})} \tag{6.169}$$

In the last example, E_B is the band energy, obtained by summing the dispersion $\varepsilon(\mathbf{k})$ over all occupied energy states, and n_F is the Fermi-Dirac occupation number. Equation (6.168) is important, since it shows that the effective band dispersion can be obtained by taking a functional derivative with respect to the occupation number. This result was used earlier in deriving the exchange energy $\Sigma_X(\mathbf{k})$.

6.3.2 Kinetic Energy

The kinetic energy is treated differently than other contributions to the energy. The reason is that it is not merely a function of the density. We give two examples of the kinetic energy of simple systems:

1. For a free electron gas, the ground-state energy is

$$E_K = \frac{3}{5} N E_F = N \frac{3\hbar^2 k_F^2}{10m} = \Omega (3\pi)^{2/3} \frac{3\hbar^2}{10m} n^{5/3} \tag{6.170}$$

where $k_F^3 = 3\pi^2 n$. The above expression is often written, for systems in which the density varies, as

$$E_K = \frac{3}{5}\int d^3 r\, n(\mathbf{r})\, E_F[n(\mathbf{r})] = (3\pi)^{2/3} \frac{3\hbar^2}{10m} \int d^3 r\, n(\mathbf{r})^{5/3} \tag{6.171}$$

In this case, one can write the kinetic energy as a function of the density. The functional derivative is

$$\frac{\delta E_K}{\delta n(r)} = (3\pi)^{2/3} \frac{\hbar^2}{2m} n(\mathbf{r})^{2/3} = \frac{\hbar^2 k_F^2[n(\mathbf{r})]}{2m} = E_F[n(\mathbf{r})] \tag{6.172}$$

The derivative brings down a factor of $\frac{5}{3}$ that cancels most of the $\frac{3}{10}$ factor in E_K. The interesting feature of the result is that the derivative gives the kinetic energy of an electron at the Fermi surface. This feature is general: the derivative of the ground-state energy, with respect to density, gives the energy of an electron at the Fermi surface. At zero temperature, this energy is the chemical potential:

$$\frac{\delta E_G}{\delta n(\mathbf{r})} = \mu \tag{6.173}$$

The chemical potential is a constant for a system in equilibrium. So must be the left-hand side of the above equation. The kinetic energy term must be large in regions where the potential energy is small, and vice versa.

2. The second simple system is the hydrogen atom. The kinetic energy of this system of one electron in the ground state is

$$E_K = \frac{\hbar^2}{2m}\int d^3r|\nabla\phi_{1s}(\mathbf{r})|^2 \tag{6.174}$$

$$n(\mathbf{r}) = |\phi_{1s}(\mathbf{r})|^2 \tag{6.175}$$

Since the eigenfunction ϕ_{1s} is real, write

$$\nabla n(\mathbf{r}) = 2\phi_{1s}\nabla\phi_{1s} \tag{6.176}$$

$$\nabla|\phi_{1s}(\mathbf{r})|^2 = \frac{|\nabla n(\mathbf{r})|^2}{4n(\mathbf{r})} \tag{6.177}$$

$$E_K = \frac{\hbar^2}{8m}\int d^3r\frac{|\nabla n(r)|^2}{n(\mathbf{r})} \tag{6.178}$$

Now the kinetic energy of the ground state is a function of $n(\mathbf{r})$ and also $\nabla n(\mathbf{r})$. In general, for most systems, the kinetic energy is more like the hydrogen atom, in that the function depends on the gradient of the density, as well as on the density. In DFT, the kinetic energy is not evaluated using a functional derivative, but is evaluated as a derivative of the eigenfunction, as in eqn. (6.174).

6.3.3 Kohn-Sham Equations

Kohn and Sham [9] took the ground-state energy theorem, and developed it into a set of equations that could be solved. These equations, with minor embellishments, are the basis of today's computer codes. Their equations are

$$H = \sum_{i=1}^{N} H_i \tag{6.179}$$

$$H_i = -\frac{\hbar^2\nabla_i^2}{2m} + V_{\text{eff}}[n(\mathbf{r})] \tag{6.180}$$

The Hamiltonian for the N-particle system is a summation of the Hamiltonians of the individual electrons. In H_i, the first term is from the kinetic energy, and the effective potential is from all other contributions to the ground-state energy. Each individual electron appears to have its own Hamiltonian H_i. It has eigenstates and eigenfunctions:

$$H_i\psi_i(\mathbf{r}) = E_i\psi_i(\mathbf{r}) = \left[-\frac{\hbar^2\nabla_i^2}{2m} + V_{\text{eff}}[n(\mathbf{r})]\right]\psi_i(\mathbf{r}) \tag{6.181}$$

It is a mistake to interpret E_i as an eigenvalue of the electron, or $\psi_i(\mathbf{r})$ as an electron eigenfunction. They are called Kohn-Sham eigenfunctions and eigenvalues, since they are solutions of the Kohn-Sham equations. The electron density is

$$n(\mathbf{r}) = \sum_i |\psi_i(\mathbf{r})|^2 \tag{6.182}$$

where the summation over i is over all occupied electron states. The two equations (6.181) and (6.182) describe a self-consistent procedure. One makes an initial guess at the density $n(\mathbf{r})$ and uses it to find the eigenfunctions $\psi_i(\mathbf{r})$. These eigenfunctions are used to calculate a refined density. Repeat the cycle until convergence to self-consistency is obtained.

The full Hamiltonian for a system of many electrons is

$$H = \sum_{i=1}^{N}\left[-\frac{\hbar^2\nabla_i^2}{2m} + U(\mathbf{r}_i)\right] + \frac{1}{4\pi\varepsilon_0}\sum_{i>j}^{N}\frac{e^2}{|\mathbf{r}_i - \mathbf{r}_j|} \tag{6.183}$$

The kinetic energy term has been discussed. The second term is the interaction $U(\mathbf{r}_i)$ of an electron with a fixed potential. In solids, it is a periodic array of nuclear charge. In an atom or small molecule, it is the nonperiodic array of nuclear charge. For large systems, the nuclear potential is replaced by a pseudopotential. The terms in the effective potential in eqn. (6.180) are

$$V_{\text{eff}}[n(\mathbf{r})] = U(\mathbf{r}) + V_H[n(\mathbf{r})] + V_X[n(\mathbf{r})] + V_C[n(\mathbf{r})] \tag{6.184}$$

$$V_H(\mathbf{r}) = \frac{e^2}{4\pi\varepsilon_0}\int d^2r' \frac{n(\mathbf{r}')}{|r - r'|} \tag{6.185}$$

The first term is the ion potential $U(\mathbf{r})$. Its contribution to the ground-state energy is

$$E_U = \int d^3r\, n(\mathbf{r})U(\mathbf{r}), \quad \frac{\delta E_U}{\delta n(\mathbf{r})} = U(\mathbf{r}) \tag{6.186}$$

The second is the Hartree potential $V_H(\mathbf{r})$, which is from the electron–electron interactions:

$$E_H = \frac{e^2}{8\pi\varepsilon_0}\int d^3r \int d^3r' \frac{n(\mathbf{r})n(\mathbf{r}')}{|r - r'|}, \quad \frac{\delta E_H}{\delta n(\mathbf{r})} = V_H(\mathbf{r}) \tag{6.187}$$

Note that $V_H(\mathbf{r})$ is not a function of the density $n(\mathbf{r})$. Instead, it is a *functional* of the density. The last two terms in eqn. (6.184) are the potential energy from exchange (V_X) and correlation (V_C). They are discussed next.

6.3.4 Exchange and Correlation

The exchange energy contributes to the self-energy of an electron $\Sigma_X(k)$, and to the ground-state energy E_X of a system of electrons. It was given earlier in eqns. (6.71) and (6.72):

$$\Sigma_X(k) = -\frac{e^2}{\varepsilon_0}\int \frac{d^3p}{(2\pi)^3}\frac{n_F(\varepsilon_p)}{|\mathbf{p}-\mathbf{k}|^2} \tag{6.188}$$

$$E_X = -\frac{e^2}{2\varepsilon_0}\Omega\int\frac{d^3k}{(2\pi)^3}\int\frac{d^3p}{(2\pi)^3}\frac{n_F(\varepsilon_p)n_F(\varepsilon_k)}{|\mathbf{p}-\mathbf{k}|^2} \tag{6.189}$$

$$= \frac{1}{2}\Omega\sum_s\int\frac{d^3k}{(2\pi)^3}n_F(\varepsilon_k)\Sigma_X(k)$$

At zero temperature, the Fermi-Dirac occupation numbers $n_F(\varepsilon_k)$ are step functions that restrict the values of $k < k_F$, $p < k_F$. The integrals were done earlier:

$$\Sigma_X(k) = -\frac{e^2k_F}{4\pi^2\varepsilon_0}S\left(\frac{k}{k_F}\right) \tag{6.190}$$

$$E_X = -N\frac{3e^2k_F}{(4\pi)^2\varepsilon_0} \tag{6.191}$$

The last equation, for the ground-state energy from exchange, can be written using $k_F = (3\pi^2n)^{1/3}$ as

$$E_X = -\frac{3e^2(3\pi^2)^{1/3}}{(4\pi)^2\varepsilon_0}\int d^3rn(\mathbf{r})^{4/3} \tag{6.192}$$

The functional derivative is

$$\frac{\delta E_X}{\delta n(\mathbf{r})} = -\frac{e^2}{4\pi^2\varepsilon_0}[3\pi^2n(\mathbf{r})]^{1/3} = \Sigma_X(k_F) \tag{6.193}$$

The effective potential from exchange is the self-energy of the electron, evaluated at the Fermi energy.

Equation (6.193) was first used by John Slater [16]. He wanted to include exchange as a local potential function, and used the free electron formula $\Sigma = -e^2k_F/(4\pi^2\varepsilon_0)$ with a local approximation for the Fermi wave vector $k_F(\mathbf{r}) = [3\pi^2n(\mathbf{r})]^{1/3}$. His potential was

$$V_{X\alpha}(\mathbf{r}) = -\frac{\alpha e^2}{4\pi^2\varepsilon_0}[3\pi^2n(\mathbf{r})]^{1/3} \tag{6.194}$$

He was unsure of the prefactor, so he added a dimensionless constant α, which was varied to obtain good results. His method was called $X\alpha$. He used it, long before Kohn invented DFT, to obtain some early band structures of solids, and electrons states in atoms and molecules. After the invention of DFT, it was recognized that Slater's $X\alpha$ method was Kohn's exchange term, with $\alpha = 1$.

The *correlation energy* is the additional energy terms, from electron–electron interactions, beyond Hartree-Fock. The Hartree-Fock terms come from first-order perturbation theory. Correlation comes from all other terms in the perturbation expansion. There is a correlation contribution to the ground-state energy E_C, and a correlation contribution to the self-energy $\Sigma_C(k)$ of the electron. These two energies are also related by a functional derivative:

$$V_C(\mathbf{r}) = \frac{\delta E_c}{\delta n(\mathbf{r})} = \Sigma_C(k_F) \tag{6.195}$$

The calculation of the correlation energy has been a major activity in many-body theory during the past half-century. Usually the results are listed for E_C as a function of the electron density. The density is given by the dimensionless constant r_s. Recall its relation to the electron density:

$$r_s^3 = \frac{3}{4\pi a_0^3 n(\mathbf{r})} \tag{6.196}$$

Some correlation functions derived by various investigators are, in Rydberg energy:

$$\begin{aligned}
\text{Nozieres and Pines [12]} \quad E_C &= -0.115 + 0.0311 \ln(r_s) \\
\text{Vashista and Singwi [17]} \quad E_C &= -0.112 + 0.0335 \ln(r_s) - \frac{0.027}{0.1 + r_s} \\
\text{Gordon and Kim [2]} \quad E_C &= -0.12312 + 0.03796 \ln(r_s) \\
\text{Perdew and Zunger [13]} \quad E_C &= \frac{-0.1423}{1 + 1.0529\sqrt{r_s} + 0.3334 r_s}
\end{aligned}$$

The four results are very similar. Making a graph of E_C as a function of r_s, over the range of metallic densities $2 < r_s < 6$, shows the four functions are very similar. One can then express r_s as a function of $n(\mathbf{r})$, and then take the derivative to obtain the potential $V_C(\mathbf{r})$.

6.3.5 *Application to Atoms*

Kohn and Sham first applied the method to calculate the properties of atoms. The nucleus has charge Z, and assume there are N electrons in bound-state orbitals. A neutral atom has $N = Z$, but the method can be applied to ions where $N \neq Z$. The steps in the calculation are as follows:

1. Solve the following Hamiltonian for each electron orbital:

$$H\psi_i(\mathbf{r}) = E_i \psi_i(\mathbf{r}) \tag{6.197}$$

$$H = -\frac{\hbar^2\nabla^2}{2m} - \frac{Ze^2}{4\pi\varepsilon_0 r} - \frac{e^2}{4\pi\varepsilon_0}\left(\frac{3}{\pi}n(\mathbf{r})\right)^{1/3} + V_C[n(\mathbf{r})] \tag{6.198}$$

2. Calculate the density:

$$n(\mathbf{r}) = \sum_{i=1}^{N} |\psi_i(\mathbf{r})|^2 \tag{6.199}$$

3. Repeat steps (1) and (2) until the density is self-consistent.

4. Then calculate the ground-state energy $E_G(N)$ for the N-electron system. The ground-state energy is the kinetic energy E_K, the exchange energy E_X, and the correlation energy E_C:

$$E_G = E_K + E_X + E_C \tag{6.200}$$

$$E_K = \frac{\hbar^2}{2m}\sum_{i=1}^{N}\int d^3r |\nabla\psi_i(\mathbf{r})|^2 \tag{6.201}$$

The exchange energy is found from the integral in eqn. (6.192), and the correlation energy is found from a similar integral for the correlation energy.

5. Repeat the above calculation for the same atom with $N-1$ electrons, and find $E_G(N-1)$. The ionization energy is

$$E_I = E_G(N-1) - E_G(N) \tag{6.202}$$

The ionization energy is not related to the eigenvalues E_i.

This calculation gives excellent results for the ionization energy of atoms with filled shells of electrons. The agreement with experiment is several significant digits.

Atoms with partially filled *d*- or *f*-shells are magnetic, and the method is less accurate. In this case, one defines separate densities for spin-up $n_\uparrow(\mathbf{r})$ and spin-down $n_\downarrow(\mathbf{r})$ electrons, and writes coupled equations to relate these two densities. The exchange energy is only between electrons of the same spin polarization.

The method fails completely for the hydrogen atom. It has only one electron, and there is no need to add terms with exchange and correlation. The helium atom has two electrons. In its ground state, one spin is up and one is down, in a spin singlet. Here there is no need for an exchange term, but there is correlation energy. Clearly, some judgment is needed to use the method correctly.

6.3.6 Time-dependent Local Density Approximation

DFT was initially called the *local density approximation*. That is an apt moniker, since the method is indeed an approximation. TDLDA is an application of DFT (or LDA) to evaluate the polarizabilities of atoms and molecules. Consider an atom or ion with a closed atomic shell of electrons, so that it is spherically symmetric. Calculate its ground-state energy E_G using the method of Kohn and Sham, as described above.

Then apply a static electric field $\mathbf{F} = F\hat{z}$ to the atom. The ground-state energy changes due to the polarizability α and hyperpolarizability γ:

$$E_G(F) = E_G(0) + 4\pi\varepsilon_0\left[\frac{\alpha}{2}F^2 + \frac{\gamma}{4}F^4 + \cdots\right] \tag{6.203}$$

DFT can be used to calculate α and γ, which are properties of the ground-state of the atom. Remember that DFT is most accurate when calculating ground-state properties. Equation (6.203) is the ground-state energy in an electric field. Here we discuss the calculation of the polarizability α. See Mahan and Subbaswamy [11] for the calculation of the hyperpolarizability.

An atom with a closed shell of electrons has no dipole moment. It will have a dipole moment in response to the applied electric field:

$$\mathbf{p} = 4\pi\varepsilon_0\alpha\mathbf{F} = \hat{z}4\pi\varepsilon_0\alpha F = \hat{z}p \tag{6.204}$$

$$p = 4\pi\varepsilon_0\alpha F \tag{6.205}$$

The dipole moment is the integral over the charge density times a microscopic dipole ez_i:

$$p = 4\pi\varepsilon_0 \alpha F = e\int d^3rz\, n(\mathbf{r}) \tag{6.206}$$

The integral is zero without the electric field, since the charge density $n(\mathbf{r})$ is spherically symmetric. Assume the electric field F is infinitesimal. It will change each Kohn-Sham eigenfunction ψ_i by an infinitesimal amount $F\phi_i$, which causes a small change in the density:

$$\psi_i(\mathbf{r}) \rightarrow \psi_i(\mathbf{r}) + F\phi_i(\mathbf{r}) + O(F^2) \tag{6.207}$$

$$n(\mathbf{r}) \rightarrow \sum_i |\psi_i + F\phi_i|^2 = n_0(\mathbf{r}) + Fn_1(\mathbf{r}) + O(F^2) \tag{6.208}$$

$$n_0(\mathbf{r}) = \sum_i |\psi_i|^2, \quad n_1(\mathbf{r}) = \sum_i [\psi_i\phi_i^* + \phi_i\psi_i^*] \tag{6.209}$$

$$\alpha = \frac{e}{4\pi\varepsilon_0}\int d^3rzn_1(\mathbf{r}) \tag{6.210}$$

The ground-state density is $n_0(\mathbf{r})$, and the change in density induced by the infinitesimal electric field is $n_1(\mathbf{r})$. The latter function determines the polarizability.

The function ϕ_i is found by solving the Hamiltonian to order $O(F)$. This can be done by perturbation theory. A better method is to write

$$H = -\frac{\hbar^2\nabla^2}{2m} + ezF + V_{\text{eff}}(n_0 + Fn_1) \tag{6.211}$$

$$= H_0 + F\delta V + O(F^2)$$

$$H_0 = -\frac{\hbar^2\nabla^2}{2m} + V_{\text{eff}}(n_0) \tag{6.212}$$

$$\delta V = ez + n_1\left(\frac{\delta V_{\text{eff}}(n)}{\delta n}\right)_{n_0} \tag{6.213}$$

The second term in δV is a screening term. The effective potential is changed by the field.

The new Hamiltonian is solved to order $O(F)$:

$$[H_0 + F\delta V][\psi_i + F\phi_i] = [E_i + O(F^2)][\psi_i + F\phi_i] \tag{6.214}$$

The terms that depend on one power of the field are

$$[H_0 - E_i]\phi_i = -\delta V\psi_i \tag{6.215}$$

This equation is solved on the computer for $\phi_i(\mathbf{r})$. It is not an eigenvalue equation. It is an inhomgeneous differential equation. It is easier to solve than an eigenvalue equation, if the functions on the right of the equal sign are known. It describes another self-consistent procedure: the right-hand side depends on $n_1(\mathbf{r})$, which depends on all of the ϕ_i for occupied orbital states. So one solves eqn. (6.215) to obtain the ϕ_i for each occupied orbital, and then calculates $n_1(\mathbf{r})$. It is used to recalculate all of the $\phi_i(\mathbf{r})$. This cycle is repeated until self-consistency is obtained. Table 6.1 shows results for the rare gas atoms. The

Table 6.1 Experimental and calculated dipole polarizabilites of rare gas atoms, in units of cubic angstroms (Å^3)

Atom	Expt.	DFT
He	0.21	0.25
Ne	0.40	0.45
Ar	1.64	1.77
Kr	2.48	2.67
Xe	4.04	4.26

polarizability has the units of volume, and the table is in cubic angstroms. The theoretical values are consistently too high. The percentage error is 20% for helium, and drops down to 5% for xenon. DFT treats the electrons in the atom as a little electron gas. This approximation gets better with increasing atomic number, where there are more electrons. These trends are found when other quantities are calculated.

6.3.7 TDLDA in Solids

The time-dependent local density approximation (TDLDA) has been applied to the calculation of the dielectric response of covalent solids by Baroni et al. [1] This case includes most semiconductors and many insulators. After solving the Kohn-Sham equations for a solid, one has a set of Kohn-Sham eigenvalues $E_n(\mathbf{k})$ and eigenfunctions $\psi_n(\mathbf{k}, \mathbf{r})$. The electron density is

$$n(\mathbf{r}) = 2\sum_n \int_{BZ} \frac{d^3k}{(2\pi)^3} |\psi_n(\mathbf{k}, \mathbf{r})|^2 \tag{6.216}$$

where the factor of two is spin degeneracy. The application of an electric field F causes the Kohn-Sham eigenfunctions to have a first-order variation $\phi_n(\mathbf{k}, \mathbf{r})$:

$$\psi_n(\mathbf{k}, \mathbf{r}) \rightarrow \psi_n(\mathbf{k}, \mathbf{r}) + F\phi_n(\mathbf{k}, \mathbf{r}) \tag{6.217}$$

$$n(\mathbf{r}) \rightarrow n(\mathbf{r}) + Fn_1(\mathbf{r}) \tag{6.218}$$

$$n_1(\mathbf{r}) = 2\sum_n \int_{BZ} \frac{d^3k}{(2\pi)^3} [\psi_n^*(\mathbf{k}, \mathbf{r})\phi_n(\mathbf{k}, \mathbf{r}) + \phi_n^*(\mathbf{k}, \mathbf{r})\psi_n(\mathbf{k}, \mathbf{r})] \tag{6.219}$$

If $U(\mathbf{r})$ is the periodic potential of the ion cores in the crystal, then the Kohn-Sham potential is

$$V(\mathbf{r}) = U(\mathbf{r}) + \frac{e^2}{4\pi\varepsilon_0} \int d^3r' \frac{n(\mathbf{r}')}{|r - r'|} + V_{xc}[n(\mathbf{r})] \tag{6.220}$$

The last term is from exchange and correlation. The self-consistent change in potential, caused by the electric field in the $\hat{z}$ direction, is

$$V(\mathbf{r}) \rightarrow V(\mathbf{r}) + FV_{\mathrm{SCF}}(\mathbf{r}) \tag{6.221}$$

$$V_{\mathrm{SCF}}(\mathbf{r}) = ez + \frac{e^2}{4\pi\varepsilon_0}\int d^3r' \frac{n_1(\mathbf{r}')}{|\mathbf{r}-\mathbf{r}'|} + n_1(\mathbf{r})\left(\frac{\delta V_{xc}}{\delta n}\right)_{n(\mathbf{r})} \tag{6.222}$$

and the first-order eigenfunction is

$$[H - E_n(\mathbf{k})]\phi_n(\mathbf{k}, \mathbf{r}) = -V_{\mathrm{SCF}}(\mathbf{r})\psi_n(\mathbf{k}, \mathbf{r}) \tag{6.223}$$

After solving eqn. (6.223) for $\phi_n(\mathbf{k}, \mathbf{r})$, one calculates $n_1(\mathbf{r})$. Use n_1 to calculate V_{SCF} and recalculate ϕ_n. This loop is repeated until self-consistency is achieved. The polarization **P** induced by the field, and dielecctric constant ε are

$$\mathbf{P}(\mathbf{r}) = \frac{Fe\hat{z}}{\Omega}\int d^3r\mathbf{r}n_1(\mathbf{r}) \tag{6.224}$$

$$\varepsilon = \varepsilon_0 + \frac{\hat{z}\cdot\mathbf{P}}{F} \tag{6.225}$$

The integral d^3r extends over the volume Ω of the crystal. This procedure is the solid-state analogue of the TDLDA procedure for atoms.

Since the applied field gives a potential energy that is p-wave ("z"), then n_1 is also p-like. The potential term

$$\frac{e^2}{4\pi\varepsilon_0}\int d^3r' \frac{n_1(\mathbf{r}')}{|\mathbf{r}-\mathbf{r}'|} \tag{6.226}$$

is dipolar. It is the potential energy induced by the dipoles on the other atoms or unit cells. It serves as a screening term.

6.4 Dielectric Function

In cubic insulators, Coulomb's law between two electrons is

$$V_{ee}(r) = \frac{e^2}{4\pi\varepsilon r} \tag{6.227}$$

$$\frac{\varepsilon}{\varepsilon_0} = 1 + \frac{4\pi\tilde{\alpha}}{1 - \frac{4\pi\tilde{\alpha}}{3}} \tag{6.228}$$

where $\tilde{\alpha} = \alpha/\Omega_0$ is the polarizability per unit volume. The latter equation is the famous Lorenz-Lorentz dielectric function. It has a denominator with the local field correction of $4\pi\tilde{\alpha}/3$ derived by summing local dipoles in the vicinity of the charges. This dielectric function applies to a hypothetical solid composed of spherical atoms that are smaller than the size of the unit cell, so there is no charge overlap. Actual solids have atoms that do overlap. In alkali halide crystals and rare gas solids, the atoms overlap but remain nearly spherical in shape. In this case the above formula can still be used to accurately calculate the dielectric function (see Mahan and Subbaswamy [11]). However, most other insulators have some degree of covalent bonding between atoms. Their dielectric function is evaluated by another formula, which was discussed in the prior section.

The next section will derive the dielectric function of a Fermi gas of electrons. The dielectric response is not a constant, but is a function of wave vector **q** and frequency ω, and is written as $\varepsilon(q, \omega)$. It is useful to work in Fourier transform space. The Fourier transform of Coulomb's law in three dimensions is $v_q = e^2/\varepsilon_0 q^2$, and the screened interaction is

$$V_{s,ee}(q, \omega) = \frac{v_q \varepsilon_0}{\varepsilon(q, \omega)} \tag{6.229}$$

The dielectric function $\varepsilon(q, \omega)$ is derived below. It will also have a local field correction, but in a different form than in eqn. (6.228).

Homogeneous materials have two dielectric functions: longitudinal $\varepsilon_\ell(\mathbf{q}, \omega)$ and transverse $\varepsilon_t(\mathbf{q}, \omega)$. The longitudinal one is used in screening charge fluctuations and in the above equation. The transverse dielectric function arises from electromagnetic fields. In trying to find the optical wave vector **k** in a solid, one solves a formula derived from Maxwell's equations:

$$k^2 = \left(\frac{\omega}{c}\right)^2 \varepsilon_t(k, \omega) \tag{6.230}$$

The two dielectric functions can be shown to be equal at long wavelength:

$$\lim_{\mathbf{q}\to 0} \varepsilon_t(\mathbf{q}, \omega) = \lim_{\mathbf{q}\to 0} \varepsilon_\ell(\mathbf{q}, \omega) \equiv \varepsilon(\omega) \tag{6.231}$$

In crystals, the dielectric response is a tensor $\varepsilon_{\mu\nu}(\mathbf{q}, \omega)$. In cubic crystals, along major axes of symmetry, the tensor is represented by its longitudinal and tranverse parts. However, for a general wave vector in the Brillouin zone, one must use the tensor formula. The longitudinal part for dielectric screening is usually

$$\varepsilon_\ell(\mathbf{q}, \omega) = \sum_{\mu\nu} \frac{q_\mu \varepsilon_{\mu\nu}(\mathbf{q}, \omega) q_\nu}{q^2} \tag{6.232}$$

Most treatments of the electron gas use the homogenous theory.

6.4.1 Random Phase Approximation

Rather than derive the screening between two electrons, it is easier to derive the Coulomb potential from a fixed impurity. Assume there is an impurity of charge Ze at the point $\mathbf{R} = 0$. This impurity introduces a potential term U into the electron Hamiltonian:

$$U = \frac{Ze^2}{4\pi\varepsilon_0} \int \frac{d^3r}{r} \rho(\mathbf{r}) = \sum_{\mathbf{q}} U(\mathbf{q}) \rho(\mathbf{q}) \tag{6.233}$$

$$U(\mathbf{q}) = Z v_q, \quad \rho(\mathbf{q}) = \frac{1}{\Omega} \sum_{\mathbf{p}\sigma} C^\dagger_{\mathbf{p}+\mathbf{q}\sigma} C_{\mathbf{p}\sigma} \tag{6.234}$$

The response of the electron gas to this perturbation is found using the equations of motion. Instead of the equation for the entire density, just examine the equation of motion for the product of operators:

$$\frac{\partial}{\partial t}\left[C^{\dagger}_{p+q\sigma}C_{p\sigma}\right] = \frac{i}{\hbar}\left[H, C^{\dagger}_{p+q\sigma}C_{p\sigma}\right] \tag{6.235}$$

$$H = H_0 + U + V_{ee} \tag{6.236}$$

This expression has several terms.

- Since the system is oscillating with a frequency ω, the time dependence is $\exp(-i\omega t)$. The time derivative on the left produces $-i\omega$.
- The commutator of H_0 was done in the prior section and produces the difference in the band energies:

$$[H_0, C^{\dagger}_{\mathbf{p}+\mathbf{q}\sigma}C_{\mathbf{p}\sigma}] = [\xi(\mathbf{p}+\mathbf{q}) - \xi(\mathbf{p})]C^{\dagger}_{\mathbf{p}+\mathbf{q},\sigma}C_{\mathbf{p}\sigma} \tag{6.237}$$

- The commutator of the impurity potential is

$$[U, C^{\dagger}_{\mathbf{p}+\mathbf{q}\sigma}C_{\mathbf{p}\sigma}] = \frac{1}{\Omega}\sum_{\mathbf{p}'\mathbf{q}'s} U(\mathbf{q}')[C^{\dagger}_{\mathbf{p}'+\mathbf{q}'s}C_{\mathbf{p}'s}C^{\dagger}_{\mathbf{p}+\mathbf{q}\sigma}C_{\mathbf{p}\sigma} - C^{\dagger}_{\mathbf{p}+\mathbf{q}\sigma}C_{\mathbf{p}\sigma}C^{\dagger}_{\mathbf{p}'+\mathbf{q}'s}C_{\mathbf{p}'s}] \tag{6.238}$$

The commutator is evaluated in the usual manner by moving all raising operators to the left. The terms with four operators cancel and the answer is the delta functions derived from the commutes:

$$\begin{aligned}[U, C^{\dagger}_{\mathbf{p}+\mathbf{q}\sigma}C_{\mathbf{p}\sigma}] &= \frac{1}{\Omega}\sum_{\mathbf{p}'\mathbf{q}'s} U(\mathbf{q}')\{C^{\dagger}_{\mathbf{p}'+\mathbf{q}'s}[\delta_{\sigma s}\delta_{\mathbf{p}'=\mathbf{p}+\mathbf{q}} - C^{\dagger}_{\mathbf{p}+\mathbf{q}\sigma}C_{\mathbf{p}'s}]C_{\mathbf{p}\sigma} \\ &\quad - C^{\dagger}_{\mathbf{p}+\mathbf{q}\sigma}[\delta_{\sigma s}\delta_{\mathbf{p}=\mathbf{p}'+\mathbf{q}'} - C^{\dagger}_{\mathbf{p}'+\mathbf{q}'s}C_{\mathbf{p}\sigma}]C_{\mathbf{p}'s}\} \\ &= \frac{1}{\Omega}\sum_{\mathbf{q}'} U(\mathbf{q}')[C^{\dagger}_{\mathbf{p}+\mathbf{q}+\mathbf{q}'\sigma}C_{\mathbf{p}\sigma} - C^{\dagger}_{\mathbf{p}+\mathbf{q}\sigma}C^{\dagger}_{\mathbf{p}-\mathbf{q}'\sigma}]\end{aligned} \tag{6.239}$$

This expression has as its largest term $\mathbf{q}'=-\mathbf{q}$, and the other terms are ignored. It is this approximation that generated the name *random phase approximation*. Actually, phases do not have anything to do with it:

$$[U, C^{\dagger}_{\mathbf{p}+\mathbf{q}\sigma}C_{\mathbf{p}\sigma}] \approx \frac{1}{\Omega}U(-\mathbf{q})[n_{\mathbf{p}\sigma} - n_{\mathbf{p}+\mathbf{q},\sigma}] \tag{6.240}$$

where $U(-\mathbf{q}) = U(\mathbf{q})$

- The next term in the Hamiltonian is V_{ee} from electron–electron interactions. The random phase approximation (RPA) is derived ignoring this interaction. In the next section, this term is included, and it leads to the local field corrections.

The terms derived so far can be collected together:

$$-\frac{i}{\hbar}[\hbar\omega + \xi(\mathbf{p}+\mathbf{q}) - \xi(\mathbf{p})]C^{\dagger}_{\mathbf{p}+\mathbf{q}\sigma}C_{\mathbf{p}\sigma} = \frac{i}{\hbar\Omega}U(\mathbf{q})[n_{\mathbf{p}\sigma} - n_{\mathbf{p}+\mathbf{q},\sigma}] \tag{6.241}$$

and solved for the product of operators:

$$C^{\dagger}_{\mathbf{p}+\mathbf{q}\sigma}C_{\mathbf{p}\sigma} = -\frac{U(\mathbf{q})}{\Omega}\frac{n_{\mathbf{p}\sigma} - n_{\mathbf{p}+\mathbf{q},\sigma}}{\hbar\omega + \xi(\mathbf{p}+\mathbf{q}) - \xi(\mathbf{p})} \tag{6.242}$$

The change in electron density caused by the impurity is,

$$\delta\rho(\mathbf{q},\omega) = \sum_{\mathbf{p}\sigma} C^{\dagger}_{\mathbf{p}+\mathbf{q}\sigma} C_{\mathbf{p}\sigma} \tag{6.243}$$

$$= U(\mathbf{q})\, P(\mathbf{q},\omega)$$

$$P(\mathbf{q},\omega) = \frac{1}{\Omega}\sum_{\mathbf{p}\sigma} \frac{n_{\mathbf{p}\sigma} - n_{\mathbf{p}+\mathbf{q},\sigma}}{\xi(\mathbf{p}) - \hbar\omega - \xi(\mathbf{p}+\mathbf{q})} \tag{6.244}$$

The function $P(\mathbf{q},\omega)$ is called the *polarization function* of the electron gas. Some of its various properties are discussed below.

The next step is to derive the RPA dielectric function. The impurity potential $U(\mathbf{q})$ causes some electron polarization $\delta\rho(\mathbf{q},\omega)$. This polarization is the screening charge. It is the electron charge that moves toward the impurity (if it is positive), or moves away from it (if it is negative), and reduces the long-range potential. The total potential is $U(\mathbf{q})$ plus that from the charge polarization:

$$U_{\text{eff}}(\mathbf{q},\omega) = U(\mathbf{q}) + U_{\text{el}}(\mathbf{q},\omega) \tag{6.245}$$

The electronic contribution is from the charge polarization $U_{\text{el}}(\mathbf{q},\omega = v_q\delta\rho(\mathbf{q},\omega)$. Henry Ehrenreich and Morrel Cohen [3] introduced the idea of the *self-consistent field*. The electrons actually respond to $U_{\text{eff}}(q)$ rather than to the bare interaction $U(q)$:

$$U_{\text{el}}(\mathbf{q},\omega) = v_q\delta\rho(\mathbf{q},\omega) = U_{\text{eff}}(\mathbf{q},\omega)\, v_q P(\mathbf{q},\omega) \tag{6.246}$$

Solve for U_{eff} by combining the above two equations:

$$U_{\text{eff}} = \frac{U(\mathbf{q})}{1 - v_q P(\mathbf{q},\omega)} \tag{6.247}$$

$$= \frac{v_q\varepsilon_0}{\varepsilon(\mathbf{q},\omega)}, \quad \varepsilon(q,\omega)/\varepsilon_0 = 1 - v_q P(\mathbf{q},\omega)$$

The last equation is the RPA dielectric function.

6.4.2 *Properties of* P(*q*, *ω*)

The polarization function and the dielectric function are evaluated in several limits.

- Examine the limit of large frequency. In eqn. (6.244) rewrite the factor $n_{\mathbf{p}+\mathbf{q}\sigma}$ by changing summation variables to $\mathbf{p}+\mathbf{q}\to -\mathbf{p}$, which gives

$$P(\mathbf{q},\omega) = -\frac{1}{\Omega}\sum_{\mathbf{p}\sigma} n_{\mathbf{p}\sigma}\left[\frac{1}{\hbar\omega + \xi(\mathbf{p}+\mathbf{q}) + \xi(\mathbf{p})} - \frac{1}{\hbar\omega - \xi(\mathbf{p}+\mathbf{q}) + \xi(\mathbf{p})}\right] \tag{6.248}$$

Expand the denominators, assuming that $\hbar\omega \gg [\xi(\mathbf{p}+\mathbf{q}) - \xi(\mathbf{p})]$, and keep the first nonvanishing term, which goes as $O(1/\omega^2)$:

$$P(\mathbf{q},\omega) = \frac{2}{(\hbar\omega)^2\Omega}\sum_{\mathbf{p}\sigma} n_{\mathbf{p}\sigma}[\xi(\mathbf{p}+\mathbf{q}) - \xi(\mathbf{p})] + O(1/\omega^4) \tag{6.249}$$

$$= \frac{1}{m(\omega)^2\Omega}\sum_{\mathbf{p}\sigma} n_{\mathbf{p}\sigma}\left[q^2 + 2\mathbf{q}\cdot\mathbf{p}\right] + O(1/\omega^4)$$

$$= \frac{n_e q^2}{m\omega^2} + O(1/\omega^4)$$

The term with q^2 gives the number of electrons ($n_e = N_e/\Omega$), while the term with vector **p** averages to zero when doing the vector integrals:

$$\lim_{\omega\to\infty} \varepsilon(q,\omega)/\varepsilon_0 = 1 - v_q\frac{n_e q^2}{m\omega^2} = 1 - \frac{\omega_p^2}{\omega^2} \tag{6.250}$$

The dielectric function becomes the classical formula in terms of the plasma frequency. A careful treatment shows that this limit is valid whenever $\omega \gg qv_F$, where v_F is the Fermi velocity.

- The other limit has $\omega < qv_F$, which is usually taken to be small frequency. Just set $\omega = 0$. This static dielectric function is

$$\varepsilon(q,0)/\varepsilon_0 = 1 - v_q\frac{1}{\Omega}\sum_{\mathbf{p}\sigma}\frac{n_{\mathbf{p}\sigma} - n_{\mathbf{p}+\mathbf{q},\sigma}}{\xi(\mathbf{p}) - \xi(\mathbf{p}+\mathbf{q})} \tag{6.251}$$

For any dispersion, write at small wave vector

$$\xi(\mathbf{p}+\mathbf{q}) \approx \xi(\mathbf{p}) + \hbar\mathbf{q}\cdot\mathbf{v}(\mathbf{p}) + O(q^2), \quad \hbar\mathbf{v}(\mathbf{p}) = \nabla_{\mathbf{p}}\xi(\mathbf{p}) \tag{6.252}$$

$$\xi(\mathbf{p}+\mathbf{q}) - \xi(\mathbf{p}) \approx \hbar\mathbf{q}\cdot\mathbf{v}(\mathbf{p}) \tag{6.253}$$

Similarly, the numerator of $P(\mathbf{q}, 0)$ is

$$[n_{\mathbf{p}\sigma} - n_{\mathbf{p}+\mathbf{q},\sigma}] \approx n_F[\xi(\mathbf{p})] - n_F[\xi(\mathbf{p}) + \hbar\mathbf{v}\cdot\mathbf{q}] \tag{6.254}$$

$$\approx -\hbar\mathbf{q}\cdot\mathbf{v}(\mathbf{p})\left(\frac{dn_F(\xi)}{d\xi}\right)$$

In the summation over $\mathbf{p}\sigma$ in the definition of $\varepsilon(\mathbf{q}, 0)$, the factor of $\hbar\mathbf{q}\cdot\mathbf{v}$ cancels in numerator and denominator:

$$\varepsilon(q,0)/\varepsilon_0 = 1 + v_q\frac{1}{\Omega}\sum_{\mathbf{p}\sigma}\left(-\frac{dn_F(\xi)}{d\xi}\right)_{\xi(\mathbf{p})} \tag{6.255}$$

At low temperatures, the factor in parentheses becomes a delta function, and the integral is just the density of states $N(\xi)$ of the metal evaluated at the chemical potential $\xi = \mu$:

$$\left(-\frac{dn_F(\xi)}{d\xi}\right)_{\xi(\mathbf{p})} = \delta[\xi(\mathbf{p})] \tag{6.256}$$

$$\frac{1}{\Omega}\sum_{\mathbf{p}\sigma}\left(-\frac{dn_F(\xi)}{d\xi}\right)_{\xi(\mathbf{p})} = N(\mu) \tag{6.257}$$

$$\varepsilon(q,0)/\varepsilon_0 = 1 + v_q N(\mu) \tag{6.258}$$

The latter formula is a good approximation at all temperatures where $k_B T \ll \mu$. For the Sommerfeld model of plane waves,

$$N(\mu) = 2\int \frac{d^3p}{(2\pi)^3}\delta\left(\frac{\hbar^2p^2}{2m} - \mu\right) = \frac{mk_F}{\hbar^2\pi^2} \tag{6.259}$$

$$\varepsilon(q, 0)/\varepsilon_0 = 1 + \frac{q_{TF}^2}{q^2} \tag{6.260}$$

$$q_{TF}^2 = \frac{3e^2n_e}{2\varepsilon_0 E_F} \tag{6.261}$$

The result is given in terms of the Thomas-Fermi screening length q_{TF}.

The dielectric function $\varepsilon(\mathbf{q}, \omega)$ has a different form, depending whether $\omega > qv_F$ or $\omega < qv_F$.

In three dimensions, the polarization function can be evaluated analytically at zero temperature for all values of wave vector and frequency. Again, start with the expression

$$P(\mathbf{q}, \omega) = -\frac{1}{\Omega}\sum_{\mathbf{p}\sigma} n_{\mathbf{p}\sigma}\left[\frac{1}{\hbar\omega + \xi(\mathbf{p}+\mathbf{q}) - \xi(\mathbf{p})} - \frac{1}{\hbar\omega - \xi(\mathbf{p}+\mathbf{q}) + \xi(\mathbf{p})}\right] \tag{6.262}$$

$$= -2\int_{p<k_F} \frac{d^3p}{(2\pi)^3}\left[\frac{1}{\hbar\omega + \varepsilon(q) + \hbar\mathbf{p}\cdot\mathbf{v}_q} - \frac{1}{\hbar\omega - \varepsilon(q) - \hbar\mathbf{p}\cdot\mathbf{v}_q}\right]$$

where $\varepsilon(q) = \hbar^2q^2/2m$ and $\mathbf{v}_q = \hbar\mathbf{q}/m$. Choose as the $\hat{z}$ direction $\hat{q}$. Then $\mathbf{p}\cdot\mathbf{v}_q = p\nu v_q$, where $\nu = \cos(\theta)$. The integration variables are $d^3p = 2\pi d\nu p^2dp$ and the angular integral gives

$$\int_{-1}^{1} d\nu\left[\frac{1}{\hbar\omega + \varepsilon(q) + \hbar p\nu v_q} - \frac{1}{\hbar\omega - \varepsilon(q) - \hbar p\nu v_q}\right] \tag{6.263}$$

$$= \frac{1}{\hbar pv_q}\left[\ln\left|\frac{\hbar\omega + \varepsilon(q) + \hbar pv_q}{\hbar\omega + \varepsilon(q) - \hbar pv_q}\right| + \ln\left|\frac{\hbar\omega - \varepsilon(q) - \hbar pv_q}{\hbar\omega - \varepsilon(q) + \hbar pv_q}\right|\right]$$

The polarization function is

$$P(\mathbf{q}, \omega) = -\frac{1}{2\pi^2\hbar v_q}\int_0^{k_F} pdp\left[\ln\left|\frac{\hbar\omega + \varepsilon(q) + \hbar pv_q}{\hbar\omega + \varepsilon(q) - \hbar pv_q}\right| + \ln\left|\frac{\hbar\omega - \varepsilon(q) - \hbar pv_q}{\hbar\omega - \varepsilon(q) + \hbar pv_q}\right|\right] \tag{6.264}$$

The wave vector integral can be evaluated, yielding the final expression:

$$\varepsilon(q, \omega)/\varepsilon_0 = 1 + \frac{q_{TF}^2}{2q^2}\left\{1 + \frac{m^2}{2\hbar^4k_Fq^3}[F(q, \omega) + F(q, -\omega)]\right\} \tag{6.265}$$

$$F(q, \omega) = [4E_F\varepsilon(q) - (\varepsilon(q) - \hbar\omega)^2]\ln\left|\frac{\varepsilon(q) + \hbar(qv_F + \omega)}{\varepsilon(q) - \hbar(qv_F - \omega)}\right| \tag{6.266}$$

An important limit is when $\omega = 0$:

$$\varepsilon(q, 0)/\varepsilon_0 = 1 + \frac{q_{TF}^2}{2q^2}\left[1 + \frac{1}{4k_Fq}(4k_F^2 - q^2)\ln\left|\frac{2k_F + q}{2k_F - q}\right|\right] \tag{6.267}$$

When $q \ll k_F$ the bracket is two, and one gets the earlier result of Thomas-Fermi.

6.4.3 Hubbard-Singwi Dielectric Functions

John Hubbard [7] was one of the first theorists to consider how electron–electron interactions influence the dielectric function. He derived a formula that is now called the Hubbard dielectric function:

$$\varepsilon_H(q,\omega)/\varepsilon_0 = 1 - \frac{v_q P(q,\omega)}{1 + G_H(q) v_q P(q,\omega)} \tag{6.268}$$

In different published papers he suggested different forms for the Hubbard function $G_H(q)$, such as

$$\begin{aligned} G_H(q) &= \frac{1}{2}\frac{q^2}{q^2 + k_F^2} \\ &= \frac{1}{2}\frac{q^2}{q^2 + q_{\mathrm{TF}}^2} \\ &= \frac{1}{2}\frac{q^2}{q^2 + k_F^2 + q_{\mathrm{TF}}^2} \end{aligned} \tag{6.269}$$

Dielectric functions today are written precisely in the Hubbard form. Much effort has gone into choosing the best form for $G_H(q)$. Now it is known that the third choice, of the three above, is probably the best. The "H" subscript on $G_H(q)$ and ε_H is dropped, since everyone uses this same form.

Note the close analogy to the Lorenz-Lorentz form for the dielectric function. For $\omega = 0$, then $P(q,0) < 0$, so that $v_q P(q,0)$ is equivalent to $-4\pi\tilde{\alpha}$. Then, to complete the analogy, the Hubbard function $G(q)$ is the local field correction, which is $\frac{1}{3}$ for insulators.

Singwi and collaborators [17] derived many properties of $G(q)$. One of their formulas is

$$G(q) = -\frac{1}{n_e}\int \frac{d^3k}{(2\pi)^3}\frac{\mathbf{q}\cdot\mathbf{k}}{k^2}[S(\mathbf{q}-\mathbf{k}) - 1] \tag{6.270}$$

where $S(\mathbf{q}-\mathbf{k})$ is the static structure factor. Using its relation to the pair distribution function $g(r)$ gives

$$\begin{aligned} G(q) &= \frac{i}{4\pi}\int d^3r[g(r) - 1]\left(\frac{\mathbf{q}\cdot\mathbf{r}}{r^3}\right)e^{i\mathbf{q}\cdot\mathbf{r}} \\ &= q\int_0^\infty dr[1 - g(r)]j_1(qr) \end{aligned} \tag{6.271}$$

where $j_1(z)$ is a spherical Bessel function. This formula makes clear that the Hubbard factor $G(q)$ is due to the "hole" around each electron. It is a change in the screening caused by the lack of electrons near the central one. It is indeed a local correction to the electric field and to the dielectric screening.

The function $G(q,\omega)$ depends on frequency as well as wave vector. The frequency dependence is poorly understood and is rarely invoked. One can also show that Hubbard was correct in that at small values of wave vector then $G(q) \propto q^2$. Figure 6.3 shows a measurement

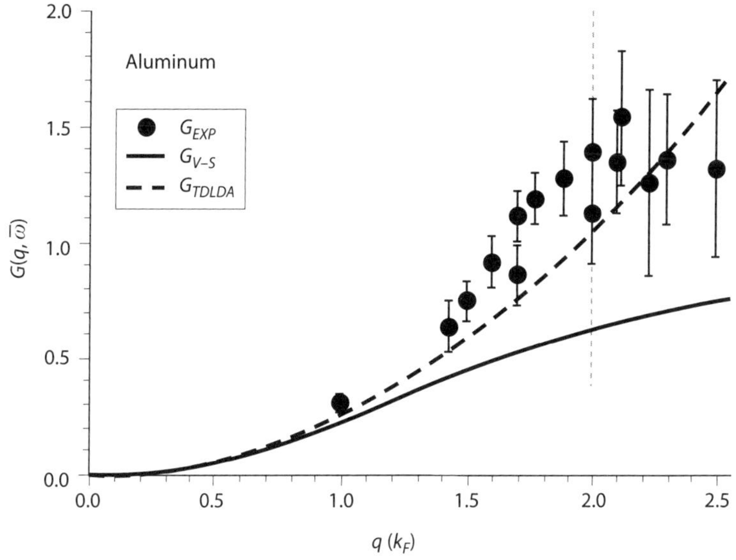

FIGURE 6.3. Measurement of $G(q)$ in metallic aluminum. By Larson et al., *Phys. Rev. Lett.* 77, 1346 (1996). Used with permission of the American Physical Society.

of $G(q)$ in aluminum, using x-ray scattering, by Larson et al. The data points are the circles with error bars. The solid and dashed lines are various theories. These measurements are made at nonzero frequency, whereas most theories are calculated at zero frequency.

6.5 Impurities in Metals

Consider the potential energy $V_s(r)$ from a charged impurity in a metal. The impurity is at an interstitial position and has a charge Z. The subscript s denotes that the Coulomb potential is screened by the dielectric function:

$$V_s(r) = Z\varepsilon_0 \int \frac{d^3q}{(2\pi)^3} e^{i\mathbf{q}\cdot\mathbf{r}} \frac{v_q}{\varepsilon(q)}, \quad v_q = \frac{e^2}{\varepsilon_0 q^2} \tag{6.272}$$

and $\varepsilon(q) = \varepsilon(q, \omega = 0)$ is the static dielectric function. The angular integrals are simple, yielding

$$V_s(r) = \frac{Ze^2}{2\pi^2 r}\int_0^\infty dq \frac{\sin(qr)}{q\varepsilon(q)} \tag{6.273}$$

This integral is simple to evaluate. Using a standard dielectric function, such as RPA, one finds that at large values of r ($k_F r \gg 1$) the potential $V_s(r)$ becomes oscillatory with a dependence $\cos(2k_F r)$. These oscillations are called *Friedel oscillations*, since Friedel first gave a simple explanation. They do not occur if one uses a Thomas-Fermi dielectric function $[\varepsilon(q) = 1 + q_{\mathrm{TF}}^2/q^2]$ but are found using the RPA or Hubbard form. At zero frequency, these functions have a nonanalytic dependence, such as

$$\ln\left|\frac{2k_F - q}{2k_F + q}\right| \tag{6.274}$$

which is the source of the oscillation.

6.5.1 Friedel Analysis

Friedel's derivation puts a single impurity in the center of a spherical metal of radius R. Crystalline effects are ignored and the electrons are treated as free.

Without the impurity, the Hamiltonian for one electron is the wave equation

$$E\psi(\mathbf{r}) = -\frac{\hbar^2\nabla^2}{2m}\psi(\mathbf{r}) \tag{6.275}$$

In spherical coordinates, the solutions are

$$\psi_{k\ell m}(r,\theta,\phi) = j_\ell(kr)\,P_\ell^{|m|}(\theta)e^{im\phi} \tag{6.276}$$

where $j_\ell(z)$ are spherical Bessel functions, and $P_\ell^{|m|}(\theta)$ are associated Legendre functions. Friedel assumed that the wave functions vanish at the surface of the sphere, $j_\ell(kR) = 0$. The asymptotic form of the spherical Bessel function is

$$\lim_{z\gg 1} j_\ell(z) = \frac{1}{z}\sin(z - \ell\pi/2) \tag{6.277}$$

The sine function vanishes at multiples of π, which gives the allowed values of wave vector:

$$n\pi = k_n R - \frac{\ell\pi}{2} \tag{6.278}$$

$$k_n = \frac{\pi}{R}\left(n + \frac{\ell}{2}\right) \tag{6.279}$$

The value of n at the chemical potential μ is n_F:

$$\mu = \frac{\hbar^2 k_F^2}{2m} = \frac{\pi^2\hbar^2}{2mR^2}\left(n_F + \frac{\ell}{2}\right)^2 \tag{6.280}$$

Repeat this calculation when there is an impurity. The Hamiltonian is now

$$E\psi(\mathbf{r}) = \left[-\frac{\hbar^2\nabla^2}{2m} + V_s(r)\right]\psi(\mathbf{r}) \tag{6.281}$$

This differential equation can be solved on the computer. Since the potential energy is spherically symmetric, the eigenfunctions can be written as

$$\psi_{k\ell m}(r,\theta,\phi) = R_\ell(kr)\,P_\ell^{|m|}(\theta)e^{im\phi} \tag{6.282}$$

$$\lim_{z\gg 1} R_\ell(z) = \frac{1}{z}\sin[z + \delta_\ell(k) - \ell\pi/2] \tag{6.283}$$

The radial function is now $R_\ell(kr)$. At large distances it can also be expressed as a sine function, whose argument contains the phase shift $\delta_\ell(kr)$. The function $R_\ell(kR)$ vanishes at the surface of the sphere. The wave vectors obey the eigenvalue equation

$$k_n R + \delta_\ell(k_n) = \pi\left[n + \frac{\ell}{2}\right] \tag{6.284}$$

How does the impurity change n_F, the maximum allowed value of n for occupied states? Differentiate the above equation with respect to wave vector:

$$\frac{dn}{dk} = \frac{R}{\pi} + \frac{d}{dk}\frac{\delta_\ell(k)}{\pi} \tag{6.285}$$

The first term on the right is the derivative if there is no impurity. The change in occupation number δn due to the impurity is from the second term:

$$\frac{d}{dk}\delta n = \frac{d}{dk}\frac{\delta_\ell(k)}{\pi} \tag{6.286}$$

$$\delta n_F = \frac{\delta_\ell(k_F)}{\pi} \tag{6.287}$$

The change in the number of states δn_F at the chemical potential is given by the phase shift divided by π. The total change in the number of electron states is obtained by summing over

- angular momentum ℓ
- the magnetic quantum number $-\ell \le m \le \ell$
- spin degeneracy, which is two

$$\delta N_F = 2\sum_{\ell, m} n_F = \frac{2}{\pi}\sum_{\ell=0}^{\infty}(2\ell + 1)\delta_\ell(k_F) \tag{6.288}$$

The phase shifts do not depend on m in a nonmagnetic metal, so their summation gives $(2\ell + 1)$.

What is the meaning of δN_F? It is the total change in the number of electrons because of the impurity. If the impurity is a cation of $Z = +2$, then two electrons are attracted to the impurity as part of its screening charge, and $\delta N_F = 2$. If the impurity is a halide ion, with $Z = -1$, one electron must leave the vicinity of the impurity to achieve charge neutrality. In general, $\Delta N_F = Z$, which gives *the Friedel sum rule*:

$$Z = 2\sum_{\ell, m} n_F = \frac{2}{\pi}\sum_{\ell=0}^{\infty}(2\ell + 1)\delta_\ell(k_F) \tag{6.289}$$

It provides an important constraint on the value of the phase shifts due to impurity scattering.

The change in electron charge are not in bound states. If the impurity has $Z > 0$, and attracts electrons, then every conduction electron spends part of its motion near the impurity. Averaging the motion of all of these electrons gives a total extra charge of Z near to the impurity. Charged impurities in a metal often do not have bound states, since the screening weakens the potential energy.

Another important quantity is the change in ground-state energy of the electrons due to the impurity. The change in energy for one set of (n, ℓ, m) is

$$\delta E = \frac{\pi^2\hbar^2}{2mR^2}\left[\left(n+\frac{\ell}{2}-\frac{\delta_\ell}{\pi}\right)^2-\left(n+\frac{\ell}{2}\right)^2\right] \tag{6.290}$$

$$\approx -\frac{\pi\hbar^2}{mR^2}\delta_\ell\left(n+\frac{\ell}{2}\right) = -\frac{\hbar^2}{mR}k_n\delta_\ell(k_n)$$

Now sum over (i) n, (ii) ℓ, (iii) m, and (iv) spin. The summation over n is changed to an integral over wave a vector using $dn = Rdk/\pi$:

$$\frac{1}{R}\sum_n f(k_n) = \int\frac{dk}{\pi}f(k) \tag{6.291}$$

and the change in ground-state energy is

$$\delta E = -\frac{2\hbar^2}{\pi m}\sum_\ell(2\ell+1)\int_0^{k_F} kdk\delta_\ell(k) \tag{6.292}$$

This expression was derived by Fumi [4]. All of the details of the screened potential are contained in the phase shifts.

The final derivation is of the charge density. The first step is to normalize the eigenfunction in spherical coordinates:

$$\int d^3r|\psi_{n\ell m}(\mathbf{r})|^2 = 1 = A^2\int_0^R r^2drj_\ell^2(k_nr)\int d\phi\sin(\theta)\,d\theta|Y_\ell^m(\theta,\phi)|^2 \tag{6.293}$$

$$\int d\phi\sin(\theta)\,d\theta|Y_\ell^m(\theta,\phi)|^2 = 1 \tag{6.294}$$

$$\int_0^R r^2drj_\ell^2(k_nr) = \frac{R}{2k_n^2} \tag{6.295}$$

$$A^2 = \frac{2k_n^2}{R} \tag{6.296}$$

The evaluation of the last integral uses the fact that $j_\ell(k_nR) = 0$. The normalization A is known, and the density is

$$n(r) = 2\sum_{n\ell m}|\psi_{n\ell m}(\mathbf{r})|^2 \tag{6.297}$$

where the factor of two is for spin degeneracy in a paramagnetic electron gas. As a check, evaluate this expression for the system with no impurity, so that

$$|\psi_{n\ell m}(r)|^2 = A^2j_\ell^2(k_nr)|Y_\ell^m(\theta,\phi)|^2 \tag{6.298}$$

$$\sum_m|Y_\ell^m(\theta,\phi)|^2 = \frac{2\ell+1}{4\pi} \tag{6.299}$$

$$\sum_{\ell} (2\ell+1) j_{\ell}^2(k_n r) = 1 \tag{6.300}$$

$$\sum_n k_n^2 = \frac{R}{\pi}\int_0^{k_F} k_n^2 dk_n = \frac{Rk_F^3}{3\pi} \tag{6.301}$$

$$n(\mathbf{r}) = \frac{4}{R}\frac{1}{4\pi}\frac{Rk_F^3}{3\pi} = \frac{k_F^3}{3\pi^2} = n_e \tag{6.302}$$

The electron density in the ground state is a constant n_e.

The density changes due to the impurity. It is easiest to calculate the change in density:

$$\delta n(\mathbf{r}) = 2A^2 \sum_{n\ell m} |Y_{\ell}^m(\theta,\phi)|^2 [R_{\ell}^2(k_n r) - j_{\ell}^2(k_n r)] \tag{6.303}$$

The summation over m is done using eqn. (6.299). The summation over n is changed to an integral $\int dk_n$. The radial eigenfunctions are evaluated in the region where $k_n r \gg 1$, and the asymptotic form is valid:

$$\delta n(\mathbf{r}) = \frac{1}{\pi^2 r^2}\sum_{\ell}(2\ell+1)\int_0^{k_F} dk_n \{\sin^2[k_n r + \delta_{\ell} - \ell\pi/2] - \sin^2[k_n r - \ell\pi/2]\} \tag{6.304}$$

Use the fact that $\sin^2(\alpha) = [1-\cos(2\alpha)]/2$ to rewrite the above formula as

$$\begin{aligned}
\delta n(\mathbf{r}) &= -\frac{1}{2\pi^2 r^2}\sum_{\ell}(2\ell+1)\int_0^{k_F} dk_n \{\cos[2k_n r + 2\delta_{\ell} - \ell\pi] - \cos[2k_n r - \ell\pi]\} \\
&= -\frac{1}{2\pi^2 r^2}\sum_{\ell}(2\ell+1)(-1)^{\ell}\int_0^{k_F} dk_n \{\cos[2k_n r + 2\delta_{\ell}(k_n)] - \cos[2k_n r]\} \\
&= \frac{1}{\pi^2 r^2}\sum_{\ell}(2\ell+1)(-1)^{\ell}\int_0^{k_F} dk_n \sin[\delta_{\ell}(k_n)]\sin[2k_n r + \delta_{\ell}(k_n)]
\end{aligned} \tag{6.305}$$

Further progress in evaluating this expression cannot be obtained without knowing how the phase shifts $\delta_{\ell}(k)$ depend on wave vector. Some examples are worked out in the homework. Clearly, the change in charge density $\delta n(\mathbf{r})$ varies with distance in an oscillatory manner. The oscillations are spherically symmetric. These oscillations can be observed in three dimensions by several methods. One is small-angle neutron scattering, while another uses nuclear magnetic resonance.

The most direct observation of Friedel oscillations is on the surface of metals. Figure 6.4 shows two measurements by Ward Plummer's group of Friedel oscillations on metal surfaces using scanning tunneling microscopes. Figure 6.4a shows the (111) surface of copper. The surface has adatoms of copper, and around each adatom are circular rings, which are Friedel oscillations in electron density.

Beryllium is an hcp metal with two electrons per atom. They would fill the conduction band, and Be would be an insulator, except for small overlaps between the conduction and valence bands, which make it a semimetal. However, on the (0001) surface, which is perpendicular to the *c*-axis, there is a large surface-state density of electrons. The surface acts very much like a two-dimensional free electron system, with negligible interaction with the three-dimensional band states of the bulk material. A surface tunneling microscope

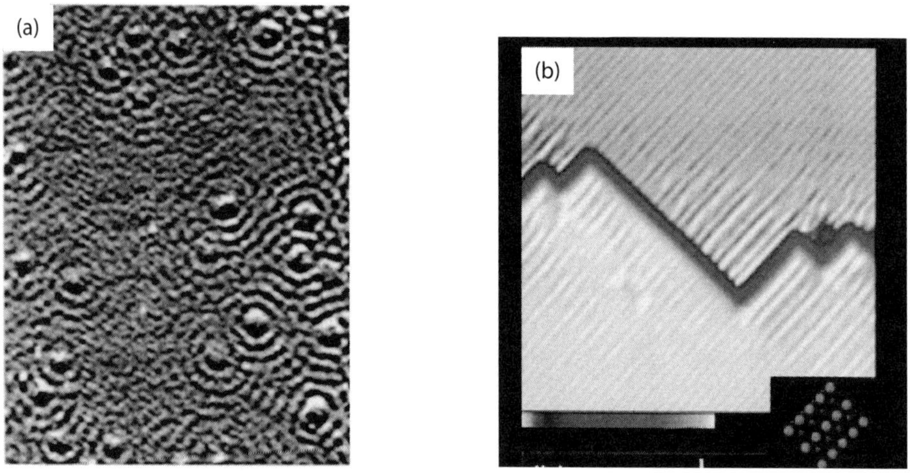

FIGURE 6.4. (a) Friedel oscillations on the (111) surface of Cu. From Petersen et al., *Phys. Rev. B* **57**, R6858 (1998). (b) Friedel oscillations in electrons density on the (0001) surface of Be near to steps. From Hofmann et al., *Phys. Rev. Lett.* **79**, 265 (1997). Both figures with permission of the American Physical Society.

(STM) of this surface observes not atoms, but the high density electron gas. Everything shows Friedel oscillations. There are circular density oscillations around isolated surface adatoms, and linear oscillations near to linear steps in atom height. Figure 6.4b shows linear oscillations adjacent to steps on the surface. The insert in the lower right shows the arrangement of surface atoms of Be.

6.5.2 RKKY Interaction

Ruderman and Kittel [15] suggested another oscillatory interaction in metals. In this case it is the spin density that is oscillating, not the charge density. Ruderman and Kittel suggested this mechanism could provide a long-range interaction between nuclear spins in metals. Following the Ruderman-Kittel paper, Kasuya [8] and Yosida [18] extended the theory to include the long-range interaction between the electronic magnetic moments, and the combined result is now called RKKY. A neutral impurity could have a magnetic moment with a local spin **S**. Magnetic moments are found in atoms from the transition metal series with unfilled *d*-shells, and from the rare-earth series with unfilled *f*-shells. The interaction is important when there are magnetic impurities in metals or in semiconductors: e.g., Mn in GaAs or Mn in Cu.

The conduction electrons have a spin $\mathbf{s}_i$ that is either up or down. The magnetic interaction between the local spin at $\mathbf{R}_0$ and the conduction spins is

$$V_{sd} = \frac{J}{\hbar^2}\sum_i \delta^3(\mathbf{r}_i - \mathbf{R}_0)\, \mathbf{S}\cdot\mathbf{s}_i \tag{6.306}$$

where the coupling constant J has dimensions of energy times volume. The notation V_{sd} is from the days when this interaction was applied to transition metal impurities in copper. The vector product of two spins is

$$\mathbf{S}\cdot\mathbf{s}_i = S_x s_{ix} + S_y s_{iy} + S_z s_{iz} \tag{6.307}$$

$$= S_z s_{iz} + \frac{1}{2}\left[S^{(+)}s_i^{(-)} + S^{(-)}s_i^{(+)}\right]$$

Denote a conduction band state as $|\mathbf{k},\uparrow\rangle$ or $|\mathbf{k}\downarrow\rangle$. The local spin has a state $|S, m_S\rangle$, where $-S < m_S < S$. Evaluate the eigenfunction of the conduction electrons using first-order perturbation theory:

$$|\mathbf{k}\uparrow\rangle = |\mathbf{k}\uparrow\rangle_0 + \sum_{\mathbf{k}',s} |\mathbf{k}'s\rangle \frac{\langle \mathbf{k}'s| V_{sd}|\mathbf{k}\uparrow\rangle_0}{E_c(\mathbf{k}) - E_c(\mathbf{k}')} \tag{6.308}$$

$$= |\mathbf{k}\uparrow\rangle_0 + \frac{J}{2\hbar\Omega^{3/2}} e^{i\mathbf{k}\cdot\mathbf{R}_0} \sum_{\mathbf{k}',s} \frac{e^{i\mathbf{k}'\cdot(\mathbf{r}-\mathbf{R}_0)}\chi_s}{E_c(\mathbf{k}) - E_c(\mathbf{k}')}\left[\langle s|\uparrow\rangle S_z + \langle s|\downarrow\rangle S^{(+)}\right]$$

$$|\mathbf{k}\downarrow\rangle = |\mathbf{k}\downarrow\rangle_0 + \frac{J}{2\hbar\Omega^{3/2}} e^{i\mathbf{k}\cdot\mathbf{R}_0} \sum_{\mathbf{k}',s} \frac{e^{i\mathbf{k}'\cdot(\mathbf{r}-\mathbf{R}_0)}\chi_s}{E_c(\mathbf{k}) - E_c(\mathbf{k}')}\left[-\langle s|\downarrow\rangle S_z + \langle s|\uparrow\rangle S^{(-)}\right] \tag{6.309}$$

The charge density for each spin configuration is

$$\rho_\sigma(\mathbf{r}) = \frac{1}{\Omega}\sum_{\mathbf{k}} n_{\mathbf{k}} \langle \mathbf{k}\sigma|\mathbf{k}\sigma\rangle \tag{6.310}$$

We retain only the first cross term: those to $O(J)$. The product $\langle \mathbf{k}\sigma|\mathbf{k}\sigma\rangle$ forces $s = \uparrow$ for spin-up, and $s = \downarrow$ for spin-down. The spin-flip terms do not contribute. The local spin has $\langle S_z\rangle = \hbar m_S$. The net spin polarization is

$$m(\mathbf{r}) = \rho_\uparrow(\mathbf{r}) - \rho_\downarrow(\mathbf{r}) = \frac{2m_S J}{\Omega^2}\sum_{\mathbf{k}} n_k e^{-i\mathbf{k}\cdot(\mathbf{r}-\mathbf{R}_0)} \sum_{\mathbf{k}'} \frac{e^{i\mathbf{k}'\cdot(\mathbf{r}-\mathbf{R}_0)}\chi_s}{E_c(\mathbf{k}) - E_c(\mathbf{k}')} \tag{6.311}$$

$$n(\mathbf{r}) = \rho_\uparrow(\mathbf{r}) + \rho_\downarrow(\mathbf{r}) = n_e + O(J^2) \tag{6.312}$$

The spin density $m(\mathbf{r})$ has oscillations to order $O(J)$. The charge density $n(\mathbf{r})$ does not have a term of order $O(J)$, but does have Friedel oscillations, which are not shown. Assume quadratic dispersion $E_c(\mathbf{k}) = \hbar^2k^2/2m$. The first integral is a simple Green's function:

$$\frac{1}{\Omega}\sum_{\mathbf{k}'} \frac{e^{i\mathbf{k}'\cdot(\mathbf{r}-\mathbf{R}_0)}\chi_s}{E_c(\mathbf{k}) - E_c(\mathbf{k}')} = \frac{2m}{\hbar^2}\frac{\cos[k(\mathbf{r}-\mathbf{R}_0)}{4\pi|\mathbf{r}-\mathbf{R}_0|} \tag{6.313}$$

Let $\tau = |\mathbf{r} - \mathbf{R}_0|$. The second integral is done in spherical coordinates:

$$2\pi\int_0^{k_F} k^2 dk \cos(k\tau)\int_0^{\pi} \sin(\theta)\, d\theta \exp[-ik\tau\cos(\theta)] \tag{6.314}$$

$$= \frac{4\pi}{\tau}\int_0^{k_F} kdk\cos(k\tau)\sin(k\tau) = \frac{2\pi}{\tau}\int_0^{k_F} kdk\sin(2k\tau)$$

$$= \frac{\pi k_F}{\tau^2}\left[\frac{\sin(2k_F\tau)}{2k_F\tau} - \cos(2k_F\tau)\right]$$

$$m(\mathbf{r}) = \frac{mJm_s k_F}{\hbar^2(2\pi)^3}\frac{1}{\tau^3}\left[\frac{\sin(2k_F\tau)}{2k_F\tau} - \cos(2k_F\tau)\right] \tag{6.315}$$

$$= n_e \frac{3Jm_S mk_F}{\pi\hbar^2} F(2k_F r) = n_e^2 \frac{9\pi Jm_S}{2E_F} F(2k_F r)$$

$$F(\theta) = \frac{1}{\theta^4}[\sin(\theta) - \theta\cos(\theta)], \quad n_e = \frac{k_F^3}{3\pi^2} \tag{6.316}$$

At large distances from the magnetic impurity, the spin density oscillates as $\cos(2k_F\tau)/\tau^3$, where $\tau = |\mathbf{r} - \mathbf{R}_0|$. The magnitude of the oscillation depends on the magnetic quantum number m_S of the local spin. Another impurity located at $\mathbf{R}_1$ will change its energy because of this polarization, and cause an effective interaction between the two magnetic impurities. Local impurities can cause both charge density oscillations and spin density oscillations, in the vicinity of the impurity.

References

1. S. Baroni, P. Giannozzi, and A. Testa, Green's function approach to linear response in solids. *Phys. Rev. Lett.* **58**, 1861–1864 (1987)
2. R. G. Gordon and Y. S. Kim, Theory for the forces between closed-shell atoms and molecules. *J. Chem. Phys.* **56**, 3122–3133 (1972)
3. H. Ehrenreich and M. H. Cohen, Self-consistent field approach to the many-electron problem. *Phys. Rev.* **115**, 786–790 (1959)
4. F. G. Fumi, Vacancies in monovalent metals. *Philos. Mag.* **46**, 1007–1020 (1955)
5. Ph. Hofmann, B. G. Briner, M. Doering, H. P. Rust, E. W. Plummer, and A. M. Bradshaw, Anisotropic Two-Dimensional Friedel Oscillations. *Phys. Rev. Lett.* **79**, 265–268 (1997)
6. P. Hohenberg and W. Kohn, Inhomogeneous electron gas. *Phys. Rev.* **136**, B864–B871 (1964)
7. J. Hubbard, Description of collective motions in terms of many body perturbation theory. *Proc. R. Soc. London Series A* **243**, 336–352 (1957)
8. T. Kasuya, A theory of metallic ferro- and antiferromagnetism on Zener's model. *Prog. Theor. Phys. (Kyoto)* **16**, 45–57 (1956)
9. W. Kohn and L. J. Sham, Self-consistent equations including exchange and correlation effects. *Phys. Rev.* **140**, A1133–A1138 (1965)
10. B. C. Larson, J. Z. Tischler, E. D. Isaacs, A. Fleszar, and A. G. Eguiluz, Inelastic X-ray scattering as a probe of many-body local field factor in metals. *Phys. Rev. Lett.* **77**, 1346–1349 (1996)
11. G. D. Mahan and K. R. Subbaswamy, *Local Density Theory of Polarizability* (Plenum, 1990)
12. P. Nozieres and D. Pines, Correlation energy of a free electron gas. *Phys. Rev.* **111**, 442–454 (1958)
13. J. P. Perdew and A. Zunger, Self-interaction correction to density–functional approximations for many-electron systems. *Phys. Rev. B* **23**, 5048–5079 (1981)
14. L. Petersen, et al., Direct imaging of the two-dimensional Fermi contour: Fourier transform of STM. *Phys. Rev. B* **57**, R6858–R6861 (1998)
15. M. A. Ruderman and C. Kittel, Indirect exchange coupling of nuclear magnetic moments by conduction electrons. *Phys. Rev.* **96**, 99–102 (1954)
16. J. C. Slater, A simplification of the Hartree-Fock method. *Phys. Rev.* **81**, 385–390 (1951)
17. P. Vashishta and K. S. Singwi, Electron correlations at metallic densities. *Phys. Rev. B* **6**, 875–887 (1972)
18. K. Yosida, Magnetic properties of Cu–Mn alloys. *Phys. Rev.* **106**, 893–898 (1957)

Homework

1. The current operator in quantum mechanics is

$$\mathbf{j}(\mathbf{r}) = \frac{e\hbar}{2mi}\left[\psi^{\dagger}(\mathbf{r})\nabla\psi(\mathbf{r}) - \nabla\psi^{\dagger}(\mathbf{r})\psi(\mathbf{r})\right] \tag{6.317}$$

Write out this operator in terms of creation and destruction operators. What is the expression when the eigenstates are plane waves?

2. Explicitly verify the equation of continuity,

$$\frac{\partial}{\partial t}\rho(\mathbf{r}, t) + \nabla\cdot\mathbf{j}(\mathbf{r}, t) = 0 \tag{6.318}$$

for a gas of free fermion particles that have

$$H = \sum_{\mathbf{k}\sigma} \varepsilon_{\mathbf{k}} c^{\dagger}_{ks} c_{ks} \tag{6.319}$$

$$\rho(\mathbf{r}, t) = \frac{1}{\Omega}\sum_{\mathbf{kq}\sigma} e^{-iq\cdot r} c^{\dagger}_{k+q,s} c_{ks} \tag{6.320}$$

$$\mathbf{j}(\mathbf{r}, t) = \frac{\hbar}{m\Omega}\sum_{\mathbf{kq}\sigma} e^{-iq\cdot r}\left(\mathbf{k} + \frac{1}{2}\mathbf{q}\right) c^{\dagger}_{k+q,s} c_{ks} \tag{6.321}$$

3. Repeat the Wigner-Seitz calculation for the energy in a spherical W-S cell using the Heine potential

$$V_{ei} = A, \quad r < r_i a_0 \tag{6.322}$$

$$= -\frac{Ze^2}{r}, \quad r_i a_0 < r < r_s a_0 \eta$$

where r_i is the ion radius in atomic units.

4. The Wigner-Seitz model for metallic hydrogen has the proton in the center of a sphere of uniform negative charge. The potential energy on the proton for a displacement r from the center is

$$V(r) = -\frac{E_{Ry}}{r_s}\left[3 - \frac{r^2}{a_B^2 r_s^2}\right] \tag{6.323}$$

Find the energy of the zero-point motion of the proton for this potential energy, assuming that each proton moves independently of the others.

5. Show that the contribution of the exchange energy to the effective mass $d\Sigma_X/dp$ is divergent as $p \to k_F$.

6. Show that the pair distribution function obeys the sum rule

$$1 = n\int d^3r[1 - g(r)] \tag{6.324}$$

7. For a three-dimensional liquid, consider a pair distribution function of the form:

$$g(r) = 0, \quad r < d \tag{6.325}$$

$$= A, \quad d < r < 2d$$

$$= 1, \quad 2d < r$$

(a) Determine A from the sum rule of the previous problem.

(b) Derive the form of $S(q)$

8. Show that in the Hartree-Fock approximation the pair distribution function is

$$S(q) = 1 - \frac{2}{N}\sum_{\mathbf{k}} n_{\mathbf{k}} n_{\mathbf{k}+\mathbf{q}} \tag{6.326}$$

Evaluate this integral to find $S(q)$.

9. Use the $S(q)$ from the prior problem to calculate the Hubbard $G_H(q)$.

10. (a) Show that in two dimensions the RPA dielectric constant is (A is area)

$$\varepsilon(q,\omega) = 1 - v_q P(q,\omega), \quad v_q = \frac{2\pi e^2}{q} \tag{6.327}$$

$$P(q,\omega) = -\frac{1}{A}\sum_{p,\sigma} \frac{n_{p\sigma} - n_{p+q,\sigma}}{\hbar\omega + \varepsilon_{p+q} - \varepsilon_p} \tag{6.328}$$

(b) Take the limit of large frequency and find the plasmon dispersion relation.

(c) Take the limit of zero frequency and find the expression for $P(q,0)$. Evaluate the integral for this expression for $q \ll k_F$.

11. Find the dispersion of the plasmon oscillations for a metal whose energy bands are given by a three-dimensional tight-binding model on the sc lattice.

12. Evaluate the Friedel oscillations in density $\delta n(\mathbf{r})$ for two cases:

- The phase shifts δ_ℓ do not depend on wave vector.
- The phase shift $\delta_\ell(k)$ has a linear dependence on wave vector:

$$\delta_\ell(k) = \Delta_\ell k \tag{6.329}$$

13. Do a Friedel analysis of a charged impurity in two dimensions. Find:

(a) The form of the Friedel sum rule.

(b) The change in ground-state energy.

(c) The change in charge density.

14. Consider a particle in one dimension near to the edge of a semi-infinite box. The edge is at $x = 0$, so the eigenfunctions are of the form $\psi(x) = \sin(kx)$. Evaluate the particle density near the edge of the box by integrating over wave vector up to the Fermi wave vector k_F. These one-dimensional Friedel oscillations are observed near step edges on the surface of Be.

15. In a ferromagnetic metal, the density of up spins is greater than the density of down spins: $n_\uparrow > n_\downarrow$. What is the magnetic susceptibility χ, which is defined as the change in magnetization δM in a small magnetic field B using the Pauli interaction

$$\delta M = \chi B \tag{6.330}$$

16. The Kohn-Sham equations in the text assume a paramagnetic metal with equal numbers of electron spins in both directions. In magnetic metals, such as iron, the conduction electron spins $n_\uparrow(\mathbf{r}) \neq n_\downarrow(\mathbf{r})$. Write out the Kohn-Sham equations for this case in the Hartree-Fock approximation.

7 Phonons

7.1 Phonon Dispersion

Phonons are among the most important excitations in condensed matter. The ions can vibrate, and a crystal of ions has collective vibrations. The description of the vibrations can be done classically or using quantum mechanics. In the quantum mechanical description, these collective motions are called phonons.

Phonons have the same frequency as the classical vibrations. Classical physics, using Newton's law of motion, is used to find the frequencies. Quantum mechanics is used to determine the amplitude of vibrations. It is the amplitude that is quantized. The initial section of this chapter employs classical physics to find force constants.

Phonon energies are reported in several different units. The frequency $\omega_\lambda(q)$ has the dimensions of radians per second. The related frequency

$$\nu_\lambda(q) = \frac{\omega_\lambda(q)}{2\pi} \tag{7.1}$$

has the units of cycles per second, which is hertz. A natural unit is terahertz (THz). For example, the optical phonon frequency in silicon at zero wave vector is 15 THz. This unit is used by those measuring phonons throughout the Brillouin zone, which can be done by either neutron or x-ray scattering from crystals.

Electron energies are usually reported in electron volts. Similarly, the phonon energy $\hbar\omega_\lambda(q)$ is often reported in milli-electron volts. The optical phonon energy in silicon is at 62 meV. Some long-wavelength phonons can be observed in Raman scattering. These optical experiments usually report the frequency in a unit that looks like a wave vector, but is actually the inverse of the wavelength of the photon with the same energy as the phonon:

$$\tilde{k} = \frac{1}{\lambda} = \frac{\nu_\lambda(q)}{c} \tag{7.2}$$

The silicon optical phonon has a frequency of 500 cm^{-1}. Accurate conversion between these units is

$$1 \text{ THz} = 4.13567 \text{ meV} = 33.3564 \text{ cm}^{-1} \tag{7.3}$$

7.1.1 *Spring Constants*

The forces between ions are caused by a combination of short-range overlap of electronic orbitals and long-range interactions due to Coulomb and van der Waals interactions. The short-range forces can be determined using density functional methods. First the ground-state energy is found for a crystal. Then displace the same atom in each unit cell a small amount Q and recalculate the ground-state energy. The increase in energy should go as $KQ^2/2$, and the calculation determines the spring constant K. Repeat this for all atoms in the unit cell, and one has a set of classical spring constants with which to calculate the vibrational frequencies. This method is called *frozen phonon.*

If one knows the potential energy between nearby atoms, such as the Van der Waals interaction, one can find the force constants directly. Define the potential between two neighboring ions at $\mathbf{R}_i$ and $\mathbf{R}_j$ as $V_{ij}(\mathbf{R}_i - \mathbf{R}_j)$. The ion positions are the equilibrium position $\mathbf{R}_i^{(0)}$ plus the ion displacement $\mathbf{Q}_i$ due to the vibrations: $\mathbf{R}_i = \mathbf{R}_i^{(0)} + \mathbf{Q}_i$. Then expand the potential energy in a Taylor series in the small displacements:

$$\begin{aligned} V_{ij}(\mathbf{R}_i - \mathbf{R}_j) &= V_{ij}(\mathbf{R}_{ij}^{(0)} + \mathbf{Q}_{ij}) \\ &= V_{ij}(\mathbf{R}_{ij}^{(0)}) + \mathbf{Q}_{ij} \cdot \nabla V_{ij}(\mathbf{R}_{ij}^{(0)}) \\ &\quad + \frac{1}{2!}(\mathbf{Q}_{ij} \cdot \nabla)^2 V_{ij}(\mathbf{R}_{ij}^{(0)}) + \cdots \end{aligned} \tag{7.4}$$

These three terms are as follows:

- The first term on the right $V_{ij}(\mathbf{R}_{ij}^{(0)})$ contributes to the ground-state binding energy of the solids, but does not contribute to the vibrational properties.
- The second term contains the force $\nabla V_{ij}(\mathbf{R}_{ij}^{(0)})$ acting on an ion from its neighbors. This term is zero when summed over all neighbors: that is the definition of equilibrium. If the net force on an ion was nonzero, then it would move until it found an equilibrium point where it was zero.
- The third term is the most important. It has two factors of the ion displacement, and therefore is similar to the potential energy for the harmonic oscillator. The features of this term are discussed below.

This method is usually accurate since the displacements $\mathbf{Q}$ are typically a small fraction of the nearest-neighbor distance $\mathbf{R}_{ij}^{(0)}$.

A *central potential* has the potential energy being a function of only $R = |\mathbf{R}|$, and does not depend on the vector properties. Then its double derivative gives

$$\frac{\partial}{\partial R_\mu} V(R) = \frac{R_\mu}{R} \frac{dV}{dR} \tag{7.5}$$

$$\frac{\partial^2}{\partial R_\mu \partial R_\nu} V(R) = \frac{\delta_{\mu\nu}}{R} \frac{dV}{dR} + \frac{R_\mu R_\nu}{R^2} \left[\frac{d^2 V}{dR^2} - \frac{1}{R} \frac{dV}{dR} \right] \tag{7.6}$$

When atoms are near to the equilibrium point, then dV/dR is nearly zero. So the largest term is the second derivative, and often only this term is retained:

$$V_{ij}(\mathbf{R}_i - \mathbf{R}_j) \approx V_{ij}(\mathbf{R}_{ij}^{(0)}) + \frac{1}{2}(\mathbf{Q}_{ij} \cdot \hat{R}_{ij})^2 K \tag{7.7}$$

$$K = \frac{d^2 V}{dR^2} \tag{7.8}$$

The potential energy between two atoms has been reduced to a single spring constant K. Potentials of this type are called *bond-directed*, since the relative displacement $\mathbf{Q}_{ij}$ has a scalar product with the bond direction $\hat{R}_{ij}$.

- In many types of crystals, a good description of the phonon modes for all values of wave vector are obtained by having spring constants K_1 between first neighbors and K_2 between second neighbors. These crystals have no long-range Coulomb forces and no covalent bonding. Examples are rare gas solids, such as argon and xenon, and simple metals. In metals, the ions have charge, but the long-range Coulomb interaction is screened by the electrons.

- In ionic solids, such as the alkali halides, the ions have a net charge q_i. When they displace from equilibrium they form a dipole $q_i^* \mathbf{Q}_i$. The *effective charge* q_i^* is usually less than the ion charge q_i. The ion motion induces some electronic polarization, which produces an electronic dipole $e_i \mathbf{Q}_i$ that is in the opposite direction to $q_i \mathbf{Q}_i$, and $q_i^* = q_i - e_i$. The *Szigeti charge* is $e^* = q_i^* / q_i$. Long-range dipole–dipole interactions are another source of potential energy between the ion displacements. It is important to include these dipolar interactions in the discussion of phonons in ionic crystals.

- Another class of crystals are those with covalent bonding, such as silicon or the carbon-based materials. Then bond bending forces are important. Bond bending is a three-atom interaction term and is not a central force. It can be included by adding a spring constant for the bond bending. For example, an excellent description of the phonons in silicon is obtained using a model proposed by Stillinger and Weber [13], which has only two spring constants: one for nearest-neighbor bond-directed interactions, and the other for bond-bending forces.

For most solids, the phonons can be well described by only a few spring constants. These constants can be found either theoretically using density functional theory, or else experimentally by measuring the phonon dispersion using neutron scattering and fitting the dispersion to a model with a few force constants. We have written phonon computer codes for numerous materials and usually get a good fit with only a few spring constants.

There is one class of solids in which the phonons cannot be calculated with a few spring constants. Crystals made up of the first three elements in the periodic table (H, He, Li) have significant quantum fluctuations due to the light mass of the ion. The phonons are not well described by the harmonic model, and anharmonic interactions are important. They are a special, interesting, and difficult case.

7.1.2 *Example: Square Lattice*

An example is presented in two dimensions. Assume there is a square lattice with one atom per unit cell, as shown in fig. 7.1. The equation of motion for bond-directed forces is

$$M\frac{\partial^2}{\partial t^2}\mathbf{Q}_j = -\sum_{\delta} K(\delta)\hat{\delta}\hat{\delta}\cdot(\mathbf{Q}_j - \mathbf{Q}_{j+\delta}) \tag{7.9}$$

where $\boldsymbol{\delta}$ are the vectors to the nearest neighbors. Assume a phonon wave $\mathbf{Q}\exp[i(\mathbf{q}\cdot\mathbf{R}_j - \omega t)]$. If including only first neighbors, with spring constant K_1, the equations for in-plane vibrations are

$$M\omega^2 Q_x = 2K_1 Q_x[1-\cos(\theta_x)], \quad \theta_x = q_x a \tag{7.10}$$

$$M\omega^2 Q_y = 2K_1 Q_y[1-\cos(\theta_y)], \quad \theta_y = q_y a \tag{7.11}$$

This solution does not have stable transverse waves. So add a second-neighbor spring constant K_2 in the directions $\hat{\delta} = (\pm 1, \pm 1)/\sqrt{2}$. The force in the x-direction is

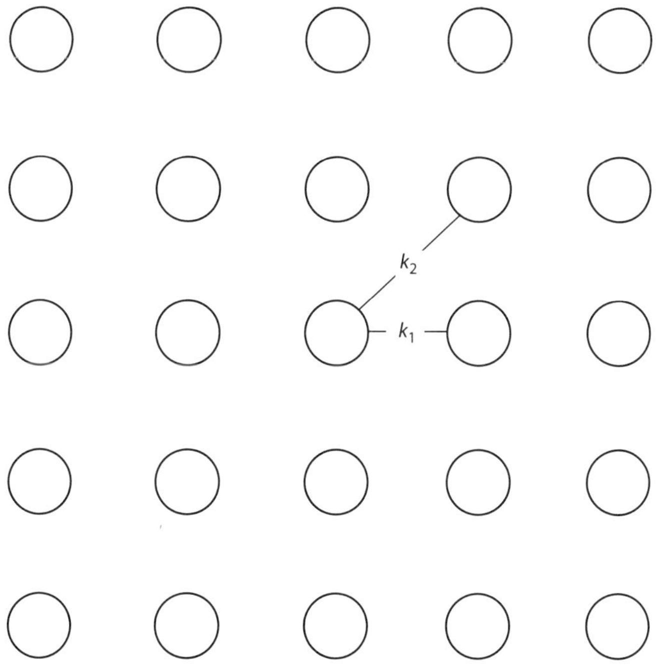

FIGURE 7.1. Square lattice in two dimensions.

$$F_x = \frac{K_2}{2}\Big[(Q_x + Q_y)(2 - e^{i(\theta_x+\theta_y)} - e^{-i(\theta_x+\theta_y)}) \tag{7.12}$$

$$+ (Q_x - Q_y)(2 - e^{i(\theta_x-\theta_y)} - e^{-i(\theta_x-\theta_y)})\Big]$$

$$F_x = 2K_2\{Q_x[1 - \cos(\theta_x)\cos(\theta_y)] + Q_y \sin(\theta_x)\sin(\theta_y)\} \tag{7.13}$$

$$F_y = 2K_2\{Q_y[1 - \cos(\theta_x)\cos(\theta_y)] + Q_x \sin(\theta_x)\sin(\theta_y)\} \tag{7.14}$$

Define $\omega_j^2 = 2K_j/M$. The *dynamical matrix* $\mathcal{D}$ has the form

$$\mathcal{D}\begin{pmatrix} Q_x \\ Q_y \end{pmatrix} = \omega^2 \begin{pmatrix} Q_x \\ Q_y \end{pmatrix} \tag{7.15}$$

$$\mathcal{D} = \begin{pmatrix} \omega_1^2(1-c_x) + \omega_2^2(1 - c_x c_y) & \omega_2^2 s_x s_y \\ \omega_2^2 s_x s_y & \omega_1^2(1-c_y) + \omega_2^2(1-c_x c_y) \end{pmatrix} \tag{7.16}$$

$$c_x = \cos(\theta_x), \quad s_y = \sin(\theta_y), \quad \text{etc.} \tag{7.17}$$

$$\omega_j^2 = \frac{2K_j}{M} \tag{7.18}$$

The vibrational frequencies are determined by setting to zero the determinant of the matrix:

$$0 = |\mathcal{D}_{\mu\nu} - \omega^2 \delta_{\mu\nu}| \tag{7.19}$$

$$\omega_\pm^2 = \omega_2^2(1 - c_x c_y) + \omega_1^2\left[1 - \frac{1}{2}(c_x + c_y)\right] \pm \frac{1}{2}\sqrt{\omega_1^4(c_x - c_y)^2 + 4\omega_2^4 s_x^2 s_y^2} \tag{7.20}$$

These formulas give the phonon dispersion: the energy bands of the phonons. They are often graphed along major symmetry directions. The eigenvectors of the matrix determine the vector displacements of the ions, which are denoted by $\hat{\xi}$. In two dimensions:

- Along the (10) direction, set $\theta_y = 0$, $s_y = 0$, $c_y = 1$ and find

$$\omega_l^2(q_x) = (\omega_1^2 + \omega_2^2)(1 - c_x) = 2(\omega_1^2 + \omega_2^2)\sin^2(\theta_x/2) \tag{7.21}$$

$$\omega_t^2(q_x) = \omega_2^2(1 - c_x) = 2\omega_2^2 \sin^2(\theta_x/2) \tag{7.22}$$

  The transverse phonon frequency is given entirely by the second-neighbor springs.

- Along the (11) direction, set $\theta_x = \theta_y$, $c_x = c_y$, $s_x = s_y$,

$$\omega_l^2(q) = \omega_1^2[1 - \cos(\theta)] + 2\omega_2^2 \sin^2(\theta) \tag{7.23}$$

$$\omega_t^2(q) = \omega_1^2[1 - \cos(\theta)] \tag{7.24}$$

Figure 7.2 shows these phonon frequncies along principal directions in the Brillouin zone. For this figure, we set $K_1 = 2K_2 \equiv 2K$, $\omega_0^2 = 2K/M$, and $\Omega = \omega/\omega_0$. This method of finding phonon modes can be appplied to actual solids.

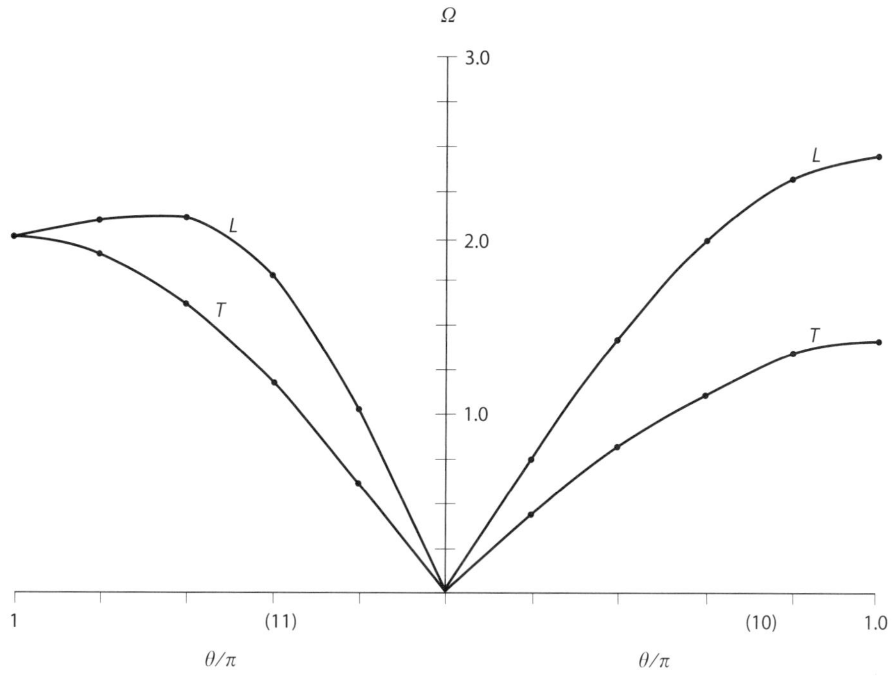

FIGURE 7.2. Phonon modes in a two-dimensional square lattice.

7.1.3 *Polar Crystals*

In polar crystals, such as alkali halides or binary semiconductors, long-range Coulomb interactions have to be included in the calculation of the phonon dispersion. When the ions vibrate, their charge creates an oscillating dipole at that ion site. Long-range dipole–dipole interactions have a large influence on the phonon frequencies.

In most materials, the optical phonons contribute to the dielectric function. This happens in all polar materials: they have some charge transfer, so that the crystal has both anions and cations. The materials in which phonons do not contribute to the dielectric function are (i) conductors, since all long-range electric fields are screened by the conduction electrons, and (ii) crystals composed of a single element, such as diamond, silicon, or argon.

Richard Zallen [18] proved the following theorem: *An elemental crystal displays symmetry-allowed, one-phonon absorption if and only if the primitive cell contains three or more atoms.* Argon, with one atom per unit cell, and silicon and diamond, with two, have no infrared active phonons at long wavelength, and phonons do not contribute to the dielectric function. However, crystals like graphite, with four carbon atoms per unit cell, do have infrared active phonons, which contribute to the dielectric function of the crystal.

7.1.4 *Phonons*

A simple model assumes that if the ion has charge e and displaces $\mathbf{Q}_j$, it creates a dipole $\mathbf{p}_j = e\mathbf{Q}_j$. Assume the phonon is a wave $\exp[i(\mathbf{q}\cdot\mathbf{R}_j - \omega t)]$. The lattice transform $T_{\mu\nu}(\mathbf{q})$ of the dipole–dipole interaction was introduced in chapter 4. The electric field on an ion, from long-range dipole interactions, is

$$E_{\mu,j} = -\frac{1}{4\pi\varepsilon_0}\sum_{\nu,j'\neq j}\phi_{\mu\nu}(\mathbf{R}_{jj'})eQ_{\nu}e^{i\mathbf{q}\cdot\mathbf{R}_{j'}} \tag{7.25}$$

$$= -\frac{e}{4\pi\varepsilon_0}\sum_{\nu}T_{\mu\nu}(\mathbf{q})Q_{\nu}e^{i\mathbf{q}\cdot\mathbf{R}_j}$$

$$\phi_{\mu\nu}(\mathbf{R}) = \frac{\delta_{\mu\nu}}{R^3} - \frac{3R_{\mu}R_{\nu}}{R^5} \tag{7.26}$$

$$T_{\mu\nu}(\mathbf{q}) = \sum_{\mathbf{R}_j\neq 0} e^{i\mathbf{q}\cdot\mathbf{R}_j}\phi_{\mu\nu}(\mathbf{R}_j) \tag{7.27}$$

This term is added to the dynamical matrix. Separate terms arise from the negative ions and from the positive ions. In binary crystals with ion charges $\pm e$, there are two separate dipole–dipole summations: $T_{(e)}$ over like ions $(++, --)$, and $T^{(i)}$ over unlike ions $(+-, -+)$. In this case the equations of motion are

$$M_+\omega^2\mathbf{Q}_j^{(+)} = \sum_{\delta}K(\delta)\hat{\delta}\hat{\delta}\cdot(\mathbf{Q}_j^{(+)} - \mathbf{Q}_{j+\delta}) + \frac{e^2}{4\pi\varepsilon_0}\left[\mathcal{T}^{(e)}\cdot\mathbf{Q}_j^{(+)} - \mathcal{T}^{(i)}\cdot\mathbf{Q}_j^{(-)}\right] \tag{7.28}$$

$$M_-\omega^2\mathbf{Q}_j^{(-)} = \sum_{\delta}K(\delta)\hat{\delta}\hat{\delta}\cdot(\mathbf{Q}_j^{(-)} - \mathbf{Q}_{j+\delta}) + \frac{e^2}{4\pi\varepsilon_0}\left[\mathcal{T}^{(e)}\cdot\mathbf{Q}_j^{(\ )} - e^2\mathcal{T}^{(i)}\cdot\mathbf{Q}_j^{(+)}\right] \tag{7.29}$$

Long-wavelength optical phonons are easily measured using Raman scattering. It is interesting to solve these equations in the limit that $\mathbf{q}\to 0$. The $\mathbf{q}$ dependence of the dipolar tensors is nonanalytic. In cubic crystals

$$\lim_{q\to 0}\mathcal{T}^{(e,i)} = \frac{4\pi}{\Omega_0}\left[\hat{q}\hat{q} - \frac{1}{3}\mathcal{I}\right] \tag{7.30}$$

where $\mathcal{I}$ is the unit tensor. Keep only the first-neighbor short-range force, and note that in cubic crystals

$$\sum_{\delta}\hat{\delta}\hat{\delta} = \frac{z}{3}\mathcal{I} \tag{7.31}$$

where z is the coordination number. The above equations simplify at long wavelength:

$$M_+\omega^2\mathbf{Q}_j^{(+)} = \frac{zK_1}{3}[\mathbf{Q}_j^{(+)} - \mathbf{Q}^{(-)}] + \frac{e^2}{\varepsilon_0\Omega_0}\left[\hat{q}\hat{q} - \frac{1}{3}\mathcal{I}\right]\cdot[\mathbf{Q}_j^{(+)} - \mathbf{Q}_j^{(-)}] \tag{7.32}$$

$$M_-\omega^2\mathbf{Q}_j^{(-)} = \frac{zK_1}{3}[\mathbf{Q}_j^{(-)} - \mathbf{Q}^{(+)}] + \frac{e^2}{\varepsilon_0\Omega_0}\left[\hat{q}\hat{q} - \frac{1}{3}\mathcal{I}\right]\cdot[\mathbf{Q}_j^{(-)} - \mathbf{Q}_j^{(+)}] \tag{7.33}$$

Divide the first equation by M_+, the second by M_-, and then subtract them. Introduce the reduced mass μ, and find

$$\frac{1}{\mu} = \frac{1}{M_-} + \frac{1}{M_+},\quad \delta\mathbf{Q} = \mathbf{Q}_j^{(+)} - \mathbf{Q}^{(-)} \tag{7.34}$$

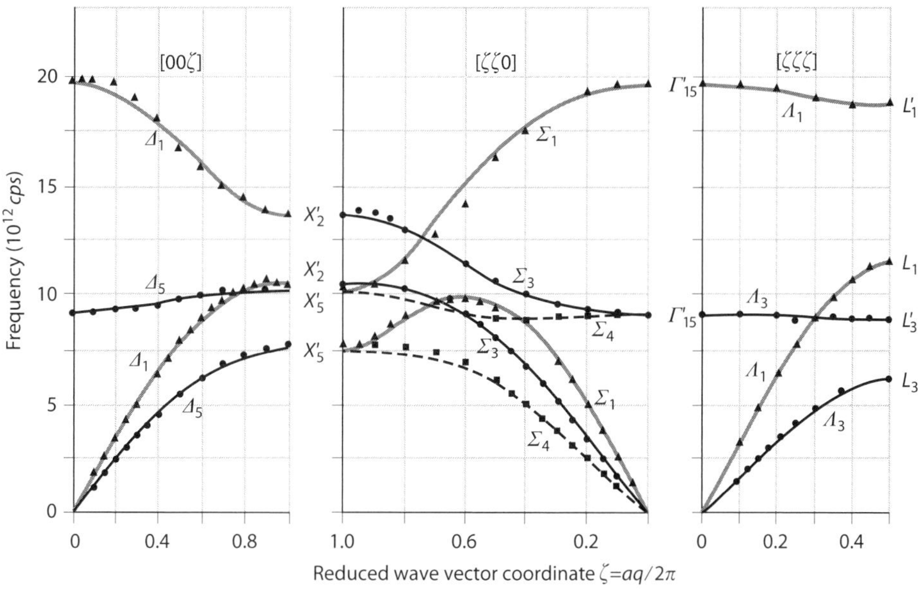

FIGURE 7.3. Phonon dispersion in LiF. Note difference between ω_{LO} and ω_{TO} at long wavelength. From Dolling et al., *Phys. Rev.* **168**, 970 (1968). Used with permission of the American Physical Society.

$$\omega^2 \delta\mathbf{Q} = \omega_0^2 \delta\mathbf{Q} + \omega_{pi}^2 \left[\hat{q}\hat{q} - \frac{1}{3}\mathcal{I} \right] \cdot \delta\mathbf{Q} \tag{7.35}$$

$$\omega_{pi}^2 = \frac{e^2}{\varepsilon_0 \mu \Omega_0}, \quad \omega_0^2 = \frac{zK_1}{3\mu} \tag{7.36}$$

The ion plasma frequency is ω_{pi}^2. The optical phonons would have a frequency ω_0 without the dipolar interaction.

- Longitudinal optical phonons have $\mathbf{q} \parallel \delta\mathbf{Q}$ and a frequency of

$$\omega_{LO}^2 = \omega_0^2 + \frac{2}{3}\omega_{pi}^2 \tag{7.37}$$

- Transverse optical phonons have $\mathbf{q} \perp \delta\mathbf{Q}$ and a frequency of

$$\omega_{TO}^2 = \omega_0^2 - \frac{1}{3}\omega_{pi}^2 \tag{7.38}$$

$$\omega_{LO}^2 - \omega_{TO}^2 = \omega_{pi}^2 \tag{7.39}$$

The longitudinal and transverse frequencies are different at long wavelength, and the difference is due to the long-range dipolar interactions. Figure 7.3 shows the measured phonon dispersion in LiF, and the LO-TO splitting is quite large. The predictions of eqn. (7.39) do not agree with the observed LO-TO splittings in actual polar crystals. There are several possible reasons for this disagreement. The first is the neglect of the dielectric screening provided by the ions in the crystal. Define $\tilde{\alpha}$ as the polarizability per unit volume of the electrons in the valence states of the ions. It has contributions from anions (α_a) and cations (α_c):

$$\tilde{\alpha} = \frac{1}{\Omega_0}(\alpha_a + \alpha_c) \equiv \frac{\alpha_e}{\Omega_0} \tag{7.40}$$

$$\varepsilon(\infty)/\varepsilon_0 = 1 + \frac{4\pi\tilde{\alpha}}{1 - 4\pi\tilde{\alpha}/3}, \quad \frac{4\pi\tilde{\alpha}}{3} = \frac{\varepsilon(\infty) - \varepsilon_0}{\varepsilon(\infty) + 2\varepsilon_0} \tag{7.41}$$

One can rewrite the dynamical equations while including the electronic polarization:

$$\omega^2 \delta\mathbf{Q} = \omega_0^2 \delta\mathbf{Q} - \frac{e}{\mu}\mathbf{E} \tag{7.42}$$

$$\mathbf{E} = -\mathcal{T} \cdot \left[\frac{e}{4\pi\varepsilon_0}\delta\mathbf{Q} + \alpha_e \mathbf{E}\right] \tag{7.43}$$

For longitudinal fields,

$$E_l = \hat{q} \cdot \mathbf{E} = -\frac{8\pi}{3\Omega_0}\left(\frac{e}{4\pi\varepsilon_0}\delta Q_l + \alpha_e E_l\right) \tag{7.44}$$

$$= -\frac{2}{3\varepsilon_0\Omega_0}\frac{e\delta Q_l}{1 + 8\pi\tilde{\alpha}/3} = -\left(\frac{\varepsilon(\infty) + 2\varepsilon_0}{3\varepsilon(\infty)}\right)\frac{2e\delta Q_l}{3\varepsilon_0\Omega_0}$$

For transverse fields,

$$E_t = \frac{4\pi}{3\Omega_0}\left(\frac{e}{4\pi\varepsilon_0}\delta Q_t + \alpha_e E_t\right) \tag{7.45}$$

$$= \frac{1}{3\varepsilon_0\Omega_0}\frac{e\delta Q_t}{1 - 4\pi\tilde{\alpha}/3} = \left(\frac{\varepsilon(\infty) + 2\varepsilon_0}{3\varepsilon_0}\right)\frac{e\delta Q_t}{3\varepsilon_0\Omega_0}$$

The two phonon frequencies are

$$\omega_{\mathrm{LO}}^2 = \omega_0^2 + \frac{2}{3}\left(\frac{\varepsilon(\infty) + 2\varepsilon_0}{3\varepsilon(\infty)}\right)\omega_{pi}^2 \tag{7.46}$$

$$\omega_{\mathrm{TO}}^2 = \omega_0^2 - \frac{1}{3}\left(\frac{\varepsilon(\infty) + 2\varepsilon_0}{3\varepsilon_0}\right)\omega_{pi}^2 \tag{7.47}$$

$$\omega_{\mathrm{LO}}^2 - \omega_{\mathrm{TO}}^2 = \omega_{pi}^2\left(\frac{\varepsilon(\infty) + 2\varepsilon_0}{3}\right)^2\frac{1}{\varepsilon(\infty)\varepsilon_0} \tag{7.48}$$

The LO-TO splitting is multiplied by the screening factor of $P(\varepsilon) = (\varepsilon + 2\varepsilon_0)^2/(9\varepsilon\varepsilon_0)$. This important factor has the value $P(\varepsilon_0) = 1$. Another value is $P(4\varepsilon_0) = 1$. Between $1 < \varepsilon(\infty)/\varepsilon_0 < 4$ it varys only slightly from unity. So the dielectric screening is not an important factor!

All of the quantities in the above expression are easily measured. The right-hand side never quite equals the left-hand side. This discrepancy is explained by assuming that the effective charge e^* on an ion when it vibrates is not an integer value. The Szigeti charge e^* is the dimensionless, effective charge on the ion, as observed in the LO-TO splitting. It is not the actual charge on the ion, but is a smaller number due to electronic polarization. It is defined as the ratio of the above two expressions:

Table 7.1 Dielectric constants, optical phonon frequencies at q = 0, and Szigeti charges for some zincblende semiconductors

Salt	$\varepsilon(0)/\varepsilon_0$	$\varepsilon(\infty)/\varepsilon_0$	ν_{TO} (THz)	ν_{LO} (THz)	ν_{pi} (THz)	e^*/e
AlP	9.80	7.54	13.16	15.02	8.67	0.72
AlAs	10.06	8.16	10.84	12.10	7.01	0.65
AlSb	12.04	10.24	9.55	10.19	5.89	0.47
GaP	11.11	9.11	10.95	12.06	7.14	0.58
GaAs	13.18	10.89	8.18	8.84	5.21	0.49
GaSb	15.69	14.44	6.70	6.98	4.20	0.32
InP	12.56	9.61	9.10	10.32	5.99	0.65
InAs	15.15	12.25	6.51	7.15	4.19	0.52
InSb	17.30	15.68	5.39	5.72	3.32	0.39

$$(e^*)^2 = \frac{\omega_{\mathrm{LO}}^2 - \omega_{\mathrm{TO}}^2}{\omega_{pi}^2} \frac{9\varepsilon(\infty)\varepsilon_0}{(\varepsilon(\infty) + 2\varepsilon_0)^2} \tag{7.49}$$

In evaluating the right-hand-side of this expression, use for e the charge of one electron. Then e^* is the effective charge on the ion for optical vibrations. Table 7.1 shows dielectric data, zone-center optical phonon frequencies, and Szigeti charges of some III–V semiconductors with the zincblende structure. The values of e^* are less than unity, which reflects the covalent bonding in these semiconductors. The experimental values in the first four columns are not known accurately, in that different experiments report slightly different numbers. The values we chose to make the table are typical.

7.1.5 *Dielectric Function*

The total dielectric function at long wavelength from electrons and phonons is

$$\varepsilon(\omega)/\varepsilon_0 = 1 + \frac{4\pi\alpha_T(\omega)/\Omega_0}{1 - 4\pi\alpha_T(\omega)/3\Omega_0} \tag{7.50}$$

$$\alpha_T = \alpha_e + \frac{e^2}{4\pi\mu\varepsilon_0} \frac{1}{\omega_0^2 - \omega^2} \tag{7.51}$$

$$\frac{4\pi\alpha_T}{\Omega_0} = 4\pi\tilde{\alpha} + \frac{\omega_{pi}^2}{\omega_0^2 - \omega^2}, \quad \omega_{pi}^2 = \frac{e^2}{\mu\Omega_0\varepsilon_0} \tag{7.52}$$

The first term in α_T is from electrons, and the second is from the optical phonons, where μ is the reduced mass of the anion and cation. This expression is the transverse dielectric function, which is the response to a transverse electric field. The same expression is obtained for the longitudinal dielectric function. This dielectric function has several features.

- It can be written as separate contributions from electrons $\varepsilon(\infty)$ and phonons

$$\varepsilon(\omega) = \varepsilon(\infty) + \left(\frac{\varepsilon(\infty) + 2\varepsilon_0}{3\varepsilon_0}\right)^2 \frac{\omega_{pi}^2 \varepsilon_0}{\omega_{TO}^2 - \omega^2} \tag{7.53}$$

$$\varepsilon(\infty)/\varepsilon_0 = 1 + \frac{4\pi\alpha_e/\Omega_0}{1 - 4\pi\alpha_e/3\Omega_0} \tag{7.54}$$

where α_e is the electronic polarizability, Ω_0 is the volume of a unit cell, and ω_{pi} is the ion plasma frequency. The second term in $\varepsilon(\omega)$ is the contribution due to phonons. It has a resonance at the frequency ω_{TO}.

- The high-frequency dielectric function is $\varepsilon(\infty)/\varepsilon_0 = n^2 - \kappa^2 \approx n^2$, which determines the refractive index in the region of no absorption. The static dielectric function is

$$\varepsilon(0)/\varepsilon_0 = 1 + \frac{4\pi\alpha_T(0)/\Omega_0}{1 - 4\pi\alpha_T(0)/3\Omega_0} \tag{7.55}$$

$$\varepsilon(0) = \varepsilon(\infty) + \left(\frac{\varepsilon(\infty) + 2\varepsilon_0}{3\varepsilon_0}\right)^2 \frac{\omega_{pi}^2 \varepsilon_0}{\omega_{\mathrm{TO}}^2} \tag{7.56}$$

- The loss function can be written as

$$\frac{1}{\varepsilon(\omega)} = \frac{1}{\varepsilon(\infty)} \frac{\omega_{\mathrm{TO}}^2 - \omega^2}{\omega_{\mathrm{LO}}^2 - \omega^2} \tag{7.57}$$

Note that its poles are at the frequency of the LO phonon.

- The above formula can be used to prove the Lyddane-Sachs-Teller relations:

$$\frac{\varepsilon(0)}{\varepsilon(\infty)} = \frac{\omega_{\mathrm{LO}}^2}{\omega_{\mathrm{TO}}^2} \tag{7.58}$$

All of the above relations can be proven using algebra. These derivations are assigned as homework.

The above relations are all valid in the limit that the photon wave vector goes to zero. For photons of high frequency in the x-ray region of the spectrum, the wave vectors will be on the order of those in the Brillouin zone $O\,(10^8/\mathrm{cm})$. In that case one should retain the wave vector dependence $\varepsilon(\mathbf{k}, \omega)$. At $\mathbf{k} \neq 0$ there are two different dielectric functions: transverse and longitudinal. The transverse dielectric function is the one that appears in Maxwell's equations, when we write $k^2 = (\omega/c)^2 \varepsilon(\mathbf{k}, \omega)$. The longitudinal dielectric function is used in screening Coulomb interactions, when we write the screened interaction as

$$V_{\mathrm{sc}}(k, \omega) = \frac{4\pi e^2}{q^2 \varepsilon(q, \omega)} \tag{7.59}$$

The longitudinal and transverse dielectric functions are identical at zero wave vector, but are different at nonzero wave vector.

7.2 Phonon Operators

7.2.1 *Simple Harmonic Oscillator*

A standard problem of quantum mechanics is the simple harmonic oscillator. It is a mass m connected to a harmonic spring of constant K. The Hamiltonian is

$$H = \frac{p^2}{2m} + \frac{K}{2}x^2 \tag{7.60}$$

with eigenvalues $E_n = \hbar\omega(n + 1/2)$, $\omega = \sqrt{K/m}$. The eigenfunctions $\phi_n(x)$ are given by Hermite polynomials.

The position x does not commute with its momentum p:

$$[x, p] = xp - px = i\hbar \tag{7.61}$$

This important postulate of quantum mechanics can be developed into an operator algebra. First define a unit of length:

$$x_0 = \sqrt{\frac{\hbar}{m\omega}} = \left(\frac{\hbar^2}{Km}\right)^{1/4} \tag{7.62}$$

Define a lowering (a) and raising ($a^{\dagger}$) operator as

$$a = \frac{1}{\sqrt{2}x_0}\left[x + \frac{ip}{m\omega}\right] \tag{7.63}$$

$$a^{\dagger} = \frac{1}{\sqrt{2}x_0}\left[x - \frac{ip}{m\omega}\right] \tag{7.64}$$

The dagger symbol (†) usually denotes Hermitian conjugate. That is its intent here, since the raising operator is the Hermitian conjugate of the lowering operator. Using the commutator of x and p to determine how these operators commute,

$$[a, a] = 0 = [a^{\dagger}, a^{\dagger}] \tag{7.65}$$

$$[a, a^{\dagger}] = \frac{1}{2x_0^2}\left[x + \frac{ip}{m\omega}, x - \frac{ip}{m\omega}\right] \tag{7.66}$$

$$= \frac{-i}{2\hbar}\{[x, p] - [p, x]\} = 1$$

The first line is trivial, since all operators commute with themselves. The lowering and raising operators do not commute, and their commutator gives one. These operators are dimensionless.

Express the Hamiltonian in terms of these operators. Note that

$$a^{\dagger}a = \frac{1}{2x_0^2}\left[x - \frac{ip}{m\omega}\right]\left[x + \frac{ip}{m\omega}\right] \tag{7.67}$$

$$= \frac{1}{2x_0^2}\left[x^2 + \frac{p^2}{m^2\omega^2} + \frac{i}{m\omega}[x, p]\right]$$

$$= \frac{1}{2x_0^2}\left[x^2 + \frac{p^2}{m^2\omega^2} - x_0^2\right]$$

$$a^\dagger a + \frac{1}{2} = \frac{1}{2x_0^2}\left[x^2 + \frac{p^2}{m^2\omega^2}\right] \tag{7.68}$$

where the commutator gave $i\hbar$. Recall that $K = m\omega^2$, so the right-hand side can be rearranged into

$$a^\dagger a + \frac{1}{2} = \frac{1}{Kx_0^2}\left[\frac{K}{2}x^2 + \frac{p^2}{2m}\right] = \frac{1}{\hbar\omega}H \tag{7.69}$$

$$H = \hbar\omega\left[a^\dagger a + \frac{1}{2}\right] \tag{7.70}$$

The Hamiltonian can be written in a simple way using the operators. The eigenfunctions have the properties, when using the Dirac notation $\phi_n = |n\rangle$, that

$$a|n\rangle = \sqrt{n}|n-1\rangle \tag{7.71}$$

$$a^\dagger|n\rangle = \sqrt{n+1}|n+1\rangle \tag{7.72}$$

Then the commutator is satisfied:

$$[a, a^\dagger]|n\rangle = a\sqrt{n+1}|n+1\rangle - a^\dagger\sqrt{n}|n-1\rangle \tag{7.73}$$

$$= (n+1-n)|n\rangle = |n\rangle$$

The Hamiltonian is

$$H|n\rangle = \hbar\omega\left[n + \frac{1}{2}\right]|n\rangle = E_n|n\rangle \tag{7.74}$$

This operator algebra for the simple harmonic oscillator will be extended and used below for many other boson systems. Another of its properties is that

$$|n\rangle = \frac{(a^\dagger)^n}{\sqrt{n!}}|0\rangle \tag{7.75}$$

where $|0\rangle$ is the ground state $n = 0$ with no excitations. The state $|n\rangle$ with n excitations is obtained from the ground state by the application of n raising operators.

Another useful identity is to solve eqns. (7.63) and (7.64) for x and p:

$$x = \frac{x_0}{\sqrt{2}}(a + a^\dagger) \tag{7.76}$$

$$p = \frac{i\hbar}{\sqrt{2}x_0}(a^\dagger - a) \tag{7.77}$$

These expressions are useful, since they show how to derive matrix elements of these operators between two states:

$$\langle n|x|\ell\rangle = \frac{x_0}{\sqrt{2}}[\langle n|a|\ell\rangle + \langle n|a^\dagger|\ell\rangle] \tag{7.78}$$

$$= \frac{x_0}{\sqrt{2}}\sqrt{n}\,[\delta_{n=\ell-1} + \delta_{n=\ell+1}]$$

For the harmonic oscillator, the matrix elements of x are nonzero only between adjacent quantum levels.

The time dependence of the operators is needed for many calculations. Consider the lowering operator. Its time derivative is

$$\frac{d}{dt}a = \frac{i}{\hbar}[H, a] = i\omega[a^\dagger a, a] = i\omega[a^\dagger, a]a = -i\omega a \tag{7.79}$$

The solution to this equation is

$$a(t) = ae^{-i\omega t} \tag{7.80}$$

$$a^\dagger(t) = a^\dagger e^{i\omega t} \tag{7.81}$$

The first equation is a solution to eqn. (7.79). The second equation is just the Hermitian conjugate of the first. These two results give the time dependence of the position operator:

$$x(t) = \frac{x_0}{\sqrt{2}}(ae^{-i\omega t} + a^\dagger e^{i\omega t}) \tag{7.82}$$

All of these formulas are used in later sections.

7.2.2 *Phonons in One Dimension*

A simple model of collective vibrations is a chain of identical atoms, with a harmonic spring K between near neighbors. The atoms vibrate only along the chain axis, as in bond-directed potentials. Assume the chain is periodic and has atoms located at $j = 1, 2, \ldots, N$ and $x_j = aj$. The periodic feature requires $Q_{N+1} \equiv Q_1$. The Hamiltonian is

$$H = \sum_{j=1}^{N}\left[\frac{P_j^2}{2M} + \frac{K}{2}(Q_j - Q_{j+1})^2\right] \tag{7.83}$$

Change variables to collective coordinates:

$$Q_n = \frac{1}{\sqrt{N}}\sum_k e^{ikan}Q_k \tag{7.84}$$

$$P_n = \frac{1}{\sqrt{N}}\sum_k e^{-ikan}P_k \tag{7.85}$$

The periodic condition with $Q_{n+N} = Q_n$ means that $\exp[ikaN] = 1$, so the values of wave vector are

$$k_\alpha = \frac{2\pi\alpha}{aN}, \quad \alpha = 0, 1, \ldots, N-1 \tag{7.86}$$

Note that if $\alpha = N-1$ then

$$\exp[ika] = \exp\left[i2\pi\left(1-\frac{1}{N}\right)\right] = \exp\left[-\frac{2\pi i}{N}\right] \tag{7.87}$$

So $\alpha = N-1$ behaves as $\alpha = -1$. Similarly, $\alpha = N-n$ behaves as $\alpha = -n$. The collective variables obey commutation relations:

$$[Q_k, P_{k'}] = i\hbar\delta_{kk'} = i\hbar\delta_{\alpha\alpha'} \tag{7.88}$$

$$[Q_n, P_m] = \frac{1}{N}\sum_{kk'}[Q_k, P_{k'}]e^{ia(kn-k'm)} \tag{7.89}$$

$$= i\hbar\frac{1}{N}\sum_k e^{iak(n-m)} = i\hbar\delta_{nm}$$

Denote the collective momentum as

$$P_k = \frac{\hbar}{i}\frac{\partial}{\partial Q_k} \tag{7.90}$$

The Hamiltonian diagonalizes when written in collective coordinates:

$$\sum_{j=1}^{N}\frac{P_j^2}{2M} = \frac{1}{2MN}\sum_{kk'}P_kP_{k'}\sum_j e^{ija(k+k')} \tag{7.91}$$

$$\frac{1}{N}\sum_j e^{ija(k+k')} = \delta_{k+k'} = \delta_{\alpha+\alpha'=N} \tag{7.92}$$

$$\sum_{j=1}^{N}\frac{P_j^2}{2M} = \sum_k\frac{P_kP_{-k}}{2M} \tag{7.93}$$

where negative wave vectors are defined as above: $\alpha = N-n \rightarrow -n$. A similar calculation for the potential energy gives

$$\frac{K}{2}\sum_j(Q_j - Q_{j+1})^2 = \frac{K}{2}\sum_j Q_j[2Q_j - Q_{j-1} - Q_{j+1}] \tag{7.94}$$

$$= K\sum_k Q_kQ_{-k}[1-\cos(ka)]$$

The frequency of the collective mode is

$$\omega^2(k) = \frac{2K}{M}[1-\cos(ka)] = \frac{4K}{M}\sin^2(ka/2) \tag{7.95}$$

$$H = \sum_k\left[\frac{P_kP_{-k}}{2M} + \frac{M\omega(k)^2}{2}Q_kQ_{-k}\right] \tag{7.96}$$

The Hamiltonian now depends only on the pair of wave vectors $(k, -k)$. The combination of P_kP_{-k} and Q_kQ_{-k} means that it is not quite the simple harmonic oscillator for each value of k. However, it can be solved easily by introducing the raising and lowering operators $(a_k^\dagger, a_k)$ for each wave vector:

$$Q_k = X_k\left(a_k + a_{-k}^\dagger\right), \quad X_k^2 = \frac{\hbar}{2M\omega(k)} \tag{7.97}$$

From the Heisenberg equation of motion,

$$\frac{d}{dt}Q_j(t) = \frac{i}{\hbar}[H, Q_j] = \frac{P_j}{M} \tag{7.98}$$

since the kinetic energy term $P_j^2/2M$ is the only term that does not commute with Q_j. Recalling the definitions of $(Q_k P_k)$,

$$P_k(t) = M\dot{Q}_{-k} \tag{7.99}$$

$$P_k = -iM\omega(k)X_k\left(a_{-k} - a_k^\dagger\right) \tag{7.100}$$

where X_k has the dimensional units of length. These definitions give the correct commutation relations:

$$[a_k, a_{k'}^\dagger] = \delta_{kk'} \tag{7.101}$$

$$\begin{aligned}[Q_k, P_{k'}] &= -\frac{i\hbar}{2}\sqrt{\frac{\omega(k')}{\omega(k)}}\left[a_k + a_{-k}^\dagger, a_{-k'} - a_{k'}^\dagger\right] \\ &= i\hbar\delta_{kk'}\end{aligned} \tag{7.102}$$

In terms of these operators, the Hamiltonian becomes

$$\begin{aligned}H &= \frac{1}{2}\sum_k \hbar\omega(k)\left[a_k^\dagger a_k + a_k a_k^\dagger\right] \\ &= \sum_k \hbar\omega(k)\left[a_k^\dagger a_k + \frac{1}{2}\right]\end{aligned} \tag{7.103}$$

The second equation is derived from the first by using the commutator $aa^\dagger = a^\dagger a + 1$. The final equation in (7.103) shows that each wave vector state does indeed act as a simple harmonic oscillator, with the frequency $\omega(k)$.

The raising and lowering operators $(a_k^\dagger, a_k)$ create or remove a phonon. The quantum mechanical solution to the collective motions has the same frequency $\omega(k)$ as the classical solution to the same problem. Quantum mechanics only changes the allowed amplitude of the atom motions. There are a discrete number of phonons n_k for each wave vector k, and the displacements can occur only in integer amounts. In thermal equilibrium, the average value for the number operator is

$$\langle a_k^\dagger a_k \rangle = n_B[\omega(k)] = \frac{1}{e^{\beta\hbar\omega(k)} - 1} \tag{7.104}$$

where again $\beta = 1/k_B T$. Phonons have no chemical potential.

The atom displacements as a function of time are given by the operator:

$$Q_n(t) = \frac{1}{\sqrt{N}}\sum_k X_k e^{ikan}\left(a_k e^{-i\omega(k)t} + a^\dagger_{-k} e^{i\omega(k)t}\right) \tag{7.105}$$

$$P_n(t) = -iM\sum_k X_k \omega(k) e^{iakn}\left(a_k e^{-i\omega(k)t} - a^\dagger_{-k} e^{i\omega(k)t}\right) \tag{7.106}$$

These operators obey the classical equations of motion (7.98).

7.2.3 *Binary Chain*

Next consider the vibrational properties of a chain of atoms in one dimension, with alternate atoms labeled A and B, as shown in fig. 7.4. They have masses M_A, M_B. The same spring of constant K is between all atoms. There are two atoms per unit cell, and N cells in the periodic chain. In each cell, let atom A be on the left, and B be on the right. The Hamiltonian is

$$H = K + V \tag{7.107}$$

$$K = \sum_{n=1}^{N}\left[\frac{P^2_{An}}{2M_A} + \frac{P^2_{Bn}}{2M_B}\right] \tag{7.108}$$

$$V = \frac{K}{2}\sum_n [(Q_{An} - Q_{Bn})^2 + (Q_{An} - Q_{B,n-1})^2] \tag{7.109}$$

$$= K\sum_n [Q^2_{An} + Q^2_{Bn} - Q_{An}(Q_{Bn} + Q_{B,n-1})]$$

Let a be the distance between atoms, so the lattice constant is $2a$. Again define collective coordinates:

$$Q_{An} = \frac{1}{\sqrt{N}}\sum_k e^{i2kan} Q_{Ak} \tag{7.110}$$

$$Q_{Bn} = \frac{1}{\sqrt{N}}\sum_k e^{ika(2n+1)} Q_{Bk} \tag{7.111}$$

The Hamiltonian can be written in terms of collective coordinates:

$$K = \sum_k \left[\frac{P_{Ak}P_{A,-k}}{2M_A} + \frac{P_{Bk}P_{B,-k}}{2M_B}\right] \tag{7.112}$$

$$V = K\sum_k [Q_{Ak}Q_{A,-k} + Q_{Bk}Q_{B,-k} - 2\cos(ka)\, Q_{Ak}Q_{B,-k}] \tag{7.113}$$

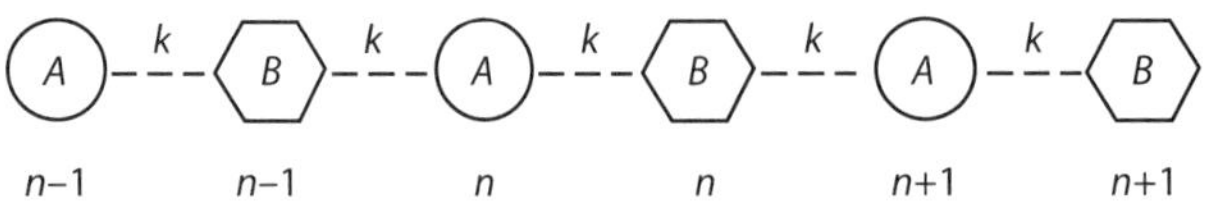

FIGURE 7.4. Binary chain with alternating masses.

The vibrational frequencies can be found using the classical equations of motion, where overdots denote time derivatives:

$$M_A \ddot{Q}_{Ak} = F_{Ak} = -\frac{\partial V}{\partial Q_{Ak}} = -2K[Q_{Ak} - \cos(ka)Q_{B,-k}] \tag{7.114}$$

$$M_B \ddot{Q}_{B,-k} = -2K[Q_{B,-k} - \cos(ka)Q_{Ak}] \tag{7.115}$$

If the normal mode has a frequency ω, then the double time derivative gives $-\omega^2$. The equations can be expressed in terms of a 2×2 dynamical matrix:

$$\omega^2 \begin{bmatrix} Q_{Ak} \\ Q_{B,-k} \end{bmatrix} = \begin{bmatrix} \frac{2K}{M_A} & -\frac{2K}{M_A}\cos(ka) \\ -\frac{2K}{M_B}\cos(ka) & \frac{2K}{M_B} \end{bmatrix} \begin{bmatrix} Q_{Ak} \\ Q_{B,-k} \end{bmatrix} \tag{7.116}$$

The frequencies are the square root of the eigenvalues of the Hamiltonian matrix:

$$0 = \omega^4 - \omega^2 2K\left(\frac{1}{M_A} + \frac{1}{M_B}\right) + \frac{4K^2}{M_A M_B}[1 - \cos^2(ka)] \tag{7.117}$$

$$\omega_\pm^2(k) = K\left(\frac{1}{M_A} + \frac{1}{M_B}\right) \pm K\sqrt{\left(\frac{1}{M_A} - \frac{1}{M_B}\right)^2 + \frac{4\cos^2(ka)}{M_A M_B}} \tag{7.118}$$

There are two branches to the phonon dispersion, as shown in fig. 7.5. The acoustical modes $\omega_-(k)$ vanish at $k=0$, while the optical modes have $\omega_+(k)^2 = 2K/\mu$ at this point, where μ is the reduced mass. The phonon wave vectors for the periodic chain are

$$k_\alpha = \frac{2\pi\alpha}{2aN}, \quad -\frac{N-2}{2} < \alpha < \frac{N}{2} \tag{7.119}$$

where the last equation assumes that N is an even integer. For large values of N, the edge of the Brillouin zone is at $\alpha = \pm N/2$ so that $ka = \pm\pi/2$. Here $\cos(ka) = 0$ and the phonon frequencies are

$$\omega_\pm^2(\pi/2a) = \frac{2K}{M_A}, \frac{2K}{M_B} \tag{7.120}$$

where ω_+ has the smaller of the two masses. In fig. 7.5 the masses are M_A, $M_B = 2M_A$. Define $\omega_0^2 = K/M_A$ and the figure shows $\Omega_\pm = \omega_\pm/\omega_0$.

The above derivation is easy since it only involves diagonalizing a 2×2 matrix. Three-dimensional crystals may have $N \sim 100$ atoms in the unit cell, and the dynamical phonon matrix has a dimension of $3N \sim 300$. Such calculations are done using a computer. The matrix in eqn. (7.116) has an undesirable feature, in that it is not symmetric. This increases the numerical complexity. However, it can be made symmetric in a simple way. Define new collective displacement amplitudes:

$$q_{Ak} = \sqrt{M_A}Q_{Ak}, \quad q_{Bk} = \sqrt{M_B}Q_{Bk} \tag{7.121}$$

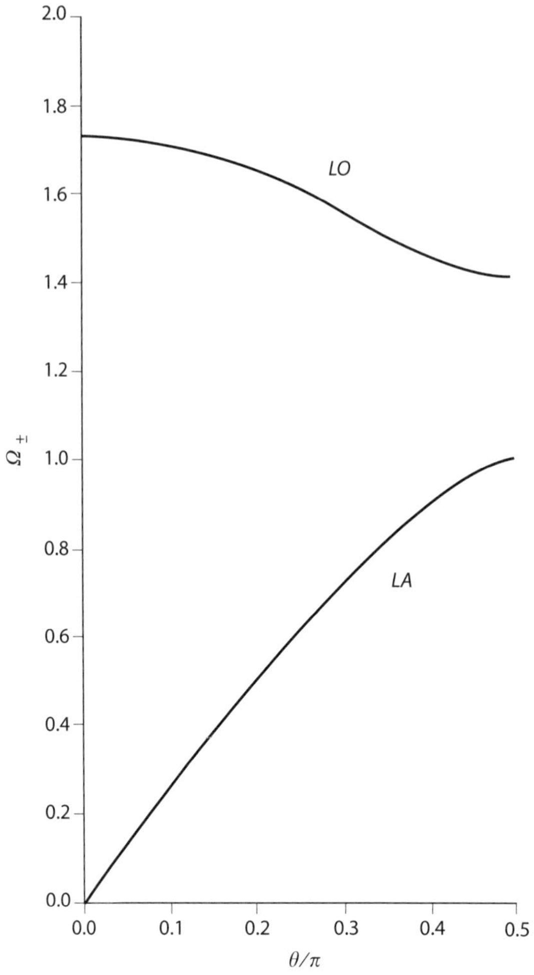

FIGURE 7.5. Two phonon branches of a binary chain.

The dynamical matrix in these variables is

$$\omega^2 \begin{bmatrix} q_{Ak} \\ q_{B,-k} \end{bmatrix} = 2K \begin{bmatrix} \dfrac{1}{M_A} & -\dfrac{\cos(ka)}{\sqrt{M_A M_B}} \\ -\dfrac{\cos(ka)}{\sqrt{M_A M_B}} & \dfrac{1}{M_B} \end{bmatrix} \begin{bmatrix} q_{Ak} \\ q_{B,-k} \end{bmatrix} \tag{7.122}$$

The dynamical matrix is now symmetric. This trick works for any number of atoms in the unit cell, even if each have a different mass. The eigenvalues (frequencies) of this matrix are identical to the original one.

Raising and lowering operators can be defined for these eigenfunctions:

$$Q_{Ak} = \sum_{\lambda=1}^{2} X_{A,k\lambda} \xi_{A,k\lambda} (a_{k\lambda} + a^\dagger_{-k\lambda}), \quad X^2_{A,k\lambda} = \frac{\hbar}{2M_A \omega_\lambda(k)} \tag{7.123}$$

$$Q_{Bk} = \sum_{\lambda=1}^{2} X_{B,k\lambda}\xi_{B,k\lambda}(a_{k\lambda} + a^{\dagger}_{-k\lambda}), \tag{7.124}$$

where λ is the polarization of the phonon: acoustical or optical. The factors $\xi_{A,k\lambda}$, $\xi_{B,k\lambda}$ are the relative displacements of each atom for that value of $(k\lambda)$, as determined by the eigenvectors of the dynamical matrix. They are normalized so that

$$\xi^2_{B,k\lambda} + \xi^2_{B,k\lambda} = 1 \tag{7.125}$$

In terms of these operators, one can write the original Hamiltonian as

$$H = \sum_{k\lambda} \hbar\omega_\lambda(k)\left[a^{\dagger}_{k\lambda}a_{k\lambda} + \frac{1}{2}\right] \tag{7.126}$$

The results for two atoms per unit cell in a linear chain can be generalized to any number of atoms in a unit cell in any dimension. Let **k** be a vector and $1 \le \lambda \le 3N_c$, where N_c is the number of atoms in the unit cell. The Hamiltonian and displacements are

$$H = \sum_{\mathbf{k}\lambda} \hbar\omega_\lambda(\mathbf{k})\left[a^{\dagger}_{\mathbf{k}\lambda}a_{\mathbf{k}\lambda} + \frac{1}{2}\right] \tag{7.127}$$

$$Q_{j\mathbf{k}} = \sum_{\lambda=1}^{3N_c} X_{j,\mathbf{k}\lambda}\xi_{j,\mathbf{k}\lambda}(a_{\mathbf{k}\lambda} + a^{\dagger}_{-\mathbf{k}\lambda}), \quad X^2_{j,\mathbf{k}\lambda} = \frac{\hbar}{2M_j\omega_\lambda(\mathbf{k})} \tag{7.128}$$

The symbol j denotes one of the N_c atoms in the unit cell.

7.3 Phonon Density of States

The phonon density of states is an important feature of the system of phonons. When averaging over the properties of the phonon system, often only the density of states is needed. The general formula for a unit cell of N_a atoms in a crystal of dimension d is

$$F(\omega) = \sum_{\lambda=1}^{N_a d} \int \frac{d^d q}{(2\pi)^d}\delta[\omega - \omega_\lambda(\mathbf{q})] \tag{7.129}$$

For example, the simple chain of identical atoms in one dimension has a density of states

$$F(\omega) = \int_{-\pi/a}^{\pi/a} \frac{dq}{2\pi}\delta[\omega - \omega_X \sin(qa/2)] \tag{7.130}$$

$$= \frac{2}{a\pi\omega_X|\cos(qa/2)|} = \frac{2}{a\pi\sqrt{\omega_X^2 - \omega^2}}$$

The density of states diverges near the edge of the Brillouin zone, when $q \to \pi/a$ and $\omega \to \omega_X$. This divergence is a van Hove singularity. Figure 7.6 shows the phonon density of states of Al and AgCl. The various peaks are Van Hove singularities, which are found in all dimensions. In Al, the high-frequency peak is from the LA phonons, while in AgCl it is from the LO phonon.

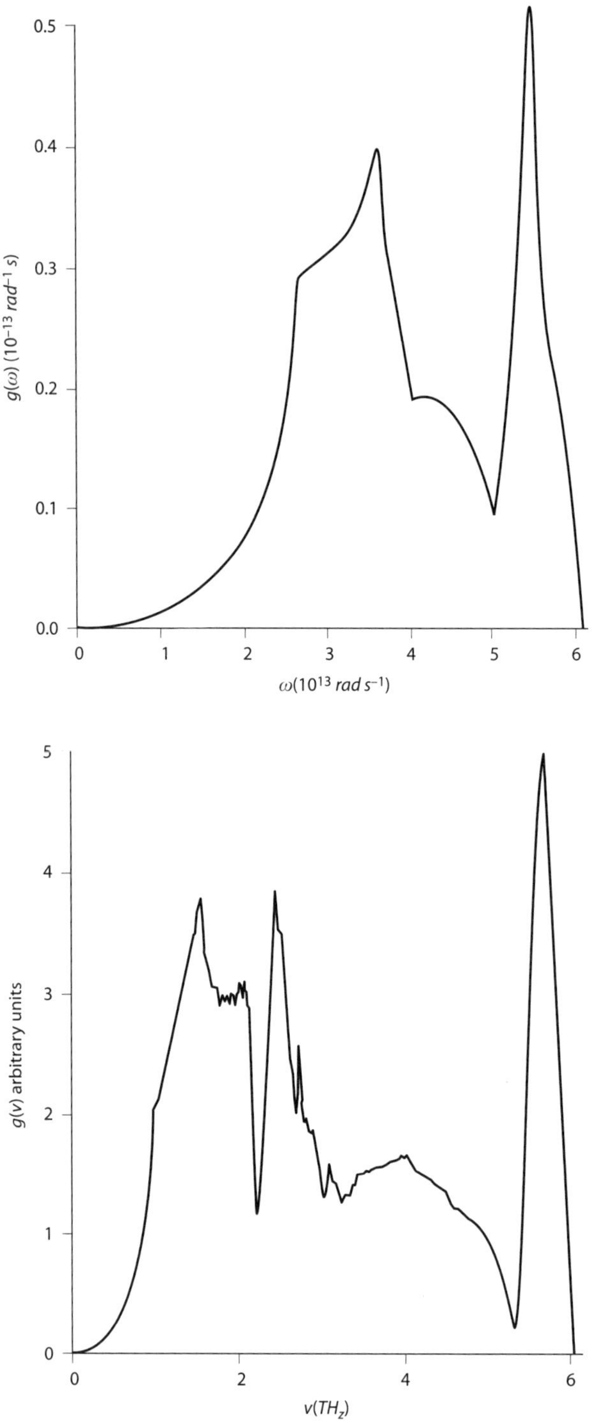

FIGURE 7.6. Phonon density of states in (a) aluminum, from Stedman et al., *Phys. Rev.* **162**, 549 (1967), and (b) AgCl, from Vijayaraghaven et al., *Phys. Rev. B* **1**, 4819 (1970). Both figures used with permission of the American Physical Society.

7.3.1 Phonon Heat Capacity

An example of using the density of phonon states is to find the average energy in the phonon system at temperature T:

$$U(T) = \sum_{\lambda} \int \frac{d^3q}{(2\pi)^3} \hbar\omega_\lambda(\mathbf{q}) \left[\frac{1}{e^{\beta\hbar\omega_\lambda(\mathbf{q})} - 1} + \frac{1}{2} \right] \tag{7.131}$$

Use the density of states $F(\omega)$ defined in eqn. (7.129):

$$U(T) = \hbar \int_0^{\omega_X} \omega d\omega F(\omega) \left[\frac{1}{e^{\beta\hbar\omega} - 1} + \frac{1}{2} \right] \tag{7.132}$$

where ω_X is the maximum phonon frequency in the Brillouin zone. It is numerically efficient to calculate the density of states $F(\omega)$ once, and then use it at each value of temperature to do the single integral to get the phonon internal energy $U(T)$. The heat capacity is

$$C(T) = \frac{dU}{dT} \tag{7.133}$$

An important expansion at high temperature occurs whenever $\beta\hbar\omega_X < 1$. Then one can expand the exponent in the occupation number:

$$\frac{1}{e^{\beta\hbar\omega} - 1} = \frac{k_B T}{\hbar\omega} - \frac{1}{2} + O(\hbar\omega/k_B T) \tag{7.134}$$

$$U(T) = k_B T \int_0^{\omega_X} d\omega F(\omega) + O(1/k_B T) \tag{7.135}$$

$$C(T) = k_B \int_0^{\omega_X} d\omega F(\omega) - O(1/k_B T^2) \tag{7.136}$$

$$= k_B dN_a \frac{\Omega_{BZ}}{(2\pi)^3} = \frac{k_B}{\Omega_0} dN_a$$

where d is the dimension. The heat capacity from phonons always goes to a constant at high temperature. Since Ω_0 is the volume of a unit cell, then Ω_0/N_a is the volume per atom.

The other important limit is at low temperature. Assume that T is quite low, so that only the low-frequency part of the density of states is important. Only the acoustical phonons contribute to the density of states at low temperature. For a solid of dimension d, there are d acoustical phonons. Acoustical phonons have a dispersion $\omega_\lambda(q) \approx C_\lambda q$ at long wavelength. The density of states is

$$\lim_{\omega \to 0} F(\omega) \propto \sum_{\lambda=1}^{d} \int dq q^{d-1} \delta[\omega - C_\lambda q] \tag{7.137}$$

$$\propto \omega^{d-1} \sum_{\lambda=1}^{d} C_\lambda^{-d}$$

$$\lim_{\omega \to 0} F(\omega) = G\omega^{d-1} \tag{7.138}$$

The low-temperature internal energy is a constant from the zero-point motion, plus a term proportional to a power of T:

$$\lim_{T\to 0} U(T) = U_0 + \hbar G \int_0^\infty d\omega \frac{\omega^d}{e^{\beta\hbar\omega} - 1} \tag{7.139}$$

where the upper limit has been replaced by infinity. Change integration variables to $x = \beta\hbar\omega$ and find

$$\lim_{T\to 0} U(T) = U_0 + \hbar G \left(\frac{k_B T}{\hbar}\right)^{d+1} \int_0^\infty \frac{x^d dx}{e^x - 1} \tag{7.140}$$

$$\lim_{T\to 0} C(T) = k_B G (d+1) \left(\frac{k_B T}{\hbar}\right)^{d} \int_0^\infty \frac{x^d dx}{e^x - 1} \tag{7.141}$$

The heat capacity from phonons at low temperature goes as $C \propto T^d$, where d is the dimension. In one dimension, such as polymers, it goes as T, while for three-dimensional crystals it goes as T^3.

The heat capacity of a solid below room temperature was measured experimentally in the nineteenth century. The fact that it became small at lower temperatures could not be explained by classical physics. Einstein first suggested that the results could be obtained by using phonons of a constant frequency, which gives an internal energy of

$$U(T) = 3N\hbar\omega_0 \left[\frac{1}{e^{\beta\hbar\omega_0} - 1} + \frac{1}{2}\right] \tag{7.142}$$

where N is the number of atoms. His proposal was one of the first applications of Planck's constant. His formula gave a heat capacity that vanished at low temperature, but did not explain the data. Today, the phrase *Einstein model* is used whenever the phonon spectrum is approximated by a single optical phonon ω_0. Debye suggested that a better model for the phonon frequencies had the three acoustical modes with constant dispersion: $\omega_\lambda(q) = C_\lambda q$ throughout the Brillouin zone. Then the internal phonon energy is similar to eqn. (7.140):

$$U(T) = U_0 + G \left(\frac{k_B T}{\hbar}\right)^{d+1} \int_0^{x_X} \frac{x^d dx}{e^x - 1} \tag{7.143}$$

$$x_X = \frac{\hbar\omega_X}{k_B T} \tag{7.144}$$

This gave better agreement with the data than Einstein's model. The first three-dimensional phonon calculation, using a spring and mass model, was done by Born and von Karman (1912, 1913) [1]. They achieved much better agreement with experiments on heat capacity.

In an earlier chapter, the low-temperature heat capacity from the electrons in a metal was shown to be $C = \gamma T$, where γ depended on the electronic density of states. Combining the contributions from electrons and phonons at low temperature gives

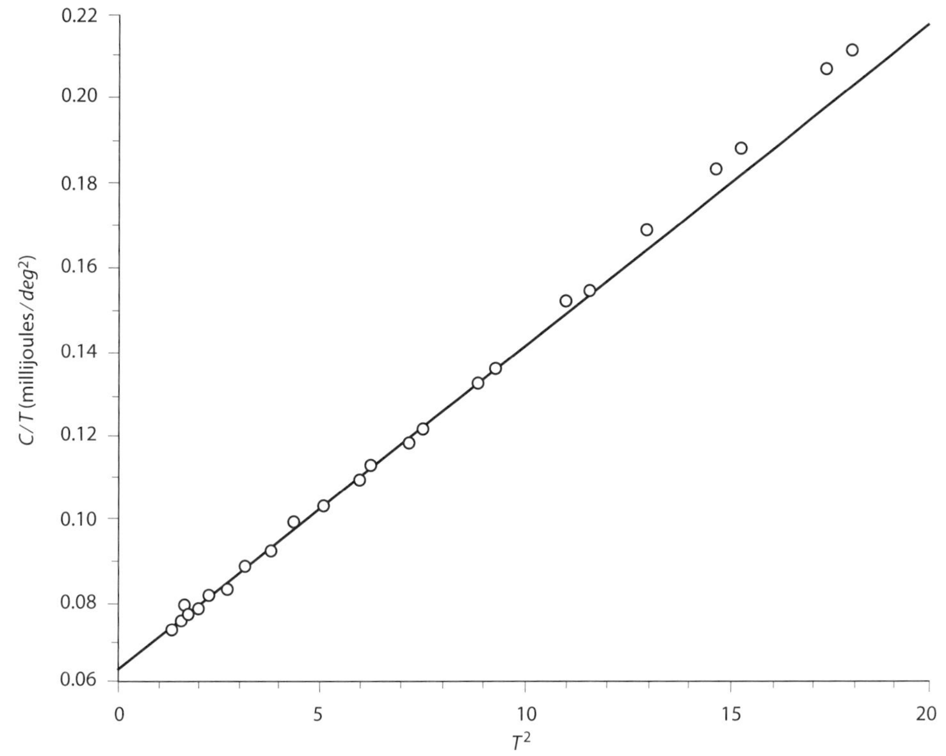

FIGURE 7.7. Heat capacity C/T vs. T^2 for metallic aluminum. From Phillips, *Phys. Rev.* **114**, 676 (1959). Used with permission of the American Physical Society.

$$C(T) = \gamma T + \xi T^3 \tag{7.145}$$

$$C(T)/T = \gamma + \xi T^2 \tag{7.146}$$

A common way to present the experimental data at low temperature is a graph of C/T vs. T^2. An example is shown in fig. 7.7 for Al. The slope gives ξ, while the zero intercept gives γ.

7.3.2 Isotopes

Most elements in the periodic table have several abundant isotopes. They have the same atomic number, but different atomic weights. Germanium is an example of a material with several abundant isotopes: ^{70}Ge (20%), ^{72}Ge (27%), ^{73}Ge (8%), ^{74}Ge (36%), ^{76}Ge (8%). Different isotopes have a different ion mass M_j. How does this affect the phonon energies? The spring constants are unaffected by the variations in mass.

Here we provide a simple example. The crystal has two isotopes. One has mass M_1 and concentration c_1, and the other has mass M_2 and concentration $c_2 = 1 - c_1$. Simplify the notation to have $c_1 \equiv c,\ c_2 \equiv 1-c$. The kinetic energy is divided into two terms, with subscripts one and two, to denote the two kinds of masses:

$$KE = \sum_{j_1=1}^{N_1} \frac{P_{j1}^2}{2M_1} + \sum_{j_2=1}^{N_2} \frac{P_{j2}^2}{2M_2} \tag{7.147}$$

$$N_1 = cN, \quad N_2 = (1-c)N \tag{7.148}$$

Since the masses enter the kinetic energy in the denominator, define an average mass $\bar{M}$ and another dimensionless constant Δ:

$$\frac{1}{\bar{M}} = \frac{c}{M_1} + \frac{1-c}{M_2} \tag{7.149}$$

$$\Delta = \bar{M}\left(\frac{1}{M_1} - \frac{1}{M_2}\right) = \frac{(M_2 - M_1)\bar{M}}{M_1 M_2} \tag{7.150}$$

Generally, $M_2 - M_1$ is one or two atomic mass units, and $\Delta \ll 1$. Solve for $M_{1,2}$ and find

$$\frac{1}{M_1} = \frac{1}{\bar{M}}[1 + (1-c)\Delta] \tag{7.151}$$

$$\frac{1}{M_2} = \frac{1}{\bar{M}}[1 - c\Delta] \tag{7.152}$$

$$KE = \sum_{j=1}^{N} \frac{P_j^2}{2\bar{M}} + \delta H \tag{7.153}$$

$$\delta H = \frac{\Delta}{2\bar{M}}\left[(1-c)\sum_{j1=1}^{N_1} P_{j1}^2 - c\sum_{j2=1}^{N_2} P_{j2}^2\right] \tag{7.154}$$

$$H = H_0 + \delta H \tag{7.155}$$

$$H_0 = \sum_{j=1}^{N} \frac{P_j^2}{2\bar{M}} + V(Q_1, Q_2, \ldots, Q_N) \tag{7.156}$$

The Hamiltonian H_0 is solved using the average mass $\bar{M}$. The term δH is the perturbation due to the mass fluctuations. It arises from the isotope fluctuations in the kinetic energy.

The ground-state energy is found for H_0 and has the usual form:

$$E_G^{(0)} = \sum_{\eta\mathbf{q}} \hbar\omega_\eta(\mathbf{q})\left(N_\eta(\mathbf{q}) + \frac{1}{2}\right) \tag{7.157}$$

where η is the polarization (LA, TA, LO, TO) and $\mathbf{q}$ is the wave vector in the Brillouin zone. The thermal occupation number is $N_\eta(\mathbf{q})$. The perturbation depends on the momentum operator:

$$\mathbf{P}_j = i\sum_{\eta\mathbf{q}} P_\eta(\mathbf{q})\hat{\xi}\left[e^{i\mathbf{q}\cdot\mathbf{R}_j}a_{\eta\mathbf{q}} - e^{-i\mathbf{q}\cdot\mathbf{R}_j}a_{\eta\mathbf{q}}^\dagger\right] \tag{7.158}$$

$$P_\eta(\mathbf{q}) = \sqrt{\frac{\hbar \bar{M}\omega_\eta(\mathbf{q})}{2N}} \tag{7.159}$$

$$P_j^2 = \sum_{\eta\mathbf{q},\eta'\mathbf{q}'} P_\eta(\mathbf{q})\, P_{\eta'}(\mathbf{q}')\, \hat{\xi}\cdot\hat{\xi}'\Big[e^{i(\mathbf{q}-\mathbf{q}')\cdot\mathbf{R}_j}(a_{\eta\mathbf{q}}a^\dagger_{\eta'\mathbf{q}'} + a^\dagger_{\eta\mathbf{q}}a_{\eta'\mathbf{q}'})$$

$$- e^{i(\mathbf{q}+\mathbf{q}')\cdot\mathbf{R}_j}a_{\eta\mathbf{q}}a_{\eta'\mathbf{q}'} - e^{i(\mathbf{q}+\mathbf{q}')\cdot\mathbf{R}_j}a^\dagger_{\eta\mathbf{q}}a^\dagger_{\eta'\mathbf{q}'}\Big] \tag{7.160}$$

Evaluate this expression using perturbation theory. First-order perturbation theory gives

$$\left\langle \frac{P_j^2}{2\bar{M}} \right\rangle = \frac{1}{4N}\sum_{\eta\mathbf{q}} \hbar\omega_\eta(\mathbf{q})\,[2N_\eta(\mathbf{q}) + 1] \tag{7.161}$$

where the only terms that contributed are those with $aa^\dagger + a^\dagger a$, and $\eta = \eta'$, $\mathbf{q} = \mathbf{q}'$. The result is the same for all sites $\mathbf{R}_j$, so averaging over the two kinds of sites gives

$$\langle \delta H \rangle = \frac{N}{4N}\sum_{\eta\mathbf{q}} \hbar\omega_\eta(\mathbf{q})\,[2N_\eta(\mathbf{q}) + 1][c(1-c) - c(1-c)] = 0 \tag{7.162}$$

The first-order energy is zero.

Next evaluate this expression in the second-order of perturbation theory. All of the terms in P_j^2 contribute. Squaring δH gives several terms, where each is a double summation over sites. Separate the terms with a single summation and those with two:

$$(\delta H)^2 = \frac{1}{4\bar{M}^2}\Big[(1-c)^2\sum_{j1} P_{j1}^4 + c^2\sum_{j2} P_{j2}^4 + (1-c)^2\sum_{j1} P_{j1}^2 \sum_{j1'\neq j_1} P_{j1'}^2$$

$$+ c^2\sum_{j2} P_{j2}^2 \sum_{j2'\neq j2} P_{j2'}^2 - 2c(1-c)\sum_{j1} P_{j1}^2 \sum_{j2} P_{j2}^2\Big] \tag{7.163}$$

When the last three terms are averaged, one gets for them

$$c^2(1-c)^2 \sum_{j=1}^{N} P_j^2 \sum_{j'\neq j}^{N} P_{j'}^2 (1-1)^2 = 0 \tag{7.164}$$

All of the double-summation terms cancel. The first two terms are evaluated in second order. All terms have the same factor in the numerator of $(\hat{\xi}\cdot\hat{\xi}')^2\hbar^2\omega_\eta(\mathbf{q})\omega_{\eta'}(\mathbf{q}')/4N^2$. The other factors are as follows:

- Terms with two raising operators create two additional phonons in the intermediate state:

$$\frac{[N_\eta(\mathbf{q}) + 1]\,[N_{\eta'}(\mathbf{q}') + 1]}{\hbar[\omega_\eta(\mathbf{q}) + \omega_{\eta'}(\mathbf{q}')]} \tag{7.165}$$

- Terms with two lowering operators remove two phonons from the intermediate state:

$$-\frac{N_\eta(\mathbf{q})\, N_{\eta'}(\mathbf{q}')}{\hbar[\omega_\eta(\mathbf{q}) + \omega_{\eta'}(\mathbf{q}')]} \tag{7.166}$$

  where the negative sign comes from the energy denominator.

- Terms with one raising operator and one lowering operator change the frequency of the phonons in the intermediate state:

$$\frac{[N_\eta(\mathbf{q})+1]N_{\eta'}(\mathbf{q}')}{\hbar[\omega_\eta(\mathbf{q})-\omega_{\eta'}(\mathbf{q}')]} \tag{7.167}$$

- The result is the same for each value of j. So the average over the two kinds of sites gives

$$N\Delta^2[c(1-c)^2+c^2(1-c)] = N\Delta^2 c(1-c) \tag{7.168}$$

The total second-order energy is

$$E_G^{(2)} = \frac{1}{4N}\Delta^2 c(1-c)\hbar \sum_{\eta\mathbf{q},\eta'\mathbf{q}'} (\hat{\xi}\cdot\hat{\xi}')^2 \omega_\eta(\mathbf{q})\omega_{\eta'}(\mathbf{q}')\left[\frac{N_\eta(\mathbf{q})+N_{\eta'}(\mathbf{q}')+1}{\omega_\eta(\mathbf{q})+\omega_{\eta'}(\mathbf{q}')}\right.$$
$$\left.+\frac{N_\eta(\mathbf{q})-N_{\eta'}(\mathbf{q}')}{\omega_\eta(\mathbf{q})-\omega_{\eta'}(\mathbf{q}')}\right] \tag{7.169}$$

The prefactor of this expression contains $\Delta^2 c(1-c)$. The factor $c(1-c) < 1/4$. The factor of $\Delta \approx \delta M/\bar{M} \ll 1$. So the energy correction is generally quite small.

The change in energy of an individual phonon mode is obtained by a functional derivative with respect to the occupation number:

$$\hbar\Omega_\eta(\mathbf{q}) = \frac{\delta E_G}{\delta N_\eta(\mathbf{q})} \tag{7.170}$$
$$= \hbar\omega_\eta(\mathbf{q})\left[1+\Delta^2 c(1-c)\hat{\xi}\cdot\mathcal{S}\cdot\hat{\xi}_\eta\right]$$

$$\mathcal{S}_\eta(\mathbf{q}) = \frac{1}{2N}\sum_{\eta'\mathbf{q}'}\omega_{\eta'}(\mathbf{q}')\hat{\xi}'\hat{\xi}'\left[\frac{1}{\omega_\eta(\mathbf{q})+\omega_{\eta'}(\mathbf{q}')}+\frac{1}{\omega_\eta(\mathbf{q})-\omega_{\eta'}(\mathbf{q}')}\right]$$
$$= \frac{\omega_\eta(\mathbf{q})}{N}\sum_{\eta'\mathbf{q}'}\omega_{\eta'}(\mathbf{q}')\frac{\hat{\xi}'\hat{\xi}'}{\omega_\eta(\mathbf{q})^2-\omega_{\eta'}(\mathbf{q}')^2} \tag{7.171}$$

$$\mathcal{S}_\eta(\mathbf{q}) = \omega_\eta(\mathbf{q})\int_0^{\omega_X}\mathcal{F}(\omega)d\omega\frac{\omega}{\omega_\eta(\mathbf{q})^2-\omega^2} \tag{7.172}$$

$$\mathcal{F}(\omega) = \frac{1}{N}\sum_{\eta'\mathbf{q}'}\delta[\omega-\omega_{\eta'}(\mathbf{q}')]\hat{\xi}'\hat{\xi}' \tag{7.173}$$

where $\mathcal{F}(\omega)$ is the tensor version of the phonon density of states. For cubic crystals the tensor is diagonal: $\mathcal{F}_{\mu\nu}(\omega) = \delta_{\mu\nu}D(\omega)$.

Including the isotopes in the ground-state energy or in the frequencies of the individual phonons makes only a small correction. Usually this conribution is ignored.

A similar expression can be obtained for the lifetime of a phonon from isotope scattering, The energy denominator $1/(\omega-\omega')$ is replaced by a delta function $2\pi\delta(\omega-\omega')$:

$$\frac{1}{\tau_\eta(\mathbf{q})} = \frac{\pi\Delta^2 c(1-c)}{N}\omega_\eta(\mathbf{q})\sum_{\eta'\mathbf{q}'}\omega_{\eta'}(\mathbf{q}')(\hat{\xi}'\cdot\hat{\xi})^2\delta[\omega_\eta(\mathbf{q})-\omega_{\eta'}(\mathbf{q}')]$$
$$= \pi\Delta^2 c(1-c)\,\omega_\eta(\mathbf{q})^2\hat{\xi}\cdot\mathcal{F}[\omega_\eta(\mathbf{q})]\cdot\hat{\xi} \tag{7.174}$$

This lifetime is quite large since the prefactors in the above expression are quite small.

7.4 Local Modes

The phonon calculations have assumed that all atoms in the chain have the same mass or, in binary chains, that the mass alternates. But solids have substitutional impurities, and they will have a different mass. The mass may vary for different isotopes, but that case was covered in the prior section.

Consider a crystal in which all atoms have the same mass M_0, except one atom that has mass M_1. The phonons in the lattice will have their maximum phonon energy given by

$$\omega_X = 2\sqrt{\frac{K}{M_0}} \tag{7.175}$$

The isotope with mass M_1 wants to have a different frequency, say $\omega_1 = 2\sqrt{K/M_1}$. If $M_1 > M_0$, then $\omega_1 < \omega_X$. The local vibrations around the impurity will be in the band of phonon frequencies. Then the local vibrations are just part of the general phonon spectrum. However, if the local mass is lighter, so that $M_1 < M_0$, then $\omega_1 > \omega_X$, and the impurity frequency is higher than any phonon mode. The local vibration can exist as a separate eigenstate. It is called a *local mode*, and is found generally in crystals when a substitutional atom has a lighter mass than the host atoms.

Let us solve classically a simple model. There is a periodic chain of $N-1$ identical atoms in one dimension, but one site L has a different mass M_1. All spring constants are alike: the chemical bonding is the same for the different isotopes. The equations of motion are

$$\omega^2 Q_j = \omega_0^2[2Q_j - Q_{j+1} - Q_{j-1}], \quad j \neq L \tag{7.176}$$

$$\omega^2 Q_L = \omega_1^2[2Q_L - Q_{L+1} - Q_{L-1}] \tag{7.177}$$

$$\omega_0^2 = \frac{K}{M_0}, \quad \omega_1^2 = \frac{K}{M_1} \tag{7.178}$$

Multiply the equation with Q_j by $\exp(i\theta j)/\sqrt{N}$, where $\theta = ka$. The lattice constant is a. Then sum over all values of j. Define the collective variable as

$$q_k = \sqrt{\frac{1}{N}}\sum_j Q_j \exp(i\theta j) \tag{7.179}$$

Equations (7.176) and (7.177) become

$$\omega^2 q_k = \omega_k^2 q_k + \sqrt{\frac{1}{N}}(\omega_1^2 - \omega_0^2)[2Q_L - Q_{L+1} - Q_{L-1}]e^{i\theta L} \tag{7.180}$$

$$\omega_k^2 = 2\omega_0^2[1-\cos(\theta)] \tag{7.181}$$

where the last terms in eqn. (7.176) are

$$\sum_j [Q_{j+1} + Q_{j-1}]e^{i\theta j} = \sum_j Q_j\{e^{i\theta(j-1)} + e^{i\theta(j+1)}\}$$

$$= 2\cos(\theta)\sum_j Q_j e^{i\theta j} \tag{7.182}$$

Solve eqn. (7.180) for q_k and take the inverse transform:

$$q_k = \sqrt{\frac{1}{N}}\frac{\omega_1^2 - \omega_0^2}{\omega^2 - \omega_k^2}[2Q_L - Q_{L+1} - Q_{L-1}]\exp(i\theta L) \tag{7.183}$$

$$Q_j = \sqrt{\frac{1}{N}}\sum_k q_k \exp(-ikaj) \tag{7.184}$$

$$Q_j = (\omega_1^2 - \omega_0^2)[2Q_L - Q_{L+1} - Q_{L-1}]\,G_{jL}(\omega^2) \tag{7.185}$$

$$G_{jL}(\omega^2) = \frac{1}{N}\sum_k \frac{\exp[ika(L-j)]}{\omega^2 - \omega_k^2} \tag{7.186}$$

In eqn. (7.185), let $j = L,\ L+1,\ L-1$ in succession. Then

$$\begin{aligned}[2Q_L - Q_{L+1} - Q_{L-1}] &= (\omega_1^2 - \omega_0^2)[2Q_L - Q_{L+1} - Q_{L-1}] \\ &\quad \times [2G_{LL} - G_{L+1,L} - G_{L-1,L}]\end{aligned} \tag{7.187}$$

Cancel the common factor of $[2Q_L - Q_{L+1} - Q_{L-1}]$ and find the equation

$$1 = (\omega_1^2 - \omega_0^2)[2G_{LL} - G_{L+1,L} - G_{L-1,L}] \tag{7.188}$$

In the limit of large N, the Green's function depends only on the difference of the positions, so simplify to

$$1 = 2(\omega_1^2 - \omega_0^2)[G_0 - G_1] \tag{7.189}$$

This equation describes the motion of vibrational states near the location of the impurity. Impurity local modes are solutions of this equation. For the chain of atoms, the two Green's function integrals can be evaluated:

$$\begin{aligned}G_0(\omega^2) &= \int_{-\pi}^{\pi}\frac{d\theta}{2\pi}\frac{1}{\omega^2 - 2\omega_0^2[1-\cos(\theta)]} \\ &= \frac{1}{\sqrt{\omega^2(\omega^2 - 4\omega_0^2)}}, \quad \omega^2 > 4\omega_0^2 \\ &= 0 \text{ if } \omega^2 < 4\omega_0^2\end{aligned} \tag{7.190}$$

$$\begin{aligned}G_1(\omega^2) &= \int_{-\pi}^{\pi}\frac{d\theta}{2\pi}\frac{\cos(\theta)}{\omega^2 - 2\omega_0^2[1-\cos(\theta)]} \\ &= \frac{1}{2\omega_0^2}\left[1 - \frac{\omega^2 - 2\omega_0^2}{\sqrt{\omega^2(\omega^2 - 4\omega_0^2)}}\right], \quad \omega^2 > 4\omega_0^2 \\ &= \frac{1}{2\omega_0^2} \text{ if } \omega^2 < 4\omega_0^2\end{aligned} \tag{7.191}$$

If $\omega^2 > 4\omega_0^2$, then $G_0 > G_1$. For eqn. (7.189) to be satisfied, then $\omega_1^2 > \omega_0^2$, which means that $M_0 > M_1$. There is a local mode of vibration whenever the impurity mass is lighter than the mass of the host ions. This feature is also found in two and three dimensions. The maximum band frequency is $\omega_k^2 = 4\omega_0^2$ is at the edge of the BZ, $\theta = \pi$. The vibrational frequency ω of the local mode is outside the normal phonon band and can exist as a separate excitation. It is an eigenstate of the Hamiltonian. Its frequency is found by solving eqn. (7.189).

- If $\omega^2 < 4\omega_0^2$ the equation has no solution, and there is no local mode. The impurity vibrates as part of the normal modes.
- When $\omega^2 > 4\omega_0^2$, there is a local mode, and the vibration of an atom ℓ away from the impurity has the amplitude

$$G_\ell = G_0(-1)^\ell e^{-\alpha|\ell|} \tag{7.192}$$

$$\omega^2 = 2\omega_0^2\left[1 + \cosh(\alpha)\right] \tag{7.193}$$

The local mode behaves as a phonon bound state.

7.5 Elasticity

Phonons of long wavelength have a wave vector $\mathbf{q} \to 0$. They are easily observed by a variety of experiments. Raman scattering can measure the frequencies of some optical phonons. The acoustic modes at long wavelength morph into ordinary sound waves. There are three sound waves in the bulk material: one longitudinal and two transverse. They are described by the theory of elasticity.

7.5.1 *Stress and Strain*

Elasticity takes the view that the material is homogeneous. The topic was developed in the nineteenth century before the discovery of atoms. Let $\mathbf{u}(\mathbf{r})$ denote the small displacement of material from equilibrium at point $\mathbf{r}$. The displacement $\mathbf{u}$ is a vector with components u_i, $(i = x, y, z)$. The *strain tensor* is defined as the symmetric derivative of the displacement:

$$\varepsilon_{ij} = \frac{1}{2}\left[\frac{\partial u_i}{\partial x_j} + \frac{\partial u_j}{\partial x_i}\right] = \varepsilon_{ji} \tag{7.194}$$

where the vector $\mathbf{r}$ has components x_i. The strain is dimensionless. The energy per unit volume in the strain field is

$$\mathcal{E} = \frac{1}{2}\sum_{ijkl} \varepsilon_{ij} C_{ij,kl} \varepsilon_{kl} \tag{7.195}$$

The *elastic constants* $C_{ij,kl}$ are a fourth rank tensor, which have units of energy per volume. The number of independent, nonzero, components depends on the symmetry of the crystal. The Voigt notation is to write this as $C_{\alpha\beta}$, where (α, β) each range from one to six according to

$$1 = (xx),\ 2 = (yy),\ 3 = (zz),\ 4 = (yz),\ 5 = (zx),\ 6 = (xy) \tag{7.196}$$

The constants are the same when the indices are reversed: $C_{\alpha\beta} = C_{\beta\alpha}$. Also define another symbol for the strain:

$$e_1 = \varepsilon_{xx},\ e_2 = \varepsilon_{yy},\ e_3 = \varepsilon_{zz},\quad e_4 = 2\varepsilon_{yz},\ e_5 = 2\varepsilon_{xz},\ e_6 = 2\varepsilon_{xy} \tag{7.197}$$

Now the energy can be written as

$$\mathcal{E} = \frac{1}{2} \sum_{\alpha,\beta=1}^{6} e_\alpha C_{\alpha\beta} e_\beta \tag{7.198}$$

The *dilation* is defined as

$$\Delta = \nabla \cdot \mathbf{u} = \sum_{\alpha=1}^{3} e_\alpha \tag{7.199}$$

It is important in defining the electron–phonon interaction.

The following are two important symmetry cases.

- Cubic crystals have three independent elastic constants, $C_{11} = C_{22} = C_{33}$, $C_{12} = C_{13} = C_{23}$, and $C_{44} = C_{55} = C_{66}$. The tensor has the form

$$\begin{bmatrix} C_{11} & C_{12} & C_{12} & 0 & 0 & 0 \\ C_{12} & C_{11} & C_{12} & 0 & 0 & 0 \\ C_{12} & C_{12} & C_{11} & 0 & 0 & 0 \\ 0 & 0 & 0 & C_{44} & 0 & 0 \\ 0 & 0 & 0 & 0 & C_{44} & 0 \\ 0 & 0 & 0 & 0 & 0 & C_{44} \end{bmatrix} \tag{7.200}$$

- Hexagonal crystals such as hcp and wurtzite have five elastic constants: C_{11}, C_{12}, C_{33}, C_{13}, C_{44}. The (x, y)-axes have the same constants, but the z-axis has different values. A sixth constant is used, which is $C_{66} = (C_{11} - C_{12})/2$. The tensor is

$$\begin{bmatrix} C_{11} & C_{12} & C_{13} & 0 & 0 & 0 \\ C_{12} & C_{11} & C_{13} & 0 & 0 & 0 \\ C_{13} & C_{13} & C_{33} & 0 & 0 & 0 \\ 0 & 0 & 0 & C_{44} & 0 & 0 \\ 0 & 0 & 0 & 0 & C_{44} & 0 \\ 0 & 0 & 0 & 0 & 0 & C_{66} \end{bmatrix} \tag{7.201}$$

The elastic tensor for other crystal groups can be found in Cady [3].

The *stress tensor* is found from Hooke's law: the stress is linear in the strain, and the constants of proportionality are the elastic constants

$$\sigma_{ij} = \sum_{kl} C_{ij,kl} \varepsilon_{kl} \tag{7.202}$$

$$\sigma_\alpha = \sum_\beta C_{\alpha\beta} e_\beta \tag{7.203}$$

where the same result is given in the two notations. The equation of elasticity is derived from Newton's second law:

$$\rho \frac{\partial^2}{\partial t^2} u_i = \sum_j \frac{\partial}{\partial x_j} \sigma_{ij} \tag{7.204}$$

where ρ is the density of the material in units of kilograms per volume. The ratio $C_{\alpha\beta}/\rho$ has the units of velocity squared. A sound wave has a displacement given by

$$\mathbf{u}(\mathbf{r}) = \mathbf{u}_0 \exp[i(\mathbf{q}\cdot\mathbf{r} - \omega t)] \tag{7.205}$$

Time derivatives give $-i\omega$, and space derivatives $\partial_i \to iq_i$. The strains are

$$e_1 = iq_x u_x, \quad e_2 = iq_y u_y, \quad e_3 = iq_z u_z \tag{7.206}$$

$$e_4 = i(q_z u_y + q_y u_z), \quad e_5 = i(q_z u_x + q_x u_z), \quad e_6 = i(q_x u_y + q_y u_x) \tag{7.207}$$

In cubic crystals the three equations that define the sound waves are

$$\rho u_x \omega^2 = u_x\left[q_x^2 C_{11} + (q_y^2 + q_z^2) C_{44}\right] + q_x(q_y u_y + q_z u_z)(C_{12} + C_{44}) \tag{7.208}$$

$$\rho u_x \omega^2 = u_y\left[q_y^2 C_{11} + (q_x^2 + q_z^2) C_{44}\right] + q_y(q_x u_x + q_z u_z)(C_{12} + C_{44}) \tag{7.209}$$

$$\rho u_z \omega^2 = u_z\left[q_z^2 C_{11} + (q_x^2 + q_y^2) C_{44}\right] + q_z(q_x u_x + q_y u_y)(C_{12} + C_{44}) \tag{7.210}$$

In hexagonal crystals the equations are

$$\rho u_x \omega^2 = u_x\left[q_x^2 C_{11} + q_y^2 C_{66} + q_z^2 C_{44}\right] + q_x[q_y u_y (C_{12} + C_{66}) + q_z u_z (C_{13} + C_{44}) \tag{7.211}$$

$$\rho u_y \omega^2 = u_x\left[q_y^2 C_{11} + q_x^2 C_{66} + q_z^2 C_{44}\right] + q_y[q_x u_x (C_{12} + C_{66}) + q_z u_z (C_{13} + C_{44}) \tag{7.212}$$

$$\rho u_z \omega^2 = u_z\left[q_z^2 C_{33} + (q_x^2 + q_y^2) C_{44}\right] + q_z(q_x u_x + q_y u_y)(C_{13} + C_{44}) \tag{7.213}$$

In every case, we can solve these three coupled equations for the dispersion of the three sound waves. There are three roots, which are the longitudinal wave and two transverse sound waves. Since $\omega(\mathbf{q}) = |\mathbf{q}| v(\hat{\mathbf{q}})$, the solution gives the angular dependence of the sound velocity. Hexagonal crystals have the feature that the three sound waves are isotropic in the basal plane (i.e., when $q_z = 0$). That is not true for cubic crystals.

In cubic crystals, simple expressions are found for the sound waves in the three principal axes.

- Along the (100) direction:

  $$v_l^2 = \frac{C_{11}}{\rho}, \quad v_{t1}^2 = v_{t2}^2 = \frac{C_{44}}{\rho} \tag{7.214}$$

  Both transverse waves have the same velocity.

- Along the (110) direction:

Table 7.2 Elastic constants in units of (10^{11} dyn/cm^2), optical phonon frequencies in THz, and piezoelectric coefficient in units of C/m^2, of some cubic semiconductors

	C_{11}	C_{12}	C_{44}	ν_{LO}	ν_{TO}	e_{14}	$\varepsilon(0)/\varepsilon_0$	$\varepsilon(\infty)/\varepsilon_0$
C	107.64	12.52	57.74	39.9	39.9	0	5.70	5.70
Si	16.01	5.78	8.00	15.7	15.7	0	11.7	11.7
Ge	12.85	4.83	6.68	9.00	9.00	0	16.0	16.0
AlP	13.20	6.30	6.15	15.0	13.2		9.8	7.5
AlAs	12.50	5.34	5.42	12.1	10.8		10.06	8.16
AlSb	8.77	4.34	4.08	10.19	9.56	0.068	11.63	10.24
GaP	14.05	6.20	7.03	12.02	10.97	−0.10	11.11	9.11
GaAs	11.81	5.32	5.94	8.75	8.06	0.16	13.18	10.89
GaSb	8.83	4.03	4.32	7.19	6.92	0.126	15.69	14.44
InP	10.22	5.76	4.60	10.34	9.11	0.042	12.40	9.61
InAs	8.33	4.53	3.96	7.28	6.57	0.045	15.13	12.25
InSb	6.47	3.65	3.02	5.70	5.34	0.071	17.12	15.68

Note. Data from Landolt-Börnstein, New Series, Vol. 41A1(α and β) (Springer-Verlag, Berlin, 2001), and Sharma et al., *J. Phys. Chem. Solids* 53, 329 (1992).

$$v_l^2 = \frac{C_{11} + C_{12} + 2C_{44}}{2\rho} \tag{7.215}$$

$$v_{t1}^2 = \frac{C_{44}}{\rho}, \quad v_{t2}^2 = \frac{C_{11} - C_{12}}{2\rho} \tag{7.216}$$

- Along the (111) direction:

$$v_l^2 = \frac{C_{11} + 2C_{12} + 4C_{44}}{3\rho} \tag{7.217}$$

$$v_{t1}^2 = v_{t2}^2 = \frac{C_{11} - C_{12} + C_{44}}{3\rho} \tag{7.218}$$

The material is isotropic whenever $C_{11} = C_{12} + 2C_{44}$. Table 7.2 shows some elastic constants, optical phonon frequencies, and the one nonzero piezoelectric constant for some group IV and III–V cubic semiconductors.

7.5.2 Isotropic Materials

Isotropic crystals have sound waves with the same velocity in all directions. This behavior is found in fluids, in some glasses, and in amorphous materials. The sound waves are given in terms of two Lamé parameters (λ, μ), which act as isotropic elastic constants.

In relation to cubic materials, $C_{11} = \lambda + 2\mu$, $\mu = C_{44}$. For a cubic material to be isotropic, it must have $C_{12} = C_{11} - 2C_{44} = \lambda$ in isotropic bodies. Then the energy density can be converted from a cubic material to be

$$\mathcal{E} = \frac{1}{2}\sum_{ijkl} \varepsilon_{ij} C_{ij,kl} \varepsilon_{kl} \tag{7.219}$$

$$= \frac{1}{2}\left[C_{11}\left(\varepsilon_{xx}^2 + \varepsilon_{yy}^2 + \varepsilon_{zz}^2\right) + 2C_{12}(\varepsilon_{xx}\varepsilon_{yy} + \varepsilon_{xx}\varepsilon_{zz} + \varepsilon_{yy}\varepsilon_{zz})\right.$$

$$\left. + 4C_{44}\left(\varepsilon_{xy}^2 + \varepsilon_{xz}^2 + \varepsilon_{yz}^2\right)\right]$$

$$= \frac{1}{2}\left[(\lambda + 2\mu)\Delta^2 + 4\mu(\varepsilon_{xy}^2 + \varepsilon_{xz}^2 + \varepsilon_{yz}^2 - \varepsilon_{xx}\varepsilon_{yy} - \varepsilon_{xx}\varepsilon_{zz} - \varepsilon_{yy}\varepsilon_{zz})\right]$$

The stress tensor is

$$\sigma_{ij} = \lambda\Delta\delta_{ij} + 2\mu\varepsilon_{ij} \tag{7.220}$$

The equation for sound waves is

$$\rho\frac{\partial^2}{\partial t^2}\mathbf{u} = \mu\nabla^2\mathbf{u} + (\lambda + \mu)\nabla(\nabla\cdot\mathbf{u}) \tag{7.221}$$

Longitudinal sound waves have $\mathbf{q} \parallel \mathbf{u}$, so that the dilation $\Delta = (\nabla \cdot \mathbf{u})$ is the major variable. Take $\nabla \cdot$ of the above equation, which gives, for waves of the phase $\exp[i(\mathbf{q}\cdot\mathbf{r} - \omega t)]$,

$$\rho\omega^2\Delta = (\lambda + 2\mu)\,q^2\Delta \tag{7.222}$$

$$\omega^2 = q^2 v_l^2, \quad v_l^2 = \frac{\lambda + 2\mu}{\rho} \tag{7.223}$$

where v_l is the longitudinal sound velocity. The last term in eqn. (7.221) is absent in transverse waves, and their velocity is

$$\omega^2 = q^2 v_t^2, \quad v_t^2 = \frac{\mu}{\rho} \tag{7.224}$$

The two transverse sound waves have the same velocity.

Instead of the Lamé parameters, the sound waves in homogeneous materials are often described in terms of other parameters:

- *Young's modulus* is defined as

$$E = \mu\frac{3\lambda + 2\mu}{\lambda + \mu} \tag{7.225}$$

- *Poisson's ratio* is

$$\sigma = \frac{\lambda}{2(\lambda + \mu)} \tag{7.226}$$

- *Modulus of hydrostatic compression* is

$$K = \lambda + \frac{2}{3}\mu \tag{7.227}$$

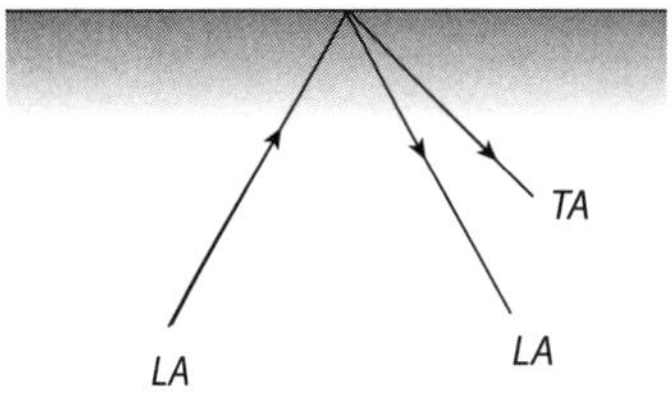

FIGURE 7.8. In two dimensions, when a longitudinal wave approaches the surface, two waves are reflected.

7.5.3 Boundary Conditions

In the harmonic approximation, phonons travel through the material without scattering. Scattering does occur when they come to a free surface or to an interface with another material.

- For the free surface, the boundary conditions are that there are no stresses at the surface. Since the stress is a tensor, then all three components of this tensor, projected on the normal vector to the surface, vanish. This gives three boundary conditions at each free surface. For example, if the free surface is in the z-direction, then

$$0 = \sigma_{zz} = \sigma_{zx} = \sigma_{zy}. \tag{7.228}$$

- For the interface between two materials, the matching conditions are twofold: (i) that all three componets of $\mathbf{u}(\mathbf{r})$ are matched at the surface, and (ii) that the three components of the stress tensor are matched at the interface.

As an example, consider a longitudinal sound wave that reflects from a free surface. An incident wave approaches the surface. Due to the boundary conditions, three waves leave the surface: one longitudinal and two transverse.

A simple example is given in two dimensions in a homogeneous material. In this case, a longitudinal wave approaches the surface, and a longitudinal and a transverse wave depart from the surface. The geometry is shown in fig. 7.8. The figure shows the (x, z)-plane. The z-direction is normal to the surface. All of the sound waves have the same wave vector parallel to the surface: $\exp(iq_x x)$.

- The incident longitudinal wave has wave vector components (q_x, q_z), and $q = \sqrt{q_x^2 + q_z^2}$, $\omega = v_l q$. Its displacement is $\hat{e}_i = (q_x, q_z)/q$.
- The reflected longitudinal wave has wave vector components $(q_x, -q_z)$, and also has $\omega = v_l q$. Its displacement is $\hat{e}_r = (q_x, -q_z)/q$.
- The reflected transverse wave has wave vector components $\mathbf{q}' = (q_x, -k_z)$, $q' = \sqrt{q_x^2 + k_z^2}$, $\omega = v_t q'$. The wave vector $k_z^2 = \omega^2/v_t^2 - q_x^2$. The displacement component is $\hat{e}_t = (k_z, q_x)/q'$, since transverse waves have $\hat{q}' \cdot \hat{e}_t = 0$.

The total displacement of the material is

$$\mathbf{u}(\mathbf{r}) = e^{iq_x x}\left[\hat{e}_i e^{iq_z z} + \hat{e}_r R e^{-iq_z z} + \hat{e}_t T e^{-ik_z z}\right] \tag{7.229}$$

Some components of strain are, at $(x = 0, z = 0)$,

$$\frac{du_x}{dx} = i\left[\frac{q_x^2}{q}(1+R) + \frac{q_x k_z}{q'}T\right] \tag{7.230}$$

$$\frac{du_z}{dz} = i\left[\frac{q_z^2}{q}(1+R) - \frac{q_x k_z}{q'}T\right] \tag{7.231}$$

$$\frac{du_x}{dz} = i\left[\frac{q_x q_z}{q}(1-R) - \frac{k_z^2}{q'}T\right] \tag{7.232}$$

$$\frac{du_z}{dx} = i\left[\frac{q_x q_z}{q}(1-R) + \frac{q_x^2}{q'}T\right] \tag{7.233}$$

$$\Delta = \frac{du_x}{dx} + \frac{du_z}{dz} = iq(1+R) \tag{7.234}$$

Transverse waves do not contribute to the dilation. The two boundary conditions for the two-dimensional problem are

$$0 = \sigma_{zz} = \lambda\Delta + 2\mu\frac{du_z}{dz} = \lambda q(1+R) + 2\mu\left[\frac{q_z^2}{q}(1+R) - \frac{q_x k_z}{q'}T\right] \tag{7.235}$$

$$0 = \sigma_{xz} = \mu\left(\frac{du_z}{dx} + \frac{du_x}{dz}\right) = \mu\left[2\frac{q_x q_z}{q}(1-R) + \frac{q_x^2 - k_z^2}{q'}T\right] \tag{7.236}$$

Solving gives the amplitudes of the two reflected waves:

$$T = -4\frac{v_t}{v_l}\frac{q_x q_z(\lambda q^2 + 2\mu q_z^2)}{D} \tag{7.237}$$

$$R = -\frac{(q_x^2 - k_z^2)(\lambda q^2 + 2\mu q_z^2) + 2\mu q_x^2 q_z k_z}{D} \tag{7.238}$$

$$D = (q_x^2 - k_z^2)(\lambda q^2 + 2\mu q_z^2) - 2\mu q_x^2 q_z k_z \tag{7.239}$$

Even this simple problem has a complicated answer.

7.5.4 Defect Interactions

Crystal defects such as vacancies and interstitials also create local strain fields. Such defects interact using long-range elastic forces. For a point defect, the lattice deformation $\mathbf{u}(\mathbf{r})$ is given in terms of the elastic constants C_{ijkl}:

$$G\frac{\partial}{\partial x_i}\delta^3(\mathbf{r}) = \sum_{jkl} C_{ijkl}\frac{\partial^2}{\partial x_j \partial x_k}u_l \tag{7.240}$$

where G is a constant parameter that is related to the pressure inside of the vacancy. It has the units of energy. Figure 7.9 shows two examples. In fig. 7.9a neighboring atoms might

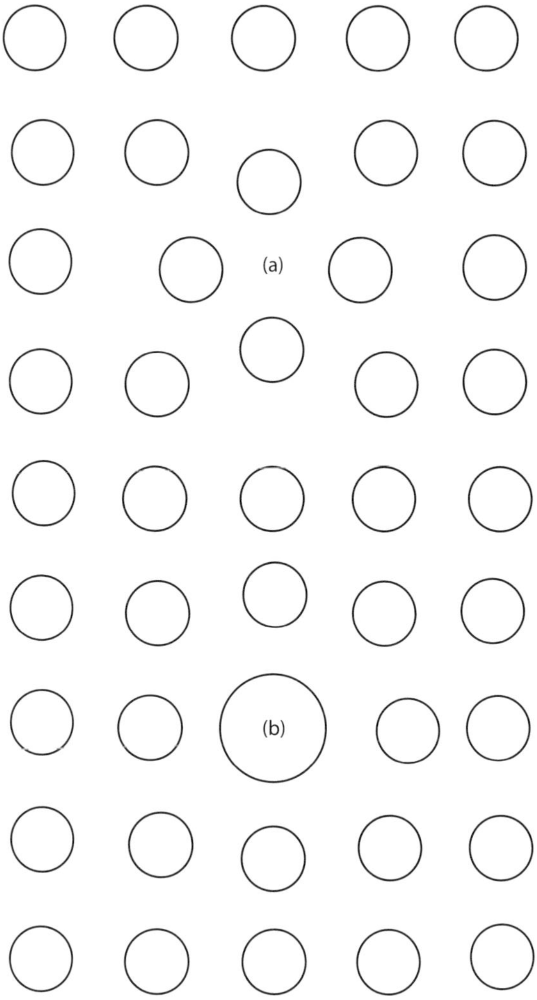

FIGURE 7.9. (a) Neighboring atoms move inward at a vacancy. (b) Neighboring atoms move outward for large substitutional impurity.

move inward at a vacancy, creating a symmetrical strain field. In fig. 7.9b, they move outward for a large substitutional impurity.

Recall that the free energy density is

$$F = \frac{1}{2}\sum_{ijkl} \varepsilon_{ij} C_{ijkl} \varepsilon_{kl} \tag{7.241}$$

The elastic constants are unchanged if various pair of indices are interchanged, so rewrite the above result as

$$F = \frac{1}{2}\sum_{ijkl} \frac{\partial u_i}{\partial x_j} C_{ijkl} \frac{\partial u_l}{\partial x_k} \tag{7.242}$$

Two vacancies at $\mathbf{r}_{1,2}$ create a strain field:

$$\mathbf{u}_T = \mathbf{u}(\mathbf{r} - \mathbf{r}_1) + \mathbf{u}(\mathbf{r} - \mathbf{r}_2) \tag{7.243}$$

The vacancy–vacancy interaction is given by the cross-term. Integrate over $\int d^3r$ to get the vacancy-induced interaction:

$$U(\mathbf{r}_1, \mathbf{r}_2) = \sum_{ijkl} \int d^3r \left[\frac{\partial}{\partial x_j} u_i(\mathbf{r} - \mathbf{r}_2) \right] C_{ijkl} \left[\frac{\partial}{\partial x_k} u_l(\mathbf{r} - \mathbf{r}_1) \right] \tag{7.244}$$

All derivatives are with respect to the variable $\mathbf{r}$. Integrate by parts:

$$U(\mathbf{r}_1, \mathbf{r}_2) = -\sum_{ijkl} \int d^3r u_i(\mathbf{r} - \mathbf{r}_2) C_{ijkl} \frac{\partial}{\partial x_j} \frac{\partial}{\partial x_k} u_l(\mathbf{r} - \mathbf{r}_1) \tag{7.245}$$

Most of the integrand can be replaced using eqn. (7.240), which generates these strains:

$$U(\mathbf{r}_1, \mathbf{r}_2) = -G \sum_i \int d^3r u_i(\mathbf{r} - \mathbf{r}_2) \frac{\partial}{\partial x_i} \delta^3(\mathbf{r} - \mathbf{r}_1) \tag{7.246}$$

Again integrate by parts, and use the delta function to eliminate the integral:

$$U(\mathbf{r}_1 - \mathbf{r}_2) = G[\nabla \cdot \mathbf{u}(\mathbf{r})]_{\mathbf{r} = \mathbf{r}_1 - \mathbf{r}_2} \tag{7.247}$$

The elastic interaction between two spherical point defects is given by the dilation $\nabla \cdot \mathbf{u}$ evaluated at the separation of the two defects. This result can also be obtained using phonon Green's functions.

As an example, evaluate this interaction in isotropic materials. The starting equation (7.240) is

$$G\nabla\delta^3(\mathbf{r}) = \mu\nabla^2\mathbf{u}(\mathbf{r}) + (\lambda + \mu)\nabla(\nabla \cdot \mathbf{u}) \tag{7.248}$$

Since the elastic interaction is given by the dilation, take the gradient of this equation, to get

$$G\nabla^2\delta^3(\mathbf{r}) = (\lambda + 2\mu)\nabla^2(\nabla \cdot \mathbf{u}) \tag{7.249}$$

Remove the factor of ∇^2 from both sides, and get

$$(\nabla \cdot \mathbf{u}) = \frac{G}{\lambda + 2\mu}\delta^3(\mathbf{r}) \tag{7.250}$$

$$U(\mathbf{r}) = G\Delta(\mathbf{r}) = \frac{G^2}{\lambda + 2\mu}\delta^3(\mathbf{r}) \tag{7.251}$$

The interaction is a delta function. The two defects interact only if they are in the same unit cell. There is no long-range elastic interaction in isotropic materials. Since crystals are not elastically isotropic, a long-range interaction exists in crystals.

An interesting example of elastic interactions is the observation of the *void lattice*. Bombarding metals with alpha radiation creates voids in the metal. The voids contain a high pressure of helium atoms—the neutral form of the alpha particles. In some metals, the voids themselves form a lattice, with a lattice constant given by tens of nanometers. The only way the voids could interact to form the lattice is through elastic interactions.

7.5.5 Piezoelectricity

The piezoelectric effect is a linear relation between stress and an electric field. The application of stress creates an internal polarization field. Conversely, the application of an electric field creates a stress. Since stress has two spatial indices (ij) and the electric field has one, the coefficient of proportionality $e_{k,ij}$ has three. The two imporant relations are

$$\sigma_{ij} = \sum_{kl} C^{E}_{ij,kl}\varepsilon_{kl} - \sum_{k} e_{k,ij}E_k \tag{7.252}$$

$$P_k = \sum_{\ell} \chi_{k\ell}E_\ell + \sum_{ij} d_{k,ij}\varepsilon_{ij} \tag{7.253}$$

where $\chi_{k\ell}$ is the dielectric response, and (e, d) are piezoelectric constants. Usually the indices are written in the Voigt notation, as $e_{k\alpha}$, $d_{k\alpha}$. The two third-rank tensors are mutual inverses. Also note the superscript E on the elastic constants. When a crystal is piezoelectric, it matters whether the elastic constants are measured at constant field E or at constant D.

A crystal must lack a center of inversion to be piezoelectric. It also must have cations and anions. Rocksalt is not piezoelectric. However, common semiconductor lattices such as zincblende and wurtzite are piezoelectric:

- Zincblende has only one nonzero piezoelectric constant, when the three indices occur in the combination of (xyz). Recall that the crystal field at an ion site has this symmetry. In the Voigt notation, the nonzero constant is e_{14}. Usually, the values are small and piezoelectricity is not a major factor in zincblende semiconductors.

- Wurtzite semiconductors have three nonzero piezoelectric constants. They are generally larger than those found in zincblende materials. Some values are shown in table 7.3. Those of ZnO are the largest.

The constants $e_{k\alpha}$ have the units of C/m^2. If Q is a charge in coulombs, and ε_0 is the dielectric permittivity of vacuum, then Q^2/ε_0 has the units of joules·meter. Similarly, the units of $e^2_{k\alpha}/\varepsilon_0$ is joules per volume. Those are the same units as the elastic constants C. Ignoring subscripts, the quantity

Table 7.3 Wurtzite piezoelectric constants in units of colombs per square meter

Crystal	e_{31}	e_{33}	e_{15}
ZnO	−0.35	−0.35	1.56
ZnS	0.09	−0.05	0.43
CdS	−0.24	0.41	−0.20
CdSe	−0.16	0.35	−0.14

$$k = \frac{e^2}{\varepsilon_0 C} \tag{7.254}$$

is dimensionless. It is called the *electromechanical couping constant*. It measures the efficiency of converting electrical energy to mechanical energy, or vice versa.

7.5.6 Phonon Focusing

Suppose there is a *hot spot* in a crystal that generates heat. Computer processors have them. Assume the heat is generated at a point source. Since heat is a gas of phonons, then phonons will radiate out in all directions from the hot spot. The mean free path of phonons is rather short, so that each phonon will divide into two other phonons by anharmonic scattering. Rather soon, the heat pulse is composed mostly of low-frequency acoustic phonons.

Do the phonons radiate outward equally in all directions? In some cases the answer is yes, but more often it is no. Phonon focusing is the name applied to the phenomenon of having the phonon energy in crystals focused along certain symmetry directions. In fact, the outward flow of acoustic phonons is usually quite anisotropic. Phonon focusing is a simple concept that is based on the elastic properties of crystals.

The effect is best illustrated by a simple example. It is easier to visualize two dimensions than three. Consider the phonons in the (x, y)-plane of a cubic crystal. The standard elastic equations for a wave of frequency ω, wave vector $\mathbf{q} = (q_x, q_y)$, and displacements (Q_x, Q_y) are

$$\rho\omega^2 Q_x = (C_{11}q_x^2 + C_{44}q_y^2)Q_x + q_x q_y(C_{12} + C_{44})\,Q_y \tag{7.255}$$

$$\rho\omega^2 Q_y = (C_{11}q_y^2 + C_{44}q_x^2)Q_y + q_x q_y(C_{12} + C_{44})\,Q_x \tag{7.256}$$

Solving this dynamical matrix gives

$$\omega_{L,T}^2 = \frac{q^2}{2\rho}\left[C_{11} + C_{44} \pm \sqrt{(C_{11} - C_{44})^2\cos^2(2\theta) + (C_{12} + C_{44})^2\sin^2(2\theta)}\,\right] \tag{7.257}$$

$$q_x = q\cos(\theta), \quad q_y = q\sin(\theta) \tag{7.258}$$

where ω_L is the plus sign, and ω_T is the minus sign. The above formula can be used to graph the frequency $\omega_{L,T}(\theta)$ as a function of the azimuthal angle θ. Note that if $C_{44} = (C_{11} - C_{12})/2$, the angular dependence vanishes and $\omega_L^2 = C_{11}/\rho$, $\omega_T^2 = C_{44}/\rho$. Group theory requires this to happen in the (x, y)-plane of hexagonal crystals, but it rarely occurs in cubic crystals.

Make a graph of the line of constant frequency as a function of angle. Invert the above formula:

$$q_{\pm}^2(\theta) = \frac{2\rho\omega^2}{C_{11} + C_{44} \pm \sqrt{(C_{11} - C_{44})^2\cos^2(2\theta) + (C_{12} + C_{44})^2\sin^2(2\theta)}} \tag{7.259}$$

For a fixed frequency, graph the wave vector as a function of the azimuthal angle. Figure 7.10 shows the result for a transverse phonon in silicon. Along the four principal axes ($\pm\hat{x}$,

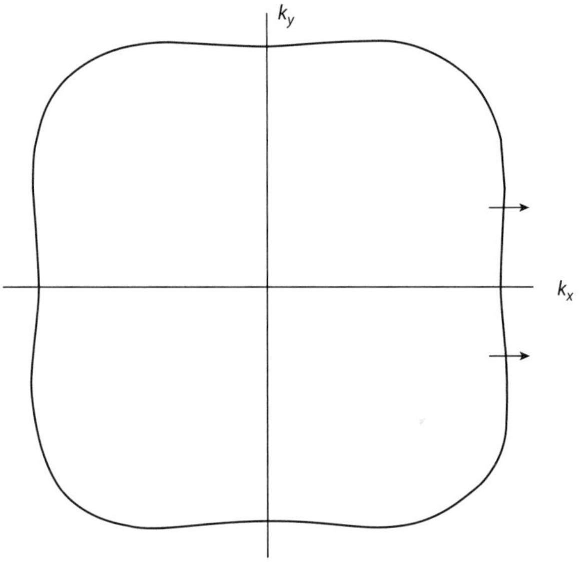

FIGURE 7.10. Line of constant frequency for a transverse phonon in the (x, y)-plane of silicon. Arrows are direction of phonon velocity.

$\pm\hat{y}$), the line is convex. This feature acts as a lens that focuses the phonons along these axes. The phonon velocity $\mathbf{v} = \nabla_q \omega(\mathbf{q})$ is perpendicular to the line of constant frequency. The direction of velocity is shown by small arrows along the $+\hat{x}$ axis. For a significant fraction of phase space (e.g., angle), the phonons velocities are focused in the direction of one of the four principal axes. A similar plot for the longitudinal phonon in Si gives almost a circle, and there is little focusing. See the review article by Jim Wolfe [17] for a further discussion.

7.6 Thermal Expansion

Most crystals expand when heated. The linear rate of expansion, in one direction r_μ, is given by α_μ

$$\alpha_\mu = \frac{1}{L_\mu}\left(\frac{dL_\mu}{dT}\right)_P \tag{7.260}$$

$$\Delta L_\mu = \alpha_\mu \Delta T L_\mu \tag{7.261}$$

Similarly, the coefficient of volume expansion β is given by

$$\Omega(T) = L_x(T)\, L_y(T)\, L_z(T) \tag{7.262}$$

$$\beta = \frac{1}{\Omega}\left(\frac{d\Omega}{dT}\right)_P \tag{7.263}$$

$$\beta\Delta T\Omega = \Omega(T+\Delta T) - \Omega(T) \tag{7.264}$$

$$= \Omega(T)[(1+\alpha_x\Delta T)(1+\alpha_y\Delta T)(1+\alpha_z\Delta T) - 1]$$

$$\beta = \alpha_x + \alpha_y + \alpha_z \tag{7.265}$$

In cubic crystals, α is independent of direction and $\beta = 3\alpha$.

There are some interesting materials that have $\alpha_\mu < 0$ in one direction. See Li et al. [8] for a review.

In most materials, α or β is a function of temperature. The temperature dependence is identical to that of the heat capacity, and $\beta(T) \propto C(T)$. This unusual relationship derives from a thermodynamic identity:

$$\beta = \frac{\gamma_T C_V}{B\Omega} \tag{7.266}$$

where

- γ_T is the Grüneisen parameter
- B is the bulk modulus

$$B = -\Omega\frac{dP}{d\Omega} \tag{7.267}$$

- C_V is the specific heat at constant volume.

Both γ_T and B are usually independent of temperature over wide ranges of T. That is the reason that $\beta(T) \propto C_V(T)$.

The coefficient of thermal expansion is due to the interatomic potential having nonlinear terms. The forces on the atoms are not perfect Hooke's law, but have cubic or quartic anharmonic terms. This nonlinearity is expressed by the Grüneisen parameter. First define the logarithmic derivative of the phonon frequencies, as a function of volume:

$$\gamma(\mathbf{q},\lambda) = -\frac{d\omega_\lambda(\mathbf{q})}{d\ln(\Omega)} \tag{7.268}$$

$$\gamma_T = \frac{\sum_{\mathbf{q}\lambda}\frac{dn_{\mathbf{q}\lambda}}{dT}\gamma(\mathbf{q},\lambda)}{\sum_{\mathbf{q}\lambda}\frac{dn_{\mathbf{q}\lambda}}{dT}\omega_\lambda(\mathbf{q})} \tag{7.269}$$

The parameter $\gamma(\mathbf{q},\lambda)$ expresses how the phonon frequencies change when the solid is compressed or expanded. This parameter is then averaged over all of the phonons to get the average parameter γ_T. This parameter is important when discussing the mean free path of phonons due to nonlinear interactions.

7.7 Debye-Waller Factor

Chapter 2 has a discussion on Bragg scattering using neutrons. Neutron scattering is also used to measure the dispersion relation $\omega_\lambda(\mathbf{q})$ of phonons in crystals. It is an example of inelastic scattering. The neutron scatters by a wave vector $\mathbf{k}_f - \mathbf{k}_i = \mathbf{q}$, and creates (or destroys) a phonon of wave vector $\mathbf{q}$. The neutron changes its kinetic energy by $E_f - E_i = \pm\hbar\omega_\lambda(\mathbf{q})$ during the scattering. The cross section for this process is the same as discussed in chapter 2.

Phonons play several roles in these experiments. In Bragg scattering, which is elastic, the earlier theory assumed that all atoms in the crystal were at their crystalline positions $\mathbf{R}_j^{(0)}$. Because of the phonon vibrations, the atoms will be at $\mathbf{R}_j(t) = \mathbf{R}_j^{(0)} + Q_j(t)$, where $Q_j(t)$ is the instantaneous displacement of the atom due to the phonons. In the theory of Bragg scattering, the factor of $\exp[i\mathbf{G}\cdot\mathbf{R}_j]$ was set equal to one. Here $\mathbf{G} = \mathbf{q}$ is the reciprocal lattice vector, which is the change in wave vector of the neutron. Now evaluate this factor using

$$e^{i\mathbf{G}\cdot\mathbf{R}_j} = e^{i\mathbf{G}\cdot(\mathbf{R}_j^{(0)}+\mathbf{Q}_j)} = e^{i\mathbf{G}\cdot\mathbf{Q}_j} \tag{7.270}$$

Since $\mathbf{Q}_j$ is an operator, the exponential function is evaluated using a *linked cluster expansion*. Expand the exponent in a power series:

$$\langle e^{i\mathbf{G}\cdot\mathbf{Q}_j}\rangle = \left\langle\left[1 + i\mathbf{G}\cdot\mathbf{Q}_j + \frac{i^2}{2!}(\mathbf{G}\cdot\mathbf{Q}_j)^2 + \cdots\right]\right\rangle \tag{7.271}$$

$$\mathbf{Q}_j = \sum_{\mathbf{k}\lambda} X(\mathbf{k}\lambda)\hat{\xi} e^{i\mathbf{k}\cdot\mathbf{R}_j^{(0)}} A_{\mathbf{k}\lambda}, \quad A_{\mathbf{k}\lambda} = a_{\mathbf{k}\lambda} + a^\dagger_{-\mathbf{k}\lambda} \tag{7.272}$$

In the expansion (7.271), the term linear in $\langle\mathbf{Q}_j\rangle = 0$, since there is an odd number of raising and lowering operators. All odd powers $\langle\mathbf{A}_j^{2n+1}\rangle = 0$ because they have an odd number of raising and lowering operators. The second-order term is nonzero:

$$\begin{aligned}\langle(\mathbf{G}\cdot\mathbf{Q}_j)^2\rangle = &\sum_{\mathbf{k}\lambda,\mathbf{k}'\lambda'} X(\mathbf{k}\lambda)X(\mathbf{k}'\lambda')(\mathbf{G}\cdot\hat{\xi})(\mathbf{G}\cdot\hat{\xi}')e^{i\mathbf{R}_j^{(0)}\cdot(\mathbf{k}-\mathbf{k}')} \\ &\times\langle(a_{\mathbf{k}\lambda} + a^\dagger_{-\mathbf{k}\lambda})(a_{\mathbf{k}'\lambda'} + a^\dagger_{-\mathbf{k}'\lambda'})\rangle\end{aligned} \tag{7.273}$$

Since $\langle|$ and $|\rangle$ have the same number of phonons in each state, the raising and lowering operators must be paired up, which forces $\mathbf{k}' = -\mathbf{k}$, $\lambda' = \lambda$:

$$\begin{aligned}\langle(a_{\mathbf{k}\lambda} + a^\dagger_{-\mathbf{k}\lambda})(a_{\mathbf{k}'\lambda'} + a^\dagger_{-\mathbf{k}'\lambda'})\rangle &= \langle a_{\mathbf{k}\lambda}a^\dagger_{-\mathbf{k}'\lambda'}\rangle + \langle a^\dagger_{-\mathbf{k}\lambda}a_{\mathbf{k}'\lambda'}\rangle \\ &= \delta_{\mathbf{k}'=-\mathbf{k}}\delta_{\lambda'=\lambda}[2N_{\mathbf{k}\lambda} + 1]\end{aligned} \tag{7.274}$$

$$N_{\mathbf{k}\lambda} = \frac{1}{e^{\hbar\omega_\lambda(\mathbf{k})/k_BT} - 1} \tag{7.275}$$

where $N_{\mathbf{k}\lambda}$ is the boson occupation number for phonons. The first nonzero term is

$$\Delta(T,\mathbf{G}) = \langle(\mathbf{G}\cdot\mathbf{Q}_j)^2\rangle = \frac{\hbar}{2MN}\sum_{\mathbf{k}\lambda}\frac{(\mathbf{G}\cdot\hat{\xi})^2}{\omega_\lambda(\mathbf{k})}[2N_{\mathbf{k}\lambda} + 1] \tag{7.276}$$

The next nonzero term has four factors. Since the raising and lowering operators must occur in pairs, the four factors are averaged in pairs:

$$\frac{1}{4!}\langle(\mathbf{G}\cdot\mathbf{Q}_j)^4\rangle = \frac{3}{24}\langle(\mathbf{G}\cdot\mathbf{Q}_j)^2\rangle^2 = \frac{1}{8}\Delta^2 \tag{7.277}$$

The factor of 3 comes from the different ways of choosing pairs for four identical factors: (12, 34), (13, 24), (14, 23). The linked cluster expansion is generating the series

$$e^{i\mathbf{G}\cdot\mathbf{Q}_j} = 1 - \frac{1}{2}\Delta + \frac{1}{8}\Delta^2 - \cdots \tag{7.278}$$

$$= \exp\left[-\frac{1}{2}\Delta\right]$$

This factor multiplies the amplitude of the scattering. The intensity of the Bragg scattering contains the square of this factor,

$$I(\mathbf{G}) = N^2 b^2 \left(\frac{2\pi\hbar}{m_n}\right)^2 e^{-\Delta(T,G)} \tag{7.279}$$

The expression $\Delta(T, \mathbf{G})$ can be evaluated on the computer using the frequencies and eigenvectors $\hat{\xi}$ of the phonons.

- At very low temperatures, the number of thermal phonon $N_{\mathbf{k}\lambda} \approx 0$. The function $\Delta(0, G) \neq 0$ due to the zero-point motions of the phonons:

$$\Delta(0, \mathbf{G}) = \frac{\hbar}{2MN} \sum_{\mathbf{k}\lambda} \frac{(\mathbf{G}\cdot\hat{\xi})^2}{\omega_\lambda(\mathbf{k})} \tag{7.280}$$

  This number is of order $O(1)$.

- At very high temperatures, where $x \equiv \hbar\omega_\lambda(\mathbf{k})/k_B T < 1$, expand

$$N = \frac{1}{e^x - 1} = \frac{1}{x} - \frac{1}{2} + O(x) \tag{7.281}$$

$$2N + 1 \approx \frac{2k_B T}{\hbar\omega_\lambda(\mathbf{q})} \tag{7.282}$$

$$\Delta(T, \mathbf{G}) = k_B T \mathbf{G}\cdot\tilde{C}\cdot\mathbf{G} \tag{7.283}$$

$$\tilde{C} = \frac{1}{MN} \sum_{\mathbf{k}\lambda} \frac{\hat{\xi}\hat{\xi}}{\omega_\lambda(\mathbf{k})^2} \tag{7.284}$$

  The tensor $\tilde{C}$ has the units of the inverse of the spring constants, so a rough estimate is $\tilde{C} \approx \mathcal{I}/K$, where $\mathcal{I}$ is the unit tensor. The intensity of the Bragg scattering falls off at high temperature because of this factor. The ion positions fluctuate in time and make a diffuse target for the x-rays, or neutrons, that are engaged in the scattering. Also note that the damping function $\exp(-\Delta)$ gets smaller for larger values of reciprocal lattice vector G.

A similar pheomenon occurs when using neutron to measure phonon spectra. Then replace G by $\mathbf{q}$. Here $\mathbf{q}$ is the wave vector of the phonon created (or destroyed) in the scattering process. The wave vector $\mathbf{k}$ is from the phonons that are jiggling the ions and reducing the intensity of the scattering cross section.

7.8 Solitons

Solitons are solitary wave solutions found in one dimension. They were first observed as wave pulses in barge canals. Modern experiments observe them as pulses of laser light traveling through materials. Most of the theoretical work on them is entirely classical: one solves the classical equations of waves in a material that has nonquadratic potentials. Several equations have been well studied.

- Korteweg and deVries [7] first derived an equation for the propagation of water waves in one-dimensional canals. Their equation is called KdV:

$$0 = \frac{\partial u}{\partial t} + au\frac{\partial u}{\partial x} + b\frac{\partial^3 u}{\partial x^3} \tag{7.285}$$

  Here a and b are constants. The variable $u(x, t)$ is the height of the water as a function of distance (x) and time (t).

- The *Toda lattice* is a model of a one-dimensional system of springs and masses [15]. The springs have a dependence on relative separation:

$$r_n = Q_{n+1} - Q_n \tag{7.286}$$

$$V(r) = \frac{a}{b}\left[e^{-br} + br - 1\right] \tag{7.287}$$

$$F(r) = -\frac{dV}{dr} = a\left[e^{-br} - 1\right] \tag{7.288}$$

  The amplitude of the force is a, while b is a length scale. At small values of r the potential goes as $V \sim r^2 + O(r^3)$. The force obeys Hooke's law at small values of (br), but goes to a constant at large positive (br), and grows exponentially at large negative values.

- A symmetric exponential potential is

$$V(r) = K[\cosh(br) - 1] \tag{7.289}$$

- Fermi, Pasta, and Ulam [5] studied a nonlinear lattice using one of the earliest computers. Their model is now called FPU, and consists of springs with quadratic and quartic components:

$$V(r) = \frac{K_2}{2}r^2 + \frac{K_4}{4}r^4 \tag{7.290}$$

These different cases are discussed below. Most of them involve Jacobian elliptic functions.

7.8.1 *Solitary Waves*

Some of the nonlinear equations have a solution for a single pulse that is going down the system. Usually this pulse is a compressive soliton: The wave makes the chain of atoms shorter. As an example, consider a semi-infinite chain of masses $j = 1, 2, 3, \ldots$ connected

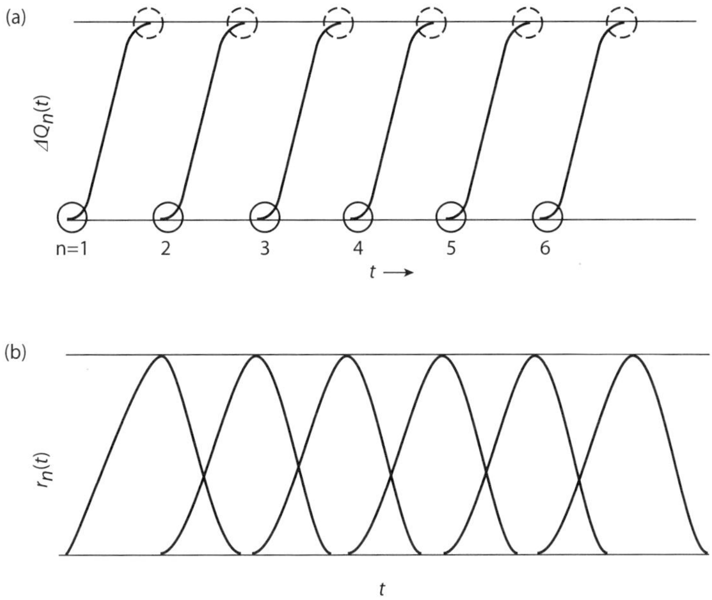

FIGURE 7.11. Motion of atoms in a compressive soliton: (a) atom motions, (b) relative displacements.

by nonlinear springs. One hits the end atom with a hammer that generates an impulse, driving the end atom toward the second one. The end atom moves to the right and stops. The second atom then moves to the right and stops. This chain reaction continues down the row of atoms, as shown in fig. 7.11a. A moving solitary pulse is found when plotting the relative displacements r_n, as shown in fig. 7.11b. An example of this behavior is found in the Toda lattice, as discussed below.

The equation of motion for atoms for the Toda lattice is

$$m\frac{\partial^2}{\partial t^2}Q_n = a\left[e^{-b(Q_n - Q_{n-1})} - e^{-b(Q_{n+1} - Q_n)}\right] \tag{7.291}$$

An exact solution for a solitary wave is

$$bQ_n(t) = \ln\left[\frac{1 + e^{2\phi_n}}{1 + e^{2\phi_{n+1}}}\right] \tag{7.292}$$

$$\phi_n = \kappa n - \omega t \tag{7.293}$$

Insert this guess into the equation of motion (7.291). The equation of motion is obeyed provided that the frequency is

$$\omega = \pm\sqrt{\frac{ab}{m}}\sinh(\kappa) \tag{7.294}$$

An important feature of this solitary wave is its stability. Toda describes the solution with two solitons on the lattice. If going in opposite directions, they pass through each other.

If going in the same direction, the faster one overtakes the slower one. They are perfectly stable, and do not break up into phonons.

The KdV equation also has a stable solution for a solitary wave. It is assigned as a homework problem.

7.8.2 Cnoidal Functions

Jacobian elliptic functions are also called cnoidal functions. They are needed for periodic solutions of the nonlinear equations.

The integral for the arcsine is

$$u = \int_0^z \frac{dt}{\sqrt{1-t^2}} = \arcsin(z), \quad z = \sin(u) \tag{7.295}$$

In a similar way, introduce another function that depends on the parameter k:

$$u = \int_0^z \frac{dt}{\sqrt{(1-t^2)(1-k^2t^2)}}, \quad z = \mathrm{sn}(u) \tag{7.296}$$

The quantity $\mathrm{sn}(u)$ is a Jacobian elliptic function. Use it to define a family of related functions:

$$\mathrm{cn}^2(u) = 1 - \mathrm{sn}^2(u) \tag{7.297}$$

$$\mathrm{dn}^2(u) = 1 - k^2\mathrm{sn}^2(u) \tag{7.298}$$

Figure 7.12 shows a graph of these functions vs. u. Note that $\mathrm{sn}(u)$ is almost flat when the function equals ± 1, while $\mathrm{cn}(u)$ has a sharper extremum.

They are periodic, with the period determined by multiples of the first elliptic integral $K(k)$:

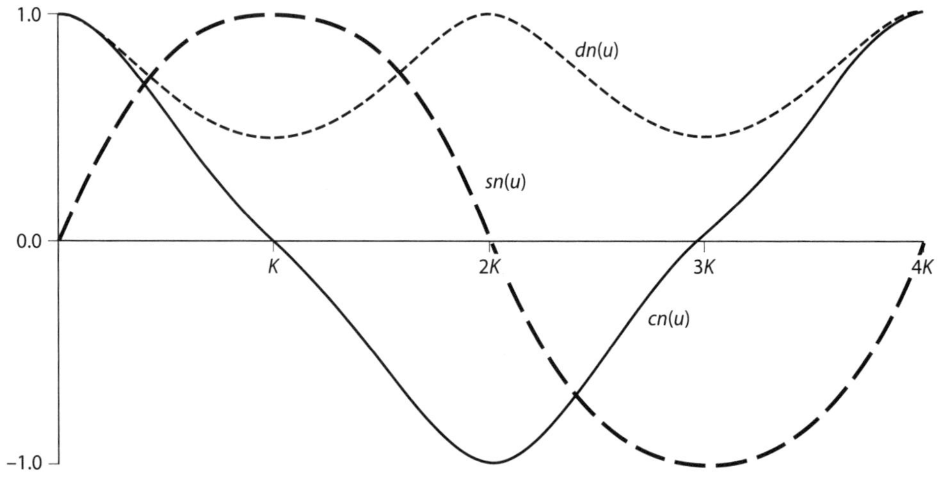

FIGURE 7.12. Jacobian elliptic functions for $k^2 = 0.8$. In this case $K = 2.2572$.

$$K(k) = \int_0^1 \frac{dt}{\sqrt{(1-t^2)(1-k^2t^2)}} = \int_0^{\pi/2} \frac{d\theta}{\sqrt{1-k^2\sin^2\theta}} \tag{7.299}$$

$$E(k) = \int_0^1 dt \sqrt{\frac{1-k^2t^2}{1-t^2}} = \int_0^{\pi/2} d\theta\sqrt{1-k^2\sin^2\theta} \tag{7.300}$$

The period of $\mathrm{dn}(u)$ is $2K(k)$, while that for $\mathrm{sn}(u)$ and $\mathrm{cn}(u)$ is $4K(k)$. Like sines and cosines, they change sign after half of a period:

$$\mathrm{cn}(u+2K) = -\mathrm{cn}(u) \tag{7.301}$$

$$\mathrm{sn}(u+2K) = -\mathrm{sn}(u) \tag{7.302}$$

The second elliptic integral $E(k)$ is used below. These functions depend on k, although that dependence is not highlighted in the notation. Further properties of these functions are given in the appendix.

7.8.3 Periodic Solutions

The nonlinear equations on a lattice also have period solutions. A related equation for the relative displacement $r_n = Q_{n+1} - Q_n$ on the Toda lattice is

$$m\frac{\partial^2}{\partial t^2} r_n = a\left[2e^{-br_n} - e^{-br_{n+1}} - e^{-br_{n-1}}\right] \tag{7.303}$$

The goal is to find periodic functions that obey this recursion differential equation. The functions sn, cn, dn do not have the correct properties. However, the proper functions are their square. The equation is satisfied by the choice of

$$e^{-br_n} = 1 + \frac{C}{a}[\mathrm{dn}^2(u_n) - D], \quad D = \frac{E(K)}{K(K)} \tag{7.304}$$

$$u_n = 2K(k)\left(\nu t \pm \frac{n}{\lambda}\right) \tag{7.305}$$

with the right choice of C, D, ν, and λ. The equation of motion is now

$$\frac{\partial^2}{\partial t^2}(br_n) = \frac{Cb}{m}[2\mathrm{dn}^2(u_n) - \mathrm{dn}^2(u_{n+1}) - \mathrm{dn}^2(u_{n-1})] \tag{7.306}$$

$$= -\frac{\partial^2}{\partial t^2}\ln\left\{\frac{C}{a}\left[\frac{a}{C} + \mathrm{dn}^2(u_n) - D\right]\right\}$$

Toda's recursion relation for the Toda lattice is

$$\mathrm{dn}^2(u+v) + \mathrm{dn}^2(u-v) - 2\mathrm{dn}^2(u) = \frac{d^2}{du^2}\ln\left[1 - k^2 sn^2(u)\, sn^2(v)\right] \tag{7].307}$$

$$= \frac{d^2}{du^2}\ln\left[\mathrm{ctn}^2(v) + \mathrm{dn}^2(u)\right]$$

$$\mathrm{ctn}(u) = \frac{\mathrm{cn}(u)}{\mathrm{sn}(u)} \tag{7.308}$$

The last identity adds a constant to the argument of the logarithm:

$$1 - k^2\mathrm{sn}^2(u)\,\mathrm{sn}^2(v) = \mathrm{sn}^2(v)\left[\frac{1}{\mathrm{sn}^2(v)} - k^2\mathrm{sn}^2(u)\right]$$

$$= \mathrm{sn}^2(v)\,[\mathrm{ctn}^2(v) + \mathrm{dn}^2(u)] \tag{7.309}$$

and the constant term $\mathrm{sn}^2(v)$ cancels from the derivative.

With the appropriate choice of constants, the second line of eqn. (7.307) is identical to the second line of (7.306). $u_n \pm v$ must be $u_{n\pm1}$, which is achieved by setting

$$v = \frac{2K(k)}{\lambda} \tag{7.310}$$

since time t is only contained in u_n, then

$$\frac{\partial^2}{\partial t^2} \to (2K\nu)^2 \frac{d^2}{du^2} \tag{7.311}$$

Equating coefficients gives

$$2K\nu = \sqrt{\frac{bC}{m}} \tag{7.312}$$

Next, consider the argument of the logarithm. For the two equations to be the same, they must be identical:

$$\mathrm{ctn}^2(v) + \mathrm{dn}^2(u) = \frac{a}{C} - D + \mathrm{dn}^2(u) \tag{7.313}$$

$$C = \frac{a}{\mathrm{ctn}^2(v) + D} \tag{7.314}$$

Since a denotes the strength of the potential, and $D = K(k)/K(k)$, $v = 2K(k)/\lambda$ are both known, then C is determined, as is the frequency of the wave:

$$\nu = \sqrt{\frac{ab}{m}} \frac{1}{2K(k)\sqrt{\mathrm{ctn}^2(v) + E/K}} \tag{7.315}$$

The wave has two free parameters. One is the wavelength λ, which is the number of lattice points in one period of the wave. The second is the parameter k, which relates to the degree of anharmonicity.

The Toda lattice is unique in that periodic solutions are available for any wavelength. Other nonlinear equations have analytic periodic solutions only for particular wavelengths. As an example, below are some solutions for zone boundary modes, where the period is two lattice constants.

- For the FPU lattice, the equation of motion for the relative displacement is

$$m\frac{\partial^2}{\partial t^2} r_n = -K_2[2r_n - r_{n+1} - r_{n-1}] - K_4\left[2r_n^3 - r_{n+1}^3 - r_{n-1}^2\right] \tag{7.316}$$

The solution has the form $r_n = q_0 \text{cn}(u_n)$, $u_n = 2Kn - \omega_2 t$. Since $\text{cn}(u \pm 2K) = -\text{cn}(u)$, alternate masses go in opposite directions, which is typical of an optical phonon. Then using the expressions for the second derivative in the appendix, one finds

$$\omega_2^2 = \frac{4}{m}\left[K_2 + q_0^2 K_4\right] \tag{7.317}$$

$$2k^2 = \frac{K_4 q_0^2}{K_2 + q_0^2 K_4} \tag{7.318}$$

where k is the parameter in $K(k)$. The frequency depends on the amplitude q_0.

- For the $\cosh(br_n)$ potential function, the equation for an atom displacement is

$$m\frac{\partial^2}{\partial t^2}Q_n = -Kb\{\sinh[b(Q_n - Q_{n+1})] + \sinh[b(Q_n - Q_{n-1})]\} \tag{7.319}$$

The period two wave has a solution

$$bQ_n = \ln\left[\frac{\text{dn}(u_n)}{\sqrt{k_1}}\right] \tag{7.320}$$

$$u_n = Kn - \omega_2 t, \quad k_1 = \sqrt{1-k^2} \tag{7.321}$$

where $\text{dn}(u)$ has a period of $2K(k)$. Note the identity

$$\text{dn}(u \pm K) = \frac{k_1}{\text{dn}(u)} \tag{7.322}$$

$$bQ_{n+1} = \ln\left[\frac{\text{dn}(u_n \pm K)}{\sqrt{k_1}}\right] = -bQ_n \tag{7.323}$$

and the frequency is

$$\omega_2 = b\sqrt{\frac{K}{mk_1}} \tag{7.324}$$

Similar periodic solutions can be found for wavelengths of four lattice constants. These periodic solutions are often unstable, and eventually decay into phonons. Only the solitary waves are stable solutions.

References

1. M. Born and Th. von Karman, Über Schwingungen in Raumgittern. *Phys. Z.* **13**, 297 (1912); **14**, 15 (1913)
2. M. Born and K. Huang, *Dynamical Theory of Crystal Lattices* (Oxford University Press, London, 1954)
3. W. G. Cady, *Piezoelectricity*, 2 vols. (Dover, New York, 1964)
4. G. Dolling, H. G. Smith, R. M. Nicklow, P. R. Vijayaraghavan, and M. K. Wilkinson, Lattice dynamics of LiF. *Phys. Rev.* **168**, 970–979 (1968)

5. E. Fermi, J. R. Pasta, and S. Ulam, Los Alamos Report No. LA-1940, Vol.2, p. 978 (1955)
6. J. R. Hardy and A. M. Karo, *The Lattice Dynamics and Statics of Alkali Halide Crystals* (Plenum, New York, 1979)
7. D. J. Korteweg and A. deVries, *Philos. Mag.* **39**, 422 (1985)
8. J. Li, A. W. Sleight, C. Y. Jones, and B. H. Toby, Trends in negative thermal expansion behavior for compounds with the delafossite structure. *J. Solid State Chem.* **178**, 285–294 (2005)
9. G. D. Mahan, *Many-Particle Physics*, 3rd ed. (Kluwer/Plenum, New York, 2001) Chap. 11
10. N. E. Phillips, Heat capacity of aluminum between 0.1 K and 4.0 K. *Phys. Rev.* **114**, 676–685 (1959)
11. S. B. Sharma, S. C. Sharma, B. S. Sharma, and S. S. Bedi, Analysis of dielectric properties of I–VII, II–VI, and III–V crystals with the ZnS structure. *J. Phys. Chem. Solids*, **53**, 329–335 (1992)
12. R. Stedman, L. Almqvist, and G. Nilsson, Phonon-frequency distributions and heat capacities of Al and Pb. *Phys. Rev.* **162**, 549–557 (1967)
13. F. H. Stillinger and T. A. Weber, Computer simulation of local order in condensed phases of silicon. *Phys. Rev. B* **31**, 5262 (1985)
14. G. P. Srivastava, *The Physics of Phonons* (Adam Hilger, New York, 1990)
15. M. Toda, *Nonlinear Waves and Solitons* (Kluwer, New York, 1989)
16. P. R. Vijayaraghavan, R. M. Nicklow, H. G. Smith, and M. K. Wilkinson, Lattice dynamics of AgCl. *Phys. Rev. B* **1**, 4819–4826 (1970)
17. J. P. Wolfe, Acoustic wavefronts in crystalline solids. *Phys. Today* 48(9) 34–40 (1995)
18. R. Zallen, Symmetry and reststrahlen in elemental crystals. *Phys. Rev.* **173**, 824 (1968)

Homework

1. The phonon Hamiltonian for the one-dimensional chain can be written as

$$H = \sum_k H_k \tag{7.325}$$

$$H_k = \frac{p_k p_{-k}}{2m} + \frac{m}{2}\omega(k)^2 q_k q_{-k} \tag{7.326}$$

where q_k and q_{-k} are independent variables. Derive the ground-state eigenfunction ψ_0 (q_k, q_{-k}):

$$H_k \psi_0 = E_k \psi_0 \tag{7.327}$$

2. Derive the vibrational frequencies of a one-dimensional chain of atoms with identical masses m. The chain has alternating spring constants (K_1, K_2).

3. An example in the text solved the problem of a one-dimensional chain with alternating masses M_A and M_B. Show that the solution may be reduced to the form

$$H = \sum_{k\lambda} \hbar\omega_{k\lambda}\left(a^\dagger_{k\lambda} a_{k\lambda} + \frac{1}{2}\right) \tag{7.328}$$

$$X_i = \sum_{k\lambda} X_{k\lambda} e^{ikR_i} \xi_i(k\lambda)(a_{k\lambda} + a^\dagger_{-k\lambda}) \tag{7.329}$$

Derive the form for the vectors $\xi_i(k\lambda)$ that give the displacements for the atoms A and B.

4. Consider the phonons in the two-dimensional square lattice. Assume that $K_1 = 2K_2$, where K_1 is the first-neighbor spring constant, and K_2 is the second-neighbor. Show that for this case a simple analytical solution is obtained for the longitudinal and transverse phonon frequencies. Also derive the eigenvectors of the dynamical matrix.

5. Derive the equation for the phonon modes in two dimensions for the pt lattice. Use only first-neighbor, bond-directed forces. Show the dispersion along the $\Gamma \rightarrow K$ direction in the Brillouin zone.

6. Solve for the phonons of a bcc metal, considering only one force contant: a bond-directed force to the first neighbor

$$M\omega^2 \mathbf{Q}_j = K\sum_{\delta=1}^{8} \hat{\delta}\hat{\delta}\cdot[\mathbf{Q}_j - \mathbf{Q}_{j+\delta}] \tag{7.330}$$

(a) Derive the cubic equation for the phonon frequencies.

(b) Give the analytical solutions for phonons going along the three principal axes: (100), (110), (111).

(c) Compare to the experimental results for lithium in S. Pal, *Phys. Rev. B* 2, 4741 (1970).

(d) What is the value of K for Li?

7. Create the analog to table 7.1 for the II–VI semiconductors with the zincblende structure.

8. Find the exact solution to

$$H = E_0 a^\dagger a + E_1(aa + a^\dagger a^\dagger) \tag{7.331}$$

where (E_0, E_1) are constants, and $(a, a^\dagger)$ are boson operators.

9. In boson systems, the thermal average of $\langle a_p^\dagger a_p \rangle$ is

$$\langle a_p^\dagger a_p \rangle = \langle n_p \rangle = \frac{1}{e^{\beta(\varepsilon_p - \mu)} - 1} \tag{7.332}$$

What is the thermal average of $\langle n_p^2 \rangle$ that arises in the energy of anharmonic phonons?

10. The dimensionless Szigeti charge e^* is defined as

$$(e^*)^2 = \frac{\mu\Omega_0\varepsilon_0}{e^2}\left[\omega_{\mathrm{LO}}^2 - \omega_{\mathrm{TO}}^2\right] \tag{7.333}$$

Numerically evaluate this charge for the three crystals in table 7.4 with the NaCl structure.

Table 7.4

Crystal	a (nm)	ν_{LO} (THz)	ν_{TO} (THz)
AgF	0.4936	9.60	5.10
AgCl	0.5550	5.88	3.17
AgBr	0.5775	4.14	2.37

11. Evaluate the phonon density of states for phonons in one dimension with frequency $\omega(q) = \omega_0 + \Delta\cos(qa)$.

12. Show that the low-temperature phonon heat capacity in the Debye model, in d-dimensions, is $C_V = \xi_d T^d$ and find ξ_1, ξ_2.

13. The heat capacity of a simple metal can be written as

$$C_V = \gamma T + \xi T^3 \tag{7.334}$$

where γ (eqn. 5.67) is from electrons and ξ (eqn. 7.141) is from phonons. Calculate the temperature T_0 in sodium where these two contributions are equal. For the speeds of sound in ξ, find them in the (110) direction.

14. This is an essay question. Given that the internal energy of a solid is defined in terms of the number of phonons,

$$U(T) = \sum_{\mathbf{q}s} \hbar\omega(\mathbf{q}, s)\, n_{\mathbf{q}s}(T), \quad n_{\mathbf{q}s}(T) = \frac{1}{e^{\hbar\omega_q/k_BT} - 1} \tag{7.335}$$

what is the smallest distance over which one can define a change in temperature? In a nanoscale device, how small can one define a change in temperature? The issue arises in thermal conductivity, where folks wish to apply Fourier's law ($\dot{\mathbf{Q}} = -\kappa\nabla T$) on nanometer distance scales.

15. Consider isotope scattering in a crystal that has three isotopes: M_j of concentration c_j, where $j = (1, 2, 3)$, and $1 = c_1 + c_2 + c_3$. Derive an expression for δH and discuss the prefactor in first- and second-order perturbation theory.

16. Solve eqn. (7.156) for the frequency of the impurity phonon mode.

17. Express the velocities of longitudinal and transverse sound in homogeneous materials in terms of Young's modulus and Poisson's ratio.

18. A three-dimensional solid of neutral atoms has an impurity complex consisting of nearest neighbors of plus and minus atoms that occupy lattice sites. The long-range Coulomb interaction between two such complexes has the form of a dipole–dipole interaction. Show that the long-range elastic interaction from LA phonons also has the form of dipole–dipole, assuming that the defects displace slightly toward each other.

19. A one-dimensional chain of springs and masses has all masses m and springs K alike except one spring is different: $K' = K + \delta K$. Derive the condition on δK to obtain a local phonon.

20. Hexagonal crystals have the feature that the velocities of sound in the basal plane ($q_z = 0$) are isotropic.

(a) Prove this statement using elasticity theory.

(b) What are the three velocities?

21. Show that in the (x, y)-plane of a cubic crystal, the direction of phonon focusing depends on whether $C_{44} > (C_{11} - C_{12})/2$ or $C_{44} < (C_{11} - C_{12})/2$. What are the focusing directions for each case?

22. Show that the following function is a solution to the KdV equation:

$$u(x, t) = \frac{A}{\cosh^2(\phi)}, \quad \phi = \xi(x - ct) \tag{7.336}$$

and relate (ξ, c) to the constants (a, b, A).

23. For the period two solution of the lattice with the $K[\cosh(qr) - 1]$ potential function, show that the energy per atom is

$$E/N = K[\cosh(q_0) - 1] \tag{7.337}$$

and determine q_0. (*Hint*: What is the amplitude of the wave?).

24. Solve the classical equation of motion for a mass M connected to a spring with potential fucntion (4.75). Use the cnodial functions. Hence, derive equation (4.78).

8 Boson Systems

Bosons are elementary particles that have integer spin. Examples are pions and photons. The helium atom of atomic weight four (^{4}He) is customarily treated as a boson. It is a composite particle of two neutrons, two protons, and two electrons. Similarly, the atom ^{3}He is treated as a fermion, since it has an odd number of them. Many collective excitations obey boson statistics, such as plasmons and phonons. In this chapter we study boson systems such as ^{4}He and spin waves (magnons).

Bosons are divided into two classes. Those that consist of real particles, such as ^{4}He, have a chemical potential μ. Those that are excitations, that can be created and destroyed, have no chemical potential. The latter group includes phonons, photons, and magnons.

8.1 Second Quantization

Raising and lowering operators for phonons were introduced in the prior chapter. They were derived starting from the commutation relation $[x, p_x] = i\hbar$. Similar raising and lowering operators are needed for boson particles. They are not based on the commutation relation between x and p_x. Instead, they are derived from the field equations. This procedure is called *second quantization*, and is now described. It was used earlier for fermions, such as electrons or holes. First quantization is derived from the operators for the harmonic oscillator, based on $[x, p_x] = i\hbar$.

For each of the bosons, assume they can be described by a Hamiltonian H that has eigenfunctions ϕ_n and eigenvalues E_n:

$$H\phi_n = E_n\phi_n \tag{8.1}$$

The most general wave function is

$$\psi(\mathbf{r}) = \sum_n a_n \phi_n \tag{8.2}$$

In discussing fermion systems in the last chapter, the symbol a_n became an operator to maintain fermion statistics. For boson systems, a_n is also an operator, but obeys a different set of statistics. These operators commute:

$$[a_n, a_{n'}] = [a_n a_{n'} - a_{n'} a_n] = 0 \tag{8.3}$$

$$\left[a_n, a^{\dagger}_{n'}\right] = \delta_{nn'} \tag{8.4}$$

$$\left[a^{\dagger}_n, a^{\dagger}_{n'}\right] = 0 \tag{8.5}$$

Raising operators commute with raising operators, lowering operators commute with lowering operators. Raising and lowering operators do not commute for the same eigenstate. These operator rules are sufficient to maintain boson properties. They have the same statistics as harmonic oscillator states:

$$a_n|\ell_n\rangle = \sqrt{\ell_n}|\ell_n - 1\rangle \tag{8.6}$$

$$a^{\dagger}_n|\ell_n\rangle = \sqrt{\ell_n + 1}|\ell_n + 1\rangle \tag{8.7}$$

where ℓ_n is the integer number of bosons in the state n. The thermodynamic average of the boson number operator is

$$\langle a^{\dagger}_n a_n\rangle = \frac{1}{e^{\beta(E_n-\mu)} - 1}, \quad \beta = \frac{1}{k_B T} \tag{8.8}$$

Second quantization is a different procedure than first quantization. For example, apply second quantization to the simple harmonic oscillator. The wave function becomes

$$\psi(x) = \sum_n b_n \phi_n(x) \tag{8.9}$$

$$\psi^{\dagger}(x) = b^{\dagger}_n \phi_n(x) \tag{8.10}$$

where the lowering and raising operators are called $(b_n, b^{\dagger}_n)$. The comutation relation is

$$\left[b_n, b^{\dagger}_m\right] = \delta_{nm} \tag{8.11}$$

Clearly, the operators $(b_n, b^{\dagger}_n)$ have no simple relationship to operators $(a, a^{\dagger})$ discussed in the previous chapter. The wave functions obey a commutation relation:

$$\begin{aligned}[\psi(x), \psi^{\dagger}(x')] &= \sum_{nm} \left[b_n, b^{\dagger}_m\right]\phi_n(x)\phi^{*}_m(x') \\ &= \sum_n \phi_n(x)\phi^{*}_n(x') = \delta(x - x')\end{aligned} \tag{8.12}$$

In second quantization the wave functions are operators that obey commutation relations.

8.2 Superfluidity

Superfluids are among the most interesting of all condensed matter systems. The entire macroscopic system is in a single, coherent quantum state. Examples of superfluids are liquid ^{4}He, liquid ^{3}He, and superconductors. Some of these are in fermion systems (^{3}He, superconductors), while others are in boson systems (^{4}He, excitons). Low-temperature superfluids are also created in dilute atomic gases trapped by standing electromagnetic potentials. This chapter discusses Bose superfluids. A later chapter discusses superconductors. Another coherent boson system is a laser, although it is not a superfluid.

8.2.1 Bose-Einstein Condensation

Einstein is famous for many predictions and discoveries. His theory of boson condensation is among the most important. The first example is for a noninteracting, spinless, boson gas in three dimensions. The particles have energy $E(k) = \hbar^2k^2/2m$. The number of particles N in a box of volume Ω is obtained by summing over all wave vector states $\mathbf{k}$:

$$N = \sum_{\mathbf{k}} \langle C_{\mathbf{k}}^{\dagger} C_{\mathbf{k}} \rangle = \sum_{\mathbf{k}} n_{\mathbf{k}} \tag{8.13}$$

$$= \sum_{\mathbf{k}} \frac{1}{e^{\beta[E(k)-\mu]} - 1} = \Omega \int \frac{d^3k}{(2\pi)^3} \frac{1}{e^{\beta[E(k)-\mu]} - 1} \tag{8.14}$$

Evaluate this integral in spherical coordinates. Since the integrand has no angular dependence, then $d^3k = 4\pi k^2 dk$. Change integration variables to $x = \beta E(k)$ and find

$$n = \frac{N}{\Omega} = \frac{1}{4\pi^2} \left(\frac{2mk_BT}{\hbar^2} \right)^{3/2} I(y) \tag{8.15}$$

$$I(y) = \int_0^\infty \frac{\sqrt{x}\,dx}{e^{x-y} - 1}, \quad y = \beta\mu \tag{8.16}$$

Consider the temperature dependence of the above formula. The particle density n is constant with temperature or varies slowly. Yet the right-hand side appears to have a large variation with temperature: there is a prefactor of $T^{3/2}$. The right-hand side can remain constant only by having $y = \beta\mu$ vary with temperature, by having the chemical potential vary $\mu(T)$.

The factor of y must be negative or else the occupation number $n_{\mathbf{k}}$ would be negative as $x \to 0$. As the temperature is lowered, the chemical potential approaches zero from below: it becomes less negative. This behavior is shown in fig. 8.1. At a temperature T_λ the chemical potential vanishes. This temperature is deduced from the above equation:

$$n = \frac{1}{4\pi^2} \left(\frac{2mk_BT_\lambda}{\hbar^2} \right)^{3/2} I(0) \tag{8.17}$$

$$I(0) = \int_0^\infty \frac{\sqrt{x}\,dx}{e^x - 1} = \Gamma(3/2)\zeta(3/2) \approx 2.315 \tag{8.18}$$

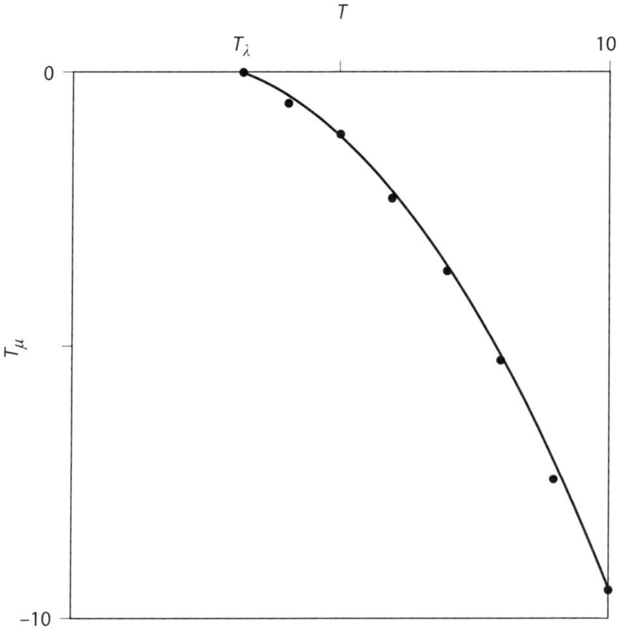

FIGURE 8.1. Chemical potential $T_\mu = \mu(T)/k_B$ as a function of temperature T. At T_λ the chemical potential vanishes.

$$k_B T_\lambda = \frac{\hbar^2}{2m}\left(\frac{4\pi^2 n}{I(0)}\right)^{2/3} \tag{8.19}$$

The last formula gives the Bose-Einstein transition temperature T_λ as a function of the particle density and mass. Using the density of $\rho = 0.146$ g/cm^3 for helium, and the helium mass, this formula predicts $T_\lambda = 3.13$ K. The experimental value is around 2.17 K. The theory is inaccurate since the excitations in helium are not plane waves.

What happens when the temperature is lowered below T_λ? It is no longer possible to satisfy eqn. (8.14). A large number of the particles begin to occupy the state with zero wave vector. The chemical potential remains at zero. So rewrite this equation as

$$N = N_0 + \sum_{\mathbf{k}\neq 0} \frac{1}{e^{\beta E(k)} - 1} = N_0 + \Omega \int \frac{d^3k}{(2\pi)^3} \frac{1}{e^{\beta E(k)} - 1} \tag{8.20}$$

$$\begin{aligned} n &= n_0 + \frac{1}{4\pi^2}\left(\frac{2mk_B T}{\hbar^2}\right)^{3/2} I(0), \quad n_0 = \frac{N_0}{\Omega} \\ &= n_0 + n\left(\frac{T}{T_\lambda}\right)^{3/2} \end{aligned} \tag{8.21}$$

The last line was derived using the definition of T_λ in eqn. (8.20). One can then solve for the density of particles n_0 in the state with $\mathbf{k} = 0$:

$$n_0 = n\left[1 - \left(\frac{T}{T_\lambda}\right)^{3/2}\right] \tag{8.22}$$

The state with $\mathbf{k} = 0$ is called the *condensate*. The Bose-Einstein transition temperature T_λ is where the system of bosons starts to become a coherent fluid by having a significant fraction of the particles occupy the same quantum state.

The above calculation is not accurate. In an actual quantum fluid, such as liquid ^{4}He, the particles have strong interactions. The low-lying excitation energies are not $E(k) = \hbar^2k^2/2m$, but are longitudinal sound waves with $E(k) = \hbar c_s k$. However, one cannot use the sound wave dispersion in the integral (8.14) for the transition temperature since that is not the proper function. Equation (8.13) is still valid in terms of the number of particles $n_\mathbf{k}$ in state $\mathbf{k}$. The question is to find this function when the excitations are sound waves. That is a complicated theoretical problem, which is discussed in the following section. Neutron scattering experiments at low temperature show that $n_\mathbf{k}$ is a Gaussian function of wave vector $n_\mathbf{k} \propto \exp[-wk^2]$, where w does not depend strongly on temperature.

Green's function theory derives a spectral function $A(\mathbf{k}, E)$, which is the probability that a He particle has wave vector $\mathbf{k}$ and energy E. The number of particles with wave vector $\mathbf{k}$ is

$$n_\mathbf{k} = \int \frac{dE}{2\pi} A(\mathbf{k}, E)\, n_B(E) \tag{8.23}$$

$$n_B(E) = \frac{1}{e^{\beta(E-\mu)} - 1} \tag{8.24}$$

where $n_B(E)$ is the Bose-Einstein distribution function in energy. The particle density of states is $N(E)$:

$$N(E) = \int \frac{d^3k}{(2\pi)^3} A(k, E) \tag{8.25}$$

$$n = \int \frac{dE}{2\pi} N(E)\, n_B(E) \tag{8.26}$$

The important quantity in the calculation of the transition temperature is the particle density of states.

An example of why the theory is poor is shown by doing the same calculation in two dimensions. Using the free particle dispersion $E(k) = \hbar^2k^2/2m$ predicts no BE condensation in two dimensions, which is a homework problem. But experimentally it does exist in films of helium.

8.2.2 Bogoliubov Theory of Superfluidity

Bogoliubov proposed a theory of superfluidity in an interacting boson gas. His theory applies to the case of weak interactions, where pairs of particles collide occasionally. It applies quite well to the Bose-Einstein condensation in dilute gases, as recently discovered in atomic physics. It is not an accurate theory for liquid helium, since there the particles are packed together and each particle is simultaneously interacting with six to eight neighboring particles at all times. However, this theory is important since it introduced many ideas to the topic of superfluids, and it does apply to superfluid gases.

Start with a Hamiltonian that has a kinetic energy term (H_0) and a particle–particle interaction term (V). It is assumed that the particles are neutral, so that the interactions are of short range:

$$H = H_0 + V \tag{8.27}$$

$$H_0 = \sum_{\mathbf{p}} \varepsilon(p)\, C^\dagger_{\mathbf{p}} C_{\mathbf{p}}, \quad \varepsilon(p) = \frac{\hbar^2 p^2}{2m} \tag{8.28}$$

$$V = \frac{1}{2\Omega} \sum_{\mathbf{pkq}} v(q)\, C^\dagger_{\mathbf{p}+\mathbf{q}} C^\dagger_{\mathbf{k}-\mathbf{q}} C_{\mathbf{k}} C_{\mathbf{p}} \tag{8.29}$$

The interaction term is similar in form to that of electron–electron interactions. An important difference is that there is no spin index, since we assume the bosons are spinless. The interaction $v(q)$ is the Fourier transform of the particle–particle interaction $V(r)$. For the interaction between two helium particles, most calculations use the Lenard-Jones 6-12 potential:

$$V(r) = 4\varepsilon\left[\left(\frac{\sigma}{r}\right)^{12} - \left(\frac{\sigma}{r}\right)^{6}\right] \tag{8.30}$$

It has a maximum well depth of ε at $r = 2^{1/6}\sigma$. The helium parameters are $\varepsilon/k_B = 10.7$ K, $\sigma = 2.648$ Å. One problem with this potential is that it lacks a Fourier transform—it is too singular near the origin. That problem can be circumvented by using a T-matrix theory and expressing the interaction in terms of phase shifts. That approach is used in crystals of cold atomic gases. However, since the theory does not apply to helium anyway, the lack of a Fourier transform is ignored. It is assumed that whatever system is described by eqn. (8.29), its interaction potential has a Fourier transform.

The term with $\mathbf{q} = 0$ is contained in the summation. It is the Hartree energy: the average energy of a particle from interacting with its neighboring particles.

Calculate the ground-state energy. The kinetic energy term is

$$E_K = \langle H_0 \rangle = \varepsilon(0)\langle C^\dagger_0 C_0 \rangle + \sum_{\mathbf{p}\neq 0} \frac{\varepsilon(p)}{e^{\beta(\varepsilon-\mu)} - 1} \tag{8.31}$$

The first term is zero since $\varepsilon(0) = 0$. The above formula is valid even when there is a condensate. Kinetic energy is contributed only by the particles not in the condensate.

Consider the ground-state energy from the particle–particle interactions. The Hartree term has $\mathbf{q} = 0$:

$$E_H = \frac{v(0)}{\Omega} N^2 \tag{8.32}$$

$$N = \sum_{\mathbf{k}} \langle C^\dagger_{\mathbf{k}} C_{\mathbf{k}} \rangle = \sum_{\mathbf{k}} \frac{1}{e^{\beta(\varepsilon(k)-\mu)} - 1} = \sum_{\mathbf{k}} n_{\mathbf{k}} \tag{8.33}$$

There is also an exchange term, derived by pairing the operators the other way:

$$E_X = \frac{1}{2\Omega}\sum_{\mathbf{pkq}} v(q)\langle C^\dagger_{\mathbf{p+q}} C_{\mathbf{k}}\rangle\langle C^\dagger_{\mathbf{k-q}} C_{\mathbf{p}}\rangle \tag{8.34}$$

$$= \frac{1}{2\Omega}\sum_{\mathbf{pq}} v(q)\, n_{\mathbf{p}} n_{\mathbf{p+q}} = \frac{1}{2}\sum_{\mathbf{p}} n_{\mathbf{p}} \Sigma_X(p)$$

$$\Sigma_X(p) = \frac{1}{\Omega}\sum_{\mathbf{q}} v(q)\, n_{\mathbf{p+q}} \tag{8.35}$$

The exchange energy has $\mathbf{q} = \mathbf{k} - \mathbf{p}$. There is no sign change from interchanging the order of the boson operators, so the exchange energy has a plus sign in front of the equation. The sign of the answer will depend on the sign of $v(\mathbf{q})$. The total ground-state energy in the Hartree-Fock approximation is

$$E_G = E_K + E_H + E_X \tag{8.36}$$

This theory applies to the fluid in the normal state.

How is the theory changed in the superfluid, when a significant fraction of the particles are in the state with $\mathbf{p} = 0$ and $\mu = 0$? Let $N_0 = n_0 \Omega$ be the number of particles in the condensate ($\mathbf{p} = 0$). The Hartree energy is the same. In evaluating the exchange energy, do not include the term with $\mathbf{q} = 0$ since that is already in the Hartree energy. Then we get

$$\Sigma_X(p) = \frac{N_0}{\Omega} v(p) + \frac{1}{\Omega}\sum_{\mathbf{q}\neq 0, -\mathbf{p}} v(q)\, n_{\mathbf{p+q}} \tag{8.37}$$

$$E_X = \frac{N_0}{2}\Sigma_X(0) + \frac{1}{2\Omega}\sum_{\mathbf{p}\neq 0}\sum_{\mathbf{q}\neq 0, -\mathbf{p}} v(q)\, n_p n_{p+q} \tag{8.38}$$

The first term comes from the condensate.

If $N_0 \gg 1$, the operators $C^\dagger_0$, C_0 become c-numbers rather than operators. These operators act on the condensate state. If N_0 is very large,

$$C_0|N_0\rangle = \sqrt{N_0}|N_0 - 1\rangle \approx \sqrt{N_0}|N_0\rangle \tag{8.39}$$

$$C^\dagger_0|N_0\rangle = \sqrt{N_0 + 1}|N_0 + 1\rangle \approx \sqrt{N_0}|N_0\rangle \tag{8.40}$$

Consider that N_0 might be 10^{20}, which makes the above expressions reasonable. So replace these operators by

$$C^\dagger_0 = \sqrt{N_0} = C_0 \tag{8.41}$$

Write out the particle–particle interaction term V by explicitly separating out all terms that have one or more of their wave vectors zero. One can have $0 = \mathbf{p} = \mathbf{k} = \mathbf{q}$, which gives

$$V = \frac{N_0^2 v(0)}{2\Omega} + \cdots \tag{8.42}$$

It is not possible to have three subscripts be zero, but not the fourth. So the next interesting terms have two wave vectors be zero. There are six ways of pairing four operators. Each term has the factor of $N_0/2\Omega$ plus the factor:

$$0 = \mathbf{k} = \mathbf{q},\ \mathbf{p} \neq 0 \text{ gives } \nu(0)\, n_p$$

$$0 = \mathbf{p} = \mathbf{q},\ \mathbf{k} \neq 0 \text{ gives } \nu(0)\, n_k$$

$$0 = \mathbf{k} = \mathbf{p},\ \mathbf{q} \neq 0 \text{ gives } \nu(q)\, C_q^\dagger C_{-q}^\dagger$$

$$\mathbf{k} = \mathbf{q},\ \mathbf{p} = -\mathbf{q} \text{ gives } \nu(q)\, C_q C_{-q}$$

$$\mathbf{k} = \mathbf{q},\ \mathbf{p} = 0 \text{ gives } \nu(q)\, C_q^\dagger C_q$$

$$\mathbf{k} = 0,\ \mathbf{p} = -\mathbf{q} \text{ gives } \nu(q)\, C_{-q}^\dagger C_{-q}$$

There are also terms with one wave vector equal to zero and three that are nonzero. Writing them out is a homework problem. The following terms have been derived:

$$V = \frac{N_0^2 \nu(0)}{2\Omega} + \frac{N_0}{\Omega} \sum_{p \neq 0} \left\{ \nu(0)\, n_p + \frac{\nu(p)}{2} \left[(C_p^\dagger + C_{-p})(C_{-p}^\dagger + C_p) - 1 \right] \right\} \tag{8.43}$$

All variables of summation have been changed to p. The last term, with -1, comes from the commutator $C_{-p}C_{-p}^\dagger - 1 = C_{-p}^\dagger C_{-p}$. The constant term with -1 is dropped.

The first term in the summation can be combined with the first term on the right. Recall that

$$N = N_0 + \sum_{p \neq 0} n_p \tag{8.44}$$

$$\frac{\nu(0)}{2\Omega} N^2 = \frac{\nu(0)}{2\Omega} \left[N_0 + \sum_{p \neq 0} n_p \right]^2 = \frac{\nu(0)}{2\Omega} \left[N_0^2 + 2N_0 \sum_{p \neq 0} n_p + \left(\sum_{p \neq 0} n_p \right)^2 \right] \tag{8.45}$$

The first two terms on the right in eqn. (8.46) contribute to the constant term:

$$V = V_0 + \frac{N_0}{2\Omega} \sum_{p \neq 0} \nu(p) (C_p^\dagger + C_{-p})(C_{-p}^\dagger + C_p) \tag{8.46}$$

$$V_0 = \frac{N^2 \nu(0)}{2\Omega} \tag{8.47}$$

The second term on the right is the Bogoliubov interaction term in the superfluid.

The easiest way to diagonalize the Hamiltonian with the above interaction is to remember that the operators obey harmonic oscillator statistics. Treat the Hamiltonian as a harmonic oscillator problem, where V is a contribution to the spring constant. Define

$$Q_p = \sqrt{\frac{1}{2\varepsilon(p)}} (C_{-p}^\dagger + C_p) \tag{8.48}$$

$$P_p = i\sqrt{\frac{\varepsilon(p)}{2}} (C_p^\dagger - C_{-p}) \tag{8.49}$$

$$[Q_{\rm p}, P_{\rm p'}] = \frac{i}{2}\delta_{\rm pp'}\{[C_{\rm p}, C_{\rm p}^\dagger] - [C_{\rm -p'}^\dagger, C_{\rm -p}]\} = i\delta_{\rm pp'} \tag{8.50}$$

The terms in the Hamiltonian can be expressed in terms of these pseudo displacements $Q_{\rm p}$ and momenta $P_{\rm p}$:

$$\varepsilon(p)\left[C_{\rm p}^\dagger C_{\rm p} + \frac{1}{2}\right] = \frac{1}{2}\left[P_{\rm p}P_{\rm -p} + \varepsilon(p)^2 Q_{\rm p}Q_{\rm -p}\right] \tag{8.51}$$

$$v(p)(C_{\rm p}^\dagger + C_{\rm -p})(C_{\rm -p}^\dagger + C_{\rm p}) = 2v(p)\varepsilon(p)\,Q_{\rm p}Q_{\rm -p} \tag{8.52}$$

The Hamiltonian is

$$H = V_0 + \frac{1}{2}\sum_{\rm p\neq 0}\left[P_{\rm p}P_{\rm -p} + E(p)^2 Q_{\rm p}Q_{\rm -p}\right] \tag{8.53}$$

$$E(p)^2 = \varepsilon(p)\left[\varepsilon(p) + 2n_0 v(p)\right] \tag{8.54}$$

where n_0 is the density of particles in the condensate. The eigenvalue is $E(p)$. Define a new set of operators $(a_{\rm p}, a_{\rm p}^\dagger)$:

$$Q_{\rm p} = \sqrt{\frac{1}{2E(p)}}\left(a_{\rm -p}^\dagger + a_{\rm p}\right) \tag{8.55}$$

$$P_{\rm p} = i\sqrt{\frac{E(p)}{2}}\left(a_{\rm p}^\dagger - a_{\rm p}\right) \tag{8.56}$$

$$H = V_0 + \frac{1}{2}\sum_{\rm p\neq 0} E(p)\left[a_{\rm p}^\dagger a_{\rm p} + \frac{1}{2}\right] \tag{8.57}$$

It is also possible to relate the new operators to the original set. Equate the two expressions for $Q_{\rm p}$ and $P_{\rm p}$ and find

$$C_{\rm -p}^\dagger + C_{\rm p} = \sqrt{\frac{\varepsilon(p)}{E(p)}}\left(a_{\rm -p}^\dagger + a_{\rm p}\right) \tag{8.58}$$

$$C_{\rm -p}^\dagger - C_{\rm p} = \sqrt{\frac{E(p)}{\varepsilon(p)}}\left(a_{\rm -p}^\dagger - a_{\rm p}\right) \tag{8.59}$$

Add and subtract these two equations:

$$C_{\rm -p}^\dagger = \frac{1}{2}\left[\sqrt{\frac{\varepsilon(p)}{E(p)}}\left(a_{\rm -p}^\dagger + a_{\rm p}\right) + \sqrt{\frac{E(p)}{\varepsilon(p)}}\left(a_{\rm -p}^\dagger - a_{\rm p}\right)\right] \tag{8.60}$$

$$C_{\rm p} = \frac{1}{2}\left[\sqrt{\frac{\varepsilon(p)}{E(p)}}\left(a_{\rm -p}^\dagger + a_{\rm p}\right) - \sqrt{\frac{E(p)}{\varepsilon(p)}}\left(a_{\rm -p}^\dagger - a_{\rm p}\right)\right] \tag{8.61}$$

The new operators are linear combinations of the original ones.

The dispersion relation $E(p)$ describes sound waves at small wave vectors. Assuming that $v(0)$ is neither zero nor infinity, then

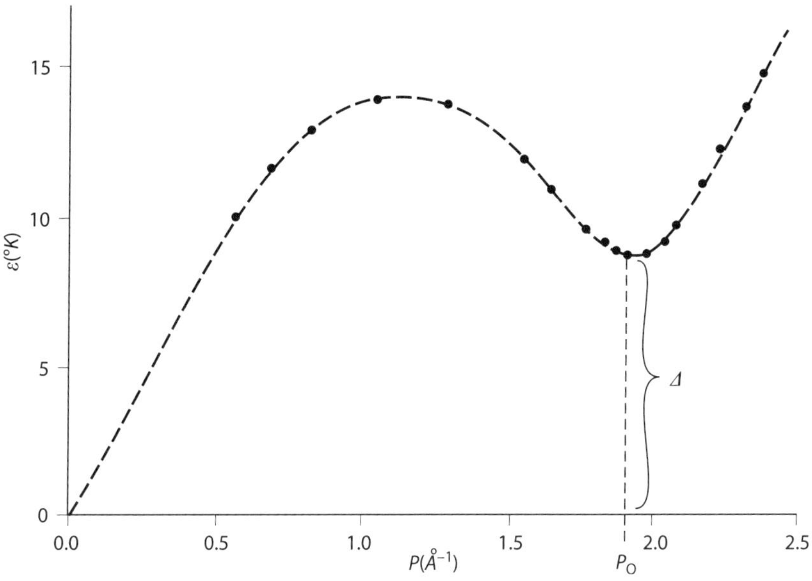

FIGURE 8.2. Dispersion of excitations in superfluid helium, as measured by neutron scattering.

$$\lim_{p \to 0} E(p) = \hbar v p, \quad v = \sqrt{\frac{n_0 v(0)}{m}} \tag{8.62}$$

where v is the velocity of these sound waves. This is the correct behavior of superfluid ^{4}He, whose dispersion is shown in fig. 8.2. The excitations at low wave vector are sound waves. There is a maximum, and a minimum around $k_0 = 2.2$ Å. This minimum is called the *roton*. Landau first deduced the need for such a minimum from thermodynamic data such as heat capacity. The speed of sound is not accurately predicted by eqn. (8.62).

A more accurate theory is found by replacing $v(q)$ with a T-matrix such as $T_{\mathbf{k}'\mathbf{k}}$, where $\mathbf{k}' = \mathbf{k} + \mathbf{q}$. For $\mathbf{q} \to 0$ then $v(0) = T_{\mathbf{kk}}$. At small values of $\mathbf{k}$ the T-matrix is approximated by its s-wave phase shift $\delta_0(k)$:

$$T_{\mathbf{kk}} \approx \frac{2\pi\hbar^2}{m}\left(\frac{\delta_0(k)}{k}\right) \tag{8.63}$$

In the limit that $k \to 0$, the phase shift is given by its scattering length a_0:

$$\lim_{k \to 0} \delta_0(k) = a_0 k \tag{8.64}$$

$$\lim_{k \to 0} T_{\mathbf{kk}} = \frac{2\pi\hbar^2 a_0}{m} \tag{8.65}$$

$$v = \frac{\hbar}{m}\sqrt{2\pi n_0 a_0} \tag{8.66}$$

This approximation is also not accurate in liquid helium, but is accurate in laser-cooled condensates of gases.

8.2.3 Off-diagonal Long-range Order

Earth-bound nature has identified several superfluids. Among these are superconductors, the low-temperature phase of liquid ^{4}He, and Bose-Einstein condensation of atoms in optical traps. Superfluids are macroscopic quantum states. The phase of all particles in the superfluid are tied together into one state that extends throughout the fluid. The phase coherence of the superfluid is similar to that found for the photons in a laser. Superfluids are unique environments in which to study macroscopic quantum phenomena.

The helium atom is found in two isotopes, which are called ^{4}He and ^{3}He. Both are found in the liquid state at very low temperatures. ^{3}He is a fermion, and ^{4}He is regarded as a boson. ^{4}He undergoes Bose-Einstein condensation at $T_\lambda = 2.17$ K. This transition is not well described by the free particle or by the Bogoliubov model, since it assumes the fluid of atoms is weakly interacting. In helium they are strongly interacting, which requires a different theory.

The modern theory of superfluidity in ^{4}He treats the liquid as a highly correlated and strongly interacting system. The representation of states in a plane-wave basis is abandoned. The one concept carried over is the condensate, or zero-momentum state, in which there is a nonzero fraction of the particles in the superfluid. How does one the define zero-momentum state in a basis set that is not plane waves? What is meant by superfluidity and Bose-Einstein condensation in a system that is strongly interacting and highly correlated? The method of doing this was introduced by Penrose [9]. Yang [12] suggested the name of *off-diagonal long-range order* (ODLRO) for the type of ordering introduced by Penrose and discussed by Penrose and Onsager [10].

ODLRO was introduced as the type of ordering for superfluids, such as Bose-Einstein condensation in ^{4}He, and electron pairs in superconductors. It is distinguished from diagonal long-range order (DLRO), which is the usual ordering one finds in crystalline solids. The distinction between these two orderings stems from the different behavior in the density matrix. These differences will be explained.

The *supersolid* state is where both DLRO and ODLRO coexist. It was suggested by Leggett [6] and discussed by Chester [1]. It may have been observed recently by Chan and Kim [4]. The nature of the order parameter in this state is being discussed.

A typical experimental system has $N \sim 10^{23}$ identical spinless bosons. The ground state is described by a many-particle wave function $\Psi_0(\mathbf{r}_1, \mathbf{r}_2, \ldots, \mathbf{r}_N)$. The subscript zero on Ψ_0 indicates it is for the ground state. There is no assumption that each atom is in the ground state (i.e., $\mathbf{k} = 0$), only that the system is in the ground-state. When the particles are strongly interacting and highly correlated, they will fluctuate between many momentum states. Correlated basis functions (CBFs) are the type of wave function most often employed in the study of the ground-state properties of ^{4}He. They have the form

$$\Psi_0(\mathbf{r}_1, \mathbf{r}_2, \ldots, \mathbf{r}_N) = L_N \exp\left[-\sum_{i>j} u(\mathbf{r}_i - \mathbf{r}_j)\right] \tag{8.67}$$

where L_N is a normalization constant, which is specified by satisfying the integral (8.69). The function $u(\mathbf{r})$ is repulsive at small distances, in order to keep the atoms apart. See Mahan [7] for a discussion of this wave function.

The ground state is the lowest possible energy state for the whole liquid. It is not a static or rigid structure, since each atom fluctuates with zero-point energy. The present estimates for ^{4}He at $T = 1$ K give the average kinetic energy per particle as 15 K and the average potential energy as -22 K, so the average binding energy is 7 K. The large value of average kinetic energy shows the large amount of zero-point motion in the fluid, even at low temperature, which comes from the quantum nature of the fluid. The classical estimate $\langle \text{K.E.} \rangle = 3k_BT/2$ is obviously inaccurate. The ground-state wave function $\Psi_0(\mathbf{r}_1, \mathbf{r}_2, \ldots, \mathbf{r}_N)$ for ^{4}He describes a system with a large amount of zero-point motion.

The square of the wave function gives the probability density for finding particles at positions $\mathbf{r}_j$ in the system, and is called the diagonal density matrix:

$$\rho_N(\mathbf{r}_1, \mathbf{r}_2, \ldots, \mathbf{r}_N) = |\Psi_0(\mathbf{r}_1, \mathbf{r}_2, \ldots, \mathbf{r}_N)|^2 \tag{8.68}$$

The subscript N indicates that it applies to N particles. ρ_N is normalized so that the integral over all coordinates gives unity:

$$1 = \int d^3r_1 \cdots d^3r_N \rho_N(\mathbf{r}_1, \mathbf{r}_2, \ldots, \mathbf{r}_N) \tag{8.69}$$

The one- and two-particle density matrix is obtained from ρ_N by integrating over all but one or two coordinate:

$$\rho_N(\mathbf{r}_1, \mathbf{r}_2) = \int d^3r_3 d^3r_4 \cdots d^3r_N \rho_N(\mathbf{r}_1, \mathbf{r}_2, \ldots, \mathbf{r}_N) \tag{8.70}$$

$$\rho_N(\mathbf{r}_1) = \int d^3r_2 d^3r_3 \cdots d^3r_N \rho_N(\mathbf{r}_1, \mathbf{r}_2, \ldots, \mathbf{r}_N) \tag{8.71}$$

$$1 = \int d^3r_1 \rho_1(\mathbf{r}_1), \quad \rho_N(\mathbf{r}) = 1/\Omega \tag{8.72}$$

The pair distribution function $g(\mathbf{r})$ is

$$\rho_N(\mathbf{r}_1, \mathbf{r}_2) = \frac{1}{\Omega^2} g(\mathbf{r}_1 - \mathbf{r}_2) \tag{8.73}$$

Penrose [9] used the idea of a general density matrix, which is defined as the product of two wave functions with different coordinates:

$$\begin{aligned} &\tilde{\rho}_N(\mathbf{r}_1, \mathbf{r}_2, \ldots, \mathbf{r}_N; \mathbf{r}_1', \mathbf{r}_2', \ldots, \mathbf{r}_N') \\ &\quad = \Psi_0^*(\mathbf{r}_1, \mathbf{r}_2, \ldots, \mathbf{r}_N) \Psi_0(\mathbf{r}_1', \mathbf{r}_2', \ldots, \mathbf{r}_N') \end{aligned} \tag{8.74}$$

This quantity is denoted by $\tilde{\rho}$, where the tilde is to distinguish it from the diagonal density matrix introduced earlier. They are identical if the two sets of coordinates in $\tilde{\rho}$ are equal:

$$\tilde{\rho}_N(\mathbf{r}_1, \mathbf{r}_2, \ldots, \mathbf{r}_N; \mathbf{r}_1, \mathbf{r}_2, \cdots, \mathbf{r}_N) = \rho_N(\mathbf{r}_1, \mathbf{r}_2, \ldots, \mathbf{r}_N). \tag{8.75}$$

The concept of ODLRO is contained in the function $\tilde{\rho}_1(\mathbf{r}_1, \mathbf{r}_1')$. It is obtained from $\tilde{\rho}_N$ when all but one set of coordinates are equal and averaged over

$$\tilde{\rho}_1(\mathbf{r}_1, \mathbf{r}_1') = \int d^3r_2 d^3r_3 d^3r_4 \cdots d^3r_N \tilde{\rho}_N(\mathbf{r}_1, \mathbf{r}_2, \ldots, \mathbf{r}_N; \mathbf{r}_1', \mathbf{r}_2, \ldots, \mathbf{r}_N) \tag{8.76}$$

$$\tilde{\rho}_1(\mathbf{r}_1, \mathbf{r}_1) = \rho_N(\mathbf{r}_1) = \frac{1}{\Omega} \tag{8.77}$$

where $\tilde{\rho}_1(\mathbf{r}_1, \mathbf{r}_1')$ becomes the diagonal density matrix ρ_N when $\mathbf{r}_1 = \mathbf{r}_1'$. In a liquid, the dependence on $\mathbf{r}_1$ and $\mathbf{r}_1'$ can only be their difference, since there is no absolute frame of reference. It is convenient to define the quantity

$$R(\mathbf{r}_1 - \mathbf{r}_1') = \Omega\tilde{\rho}_1(\mathbf{r}_1, \mathbf{r}_1') \tag{8.78}$$

$$R(0) = 1 \tag{8.79}$$

which is normalized to unity at $\mathbf{r} = 0$. $R(r)$ is the function which is important in understanding ODLRO.

Some insight into $\tilde{\rho}_1(\mathbf{r}_1, \mathbf{r}_1')$ is obtained by remembering the techniques used in the weakly interacting systems. There a one-particle state function is defined in the plane-wave representation as

$$\Phi(\mathbf{r}) = \frac{1}{\sqrt{\Omega}}\sum_{\mathbf{k}} e^{i\mathbf{k}\cdot\mathbf{r}} C_{\mathbf{k}} \tag{8.80}$$

$$C_{\mathbf{k}} = \frac{1}{\sqrt{\Omega}}\int d^3 r e^{-i\mathbf{k}\cdot\mathbf{r}}\Phi(\mathbf{r}) \tag{8.81}$$

The number of particles in state $\mathbf{k}$ is

$$n_{\mathbf{k}} = \langle C_{\mathbf{k}}^{\dagger} C_{\mathbf{k}}\rangle = \frac{1}{\Omega}\int d^3 r d^3 r' e^{-i\mathbf{k}\cdot\mathbf{r}(\mathbf{r}-\mathbf{r}')}\langle\Phi^{\dagger}(\mathbf{r})\Phi(\mathbf{r}')\rangle \tag{8.82}$$

$n_{\mathbf{k}}$ is found to be the Fourier transform of the quantity $\Phi^{\dagger}(\mathbf{r})\Phi(\mathbf{r}')$, which must be a function of $\mathbf{r} - \mathbf{r}'$. Take a particle and find its wave function at two different points $\mathbf{r}$ and $\mathbf{r}'$. This product is averaged, which is taken over the other particles and their positions. This procedure is exactly the one that was used to obtain $\tilde{\rho}_1(\mathbf{r}_1, \mathbf{r}_1')$; take a product wave function with one particle at two points $\mathbf{r}_1$ and $\mathbf{r}_1'$ and average it over the other particles and their positions. The quantity $\tilde{\rho}_1(\mathbf{r}_1, \mathbf{r}_1')$, or its equivalent $R(\mathbf{r}_1 - \mathbf{r}_2)$, is the many-body definition of the quantity $\langle\Phi^{\dagger}(\mathbf{r})\Phi(\mathbf{r}')\rangle$. $R(r)$ is quite a different function than the pair distribution function $g(r)$ and the two are not related.

The procedure for finding $n_{\mathbf{k}}$ is the same as in the free-particle case (8.82). Instead of the Fourier transform of $\Phi^{\dagger}(\mathbf{r})\Phi(\mathbf{r}')$, take the Fourier transform of the equivalent quantity $R(\mathbf{r} - \mathbf{r}')$. Before computing the Fourier transform, examine how it behaves as $r \rightarrow \infty$. For fluids at rest, so that the macroscopic occupation is in the zero-momentum state, $R(r)$ goes to a constant $R(\infty)$ as $\mathbf{r} \rightarrow \infty$. The Fourier transform is

$$\begin{aligned} n_{\mathbf{k}} &= n\int d^3 r R(r) e^{i\mathbf{k}\cdot\mathbf{r}} \\ &= nR(\infty)(2\pi)^3\delta^3(\mathbf{k}) + n\int d^3 r[R(r) - R(\infty)]e^{i\mathbf{k}\cdot\mathbf{r}} \\ &= N_0\delta_{k=0} + n\int d^3 r[R(r) - R(\infty)]e^{i\mathbf{k}\cdot\mathbf{r}} \end{aligned} \tag{8.83}$$

The density is $n = N/\Omega$, $N_0 = f_0 N$. There is a delta function term at $\mathbf{k} = 0$ whose amplitude is $nR(\infty)$. The quantity $R(\infty) = f_0$ is the fraction of particles in the zero-momentum

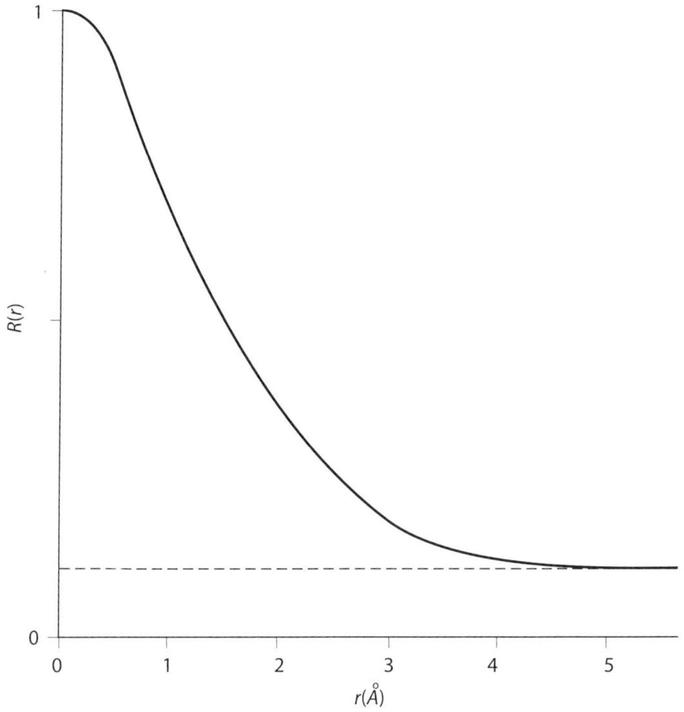

FIGURE 8.3. Single-particle density matrix as a function of separation. The solid line is the Monte Carlo calculation of McMillan which asymptotically approaches a density fraction of $f_0 = 0.11$. From *Phys. Rev.* **138**, A442 (1965). Used with permission of the American Physical Society.

state. This fraction is the "order" that exists in the off-diagonal density matrix $R(r)$. For $\mathbf{k} \neq 0$, the momentum distribution of particles $n_{\mathbf{k}}$ is found from the Fourier transform of $[R(r) - f_0]$.

Figure 8.3 shows the function $R(r)$ calculated for liquid ^{4}He by McMillan [8] using Monte Carlo techniques. He used the Lennard-Jones potential between helium atoms and found $f_0 = 0.11$. His fraction is similar to that obtained earlier by Penrose and Onsager [10], who got $f_0 = 0.08$ for a gas of hard spheres. The latter estimate is also based on a Monte Carlo calculation. There have been many calculations of these quantities. Figure 8.3 shows that $R(r)$ starts at unity and falls smoothly to its asymptotic value, which it reaches at about $r = 0.4$ nm. This distance is rather short, since it is only 1.5 atomic diameters.

Actual experimental systems have a finite size. Then there are no $\mathbf{k} = 0$ states: for a particle in a finite box of length L the lowest wave vector is of $O(\pi/L)$. A commonly used formula for the generalized particle density is

$$\rho(\mathbf{r}, \mathbf{r}') = \sum_j n_j \phi_j^*(\mathbf{r}) \phi_j(\mathbf{r}') \tag{8.84}$$

where n_j is the number of particles $n_j = \langle C_j^\dagger C_j \rangle$ in the state with quantum number j. In laser cooling of atoms, the potential that confines the atoms is often quadratic, so the eigenfunctions $\phi_j(\mathbf{r})$ are those of a harmonic oscillator. In any case, Bose-Einstein condensation is the occupation of the lowest quantum state. For the above formula, it is the occupation of the state ϕ_j which has the lowest eigenvalue.

8.3 Spin Waves

Spin waves, often called *magnons*, are a low-energy, low-temperature excitation of magnetically ordered crystals, such as ferromagnets or antiferromagnets. In two and three dimensions they are treated as bosons. The dispersion relations of these excitations can be measured using neutron scattering. They affect many low-temperature measurements of thermodynamic properties of magnetic systems, such as heat capacity and magnetization. Magnetic crystals also have phonons, so they have two sets of independent boson excitations.

Assume that in a ferromagnet all of the atoms in the crystal have the same spin $\mathbf{S}_j$. In the ground state they might all point in the same direction: they would be in the same quantum state. What is the low-energy excitation of this magnetic spin? It generally costs too much energy to flip a single spin at a single site to another quantum state. The low-energy excitations are wave-like, where each spin gets lowered according to its phase. If $S_j^{(-)}$ is a spin lowering operator for site $\mathbf{R}_j$, one might create an excitation by operating on the ground state by an operator such as

$$S_k^{(-)} = \frac{1}{\sqrt{N}}\sum_j e^{i\mathbf{k}\cdot\mathbf{R}_j} S_j^{(-)} \tag{8.85}$$

What are the properties of such operators?

First consider a magnetic crystal that is an insulator and has local spins at each atomic site. Label the site of the spin at site $\mathbf{R}_j$ by subscript j, such as $S_j^{(\pm)}$. The spin states are labeled with angular momenum J and magnetic quantum number M. The general properties of spin operators at the same site are

$$S_j^{(\pm)} = S_j^{(x)} \pm iS_j^{(y)} \tag{8.86}$$

$$S_j^{(z)}|J, M\rangle = \hbar M|J, M; j\rangle \tag{8.87}$$

$$S_j^{(+)}|J, M\rangle = \hbar\sqrt{J(J+1) - M(M+1)}|J, M+1; j\rangle \tag{8.88}$$

$$S_j^{(-)}|J, M\rangle = \hbar\sqrt{J(J+1) - M(M-1)}|J, M-1; j\rangle \tag{8.89}$$

Some commutation relations between raising and lowering operators at the same site are

$$\left[S_j^{(+)}, S_j^{(-)}\right] = 2\hbar S_j^{(z)} \tag{8.90}$$

$$\left[S_j^{(z)}, S_j^{(+)}\right] = \hbar S_j^{(+)} \tag{8.91}$$

When two operators are commuted, one finds another operator. This creates a problem when trying to define collective coordinates. When boson operators are commuted, one gets a constant. When fermion operators are anticommuted, one also finds a constant. Having a commutator produce another operator means that spins are neither bosons nor fermions.

Consider the case for spin-1/2 states. Their raising and lowering operators commute at different sites, since they are in different Hilbert spaces:

$$\left[S_j^{(+)}, S_{j'}^{(-)}\right] = 0 \;\; \text{if } j \neq j' \tag{8.92}$$

However, on the same site they anticommute. For a state $|\pm\frac{1}{2}; j\rangle$ at a site $\mathbf{R}_j$ these operators give

$$\left[S_j^{(+)} S_j^{(-)} + S_j^{(-)} S_j^{(+)}\right]|\tfrac{1}{2}; j\rangle = \hbar\, S_j^{(+)}|-\tfrac{1}{2}; j\rangle = \hbar^2|\tfrac{1}{2}; j\rangle \tag{8.93}$$

$$\left[S_j^{(+)} S_j^{(-)} + S_j^{(-)} S_j^{(+)}\right]|-\tfrac{1}{2}; j\rangle = \hbar\, S_j^{(-)}|\tfrac{1}{2}; j\rangle = \hbar^2|-\tfrac{1}{2}; j\rangle \tag{8.94}$$

Since the Hilbert space at site j is spanned by the two states $|\pm\frac{1}{2}; j\rangle$, the anticommutator $\{S_j^{(+)}, S_j^{(-)}\}$ is equivalent to multiplying by $\hbar^2$. The problem arises in trying to make collective operators such as eqn. (8.84). Since the operators $S_j^{(-)}$ commute on different sites and anti-commute on the same site, the collective operators have inconsistent commutation relations. The case of spin-1/2 is just one example that spin operators are unique.

The resolution to this problem depends on the dimension. In one dimension, the spin operators act as fermions, while in two and higher dimensions they are approximate bosons. These two cases are discussed separately.

8.3.1 Jordan-Wigner Transformation

In one dimension the spin-1/2 operators can be made into exact fermion operators. This transformation was discovered by Jordan and Wigner [3]. In one dimension, the spins are aligned along a chain. A new set of operators are defined: the spin raising and lowering operators are multiplied by a phase factor that is dependent on spin site:

$$d_j = e^{i\phi_j} S_j^{(-)}/\hbar \tag{8.95}$$

$$d_j^\dagger = e^{-i\phi_j} S_j^{(+)}/\hbar \tag{8.96}$$

$$d_j^\dagger d_j = \frac{1}{\hbar^2} S_j^{(+)} S_j^{(-)} = S(S+1) - \frac{1}{\hbar^2} S_j^{(z)2} + \frac{1}{\hbar} S_j^{(z)} = \frac{1}{2} + \frac{1}{\hbar} S_j^{(z)} \tag{8.97}$$

The phase factor ϕ_j is chosen to be π times an operator that measures the number of spin-up operators to the left of site j:

$$j > 1:\; \phi_j = \pi \sum_{n=1}^{j-1} \left(\frac{1}{2} + \frac{1}{\hbar} S_n^{(z)}\right) = \pi \sum_{n=1}^{j-1} d_n^\dagger d_n \tag{8.98}$$

The chain is numbered from one end, say the left, with site indices $j = 1, 2, 3\ldots$. Set $\phi_1 = 0$. The $d^\dagger$ operators are interpreted as creating fermion particles, and $d_j^\dagger d_j = n_j$ is the number operator for each site. The phase ϕ_j is π times the number of such fermionss to the left of the site j. The phase factor commutes with $S_j^{(-)}$ and $S_j^{(+)}$, since the operator ϕ_j

is the number to the left and does not involve the number operator on the same site. On the same site, these operators anticommute,

$$\{d_j, d_j^\dagger\} = \frac{1}{\hbar^2}\{S_j^{(-)}, S_j^{(+)}\} = 1 \tag{8.99}$$

since that is the property of the spin operators themselves. On different sites, they also anticommute, with the help of this new phase factor. By taking the anticommutator,

$$\hbar^2\{d_\ell, d_m^\dagger\} = e^{i\phi_\ell}S_\ell^{(-)}e^{-i\phi_m}S_m^{(+)} + e^{-i\phi_m}S_m^{(+)}e^{i\phi_\ell}S_\ell^{(-)} \tag{8.100}$$

Since $\ell \neq m$, assume that $\ell > m$. The phase factor ϕ_m then commutes with $S_\ell^{(-)}$, but ϕ_ℓ does not commute with $S_m^{(+)}$. The anticommutator is then

$$\hbar^2\{d_\ell, d_m^\dagger\} = e^{i\phi_\ell - i\phi_m}\left[S_\ell^{(-)}S_m^{(+)} + e^{-i\pi n_m}S_m^{(+)}e^{i\pi n_m}S_\ell^{(-)}\right] \tag{8.101}$$

The right-hand term contains the operator combination

$$e^{-i\pi(1/2+S^{(z)})}S_m^{(+)}e^{i\pi(1/2+S^{(z)})} = e^{-i\pi}S_m^{(+)} = -S_m^{(+)} \tag{8.102}$$

The $S^{(+)}$ operator must always raise the magnetic quantum number m by unity, so that $S^{(z)}$ on the left always measures one integer higher in value than the same operator on the right of $S^{(+)}$. One gets an extra phase factor of $-i\pi$, which changes the sign of the term. This phase factor ϕ_ℓ was chosen to produce this sign change. The anticommutator of the d operators is equal to the commutator $[S^{(+)}, S^{(-)}]$ for different sites, which is zero:

$$\{d_\ell, d_m^\dagger\} = 0 \qquad \text{for } \ell \neq m \tag{8.103}$$

The d operators anticommute for different sites. For the same site the anticommutation gives unity. They are pure fermion operators and obey fermion statistics. One can define operators for collective states:

$$d_k = \frac{1}{\sqrt{N}}\sum_j e^{ikaj}d_j \tag{8.104}$$

$$d_k^\dagger = \frac{1}{\sqrt{N}}\sum_j e^{-ikaj}d_j^\dagger \tag{8.105}$$

The collective operators also anticommute:

$$\{d_k, d_{k'}^\dagger\} = \frac{1}{N}\sum_{j\ell} e^{ia(kj-k'\ell)}\{d_j, d_\ell^\dagger\} \tag{8.106}$$

$$= \frac{1}{N}\sum_{j\ell} e^{ia(kj-k'\ell)}\delta_{j\ell} = \delta_{kk'}$$

In thermal equilibrium the number operator is

$$n_k = d_k^\dagger d_k = \frac{1}{e^{\beta\varepsilon(k)}+1} \tag{8.107}$$

where $\varepsilon(k)$ is the dispersion relation for the spin wave. There is no chemical potential for spin waves.

In one dimension the XY model is

$$H_{XY} = -J\sum_j \left(S_j^{(+)} S_{j+1}^{(-)} + S_j^{(+)} S_{j-1}^{(-)}\right) \tag{8.108}$$

It may be transformed into a Hamiltonian in terms of the d operators:

$$\begin{aligned} H_{XY} &= -J\sum_j \left(e^{i\phi_j} d_j^\dagger e^{-i\phi_{j+1}} d_{j+1} + e^{i\phi_j} d_j^\dagger e^{-i\phi_{j-1}} d_{j-1}\right) \\ &= -J\sum_j \left(d_j^\dagger e^{-i\pi n_j} d_{j+1} + d_j^\dagger e^{-i\pi n_{j-1}} d_{j-1}\right) \\ &= -J\sum_j \left(d_j^\dagger d_{j+1} + d_j^\dagger d_{j-1}\right) \end{aligned} \tag{8.109}$$

which is just the tight-binding model for fermions. The phase factors vanish because in the first term n_j is zero if it precedes a raising operator, and in the second term it is zero if it follows a lowering operator. Next the Hamiltonian is changed to collective coordinates:

$$H = -2J\sum_k \cos(ka)\, d_k^\dagger d_k, \quad \varepsilon(k) = -2J\cos(ka) \tag{8.110}$$

The XY-model is solved exactly in one dimension, using the Jordan-Wigner transformation. The Hamiltonian becomes a simple fermion problem. The exact partition function is

$$Z = \mathrm{Tr} e^{-\beta H} = \Pi_k \left(1 + e^{-\varepsilon(k)\beta}\right) \tag{8.111}$$

The Jordan-Wigner transformation shows that in one dimension the spin-1/2 operators may be represented exactly as fermions. This result is not valid for higher dimensions. No one has been able to find an equivalent transformation for two or three dimensions. Indeed, most approximate analyses assume that in two or three dimensions the spin excitations behave approximately as bosons rather than as fermions.

8.3.2 Holstein-Primakoff Transformation

In two and three dimensions, the spin operators are treated as approximate bosons using a transformation introduced by Holstein and Primakoff [2]. Let $(a_j, a_j^\dagger)$ be boson raising and lowering operators for the spin site $\mathbf{R}_j$. They obey the usual boson statistics:

$$a_j|n_j\rangle = \sqrt{n_j}\,|n_j - 1\rangle \tag{8.112}$$

$$a_j^\dagger|n_j\rangle = \sqrt{n_j + 1}\,|n_j + 1\rangle \tag{8.113}$$

For a spin S, the number of allowed values of M is $2S + 1$. Therefore, we allow $2S + 1$ values of n_j at each site: $n_j = 0, 1, 2, \ldots, 2S$. The value of $n_j = S - M$ is the change in the magnetic quantum number M below the ground-state value of $S = M$. The boson operators are related to the spin operators by

$$S_j^{(+)} = \hbar\sqrt{2S - a_j^\dagger a_j}\, a_j \tag{8.114}$$

$$S_j^{(-)} = \hbar a_j^\dagger \sqrt{2S - a_j^\dagger a_j} \tag{8.115}$$

$$S_j^{(z)} = \hbar\left(S - a_j^\dagger a_j\right) \tag{8.116}$$

The boson $(a_j^\dagger)$ is created using the spin lowering operator. The ground state has all spins in the same quantum state, which we take as $M = S$. The boson excitation is created by lowering the value of M.

The Holstein-Primakoff transformation is nonlinear. It has the virtue of correctly giving all commutation relations for spin operators. One example is

$$\left[S_j^{(z)}, S_j^{(+)}\right] = \hbar S_j^{(+)} \tag{8.117}$$

$$\hbar^2\left[S - a^\dagger a, \sqrt{2S - a^\dagger a}\, a\right]|n\rangle = \hbar^2\left[(S - a^\dagger a)\sqrt{2S - a^\dagger a}\, a - \sqrt{2S - a^\dagger a}\, a(S - a^\dagger a)\right]|n\rangle \tag{8.118}$$

Evaluating the right-hand side gives

$$= \hbar^2\left[(S - n + 1)\sqrt{2S - n + 1} - \sqrt{2S - n + 1}\,(S - n)\right]\sqrt{n}\,|n - 1\rangle$$

$$= \hbar^2\sqrt{2S - n + 1}\sqrt{n}\,|n - 1\rangle = \hbar S_j^{(+)} \tag{8.119}$$

Similar considerations prove that all of the commutation relations work among the spin-1/2 operators. The Holstein-Primakoff transformation is exact.

8.3.3 Heisenberg Model

The Holstein-Primakoff transformation is used to solve for the low-temperature properties of the Heisenberg model. It is a model of magnetism in an insulator. Each atomic site has a spin $\mathbf{S}_j = (S_j^{(x)}, S_j^{(y)}, S_j^{(z)})$ that interacts only with spins on the nearest-neighbor sites. The interaction is through electron exchange, caused by the overlap of the electronic orbitals on neighboring atoms. The customary way of writing the Hamiltonian is

$$H = -\frac{J}{\hbar^2}\sum_{j,\delta} \mathbf{S}_j \cdot \mathbf{S}_{j+\delta} \tag{8.120}$$

The constant J has the units of energy. The minus sign in front is chosen since $J > 0$ in most cases. The lowest energy state when $J > 0$ is to have all spin pointing in the same direction, which is called *up*. This is the ferromagnetic arrangement. Spin waves are the low-energy excitations of the ferromagnetically ordered spin system. If $J < 0$, the ground state has alternate spins pointing *up* and *down*. This latter arrangement is antiferromagnetic.

The spin $\mathbf{S}_j$ is at site $\mathbf{R}_j$, while $\mathbf{S}_{j+\delta}$ is at neighboring site $\mathbf{R}_j + \delta$. Write the interaction in terms of spin raising and lowering operators, and then make the Holstein-Primakoff transformation. Assume that $S = \frac{1}{2}$.

$$H = -\frac{J}{\hbar^2}\sum_{j\delta}\left[S_j^{(x)}S_{j+\delta}^{(x)} + S_j^{(y)}S_{j+\delta}^{(y)} + S_j^{(z)}S_{j+\delta}^{(z)}\right]$$

$$= -\frac{J}{\hbar^2}\sum_{j\delta}\left[\frac{1}{2}\left(S_j^{(+)}S_{j+\delta}^{(-)} + S_j^{(-)}S_{j+\delta}^{(+)}\right) + S_j^{(z)}S_{j+\delta}^{(z)}\right]$$

$$= -J\sum_{j\delta}\left[\left(\frac{1}{2} - n_j\right)\left(\frac{1}{2} - n_{j+\delta}\right) + \frac{1}{2}\sqrt{1-n_j}\, a_j a_{j+\delta}^\dagger \sqrt{1-n_{j+\delta}}\right.$$

$$\left. + \frac{1}{2} a_j^\dagger \sqrt{1-n_j}\sqrt{1-n_{j+\delta}}\, a_{j+\delta}\right] \tag{8.121}$$

Assume a low temperature, where there is magnetic ordering. The ground-state has all of the spins in the up direction. In spin language, the ground state of a chain of N spins is

$$|g\rangle = |\uparrow_1, \uparrow_2, \uparrow_3, \dots, \uparrow_N\rangle \tag{8.122}$$

In the language of bosons, all sites have no boson excitations, since spin-up is $n = 0$:

$$|g\rangle = |0_1, 0_2, 0_3, \dots, 0_N\rangle \tag{8.123}$$

At low temperature, some of the spins are flipped due to thermal excitations and some of the boson states are excited. The net spin polarization of a state $|O\rangle$ is

$$N_\uparrow - N_\downarrow = \frac{2}{\hbar}\sum_j S_j^{(z)}|O\rangle = \sum_j \left[1 - 2a^\dagger a_j\right]|O\rangle \tag{8.124}$$

At low temperature the density of such excitations is small. The boson excitation operator $n_j = a_j^\dagger a_j$ will be small, and n_j^2 will be very small. At low temperature, it is accurate to linearize eqn. (8.121), and keep only terms that are bilinear in the boson operators. Terms with four boson operators are small and ignored:

$$H \approx E_0 + \frac{J}{2}\sum_{j\delta}\left[n_j + n_{j+\delta} - a_j a_{j+\delta}^\dagger - a_j^\dagger a_{j+\delta}\right] \tag{8.125}$$

$$E_0 = -\frac{zJ}{4}N \tag{8.126}$$

where E_0 is the ground-state energy of the Heisenberg model. The constant z is the number of nearest neighbors and is called the *coordination number*. For a simple chain in one dimension it is two, for a simple cubic lattice it is six, and for a face-centered cubic lattice it is twelve.

Go to collective states, as was done for the XY-model, and find

$$H \approx E_0 + zJ\sum_{\mathbf{k}}(1 - \gamma_{\mathbf{k}})\, a_{\mathbf{k}}^\dagger a_{\mathbf{k}} \tag{8.127}$$

$$\gamma_{\mathbf{k}} = \frac{1}{z}\sum_{\delta=1}^{z} e^{i\mathbf{k}\cdot\boldsymbol{\delta}}, \quad \varepsilon(\mathbf{k}) = zJ(1 - \gamma_{\mathbf{k}}) \tag{8.128}$$

The spin waves have an excitation energy $\varepsilon(\mathbf{k})$. For any lattice with cubic symmetry, the dispersion at small wave vector is

$$\lim_{\mathbf{k}\to 0}\gamma_{\mathbf{k}} \to \frac{1}{z}\sum_{\delta=1}^{z}\left[1 + i\mathbf{k}\cdot\boldsymbol{\delta} + \frac{1}{2}(i\mathbf{k}\cdot\boldsymbol{\delta})^2 + \cdots\right] \tag{8.129}$$

$$= 1 - \frac{(k\delta)^2}{2d} + O(k^4), \quad \sum_{\delta}\hat{\delta}\hat{\delta} = \frac{z}{d}\mathcal{I}$$

$$\lim_{\mathbf{k}\to 0}\varepsilon(\mathbf{k}) = uk^2, \quad u = \frac{zJ\delta^2}{2d} \tag{8.130}$$

where d is the dimension. The dispersion is quadratic. The net magnetization as a function of temperature is

$$M = \frac{N_\uparrow - N_\downarrow}{N} = \frac{1}{N}[1 - 2a^\dagger_{\mathbf{k}}a_{\mathbf{k}}] \tag{8.131}$$

$$= 1 - 2\Omega_0\int\frac{d^3k}{(2\pi)^3}\frac{1}{e^{\beta\varepsilon(k)} - 1}$$

where Ω_0 is the volume of the unit cell. At low temperature, only spin waves that have long wavelength are excited, and we can approximate the dispersion as quadratic. Evaluate the integral in spherical coordinates ($d^3k = 4\pi k^2 dk$), change integration variables to $x = \beta uk^2$, and find

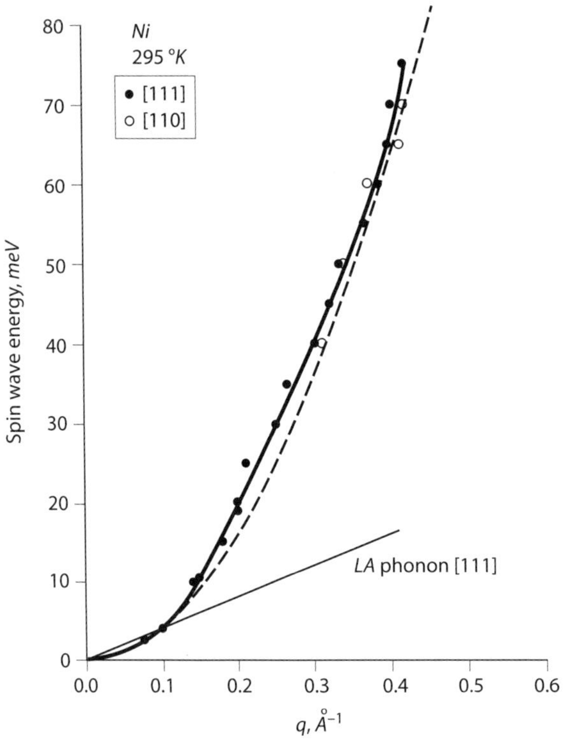

FIGURE 8.4. Dispersion of spin waves in ferromagnetic nickel in two directions in the crystal. They have quadratic dispersion at low temperatures. Data from Minkiewicz et al., *Phys. Rev.* **182**, 624 (1969). Used with permission of the American Physical Society.

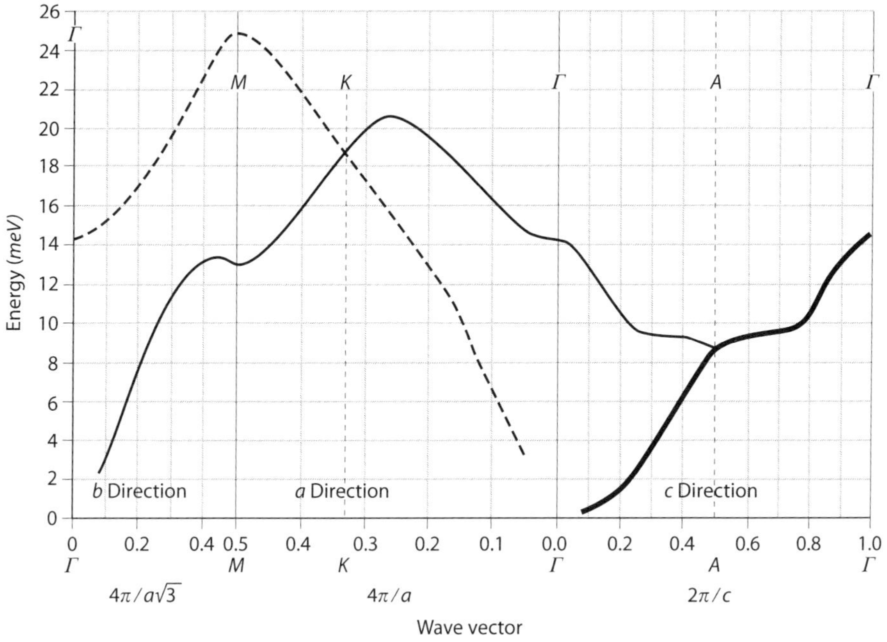

FIGURE 8.5. Spin wave dispersion in gadolinium in three principal directions of the hexagonal crystal. There are usually two spin wave modes in each direction. From Koehler et al., *Phys. Rev. Lett.* **24**, 16 (1970). Used with permission of the American Physical Society.

$$M = 1 - CT^{3/2} \tag{8.132}$$

$$C = \frac{\Omega_0}{2\pi^2}\left(\frac{k_B}{u}\right)^{3/2}\int_0^\infty \frac{\sqrt{x}\,dx}{e^x - 1} \tag{8.133}$$

The magnetization decreases as $T^{3/2}$, which agrees with experiments on ferromagnetics. This temperature dependence is predicted correctly by spin wave theory, but not by the mean field theory of magnetism. Figure 8.4 shows the spin wave dispersion in ferromagnetic nickel at low temperatures, as measured by neutron scattering by Minkiewicz et al. [11]. Spin waves are excitations of the electronic system, while neutrons scatter from the nucleus of the atoms. In an atom, the nuclear spin and electronic spins interact through the hyperfine interaction. The neutron interacts with the nuclear spin, which interacts with the spin waves through this hyperfine interaction. Figure 8.5 shows the spin waves in gadolinium. It has a hexagonal lattice and is a ferromagnet below $T_c = 300$ K. There are two spin wave modes for each wave vector. Data from Koehler et al. [5].

References

1. G. V. Chester, Speculations on Bose-Einstein condensation and quantum crystals. *Phys. Rev. A* **2**, 256–258 (1970)
2. T. Holstein and H. Primakoff, Field dependence of the intrinsic domain magnetization of a ferromagnet. *Phys. Rev.* **58**, 1098–1113 (1940)

3. P. Jordan and E. Wigner, Über das Paulische Äquivalenzverbot. *Z. Phys.* **47**, 631–651 (1928)
4. E. Kim and M.H.W. Chan, Observation of superflow in solid helium. *Science* **305**, 1941–1944 (Sept. 24, 2004)
5. W. C. Koehler, H.R. Child, R. M. Nicklow, H. G. Smith, R. M. Moon, and J. W. Cable, Spin wave dispersion in gadolinium. *Phys. Rev. Lett.* **24**, 16–18 (1970)
6. A. J. Leggett, Can a solid be "superfluid"? *Phys. Rev. Lett.* **25**, 1543–1546 (1970)
7. G. D. Mahan, *Many-particle Physics*, 3rd ed. (Kluwer/Plenum, New York, 2001) Chap. 11
8. W. L. McMillan, Ground state of liquid ^{4}He. *Phys. Rev.* **138**, A442 (1965)
9. O. Penrose, On the quantum mechanics of helium. *Philos. Mag.* **42**, 1373–1377 (1951)
10. O. Penrose and L. Onsager, Bose-Einstein condensation and liquid helium. *Phys. Rev.* **104**, 576–584 (1956)
11. V. J. Minkiewicz, M. F. Collins, R. Nathans, and G. Shirane, Critical and spin-wave fluctuations in nickel by neutron scattering. *Phys. Rev.* **182**, 624–631 (1969)
12. C. N. Yang, Concept of off-diagonal long-range order and quantum phases of liquid He and of superconductors. *Rev. Mod. Phys.* **34**, 694–704 (1962)

Homework

1. For bosons, the thermal average of the occupation number n is just the Bose-Einstein distribution function. What is the thermal average of n^2?

2. Derive the Bose-Einstein theory in two dimensions using quadratic dispersion for the particles. Show that the chemical potential can never ever be zero, except at zero temperature.

3. In the Bogoliubov theory of Bose-Einstein condensation, write out all of the terms in the particle–particle interaction that have only *one* wave vector be zero (a_0, $a_0^\dagger$) while three are nonzero. These are scattering terms.

4. In the Bogoliubov theory of helium, evaluate $n_k = a_k^\dagger a_k$ in the ground state at zero temperature.

5. Consider a system of N spinless bosons in a one-dimensional box of length L. Assume periodic boundary conditions. The Hamiltonian is

$$H = \sum_k E(k)\, a_k^\dagger a_k + \frac{V_0}{2L}\sum_{kpq} a_{k+q}^\dagger a_{p-q}^\dagger a_p a_k \tag{8.135}$$

   Calculate the ground-state energy per particle, including Hartree and Fock terms. There is no Bose-Einstein condensation in 1D.

6. The XY-model in a magnetic field in one dimension has the Hamiltonian

$$H = -\sum_j \left[J\left(S_j^{(+)}S_{j+1}^{(-)} + S_j^{(+)}S_{j-1}^{(-)}\right) + \Delta S_j^{(z)}\right] \tag{8.136}$$

   where $\Delta = \mu B/\hbar$ is the magnetic energy. Find the exact eigenvalues using the Jordan-Wigner transformation.

7. Use spin commutation relations to evaluate the commutator $[S^{(-)}, S^{(+)}] = ?$. Then obtain the same result using the boson operators from the Holstein-Primakoff transformation.

8. Derive the dispersion relation for magnons in an antiferromagnet on a bipartite lattice in three dimensions.

9. Use spin wave theory to derive the low-temperature heat capacity for a three-dimensional ferromagnetic.

10. Consider a bipartite lattice with spins S_A and S_B on alternate sites. The system is ferrimagnetic: the ordered arrangement has S_A all up and S_B all down. What is the spin wave spectrum if $S_A \neq S_B$?

9 Electron–Phonon Interactions

The electron–phonon interaction is important in all condensed matter systems where electronic conduction is important. It determines most transport coefficients, such as electrical conductivity and the Seebeck coefficient. It determines many optical properties. The interaction is treated similarly in semiconductors and insulators, but differently in metals. However, in all cases the interaction has the form

$$V_{ep} = \frac{1}{\sqrt{\Omega}} \sum_{\mathbf{k},\mathbf{q},\eta} M_\eta(\mathbf{k}+\mathbf{q},\mathbf{k})\, A_{\mathbf{q},\eta}\, C^\dagger_{\mathbf{k}+\mathbf{q},\sigma} C_{\mathbf{k}\sigma} \tag{9.1}$$

$$A_{\mathbf{q},\eta} = a_{\mathbf{q},\eta} + a^\dagger_{-\mathbf{q},\eta} \tag{9.2}$$

The operators $C^\dagger_{\mathbf{k}+\mathbf{q},\sigma} C_{\mathbf{k},\sigma}$ describe scattering an electron from the band state $\mathbf{k}$ to $\mathbf{k}+\mathbf{q}$, while destroying $(a_{\mathbf{q},\eta})$ or creating $(a^\dagger_{-\mathbf{q},\eta})$, a phonon in band η with wave vector $\mathbf{q}$. The task at hand is to derive expressions for the matrix element $M_\eta(\mathbf{k}+\mathbf{q},\mathbf{k})$ for this process. Quite often it depends only on the difference of the electron wave vectors, and is written in the simpler form $M_\eta(\mathbf{q})$.

9.1 Semiconductors and Insulators

This section discusses the electron–phonon interaction in semiconductors or insulators. In these crystals most electrons have energies near the bottom of the conduction band, and holes have energies near the top of valence band. The particle kinetic energy is similar to the thermal energy:

$$\varepsilon(k) = \frac{\hbar^2 k^2}{2m^*} \sim k_B T \tag{9.3}$$

$$k = \sqrt{\frac{2m^* k_B T}{\hbar^2}} = \frac{1}{a_0}\sqrt{\frac{m^*}{m_e}}\sqrt{\frac{k_B T}{E_{Ry}}} \tag{9.4}$$

The symbol a_0 denotes the Bohr radius, m^* an effective mass, while $E_{Ry} = 13.6$ eV is the Rydberg of energy. In GaAs, where $m^*/m_e = 1/16$, the above value at room temperature is $k \sim 0.02 \times 10^8$/cm. This value should be compared with the edge of the Brillouin zone, which is typically $k_x \sim \pi/a \sim 10^8$/cm. One finds that $k/k_x \ll 1$. The wave vectors of interest for thermally excited electrons are only a small fraction of the maximum values in the Brillouin zone.

The second part of the argument is that when an electron scatters a phonon, its initial wave vector k_i gets changed into a final wave vectors k_f according to

$$\varepsilon(k_f) = \varepsilon(k_i) \pm \hbar\omega(q) \tag{9.5}$$

Since the phonon energies are similar to thermal energies, then k_f is of the same size of k_i. The three wave vectors (k_i, k_f, q) are all very small. The scattering is only by phonons with long wavelength: those that are near to the center of the Brillouin zone. This insight allows us to make some useful approximations:

1. Use the Debye model for acoustical phonons: $\omega_\eta(q) = c_\eta q$. The symbol η denotes phonon mode: TA (transverse acoustical) or LA (longitudinal acoustical).

2. Use the Einstein model for optical phonons: $\omega_\eta(q) = \omega_\eta(0) \equiv \omega_0$.

These approximations make the integrals over wave vector easy in many cases. Analytical answers can be obtained for most expressions. The long-wavelength approximation can also be applied to the matrix element, as described below. In semiconductors and insulators with multivalley band minimum, phonon scattering between valleys is important. These processes involve phonons of large wave vector. Then the actual phonon energies must be used.

9.1.1 Deformation Potentials

The deformation potential interaction between electrons and acoustical phonons is very important in semiconductors and insulators. If one applies hydrostatic pressure to a solid, the lattice constant gets smaller. All of the energy bands change their value, and at small pressures this change is proportional to the change in lattice constant. An acoustical phonon is a local compression or dilation of the lattice, and therefore moves the local electron energy bands up or down.

Phenomenological arguments are used to derive a Hamiltonian for this interaction. A formula is borrowed from fluid dynamics: the equation of continuity

$$0 = \frac{\partial\rho}{\partial t} + \nabla\cdot(\mathbf{v}\rho) = \frac{\partial\rho}{\partial t} + \rho\nabla\cdot\mathbf{v} + \mathbf{v}\cdot\nabla\rho \tag{9.6}$$

The last term on the right is usually the smallest and is neglected. Replace the fluid velocity $\mathbf{v}$ with the time derivative of the atom displacement $\mathbf{Q}$. Canceling the two time derivatives gives

$$\frac{\delta\rho}{\rho} = -\nabla \cdot \mathbf{Q} = -\Delta \tag{9.7}$$

where Δ is the dilation. The fractional change in volume is proportional to the dilation $\nabla \cdot \mathbf{Q}$ and the shift in band energy is proportional to the same quantity. The constant of proportionality is called the *deformation potential constant D*. The Hamiltonian as a function of $\mathbf{r}$ is

$$H_{ep}(\mathbf{r}) = D\nabla \cdot \mathbf{Q}(\mathbf{r}) \tag{9.8}$$

$$\mathbf{Q}(\mathbf{r}) = \sum_{\mathbf{q},\eta} \frac{X(\mathbf{q},\eta)}{\sqrt{N}} \hat{\xi}(\mathbf{q},\eta) e^{i\mathbf{q}\cdot\mathbf{r}} A_{\mathbf{q},\eta} \tag{9.9}$$

$$X(\mathbf{q},\eta)^2 = \frac{\hbar}{2M\omega_\eta(q)} \tag{9.10}$$

$$H_{ep}(\mathbf{r}) = iD\sum_{\mathbf{q},\eta} \frac{X(\mathbf{q},\eta)}{\sqrt{N}} \mathbf{q}\cdot\hat{\xi}(\mathbf{q},\eta) e^{i\mathbf{q}\cdot\mathbf{r}} A_{\mathbf{q},\eta} \tag{9.11}$$

The interaction is cast into the form of eqn. (9.1) by integrating the above expression with the electron density operator:

$$V_{ep} = \int d^3r \rho(\mathbf{r}) H_{ep}(\mathbf{r}) \tag{9.12}$$

$$\rho(\mathbf{r}) = \frac{1}{\Omega}\sum_{\mathbf{q}} e^{i\mathbf{q}\cdot\mathbf{r}} \sum_{\mathbf{k}\sigma} C^\dagger_{\mathbf{k}+\mathbf{q},\sigma} C_{\mathbf{k}\sigma} \tag{9.13}$$

The resulting matrix element is

$$M_\eta(q) = iD\sqrt{\Omega_0}\,\mathbf{q}\cdot\hat{\xi}(\mathbf{q},\eta) X(\mathbf{q},\eta) \tag{9.14}$$

where $\Omega_0 = \Omega/N$ is the volume of the unit cell in the crystal.

Since the interaction is used only for phonons of long wavelength, the expression can be simplified. At long wavelength, longitudinal acoustical phonons (LA) have their atomic displacement along the direction of wave vector ($\hat{\xi} \parallel \hat{q}$) and TA phonons have the atomic displacement perpendicular to the wave vector ($\hat{\xi} \perp \hat{q}$). The factor of $\mathbf{q}\cdot\hat{\xi}(\mathbf{q},\eta)$ is nonzero only for LA phonons. The deformation potential interaction is only between electrons and LA phonons for an electron in a simple band.

For holes in the valence band, the deformation potential interaction can scatter holes between the different bands: light hole, heavy hole, and split-off bands. In some cases, TA phonons can do the scattering. In zincblende materials the deformation interaction for holes in the fourfold degenerate $p_{3/2}$ valence band was given by Kleiner and Roth [6]:

$$V_{ep} = D_d^v \Delta + \frac{2}{3} D_u \left[\left(J_x^2 e_{xx} + J_y^2 e_{yy} + J_z^2 e_{zz} - \frac{1}{3}\Delta J^2\right]\right.$$
$$+ \frac{2}{3} D_u' [\{J_x, J_y\} e_{xy} + \text{c.p.}] \tag{9.15}$$

$$\Delta = e_{xx} + e_{yy} + e_{zz} = \nabla \cdot \mathbf{Q} \tag{9.16}$$

where $e_{\alpha\beta}$ are strain components, and the symbol D denotes different deformation potential constants. The symbols J_μ denote $J=\frac{3}{2}$ angular momentum matrices. The first term $D_d^v\Delta$ is the usual form that acts on LA phonons. The other terms can act on TA phonons as well.

The phonon energy in $X(\mathbf{q},\eta)$ can be written $\omega_\eta(q)=c_{LA}q$, so that the matrix element at long wavelength is

$$M_{LA}(q) = C\sqrt{q}, \quad C = D\sqrt{\frac{\hbar\Omega_0}{2Mc_{LA}}} \tag{9.17}$$

This result will be used in numerous calculations.

9.1.2 Fröhlich Interaction

This interaction is between electrons and optical phonons in polar materials. Polar materials are those with ions of different valence, such as GaAs or BaF_2. All insulators are polar except those composed of a single element.

The simplest case is a binary solid such as NaCl or GaAs. Alternate atoms are anions or cations. For an optical phonon at long wavelength, two nearest neighbors oscillate against each other, which creates an oscillating dipole moment. In terms of the effective charge e^* on the ions,

$$\mathbf{p}(t) = e^*[\mathbf{Q}_+(t) - \mathbf{Q}_-(t)] \tag{9.18}$$

This oscillating dipole creates a potential a distance $\mathbf{r}$ away of

$$\phi(\mathbf{r}) = \frac{\mathbf{p}\cdot\mathbf{r}}{4\pi\varepsilon_0 r^3} \tag{9.19}$$

The Fourier transform of this potential is

$$\tilde{\phi}(\mathbf{q}) = \frac{1}{\varepsilon_0\Omega_0}\frac{\mathbf{p}_q\cdot\hat{q}}{q} \tag{9.20}$$

$$\mathbf{p}_q = e^* X_0(\hat{\xi}_+ - \hat{\xi}_-)A_q \tag{9.21}$$

There are two possibilities:

1. Longitudinal optical (LO) phonons have the atom displacements along the direction of wave vector, and $\hat{\xi}_\pm \parallel \hat{q}$, and also $\hat{\xi}_+ = -\hat{\xi}_-$.

2. Transverse optical phonons have the atomic displacements perpendicular to the wave vector. In this case $\hat{q}\cdot(\hat{\xi}_+ - \hat{\xi}_-) = 0$ and there is no interaction.

The interaction is only between LO phonons and the electrons and holes. Write the matrix element as

$$M_{LO}(q) = \frac{M_0}{q}, \quad M_0^2 = \frac{e^2\hbar\omega_{LO}}{2}\left(\frac{1}{\varepsilon(\infty)} - \frac{1}{\varepsilon(0)}\right) \tag{9.22}$$

The formula for M_0^2 is derived in the last section of this chapter. The two symbols $\varepsilon(\infty)$, $\varepsilon(0)$ are the dielectric constants at high and low frequency. They are the same ones used earlier in the dielectric function for polar materials. The LO phonon frequency ω_{LO} is a constant at long wavelength.

The lack of interaction with TO phonons is an important feature of the answer. Further insight into this result is provided by examining Maxwell's equations:

$$\nabla \times \mathbf{E} = -\frac{\partial}{\partial t}\mathbf{B} \tag{9.23}$$

$$\nabla \times \mathbf{B} = \mu_0 \frac{\partial}{\partial t}(\varepsilon_0 \mathbf{E} + \mathbf{P}) \tag{9.24}$$

These equations are solved by treating the polarization $\mathbf{P}$ as due to the phonon wave:

$$\mathbf{P} = \mathbf{P}_0 e^{i(\mathbf{q}\cdot\mathbf{r} - \omega_0 t)} \tag{9.25}$$

$$\mathbf{q} \times (\mathbf{q} \times \mathbf{E}) = -\frac{\omega_0^2}{c^2}\left(\mathbf{E} + \frac{1}{\varepsilon_0}\mathbf{P}\right) \tag{9.26}$$

This equation is solved for the two types of phonons:

1. For LO phonons, both $\mathbf{E}$, $\mathbf{P}$ are parallel to $\hat{q}$ so that $\mathbf{q} \times \mathbf{E} = 0$. The solution is

$$\mathbf{E} = -\frac{1}{\varepsilon_0}\mathbf{P} \tag{9.27}$$

which is the electric field for the Fröhlich interaction.

2. For TO phonons, the solution has the form

$$\mathbf{E} = \frac{\omega_0^2}{q^2c^2 - \omega_0^2}\frac{1}{\varepsilon_0}\mathbf{P} \approx \frac{\omega_0^2}{q^2c^2}\frac{1}{\varepsilon_0}\mathbf{P} \ll \frac{1}{\varepsilon_0}\mathbf{P} \tag{9.28}$$

The electric fields from TO phonons are very small due to the factor of the square of the speed of light in the denominator. Any transverse polarization wave wants to become a photon, rather than a phonon, and the effective electric field becomes very small. This does not occur for longitudinal fields, which is why the interaction is large only in this case.

9.1.3 *Piezoelectric Interaction*

Many crystals are piezoelectric, and they have an electron–phonon interaction due to this phenomenon. The phenomenon is simple: squeezing on the crystal produces an electric field. The requirement for being piezoelectric is to (i) be polar, and (ii) lack a center of inversion. Most II–V and II–VI semiconductors are piezoelectric. Those with the wurtzite structure, such as ZnO and CdS, have a strong piezoelectric interaction. The titanates are another class of crystals with a strong piezoelectric interaction in their ferroelectric phases.

The piezoelectric interaction is between electrons, or holes, and acoustical phonons. The acoustical phonons make a local dilation, which creates an electric field E_μ that is proportional to the strain ε_{ij}. The strain is the symmetric derivative with respect to phonon displacement. The constant of proportionality is the piezoelectric constant—a third rank tensor $e_{\mu,ij}$:

$$E_\mu = \sum_{ij} e_{\mu,ij}\varepsilon_{ij} \tag{9.29}$$

$$\varepsilon_{ij} = \frac{1}{2}\left(\frac{dQ_i}{dr_j} + \frac{dQ_j}{dr_i}\right) \tag{9.30}$$

$$= i\sum_{\mathbf{q},\eta} \frac{X(\mathbf{q},\eta)}{2\sqrt{N}} e^{i\mathbf{q}\cdot\mathbf{r}}[\xi_i q_j + \xi_j q_i] A_{\mathbf{q},\eta}$$

Apply the argument of the last section to conclude that the electric fields are longitudinal. For acoustical phonons at long wavelength, the ratio

$$\frac{\omega_\eta(q)^2}{q^2c^2} = \left(\frac{c_\eta}{c}\right)^2 \approx 10^{-10} \tag{9.31}$$

so that transverse electric fields are completely suppressed. The potential energy arising from the longitudinal component of the above electric field is

$$\phi(q) = \frac{1}{2q^2}\sum_{\mu,ij} q_\mu e_{\mu,ij}[q_i Q_j(q) + q_j Q_i(q)] \tag{9.32}$$

The third rank tensor $e_{\mu,ij}$ has nonzero components depending on crystal symmetry. The interaction is with both TA and LA phonons, and is quite anisotropic. Usually the matrix element is treated as a constant, and the value is obtained by averaging over all angles:

$$\langle M^2\rangle = \int \frac{d\Omega_q}{4\pi}|\phi(q)|^2 \equiv \frac{C_\eta^2}{q} \tag{9.33}$$

The potential $\phi(q)$ has a factor of q^{-2} in front, which is canceled dimensionally by the other two factors of $q_\mu q_{i,j}$ inside the summation. $\phi(q)$ is largely independent of the magnitude of wave vector. The only q-dependence is from the factor of phonon frequency in $X(\mathbf{q},\eta)$. The matrix element is

$$M_\eta(q) = \frac{C_\eta}{\sqrt{q}} \tag{9.34}$$

where the values of C_η depend on the angular averaging, and is different for TA and LA phonons.

9.1.4 Tight-binding Models

Another form of the electron–phonon interaction is found in system where the electronic energy bands are found using a tight-binding model. The simplest model is where each atomic site has one orbital, which overlaps weakly with its nearest neighbors. Then the tight-binding Hamiltonian for electrons has the form

$$H = \sum_{j\sigma} \left[E_0 C^{\dagger}_{j\sigma} C_{j\sigma} + \sum_{\delta} E_1(\delta) C^{\dagger}_{j+\delta,\sigma} C_{j\sigma} \right] \tag{9.35}$$

The first term is the onsite interaction, and the second term provides band dispersion from hopping to neighbors.

The hopping energy $E_1(\delta)$ depends on the distance to the neighbors. If the atoms are vibrating from phonons, this distance is modulated, which provides the interaction

$$E_1(\boldsymbol{\delta} + \mathbf{Q}_{j+\delta} - \mathbf{Q}_j) \approx E_1(\delta) + (\mathbf{Q}_{j+\delta} - \mathbf{Q}_j) \cdot \nabla_{\delta} E_1(\delta) \tag{9.36}$$

The second term provides an electron–phonon interaction. This form of phonon modulated hopping is widely used in organic compounds, including various forms of carbon.

9.1.5 *Electron Self-energies*

In semiconductors and insulators, the electron–phonon interaction can affect properties of the conduction electrons, and the valence band holes. It can change their band energy, and their effective mass. The electron–phonon interaction is relatively weak, so its influence is calculated using perturbation theory. The first-order perturbation $\langle V_{ep} \rangle = 0$ since V_{ep} has an odd number of phonon raising or lowering operators. The most important contribution is found using second-order perturbation theory. Figure 9.1 denotes the self-energy of the electron. An electron in state $\mathbf{k}$ emits a phonon and scatters to state $\mathbf{k} + \mathbf{q}$. Later it reabsorbs the same phonon. It is a contribution from the second-order of perturbation theory:

$$\Sigma(\mathbf{k}) = \frac{1}{\Omega} \sum_{\mathbf{q},\eta} |M_{\eta}(\mathbf{q})|^2 \left\{ \frac{N_{\eta}(\mathbf{q}) + 1 - n_F(\varepsilon(\mathbf{k}+\mathbf{q}))}{\varepsilon(\mathbf{k}) - \varepsilon(\mathbf{k}+\mathbf{q}) - \hbar\omega_{\eta}(\mathbf{q})} + \frac{N_{\eta}(\mathbf{q}) + n_F(\varepsilon(\mathbf{k}+\mathbf{q}))}{\varepsilon(\mathbf{k}) - \varepsilon(\mathbf{k}+\mathbf{q}) + \hbar\omega_{\eta}(\mathbf{q})} \right\} \tag{9.37}$$

$$N_{\eta}(\mathbf{q}) = \frac{1}{e^{\hbar\omega_{\eta}(\mathbf{q})/k_BT} - 1}, \quad n_F(\varepsilon(\mathbf{k}+\mathbf{q})) = \frac{1}{e^{[\varepsilon(\mathbf{k}+\mathbf{q}) - \mu]/k_BT} + 1} \tag{9.38}$$

where $(\mathbf{q}, \eta)$ are the wave vector and polarization (e.g., TA, LA, TO, LO) of the phonon, $N_{\eta}(\mathbf{q})$ is the phonon occupation number, and $n_F(\varepsilon(\mathbf{k}+\mathbf{q}))$ is the electron occupation number in the intermediate state. The first term is from the electron emitting a phonon in the intermediate state. The second term is from the electron absorbing an existing phonon

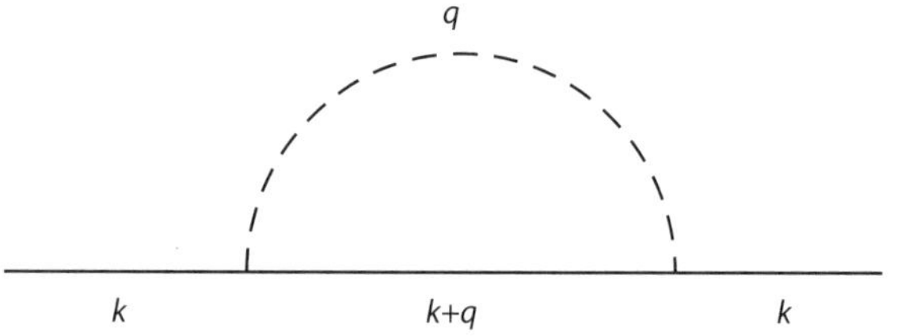

FIGURE 9.1. Feynman diagram of electron self-energy from phonons. Solid line is an electron, dashed line is a phonon.

to go to the intermediate state. This expression is evaluated for a number of cases. The lifetime of an electron from scattering by phonons is

$$\frac{\hbar}{\tau(\mathbf{k})} = \frac{2\pi}{\Omega}\sum_{\mathbf{q},\eta}|M_\eta(\mathbf{q})|^2\{[N_\eta(\mathbf{q}) + 1 - n_F(\varepsilon(\mathbf{k}+\mathbf{q}))]\delta[\varepsilon(\mathbf{k}) - \varepsilon(\mathbf{k}+\mathbf{q}) - \hbar\omega_\eta(\mathbf{q})]$$
$$+ [N_\eta(\mathbf{q}) + n_F(\varepsilon(\mathbf{k}+\mathbf{q}))]\delta[\varepsilon(\mathbf{k}) - \varepsilon(\mathbf{k}+\mathbf{q}) + \hbar\omega_\eta(\mathbf{q})]\} \tag{9.39}$$

The first term is from emitting a phonon, the second term is from absorbing an existing phonon. The factor of $n_F(\xi)$ is very important in metals. In semiconductors or insulators, we often consider the lifetime of a single electron in the conduction band, in which case $n_F(\xi) = 0$. In this case the expression for the lifetime simplifies to

$$\frac{\hbar}{\tau(\mathbf{k})} = \frac{2\pi}{\Omega}\sum_{\mathbf{q},\eta}|M_\eta(\mathbf{q})|^2\{[N_\eta(\mathbf{q}) + 1]\delta[\varepsilon(\mathbf{k}) - \varepsilon(\mathbf{k}+\mathbf{q}) - \hbar\omega_\eta(\mathbf{q})]$$
$$+ N_\eta(\mathbf{q})\delta[\varepsilon(\mathbf{k}) - \varepsilon(\mathbf{k}+\mathbf{q}) + \hbar\omega_\eta(\mathbf{q})]\} \tag{9.40}$$

These formulas are evaluated in the homework assignments.

Temperature Variation of Energy Gap

Semiconductors and insulators have an energy gap between the conduction and valence bands. Experimentally it is found that the band-gap energy $E_G(T)$ decreases with increasing temperature. The result for CdTe is shown in fig. 9.2. This phenomenon is due to the electron–phonon interaction. For an electron in the conduction band of an otherwise empty band, then $n_F(\varepsilon(\mathbf{k}+\mathbf{q})) = 0$. At high temperature use the expansion

$$N_\eta(\mathbf{q}) = \frac{k_B T}{\hbar\omega_\eta(\mathbf{q})} - \frac{1}{2} + O\left(\frac{\hbar\omega}{k_B T}\right) \tag{9.41}$$

The terms with $\frac{1}{2}$ are small compared to the first term. The bottom of the band is assumed to be at $\mathbf{k} = 0$. The change in energy with temperature of the minimum of the conduction band is

$$\Sigma(0) = -k_B T\Omega_0\sum_\eta\int\frac{d^3q}{(2\pi)^3}\frac{M_\eta(\mathbf{q})^2}{\hbar\omega_\eta(\mathbf{q})}\left[\frac{1}{\varepsilon(\mathbf{q}) + \hbar\omega_\eta(\mathbf{q})} + \frac{1}{\varepsilon(\mathbf{q}) - \hbar\omega_\eta(\mathbf{q})}\right] \tag{9.42}$$

The minus sign denotes that the energy of the electron in the conduction band is decreasing with temperature. A similar expression applies to holes in the valence band. The valence band rises in energy, while the conduction band falls, so the energy gap decreases. The dependence is indeed linear in temperature for higher temperatures, above the Debye temperature.

Effective Mass

The electron–phonon interaction makes an important contribution to the effective mass of electrons in semiconductors and insulators. Again set the electron occupation number n_F to zero in eqn. (9.37). Expand the energy denominator in powers of the electron velocity: the energy denominators are

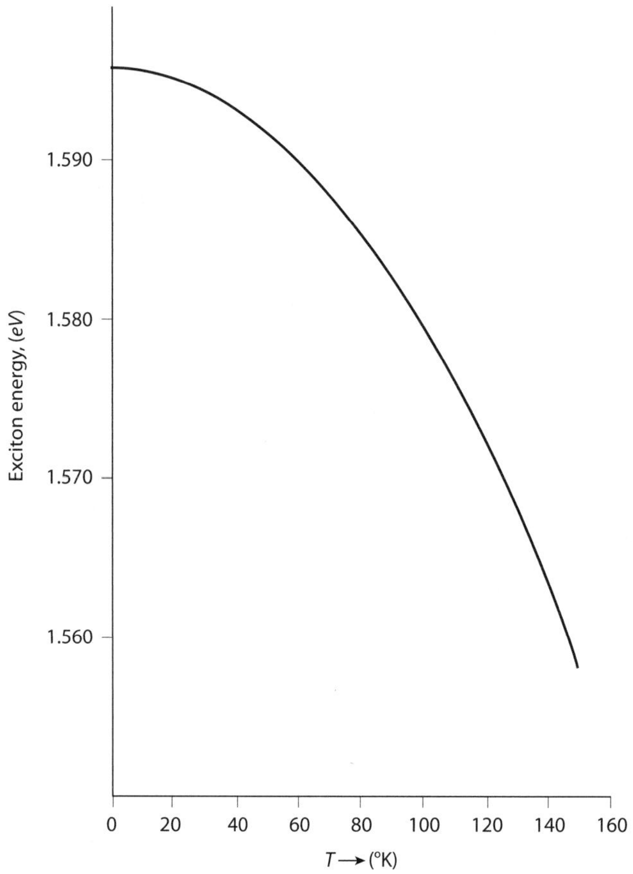

FIGURE 9.2. Energy gap variation with temperature in CdTe. From Mahan, *J. Phys. Chem. Solids* **26**, 751 (1965). Used with permission from Elsevier.

$$\frac{1}{\varepsilon(\mathbf{q}) + \hbar \mathbf{v}_k \cdot \mathbf{q} \pm \hbar\omega_\eta(\mathbf{q})} = \frac{1}{\varepsilon(\mathbf{q}) \pm \hbar\omega_\eta(\mathbf{q})}\left[1 - \frac{\hbar \mathbf{v}_k \cdot \mathbf{q}}{\varepsilon(\mathbf{q}) \pm \hbar\omega_\eta(\mathbf{q})} + \frac{(\hbar \mathbf{v}_k \cdot \mathbf{q})^2}{[\varepsilon(\mathbf{q}) \pm \hbar\omega_\eta(\mathbf{q})]^2}\right] \tag{9.43}$$

The first correction term is proportional to the vector $\mathbf{q} \cdot \mathbf{v}_k$ and averages to zero when evaluating the angular integrals. In cubic crystals, the second angular term gives an average $\langle(\mathbf{q} \cdot \mathbf{v}_k)^2\rangle = (\frac{1}{3}) v_k^2 q^2$. The self-energy function can be written as

$$\Sigma(\mathbf{k}) = \Sigma(0) - B\varepsilon(k) + \cdots \tag{9.44}$$

$$\Sigma(0) = -\sum_\eta \int \frac{d^3q}{(2\pi)^3} M_\eta(q)^2 \left[\frac{N_\eta(\mathbf{q}) + 1}{\varepsilon(\mathbf{q}) + \hbar\omega_\eta(\mathbf{q})} + \frac{N_\eta(\mathbf{q})}{\varepsilon(\mathbf{q}) - \hbar\omega_\eta(\mathbf{q})}\right] \tag{9.45}$$

$$B = \frac{4}{3}\sum_\eta \int \frac{d^3q}{(2\pi)^3} \varepsilon(q) M_\eta(q)^2 \left[\frac{N_\eta(\mathbf{q}) + 1}{[\varepsilon(\mathbf{q}) + \hbar\omega_\eta(\mathbf{q})]^3} + \frac{N_\eta(\mathbf{q})}{[\varepsilon(\mathbf{q}) - \hbar\omega_\eta(\mathbf{q})]^3}\right] \tag{9.46}$$

The total energy is

$$E(\mathbf{k}) = \varepsilon(\mathbf{k}) + \Sigma(\mathbf{k}) = \Sigma(0) + \varepsilon(\mathbf{k})(1 - B) + \cdots \tag{9.47}$$

$$= \Sigma(0) + \frac{\hbar^2 k^2}{2m^*} + \cdots$$

$$m^* = \frac{m_c}{1 - B} \tag{9.48}$$

The effective mass m^* of the electron in the band is determined by the band curvature m_c and the electron–phonon correction $(1 - B)$.

As an example, consider the polar interaction with optical phonons. Then $\omega_\eta(\mathbf{q}) = \omega_0$ and $N_\eta(\mathbf{q}) = N_0$ are both constants. The matrix element is $M^2 = C^2/q^2$. For this case, the second term in brackets in the above two equations gives a zero integral. The answer comes only from the first term:

$$\Sigma(0) = -\alpha[N_0 + 1]\hbar\omega_0 \tag{9.49}$$

$$\alpha = \frac{1}{\pi a_0^* q_0}\left(\frac{\varepsilon(0)}{\varepsilon(\infty)} - 1\right) \tag{9.50}$$

$$a_0^* = a_0 \frac{\varepsilon(0)}{\varepsilon_0}\frac{m}{m_c}, \quad q_0 = \sqrt{\frac{2m_c\omega_0}{\hbar^2}} \tag{9.51}$$

$$B = \frac{\alpha}{6}[N_0 + 1] \tag{9.52}$$

where a_0 is the atomic Bohr radius, and a_0^* is the effective one, which includes corrections for the band mass and dielectric constant. The effective mass of the electron in the band is increased by the factor $1/(1 - \alpha/6)$. The polaron constant α is usually of order $O(1)$ or less in most polar semiconductors. The notable exceptions are the silver and copper halides, where $\alpha \sim 2$–3. There the effective mass can be twice the band mass.

Wherever the electron is located in the crystal, it interacts locally with the nearby ions. The electrons cause the ions to displace a small amount, due to the interaction with the electron. As the electron moves through the crystal, it must drag with it this local ion displacement. That physical process accounts for the increase in mass.

9.2 Electron–Phonon Interaction in Metals

The usual form for the electron–phonon interaction in metals is the screened ion interaction. In its simplest form, the interaction between an electron (of charge e) and an ion (of charge Ze), is

$$V_{ei}(r) = -\frac{Ze^2}{4\pi\varepsilon_0 r} \tag{9.53}$$

This form assumes the ion is a point charge. Better models include the feature that the ion has a finite size, and the interaction between the electron and the ion is a pseudopotential.

The electron–phonon interaction comes by letting the atom move a distance **Q** and finding the change in energy:

$$H_{ep} = -\frac{Ze^2}{4\pi\varepsilon_0}\left(\frac{1}{|\mathbf{r}+\mathbf{Q}|} - \frac{1}{r}\right) \approx \frac{Ze^2}{4\pi\varepsilon_0}\frac{\mathbf{Q}\cdot\mathbf{r}}{r^3} \tag{9.54}$$

Taking the Fourier transform of the dipole interaction gives a matrix element:

$$M_\eta(q) = \frac{Ze^2}{q^2\varepsilon(q)} i\mathbf{q}\cdot\hat{\xi}(\mathbf{q},\eta)\sqrt{\Omega_0}X(\mathbf{q},\eta) \tag{9.55}$$

The formula is widely applied to simple metals. A better choice is to use a pseudopotential and its Fourier transform.

9.2.1 λ

The dimensionless number that gives the strength of the electron–phonon interaction in metals is called lambda (λ). The electron–phonon interaction in a metal is treated very differently compared to the previous section on semiconductors. First, many transport and other properties of the metal are determined by the scattering of electrons that are within a thermal energy $\sim k_BT$ of the Fermi surface. The electrons have kinetic energy of many electron volts. These energies are much larger than the energies of the phonons. This difference has an advantage: often the energy of the phonon may be neglected in doing integrals.

Secondly, the electrons scatter from a point $\mathbf{k}_i$ on the Fermi surface to another point $\mathbf{k}_f = \mathbf{k}_i + \mathbf{q}$. Both $\mathbf{k}_i$ and $\mathbf{k}_f$ are near to the Fermi surface, so the phonon wave vectors **q** can be large. Typically, the scattering involves all of the phonons in the Brillouin zone.

Many integrals are over all of the phonon modes of some function of the frequency. So examine the following integral:

$$\langle G(\mathbf{k})\rangle = \sum_\eta \int \frac{d^3q}{(2\pi)^3} M_\eta(\mathbf{q})^2 G[\omega_\eta(\mathbf{q})]\delta[\varepsilon(\mathbf{k}) - \varepsilon(\mathbf{k}+q)] \tag{9.56}$$

The summation is over all phonon polarizations (here denoted by η). The three-dimensional wave vector integral goes over the Brillouin zone. The phonon frequency is $\omega_\eta(\mathbf{q})$. The last delta function of energy conservation means that only those wave vectors are included that scatter an electron from one point on the Fermi surface to another.

The function $\alpha^2F_\mathbf{k}(\omega)$ was introduced by McMillan [8]. It contains all of the useful information regarding the electron–phonon interaction in metals. The symbol suggests that it is a product of two functions, but it is actually only a single function:

$$\alpha^2F_\mathbf{k}(\omega) = \frac{1}{\hbar}\sum_\eta \int \frac{d^3q}{(2\pi)^3} M_\eta(\mathbf{q})^2 \delta[\omega - \omega_\eta(\mathbf{q})]\delta[\varepsilon(\mathbf{k}) - \varepsilon(\mathbf{k}+\mathbf{q})] \tag{9.57}$$

$$\langle G(\mathbf{k})\rangle = \hbar\int_0^{\omega_M} d\omega G(\omega)\alpha^2F_\mathbf{k}(\omega) \tag{9.58}$$

where ω_M is the maximum phonon frequency.

The symbol $F(\omega)$ denotes the density of states of the phonons:

$$F(\omega) = \lambda \int \frac{d^3q}{(2\pi)^3} \delta[\omega - \omega_\lambda(\mathbf{q})] \tag{9.59}$$

The function $\alpha_k^2 F(\xi, \omega)$ has these factors plus several others. The symbol α^2 is to denote the matrix element $|M_\eta(\mathbf{q})|^2$, which modulates the density of states. The other important factor is the delta function for electronic energy conservation $\delta[\xi_\mathbf{k} - \xi_{\mathbf{k}+\mathbf{q}}] = \delta[\varepsilon_\mathbf{k} - \varepsilon_{\mathbf{k}+\mathbf{q}}]$. The initial wave vector $\mathbf{k}$ is equal to, or very close to, the Fermi wave vector k_F. Similarly, the energy $\xi_\mathbf{k} = \varepsilon_\mathbf{k} - \mu$ is small, so the electron energy ε is near the Fermi energy μ. The constraint $\xi_\mathbf{k} = \xi_{\mathbf{k}+\mathbf{q}}$ requires that the final electron energy after absorbing or emitting the phonon is also at or near the Fermi surface. So the factor $\delta[\xi_\mathbf{k} - \xi_{\mathbf{k}+\mathbf{q}}]$ selects events where an electron is scattered from one point on the Fermi surface $\mathbf{k}$ to another point $\mathbf{k} + \mathbf{q}$.

Many theoretical calculations have been performed on many metals to obtain McMillan's function. It is found that it is a strong function of the phonon energy ω, but a weak function of the electron energy ξ. This function is used for electrons near the chemical potential and, over the scale of the phonon energy (milli-electron volts), the electron density of states does not change rapidly. So the energy dependence of this function is usually ignored.

An inspection of eqn. (9.57) shows that $\alpha^2 F_\mathbf{k}(\omega)$ is dimensionless. A dimensionless constant is obtained by taking the integral

$$\lambda = 2 \int_0^{\omega_M} \frac{d\omega}{\omega} \alpha^2 F_\mathbf{k}(\omega) \tag{9.60}$$

which is the definition of λ. It actually depends on the starting point $\mathbf{k}$. Usually an average value is obtained by averaging it over the Fermi surface.

As an example of using $\alpha^2 F_\mathbf{k}(\omega)$, consider the lifetime $\tau(\mathbf{k})$ of an electron of wave vector $\mathbf{k}$ due to scattering by phonons:

$$\begin{aligned} \frac{1}{\tau(k)} = \frac{2\pi}{\hbar} \sum_\eta \int \frac{d^3q}{(2\pi)^3} M_\eta(q)^2 \\ \times \{[N_\eta(\mathbf{q}) + 1 - n_F[\varepsilon(\mathbf{k}+\mathbf{q})]]\delta(\varepsilon(\mathbf{k}) - \varepsilon(\mathbf{k}+\mathbf{q}) + \hbar\omega_\eta(\mathbf{q})) \\ + [N_\eta(\mathbf{q}) + n_F[\varepsilon(\mathbf{k}+\mathbf{q})]]\delta(\varepsilon(\mathbf{k}) - \varepsilon(\mathbf{k}+\mathbf{q}) - \hbar\omega_\eta(\mathbf{q}))\} \end{aligned} \tag{9.61}$$

Most of the factors in this expression are familiar from Fermi's golden rule for the transition rate in quantum mechanical systems. The various occupation factors are derived in a later chapter on the Boltzmann equation. Equation (9.61) can be expressed in terms of $\alpha^2 F_\mathbf{k}(\omega)$. First, use the delta function for energy conservation to replace the argument of $n_F[\varepsilon(\mathbf{k}+\mathbf{q})] = n_F[\varepsilon(\mathbf{k}) \pm \hbar\omega_\eta(\mathbf{q})]$. Next drop the phonon energy in the delta functions. The remaining expression is identical to the McMillan function, so

$$\frac{1}{\tau(\mathbf{k})} = 2\pi \int_0^{\omega_M} d\omega \alpha^2 F_\mathbf{k}(\omega)[2n_B(\omega) + 1 + n_F(\varepsilon(k) - \hbar\omega) - n_F(\varepsilon(k) + \hbar\omega)] \tag{9.62}$$

At high temperatures, expand these terms to get a simple expression. The Bose occupation factors can be evaluated using

$$n_B(\omega) \approx \frac{k_BT}{\hbar\omega} - \frac{1}{2} \tag{9.63}$$

$$2n_B(\omega) + 1 \approx 2\frac{k_BT}{\hbar\omega} + O\left(\frac{\hbar\omega}{k_BT}\right) \tag{9.64}$$

The fermion occupation factors can be evaluated using

$$n_F(\varepsilon(k) - \hbar\omega) - n_F(\varepsilon(k) + \hbar\omega) \approx 2\hbar\omega\left(-\frac{dn_F(\varepsilon)}{d\varepsilon}\right) \tag{9.65}$$

which gives for the lifetime

$$\frac{1}{\tau(\mathbf{k})} = \frac{2\pi k_BT\lambda}{\hbar} + 2\pi\hbar\lambda\langle\omega^2\rangle\left(-\frac{dn_F(\varepsilon)}{d\varepsilon}\right) \tag{9.66}$$

$$\lambda\langle\omega^2\rangle = 2\int_0^{\omega_M} \omega d\omega \alpha^2 F_{\mathbf{k}}(\omega) \tag{9.67}$$

The factor of $\lambda\langle\omega^2\rangle$ is another moment of $\alpha^2F(\omega)$. At high temperatures, the first term in $1/\tau$ is much larger than the second one, and $1/\tau \propto k_BT$. A measurement of the electrical resistivity at high temperatures often provides a direct measurement of λ as the coefficient of $\rho(T) = m/n_e e^2\tau \propto k_BT$.

The matrix element for the electron–phonon interaction contains the ion mass M. The ratio of the electron mass divided by the ion mass is very small: $m_e/M \ll 1$. This small ratio forms the dimensionless coupling constant in the perturbation expansion for the electron–phonon interaction. The square of the matrix element always contains the factor of

$$X(\mathbf{q}, \eta)^2 = \frac{\hbar}{2M\omega_\eta(q)} \tag{9.68}$$

For a simple metal with a simple acoustic phonon band,

$$\omega_\eta(q) = 2\sqrt{\frac{K}{M}}\sin(qa/2) \tag{9.69}$$

$$X(\mathbf{q}, \eta)^2 = \frac{\hbar}{4\sqrt{KM}\sin(qa/2)} = \sqrt{\frac{m_e}{M}}\frac{\hbar}{4\sqrt{Km_e}\sin(qa/2)} \tag{9.70}$$

The first factor is a dimensionless coupling constant:

$$g = \sqrt{\frac{m_e}{M}} \tag{9.71}$$

The second factor has the dimensions of (length)^2 and contains purely electronic factors, so its lengths must be on the scale of a Bohr radius.

For a metal such as potassium with an atomic mass of 39 AMU, the ratio is

$$g \sim 0.0037 \tag{9.72}$$

which is indeed a small number. If one multiplies this ratio by a typical electronic energy (13.6 eV), then one gets an energy of 51 meV, which is a typical phonon energy. So the self-energy of an electron from the electron–phonon interaction has the same energy scale as the phonon energy itself. These conclusions apply to processes involving a single phonon.

One might consider high-order processes involving two phonons. Such self-energy terms have an effective coupling of $g^2 \sim 10^{-5}$. This number is very small, and two phonon events make a negligible contribution to the energy of an electron in most solids. This argument is changed in solids made from very light elements, such as hydrogen and helium, since then the ion mass is small and g is larger.

9.2.2 Phonon Frequencies

In a semiconductor or an insulator, the density of conduction electrons is usually very small: typically $n_e \sim 10^{18}/\text{cm}^3$ in semiconductors, and much less in insulators. The total mass of the electrons is very small, and has negligible influence on the phonon frequency. The electrons may provide some screening of the ion–ion interaction. The screening becomes inportant only when the plasma frequency of the electrons becomes comparable to the frequencies of the optical phonons.

The situation is very different in metals. The electrons have a much higher density ($n_e \sim 10^{23}/\text{cm}^3$) and provide significant screening of the ion–ion interaction.

Imagine a solid composed of only positive ions without any electrons. It would want to fly apart due to the Coulomb repulson of the ions. If you could manage to hold it together, say by the application of pressure, the ions would vibrate at the ion plasma frequency. There would not be any acoustical phonon modes. Now, add the electrons. Their negative charge cancels the Coulomb repulsion, and the metal is stable. The electrons screen the ion–ion interactions, which makes the interaction between ions be short range. The short-range potential leads to acoustical phonons. Clearly, the electrons have a dramatic influence on the vibrational properties of the metal. They are not a small perturbation: they are the entire answer.

Consider a metal made up of protons and electrons. It is called "metallic hydrogen." The reader should be cautioned that actual solid hydrogen is an insulator composed of neutral H_2 molecules. The phrase "metallic hydrogen" refers to a theoretical concept rather than a physical reality. The proton has a charge e and an interaction of $e^2/4\pi\varepsilon_0 R_{ij}$. The Fourier transform of this potential is

$$V_{B,ii}(q) = \frac{e^2}{\varepsilon_0 q^2} \equiv v_q, \quad \text{bare} \tag{9.73}$$

$$V_{S,ii}(q) = \frac{e^2}{q^2 \varepsilon(q)}, \quad \text{screened} \tag{9.74}$$

The dielectric function $\varepsilon(q,\omega)$ is due to electrons. Usually one uses RPA or some similar form. It is evaluated at zero frequency, since phonon energies are miniscule compared to plasmon frequencies. This method of screening is valid only for point charges.

Another hypothetical metal has point nuclei of alpha particles. This "metallic helium" also does not exist. So the two possible metals, whose nuclei could be points, prefer to be nonmetals.

The third element in the periodic table is lithium, which is a metal. However, the ion has a closed shell of $(1s)^2$ electrons, so it donates only one electron to the metallic band.

For this case, and all other metal ions, it is best to represent the electron–ion interaction $V_{ei}(r)$ by a pseudopotential which accounts for the electrons bound in the ion core. This pseudopotential is Coulombic at large distance, but is a constant at small distance. Its Fourier transform is $V_{ei}(q)$. The total ion–ion interaction is

$$V_{ii}^{(\text{eff})}(q) = V_{ii}(q) + \frac{V_{ei}^2(q)P(q)}{\varepsilon(q)} \tag{9.75}$$

$$P(q) = 2\int \frac{d^3k}{(2\pi)^2} \frac{n_F(E(k)) - n_F(E(k+q))}{E(k) - E(k+q)} \tag{9.76}$$

$$\varepsilon(q) = 1 - v_q P(q) \tag{9.77}$$

Here $P(q)$ is the standard polarization operator of the electron gas, which is evaluated here at zero frequency.

The first term is the direct interaction between the ions. The second is the potential due to electron–ion interaction. The ion interacts with the electrons (V_{ei}), which polarizes the electrons (P/ε), and this polarization acts back on the ion system (V_{ei}). For the special case that all of the ions are points, and $V_{ii} = v_q = V_{ei}$, the interaction gives the prior result:

$$V_{ii}^{(\text{eff})}(q) = v_q\left[1 + \frac{v_q P}{1 - v_q P}\right] = \frac{v_q}{\varepsilon(q)} \tag{9.78}$$

Another way to derive this result is to consider the self-energy of a phonon due to the electron–phonon interaction:

$$V_{ep} = \frac{1}{\sqrt{\Omega}} \sum_{\mathbf{q},\eta} M_\lambda(\mathbf{q}) A_{\mathbf{q},\eta} \rho(\mathbf{q}) \tag{9.79}$$

$$\rho(\mathbf{q}) = \sum_{\mathbf{k}\sigma} C^\dagger_{\mathbf{k}+\mathbf{q},\sigma} C_{\mathbf{k}\sigma} \tag{9.80}$$

The Feynman diagram for this process is shown in fig. 9.3. The first diagram shows a dashed line (phonon), creating a closed loop of electrons (solid lines), which remakes the phonon. This gives a phonon self-energy of

$$\Pi(\mathbf{q}) = M_\lambda(\mathbf{q})^2 P(\mathbf{q}) \tag{9.81}$$

The second line of diagrams shows the influence of electron–electron interactions, which are the vertical dashed lines. Each interaction introduces another factor of the polarization. One gets a series of terms in $(v_q P)^n$, which are summed to produce the RPA dielectric function $\varepsilon(q)$:

$$M^2 P[1 + (vP) + (vP)^2 + \cdots] = \frac{M^2 P}{\varepsilon} \tag{9.82}$$

Since $M \propto V_{ei}$, we have just derived the second term in eqn. (9.75).

9.2.3 *Electron–Phonon Mass Enhancement*

An important feature of the electron–phonon interaction is that it lowers the energy of the electron when it is in motion. The Coulomb interaction between the electrons and its

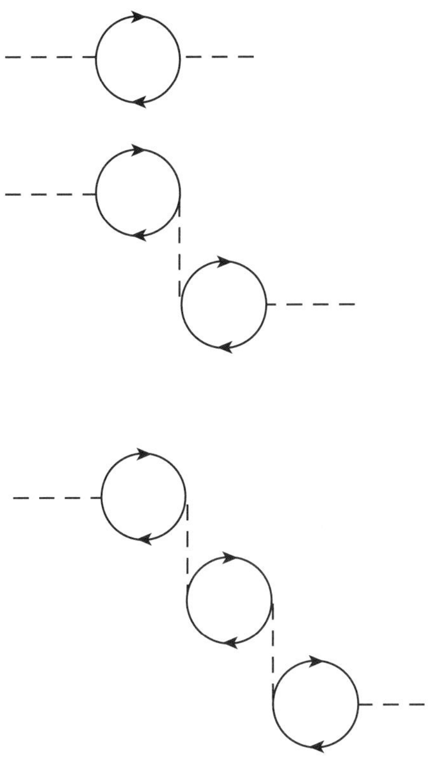

FIGURE 9.3. Phonon self-energy. Dashed lines are phonons, solid lines are electrons. Electron closed loop is $P(q)$. Vertical dashed lines are electron–electron interaction.

surrounding ions causes a local displacement of the atoms near to the electron. As the electron moves, it moves this deformation along with it. Moving this extra bit of energy results in a heavier mass for the electron: the electron–phonon interaction causes the electron mass to be heavier.

Second-order perturbation theory can be used to calculate change of electron energy due to the electron–phonon interaction. Call this contribution $\Sigma(\mathbf{k}, E)$, where E is the energy of the electron:

$$\Sigma(\mathbf{k}, E) = \sum_{\eta} \int \frac{d^3q}{(2\pi)^3} |M_{\eta}(q)|^2 \left[\frac{N_{\eta}(\mathbf{q}) + 1 - n_F(\xi_{k+q})}{E - \xi_{k+q} - \hbar\omega_{\eta}(q)} + \frac{N_{\eta}(q) + n_F(\xi_{k+q})}{E - \xi_{k+q} + \hbar\omega_{\eta}(q)} \right] \tag{9.83}$$

The first term corresponds to the virtual emission of a phonon, and the second term is from the virtual absorption of a phonon. There are two kinds of perturbation theory in quantum mechanics: (i) Rayleigh-Schrödinger theory and (ii) Brillouin-Wigner theory. Rayleigh-Schrödinger theory sets $E = \xi_k$ and is called "on the mass shell"; Brillouin-Wigner theory leaves E as a separate variable, and is called "off the mass shell." At the moment leave E unspecified, since it can be set $E=\xi_k$ later. The energy E is measured from the chemical potential.

This expression is divided into two terms:

$$\Sigma = \Sigma^{(a)} + \Sigma^{(b)} \tag{9.84}$$

$$\Sigma^{(a)}(\mathbf{k}, E) = \sum_\eta \int \frac{d^3q}{(2\pi)^3} |M_\eta(q)|^2 \left[\frac{N_\eta(q)+1}{E-\xi_{k+q}-\hbar\omega_\eta(q)} + \frac{N_\eta(q)}{E-\xi_{k+q}+\hbar\omega_\eta(q)} \right] \tag{9.85}$$

$$\Sigma^{(b)}(\mathbf{k}, E) = \sum_\eta \int \frac{d^3q}{(2\pi)^3} |M_\eta(q)|^2 n_F(\xi_{k+q}) \times \left[\frac{1}{E-\xi_{k+q}-\hbar\omega_\eta(q)} - \frac{1}{E-\xi_{k+q}+\hbar\omega_\eta(q)} \right] \tag{9.86}$$

- In an insulator or a semiconductor with few electrons in the conduction band, $n_F(\xi)=0$ and $\Sigma^{(b)}=0$. Then only $\Sigma^{(a)}$ is evaluated. This case was discussed in the previous section.
- In metals, both terms $\Sigma^{(a,b)}$ are retained. The term $\Sigma^{(a)}$ gives an overall temperature dependence to the energy of the electron. Since this expression contains no occupation factors for the electron, nothing special happens to the function at the chemical potential.
- In metals, the term $\Sigma^{(b)}$ has interesting physics, since it affects only those electrons at the Fermi surface. Evaluate the wave vector coordinates in spherical coordinates:

$$d^3q = q^2 dq d\phi d\nu, \quad \nu = \cos(\theta) \tag{9.87}$$

The angle θ is between the vectors $\mathbf{k}$ and $\mathbf{q}$. Then define a new wave vector $p^2 = (\mathbf{k}+\mathbf{q})^2$ and change integration variables from ν to p and then to ξ_p:

$$p^2 = k^2 + q^2 + 2kq\nu, \quad pdp = kqd\nu \tag{9.88}$$

$$\int d^3q n_F(\xi_{k+q}) f(\xi_{k+q}) = \frac{1}{k}\int qdqd\phi \int pdp n_F(\xi_p) f(\xi_p)$$
$$= \frac{1}{v_k}\int qdqd\phi \int d\xi n_F(\xi) f(\xi) \tag{9.89}$$

The remaining factors in the integrand are exactly those that define $\alpha^2F(\omega)$. The self-energy can be written at zero temperature as

$$\Sigma^{(b)}(\mathbf{k}, E) = \hbar \int_0^{\omega_D} d\omega \alpha^2 F(\omega) \int_{-\mu}^0 d\xi \left[\frac{1}{E-\xi+\hbar\omega} - \frac{1}{E-\xi-\hbar\omega} \right]$$
$$= -\hbar \int_0^{\omega_D} d\omega \alpha^2 F(\omega) \ln\left| \frac{E+\hbar\omega}{E-\hbar\omega} \right| \tag{9.90}$$

A simple expression is found for the self-energy. It always has the characteristic shape shown in fig. 9.4 for a metal with $\lambda=1$. The value of the self-energy is typically a few

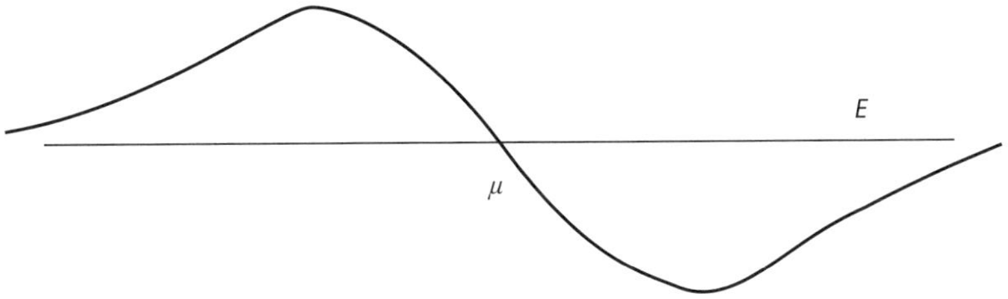

FIGURE 9.4. Typical electron–phonon self-energy in a metal for electrons near the Fermi surface at $E = 0$.

millivolts, which is also the size of phonon energies. $\Sigma(\mathbf{k}, E)$ is a function of wave vector and energy. The wave vector dependence is weak, and typically

$$\frac{1}{v_k}\frac{\partial \Sigma(\mathbf{k}, E)}{\partial k} \sim \frac{\Sigma}{E_F} \ll 1 \tag{9.91}$$

An interesting result is obtained when taking a derivative of this function with respect to E at $(E = 0)$:

$$\frac{\partial \Sigma(\mathbf{k}, E)}{\partial E} = -\hbar \int_0^{\omega_D} d\omega \alpha^2 F(\omega) \left[\frac{1}{E + \hbar\omega} - \frac{1}{E - \hbar\omega}\right] \tag{9.92}$$

$$\left(\frac{\partial \Sigma(\mathbf{k}, E)}{\partial E}\right)_{E=0} = -2\int_0^{\omega_D} \frac{d\omega}{\omega} \alpha^2 F(\omega) = -\lambda \tag{9.93}$$

The self-energy, when graphed as a function of energy, has a slope at the Fermi surface that is negative and equal in magnitude to the electron–phonon coupling constant λ. This derivative makes a significant contribution to the effective mass of the electrons at the Fermi surface.

A formula for the effective mass m^* is derived by writing the kinetic energy of the electron as

$$E(k) = \frac{\hbar^2 k^2}{2m^*} = \frac{\hbar^2 k^2}{2m_B} + \Sigma(k, E) \tag{9.94}$$

$$\varepsilon_k = \frac{\hbar^2 k^2}{2m} \tag{9.95}$$

where m_B is the *band mass* that results from the curvature of the energy bands, and m is the electron mass in vacuum. A formula for the effective mass is obtained by taking a derivative of E with respect to ε_k:

$$\frac{m}{m^*} = \frac{dE}{d\varepsilon_k} = \frac{m}{m_B} + \frac{\partial \Sigma}{\partial \varepsilon_k} + \frac{\partial \Sigma}{\partial E}\frac{\partial E}{\partial \varepsilon_k} \tag{9.96}$$

The term $\partial\Sigma/\partial\varepsilon_k$ is negligible. So the above expression simplifies to

$$\frac{m}{m^*} = \frac{m/m_B}{1 + \lambda} \tag{9.97}$$

$$\frac{m^*}{m_B} = 1 + \lambda \tag{9.98}$$

For those electrons whose energy is right at the Fermi surface, their effective mass is increased from the band mass by the factor of $1 + \lambda$. Values of λ for typical metals are as follows:

- Monovalent metals such as Na and K have $\lambda \sim 0.1$.
- Divalent metals such as Be, Mg, and Zn have $\lambda \sim 0.4$.
- Trivalent metals such as Ga and In have $\lambda \sim 1.0$.
- Pb has $\lambda = 1.5$.

For multivalent metals the correction to the effective mass is sizable.

Earlier chapters discussed the contribution to the energy of the electron from electron–electron interactions. The contributions of electron–electron and electron–phonon interactions are compared here:

Interaction	Self-energy	Mass Contribution
Electron–electron	Large	Small
Electron–phonon	Small	Large

Electron–electron interactions make a large correction to the energy through screened exchange, but this term is not very dependent on wave vector, and so makes a small contribution to the effective mass. The phonon contribution to the energy of the electron is small in value, but it makes a large contribution to the effective mass of those electrons at the Fermi surface.

9.3 Peierls Transition

The Peierls transition is found in one-dimensional chains of atoms. It occurs in carbon polymers and other molecular chain systems. The most obvious ground state has the carbon atoms in the chain spaced equally. Peierls noted [9] that this arrangement is unstable and the system prefers to dimerize: alternate pairs of atoms move closer together. This bond alternation is usually found in carbon chains: chemists call it alternate single and double bonds. Physicists call it a Peierls transition. An example is shown in fig. 9.5.

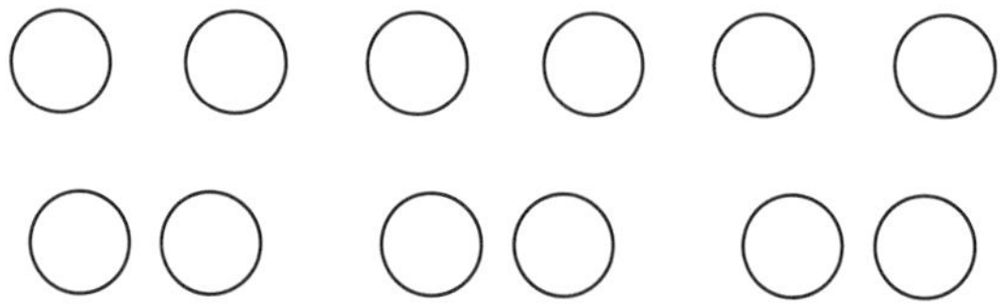

FIGURE 9.5. Top line shows atoms equally spaced. Lower line shows them after dimerization.

Begin by considering a simple model of a chain of atoms that are equally spaced. The Hamiltonian has a phonon term, an electron term, and the electron–phonon interaction:

$$H = H_p + H_e + H_{ep} \tag{9.99}$$

$$H_p = \sum_{n=1}^{N} \left[\frac{P_n^2}{2M} + \frac{K}{2}(Q_n - Q_{n+1})^2 \right] \tag{9.100}$$

$$H_e = \sum_{ns} \left[E_0 C_{ns}^{\dagger} C_{ns} - \frac{t}{2}\left(C_{n+1,s}^{\dagger} C_{ns} + C_{ns}^{\dagger} C_{n+1,s} \right) \right] \tag{9.101}$$

where t is the matrix element for an electron hopping to its neighbor site. Both of these Hamiltonians can be solved in terms of collective coordinates:

$$H_p = \sum_q \hbar\omega(q)\left[a_q^{\dagger} a_q + 1/2\right], \quad \omega(q) = 2\sqrt{\frac{K}{M}} \sin(qa/2) \tag{9.102}$$

$$H_e = \sum_{ks} \varepsilon(k)\, C_{ks}^{\dagger} C_{ks}, \quad \varepsilon(k) = E_0 - t\cos(ka) \tag{9.103}$$

where t has a negative sign in front in order that the minimum electron energy is at $k = 0$. If t had a plus sign, the minimum energy would be at the band edge $ka = \theta = \pi$, and the Peierls theory is unchanged. The top curve in fig. 9.6 shows the energy bands of the electrons when the atoms are equally spaced. It is a graph of eqn. (9.103). The lower curve shows the energy bands after dimerization.

The dimerization process is the static movement of alternate atoms in a different direction. Say the atoms with even numbers $(n = 2n')$ move to the right $Q_{2n'} = q$, while those with odd numbers $(n = 2n' + 1)$ move to the left $(Q_{2n'+1} = -q)$. The atoms are now paired. The Wigner-Seitz cell has doubled in size, and the Brillouin zone is half as large. The new zone edge is at $ka = \pi/2$.

This movement stretches the phonon spring constants, and increases the phonon potential energy by

$$\delta V_p = 2KNq^2 \tag{9.104}$$

The dimerization must lower the total energy in the electron system by an amount greater than δV_p so that the displacements lower the total energy of the system.

An important assumption in the Peierls transition is that the band of electrons is exactly half-full. Each atom contributes one electron to the system. Since each band state has room for a spin-up and a spin-down electron, half of the band states are occupied in a paramagnetic system. The dimerization changes the electron energy bands as shown in fig. 9.6b. All of the occupied electron states decrease in energy, which could lower the total energy of the system.

Consider how the electron states are changed by dimerization. The hopping term $t(a)$ depends on the distance a to the neighbor. After dimerization, it changes to

$$t(a \pm 2q) \approx t(a) \pm 2qt'(a) \tag{9.105}$$

where $t' = dt/da$. The electron Hamiltonian after dimerization is

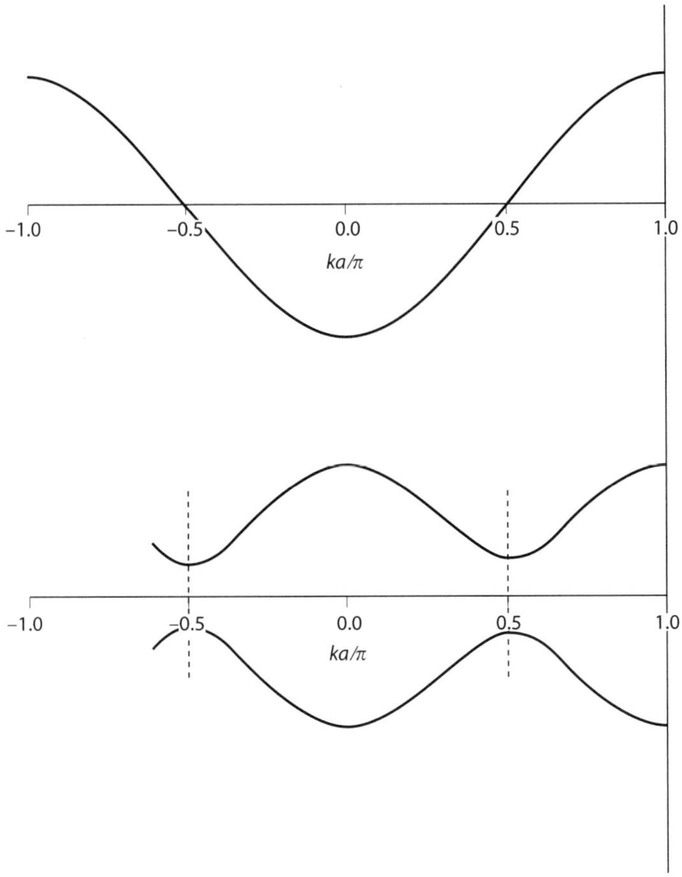

FIGURE 9.6. The top curve is the electronic energy band when atoms are equally spaced. The lower curves are the energy bands after dimerization.

$$H_e = \sum_{n=0}^{N/2} \Big[E_0\left(C^\dagger_{2n}C_{2n} + C^\dagger_{2n+1}C_{2n+1}\right) - \frac{1}{2}(t - 2qt')\left(C^\dagger_{2n}C_{2n+1} + C^\dagger_{2n+1}C_{2n}\right)$$
$$-\frac{1}{2}(t + 2qt')\left(C^\dagger_{2n}C_{2n-1} + C^\dagger_{2n-1}C_{2n}\right)\Big] \tag{9.106}$$

There are now two sites per unit cell. Call the even numbered sites A and the odd numbered sites B. Do a wave vector transform and find

$$H_e = \sum_k \{E_0\left(C^\dagger_{Ak}C_{Ak} + C^\dagger_{Bk}C_{Bk}\right)$$
$$- t\cos(\theta)\left(C^\dagger_{Ak}C_{Bk} + C^\dagger_{Bk}C_{Ak}\right) + 2iqt'\sin(\theta)\left(C^\dagger_{Ak}C_{Bk} - C^\dagger_{Bk}C_{Ak}\right)\} \tag{9.107}$$

where $\theta = ka$ and ranges in value $-\pi/2 < \theta < \pi/2$. This Hamiltonian can also be rewritten as a 2×2 matrix:

$$H_{ek} = \begin{pmatrix} E_0 & -E_1 \\ -E_1^* & E_0 \end{pmatrix}, \quad E_1 = t\cos(\theta) - 2iqt'\sin(\theta) \tag{9.108}$$

The eigenvalues are

$$E_{\pm}(\theta) = E_0 \pm |E_1| = E_0 \pm \sqrt{t^2 \cos^2(\theta) + (2qt')^2 \sin^2(\theta)} \tag{9.109}$$

The lower band in fig. 9.6b is $E_-(\theta)$. The band edge is at $\theta = \pm\pi/2$, where there is a band gap:

$$E_g = E_+ - E_- = 4|qt'| \tag{9.110}$$

How much electronic energy is gained by the dimerization? The change in energy is

$$\Delta E_e = -2\sum_k \left[\sqrt{t^2 \cos^2(\theta) + (2qt')^2 \sin^2(\theta)} - t\cos(\theta)\right] \tag{9.111}$$

where the factor of two is for spin degeneracy. Change to an integral over wave vector:

$$\sum_k = \frac{L}{2\pi}\int_{-\pi/2a}^{\pi/2a} dk = \frac{N}{\pi}\int_0^{\pi/2} d\theta \tag{9.112}$$

where $\theta = ka$ and $L = Na$. Under the square root change $\cos^2(\theta) = 1 - \sin^2(\theta)$ and get

$$\Delta E_e = -N\frac{2t}{\pi}\int_0^{\pi/2} d\theta\left[\sqrt{1 - m\sin^2(\theta)} - \cos(\theta)\right] \tag{9.113}$$

$$m = 1 - m_1, \quad m_1 = (q/q_0)^2, \quad q_0 = t/2t' \tag{9.114}$$

The integral is expressed in terms of an elliptic integral:

$$\Delta E_e = -N\frac{2t}{\pi}[E(m) - 1] \tag{9.115}$$

For small values of m_1 the integral has the expansion

$$E(m) = 1 + m_1\left[a_1 - \frac{1}{4}\ln(m_1)\right] + O\left(m_1^2\right), \quad a_1 = 0.463 \tag{9.116}$$

Only the terms that are shown are kept. The total energy from electrons and phonons is

$$\Delta E = Nm_1[\alpha + \beta\ln(m_1)] \tag{9.117}$$

$$\alpha = 2Kq_0^2 - \frac{2ta_1}{\pi}, \quad \beta = \frac{t}{2\pi} \tag{9.118}$$

Treat this derivation as a variational calculation, and vary q (actually m_1) to find the minimum energy:

$$\frac{d\Delta E}{dm_1} = N[\alpha + \beta + \beta\ln(m_1)] = 0 \tag{9.119}$$

$$\ln(m_1) = -\frac{\alpha + \beta}{\beta} \tag{9.120}$$

$$\Delta E(m_1) = -N\beta\exp\left[-\frac{\alpha + \beta}{\beta}\right] \tag{9.121}$$

Since β is positive, the system lowers its energy. The other coefficient α could have either sign. This total energy is for the chain of N atoms. Divide by N to get the change in energy per atomic site.

The dimerized state has the lowest energy regardless of the values of α and β, as long as β is positive. Actually, if t is negative, one can redo the theory. The band minimum is now at the zone edge, but the dimerization still occurs. So all one-dimensional chains of atoms or molecules with half-filling are unstable to a Peierls transition. Actual carbon chains are always dimerized.

9.4 Phonon-mediated Interactions

In field theory one learns that forces that act at a distance between two particles are due to the exchange of a boson particle between them. Examples are Coulomb forces (photon exchange) and nuclear forces (pion exchange). In a similar way, there is a force between electrons due to the exchange of phonons. Since the matrix element for electron–phonon interactions is typically small (due to the ratio of $\sqrt{m/M}$), the phonon-mediated interaction is small at all distance scales. At short distances the electron–electron interactions dominate. However, Coulomb interactions are effectively screened to zero at large distances. Phonon-mediated interactions are not screened, since the phonons can travel long distances. At large distances the phonon-mediated interaction is the most important. An earlier section showed that elasticity theory predicted an interaction between static defects of

$$U(\mathbf{r}) = G\nabla \cdot \mathbf{u}(\mathbf{r}) \tag{9.122}$$

where G is a constant, and $\mathbf{u}(\mathbf{r})$ is the displacement due to the defects. This result is extended to other particles that interact by elastic fields.

For example, when the author was a student he was taught that ferroelectric transitions were due to dipole–dipole Coulomb forces. When ions displace, they make a dipole moment, and collective motions are driven by dipolar interactions. Now it is known that such ion displacements also create a strain field, and dipoles also interact by straining the lattice. This latter phenomenon is now called a phonon-mediated interaction. Some ferroelectric transitions are driven by collective elastic strains.

9.4.1 *Fixed Electrons*

A problem that can be solved exactly is the potential between two electrons that are each attached to a different defect. The defect state has a fixed energy and binds the electron to that spot. In this case write the electron–phonon interaction as

$$H = H_0 + V_{ep} \tag{9.123}$$

$$H_0 = \sum_q \hbar\omega_q\left[a_q^{\dagger}a_q + 1/2\right] \tag{9.124}$$

$$V_{ep} = \sum_q M(q)(a_q + a^\dagger_{-q})J_q \tag{9.125}$$

$$J_q = n_1 e^{i\mathbf{q}\cdot\mathbf{R}_1} + n_2 e^{i\mathbf{q}\cdot\mathbf{R}_2}, \quad n_j = C^\dagger_j C_j \tag{9.126}$$

The Hamiltonian can be solved exactly by completing the square: first note that the interaction term can be written as $a_q J_q + a^\dagger_q J^*_q$ and the total Hamiltonian is

$$H = \sum_q \hbar\omega_q \left\{ \left[\left(a^\dagger_q + \frac{M(q)}{\hbar\omega_q} J_q \right)\left(a_q + \frac{M(q)}{\hbar\omega_q} J^*_q \right) + 1/2 \right] - \frac{M^2(q)}{\hbar\omega_q} |J_q|^2 \right\} \tag{9.127}$$

The first term suggests defining new raising and lowering operators:

$$A_q = a_q + \frac{M(q)}{\hbar\omega_q} J^*_q \tag{9.128}$$

$$A^\dagger_q = a^\dagger_q + \frac{M(q)}{\hbar\omega_q} J_q \tag{9.129}$$

$$H = \sum_q \hbar\omega_q [A^\dagger_q A_q + 1/2] + \delta H \tag{9.130}$$

The last term gives the effective interaction:

$$\delta H = -E_0[n_1 n_1 + n_2 n_2] + n_1 n_2 \tilde{V}(\mathbf{R}_{12})] \tag{9.131}$$

$$E_0 = \sum_q \frac{|M(q)|^2}{\hbar\omega_q} \tag{9.132}$$

$$\tilde{V}(\mathbf{R}) = -2\sum_q \frac{|M(q)|^2}{\hbar\omega_q} \cos(\mathbf{q}\cdot\mathbf{R}) \tag{9.133}$$

Equation (9.133) is the final formula. It is the exact interaction between two fixed particles due to exchange of phonons. The classical picture is that one electron interacts with the surrounding atoms, and creates a local strain field. This strain field then affects the energy of the other fixed electron. The quantum picture is that they interact due to the virtual exchange of phonons. Both models, classical and quantum, give the same formula. The long-wavelength part of the interaction is given by elasticity theory. The strain energy E_0 is required to put a single defect into the crystal.

An important example is provided by the Fröhlich interaction with optical phonons in polar materials:

$$M(q) = \frac{C}{q}, \quad \omega(q) = \omega_{\mathrm{LO}} \tag{9.134}$$

$$\tilde{V}(\mathbf{R}) = -2\frac{C^2}{\hbar\omega_{\mathrm{LO}}} \int \frac{d^3q}{(2\pi)^3} \frac{e^{i\mathbf{q}\cdot\mathbf{R}}}{q^2} = -\frac{C^2}{2\pi\hbar\omega_{\mathrm{LO}}} \frac{1}{R} \tag{9.135}$$

The integral is just the inverse Fourier transform for the Coulomb potential. The interaction between the electrons bound to the defects has the form of Coulomb's law. The above result is a contribution to the dielectric screening. The dielectric constant for polar materials is

$$\varepsilon(\omega) = \varepsilon(\infty) + \frac{\varepsilon(0) - \varepsilon(\infty)}{1 - \omega^2/\omega_{\mathrm{TO}}^2} \tag{9.136}$$

The first term on the right is from electronic polarization. The second term in the dielectric function is the contribution from phonons. At high frequencies ($\omega \gg \omega_{\mathrm{TO}}$) the last term is negligible, and $\varepsilon \approx \varepsilon(\infty)$. Then the Coulomb interaction between two charges is

$$V = \frac{e^2}{4\pi\varepsilon(\infty)R} \tag{9.137}$$

At $\omega = 0$ then $\varepsilon = \varepsilon(0)$ and the interaction between two charges is

$$V' = \frac{e^2}{4\pi\varepsilon_0 R} \tag{9.138}$$

The electron–phonon contribution to the screening is the difference between these two formulas:

$$V' - V = -\frac{e^2}{4\pi R}\left(\frac{1}{\varepsilon(\infty)} - \frac{1}{\varepsilon(0)}\right) \tag{9.139}$$

Set this equal to the phonon-mediated interaction in eqn. (9.135), which provides the definition of C^2:

$$C^2 = \frac{e^2}{2}\hbar\omega_{\mathrm{LO}}\left(\frac{1}{\varepsilon(\infty)} - \frac{1}{\varepsilon(0)}\right) \tag{9.140}$$

In many semiconductors, these two dielectric constants have different values: often $\varepsilon(0)$ is over twice as large as $\varepsilon(\infty)$.

9.4.2 Dynamical Phonon Exchange

Electrons can interact by phonon exchange even when they are not fixed. The Green's function for phonons is

$$D_\lambda(q, \omega) = \frac{2\omega_{q\lambda}}{\omega^2 - \omega_{q\lambda}^2} \tag{9.141}$$

$$V_p(q, \omega) = \sum_\lambda |M_\lambda(q)|^2 D_\lambda(q, \omega) \tag{9.142}$$

The frequency is the change in energy of the electron: $\hbar\omega = \varepsilon(k) - \varepsilon(\mathbf{k} + \mathbf{q})$. The Feynman diagram is shown in fig. 9.7. When two electrons interact by a Coulomb interaction, the interaction is shown as a dotted line. When they interact by exchanging a phonon, the interaction line is dashed. The Coulomb interaction is instantaneous in time. It is drawn vertically, since time increases horizontally. The phonon-mediated interaction takes some time, so it is drawn at an angle.

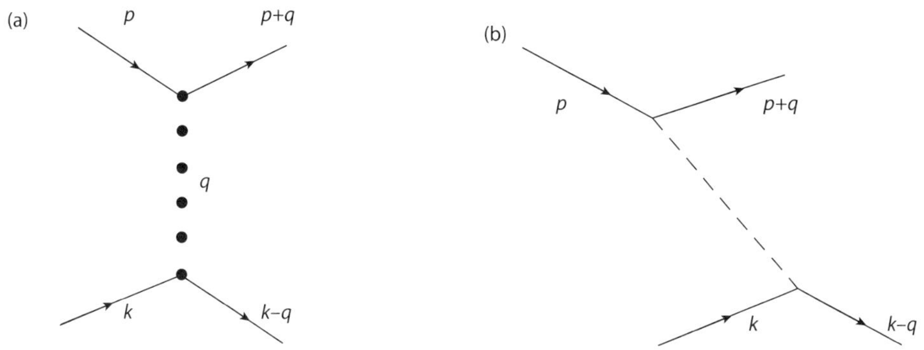

FIGURE 9.7. Solid lines are electrons, dotted line is Coulomb interaction, and dashed line is phonon-mediated interaction between two electrons.

If the electrons are fixed and cannot change their energy, then $\omega = 0$ and the formula is the previous exact expression for the phonon-mediated interaction. The above form for V_p is not exact at nonzero frequencies, but is akin to second-order perturbation theory. However, higher orders of perturbation theory can be neglected, so the above formula is all that is ever used.

Since the two interactions between electrons, Coulomb and phonon-mediated, both have a similar Feynman diagram, it makes sense to add them and treat them together:

$$W(q, \omega) = v(q) + V_p(q, \omega), \quad v(q) = \frac{e^2}{\varepsilon_0 q^2} \tag{9.143}$$

This combined interaction is then evaluated in the random phase approximation (RPA) to include screening:

$$W_{sc}(q, \omega) = \frac{W}{1 - WP} \tag{9.144}$$

$$P(q, \omega) = \frac{1}{\Omega} \sum_{k\sigma} \frac{n_k - n_{k+q}}{\hbar\omega + \varepsilon(\mathbf{k}) - \varepsilon(\mathbf{k}+\mathbf{q})} \tag{9.145}$$

where P is the usual polarization function for the electron gas. Separate this formula into screened electron–electron interactions, and a screened phonon-mediated interaction. Such separation is reasonable since the two interactions occur on very different frequency scales: the electrons respond at the plasma frequency, which is a hundred times higher than a phonon frequency.

The screened Coulomb interaction $v_{sc}(q)$ is defined as before. The screened phonon interaction is everything else. Denote $v = v(q)$, $V = V_p$:

$$v_{sc}(q) = \frac{v(q)}{\varepsilon(q, \omega)}, \quad \varepsilon(q, \omega) = 1 - v(q)\, P(q, \omega) \tag{9.146}$$

$$V_{ep,sc}(q, \omega) = W_{sc}(q, \omega) - v_{sc}(q) \tag{9.147}$$

$$= \frac{v+V}{1-(v+V)P} - \frac{v}{1-vP}$$

$$= \frac{V}{(1-vP)[1-(v+V)P]}$$

$$V_{p,sc}(q,\omega) = \frac{V_p}{\varepsilon^2}\frac{1}{1-V_pP/\varepsilon} \tag{9.148}$$

The above formula is the most general expression. It simplifies for the case there is only one phonon band of interest, and there is no summation over λ:

$$V_{ep,sc}(q,\omega) = \frac{|M(q)|^2}{\varepsilon(q)^2}\frac{2\omega(q)}{\omega^2-\Omega(q)^2} \tag{9.149}$$

$$\Omega(q)^2 = \omega(q)^2 + 2\omega(q)|M(q)|^2\frac{P(q)}{\varepsilon(q)} \tag{9.150}$$

The frequency $\Omega(q)$ is the actual phonon frequency in the metal. The interaction from the bare ion–ion interaction gives $\omega(q)^2$ in the first term. The second term gives the influence of the electron–ion interaction. Recall that the electron–phonon matrix element $M(q)$ is from the electron–ion interaction.

We prefer to write this result as

$$V_{p,sc} = |\tilde{M}(q)|^2\tilde{D}_\lambda(q,\omega) \tag{9.151}$$

$$\tilde{M}(q) = \frac{M(q)}{\varepsilon(q)}\sqrt{\frac{\omega(q)}{\Omega(q)}} \tag{9.152}$$

$$\tilde{D}(q,\omega) = \frac{2\Omega(q)}{\omega^2-\Omega(q)^2} \tag{9.153}$$

The matrix element contains a factor of the square root of the ratio of the two phonon frequencies. This factor is to account for the phonon frequency in the matrix element:

$$\tilde{M}(q) = i\mathbf{q}\cdot\hat{\xi}\tilde{X}(q)\frac{V_{ei}}{\varepsilon(q)} \tag{9.154}$$

$$\tilde{X}(q) = \sqrt{\frac{\hbar}{2MN\Omega(q)}} \tag{9.155}$$

The factor of $X(q)$ depends on the phonon frequency $\omega(q)$. It is replaced by the actual frequency $\Omega(q)$. The dielectric function $\varepsilon(q)$ screens the electron–ion interaction. This expression is quite reasonable. The phonon-mediated interaction is screened by electron–electron interactions. This phenomenon changes the phonon frequencies to $\Omega(q)$ and changes the matrix element to $\tilde{M}(q)$. However, the phonon-mediated interaction still has the same basic form and can still act over large distances. The screening from electron–electron interaction changes the form of the interaction, but does not eliminate it at large distances.

The idea that electrons can interact at large distances by exchanging phonons is very important. It explains superconductivity in the elemental metals such as Pb, Sn, and Al.

The phonon-mediated interaction can be attractive and can cause two electrons to bind into pairs. These pairs are the cause of superconductivity in the simple metals.

9.5 Electron–Phonon Effects at Defects

There are a variety of defects in solids, which are discussed in the next chapter. Many of them have pronounced effects due to the electron-phonon interaction. Two important examples are discussed below: F-centers and the Jahn-Teller effect.

9.5.1 *F-Centers*

Wide band-gap insulators are colorless when purified. When they contain a significant density of defects, they often have a bright color. The color is due to lattice defects. Impurities such as rare earth and transition metal atoms often have internal electronic transitions in the visible range of frequencies. These systems are technologically important as phospors. The German word for color is *farbe*, and farbe center is shortened to F-center.

A typical feature of these defect states is that the optical absorption and emission spectra are widely separated in frequency. This is called the *Franck-Condon effect*. A typical example is shown in fig. 9.8 for the F-center in KBr. The figure shows the absorption has a Gaussian line shape. The energy width of the Gaussian depends on temperature. In alkali halides, the F-center is due to an anion vacancy, which traps an electron. For more information see Brown [1].

The temperature dependence suggests the linewidth is due to phonons. The two processes are an electronic transition of energy ΔE plus the creation of numerous phonons:

- In absorption, $\hbar\omega = \Delta E+$ the energy of numerous phonons.
- In emission, $\Delta E = \hbar\omega+$ the energy of numerous phonons.

Since ΔE is fixed, the average photon energy $\hbar\omega$ is much greater in absorption than in emission.

There is a very simple model that contains the correct physics. The electron is initially in the state i and has an optical transition to the state f. The motion of all of the ions in the vicinity of the defect is represented by a single coordinate Q. The effective Hamiltonian is

$$H = H_e + H_p + H_{ep} \tag{9.156}$$

$$H_p = \frac{P^2}{2M} + \frac{M\omega_0^2}{2} Q^2 \tag{9.157}$$

$$H_e = E_i C_i^\dagger C_i + E_f C_f^\dagger C_f \tag{9.158}$$

$$H_{ep} = Q\,[M_i C_i^\dagger C_i + M_f C_f^\dagger C_f] \tag{9.159}$$

and $\Delta E = E_f - E_i$. There are two possibilities:

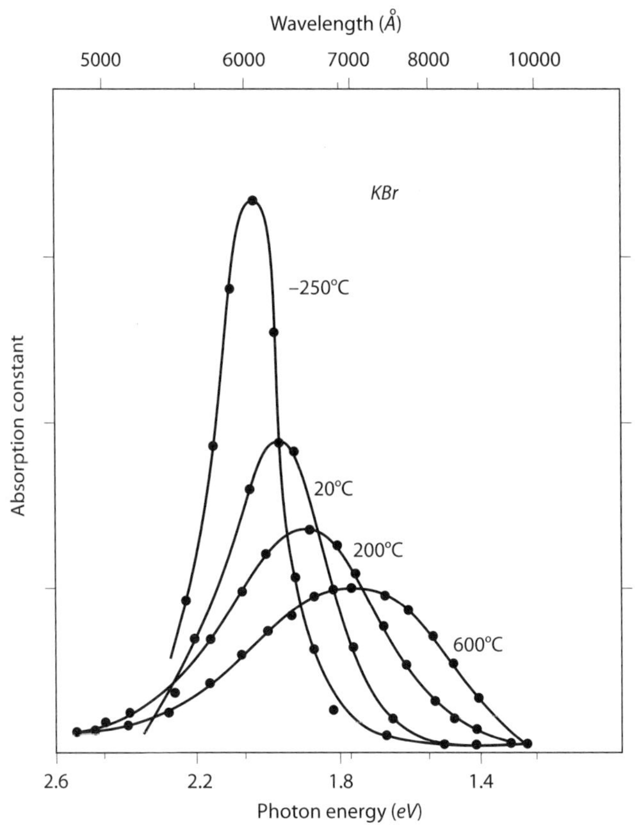

FIGURE 9.8. Optical absorption of an F-center in KBr at different temperatures. From Brown [1].

- If the electron is in the state i, the phonon state is a harmonic oscillator centered about Q_i:

$$H_{pi} = \frac{P^2}{2M} + \frac{M\omega_0^2}{2}(Q - Q_i)^2 - \frac{M\omega_0^2}{2}Q_i^2 \tag{9.160}$$

$$Q_i = -\frac{M_i}{M\omega_0^2}, \quad \psi_i(Q) = \phi_n(Q - Q_i) \tag{9.161}$$

$$\Delta_i = \frac{M\omega_0^2}{2}Q_i^2 \tag{9.162}$$

The phonon eigenfunction is ϕ_n which is the usual harmonic oscillator state. The self-energy is $-\Delta_i$.

- If the electron is in the final state j, the formulas are similar, with M_f instead of M_i, and Q_f instead of Q_i.

The important assumption is that $M_f \neq M_i$. Figure 9.9 shows the phonon potential energy diagrams of the initial and final electronic states. The final state has been shifted vertically upward by the electronic energy ΔE. $Q_i < 0$ and $Q_f > 0$ are shown with opposite signs to

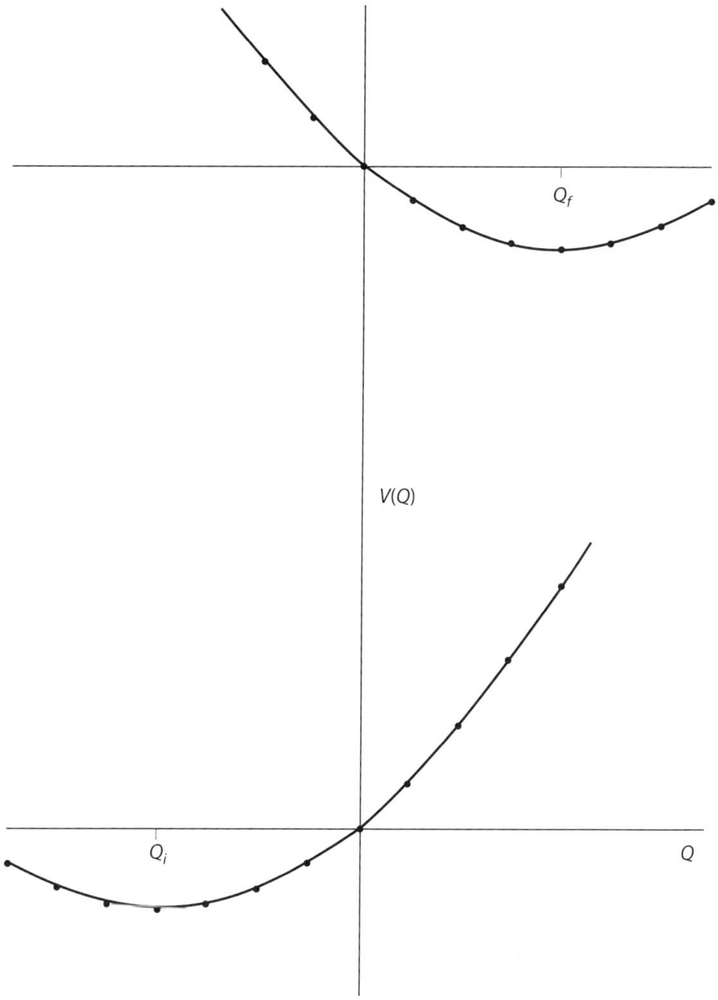

FIGURE 9.9. Configurational coordinate diagram of initial- and final-state phonon potential energies.

emphasize the possible difference in these two values. Such a figure is called a *configurational coordinate diagram.*

The optical absorption process has a matrix element

$$M_{if} = \langle f|p \cdot A|i\rangle I_{nm} \tag{9.163}$$

$$I_{nm}(Q_i - Q_f) = \int dQ\phi_n(Q - Q_f)\phi_m(Q - Q_i) \tag{9.164}$$

$$\hbar\omega = \Delta E - \Delta_f + \Delta_i + \hbar\omega_0(n - m) \tag{9.165}$$

The first term in M_{if} is the electronic matrix element of the $p \cdot A$ interaction. The second factor is the overlap between the phonon harmonic oscillator states. Square the matrix element and sum over all possible values of (n, m). The initial state m must be weighted by its thermal probability $\exp[-m\hbar\omega_0/k_BT]$. The real part of the optical conductivity is

$$\sigma(\omega) = \frac{2\pi C}{\omega}\sum_{nm} e^{-\beta m\hbar\omega_0} I_{nm}^2 \delta[\hbar\omega - \Delta E + \Delta_f - \Delta_i - \hbar\omega_0(n-m)] \qquad (9.166)$$

Replace the delta function by an integral over time. Then analytically sum the double series. The final expression is

$$\sigma(\omega) = \frac{C}{\omega}\int_{-\infty}^{\infty} dt \exp\{it[\hbar\omega - \Delta E + \Delta_f - \Delta_i] - \Phi(t)\} \qquad (9.167)$$

$$\Phi(t) = g[(N_0+1)(1-e^{-i\omega_0 t}) + N_0(1-e^{i\omega_0 t})] \qquad (9.168)$$

$$g = \frac{\hbar}{2M\omega_0}(Q_i - Q_f)^2, \quad N_0 = \frac{1}{e^{\hbar\omega_0/k_BT}-1} \qquad (9.169)$$

where g is a dimensionless coupling constant. Evaluating this expression for large values of $g > 10$ does give Gaussian line shapes in agreement with the observed absorption. The expression for emission is similar. Acoustical phonons with energy $\hbar\omega(q)$ can also be included by having all functions depend on q and summing over this wave vector.

9.5.2 *Jahn-Teller Effect*

The Jahn-Teller effect is often found in impurity complexes in crystals. It is based on a pair of theorems:

- The first theorem is due to Kramers [4], and asserts that all states of a single electron are at least doubly degenerate in the absence of a magnetic field. Usually the degeneracy is due to spin: spin-up and spin-down have the same energy states. However, there are cases involving the spin–orbit interaction where states $\phi(\mathrm{k},\mathbf{r})\sigma$ are degenerate with $\phi(-\mathbf{k},\mathbf{r})\bar{\sigma}$, where $\bar{\sigma} = -\sigma$.

- The Jahn-Teller theorem is well stated by Ham [3]: "Every nonlinear molecule or crystal defect that has orbital electronic degeneracy when the nuclei are in a symmetrical configuration is unstable with respect to at least one asymmetric distortion of the nuclei which lifts the degeneracy."

The Kramers theorem states that all states of a single electron must be doubly degenerate. The Jahn-Teller theorem states that if a localized electron state has a larger degeneracy than two, the orbitals distort and there is a lower energy state with double degeneracy.

The Jahn-Teller theorem is based on group theory. Given an electronic state i that couples to a group of local vibrations Q_α, the matrix element for a linear distortion is

$$V = \sum_{i\alpha} M_{i\alpha} Q_\alpha C_i^\dagger C_i \qquad (9.170)$$

$$H_e = E_0 \sum_i^n C_i^\dagger C_i \qquad (9.171)$$

The theorem is that there is always at least one state (i) out of the n degenerate energy levels and one vibrational mode Q_α for which the matrix element $M_{i\alpha} \neq 0$. Then that mode can relax and give the lower energy configuration $-\Delta_{i\alpha}$:

$$\Delta_{i\alpha} = \frac{M_{i\alpha}^2}{\hbar\omega_\alpha} \tag{9.172}$$

The Jahn-Teller effect breaks symmetry. The Peierls transition is a collective Jahn-Teller transition.

References

1. F. C. Brown, Color centers in alkali halides. Chapter 11 in *The Physics of Solids* (Benjamin, New York, 1967)
2. G. Grimvall, *The Electron-Phonon Interaction in Metals* (North Holland, Amsterdam, 1981)
3. F.S. Ham, The Jahn-Teller effect: a retrospective view. *J. Luminescence* **85**(4), 193–197 (2000)
4. H. A. Kramers, *Koninkl. Ned. Acad. Witenschap. Proc.* **33**, 959 (1930)
5. H. A. Jahn and E. Teller, Stability of polyatomic molecules in degenerate electronic states: orbital degeneracy. *Proc. R. Soc. London, Ser. A* **161**, 220–235 (1937)
6. W. H. Kleiner and L. M. Roth, Deformation potential in germanium from optical absorption lines for exciton formation. *Phys. Rev. Lett.* **2**, 334–336 (1959)
7. G. D. Mahan, Temperature dependence of the band gap in CdTe. *J. Phys. Chem. Solids* **26**, 751–756 1965)
8. W. L. McMillan, Transition temperatures of strong-coupled superconductors. *Phys. Rev.* **167**, 331–344 (1968)
9. R. E. Peierls, *Quantum Theory of Solids* (Clarendon Press, Oxford, UK, 1955)
10. R. T. Shuey and H. U. Beyeler, The elastic fields and interactions of point defects in isotropic and cubic media. *A. Angew. Math. Phys.* **19**, 278–300 (1968)
11. J. M. Ziman *Electrons and Phonons* (Clarendon Press, Oxford, UK, 1962)

Homework

1. The electron self-energy from the Fröhlich polaron at zero temperature is the integral

$$\Sigma_1(k) = C^2 \int \frac{d^3q}{q^2(2\pi)^3} \frac{1}{\varepsilon_k - \varepsilon_{k+q} - \hbar\omega_0} \tag{9.173}$$

Evaluate this integral and show that it can be written for small values of k as

$$\Sigma_2(k) = -U\frac{\arcsin(y)}{y}, \quad y = k/q_0 \tag{9.174}$$

Determine the constants (U, q_0).

2. The lifetime of the electron from the electron–phonon Fröhlich interaction is

$$\frac{1}{\tau(\varepsilon_k)} = \frac{2\pi C^2}{\hbar}\int \frac{d^3q}{q^2(2\pi)^3}[(N_0+1)\delta(\varepsilon_k - \varepsilon_{k+q} - \hbar\omega_0) + N_0\delta(\varepsilon_k - \varepsilon_{k+q} + \hbar\omega_0)] \tag{9.175}$$

Evaluate this integral for the case that $\varepsilon_k = 0$.

3. Evaluate the integral for the lifetime of the electron from deformation potential interaction with acoustical phonons. Use the Debye model and the high-temperature limit for the phonon occupation number. Evaluate for $k \neq 0$.

4. Show that at zero temperature, using first-order pertubation theory, the electron–phonon interaction gives a wave function for the electron in a semiconductor as

$$\psi(\mathbf{p}, \mathbf{r}) = e^{i\mathbf{r}\cdot\mathbf{p}}\left[1 + \frac{1}{\sqrt{\Omega}}\sum_{\mathbf{q}} e^{i\mathbf{q}\cdot\mathbf{r}} \frac{M(\mathbf{q})\, a_q^{\dagger}|0\rangle}{\varepsilon_p - \varepsilon_{p+q} - \hbar\omega(q)}\right] \tag{9.176}$$

Define the number of phonons in the polaron cloud as

$$N_u = \left\langle \psi \middle| \sum_{\mathbf{q}'} a_{q'}^{\dagger} a_{q'} \middle| \psi \right\rangle \tag{9.177}$$

Evaluate this expression at $p = 0$, $T = 0$ for the (i) polar interaction, and (ii) deformation potential interaction.

5. Consider a metal that has a simple form of the electron–phonon interaction:

$$\alpha^2 F(\omega) = \begin{cases} \left(\dfrac{\omega}{\omega_D}\right) & 0 < \omega < \omega_D \\ 0 & \omega_D < \omega \end{cases} \tag{9.178}$$

(a) What is λ?

(b) Derive the real part of the self-energy of the electron at zero temperature.

6. Five bonus points for making a good graph of the result in part (b) above.

7. Another electron–phonon function in metals is defined as

$$\lambda\langle\omega^2\rangle = 2\int_0^{\omega_D} \omega d\omega \alpha^2 F(\omega) \tag{9.179}$$

Show that this quantity is actually independent of phonon properties.

8. Derive the phonon-mediated interaction between two fixed electrons using the deformation potential interaction for LA phonons. A simple answer is obtained when using the Debye model for the phonon dispersion.

9. Use the piezoelectric interaction to acoustical phonons to calculate the effective interaction potential between two fixed electrons. (*Hint*: This interaction contributes to the dielectric function.)

10. The Gaussian line shape in F-center absorption is easily derived from eqn. (9.167). At large values of g, only the short time response is needed for $\Phi(t)$. Show that one can expand

$$\lim_{t\to 0} \Phi(t) = -iat - bt^2 \tag{9.180}$$

Find the expressions for (a, b). Show that this form gives a Gaussian line shape, and identify the temperature dependence of the width.

10 Extrinsic Semiconductors

10.1 Introduction

After the invention of the transistor, and later the integrated circuit, much scientific effort was applied to understanding the properties of semiconductors. After this goal was largely accomplished, condensed matter physics shifted to a historical phase where unusual metals (e.g., heavy fermions, Kondo insulators, high-temperature superconductors) were the fashionable research topics. Recently, with the emphasis on nanostructures and nano-physics, interest in semiconductors has revived. This chapter discusses the interesting properties of impure semiconductors. An intrinsic semiconductor is one that is pure and has no defects. That ideal can never be met: all semiconductors have impurities and/or defects. They are called *extrinsic*. It is these blemishes that give semiconductors their interesting physical properties and makes them useful for devices.

Atoms that have an extra valence electron can be put in the semiconductor as a substitutional impurity. An example is arsenic in silicon. These are called *donors*, and the extra electron will be bound to the donor impurity at low temperature. However, the binding is weak, and at higher temperatures the extra electron will be in the conduction band and is quite mobile. The properties of the conduction band are quite important for understanding the electrical properties of semiconductors with donors. Similarly, one can intentionally add an atom that has one fewer valence electron, such as gallium to silicon. This *acceptor* will create a *hole* in the valence band, which is an absence of an electron. This hole will also be quite mobile at higher temperatures, and the electrical properties of a semiconductor with acceptors is determined by the highest energy states in the valence bands. In an insulator, by constrast, the holes and electrons are strongly bound to the impurities and are not usually mobile at room temperature.

10.1.1 *Impurities and Defects in Silicon*

Silicon is probably the most important and most investigated semiconductor. Its defects and impurities determine many device characteristics. It has four valence electrons and is among the group IV elements in the periodic table. The elements in group V have one more valence electron and act as donors when they are a substitutional impurity in silicon. The donor binding energies are

Element	E_B (eV)
N	0.045
P	0.045
As	0.054
Sb	0.043

All except arsenic have nearly the same binding energy.

Similarly, the group III elements in the periodic table have one less valence electron than Si. They act as acceptors of electrons. When they are a substitutional impurity in silicon they have the following acceptor binding energies:

Element	E_B (eV)
B	0.045
Al	0.069
Ga	0.071
In	0.155

These numbers have less consistency. The acceptor is a more complicated defect state, due to the degeneracy of the valence band.

Silicon is the best material for integrated circuits since it has the lowest density of defects. George D. Watkins devoted much of his scientific career to understanding the most common defects in silicon. Thanks to Watkins, more is known about defects in silicon than for any other material. They are rarely simple! The following list is intended to be illustrative, but neither universal nor complete. Other materials have different types of defects.

- The easiest defect to visualize is a silicon vacancy: a silicon atom is missing. This defect is complicated, and can exist in five different charge states: V^{++} V^{+}, V^{0}, V^{-}, V^{-}. The five different states arise from five different ways the four dangling bonds can pair up and the bonding atoms can relax.

  The vacancy is a *negative U system.* Two electrons bound to the vacancy (V^{--}) have a larger binding energy than one electron (V^{-}), despite the Coulomb repulsion between the electrons. The additional binding energy for the second electron comes from polarons effects: recall that the polaron binding energy of a localized electron is

$$E_B = -\sum_{q} \frac{M(\mathbf{q})^2}{\hbar\omega(q)} \tag{10.1}$$

If $M(\mathbf{q}) \propto Z$, where Z is the number of electrons, then the polaron binding of two electrons is four times that of one electron. The polaron effects compensate for the Coulomb repulsion. In the language of the Hubbard model, the onsite energy U for the second electron is negative.

- The other intrinsic defect is an interstitial atom of silicon. They are rarely found alone. Apparently, the interstials diffuse rapidly and locate another defect. They join and form a complex of impurity plus interstital. If the defect is a vacancy, the silicon interstitial hops into the vacant site, and the crystal arrangement is restored.
- A boron atom may occupy a silicon site, where it acts as an acceptor. It may occupy an interstitial site, where it is another example of negative U. It can accept one or two electrons, and the second electron is more bound than the first. The one-electron state is 0.15 eV below the conduction band, while the two-electron state is 0.45 eV below E_C.
- When carbon, nitrogen, or oxygen is a substitional impurity in silicon, it relaxes to an "off-center" position. It does not sit at the silicon site, but breaks symmetry and moves away. There is a Jahn-Teller effect between the electronic and vibrational coordinates.
- Carbon can also occupy an interstitial position and acts as an acceptor. This fact is surprising, since carbon has the same number of bonding electrons as silicon.
- The 3d transition metal atoms are found in interstitial sites. They act as deep levels, which means the electron states are near to mid gap. These states are atomic in nature and are not described by effective mass theory.

The above list is not exhaustive, but is intended to illustrate the different kinds of behavior. Often defect complexes are composed of several items, such as a vacancy plus an impurity. Landolt-Börnstein gives a complete list of the elements and their type of impurity state in silicon.

10.1.2 Donors

Each silicon atom has four first neighbors in a tetrahedral arrangement. When a donor atom such as P is substituted for Si, four of the P valence electrons participate in the tetrahedral bonding. The fifth P valence electron is excluded from this arrangement. It is bound to the P atom in a bound state that extends over many unit cells.

First consider a semiconductor with the conduction minimum at $k = 0$ (Γ-point). GaAs and CdTe are examples. They have an isotropic effective mass m_c. In these crystals the donor is well described by the solution to the effective Hamiltonian:

$$E\psi(\mathbf{r}) = \left[E_{c0} - \frac{\hbar^2\nabla^2}{2m_c} - \frac{e^2}{4\pi\varepsilon(0)\, r} \right] \psi(\mathbf{r}) \tag{10.2}$$

Two notable features of this Hamiltonian are as follows:

1. The effective band mass m_c determines the kinetic energy, since the motion of the electron in the conduction band is constrained to be on this dispersion curve.

2. The attractive Coulomb interaction between the donor and the electron has the donor with a charge of plus one. The other four electrons in the valence band of the donor contribute to the σ-bonding and form a neutral arrangement. The potential energy is screened by the static dielectric function $\varepsilon(0)$. This choice of dielectric function is not always accurate. In polar semiconductors, such as CdTe, the optical phonons contribute to the dielectric function. If the donor binding energy has a numerical value similar to the optical phonon energy, then polaron effects must be included in the calculation of the donor binding energy.

The donor Hamiltonian is identical in form to that of the hydrogen atom. In this case the donor energy is

$$E_n = E_{c0} - \frac{E^*_{\text{Ry}}}{n^2} \tag{10.3}$$

$$E^*_{\text{Ry}} = \frac{m_c e^4}{32\pi^2 \varepsilon^2(0)\hbar^2} = E_{\text{Ry}}\left(\frac{m_c}{m_e}\right)\frac{\varepsilon_0^2}{\varepsilon^2(0)} \tag{10.4}$$

where $E_{\text{Ry}} = 13.60$ eV is the binding energy of the hydrogen atom, and m_e is the mass of a free electron. The effective Rydberg of the donor E^*_{Ry} is significantly reduced by the small effective mass $m_c \ll m_e$ and the large dielectric constant $\varepsilon(0)/\varepsilon_0 \gg 1$. For example, in GaAs, at room temperature, the values are

$$\left(\frac{m_c}{m_e}\right) = 0.0636, \quad \varepsilon(0) = 12.8\varepsilon_0 \tag{10.5}$$

$$E^*_{\text{Ry}} = 5.28 \text{ meV}, \quad a^*_B = a_B\varepsilon\left(\frac{m_e}{m_c}\right) = 10.7 \text{ nm} \tag{10.6}$$

where $a_B = 0.05292$ nm is the Bohr radius of the hydrogen atom, and a^*_B is the Bohr radius of the donor. The effective Bohr radius is quite large and shows that the bound electron wanders over many unit cells during its orbit. This observation justifies the use of the effective mass approximation. The calculated donor binding energy is in good agreement with observed donors in GaAs. The following are some experimental values for the impurity atom and the element it substitutes for in GaAs:

Element	Site	E^*_{Ry} (meV)
S	As	5.87
Se	As	5.79
C	Ga	5.91
Ge	Ga	5.88
Sn	Ga	5.82

The calculated value (5.3 meV) is about 10% low compared to experimental results. (5.8–5.9 meV). Most of this difference is due to central cell corrections. When the bound electron gets to the unit cell that contains the donor atom, the potential energy is no longer a simple Coulomb form, but has a pseudopotential form.

The calculation of the donor binding energy in silicon or germanium is more complicated due to the multivalley and ellipsoidal nature of the band dispersion. The donor Hamiltonian is

$$\left[H - \frac{e^2}{4\pi\varepsilon(0)r}\right]\psi_D(\mathbf{r}) = E\psi_D(\mathbf{r}) \tag{10.7}$$

where H is the Hamiltonian of the perfect crystal. This Hamiltonian gives the energy bands

$$H\psi_n(\mathbf{k}, \mathbf{r}) = E_n(\mathbf{k})\,\psi_n(\mathbf{k}, \mathbf{r}) \tag{10.8}$$

In solving the donor Hamiltonian, it is assumed the electron is in the conduction band, and these band states are used for expanding the solution:

$$\psi_D(\mathbf{r}) = \sum_{\mathbf{k}} a(\mathbf{k})\,\psi_c(\mathbf{k}, \mathbf{r}) \tag{10.9}$$

$$E\psi_D(\mathbf{r}) = \left[H - \frac{e^2}{4\pi\varepsilon(0)r}\right]\psi_D(\mathbf{r}) = \sum_{\mathbf{k}} a(\mathbf{k})\left[E_c(\mathbf{k}) - \frac{e^2}{4\pi\varepsilon(0)r}\right]\psi_c(\mathbf{k}, \mathbf{r}) \tag{10.10}$$

Let $E_c(0) = E_{c0}$ be the conduction band minimum, and $\varepsilon_c(\mathbf{k})$ be the band dispersion, which is zero at the band minimum: $E_c(\mathbf{k}) = E_{c0} + \varepsilon_c(\mathbf{k})$. Also write $E = E_{c0} - E_D$, and cancel E_{c0} from both sides of the equation:

$$0 = \sum_{\mathbf{k}} a(\mathbf{k})\left[\varepsilon_c(\mathbf{k}) + E_D - \frac{e^2}{4\pi\varepsilon(0)r}\right]e^{i\mathbf{k}\cdot\mathbf{r}}u_c(\mathbf{k}, \mathbf{r}) \tag{10.11}$$

In the effective mass approximation, the electron orbit extends over many unit cells. It is extended in real space, which means that it contains significant contributions from small wave vectors. The cell-periodic term $u_c(\mathbf{k},\mathbf{r})$ has only large Fourier coefficients, with wave vectors given by reciprocal lattice vectors. At small values of wave vector it is treated as being independent of wave vector and is evaluated at the wave vector of the band minimum $\mathbf{k}_0$:

$$0 = u_c(\mathbf{k}_0, \mathbf{r})\sum_{\mathbf{k}} a(\mathbf{k})\left[\varepsilon_c(\mathbf{k}) + E_D - \frac{e^2}{4\pi\varepsilon(0)r}\right]e^{i\mathbf{k}\cdot\mathbf{r}} \tag{10.12}$$

For semiconductors with a band minimum at the Γ-point, $\mathbf{k}_0 = 0$ the above expression is

$$0 = \left[\varepsilon\left(\frac{\hbar\nabla}{i}\right) + E_D - \frac{e^2}{4\pi\varepsilon(0)r}\right]\phi_D(\mathbf{r}) \tag{10.13}$$

$$\phi_D(\mathbf{r}) = \sum_{\mathbf{k}} a(\mathbf{k})\,e^{i\mathbf{k}\cdot\mathbf{r}} \tag{10.14}$$

Equation (10.13) justifies the donor Hamiltonian (10.2) for GaAs. Silicon has six equivalent band minima along the (100) directions. First consider one band minimum, say $\mathbf{k}_0 = k_0\hat{z}$. The band dispersion near the minimum is ellipsoidal:

$$\varepsilon_c(\mathbf{k}) = \frac{\hbar^2}{2m_t}(k_x^2 + k_y^2) + \frac{\hbar^2}{2m_l}(k_z - k_0)^2 \tag{10.15}$$

Multiply eqn. (10.12) by $\exp(-i\mathbf{k}_0 \cdot \mathbf{r})$ and then get the effective donor Hamiltonian:

$$0 = \left[-\frac{\hbar^2}{2m_t}\left(\frac{\partial^2}{\partial x^2} + \frac{\partial^2}{\partial y^2}\right) - \frac{\hbar^2}{2m_l}\frac{\partial^2}{\partial z^2} - \frac{e^2}{4\pi\varepsilon(0)r} \right]\phi_D(\mathbf{r}) \tag{10.16}$$

$$\phi_D(\mathbf{r}) = \sum_{\mathbf{k}} a(\mathbf{k})\, e^{i(\mathbf{k}-\mathbf{k}_0)\cdot\mathbf{r}} \tag{10.17}$$

The anisotropy in the effective masses ($m_t \neq m_l$) makes it difficult to solve this Hamiltonian analytically. An approximate solution can be obtained using a variational approach with the trial function:

$$\phi_D(\mathbf{r}) = A \exp\left[-\frac{b}{a_0}\sqrt{z^2 + \alpha\rho^2} \right] \tag{10.18}$$

where (b, α) are variational parameters. Using $\varepsilon(0) = 12.1\varepsilon_0$, $m_l = 0.916$ m, and $m_t = 0.19$ m, this function predicts a binding energy of $E_D = 27.6$ meV, which is far too small compared to experimental values of 43–45 meV.

Ning and Sah [15] explained that the increased donor energy in silicon is provided by the multiple band minima. They calculated the binding energy, including the valley–orbit splitting, and found very good agreement with experiments.

10.1.3 Statistical Mechanics of Defects

At zero kelvins, a crystal could be ideal, with every atom at a lattice site and all sites occupied. At nonzero temperatures, it is probable that intrinsic defect occur. Although defect states have a higher energy than the ground state, the entropy of the many possible sites make defects likely.

Two types of intrinsic defects are as follows.

- *Schottky defects are vacancies.* A lattice site is missing an atom. Often vacancies diffuse into the crystal from the surface. An atom, originally near the surface, hops onto the surface, leaving behind a vacant site. As more atoms hop to the surface, a new atomic layer is formed. Let n be the number of vacancies, N be the number of atoms, and $N + n$ be the number of sites, including the new ones on the surface. The energy required to create the vacancy is $E_V > 0$. The partition function for these arrangements is

$$Z = \sum_n \frac{(N+n)!}{n!N!} e^{-n\Delta}, \quad \Delta = \frac{E_V}{k_B T} \tag{10.19}$$

The binomial coefficient

$$\binom{N+n}{n} = \frac{(N+n)!}{n!N!} \tag{10.20}$$

is the number of ways of arranging n vacancies on $N+n$ sites. The summation is dominated by its largest term. Using $n! \approx \sqrt{2\pi}\, n^n e^{-n}$, the most likely value of n is

$$n \approx Ne^{-\Delta} \tag{10.21}$$

where it is assumed that $n \ll N$. As an example, if $E_V = 0.30$ eV, and $k_BT = 0.025$ eV at room temperature, the fraction of sites with vacancies is $\exp(-12) = 6$ ppm (parts per million), which is a sizable number.

- *Frenkel defects are interstitials.* Atoms are found at locations that are not lattice sites. Again, one can think of interstitials diffusing in from the surface. Let N' be the number of interstitial sites, and n be the number of atoms at interstitial sites. The energy to occupy an interstitial site is $E_I > 0$. The number of interstitial atoms is calculated as for vacancies, and is

$$n \approx N'e^{-E_I/k_BT} \tag{10.22}$$

Now consider the possibility of making n Schottky-Frenkel pairs. The vacancy is formed by the creation of an interstitial. The partition function for this process is

$$Z = \sum_n \frac{(N)!}{n!(N-n)!} \frac{(N')!}{n!(N'-n)!} e^{-n\Delta}, \quad \Delta = \frac{E_{VI}}{k_BT} \tag{10.23}$$

The first binomial coefficient is the number of ways of making n vacancies, and the second coefficient is the number of ways of making n interstitials. Again find the term in the summation with the maximum value, which has a value of n

$$n \approx \sqrt{NN'}\, e^{-\Delta/2} \tag{10.24}$$

In ionic crystals, Schottky defects are often found in pairs, with equal numbers of anion and cation sites vacant. This equality maintains charge neutrality. Coulomb's law also dictates that the anion and cation vacancies are found in pairs. If they each have a charge e and are one lattice site a away, then they have a Coulomb interaction energy

$$V_{CA} = -\frac{e^2}{4\pi a\varepsilon(0)} \approx -1 \text{ eV} \tag{10.25}$$

This large Coulomb binding prevents the two vacancies from diffusing away from each other.

Silver ions, as cations, like to diffuse through interstitial sites. Crystals such as silver halides or silver chalcogenides often have many interstitial–vacancy pairs, where the vacancies are of the Ag sites. Copper cations also diffuse, but seldom as easily as silver.

Since crystals are grown at nonzero temperatures, they are grown with defects in place. When the crystal is cooled, the defects are no longer in thermal equilibrium. They try to leave by diffusing to the surface. The diffusion coefficient is very temperature dependent, and gets much smaller at low temperatures. This process is slow, and all of the defects are never eliminated. All crystals have defects.

10.1.4 n–p *Product*

Let n_e denote the density of conduction electrons in the semiconductor, while n_h denotes the density of holes. Each of them is given by an integral over the Fermi-Dirac distribution function. For electrons it is

$$n_e = 2N_{c0}\int \frac{d^3k}{(2\pi)^3}\frac{1}{e^{\beta[E_c(k)-\mu]}+1} \tag{10.26}$$

$$E_c(k) = E_{c0} + \varepsilon(\mathbf{k}) \tag{10.27}$$

where N_{c0} is the number of equivalent conduction band minima. It is assumed the chemical potential μ is in the energy gap, between the conduction and valence bands. Then $\beta(E_{c0}-\mu)\gg 1$ and we can expand the denominator of the distribution function:

$$n_e = 2N_{c0}e^{-\beta(E_{c0}-\mu)}\int \frac{d^3k}{(2\pi)^3}\exp[-\beta\varepsilon(\mathbf{k})] \tag{10.28}$$

$$= N_c(T)\,e^{-\beta(E_{c0}-\mu)}$$

$$N_c(T) = 2N_{c0}\sqrt{m_x m_y m_z}\left(\frac{k_BT}{2\pi\hbar^2}\right)^{3/2} \tag{10.29}$$

$$\varepsilon(\mathbf{k}) = \frac{\hbar^2}{2}\left[\frac{k_x^2}{m_x}+\frac{k_y^2}{m_y}+\frac{k_z^2}{m_z}\right] \tag{10.30}$$

The band dispersion $\varepsilon(\mathbf{k})$ has a general form that includes all sorts of possible ellipsoids. A similar calculation for the valence band gives the density of holes. It is assumed the three bands are each isotropic and have masses: m_{lh} (light hole), m_{hh} (heavy hole), and m_s (split-off band):

$$n_h = N_v(T)\,e^{\beta(E_{v0}-\mu)} \tag{10.31}$$

$$N_v(T) = 2N_{v0}\left[(m_{\text{lh}})^{3/2} + (m_{\text{hh}})^{3/2} + e^{-\beta\Delta}(ms)^{3/2}\right]\left(\frac{k_BT}{2\pi\hbar^2}\right)^{3/2} \tag{10.32}$$

Note the sign of the exponent in eqn. (10.31). When the chemical potential is in the energy gap, then $\mu > E_{v0}$. The *n–p product* is obtained by multiplying together the density of electrons and holes:

$$n_e n_h = n_i^2 = N_c(T)\,N_v(T)\exp[-\beta E_G] \tag{10.33}$$

The chemical potential has canceled, and the energy gap comes from $E_G = E_{c0} - E_{v0}$. This important formula shows that the product of electron and hole densities depends on temperature and not on the doping density of donors or acceptors. If the semiconductor is very pure and has no intentional impurities, then $n_e = n_h$, since electrons and holes are made in pairs by thermally exciting an electron from the valence to the conduction band. This intrinsic carrier density is

$$n_e = n_h = n_i = \sqrt{N_c(T)\,N_v(T)}\,\exp\left[-\frac{\beta E_G}{2}\right] \tag{10.34}$$

In Si at room temperature this density is about 10^{16} electrons per cubic centimeters.

If the semiconductor is intentionally doped with donors at a density of n_D, they are usually ionized at room temperature. Then the density of conduction electrons is $n_e \sim n_D$, and the density of holes, as the minority carrier, is $n_h \sim n_i^2/n_D$.

10.1.5 Chemical Potential

The chemical potential μ of a crystal is an important quantity. It has two possible definitions.

- The energy to add an electron to the crystal: $\mu = E_{N+1} - E_N$.
- The energy to remove an electron from the crystal: $\mu = E_N - E_{N-1}$.

In a metals these two definitions give the same chemical potential, since all electrons are added or removed at the Fermi surface. In a perfect insulator, they give different energies. An added electron must go into the conduction band, while an electron is removed from the valence band. These energies differ by the band gap.

Actual insulators are not perfect, due to defects. The defects determine the value of the chemical potential. Since each crystal has a different collection of defects, the chemical potential has a different value in every insulator crystal: it is different even in different pieces of the same nominal material. Often the chemical potential is located in the energy gap, between the conduction and valence bands. When determined by defects, it is very dependent on temperature.

Figure 10.1 shows semiconductor energy bands. There is a conduction band of energy E_{c0} and a valence band of energy E_{v0}, where the energy gap is $E_G = E_{c0} - E_{v0}$. There is also a donor level at energy $E_{c0} - E_D$. The donor of density $[N_D]$ has a single positive charge. Square brackets denote concentrations. It can be neutral (concentration $[N_D^*]$), when it binds one electron, or ionized (concentration $[N_D']$). Write rate processes using the language of chemical reactions.

- Ionizing a donor is

$$e + N_D' \rightarrow N_D^* \tag{10.35}$$

so the concentrations are

$$n[N_D'] = [N_D^*]K_D, \quad K_D = N_c e^{-\beta E_D} \tag{10.36}$$

$$N_c(T) = 2\left(\frac{m_c k_B T}{2\pi\hbar^2}\right)^{3/2} \tag{10.37}$$

where n is the concentration of electrons, and N_c is the density of states of the conduction band. The factor of two is spin degeneracy. The total concentration of donors $[N_D] = [N_D'] + [N_D^*]$ is fixed, so the above equation gives

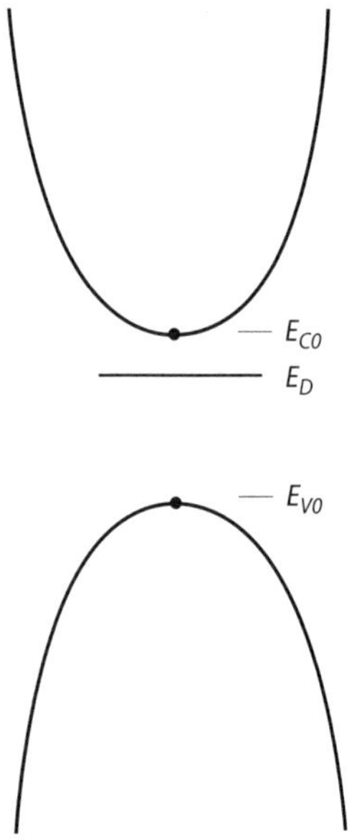

FIGURE 10.1. Energy level diagram showing conduction and valence bands and the donor level.

$$n[N_D'] = K_D\{[N_D] - [N_D']\} \tag{10.38}$$

$$[N_D'] = \frac{K_D[N_D]}{n + K_D} \tag{10.39}$$

The density of holes (p) in the valence band is given by the n–p product:

$$np = K_{np} = N_c N_v e^{-\beta E_G} \tag{10.40}$$

- The last equation is charge neutrality: the negative charges are electrons, and the positive charges are holes and ionized donors:

$$n = p + [N_D'] = \frac{K_{np}}{n} + \frac{K_D[N_D]}{n + K_D} \tag{10.41}$$

Multiply by the two denominators, and get a cubic equation for the electron density

$$0 = n^3 + n^2 K_D - n\{K_{np} + [N_D]K_D\} - K_D K_{np} \tag{10.42}$$

This equation can be solved to obtain the electron density $n(T)$. All of the coefficients vary with temperature.

So far the chemical potential has not entered the calculation. A chemical potential can be defined by assuming that n is also given by the equation

$$n = 2\int \frac{d^3k}{(2\pi)^3} \frac{1}{e^{\beta[E_c(k)-\mu]} + 1} \approx N_c(T)e^{\beta(\mu - E_{c0})} \tag{10.43}$$

$$\mu \approx E_{c0} + k_B T \ln\left[\frac{n}{N_c}\right] \tag{10.44}$$

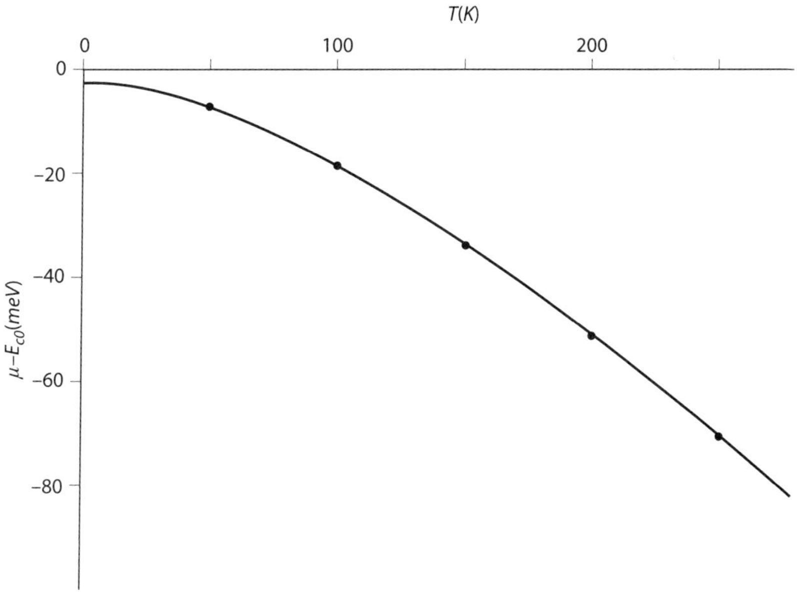

FIGURE 10.2. Chemical potential in GaAs as a function of temperature for 10^{16} donors per cubic centimeter.

Then, knowing $n(T)$, we can deduce $\mu(T)$. The chemical potential depends on the density and energy levels of various defect states, and on the parameters of the conduction and valence bands. Generally, for n-type semiconductors the chemical potential is near the conduction band edge, while for p-type semiconductors it is near the valence band edge. Figure 10.2 shows a calculation of $\mu(T)$ in GaAs. The vertical axis is $\mu - E_{c0}$ in meV. As the temperature increases, the chemical potential has increasingly negative values.

10.1.6 Schottky Barriers

Silicon, germanium, and most binary semiconductors have a surface electronic state. The energy of this state is usually between the conduction and valence bands, in the forbidden energy gap. Often it is one-third of the way between the valence and conduction bands:

$$E_S = E_{c0} - E_B, \quad E_B \approx \tfrac{2}{3} E_G \tag{10.45}$$

The actual value will depend on the choice of surface plane for the crystal. Table 10.1 shows experimental values of the barrier heights for Au and Al on the surface of silicon and gallium arsenide. V_{Bn} and V_{Bp} denote the values for electrons and holes, respectively. The sum of these two values is approximately the energy gap.

An oversimplified viewpoint is that the surface state is due to dangling bonds. When the crystal is cleaved to create a surface, the atoms on the surface have some of their bonding orbitals without any other atoms for bonding. An electron can come and be attached to these bonds. So the surface has dangling bond states, which hold electrons. The reason that this model is oversimplified is that the fraction of surface atoms with attached

Table 10.1 Schottky barrier heights for gold and aluminum contacts on silicon and gallium arsenide

Semiconductor	Metal	E_G	V_{Bn}	V_{Bp}
Si	Au	1.12	0.80	0.34
Si	Al	1.12	0.72	0.58
GaAs	Au	1.42	0.90	0.42
GaAs	Al	1.42	0.80	

Note. All energies in electron volts. V_{Bn} is the barrier height for n-type material, while V_{Bp} is the barrier for holes in p-type material.

electrons is actually quite small. This estimate is provided below. Because the surface state traps electrons, the chemical potential at the surface of the semiconductor is pinned to the surface state energy. Since the chemical potential in the bulk of the semiconductor usually has a very different value, as discussed earlier, the energy bands must bend near the surface. Figure 10.3 shows the band bending for n- and p-type semiconductors. This band bending causes a space-charge region near the surface, which is called the *depletion region.*

Our analysis is done for n-type semiconductors, but an identical analysis applies to p-type. Usually, the bulk of the semiconductor is charge neutral. There are as many conduction electrons n as ionized donors. The bands bend upward by an energy that is much greater than the thermal energy k_BT. So the band bending acts as a repulsive potential for the conduction electrons, and they are repulsed from this region. In the depletion region, the ionized donors provide a space charge of density N_D. The potential energy is found from Poisson's equation:

$$\frac{d^2}{dz^2}\phi(z) = \frac{e}{\varepsilon(0)} N_D \tag{10.46}$$

$$\phi(z) = \frac{e}{2\varepsilon(0)} N_D (z_0 - z)^2 \tag{10.47}$$

$$e\phi(z=0) = E_B = \frac{e^2}{2\varepsilon(0)} N_D (z_0)^2 \tag{10.48}$$

The band bending is parabolic, starting from the interior point z_0, as shown in fig. 10.3. The amount of surface charge per unit area is $n_s = N_D z_0$. It is balanced by the electron density in the surface states:

$$n_s = N_D z_0 = \sqrt{\frac{2\varepsilon(0)\, E_B N_D}{e^2}} \tag{10.49}$$

A typical set of values is $\varepsilon(0)/\varepsilon_0 = 10$, $E_B = 0.3$ eV, $N_D = 10^{18}$ donors/cm^3. These values give $n_s \approx 1.8\times10^{12}$/cm^2. Since the number of surface atoms is 10^{15}/cm^2, only about one out of a thousand surface atoms has a bound electron in the surface state. The width of the depletion region is z_0=18 nm.

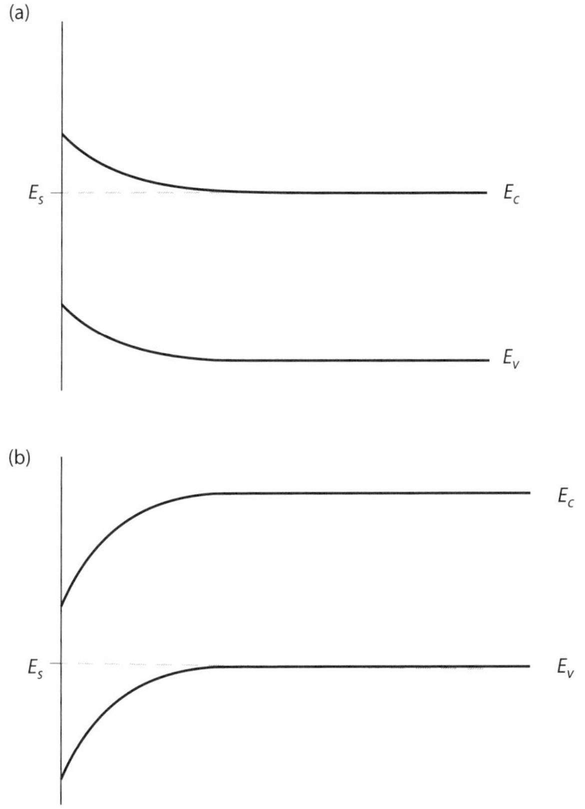

FIGURE 10.3. Surface band bending for (a) *n*-type and (b) *p*-type semiconductors.

At room temperature, silicon is good conductor, with a high value for the electron mobility. If one takes this material, grinds it up into small crystallites, and makes a new solid composed of polycrystalline material, the resistivity is increased by a factor of one thousand. Each crystalline grain has a high mobility. But at the interface between two grains, called the grain boundary, there is a *double-depletion region*. Each grain has a Schottky barrier, which forms an obstacle to the flow of electrons. Most semiconductors have the same property. ZnO is an exception, since it has no surface state and does not form Schottky barriers. Figure 10.4a shows a double Schottky barrier between two identical semiconductors.

Modern surface science, with high vacuum chambers, makes it easy to evaporate a thick metal film on the surface of the semiconductor. Each metal is characterized by a work function W, which is the energy difference between the chemical potential and the vacuum energy. It is the energy required to remove an electron from the metal and take it far away. The question is how the surface-state energy $E_s(W)$ depends on the work function of the metal after the evaporation. The slope of a graph of $E_s(W)$ vs. W is called S. It is one of the quantities graphed in the Mead plot in fig. 3.6. This graph showed that semiconductors have a very small value of S. The surface state energy E_s is independent of the choice of the metal.

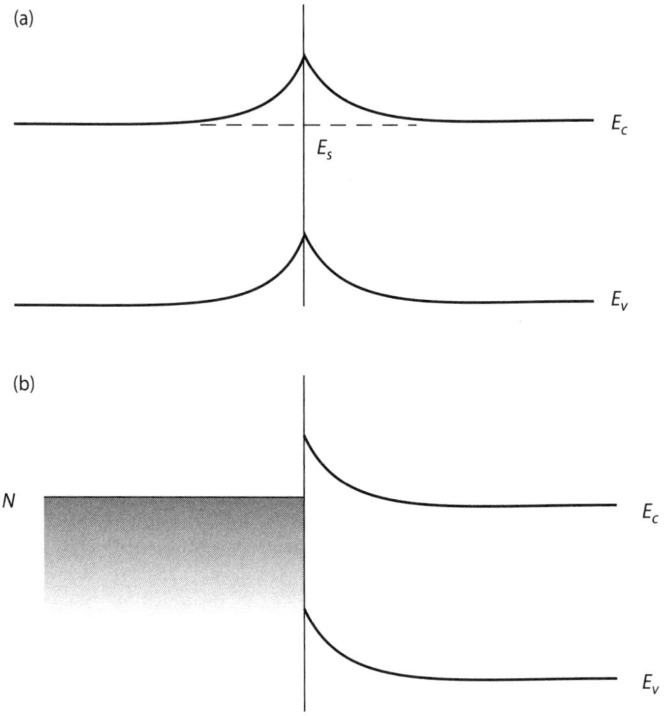

FIGURE 10.4. (a) Double Schottky barrier at a grain boundary between two identical semiconductors. (b) Metal evaporated on an *n*-type semiconductor.

The depletion region acts as a capacitor. Suppose one applies a voltage V_a between the semiconductor and the metal. The applied voltage creates a difference between the values of the chemical potential for the metal μ_m and semiconductor μ_s:

$$eV_a = \mu_s - \mu_m \tag{10.50}$$

This voltage difference occurs in the depletion region. In all of the above formula, replace E_B by $E_B - eV_a$. The width z_0 becomes a function of voltage:

$$z_0 = \sqrt{2\frac{\varepsilon(0)}{N_D e^2}|E_B - eV_a|} \tag{10.51}$$

The capacitance C of a junction is proportional to the area and inversely proportional to the width. So a plot of $C^{-2} \propto z_0^2$ as a function of V_a gives a straight line, whose intercept is the barrier height E_B. These capacitance graphs are one way of measuring barrier heights.

This device, of metal on semiconductor, is the *Schottky barrier*. It is a diode. At high temperatures, electron flow from metal to semiconductors is thermally activated over the potential barrier. At low temperatures, thermal activation is unlikely, and the current is carried by electron tunneling. The tunneling features of this device are evaluated in the next chapter.

10.2 Localization

Localization is an important phenomenon in condensed matter physics. The theory mixes two difficult ingredients: electron–electron interactions and disorder. The theory has evolved for the past thirty years, and is still evolving. Generally, the theory is well understood in one dimension, where every electron state is localized in the presence of disorder. Similarly, in three dimensions some states are localized, while others are conducting. The controversial case is two dimensions. An early prediction that all states are localized has been tested by experiments in the laboratory and experiments using computer simulations. Now it is believed that two-dimensional, extrinsic, crystalline semiconductors can have metallic conduction.

10.2.1 *Mott Localization*

For an isolated donor the binding energy of an electron is the effective Rydberg E^*_{Ry}. As more donors are added to the semiconductor, which increases the density n_e of electrons in the conduction band, the interaction between the conduction electrons and the charged donor gets screened by electron–electron interactions. The screening weakens the interaction between the donor and the electron and lowers the binding energy. Adding more electrons eventually causes the donor binding energy to vanish—the potential is too weak to have a bound state. This vanishing of the binding energy only happens in three dimensions. Figure 10.5 shows an early report on the hole binding energy to acceptors, in silicon, as a function of acceptor concentration. The binding energy vanishes around $n_c \sim 10^{18}$ cm^{-3}. A few data points are shown for donors.

If all of the electrons are no longer bound to the donors, they are free to roam the crystal. The semiconductor has become metallic: here metallic is defined as having nonzero conductivity at zero temperature. There is a critical concentration of donors n_c such that

- For $N_D > n_c$ the semiconductor is conducting at low temperature.
- For $N_D < n_c$ the semiconductor is insulating at low temperature.

Mott was the first to discuss this transition. The process of going from metallic to insulating was originally called *Mott localization*. It is one example of a *metal–insulator transition*. Other mechanisms for the metal–insulator transition and other mechanisms for localization are discussed in later chapters. Nowadays the nomenclature has changed and this type of localization is usually called *Anderson localization*.

One way to calculate this phenomenon is to use the Thomas-Fermi model for the dielectric function, which gives the screened potential energy of

$$V(r) = -\frac{e^2}{4\pi\varepsilon r} e^{-q_{\mathrm{TF}} r} = -\frac{e^2}{4\pi\varepsilon r} e^{-\lambda r/a_0^*} \tag{10.52}$$

$$\lambda = q_{\mathrm{TF}} a_0^* = \frac{\sqrt{12\pi}}{(3\pi^2)^{1/3}} \left[n_e a_0^{*3} \right]^{1/6} \tag{10.53}$$

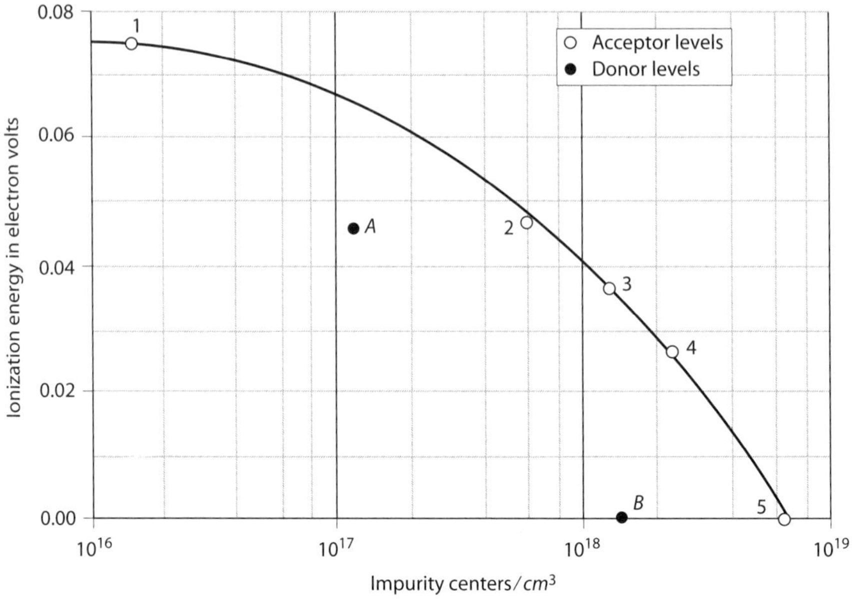

FIGURE 10.5. Donor and acceptor binding energy in silicon as a function of impurity concentration. Data from Pearson and Bardeen, *Phys. Rev.* **75**, 865 (1949). Empty circles are acceptor data, filled circles are donor data. Used with permission of the American Physical Society.

where a_0^* is the effective Bohr radius of the isolated donor. The bound states of this potential can be solved on the computer. In dimensionless form, with $\tau = r/a_0^*$, the eigenvalue equation for *s*-states is

$$0 = \left[\frac{d^2}{d\tau^2} + \frac{2}{\tau}\frac{d}{d\tau} + \frac{2}{\tau}e^{-\lambda\tau} - \varepsilon\right]\phi(\tau) \tag{10.54}$$

$$\varepsilon = \frac{|E|}{E^*_{\text{Ry}}} \tag{10.55}$$

The solution is shown in fig. 10.6.

The binding energy vanishes at the critical value of $\lambda_c = 1.19$. Using eqn. (10.53) gives

$$n_c^{1/3}a_0^* = 0.36 \tag{10.56}$$

Mott did a better calculation and found the better value $n_c^{1/3}a_0^* = 0.24$. Our value is an approximation, since at nonzero value of λ the binding energy decreases and bound orbits become larger in space and encompass more donor sites.

Mott localization is found in all semiconductors as a function of the increased number of donors or acceptors. For example, in GaAs, with $a_0^* = 10.7$ nm, then $n_c = 1.3\times10^{16}$ donors/cm^3. It is also found in nonsemiconductor systems. They all have the feature of randomness in the donor sites. Examples are alloys of tungsten and sodium–ammonia solutions [18]. Figure 10.7 shows data compiled by Edwards and Sienko [5] that shows the critical density of localization for many different experimental systems. The solid line is $n_c^{1/3}a_0^* = 0.26$. Mott localization is a widespread phenomenon.

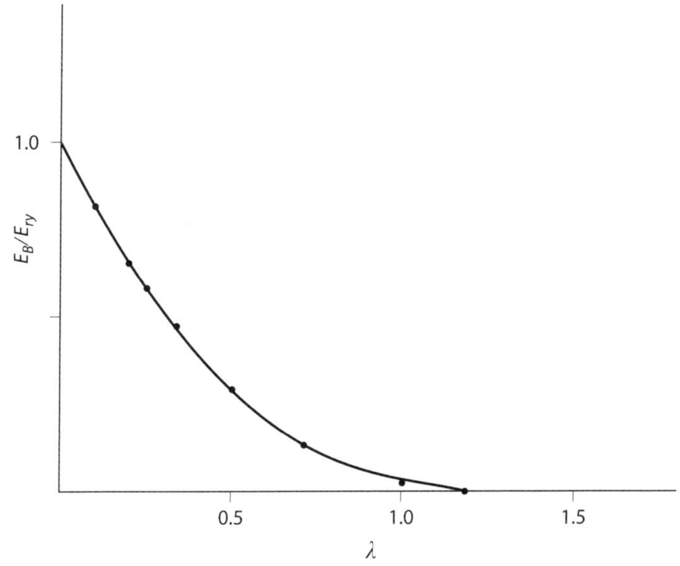

FIGURE 10.6. Eigenvalue $\varepsilon = |E|/E^*_{\mathrm{Ry}}$ of screened Coulomb interaction as a function of λ.

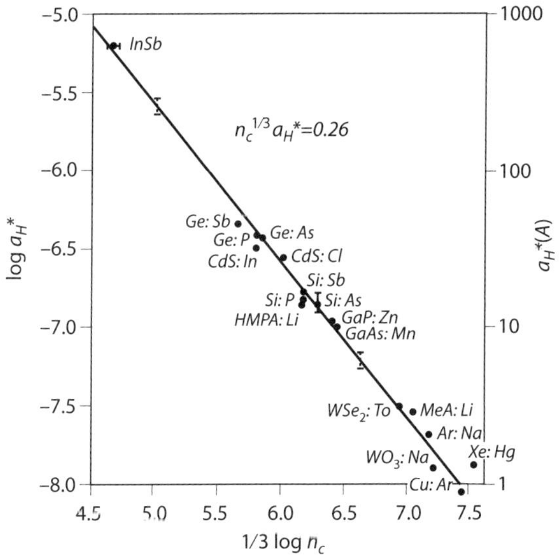

FIGURE 10.7. Critical density for Mott localization. The solid line is $n_c^{1/3} a_0^* = 0.26$. From Edwards and Sienko, *Phys. Rev. B* **17**, 2575 (1978). Used with permission of the American Physical Society.

There are many interesting experimental features of the metal–insulator transition in semiconductors. At low temperature, near to the critical concentration, the electrical resistivity can be scaled as

$$\rho(T, n) = \rho_0 f[T/T_0(n)], \quad \rho_0 = \frac{h}{e^2} \tag{10.57}$$

where $f(x)$ is a universal function, $T_0(n) \sim (n - n_c)^\alpha$ depends only on concentration n with respect to the critical concentration n_c. Figure 10.8 shows low-temperature electrical resistivity of a Si MOSFET. The two curves represent insulating and conducting systems as the temperature goes to zero. A MOSFET is a two-dimensional system.

Figure 10.9 shows a typical MOSFET (metal-oxide semiconductor field effect transistor) geometry. Current flows from the source (S) to the drain(D). The two-dimensional density n of electrons in the channel is controlled by the gate voltage, as shown in fig. 10.9b. Denote N_D as the two-dimensional density of donors in the channel. If $n > N_D$, then all donors have an electron and are neutral. The remaining electrons can conduct well since the cross section for scattering from neutral donors is small. If $n < N_D$, all electrons are bound to donors and the system is insulating. The case where $n = N_D$ is the crossover point between these two limiting behaviors. The case $n_e = N_D$ is the Mott system, and the Kravchenko and Sarachik data [11] does not reveal whether this system is conducting or insulating in 2D.

10.2.2 Anderson Localization

Anderson first discussed the localization of an electron on a crystal lattice when the site energy had fluctuations. The material is a crystal. The cnergy on site $\mathbf{R}_j$ is

$$E_j = E_0 + V\eta_j \tag{10.58}$$

Anderson took η_j as a random variable, with equal probability of having any value in the range $1 > \eta_j > -1$. He showed that if V/E_0 had a large enough value, the electrons states become localized. Note that his model was for a single electron, and contained no electron–electron interactions. The material becomes an insulator, although there is no gap in the density of states.

Anderson's calculation was analytic. Today most theoretical investigations of this topic are done on the computer. Figure 10.10 shows the electrical conductivity of a Bethe lattice as a function of V by Girvin and Jonson [8]. They normalized the energy scale to $4t$, where t is the hopping term in a tight-binding model. The Bethe lattice is a lattice with only one path between each pair of sites, which makes the calculation easier. At a critical value of $V \sim 5.5$, the conductivity drops by 10^{-35} because of Anderson localization.

Anderson's theory of localization applies to a crystal, while that of Mott applies to a disordered array of conducting sites. Today the modern usage of these names is just the opposite. Anderson localization is the name when the system has disorder, while Mott localization is the name when the system is crystalline. The metal–insulator transition in a doped semiconductor is now called Anderson localization. The metal–insulator transition in transition-metal oxides is an example of Mott localization.

10.2.3 Weak Localization

Another type of localization is called *weak localization*. Consider the experiment of a coherent beam of light shining on the planar surface of a disordered material. The disorder

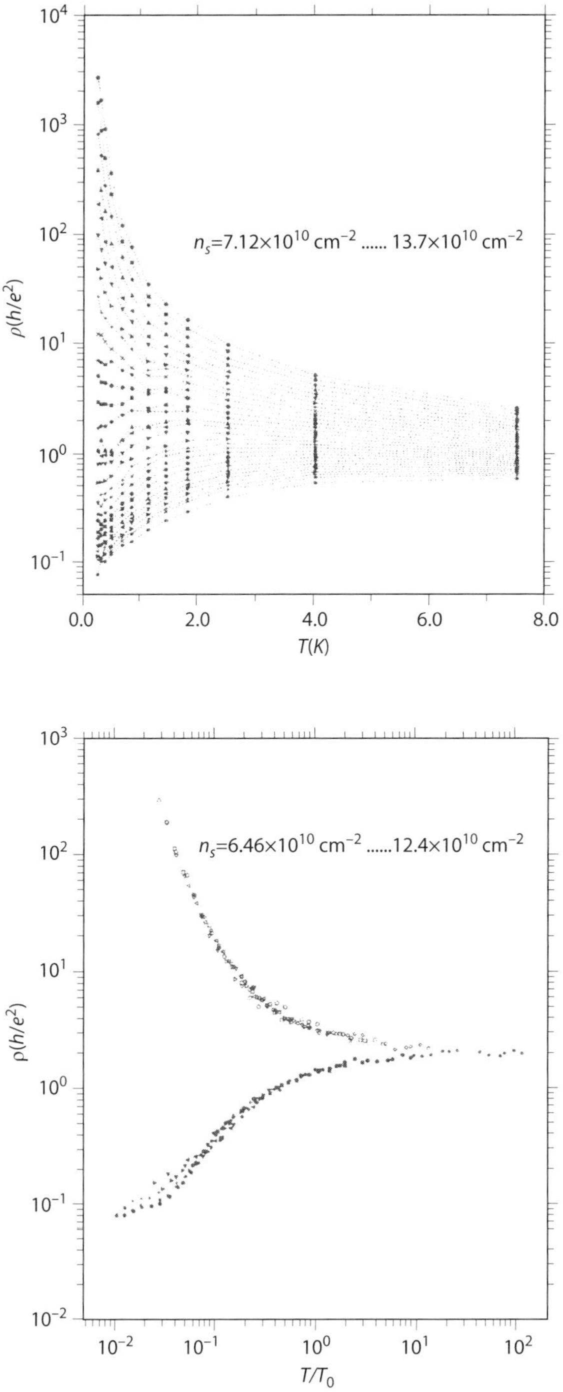

FIGURE 10.8. Low-temperature electrical resistivity of Si MOSFET. (a) Different lines are different electron densities n_e, which is controlled by a gate voltage. (b) The curves in (a) can be rescaled with a temperature T_0 to fit universal behavior. Data from Kravchenko et al., *Phys. Rev. B* **51**, 7038 (1995). Used with permission of the American Physical Society.

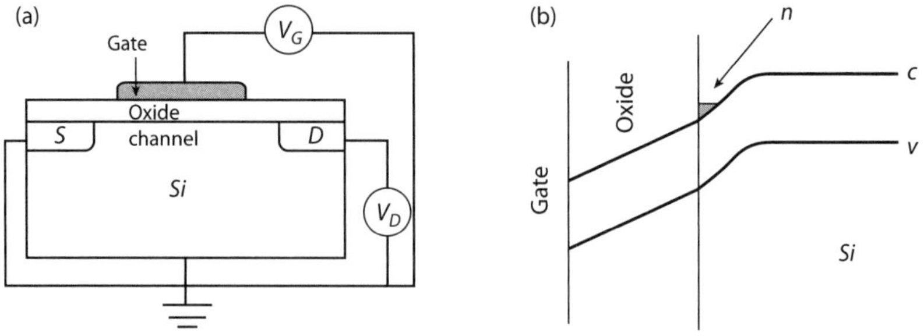

FIGURE 10.9. Metal-oxide semiconductor field effect transistor(MOSFET). (a) Current flows from source (S) to drain (D). (b) The density of electrons n in the channel is controlled by a gate voltage.

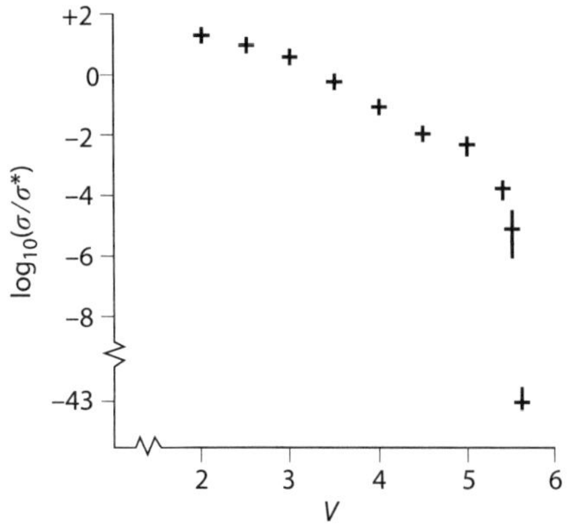

FIGURE 10.10. Conductivity of a Bethe lattice of connectivity $K = 2$ as a function of disorder V. Note the broken scales. From Girvin and Jonson, *Phys. Rev. B* **22**, 3583 (1980). Used with permission of the American Physical Society.

forces the incoming photons to diffuse through the material, diffusion being a form of random walk as the photon scatters from one atom to the next. Some of the incoming photons scatter and eventually get emitted from the original surface and contribute to the reflected light. A special case is those photons that get emitted coparallel to the incoming photons. These photons will constructively interfere with the photon path going exactly in the opposite direction. This effect is largest at normal incidence. A measurement as a function of angle shows a large emitted intensity normal to the surface, which is a peak of few degrees in width. This peak is the constructive interference between the photons going on a path in the two directions. This phenomenon is weak localization.

Strong localization is thought to be caused by the strong scattering of a particle from multiple sites. Thus, weak localization should be a precurser to strong localization. However, this relationship is still ill-defined.

10.2.4 Percolation

Percolation theory investigates another type of metal–insulator transition. Say that one has a lattice of atoms that make a metal, such as copper. Suppose one starts removing the copper atoms, one-by-one, in a random fashion. The crystal structure is maintained by replacing copper by an insulating atom such as argon. Define p $(1 \geq p \geq 0)$ as the fraction

Table 10.2 Critical thresholds for removing atoms (*A*) or bonds (*B*) for two- and three-dimensional lattices

Lattice	Z	$p_c(A)$	$p_c(B)$
2D			
pt	6	1/2	0.33
sq	4	0.59	1/2
hc	3	0.70	0.63
3D			
fcc	12	0.20	0.11
bcc	8	0.25	
sc	6	0.31	1/4
dia	4	0.43	0.39

Note: dia denotes the diamond lattice. Fractions are analytical results, decimals are from computer simulations.

of sites occupied by the copper atoms. As one removes the copper atoms, the solid becomes less conducting. There is a critical value p_c of the occupancy where the conductivity goes to zero. This value depends on the structure and on the number of first neighbors z. Table 10.2 shows some results for removing atoms under the column $p_c(A)$. Fractions are analytical results, while decimals are computer studies. For three-dimensional cubic lattices, the critical fractions are all less than one-half.

In Anderson localization, each site had an electron energy of $E_0 + V\eta_j$, where η_j was a random number $-1 \leq \eta_j \leq 1$. Suppose there is a lattice in which $\eta_j = +1$ on half of the sites, chosen randomly, and $\eta_j = -1$ on the other half. Half of the atomic sites have energy $E_- = E_0 - V$ and the other half have energy $E_+ = E_0 + V$. If those with E_- are copper and those with E_+ are argon, this lattice conducts quite well, since $p = 0.5 > p_c$. Having $\eta_j = \pm 1$ is not a suitable model of Anderson localization.

A related percolation process is, instead of removing atoms, to remove bonds that connect the atoms. The critical fraction of bonds that give the onset of zero conductivity is shown in the last column of table 10.2. These values are slightly lower than the critical concentration of atoms.

Substitional random alloys are systems with typically two kinds of atoms that are located randomly on a lattice. A band structure can be calculated for this alloy by representing the electron–ion potential energy by a suitable average of the potential of the separate atoms. The best method of doing this average was invented by Bruggeman [3], and is called the *coherent potential approximation* (CPA).

Scott Kirkpatrick [9] showed that CPA accurately predicts the effective resistance of networks of random resistors. Suppose one has a lattice (network) in which the atoms are conducting points and the bonds are resistors. Suppose there are two different resistors: R_1 of concentration p, and R_2 of concentration $1 - p$. What is the average resistance $\bar{R}$ of the network? A crude guess would be

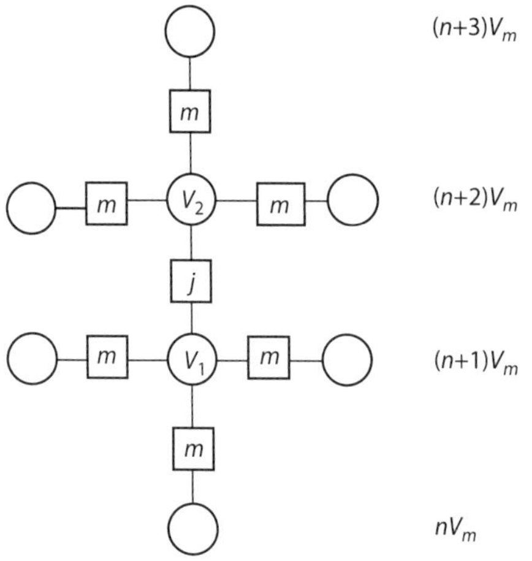

FIGURE 10.11. CPA analysis of resistor networks. Circles denote nodes, and the boxes denote resistors. All resistors have conductance σ_m except the one in the center. All circles have the average voltage nV_m, except the two marked $V_{1,2}$. Current flows upward.

$$\bar{R} \overset{?}{=} pR_1 + (1-p)\,R_2 \tag{10.59}$$

This choice gives poor results, as shown below. Instead, use CPA.

Instead of resistance, it is more convenient to talk about conductance $\sigma_j = 1/R_j$. The current between two lattice points (nodes) is $I = \sigma_j \Delta V$. Define σ_m as the average conductance of the system, and V_m as the average voltage difference between two nodes, along the direction of the current. Then $I_m = \sigma_m V_m$. Figure 10.11 shows a region of this network with a conductor σ_j (j = 1, 2) in the center. All other conductors, denoted by boxes, have the average value σ_m. The circles denote nodes, and empty circles have their average value nV_m, depending on the line n of atoms. Since only voltage differences are important, set $n = 0$. The nodes adjacent to the conductance σ_j have voltages $V_{1,2}$. A circuit analysis is done on these two nodes, using Kirchhoff's law that all currents coming into a node i sum to zero:

$$0 = \sum_j \sigma_{ij}(V_j - V_i) \tag{10.60}$$

where σ_{ij} is the conductance of the resistor between nodes i and j. For the two node points $V_{1,2}$ this law gives

$$0 = \sigma_j(V_1 - V_2) + \sigma_m[V_1 + 2(V_1 - V_m)] \tag{10.61}$$

$$0 = \sigma_j(V_2 - V_1) + \sigma_m[V_2 - 3V_m + 2(V_2 - 2V_m)] \tag{10.62}$$

The last term in each equation has a factor of two. This term is for current coming in from the sides. In a two-dimensional sq lattice there are two side neighbors. In a three-dimensional sc lattice there are four side neighbors. So replace the two by $z - 2$, where z is the coordination number:

$$0 = \sigma_j(V_1 - V_2) + \sigma_m[V_1 + (z-2)(V_1 - V_m)] \tag{10.63}$$

$$0 = \sigma_j(V_2 - V_1) + \sigma_m[V_2 - 3V_m + (z-2)(V_2 - 2V_m)] \tag{10.64}$$

These equations are solved for $V_{1,2}$:

$$V_1 = \frac{V_m}{D}[3\sigma_j + (z-2)\sigma_m] \tag{10.65}$$

$$V_2 = \frac{V_m}{D}[3\sigma_j + (2z-1)\sigma_m] \tag{10.66}$$

$$D = 2[\sigma_j + \lambda\sigma_m], \quad \lambda = \frac{1}{2}(z-1) \tag{10.67}$$

$$V_2 - V_1 = \frac{\sigma_m V_m (z+1)}{2(\sigma_j + \lambda\sigma_m)} \tag{10.68}$$

The voltage difference $V_2 - V_1$ should be equal, on the average, to the average difference V_m. It is if $\sigma_j = \sigma_m$. Denote this difference as δV_j:

$$\delta V_j = V_2 - V_1 - V_m = V_m \frac{\sigma_m - \sigma_j}{\sigma_j + \lambda\sigma_m} \tag{10.69}$$

There are two kinds of conductors. The value for σ_m is chosen by making this difference δV_j average to zero:

$$0 = \sum_{j=1}^{2} p_j \delta V_j = p\delta V_1 + (1-p)\delta V_2 \tag{10.70}$$

Plugging in the formula for δV_j gives

$$0 = V_m\left[p\frac{\sigma_m - \sigma_1}{\sigma_1 + \lambda\sigma_m} + (1-p)\frac{\sigma_m - \sigma_2}{\sigma_2 + \lambda\sigma_m}\right] \tag{10.71}$$

There is a quadratic equation for σ_m:

$$\sigma_m = \frac{1}{2\lambda}\left[B + \sqrt{B^2 + 4\lambda\sigma_1\sigma_2}\right] \tag{10.72}$$

$$B = \sigma_1[p(1+\lambda) - 1] + \sigma_2[\lambda - p(1+\lambda)] \tag{10.73}$$

This formula has some useful limits:

- If $\sigma_1 = \sigma_2 = \sigma$, then $\sigma_m = \sigma$.
- If $p = 0$, then $\sigma_m = \sigma_2$; if $p = 1$, then $\sigma_m = \sigma_1$.

Kirkpatrick used a slightly different way of averaging over the local conductance. He gets the same formula, but with $\lambda = z/2 - 1$ instead of $\lambda = (z-1)/2$. Some of his computer simulations are shown in fig. 10.12 for the 2D sq network. The points are computation, while the line is the CPA theory with his λ, which fits the simulations quite well.

Examine the case where one of the conductances is zero, say $\sigma_2 = 0$. The above formula gives

$$\sigma_m = \frac{1}{2\lambda}[B + |B|] \tag{10.74}$$

$$B = \sigma_1[p(1+\lambda) - 1] \tag{10.75}$$

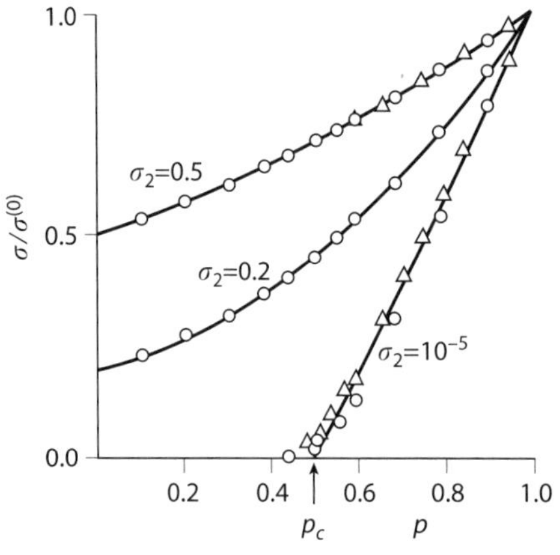

FIGURE 10.12. Average conductance of a two-dimensional resistor network on a square lattice. One resistor of concentration p has $\sigma_1 = 1$. The other resistor of concentration $1-p$ has conductance σ_2 as shown. The percolation threshold is $p = \frac{1}{2}$. An original figure using data from Kirkpatrick, *Rev. Mod. Phys.* **45**, 574 (1973).

so $\sigma_m = 0$ when $B < 0$, and $\sigma_m = B/\lambda \neq 0$ when $B > 0$. This predicts that the critical value of concentration is

$$p_c = \frac{1}{1+\lambda} \tag{10.76}$$

Using Kirkpatrick's formula for $\lambda = z/2 - 1$ gives

$$p_c = \frac{2}{z} \tag{10.77}$$

This prediction gives almost perfect agreement with the values $p_c(B)$ in table 10.2 for two-dimensional lattices: $p_c = \frac{1}{3}$ for pt $(n = 6)$, $p_c = \frac{1}{2}$ for sq $(z = 4)$, and $p_c = \frac{2}{3}$ for hc $(z = 3)$. It also predicts that $\sigma_m(p)$ is a linear function of p:

$$\sigma_m(p) = \sigma_1 \frac{p - p_c}{1 - p_c} \Theta(p - p_c) \tag{10.78}$$

However, it does not work well for three-dimensional lattices.

10.3 Variable Range Hopping

Semiconductors have a nonzero electrical conductivity at nonzero temperatures below the Mott transition. The conductivity is usually found to have a temperature dependence given by an exponent that is a fractional power of the inverse temperature:

$$\sigma(T) \approx \sigma_0 \exp\left[-\left(\frac{T_0}{T}\right)^{\nu}\right], \quad \nu = \frac{1}{2}, \frac{1}{4} \tag{10.79}$$

The fractional exponent is sometimes found to be one-half, and is sometimes one-quarter. Both cases are the result of variable range hopping.

Mott first gave the argument that predicted the dependence $T^{-1/4}$. The electrons are localized on donor sites, and each donor site has a slightly different energy E_i, depending on the number and ionization of nearby neighboring donors. The hopping to the neighbor is due to the small overlap of the neighboring eigenfunctions. If an eigenfunction goes as $\exp(-\alpha R)$, then the mutual overlap goes as $\exp(-2\alpha R)$. If an electron hops, it has to change its energy by ΔE, which may depend on how far it hops: write $\Delta E(R)$. The total probabiity of hopping is then

$$P(R) = P_0 \exp\left[-2\alpha R - \frac{\Delta E(R)}{k_B T}\right] \tag{10.80}$$

Mott assumed that the density of states $N(E)$ was a constant near the chemical potential. Since the density of states has the dimensions of number of states per energy per volume, the characteristic energy in a volume $4\pi R^3/3$ is

$$\Delta E \approx \frac{3}{4\pi R^3 N(E)} \tag{10.81}$$

$$P(R) = P_0 \exp[-W(R)] \tag{10.82}$$

$$W(R) = 2\alpha R + \frac{3}{4\pi R^3 N(\mu) k_B T} \tag{10.83}$$

The most likely hops are those that minimize $W(R)$, so take

$$0 = \frac{dW}{dR} = 2\alpha - \frac{9}{4\pi R_0^4 N(\mu) k_B T} \tag{10.84}$$

$$R_0 = \left(\frac{9}{8\pi\alpha N(\mu) k_B T}\right)^{1/4} \tag{10.85}$$

$$W(R_0) = \left(\frac{T_0}{T}\right)^{1/4} \tag{10.86}$$

$$k_B T_0 = \frac{(8\alpha)^3}{9\pi N(\mu)} \tag{10.87}$$

Some experimental data show this dependence. Figure 10.13 shows typical electrical conductivity of semiconductors that exhibit variable range hopping

10.4 Mobility Edge

In three dimensions some electronic states are localized while others are conducting. The localized states are usually near the edges of the energy bands. Mott proposed that the density of states is continuous at the energies E_c where the transition occurs between localized and conducting states.

Mott also proposed that the mobility σ of an electron is nonzero at the mobility edge. That seems to be true at room temperature. However, at low temperature, and particularly in the limit of zero temperature, the mobility vanishes at the metal–insulator transition

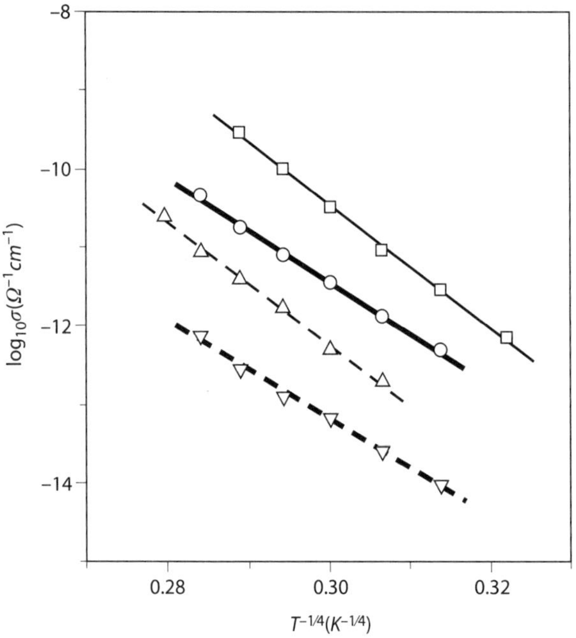

FIGURE 10.13. Electrical conductivities in zinc vanadate semiconducting glasses showing variable range hopping. A plot of log(σ) vs. $T^{-1/4}$ for different mixtures of V_2O_5 and ZnO. From Ghosh et al. [7]. (Used with permission of the American Institute of Physics.)

point. Figure 10.14 shows the electrical conductivity $\sigma(T)$, in the limit that $T \to 0$, in boron doped silicon, as a function of boron concentration. The transition seems to go as $\sigma(0) \sim \sqrt{n - n_c}$, where the critical concentration n_c for the metal–insulator transition is around $n_c \sim 4.1 \cdot 10^{18}/\text{cm}^3$.

The zero temperature conductivity in disordered systems is always found to go as

$$\sigma(0) \propto (n - n_c)^s \tag{10.88}$$

The exponent s is one-half in uncompensated samples, in multivalley semiconductors, as shown in fig. 10.14. For compensated samples, the exponent is $s = 1$. Amorphous materials usually have $s = 1$.

At zero temperature, the mobility is zero at the critical concentration. However, at nonzero temperatures, the mobility is nonzero. It is also nonzero when plotted as a function of electron energy. In that case Mott is correct: the mobility at room temperature in three dimensions is zero for $E < E_c$ and nonzero for $E > E_c$.

10.5 Band Gap Narrowing

The energy gap of a semiconductor depends on the density of impurities. A specific case is how the silicon band gap varies with the concentration of a donor such as phosphorus.

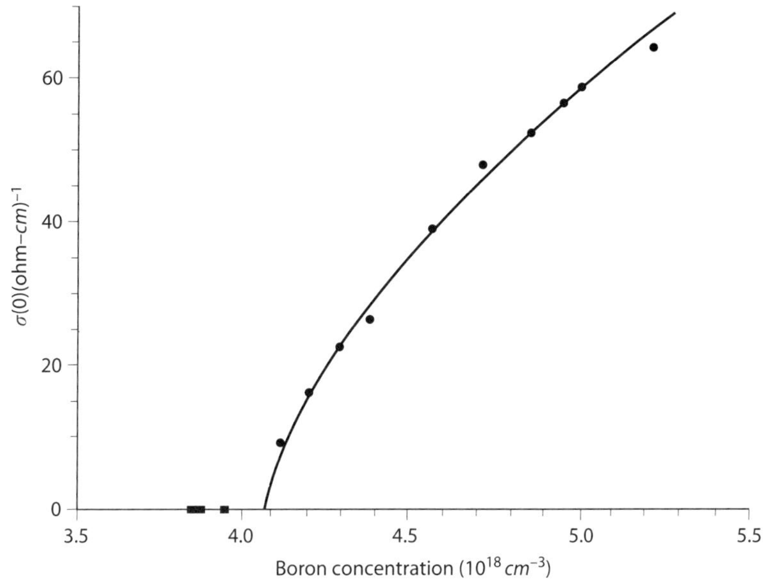

FIGURE 10.14. Zero-temperature electrical conductivity $\sigma(0)$ as a function of boron concentration in silicon. Filled circles are metallic samples and filled squares are insulators. From Dai et al., *Phys. Rev. Lett.* **66**, 1914 (1991). Used with permission of the American Physical Society.

When adding the impurities, one adds conduction electrons. At sufficiently high densities of impurities the conduction electrons are metallic, since the potential energy from the donors is screened by electron–electron interactions. There are several definitions of band gap, as shown in fig. 10.15.

- The band gap E_1 is between the top of the valence band and the bottom of the conduction band. The value is smaller than in the undoped semiconductor, as explained below.

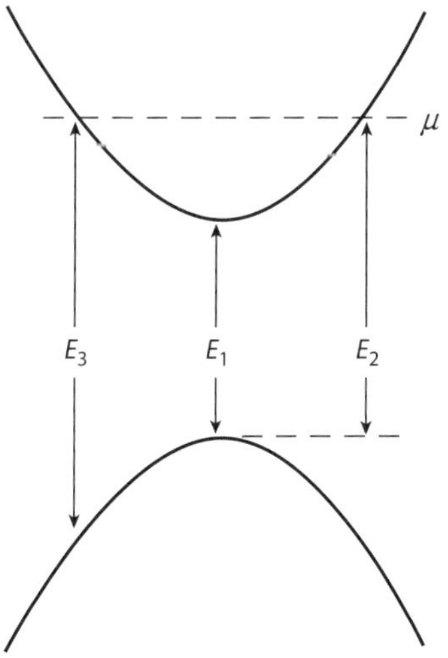

FIGURE 10.15. Three energy gaps for a heavily doped *n*-type semiconductor.

- The second E_2 is the thermal gap: how much energy it takes to remove an electron from the valence band and to insert it into the conduction band. The only available states in the conduction band are at the chemical potential μ.

- The third gap E_3 is the energy gap measured in a direct interband optical transition between the two bands.

In general, $E_3 > E_2 > E_1$.

The self-energy terms that contribute to these gaps are determined by electron–electron and electron–impurity interactions. In each case the self-energy $\Sigma(k)$ of an electron is calculated as a function of its wave vector. When discussing the gap E_1, these self-energies are evaluated at $k = 0$. When discussing E_2 and E_3, these self-energies are evaluated at the Fermi surface $k = k_F$.

1. The chemical potential μ increases in value above the bottom of the conduction band, as the conduction electrons fill up the energy states at the bottom of the conduction band. The increase in the chemical potential is called the *Burstein shift*. The density of conduction electrons at zero temperature is

$$n_e = N_c \frac{k_F^3}{3\pi^2} \tag{10.89}$$

where N_c is the number of conduction band minima. It is assumed the energy ellipsoids are spherical, and k_F is the Fermi wave vector for one of the N_c equivalent ellipsoids. The increase in chemical potential is

$$\mu = E_{c0} + \frac{\hbar^2 k_F^2}{2m_c} = E_{c0} + \frac{\hbar^2}{2m_c}\left(\frac{3\pi^2 n_e}{N_c}\right)^{2/3} \tag{10.90}$$

This energy term contributes to E_2 and E_3 but not to E_1. In silicon, where $N_c = 6$, and in germanium, where $N_c = 4$, the Burstein shift is greatly reduced because the conduction electrons get distributed between the many equivalent band minima.

2. The electron–donor interaction contributes to the self-energy of the conduction electrons. One theory is to use first- and second-order perturbation theory:

$$\Sigma_{ei}(k) = N_D\left[\tilde{v}(0) - \int \frac{d^3k'}{(2\pi)^3} \frac{\tilde{v}(k-k')^2}{\varepsilon(k) - \varepsilon(k')}\right] \tag{10.91}$$

The matrix element is the Fourier transform $\tilde{v}(q)$ of the screened electron–donor interaction:

$$V(r) = -\frac{e^2}{4\pi\varepsilon(0)\, r} e^{-Qr} \tag{10.92}$$

$$\tilde{v}(q) = -\frac{e^2}{\varepsilon(0)} \frac{1}{Q^2 + q^2} \tag{10.93}$$

$$Q^2 = \frac{3e^2 n_e}{2E_F \varepsilon(0)} = \frac{12\pi}{(3\pi^2)^{3/2}} \frac{(n_e N_c^2)^{1/3}}{a_0} \tag{10.94}$$

where Q is the Thomas-Fermi screening wave vector, and $a_0 = 4\pi\hbar^2\varepsilon(0)/e^2 m_c$ is the Bohr radius of the donor. The integrals can all be evaluated, giving

$$\Sigma_{ei}(k) = -N_D\left[\frac{2E_F}{3n_e} + \frac{4\pi}{1+4k^2/Q^2}\frac{m_c e^4}{\hbar^2 Q^3 \varepsilon_0^2}\right] \tag{10.95}$$

The donors are all ionized in the metallic state, so that $N_D = n_e$:

$$\Sigma_{ei}(k) = -\frac{2}{3}E_F - \frac{E_D}{N_c}\sqrt{n_e a_0^3}\,\frac{3\pi^3}{(12\pi)^{3/2}}\frac{1}{1+(2k/Q)^2} \tag{10.96}$$

where E_D is the binding energy of an electron bound to a single donor in an empty band. This self-energy is evaluated at $k=0$ for E_1, and at $k=k_F$ for the other energy gaps. The first term $-2E_F/3$ cancels out much of the Burstein shift.

The above formula is correct for the case that there are a few impurities in a metallic environment. The donor-doped semiconductor is a special system in that the number of donor impurities equals the number of conduction electrons. For this system, we have been double counting electrons in the conduction band. First, we state they form a uniform electron gas with a Fermi wave vector k_F. Next we state they are screening the electron–donor interaction, with one electron of screening charge for each donor. That means they spend most of their time in the vicinity of the donors, and do not form a uniform electron gas.

A better model assumes that the electrons spend all of their time screening the donors. Use the Wigner-Seitz method to compute the energy of the electron. The electron density n_e from the potential function (10.92) is

$$\nabla^2 V = \frac{e^2}{\varepsilon(0)} n_e(r) \tag{10.97}$$

$$n_e(r) = \frac{Q^2}{4\pi r} e^{-Qr} \tag{10.98}$$

The ground-state energy is the summation of the electron–donor energy E_{eD} and electron–electron energy E_{ee}:

$$E_{eD} = -\frac{e^2}{4\pi\varepsilon(0)}\int \frac{d^3r}{r} n_e(r) = \frac{e^2 Q}{4\pi\varepsilon(0)} \tag{10.99}$$

$$E_{ee} = \frac{e^2}{8\pi\varepsilon(0)}\int d^3r_1 \int d^3r_2 \frac{n_e(r_1)\,n_e(r_2)}{|\mathbf{r}_1 - \mathbf{r}_2|} = \frac{e^2 Q}{16\pi\varepsilon(0)} \tag{10.100}$$

$$E_T = E_{eD} + E_{ee} = -\frac{3e^2 Q}{16\pi\varepsilon(0)} \tag{10.101}$$

This formula is the total potential energy in the system of conduction electrons and donors. It is converted to a self-energy for a single electron by taking the functional derivative with respect to the number of electrons. Since the Thomas-Fermi screening wave vector $Q \propto n_e^{1/6}$, then

$$\Sigma_{ei} = -\frac{e^2 Q}{32\pi\varepsilon(0)} \tag{10.102}$$

This formula is considered more accurate than eqn. (10.95) and will be used in the numerical example.

3. The exchange part of the electron–electron interactions causes the bottom of the conduction band to lower in energy. This effect lowers the value of the conduction-band energy E_{c0}, and lowers energy gap. The exchange self-energy for an electron $\Sigma_X(k)$ depends on wave vector. At the bottom of the band it is

$$\Sigma_X(0) = -\frac{e^2 k_F}{2\pi^2 \varepsilon(0)} = -\frac{e^2}{2\pi^2 \varepsilon(0)}\left(\frac{3\pi^2 n_e}{N_c}\right)^{1/3} \tag{10.103}$$

This self-energy contributes to E_1. The exchange energy at the chemical potential is

$$\Sigma_X(k_F) = -\frac{e^2 k_F}{4\pi^2 \varepsilon(0)} = -\frac{e^2}{4\pi^2 \varepsilon(0)}\left(\frac{3\pi^2 n_e}{N_c}\right)^{1/3} \tag{10.104}$$

This contributes to E_2 and E_3.

4. The holes in the valence band also interact with the donor impurities. This changes the energy E_{v0} of the valence band maxima. For example, the hole–donor interaction is repulsive. The hole interaction with the donors due to the screened interaction is

$$\Sigma_{hi}(k) = \frac{2}{3}E_F - \frac{E_A}{N_c}\sqrt{n_e a_0^3}\,\frac{3\pi^3}{(12\pi)^{3/2}}\,\frac{1}{1+(2k/Q)^2} \tag{10.105}$$

where E_A is the acceptor binding energy that is calculated using the hole effective mass m_v. The quantities E_F, Q, and N_c are the Fermi energy, the screening wave vector, and the number of valleys of the electrons in the conduction band. These two contributions tend to cancel, and are usually omitted.

The terms that contribute to the first two energy gaps are

$$E_1 = -\frac{e^2 k_F}{2\pi^2 \varepsilon(0)} - \frac{e^2 Q}{32\pi\varepsilon(0)} \tag{10.106}$$

$$E_2 = \frac{\hbar^2 k_F^2}{2m} - \frac{e^2 k_F}{4\pi^2 \varepsilon(0)} - \frac{e^2 Q}{32\pi\varepsilon(0)} \tag{10.107}$$

E_1 has negative values, which means this band gap gets smaller with increased donor concentration. It is useful to write the energy in dimensionless form, using the donor binding energy E_D as the reference. Also let $\xi = n_e^{1/3} a_0$, where a_0 is the Bohr radius of the donor:

$$E_1 = -E_D\left[\frac{4}{\pi}(k_F a_0) + \frac{Q a_0}{4}\right] \tag{10.108}$$

$$E_2 = E_D\left[(k_F a_0)^2 - \frac{2}{\pi}(k_F a_0) - \frac{Q a_0}{4}\right] \tag{10.109}$$

$$(k_F a_0) = \left(\frac{3\pi^2}{N_c}\right)^{1/3}\xi, \quad Q a_0 = 2\left(\frac{3}{\pi}\right)^{1/6}(N_c)^{1/3}\sqrt{\xi} \tag{10.110}$$

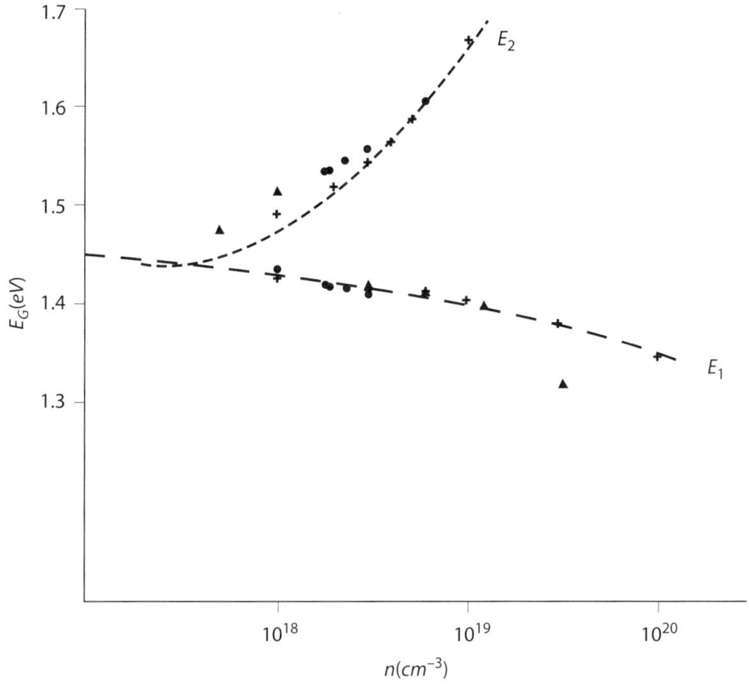

FIGURE 10.16. Band gap energies as a function of donor impurities in GaAs. Line is theory. Points are experimental. From Yao and Compaan [20] and Borghs et al. [2].

In the metallic state, the values of $\xi > \xi_c = 0.24$ start at the Mott criteria. At this value, E_2 is negative, as the potential energy terms are larger than the kinetic energy. As the donor density is increased and ξ increases, the kinetic energy term becomes large and E_2 becomes positive. Figure 10.16 shows the above theory, as the solid line, compared with experimental points as measured optically in absorption and emission. The agreement is quite good.

References

1. P. W. Anderson, The absence of diffusion in certain random lattices. *Phys. Rev.* **109**, 1492–1504 (1958)
2. G. Borghs, K. Bhattacharyya, K. Deneffe, P. Van Mieghem, and R. Mertens, Band-gap narrowing in highly doped *n*-type and *p*-type GaAs studied by photoluminescence spectroscopy. *J. Appl. Phys.* **66**, 4381–4386 (1989)
3. D.A.G. Bruggeman, Calculation of different physical constants of heterogeneous substances. *Ann. Phys.* (*Leipzig*) **24**, 636 (1935)
4. P. Dai, Y. Zhang, and M.P. Sarachik, Critical conductivity exponent for Si:B. *Phys. Rev. Lett.* **66**, 1914–1917 (1991)
5. P. P. Edwards and M. J. Sienko, Universality aspects of the metal–nonmetal transition in condensed media. *Phys. Rev. B* **17**, 2575–2581 (1978)
6. H. L. Frisch, J. M. Hammersley, and D.J.A. Welsh, Monte Carlo estimates of percolation probabilities for various lattices. *Phys. Rev.* **126**, 949–951 (1962)

7. A. Ghosh, S. Bhattacharya, D. P. Bhattacharya, and A. Ghosh, Hopping conduction in zinc vanadate semiconducting glasses. *J. Appl. Phys.* **103**, 083703 (2008)
8. S. M. Girvin and M. Jonson, Dynamical electron–phonon interaction and conductivity in strongly disordered metal alloys. *Phys. Rev. B* **22**, 3583–3597 (1980)
9. S. Kirkpatrick, Percolation and conduction. *Rev. Mod. Phys.* **45**, 574–588 (1973)
10. S. V. Kravchenko, W. E. Mason, G. E. Bowker, J. E. Furneaux, V. M. Pudalov, and M. D'lorio, Scaling of an anomalous metal–insulator transition in a two-dimensional system in silicon at $B = 0$. *Phys. Rev. B* **51**, 7038–7045 (1995)
11. S. V. Kravchenko and M. P. Sarachik, Metal–insulator transition in 2D electron systems. *Rep. Prog. Phys.* **67**, 1–44 (2004)
12. H. Landolt and R. Börnstein, New Series, Group III, Vol. 17a (Springer-Verlag, New York, 1982)
13. N. F. Mott, On the transition to metallic conduction in semiconductors. *Can. J. Phys.* **34**, 1356–1368 (1956)
14. N. F. Mott, *Metal–Insulator Transitions*, 2nd. ed. (Taylor & Francis, London, 1990)
15. T. H. Ning and C. T. Sah, Multivalley effective-mass approximation for donor states in silicon. *Phys. Rev. B* **4**, 3468–3481 (1971)
16. G. L. Pearson and J. Bardeen, Electrical properties of pure silicon and silicon alloys containing boron and phosphorus. *Phys. Rev.* **75**, 865–883 (1949)
17. H. Scher and R. Zallen, Critical density in percolation processes. *J. Chem. Phys.* **53**, 3759–3761 (1970)
18. G. A. Thomas, Asymmetry in the metal–ammonia phase diagram. *J. Phys. Chem.* **88**, 3749–3751 (1984)
19. G. D. Watkins, Intrinsic defects in silicon. *Mater. Sci. Semicond. Process.* **3**, 227–235 (2000)
20. H. Yao and A. Compaan, Plasmons, photoluminescence, and band-gap narrowing in very heavily doped *n*-GaAs. *Appl. Phys. Lett.* **57**, 147–149 (1990)

Homework

1. Neglecting spin–orbit splitting of the valence band, the $k \cdot p$ theory gave for a band dispersion

$$[E_{c0} - E][E_{v0} - E] = \hbar^2 w^2 k^2 \tag{10.111}$$

In the presense of a donor impurity, in the effective mass approximation, the equation for the bound-state energy is the solution to the equation

$$0 = \left\{ \hbar^2 w^2 \nabla^2 + \left[E_{c0} - E - \frac{e^2}{\varepsilon r} \right] \left[E_{v0} - E - \frac{e^2}{\varepsilon r} \right] \right\} \psi(\mathbf{r}) \tag{10.112}$$

This equation has a strong formal resemblance to the Klein-Gordon equation for the hydrogen atom, and can be solved the same way. Show that the eigenvalues are

$$E_n = \bar{E} + \frac{E_G/2}{\sqrt{1 + s^2/n^2}}, \quad \bar{E} = \frac{1}{2}(E_{c0} + E_{v0}) \tag{10.113}$$

and define s and n.

2. Derive an expression for the chemical potential $\mu(T)$ for a semiconductor that does not have any donors, acceptors, or other defects.

3. Derive an equation for the density of holes p in a semiconductor, in terms of the density of acceptors $[N_A]$ and other parameters, assuming the acceptors are the only important impurity.

4. Figure 10.17 shows a double Schottky barrier at the grain boundary between two identical grains of the same n-type semiconductor. The donor density is n_D, the dielectric constant is ε_0, and the barrier height is E_B.

   (a) How much surface charge density σ is stored in the surface states?

   (b) An applied voltage V_a is put between the two grains. The right side goes into forward bias V_R, while the left side goes into reverse bias V_L, where $V_a = V_R + V_L$. Derive formulas for $V_{L,R}(V_a)$ assuming that σ does not change. What is the flat band voltage?

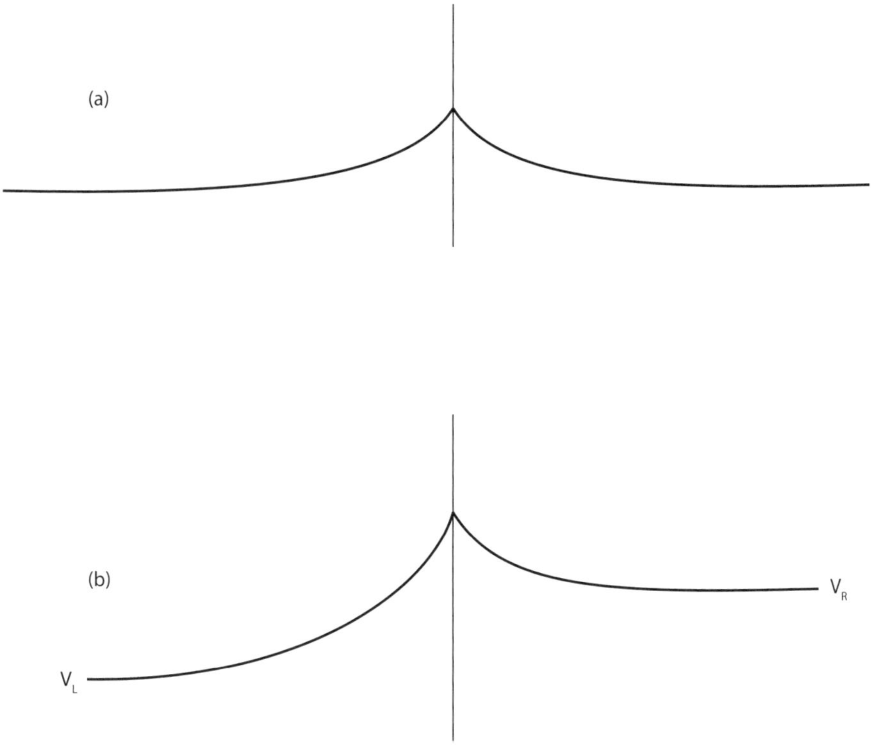

FIGURE 10.17. Double Schottky barrier with (a) zero bias and (b) positive bias.

5. Use CPA theory to derive the formula for σ_m in percolation theory for the hc lattice in 2D. Align the electric field, and current flow, to be in the zigzag direction.

6. Use CPA theory to derive the formula for σ_m in percolation theory for the pt lattice in 2D. Align the electric field and current flow to be in the straight line direction.

11 Transport Phenomena

Solids transport electricity, heat, and light. This chapter describes the transport of electricity and heat. The transport of electricity is described by the current density $\mathbf{J}(\mathbf{r},t)$, which has units of amperes per area. The transport of heat is described by the heat current $\mathbf{J}_Q(\mathbf{r},t)$, which has the units of watts per area. The theory is based on the Boltzmann equation, which is derived in the next section. The transport of light is not discussed in this chapter. However, light diffuses in disordered materials, and that process is also described by a Boltzmann equation for photons.

11.1 Introduction

Students of physics are taught in quantum mechanics that the current is given by

$$\mathbf{j}(\mathbf{r},t) = \frac{e\hbar}{2mi}\left[\psi^{*}\nabla\psi - \psi\nabla\psi^{*}\right] \tag{11.1}$$

The word "current" has two possible definitions. The above formula has a factor of the charge e in front, so that its units are amperes per area, suitable for electrical current. However, current can also be defined for neutral particles. Use the above formula without the e and the units are particles per area per unit time. This latter formula is used in fluid flow.

This formula suggests that the current can be found from the wave function. That is not true for normal systems. The flow of current is a nonequilibrium state, which cannot be defined by a wave function $\psi(\mathbf{r},t)$. A wave function for a plane wave is written in terms of operators:

$$\psi(\mathbf{r},t) = \frac{1}{\sqrt{\Omega}}\sum_{\mathbf{p}\sigma} C_{\mathbf{p}\sigma}e^{i\mathbf{p}\cdot\mathbf{r}} \tag{11.2}$$

$$\mathbf{j}(\mathbf{r}, t) = \sum_{\mathbf{p}\sigma} \frac{e\hbar\mathbf{p}}{m\Omega} C^{\dagger}_{\mathbf{p}\sigma} C_{\mathbf{p}\sigma} \tag{11.3}$$

where Ω is the volume of the system. If one replaces the operators by their thermodynamic average, then

$$\langle C^{\dagger}_{\mathbf{p}\sigma} C_{\mathbf{p}\sigma} \rangle = n_F(\xi_p) = \frac{1}{e^{\beta\xi} + 1} \tag{11.4}$$

$$\mathbf{j}(\mathbf{r}, t) = \sum_{\mathbf{p}\sigma} \frac{e\hbar\mathbf{p}}{m\Omega} n_F(\xi_p) = 0 \tag{11.5}$$

Now the current is zero, since the angular average of the vector $\mathbf{p}$ is zero. This result is correct, since there are no currents in equilibrium.

When currents are flowing, the average $\langle C^{\dagger}_{\mathbf{p}\sigma} C_{\mathbf{p}\sigma} \rangle$ is no longer given by its equilibrium value, and one can get a nonzero result for the integral. The method of finding this function uses the Boltzmann equation, as discussed below.

If the current flows in response to a static electric field $\mathbf{E}$, then in a cubic crystal the current is given by

$$\mathbf{j} = \sigma \mathbf{E}, \quad \sigma = \frac{n_e e^2 \tau_t}{m} \tag{11.6}$$

where τ_t is the lifetime of the electrons as found by solving the Boltzmann equation. The lifetime of a particle is not part of $\psi(\mathbf{r}, t)$, which also suggests that the quantum formula is not sufficient for finding the current. The electrical conductivity σ has the units of siemens per meter.

All of the above statements do not apply to superfluids such as superconductors or superfluid ^{4}He. If one applies a static magnetic field $\mathbf{B}$ to a superconductor, the Meissner effect occurs when the superconductor creates currents to reduce the magnetic field inside of the material. These currents are part of the ground-state wavefunction of the superconductor in a magnetic field, and one can indeed evaluate them using eqn. (11.1). The wave function of the superfluid is usually found using Ginzburg-Landau theory. Superfluids are very special cases, since all particles are in a coherent state, which is why their study is so interesting.

11.2 Drude Theory

Drude derived a classical formula for the motion of electrons in solids. It predated quantum mechanics and is based on Newton's second law of motion. If r is the position of an electron of velocity $\mathbf{v}$, then

$$m^* \frac{d}{dt} \mathbf{v} = F - \frac{m^* \mathbf{v}}{\tau} \tag{11.7}$$

$$\mathbf{F} = e[\mathbf{E} + \mathbf{v} \times \mathbf{B}] \tag{11.8}$$

where $\mathbf{F}$ is the force on the electron from electric ($\mathbf{E}$) and magnetic ($\mathbf{B}$) fields, and m^* is the effective mass. The last term in eqn. (11.7) introduces a phenomenological relaxation time τ, which accounts for the damping of the electron. At first one thinks this formula cannot be useful, since quantum mechanics is required to describe the motion of electrons. Also, in a metal, the electrons have a wide range of velocities within the occupied Fermi sea. The above application of Newton's law seems too simple to be accurate. In fact, it works quite well for many situations. Several examples of this formula are discussed.

1. Assume there is a constant dc electric field $\mathbf{E} = \hat{x}E_0$. There is no time dependence and no magnetic field, so the solution is

$$v_x = \mu E_0, \quad \mu = \frac{e\tau}{m^*} \tag{11.9}$$

The quantity μ is the *mobility* of the electron. It gives the average velocity of an electron in response to the electric field.

2. Assume the above electric field is suddenly terminated, say at $t = t_0$. then for $t > t_0$ the solution is

$$v_x = \mu E_0 e^{-(t-t_0)/\tau} \tag{11.10}$$

The relaxation time gives the decay of the average velocity of the electrons.

3. Assume there is an ac electric field $\mathbf{E} = E_0 \exp(-i\omega t)$ acting on the electron. Then the time derivative gives $\dot{\mathbf{v}} = -i\omega\mathbf{v}$ and the above equation has a solution

$$m^*\mathbf{v}\left(-i\omega + \frac{1}{\tau}\right) = e\mathbf{E} \tag{11.11}$$

$$\mathbf{v}(t) = \frac{e\tau}{m^*}\mathbf{E}(t)\frac{1}{1 - i\omega\tau} \tag{11.12}$$

The electrical current is $\mathbf{J} = n_e e\mathbf{v}$ and the electrical conductivity as a function of frequency is

$$\sigma(\omega) = \frac{\sigma_0}{1 - i\omega\tau}, \quad \sigma_0 = \frac{n_e e^2\tau}{m^*} \tag{11.13}$$

This form for the conductivity is called the *Drude formula*. Since it is based on a classical theory, it should appear to be invalid for electrons in metals. In fact, the formula is quite accurate and is a good description of many simple metals. It describes the infrared response of metals very well.

One reason the Drude formula is accurate is that it can also be derived from the Boltzmann transport equation (11.22). This derivation is in a later section. Earlier the Drude theory was used to discuss the quantum Hall effect.

11.3 Bloch Oscillations

The acceleration of an electron in a crystal is not given by the derivative of the velocity. The exact formula is

$$\hbar \frac{d}{dt}\mathbf{k} = \mathbf{F} \tag{11.14}$$

The time derivative of the momentum $\mathbf{p} = \hbar\mathbf{k}$ equals the force. For nearly free electrons, with $\hbar\mathbf{k} = m\mathbf{v}$, this formula is the same as Drude's formula. However, when the energy band is given by the tight-binding model, the predictions are quite different.

Consider the motion of a single electron in a conduction band. Keep the discussion simple by having it be a tight-binding band in one dimension:

$$\varepsilon(k) = w[1 - \cos(ka)] \tag{11.15}$$

The motion of the electron is determined by

$$\hbar\dot{k} = F = eE \tag{11.16}$$

$$v(t) = \frac{1}{\hbar}\frac{d\varepsilon(k)}{dk} = \frac{wa}{\hbar}\sin(ka) \tag{11.17}$$

where E is the electric field along the direction of motion. If the particle does not scatter and the electric field is a constant, the wave vector increases as

$$k(t) = \frac{eEt}{\hbar} \tag{11.18}$$

$$v(t) = \frac{wa}{\hbar}\sin(\omega t), \quad \omega = \frac{eEa}{\hbar} \tag{11.19}$$

$$x(t) = \int_0^t dt' v(t') = \frac{wa}{\hbar\omega}[1 - \cos(\omega t)] \tag{11.20}$$

The motion is more complicated than suggested by these simple equations. The value of $k(t)$ is limited to stay within the first BZ. After reaching the value $k = \pi/a$ at the edge of the Brillouin zone, the electron Bragg reflects by the reciprocal latttice vector $G = -2\pi/a$. The electron finds itself at the wave vector $k = -\pi/a$ at the other side of the Brillouin zone. Then it continues increasing as before.

The above equations show that the particle motion $x(t)$ is periodic in real space. These periodic oscillations are called the *Bloch oscillations*. They are rarely observed, since most particles have a finite mean free path. After gaining energy from the electric field, the particle loses some of that energy by emitting a phonon. While emitting the phonon, the particle's wave vector is returned to a small value, so the value of k never reaches the edge of the Brillouin zone. The periodic motion is now different: the particle accelerates in the field, emits a phonon, accelerates in the field, emits a phonon, etc.

The Bloch oscillation can be observed only if $\omega\tau > 1$, $\omega = eEa/\hbar$. This can be achieved by having a laser pulse generate a strong electric field over a short period of time.

The best observation of Bloch oscillations was in optical lattices by Ben Dahan et al. [7] Since the atoms are far apart, the lifetime of the atoms is quite long, and Bloch oscillations of atoms was observed. The three curves in fig. 11.1 represent three different laser intensities, which give different depths of the optical potential. The constant forces on the cesium atoms were generated using beating of different laser beams. Gravitation forces could also be used.

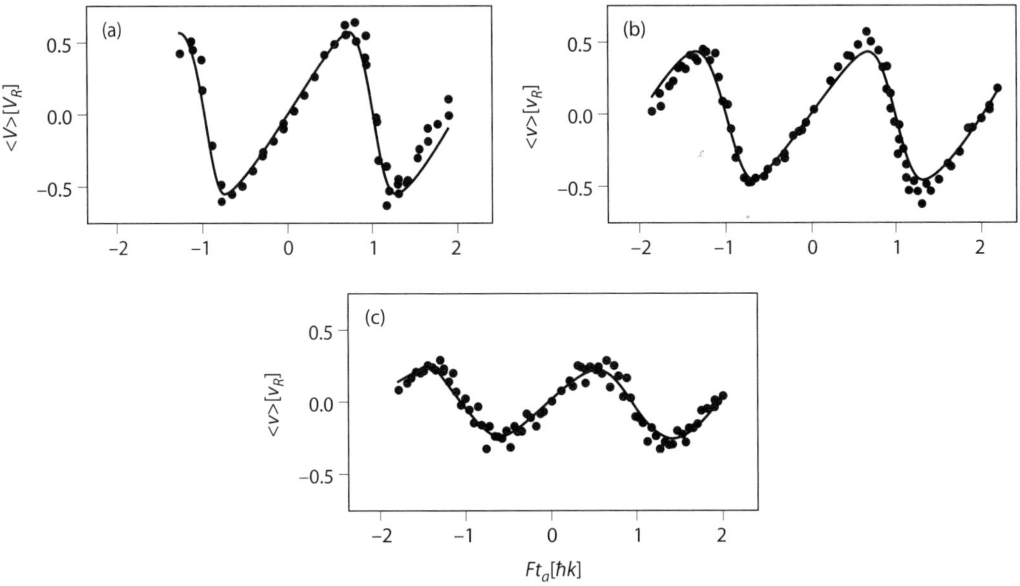

FIGURE 11.1. Bloch oscillations of cesium atoms. From Ben Dahan et al., *Phys. Rev. Lett.* **76**, 4508 (1996). Used by permission of the American Physical Society.

If the band is full of electrons, as are valence bands, the electron cannot emit a phonon, since there is no empty final state. Since the change in wave vector in eqn. (11.18) is independent of wave vector, all electrons change their wave vectors together: they all flow along the band at the same speed. When getting to the band edge, they Bragg reflect to $-\pi/a$, and keep marching forward as before. Half of the particles are going to the left and half to the right, so there is no net current. Another physical picture for the same system of a filled band is that the electrons do not move at all: they cannot change wave vector since all other states are filled. This alternate interpretation gives the same physical result: there is no current. Which of these two pictures is correct? Both interpretations have their advocates.

11.4 Boltzmann Equation

This section summarizes the derivation of the Boltzmann equation as applied to transport in solids. Start with the distribution function $f(\mathbf{r}, \mathbf{k}, t)$, which is the probability that an electron has wave vector $\mathbf{k}$ at time t at position $\mathbf{r}$. The distribution is a semiclassical concept since position and momentum $\mathbf{p} = \hbar\mathbf{k}/m$ are specified at the same time. For a noninteracting system, the distribution is that of Fermi-Dirac:

$$f^{(0)}(\mathbf{r}, \mathbf{k}, t) = \frac{1}{e^{\beta\xi(k)} + 1}, \quad \xi(k) = \varepsilon(k) - \mu \tag{11.21}$$

For transport add a perturbation to the system, such as an electric field $\mathbf{E}$ or a temperature gradient ∇T, and determine how the distribution is changed away from equilibrium.

The time variation of $f(\mathbf{r}, \mathbf{k}, t)$ is found hydrodynamically by letting each variable change slightly with time. The time change of $\mathbf{r}$ is $\mathbf{v}$, while the time change of a wave vector is the force on the particle $\hbar\dot{\mathbf{k}} = \mathbf{F}$:

$$0 = \frac{\partial f}{\partial t} + \nabla_r f \cdot \mathbf{v} + \frac{1}{\hbar}\nabla_k f \cdot [e\mathbf{E} + \mathbf{v}\times\mathbf{B}] + \left(\frac{\partial f}{\partial t}\right)_S \tag{11.22}$$

where the last term is from scattering. In steady-state current flow, the total time variation should be zero, which is why the total is zero. The first term is neglected since steady current is assumed.

Now write that $f = f^{(0)} + \delta f$, where δf is the change in the distribution due to the electric fields or temperature gradients. Retain in eqn. (11.22) terms that are first order in the perturbation: they depend on either δf, $\mathbf{E}$, or else ∇T. The evaluation of the derivatives $\nabla_r f$ and $\nabla_k f$ requires evaluating the derivative of $f^{(0)}$. The second and third terms contain

$$\nabla_r f = \nabla\left(\frac{1}{e^{\beta(r)\xi} + 1}\right) = \frac{\xi}{T}\nabla T\left(-\frac{df^{(0)}}{d\xi}\right) \tag{11.23}$$

$$\nabla_k f = \hbar\mathbf{v}_k \frac{df^{(0)}}{d\xi} \tag{11.24}$$

The magnetic field term drops out since it is part of $\mathbf{v}\cdot(\mathbf{v}\times\mathbf{B}) = 0$. Write the remaining terms as

$$\left(\frac{\partial f}{\partial t}\right)_S = \mathbf{v}_k \cdot \boldsymbol{\Lambda} \tag{11.25}$$

$$\boldsymbol{\Lambda} = \left[e\mathbf{E} - \frac{\xi_k}{T}\nabla T\right]\left(-\frac{df^{(0)}}{d\xi}\right) \tag{11.26}$$

The distribution function for electrons is written as

$$f(\mathbf{k}) = f^{(0)}(\xi_k) + \delta f(\mathbf{k}) = f^{(0)}(\xi_k) + C(\xi_k)\mathbf{v}_k \cdot \boldsymbol{\Lambda} \tag{11.27}$$

The change in the distribution function δf is also proportional to $\mathbf{v}_k \cdot \boldsymbol{\Lambda}$, and multiplied by an unknown function $C(\xi_k)$ that needs to be determined.

The scattering term is also determined by the golden rule of quantum mechanics. Consider first the case of an electron scattering from random impurities of density n_i. The exact scattering is described by a T-matrix:

$$\left(\frac{\partial f}{\partial t}\right)_S = \frac{2\pi}{\hbar} n_i \int \frac{d^3k'}{(2\pi)^3} |T_{kk'}|^2 \delta(\xi_k - \xi_{k'})[f(\mathbf{k}) - f(\mathbf{k}')] \tag{11.28}$$

$$= \frac{2\pi}{\hbar} n_i \int \frac{d^3k'}{(2\pi)^3} |T_{kk'}|^2 \delta(\xi_k - \xi_{k'})[\delta f(\mathbf{k}) - \delta f(\mathbf{k}')]$$

The last bracket contains the difference of the two distribution functions. The equilibrium terms cancel whenever $\xi_k = \xi_{k'}$, as required by the delta function for energy. This delta function also forces $|\mathbf{k}| = |\mathbf{k}'|$, so the two velocities $\mathbf{v}_k$ and $\mathbf{v}_{k'}$ differ only in their direction.

The angular variables must be treated carefully. Define the $\hat{z}$ direction as $\hat{k}$. The angles (θ, θ') are defined by

$$\cos(\theta') = \hat{k} \cdot \hat{k}' \tag{11.29}$$

$$\cos(\theta) = \hat{k} \cdot \hat{\Lambda} \tag{11.30}$$

$$\hat{\Lambda} \cdot \hat{k}' = \cos(\theta)\cos(\theta') + \sin(\theta)\sin(\theta')\cos(\phi) \tag{11.31}$$

The latter identity is the law of cosines. The azimuthal angle ϕ does not appear anywhere else in the integrand. The *T*-matrix $T_{kk'}$ depends only on θ'. So when integrating over $d\phi$, the term in $\cos(\phi)$ averages to zero. What is left is

$$\begin{aligned}\int_0^{2\pi} \frac{d\phi}{2\pi} [\delta f(\mathbf{k}) - \delta f(\mathbf{k}')] &= C(\xi_k)\Lambda(\xi_k)\, v_k \cos(\theta)\,[1 - \cos(\theta')] \\ &= C(\xi_k)\Lambda(\xi_k) \cdot \mathbf{v}_k [1 - \cos(\theta')] \\ &= [f(\mathbf{k}) - f^{(0)}(\xi_k)][1 - \cos(\theta')]\end{aligned} \tag{11.32}$$

The factors in front of the bracket $[1 - \cos(\theta')]$ can be removed from the integrand. The remaining parts of the integral define the *transport lifetime* τ_t:

$$\left(\frac{\partial f}{\partial t}\right)_S = \frac{f - f^{(0)}}{\tau_t(\xi_k)} \tag{11.33}$$

$$\frac{1}{\tau_t(\xi_k)} = \frac{2\pi}{\hbar} n_i \int \frac{d^3k'}{(2\pi)^3} |T_{kk'}|^2 \delta(\xi_k - \xi_{k'})\,[1 - \cos(\theta')] \tag{11.34}$$

$$\frac{1}{\tau(\xi_k)} = \frac{2\pi}{\hbar} n_i \int \frac{d^3k'}{(2\pi)^3} |T_{kk'}|^2 \delta(\xi_k - \xi_{k'}) \tag{11.35}$$

The usual lifetime τ is the average time between scattering events. The transport lifetime τ_t has an additional factor of $[1 - \hat{k} \cdot \hat{k}']$ that measures how much one scatters. Large-angle scattering events make this factor larger and contribute more to the resistivity. Combine eqns. (11.25) and (11.33) to obtain

$$\frac{f - f^{(0)}}{\tau_t(\xi_k)} = \mathbf{v}_k \cdot \mathbf{\Lambda} \tag{11.36}$$

$$\delta f = \tau_t(\xi_k)\mathbf{v}_k \cdot \mathbf{\Lambda}, \quad C(\xi_k) = \tau_t(\xi_k) \tag{11.37}$$

The latter identity gives the change in the distribution function of electrons due to the application of a small electric field or a small temperature gradient.

The first term in the Boltzmann equation contains the time derivative. The equilibrium part of $f(\mathbf{k}, t)$ does not depend on time. If the driving term Λ oscillates with a frequency ω, the first term gives

$$\frac{\partial f}{\partial t} = \frac{\partial(\delta f)}{\partial t} = -i\omega\delta f \tag{11.38}$$

Now write the Boltzmann equation as

$$0 = -i\omega\delta f - \mathbf{v}\cdot\mathbf{\Lambda} + \left(\frac{\partial f}{\partial t}\right)_S \tag{11.39}$$

From eqn. (11.33),

$$\left(\frac{\partial f}{\partial t}\right)_S = \frac{\delta f}{\tau_t} \tag{11.40}$$

Combining the above two equations gives

$$\delta f = \frac{\tau_t \mathbf{v}\cdot\mathbf{\Lambda}}{1 - i\omega\tau_t} \tag{11.41}$$

The ac conductivity is the Drude result, eqn. (11.13). The Drude formula for the frequency dependence of the electrical conductivity can be derived from the Boltzmann equation. That puts the Drude formula on a much firmer theoretical basis, and explains why it is valid even in quantum systems.

11.5 Currents

The electrical current density $\mathbf{J}$ (units: amperes/area) and heat current density $\mathbf{J}_Q$ (units: watts/area) are defined as

$$\mathbf{J} = 2e\int \frac{d^3k}{(2\pi)^3}\mathbf{v}_k f(\mathbf{k}) \tag{11.42}$$

$$\mathbf{J}_Q = 2\int \frac{d^3k}{(2\pi)^3}\mathbf{v}_k \xi_k f(\mathbf{k}) \tag{11.43}$$

The factor of two is for spin degeneracy. A paramagnetic arrangement is assumed, with equal numbers of up and down spins. The electrical current $\mathbf{J}$ is defined in an obvious way as the density of particles times their velocity, times the charge e. The thermal current $\mathbf{J}_Q$ has an additional factor of $\xi_k = \varepsilon_k - \mu$, where ε_k is the energy of the particle in the band. It does not have a factor of e. Heat currents are evaluated with respect to the chemical potential μ. If $\xi_k > 0$ (so that $\varepsilon_k > \mu$) the particle is said to be "hot," and to carry excess energy. If $\xi_k < 0$ (so that $\varepsilon_k < \mu$) the particle is "cold," and carries less than its share of heat. Interaction terms, such as electron–electron or electron–phonon, may add addditional terms to the heat current.

11.5.1 *Transport Coefficients*

The two integrals (11.42, 11.43) are zero in equilibrium $f = f^{(0)}$, since equilibrium is defined as the state with no currents flowing! The currents are provided by δf:

$$\mathbf{J} = 2e\int \frac{d^3k}{(2\pi)^3}\mathbf{v}_k \delta f(\mathbf{k}) \tag{11.44}$$

$$\mathbf{J}_Q = 2\int \frac{d^3k}{(2\pi)^3} \mathbf{v}_k \xi_k \delta f(\mathbf{k}) \tag{11.45}$$

Using the expression (11.37) gives

$$\mathbf{J} = 2e\int \frac{d^3k}{(2\pi)^3} \tau_t(k)\, \mathbf{v}_k \mathbf{v}_k \cdot \left[e\mathbf{E} - \frac{\xi_k}{T}\nabla T\right]\left(-\frac{df^{(0)}}{d\xi_k}\right) \tag{11.46}$$

$$\mathbf{J}_Q = 2\int \frac{d^3k}{(2\pi)^3} \xi_k(\mathbf{k})\, \tau_t(k)\, \mathbf{v}_k \mathbf{v}_k \cdot \left[e\mathbf{E} - \frac{\xi_k}{T}\nabla T\right]\left(-\frac{df^{(0)}}{d\xi_k}\right) \tag{11.47}$$

The brackets contain the factor $\mathbf{\Lambda}$. In cubic crystals, symmetry dictates that the currents are in the direction of $\hat{\Lambda}$. Then the angular integrals are $(\mathbf{v}_k \cdot \hat{\Lambda})^2 = v_k^2/3$. The integrals for the two currents are evaluated in terms of three coefficients: the electrical conductivity σ, the Seebeck coefficient S, and the electron part of the thermal conductivity K_e':

$$\mathbf{J} = \sigma[\mathbf{E} - S\nabla T] \tag{11.48}$$

$$\mathbf{J}_Q = \sigma S T \mathbf{E} - K_e' \nabla T \tag{11.49}$$

$$\sigma = \frac{2e^2}{3}\int \frac{d^3k}{(2\pi)^3} \tau_t(k)\, v_k^2 \left(-\frac{df^{(0)}}{d\xi_k}\right) \tag{11.50}$$

$$\sigma S = \frac{2e}{3T}\int \frac{d^3k}{(2\pi)^3} \tau_t(k)\, \xi_k v_k^2 \left(-\frac{df^{(0)}}{d\xi_k}\right) \tag{11.51}$$

$$K_e' = \frac{2}{3T}\int \frac{d^3k}{(2\pi)^3} \tau_t(k)\, \xi_k^2 v_k^2 \left(-\frac{df^{(0)}}{d\xi_k}\right) \tag{11.52}$$

The three coefficients differ in their powers of ξ_k^n in the integrand and the powers of charge e in the prefactor. The second integral is actually σS. Move the conductivity σ to the right-hand side, and the Seebeck S is the ratio of two integrals.

The electrical current is given by eqn. (11.48), where S is the Seebeck coefficient. Consider a bar of material of length L. Assume one end is hotter, so there is a temperature gradient $\nabla T = \Delta T/L$ between the ends, but it is insulated so no current flows $\mathbf{J} = 0$. The above formula shows there is an electric field E and voltage $\Delta V/L = -E$ between the ends. Cancel the length and find

$$S = -\frac{\Delta V}{\Delta T} \tag{11.53}$$

The temperature gradient induces a voltage difference. The Seebeck coefficient has the units of volts per degree.

The three integrals for (σ, S, K_e) are the starting point for many calculations. One starts with an interaction, such as the electron–phonon interaction, that scatters electrons. Use it to calculate the transport lifetime $\tau_t(k)$. Using the result for the lifetime, one can calculate the three transport coefficients.

Before leaving this topic, it is useful to rearrange eqn. (11.48) and put the result in the heat current equation:

$$\sigma \mathbf{E} = \mathbf{J} + \sigma S \nabla T \tag{11.54}$$

$$\mathbf{J}_Q = ST(\mathbf{J} + \sigma S \nabla T) - K' \nabla T = ST\mathbf{J} - K \nabla T \tag{11.55}$$

$$K = K' - \sigma T S^2 \tag{11.56}$$

Here the thermal conductivity includes both electron and phonon contributions. The thermal conductivity K' is the value measured with no electric field in the sample (no voltage difference), while K is that measured with no electrical current in the sample. These two values will be different in a material with a large value of $\sigma S^2 T$. Rewrite the above equation as

$$K = K'[1 - Z'T], \quad Z' = \frac{\sigma S^2}{K'} \tag{11.57}$$

$$Z = \frac{\sigma S^2}{K} = \frac{Z'}{1 - Z'T} \tag{11.58}$$

The quantity Z is called the *figure of merit* for thermoelectric materials. The quantity Z' has a similar form. The product $Z'T < 1$ is required by the fact that $K > 0$. There is no such requirement on ZT. In fact, materials are found at high temperatures ($T \sim 700$ K) that have $ZT \sim 2$.

Equation (11.55) shows that one can drive heat flow with an electrical current in a material with a large Seebeck coefficient. This phenomenon is the basis of thermoelectric refrigerators, where heat is driven from the cold to the hot side of the device by an electrical current.

11.5.2 *Metals*

The three transport coefficients are usually easy to evaluate in a metal. Two conditions must be met: (i) the density of states $N(\varepsilon)$ must be a smooth function of energy near to the chemical potential, on the energy scale of several $k_B T$; and (ii) the value of the chemical potential μ must be much larger than the thermal energy $k_B T$.

The derivation is based on several ideas.

1. The lifetime τ_t is a constant.

2. Having done the angular integrals previously, the integral over wave vector can be replaced by the density of states $N(\varepsilon)$:

$$N(\varepsilon) = \int \frac{d^3k}{(2\pi)^3} \delta[\varepsilon - E(\mathbf{k})] \tag{11.59}$$

$$\int \frac{d^3k}{(2\pi)^3} F(\xi) = \int N(\varepsilon) d\xi F(\xi), \quad \xi = \varepsilon - \mu \tag{11.60}$$

3. The density of states can be expanded in a Taylor series around the chemical potential: $\varepsilon = \mu + \xi$ so that

$$N(\varepsilon) = N(\mu) + \xi N(\mu)' + O(\xi^2) \tag{11.61}$$

where prime denotes derivative.

4. Since $\beta\mu \gg 1$, the integral of $d\xi$ can be extended to the range $-\infty < \xi < \infty$.
5. The derivative of the occupation function is exactly a symmetric function of ξ:

$$\left(-\frac{df^{(0)}(\xi)}{d\xi}\right) = \beta\frac{e^{\beta\xi}}{(e^{\beta\xi}+1)^2} = \frac{\beta}{(e^{-\beta\xi}+1)(e^{\beta\xi}+1)} \tag{11.62}$$

$$= \frac{\beta}{2[1+\cosh(\beta\xi)]}$$

6. The three important integrals are

$$\int_{-\infty}^{\infty} d\xi\left(-\frac{df^{(0)}(\xi)}{d\xi}\right) = 1 \tag{11.63}$$

$$\int_{-\infty}^{\infty} \xi d\xi\left(-\frac{df^{(0)}(\xi)}{d\xi}\right) = 0 \tag{11.64}$$

$$\int_{-\infty}^{\infty} \xi^2 d\xi\left(-\frac{df^{(0)}(\xi)}{d\xi}\right) = \frac{\pi^2}{3}(k_BT)^2 \tag{11.65}$$

The second is zero because the integrand is an odd function of ξ.

These approximations, along with the three integrals, allow the evaluation of all three transport coefficients in a metal. All of the integrands contain the additional factor of $v_k^2 = (2/m)(\mu + \xi)$. Also assume that the lifetime $\tau_t(\varepsilon)$ is a function of the energy. Then define $\Sigma(\varepsilon)=\tau_t(\varepsilon)\varepsilon$. Also assume this latter function can be expanded around the chemical potential $\Sigma(\mu+\xi) \approx \Sigma(\mu) + \xi\Sigma'(\mu)$.

1. The electrical conductivity is

$$\sigma = \frac{4e^2}{3m}\int d\xi[N(\mu) + \xi N'][\Sigma(\mu) + \xi\Sigma']\left(-\frac{df^{(0)}(\xi)}{d\xi}\right) \tag{11.66}$$

$$= \frac{4e^2}{3m}[\Sigma(\mu)N(\mu) + O(T^2)]$$

2. The Seebeck coefficient is

$$\sigma S = \frac{4e}{3mT}\int d\xi\xi[N(\mu) + \xi N'][\Sigma(\mu) + \xi\Sigma']\left(-\frac{df^{(0)}(\xi)}{d\xi}\right) \tag{11.67}$$

$$= \frac{4e}{3mT}\frac{\pi^2}{3}(k_BT)^2[N\Sigma' + \Sigma N']$$

$$= \frac{4e}{3m}(k_B^2T)\frac{d}{d\mu}[\Sigma(\mu)N(\mu)] \tag{11.68}$$

3. The electron part of the thermal conductivity is

$$K_e = \frac{4}{3mT}\int d\xi\xi^2[N(\mu) + \xi N'][\Sigma(\mu) + \xi\Sigma']\left(-\frac{df^{(0)}(\xi)}{d\xi}\right) \tag{11.69}$$

$$= \frac{4}{3mT}\frac{\pi^2}{3}(k_BT)^2[\Sigma(\mu)N(\mu)]$$

There are several important relationships among these expressions.

1. Divide the two results K_e/σ. Most of the factors cancel. Those that remain are

$$K_e = L_0\sigma T, \quad L_0 = \frac{\pi^2}{3}\left(\frac{k_B}{e}\right)^2 \tag{11.70}$$

The constant L_0 is called the *Lorenz number* and has the value of $L_0 = 2.45\times10^{-8}(\mathrm{V/K})^2$. This important expression is called *the Wiedemann-Franz law*. It states that the electron part of the thermal conductivity K_e is directly proportional to the electrical conductivity. This formula is widely used in practice. Most measurements of the quantity K_e, in fact, measure σ and use the Wiedemann-Franz law.

2. To get the Seebeck coefficient, one must divide the right-hand-side of eqn. (11.68) by the result for the electrical conductivity:

$$S = \frac{\pi^2}{3}\left(\frac{k_B}{e}\right)k_BT\frac{d}{d\mu}\ln[\Sigma(\mu)N(\mu)] \tag{11.71}$$

This important expression is called the *Mott formula* for the thermopower. It shows that the Seebeck coefficient in metals should be linear in temperature. Experiments show $S \sim T$ over wide ranges of temperature. It breaks down at low temperature because of phonon drag. The size of S in metals is then

$$S \sim \frac{k_B^2T}{e\mu} \sim 1\mu V/K \tag{11.72}$$

which is quite small.

The total thermal conductivity of a metal is the combination of phonon and electron terms. At high temperature they can be treated as independent: $K = K_p + K_e$. At very low temperature, the phonons follow the electrons, which leads to *phonon drag*. Then the two contributions are no longer independent. See Bailyn [4] for a full discussion of phonon drag.

Figure 11.2 shows the Seebeck coefficient as a function of temperature for Ag and Cu. Data for Au are similar. The result is linear at high temperature, in agreement with the Mott formula. The peaks at low temperature are due to phonon drag. The values are small. Thermoelectric devices are made from semiconductors that have values in the range of $S \sim 200 - 300\ \mu\mathrm{V/K}$.

3. The thermal conductivity can also be written as

$$K_e = D_tC_e \tag{11.73}$$

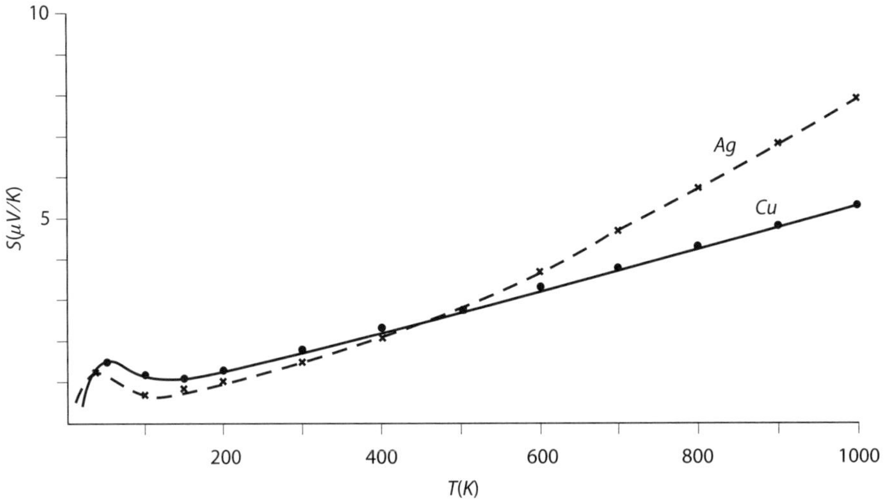

FIGURE 11.2. Seebeck coefficient of Ag and Cu as a function of temperature. The peaks at low temperature are due to phonon drag. The linear behavior at high temperature is an example of the Mott formula. Replotted data from Cusack and Kendall, *Proc. R. Soc. (London)* **72**, 898 (1958).

$$D_t = \frac{1}{3} v_F^2 \tau_t (k_F) \tag{11.74}$$

$$C_e = \frac{2\pi^2}{3} k_B^2 TN(\mu) \tag{11.75}$$

where D_T is a diffusion coefficient, and C is the heat capacity of a metal. The heat capacity was derived in chapter 5. The thermal conductivity is always equal to the diffusion coefficient multiplied by the heat capacity. Below are written (i) the equation of continuity for electrons, and (ii) the equation of continuity for energy:

$$0 = \frac{\partial \rho}{\partial t} + \nabla \cdot \mathbf{J} \tag{11.76}$$

$$W = \frac{\partial U}{\partial t} + \nabla \cdot \mathbf{J}_E \tag{11.77}$$

The units of the last equation are watts per unit volume. A typical source term W on the left could be the Joule heating due to an electrical current $W = \rho J^2$. The internal energy $U(T)$ is a function of temperature, so $\partial U/\partial t = (\partial U/\partial T)(\partial T/\partial t)$, and $\partial U/\partial T = C$, the heat capacity. Similarly, the energy current from heat flow, neglecting thermoelectric effects, is $\mathbf{J}_E = -K\nabla T$:

$$W = C\frac{\partial T}{\partial t} - \nabla\left[K(T)\nabla T\right] \tag{11.78}$$

Divide by C and get the equation for thermal diffusion:

$$\frac{W}{C} = \frac{\partial T}{\partial t} - \frac{1}{C}\nabla\left[K(T)\nabla T\right] \tag{11.79}$$

where K/C equals the diffusion coefficient. The diffusion coefficient is always written

$$D = \frac{1}{3}\bar{v}^2\bar{\tau} \tag{11.80}$$

where $\bar{v}$ is the average velocity, and $\bar{\tau}$ is the average lifetime. For electrons in a metal, the factors are the Fermi velocity, and lifetime at the Fermi surface. For a classical gas of free particles,

$$\frac{m}{2}\bar{v}^2 = \frac{3}{2}k_BT \tag{11.81}$$

$$D = \frac{k_BT\tau_t}{m} \tag{11.82}$$

This latter equation is the *Einstein relation* for the diffusion coefficient.

The average mean free path is $\bar{\ell} = \bar{v}\bar{\tau}$, so the diffusion coefficient is sometimes written as $D = \bar{v}\bar{\ell}/3$. All of the above relations are also valid in two dimensions, with the small change that the factor of $\frac{1}{3}$ becomes $\frac{1}{2}$.

11.5.3 Semiconductors and Insulators

Metals are rather loosely defined as materials with a chemical potential that is above the band minimum by several thermal energies: $\mu - E_{c0} \gg k_BT$. Insulators are materials with a low electrical conductivity because there are few carriers in the conduction band. The low conductivity occurs when the chemical potential is well below the energy band minimum:

$$\mu - E_{c0} \ll -k_BT \tag{11.83}$$

An intrinsic semiconductor is an insulator, with the chemical potential located between the conduction band minimum and the valence band maximum. This case is discussed in this section. Of course, semiconductors can be doped with donor or acceptor impurities. Donors create conduction band electrons. If the donor doping has a high concentration N_D, the chemical potential can be above the conduction band minimum.

For an intrinsic semiconductor or an insulator, assume eqn. (11.83) is valid. An electron in the conduction band has an energy

$$E_c(k) = E_{c0} + \varepsilon(k), \quad \varepsilon(k) = \frac{\hbar^2k^2}{2m_c} \tag{11.84}$$

$$E_c(k) - \mu = E_{c0} - \mu + \varepsilon(k) \gg k_BT \tag{11.85}$$

where the last inequality follows from eqn. (11.83). The above expression appears in the formula for $f^{(0)}(k)$, which permits the following approximation:

$$f^{(0)}(k) = \frac{1}{e^{\beta(E_c(k)-\mu)} + 1} \approx e^{\beta(\mu - E_{c0})}e^{-\beta\varepsilon(k)} \tag{11.86}$$

The distribution has become Maxwell-Boltzmann. The conduction band electron density is n, which is a constant at high temperature. It is determined by the donor density. An expression for the chemical potential is derived by normalizing the distribution:

$$n = 2\int \frac{d^3k}{(2\pi)^3} f^{(0)}(k) = 2e^{\beta(\mu - E_{c0})} \int \frac{d^3k}{(2\pi)^3} e^{-\beta\varepsilon(k)} \tag{11.87}$$

$$= e^{\beta(\mu - E_{c0})} N_c, \quad N_c = 2\left(\frac{m_c k_B T}{2\pi\hbar^2}\right)^{3/2}$$

$$\mu = E_{c0} + k_B T \ln\left[\frac{n}{N_c}\right] \tag{11.88}$$

Since $n/N_c < 1$, the argument of the logarithm is small and the logarithm is a negative number, which puts the chemical potential below the conduction band.

The above expression for $f^{(0)}(k)$ can be used in solving for the transport coefficients using the Boltzmann equation. Now all of the integrals over d^3k have a simple Gaussian form. Also,

$$-\frac{df^{(0)}}{d\varepsilon_k} = \beta f^{(0)} \tag{11.89}$$

Assume the lifetime $\tau_t(k)$ is independent of k. The case that it depends on k is treated in the problem assignments.

1. The integral for the conductivity contains a factor of $v_k^2 = 2\varepsilon_k/m$:

$$\sigma = \frac{4e^2 n_e \tau_t}{3mN_c} \beta \int \frac{d^3k}{(2\pi)^3} \varepsilon(k) e^{-\beta\varepsilon(k)} \tag{11.90}$$

$$= \frac{e^2 n\tau_t}{m}$$

When the lifetime depends on k the integral has a slightly different value.

2. The expression for the Seebeck has an extra factor of $\xi_k = E_{c0} + \varepsilon_k - \mu$. Since $E_{c0} - \mu$ is a constant, its integral is identical to the one for the conductivity:

$$S = \frac{1}{eT}(E_{c0} - \mu) + \delta S \tag{11.91}$$

$$E_{c0} - \mu = -k_B T \ln\left(\frac{n}{N_c}\right) \tag{11.92}$$

where the latter relation is from the Maxwell-Boltzmann distribution. The integral for δS has an additional factor of the kinetic energy, and gives

$$\delta S = \frac{4k_B}{3eN_c} \int \frac{d^3k}{(2\pi)^3} [\beta\varepsilon(k)]^2 e^{-\beta\varepsilon(k)} \tag{11.93}$$

$$= \frac{5k_B}{2e}$$

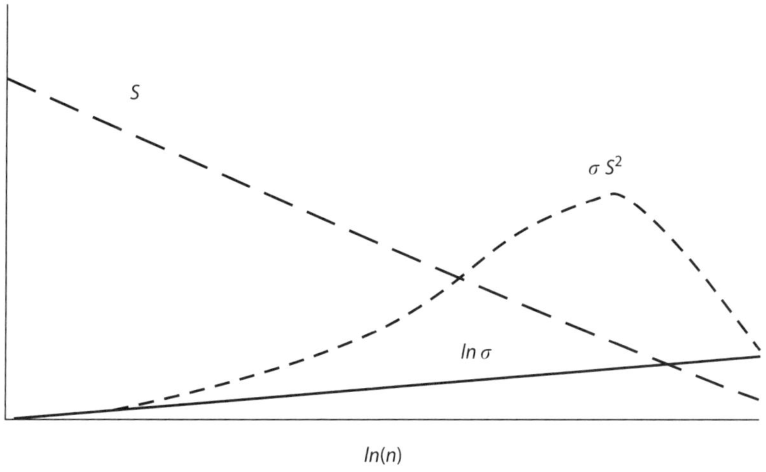

FIGURE 11.3. A graph of the Seebeck coefficient and conductivity as a function of n in a typical semiconductor. The Seebeck falls with increasing n, while the conductivity increases. The power factor σS^2 has a maximum around $n \sim 10^{19}$ electrons/cm^{-3}.

$$S = \frac{k_B}{e}\left[\frac{5}{2} - \ln\left(\frac{n}{N_c}\right)\right] \tag{11.94}$$

The Seebeck coefficient of insulators or intrinsic semiconductors can have a very high value due to the last term on the right. If the density n is quite small, then S is quite large. Insulators such as table salt (NaCl) have a very high Seebeck coefficient, which varies from sample to sample due the variation in electron density. Experiments on insulators show that $S \sim -\ln(n)$.

The efficiency of thermoelectric devices depends on the figure of merit:

$$Z = \frac{\sigma S^2 T}{K_e + K_p} \tag{11.95}$$

The numerator contains the *power factor* $P = \sigma S^2$. Figure 11.3 shows how the power factor varies with electron concentration n. It peaks at densities around 10^{19}/cm^3. This density is easily achievable by adding impurities to semiconductors, and explains why most thermoelectric devices are made from semiconductors.

11.6 Impurity Scattering

All crystals have impurities. The scattering of electrons by impurities is an important contribution to the resistivity of all solids. At very low temperatures, the phonons are frozen out and scattering by phonons vanishes. Then the impurities provide the major contribution to the electrical resistance. Two important theories of the scattering by impurities are evaluated in this section.

11.6.1 Screened Impurity Scattering

Harvey Brooks and Conyers Herring first derived the lifetime of the electron for scattering from screened, ionized impurities in a semiconductor. Assume a semiconductor is intentionally doped to be *n*-type, with a high concentration N_D of donors. Then the density of electrons is identical to the density of donors, $n_e \approx N_D$, since they are all ionized in a high-density system at room temperature. The lifetime of the electron for impurity scattering is

$$\frac{1}{\tau_t(k)} = \frac{2\pi N_D}{\hbar}\int \frac{d^3k'}{(2\pi)^3}|T_{\mathbf{kk}'}|^2\delta[\varepsilon(k) - \varepsilon(k')][1 - \cos(\theta)] \tag{11.96}$$

where θ is the angle between the two wave vectors $\mathbf{k}$ and $\mathbf{k}'$. The first Born approximation is used for the *T*-matrix, which is the Fourier transform of the potential. The latter is the Coulomb potential with Thomas-Fermi screening:

$$T_{\mathbf{kk}'} \approx v(\mathbf{k} - \mathbf{k}') \tag{11.97}$$

$$v(q) = \frac{1}{4\pi\varepsilon(0)}\int d^3r e^{i\mathbf{q}\cdot\mathbf{r}}\frac{e^2}{r}e^{-Qr}, \quad Q^2 = \frac{3e^2 n_e}{2E_F\varepsilon(0)} \tag{11.98}$$

$$v(q) = \frac{e^2}{\varepsilon(0)}\frac{1}{Q^2 + q^2} \tag{11.99}$$

$$v(\mathbf{k} - \mathbf{k}') = \frac{e^2}{\varepsilon(0)}\frac{1}{Q^2 + 2k^2[1 - \cos(\theta)]} \tag{11.100}$$

The last equation uses the fact that $k' = k$ because of the delta function for energy conservation. Evaluate the wave vector and angular integrals in the formula for the lifetime, and find

$$\frac{1}{\tau_t(k)} = \frac{N_D e^4 m_c}{4\pi\varepsilon(0)^2\hbar^3 k^3}\left[\ln[1 + \lambda(k)] - \frac{\lambda(k)}{1 + \lambda(k)}\right] \tag{11.101}$$

$$\lambda(k) = \frac{4k^2}{Q^2} = \pi(k_F a_0)\left(\frac{k}{k_F}\right)^2 \tag{11.102}$$

where the last relation used the definition $Q^2 = 3e^2 n/(2\varepsilon(0)E_F)$, and a_0 is the Bohr radius of the donor. Use the relation $N_D = n_e = k_F^3/(3\pi^2)$ to rewrite the prefactor. Most of the other factors give the donor binding energy:

$$E_D = \frac{e^4 m_c}{32\pi^2\hbar^2\varepsilon(0)^2} \tag{11.103}$$

$$\frac{1}{\tau_t(k)} = \frac{8E_D}{3\pi\hbar}\left(\frac{k_F}{k}\right)^3\left[\ln[1 + \lambda(k)] - \frac{\lambda(k)}{1 + \lambda(k)}\right] \tag{11.104}$$

$$\frac{1}{\tau_t(k_F)} = \frac{8E_D}{3\pi\hbar}\left[\ln[1 + \lambda(k_F)] - \frac{\lambda(k_F)}{1 + \lambda(k_F)}\right] \tag{11.105}$$

This last result is the *Brooks-Herring formula* for the electron lifetime in a heavily doped semiconductor. The electron mobility is $\mu = e\tau_t(k_F)/m_c$. The density of donors, or of electrons, has canceled from the prefactor. There is still a weak dependence on density in the factor $\lambda(k_F) = \pi k_F a_0$. The electron lifetime is mainly related to the binding energy of the donor $\tau_t \sim \hbar/E_D$.

11.6.2 T-matrix Description

The lifetime for scattering on an electron by an impurity of concentration n_i is given by the golden rule:

$$\frac{1}{\tau(k)} = \frac{2\pi n_i}{\hbar}\int \frac{d^3k'}{(2\pi)^3}|T_{\mathbf{kk}'}|^2 \delta[\varepsilon(\mathbf{k}) - \varepsilon(\mathbf{k}')] \tag{11.106}$$

This formula, which lacks the factor of $1 - \cos(\theta)$, is the average time τ between scattering events. It can also be written in terms of the particle velocity v_k and scattering cross section $\sigma(k)$ as

$$\frac{1}{\tau(k)} = n_i v_k \sigma(k) \tag{11.107}$$

$$\sigma(k) = \frac{2\pi}{\hbar v_k}\int \frac{d^3k'}{(2\pi)^3}|T_{\mathbf{kk}'}|^2 \delta[\varepsilon(\mathbf{k}) - \varepsilon(\mathbf{k}')] \tag{11.108}$$

The scattering is elastic, so the magnitudes of the two wave vectors are equal: $k = k'$. The two wave vectors differ only in their directions, and make a relative angle $\cos(\theta) = \hat{k}\cdot\hat{k}'$. In this case the exact *T*-matrix can be expressed in terms of the phase shifts $\delta_\ell(k)$ for scattering, where ℓ is angular momentum:

$$T_{\mathbf{kk}'} = T(k,\theta) = \frac{2\pi\hbar^2}{mk}\sum_{\ell=0}^{\infty}(2\ell + 1)e^{i\delta_\ell}\sin[\delta_\ell(k)]P_\ell(\cos\theta) \tag{11.109}$$

The integral $d^3k' = k'^2 dk' d\phi \sin(\theta) d\theta$. The integrals $dk'd\phi$ are evaluated easily: the integral over $d\phi$ gives 2π, while the integral over dk' eliminates the delta function for energy conservation

$$\int k'^2 dk' \delta[\varepsilon(\mathbf{k}) - \varepsilon(\mathbf{k}')] = \frac{mk}{\hbar^2} \tag{11.110}$$

$$\sigma(k) = 2\pi\left(\frac{m}{2\pi\hbar^2}\right)^2\int_0^\pi \sin(\theta)\,d\theta |T(k,\theta)|^2 \tag{11.111}$$

The factor of $|T(k,\theta)|^2$ creates a double summation over angular momentum, with the double factor of $P_\ell(\cos\theta)P_m(\cos(\theta)$. The angular integrals give zero unless the Legendre functions have identical index ($\ell = m$):

$$\delta_{\ell m}\frac{2}{2m+1} = \int_0^\pi \sin(\theta)\,d\theta P_\ell(\cos\theta)\,P_m(\cos(\theta) \tag{11.112}$$

$$\sigma(k) = \frac{4\pi}{k^2}\sum_{\ell}(2\ell + 1)\sin^2[\delta_\ell(k)] \tag{11.113}$$

The lifetime of an electron from impurity scattering is given in terms of the phase shifts. The same phase shifts were discussed earlier for the Friedel sum rule.

The electrical resistivity is given in terms of the transport lifetime τ_t:

$$\frac{1}{\tau_t(k)} = n_i v_k \sigma_t(k) \tag{11.114}$$

$$\sigma_t(k) = 2\pi\left(\frac{m}{2\pi\hbar^2}\right)^2 \int_0^\pi \sin(\theta)d\theta|T(k,\theta)|^2[1 - \cos(\theta)] \tag{11.115}$$

It has an additional factor of $[1-\cos(\theta)]$ in the angular integrals. The angular integral over the T-matrix gives a different answer compared to $\sigma(k)$:

$$\sigma_t(k) = \frac{4\pi}{k^2}\sum_{\ell=0}(\ell + 1)\sin^2[\delta_\ell(k) - \delta_{\ell+1}(k)] \tag{11.116}$$

In most cases the phase shifts are nonzero for a few small values of ℓ. The calculation by phase shifts is a more accurate procedure than using the first Born approximation, as was done for the Brooks-Herring formula. Phase shifts for atoms in electron gases are calculated in ref. [24].

The above cross section can be measured experimentally. If an impurity is in a metal, an electrical current exerts a force on the impurity ion. Eventually, over a long time, a dc current will encourage the diffusion of the impurity in the direction of the flow of electrons. This phenomenon is called *electromigration*. A related experiment is to inject the impurity into the metal and to determine how fast it slows down. This measurement is called the *stopping power*. It defines a coefficient Q, whose units are inverse area, which is the constant of proportionality between the particle velocity and its rate of energy loss:

$$\frac{dE}{dx} = \hbar v Q \tag{11.117}$$

$$Q = k_F n_e \sigma_t(k_F) \tag{11.118}$$

The stopping power is directly proportional to the transport cross section of electrons on the Fermi surface when scattering from the impurity. Since the electron density n_e and Fermi wave vector are known, a measurement of Q is a measurement of σ_t. Figure 11.4 shows measurements and theory of the stopping power of ions of atomic number Z in aluminum metal. The theory calculates the phase shifts, which obey the Friedel sum rule. The peak around $Z \approx 6$ is due to the bound states in the partially filled p-orbital.

11.6.3 Mooij Correlation

Most of the prior discussion concerned good metals, where the mean free path of electrons near the Fermi surface is many unit cells. In random alloys or amorphous solids the mean free path of electrons is rather short. Mooij measured the temperature dependence

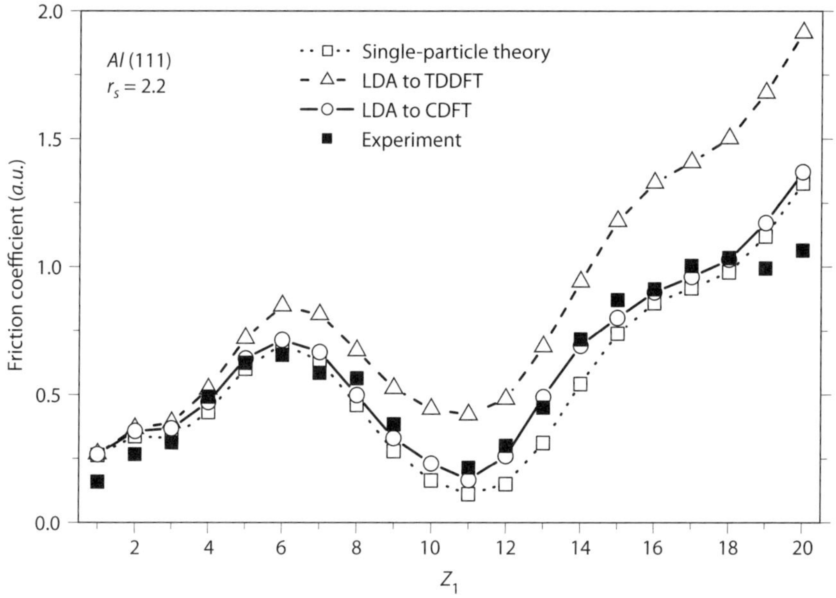

FIGURE 11.4. Stopping power of ions of atomic number Z in aluminum. Black squares are experiment, circles are free electron theory. Taken from Nazarov et al., *Phys. Rev. B* **76**, 205103 (2007). Used with permission of the American Physical Society.

of the resistivity for many such materials. Figure 11.5a shows the resistivity of a good metal, such as copper, where the resistivity increases with increasing temperature. The temperature dependence is due to scattering of the electrons by phonons, as discussed in the next section. Mooij defined the constant α as the temperature derivative of the resistivity at room temperature:

$$\alpha = \left(\frac{1}{\rho}\frac{d\rho}{dT}\right)_{T=300} \tag{11.119}$$

For copper, α is a positive number and ρ is relatively small. Figure 11.5b shows the resistivity $\rho(T)$ for a random alloy. The resistivity actually decreases with increasing temperature. For this material, α is negative and ρ is fairly high. Mooij made measurements on about 200 materials, and made the graph in fig.11.5c of α vs. ρ, all at room temperature. Each dot represents one material. On the left, with small ρ and positive α, are the good conductors such as copper. On the right, at large ρ and negative α, are the poor conductors, as found in random alloys and amorphous materials. The data points form a band of values, which cross $\alpha = 0$ around $\rho^* = 150\ \mu\Omega$ cm. Mooij suggested this value of resistivity has a special meaning. Since his work, many other data points have been added to this curve. They scatter somewhat more widely than found in Mooij's graph. Yet the trend is unmistakable.

Several explanations have been proposed for negative values of α. One of them starts with the idea that the random alloys are near the limit of Anderson localization. Most of the electrons are in localized states. Then, as the temperature is increased, conduction is allowed by variable range hopping.

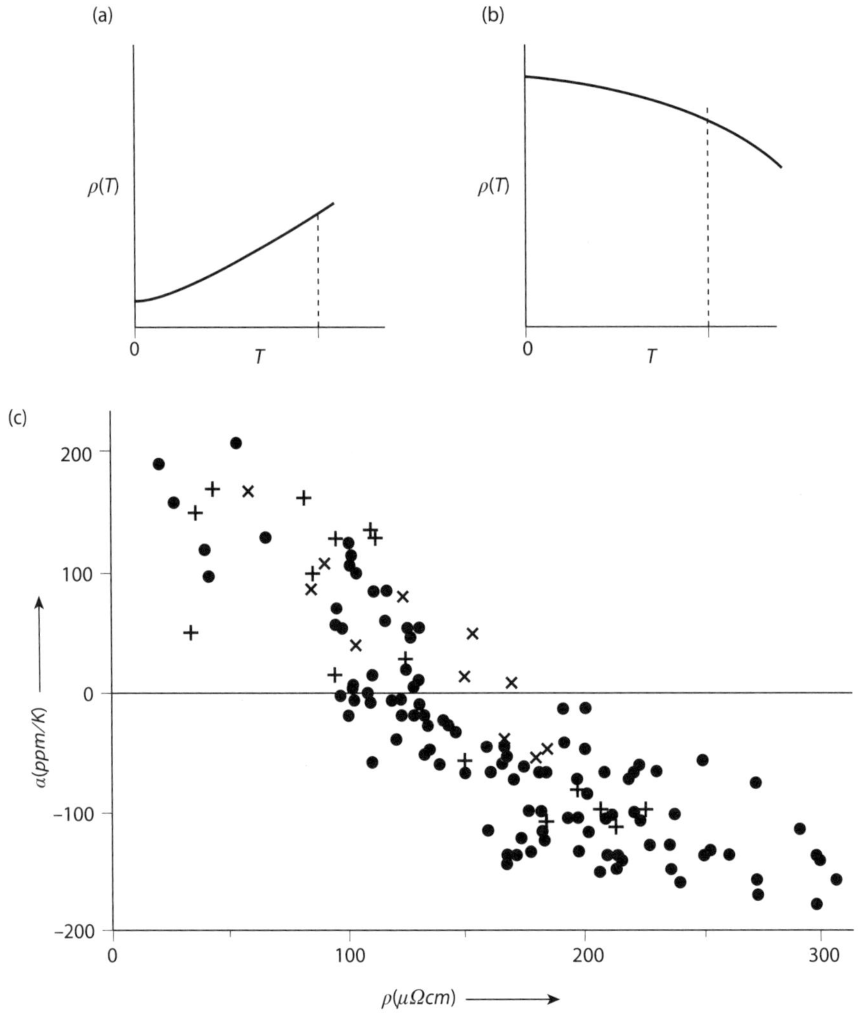

FIGURE 11.5. (a) Resistivity $\rho(T)$ of a "good" metal, where the resistivity increases with temperature. (b) Resistivity of a "bad" metal, where the resistivity decreases with temperature. (c) The Mooij correlation, showing α vs. ρ at room temperature. From Mooij, *Phys. Status Solidi A* **17**, 51 (1973). Used with permission of Wiley-VCH Verlag.

11.7 Electron–Phonon Interaction

The electron–phonon interaction is another very important contribution to the lifetime of the electron. For intrinsic solids without impurities or defects, it is the only important scattering contribution. At room temperature the scattering is usually dominated by high-frequency optical phonons. However, at low temperature they are frozen out and the scattering is dominated by long-wavelength acoustical phonons.

The electron–phonon scattering is very different in insulators or semiconductors than it is in metals.

1. In semiconductors, the conduction electrons are within a thermal energy $\varepsilon(k) \sim k_B T$ of the bottom of the conduction band. These electrons have values of wave vector close to the the band minimum. When a phonon is emitted or absorbed, energy conservation requires that the final electron state is also near the bottom of the band. So the phonons have small values of wave vector. Consequently, in discussing acoustical phonons use the Debye model ($\omega(q) = c_s q$), and in discussing optical phonons use an Einstein model ($\omega(q) = \omega_0$).

2. In metals, an electron near the Fermi surface initially has a wave vector $|k| \sim k_F$, which usually has a large value. When it emits or absorbs a phonon, it remains in energy near the Fermi surface, since phonon energies are small. Then its final wave vector is also close in value to $|\mathbf{k} + \mathbf{q}| \sim k_F$. In this case, all values of phonon wave vector are allowed that are (i) within the Brillouin zone, and (ii) obey $0 < q < 2k_F$. One includes in this process (i) all phonons, (ii) at all wavevectors, and (iii) with all polarizations.

Below are some basic ideas of the scattering, which apply to both metals and semiconductors.

11.7.1 *Lifetime*

The basic process has an electron initially in the state $\mathbf{k}$, with energy $\varepsilon(\mathbf{k})$, scattering to a final state $\mathbf{k} + \mathbf{q}$, with energy $\varepsilon(\mathbf{k} + \mathbf{q})$, by one of two processes: (i) the absorption of a phonon of wave vector $\mathbf{q}$, or (ii) the emission of a phonon of wave vector $-\mathbf{q}$. The scattering term has the form

$$\left(\frac{\partial f}{\partial t}\right)_S = \frac{2\pi}{\hbar} \sum_\lambda \int \frac{d^3q}{(2\pi)^3} |M_\lambda(\mathbf{q})|^2 \tag{11.120}$$

$$\times \{\delta[\varepsilon(\mathbf{k}) - \varepsilon(\mathbf{k}+\mathbf{q}) + \hbar\omega_\lambda(\mathbf{q})] H_a(\mathbf{k}, \mathbf{q})$$

$$+ \delta[\varepsilon(\mathbf{k}) - \varepsilon(\mathbf{k}+\mathbf{q}) - \hbar\omega_\lambda(\mathbf{q})] H_e(\mathbf{k}, \mathbf{q})\}$$

The first term includes the absorption of a phonon, and the second the emission of a phonon. The important occupation factors are contained in $H_{a,e}(\mathbf{k}, \mathbf{q})$.

1. For phonon absorption, the initial state has an electron in k $[f(\mathbf{k})]$, a phonon in $(\lambda, \mathbf{q})$ $[N_q]$, and an unoccupied electron state in $\mathbf{k} + \mathbf{q}$ $[1 - f(\mathbf{k} + \mathbf{q})]$:

$$H_a = f(\mathbf{k}) N_q[1 - f(\mathbf{k}+\mathbf{q})] - [1 - f(\mathbf{k})][N_q + 1] f(\mathbf{k}+\mathbf{q}) \tag{11.121}$$

The first term describes an electron in $\mathbf{k}$ absorbing a phonon and going to $\mathbf{k} + \mathbf{q}$. The second term describes the reverse reaction where an electron in $\mathbf{k} + \mathbf{q}$ emits a phonon and goes to $\mathbf{k}$. In equilibrium, these two processes are in balance and cancel. That is shown by using the properties of Fermi-Dirac and boson distribution functions [$x = \beta\varepsilon(\mathbf{k})$, $y = \beta\varepsilon(\mathbf{k} + \mathbf{q})$, $z = \beta\hbar\omega(\mathbf{q})$]:

$$1 - f^{(0)}(\mathbf{k}) = 1 - \frac{1}{e^x + 1} = \frac{e^x}{e^x + 1} = e^x f^{(0)}(\mathbf{k}) \tag{11.122}$$

$$N_q + 1 = \frac{1}{e^z - 1} + 1 = \frac{e^z}{e^z - 1} = e^z N_q \tag{11.123}$$

Therefore, express H_a in equilibrium as

$$H_a = f^{(0)}(\mathbf{k}) f^{(0)}(\mathbf{k}+\mathbf{q})\, N_q \left[e^{\beta\varepsilon(\mathbf{k}+\mathbf{q})} - e^{\beta[\varepsilon(\mathbf{k}) + \hbar\omega(q)]}\right] \tag{11.124}$$

The two terms in brackets are equal and cancel: they are equal because of the delta function for energy conservation requires that $\varepsilon(\mathbf{k}+\mathbf{q}) = \varepsilon(\mathbf{k}) + \hbar\omega(\mathbf{q})$.

2. For phonon emission, the initial state has an electron in $\mathbf{k}$ $[f(\mathbf{k})]$. Creating a phonon has a factor of $N_q + 1$, so that

$$H_e = f(\mathbf{k})(N_q + 1)[1 - f(\mathbf{k}+\mathbf{q})] - [1 - f(\mathbf{k})]\, N_q f(\mathbf{k}+\mathbf{q}) \tag{11.125}$$

This factor is also zero in equilibrium.

During transport the system is not in equilibrium. The factors $H_{a,e}$ are not zero, and the above expression (11.120) is not zero.

Since the integral in eqn. (11.120) is over d^3q, the factor of $f(\mathbf{k})$ can be taken out of the integral. Its coefficient defines the lifetime for an electron scattering from phonons. Write eqn. (11.120) as

$$\left(\frac{\partial f}{\partial t}\right)_S = \frac{f(\mathbf{k})}{\tau(\mathbf{k})} - \text{other terms} \tag{11.126}$$

$$\frac{1}{\tau(\mathbf{k})} = \frac{2\pi}{\hbar} \sum_\lambda \int \frac{d^3q}{(2\pi)^3} |M_\lambda(\mathbf{q})|^2 \tag{11.127}$$

$$\times \{\delta[\varepsilon(\mathbf{k}) - \varepsilon(\mathbf{k}+\mathbf{q}) + \hbar\omega_\lambda(\mathbf{q})][N_q + f(\mathbf{k}+\mathbf{q})]$$

$$+ \delta[\varepsilon(\mathbf{k}) - \varepsilon(\mathbf{k}+\mathbf{q}) - \hbar\omega_\lambda(\mathbf{q})][N_q + 1 - f(\mathbf{k}+\mathbf{q})]\}$$

The lifetime $\tau(k)$ measures how often an electron gets scattered by a phonon and leaves the state $f(\mathbf{k})$. The "other terms" in the above equation are the events where an electron is scattered back into the state $f(\mathbf{k})$.

In eqn. (11.120), replace

$$f(\mathbf{k}) = f^{(0)}(k) + \delta f(\mathbf{k}) \tag{11.128}$$

$$f(\mathbf{k}+\mathbf{q}) = f^{(0)}(|\mathbf{k}+\mathbf{q}|) + \delta f(\mathbf{k}+\mathbf{q}) \tag{11.129}$$

$$N_q = N_q \tag{11.130}$$

It is assumed the phonon system is in equilibrium. At low temperatures, this assumption is invalid. Then one has to add a term δN_q and write coupled equations for the phonon and electron distribution functions. This coupling is the origin of phonon drag. However, at room temperatures, the electrons dissipate energy to the phonons, and the phonons stay in thermal equilibrium by dissipating energy among the various phonon states using anharmonic interactions. See Bailyn [4] for a discussion of phonon drag.

With the above assumptions, the factors $H_{a,e}$ can be simplified. Keep only first-order terms, and ignore those with two factors of δf:

$$H_a = \delta f(\mathbf{k})[N_q + f^{(0)}(|\mathbf{k}+\mathbf{q}|)] - \delta f(\mathbf{k}+\mathbf{q})[N_q + 1 - f^{(0)}(k)] \tag{11.131}$$

$$H_e = \delta f(\mathbf{k})[N_q + 1 - f^{(0)}(|\mathbf{k}+\mathbf{q}|)] - \delta f(\mathbf{k}+\mathbf{q})[N_q + f^{(0)}(\mathbf{k})] \tag{11.132}$$

$$\left(\frac{\partial f}{\partial t}\right)_S = \frac{\delta f(\mathbf{k})}{\tau(\mathbf{k})} - \left(\frac{\partial f}{\partial t}\right)_{\text{back}} \tag{11.133}$$

$$\begin{aligned}\left(\frac{\partial f}{\partial t}\right)_{\text{back}} &= \frac{2\pi}{\hbar}\sum_\lambda \int \frac{d^3q}{(2\pi)^3}|M_\lambda(\mathbf{q})|^2 \delta f(\mathbf{k}+\mathbf{q}) \\ &\quad \times \{\delta[\varepsilon(\mathbf{k}) - \varepsilon(\mathbf{k}+\mathbf{q}) + \hbar\omega_\lambda(\mathbf{q})][N_q + 1 - f^{(0)}(k)] \\ &\quad + \delta[\varepsilon(\mathbf{k}) - \varepsilon(\mathbf{k}+\mathbf{q}) - \hbar\omega_\lambda(\mathbf{q})][N_q + f^{(0)}(k)]\}\end{aligned} \tag{11.134}$$

During this scattering the electron changes its energy by the energy of the phonon. The scattering is inelastic. Boltzmann's equation becomes an integral equation, which must be solved numerically. This has been done a few times. Usually one replaces eqn. (11.133) by an approximation such as

$$\left(\frac{\partial f}{\partial t}\right)_S = \frac{\delta f(\mathbf{k})}{\tau_t(\mathbf{k})} \tag{11.135}$$

where τ_t is the transport lifetime for an electron scattering by a phonon. It is found from the usual lifetime by adding a factor of $(1 - \cos\theta)$ to the integral. For the case of metals, the initial electron wave vector $\mathbf{k}$ and the final wave vector $\mathbf{k}' = \mathbf{k} + \mathbf{q}$ are both nearly equal to the Fermi wave vector. Assuming that $k = k'$, then

$$1 - \cos(\theta) = 1 - \hat{k}\cdot\hat{k}' = 1 - \frac{\mathbf{k}\cdot(\mathbf{k}+\mathbf{q})}{k^2} = -\frac{\mathbf{k}\cdot\mathbf{q}}{k^2} \tag{11.136}$$

$$\begin{aligned}\frac{1}{\tau_t(\mathbf{k})} &= \frac{2\pi}{\hbar}\sum_\lambda \int \frac{d^3q}{(2\pi)^3}|M_\lambda(\mathbf{q})|^2 \left(-\frac{\mathbf{k}\cdot\mathbf{q}}{k^2}\right) \\ &\quad \times \{\delta[\varepsilon(\mathbf{k}) - \varepsilon(\mathbf{k}+\mathbf{q}) + \hbar\omega_\lambda(\mathbf{q})][N_q + f(\mathbf{k}+\mathbf{q})] \\ &\quad + \delta[\varepsilon(k) - \varepsilon(\mathbf{k}+\mathbf{q}) - h\omega_\lambda(\mathbf{q})][N_q + 1 - f(\mathbf{k}+\mathbf{q})]\}\end{aligned} \tag{11.137}$$

This latter expression is used in many calculations. Then one can solve Boltzmann's equation, as was done for impurity scattering, and get the same expressions for the three transport coefficients: σ, S, K_e.

11.7.2 *Semiconductors*

For semiconductors, it is a poor approximation to use eqn. (11.135). Instead, one must numerically solve the integral equation for the electron energy. An efficient numerical method was give by Rode [24]. If the density of carriers is dilute, then one can ignore the factor of $f^{(0)}(|\mathbf{k}+\mathbf{q}|)$ in the definition of the relaxation time:

$$\frac{1}{\tau(\mathbf{k})} = \frac{2\pi}{\hbar}\sum_{\lambda}\int\frac{d^3q}{(2\pi)^3}|M_{\lambda}(\mathbf{q})|^2 \tag{11.138}$$

$$\times\{N_q\delta[\varepsilon(\mathbf{k}) - \varepsilon(\mathbf{k}+\mathbf{q}) + \hbar\omega_{\lambda}(\mathbf{q})]$$

$$+[N_q+1]\delta[\varepsilon(\mathbf{k}) - \varepsilon(\mathbf{k}+\mathbf{q}) - \hbar\omega_{\lambda}(\mathbf{q})]\}$$

The first term is due to the absorption of a phonon, while the second term is due to the emission of a phonon.

In silicon, the optical phonon has a large energy of about 15 THz (62 meV). This is much larger than the thermal energy at room temperature. In a Maxwell-Boltzmann distribution, most of the electrons have $\varepsilon(k) < \hbar\omega_0$. In that case, phonon emission cannot occur. The only process is the absorption of a phonon. The above expression can be written using an Einstein model for the optical phonons:

$$\frac{1}{\tau_0} = N_0 w, \quad N_0 = \frac{1}{e^{\beta\hbar\omega_0} - 1} \tag{11.139}$$

$$w = \frac{2\pi}{\hbar}\sum_{\lambda}\int\frac{d^3q}{(2\pi)^3}|M_{\lambda}(\mathbf{q})|^2\delta[\varepsilon(\mathbf{k}) - \varepsilon(\mathbf{k}+\mathbf{q}) + \hbar\omega_0] \tag{11.140}$$

Then the electrical conductivity of silicon behaves as

$$\sigma = \frac{n_e e^2 \tau_0}{m_c} = \frac{n_e e^2}{m_c w}\left(e^{\beta\hbar\omega_0} - 1\right) \tag{11.141}$$

Besides scattering by long-wavelength optical phonons, electrons can have intervalley scattering. This requires phonons of large wave vector and large energy, and has a formula of similar form.

The conductivity depends on the electron density n_e, which usually varies with temperature and depends on the impurity content of the material. A better experimental quantity is the *mobility*, which is called μ and is defined as

$$\mu = \frac{e\tau_t}{m_c} = \frac{e}{m_c w}\left(e^{\beta\hbar\omega_0} - 1\right) \tag{11.142}$$

The mobility gives the average velocity of the electrons $\langle v\rangle = \mu E$ in response to an electric field E. At low temperatures, the mobility will rise exponentially. Indeed, it does, as shown in fig. 11.6. The dramatic rise at low temperature is due to the declining number of optical phonons that are available to scatter the electrons. At low temperature the mobility is limited by scattering by acoustical phonons.

11.7.3 Saturation Velocity

At small values of electric field, the average drift velocity of electrons is proportional to the electric field. As the electric field is increased to a high value, the average drift velocity

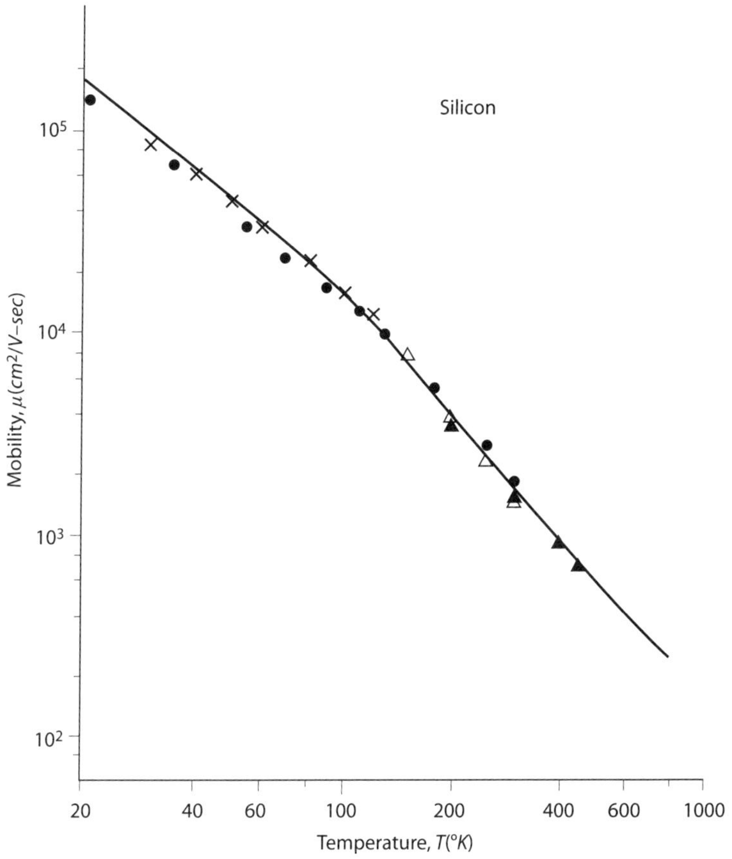

FIGURE 11.6. Mobility of pure silicon as a function of temperature. The rise at low temperature is due to the declining number of optical phonons. Solid line is theory, points are experimental. From Rode, *Phys. Status Solidi B* **53**, 245 (1972). Used with permission of Wiley-VCH Verlag.

goes to a constant value, which does not depend on the field. A typical experimental curve is shown in fig. 11.7 for silicon. For low electric fields, the electrons accelerate slowly. They are scattered by acoustical phonons, which maintains them at a small average velocity. But at high electric fields, the electrons gain energy rapidly. The scattering by acoustical phonons is too slow. The electron has a strong scattering by the creation of optical phonons, but the electron has to acquire enough kinetic energy to emit the phonon, which in silicon has an energy of $\hbar\omega_0 = 62$ meV. This process can be described by a simple model in one dimension, as shown in fig. 11.8. Figure 11.8a shows a band with quadratic dispersion. The horizontal line is at the energy of the optical phonon $\hbar\omega_0$. Using the formula that $k(t) = eEt/\hbar$, the particle gains energy at the rate

$$E(t) = \frac{1}{2m^*}(eEt)^2 \tag{11.143}$$

$$E(t_0) = \frac{1}{2m^*}(eEt_0)^2 = \hbar\omega_0 \tag{11.144}$$

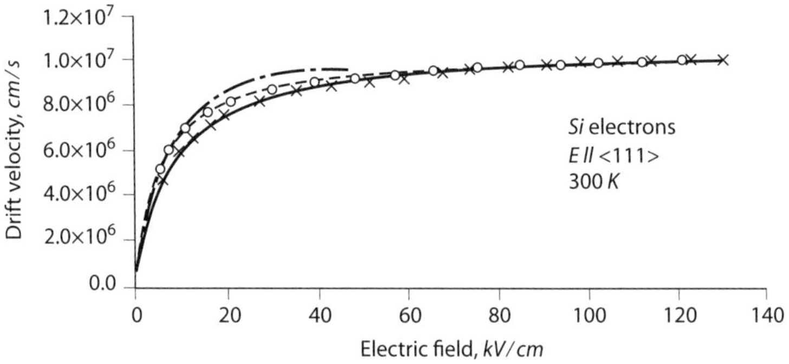

FIGURE 11.7. Drift velocity in silicon as a function of electric field. From Smith et al., *Appl. Phys. Lett.* **37**, 797 (1980). Used with permission of the American Institute of Physics.

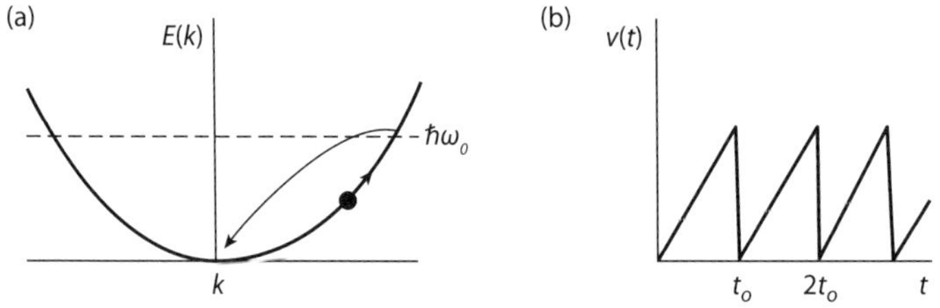

FIGURE 11.8. (a) A particle accelerates in the electric field until it has enough energy to emit an optical phonon. After emitting the phonon, it starts over again at $k = 0$. (b) Velocity $v(t)$ has a sawtooth shape.

At the time t_0 the electron has enough energy to emit the optical phonon. It returns to zero wave vector and starts over again. The velocity $v(t)$ is

$$v(t) = \frac{eEt}{m^*} \tag{11.145}$$

during one acceleration period. Its actual behavior is shown in fig. 11.8b. It has a sawtooth shape as a function of time. The saturation velocity is the average of the sawteeth:

$$v_s = \langle v \rangle = \frac{eEt_0}{2m^*} = \sqrt{\frac{\hbar\omega_0}{2m^*}} \tag{11.146}$$

where eqn. (11.144) is used to define eEt_0. The saturation velocity depends on the energy of the optical phonon and the effective mass of the electron or hole. This formula gives the correct magnitude, which is $1\text{–}2\times10^7$ cm/s in various semiconductors.

This simple model, in one dimension, applies equally well to motion in two and three dimensions. When an electron emits the optical phonon, it returns to the lowest energy state, which is the minimum of the energy band; e.g., in GaAs it is at $k = 0$. Then its subsequent acceleration is one dimensional, in the direction of the electric field.

11.7.4 Metals

Bill McMillan introduced a function $\alpha^2F(\omega)$ for the electron–phonon interaction in metals, which was discussed in chapter 9. A similar function is needed for transport theory, where the factor of $[1-\cos(\theta)] \approx -\mathbf{k}\cdot\mathbf{q}/k^2$ is added to the integrand. It has subscript of t:

$$\alpha_t^2 F(\xi,\omega) = \sum_\lambda \int \frac{d^3q}{(2\pi)^3} |M_\lambda(\mathbf{q})|^2 \left(-\frac{\mathbf{k}\cdot\mathbf{q}}{k^2}\right) \delta[\omega - \omega_\lambda(\mathbf{q})] \delta[\xi - \xi_{\mathbf{k}+\mathbf{q}}] \tag{11.147}$$

The formula for the transport lifetime of the electron from the electron–phonon interaction can be rewritten as

$$\begin{aligned}\frac{1}{\tau_t(\mathbf{k},\xi)} &= 2\pi \int_0^{\omega_D} d\omega \alpha_t^2 F(\omega) \int d\xi' \{\delta(\xi - \xi' + \hbar\omega)[n_B(\omega) + n_F(\xi')] \\ &\quad + \delta(\xi - \xi' - \hbar\omega)[n_B(\omega) + 1 - n_F(\xi')]\} \\ &= 2\pi \int_0^{\omega_D} d\omega \alpha_t^2 F(\omega)[2n_B(\omega) + 1 + n_F(\xi + \hbar\omega) - n_F(\xi - \hbar\omega)]\end{aligned} \tag{11.148}$$

The lifetime is expressed quite compactly in terms of the McMillan function. The major effort is in the calculation of $\alpha_t^2 F$. It requires a computer program that finds the energy bands in the metal, the phonons in the metal, and the matrix elements!

The above formula is simple at higher temperatures. The phonon occupation numbers are

$$2n_B(\omega) + 1 \approx 2\frac{k_BT}{\hbar\omega} \gg 1 \tag{11.149}$$

The electron occupation numbers can be neglected. The lifetime is

$$\frac{1}{\tau_t(\mathbf{k},\xi)} \approx \frac{2\pi k_B T \lambda_t}{\hbar} \tag{11.150}$$

$$\lambda_t = 2\int_0^{\omega_D} \frac{d\omega}{\omega} \alpha_t^2 F(\omega) \tag{11.151}$$

The inverse lifetime is proportional to the temperature, and the constant of proportionality is determined by λ_t. The electrical resistivity from electron–phonon interactions is

$$\rho(T) = \frac{m}{ne^2\tau_t} = \frac{2\pi m\lambda_t}{ne^2\hbar} k_B T \tag{11.152}$$

The electron–phonon couping constant λ_t can be measured as the coefficient of $d\rho/dT$ at high temperature. Figure 11.9 shows the temperature dependence of the resistivity $\rho(T)$ for a typical metal. Impurity scattering gives a constant term ρ_0. The temperature dependence is provided by the scattering of electrons by phonons. The resistivity is linear at higher temperatures, and the slope is given by λ_t. Actual raw data show a quadratic dependence on T at high temperature, due to the thermal expansion of the solid. The linear

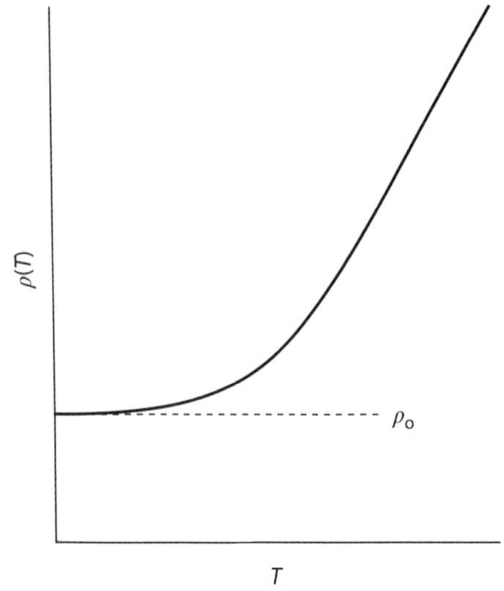

FIGURE 11.9. Temperature dependence of the electrical resistivity in metals. Impurity scattering gives a constant term ρ_0. The temperature dependence is provided by the scattering of electrons by phonons.

behavior is found when the measurements are corrected to give a result for constant volume.

Experiments and theoretical calculations show that the two constants λ, λ_t are similar in value. Often they are treated as identical and used interchangeably.

The resistivity contributions from impurity scattering and from phonons is additive. This feature is called *Matthiessen's rule*. The idea is that one can just add the scattering rates from impurity scattering τ_{ei} and phonons τ_{ep}:

$$\rho = \frac{m}{ne^2}\left(\frac{1}{\tau_{ei}} + \frac{1}{\tau_{ep}}\right) \tag{11.153}$$

Actually, the rule is not well based theoretically, since Boltzmann equation states that one must average τ, not $1/\tau$. Nevertheless, the rule is widely observed.

11.7.5 Temperature Relaxation

In thermal equilbrium, the temperature of the electron system (T_e) and the phonon system (T_p) are equal. It is possible to perturb the system so that these two temperatures are temporarily different. For example, sending a high-powered laser pulse at a metal will heat the electrons, since the electrons absorb the laser energy. Then the electron system is hotter than the phonon system. The two systems will rapidly come to thermal equilibrium with a relaxation time τ_{ep}.

Phil Allen [1] first calculated this relaxation time. The derivation of his result starts with eqn. (11.120) for the scattering time of the distribution. In evaluating this expression, keep in mind that $T_e \neq T_p$, so the factors $H_{a,e} \neq 0$. The formula is derived for a metal with a chemical potential, and uses the McMillan function α^2F to rewrite this expression as $[\varepsilon \equiv \varepsilon(\mathbf{k}), \varepsilon' = \varepsilon(\mathbf{k}+\mathbf{q})]$:

$$\left(\frac{\partial f}{\partial t}\right)_s = 2\pi \int_0^{\omega_D} d\omega \alpha^2 F(\omega) \int d\varepsilon' [\delta(\varepsilon - \varepsilon' + \hbar\omega) H_a + \delta(\varepsilon - \varepsilon' - \hbar\omega) H_e] \quad (11.154)$$

$$H_a(\varepsilon, \varepsilon', \omega) = n_F(\varepsilon) n_B(\omega) [1 - n_F(\varepsilon')] - [1 - n_F(\varepsilon)][n_B(\omega) + 1] n_F(\varepsilon') \quad (11.155)$$

$$H_e(\varepsilon, \varepsilon', \omega) = n_F(\varepsilon)(n_B(\omega) + 1)[1 - n_F(\varepsilon')] - [1 - n_F(\varepsilon)] n_B(\omega) n_F(\varepsilon') \quad (11.156)$$

where $n_{F,B}$ are the Fermi-Dirac and Bose-Einstein occupation functions. It is assumed these distributions are isotropic in wave vector space, and nonequilibrium is due to the difference in the two temperatures.

The important quantity is the time rate of change of the electron energy. When the electrons adsorb a phonon, add a factor of $\hbar\omega$: when they emit a phonon, subtract this energy. The above expression is changed to

$$\left(\frac{\partial E}{\partial t}\right)_s = 2\pi\hbar \int_0^{\omega_D} \omega d\omega \alpha^2 F(\omega) \int d\varepsilon' [\delta(\varepsilon - \varepsilon' + \hbar\omega) H_a - \delta(\varepsilon - \varepsilon' - \hbar\omega) H_e] \quad (11.157)$$

Performing the integral over $d\varepsilon'$ gives

$$\left(\frac{\partial E}{\partial t}\right)_s = 2\pi\hbar \int_0^{\omega_D} \omega d\omega \alpha^2 F(\omega) [H_a(\varepsilon, \varepsilon + \hbar\omega, \omega) - H_e(\varepsilon, \varepsilon - \hbar\omega, \omega)] \quad (11.158)$$

The next step is to average this distribution over all electron energies ε. Multiply by the density of electron states $2N(\varepsilon)$, where the factor of two is for spin degeneracy. The resulting integral is the change in internal energy U in terms of the heat capacity C:

$$\frac{dU}{dt} = C\frac{dT_e}{dt} = 2\int d\varepsilon N(\varepsilon) \left(\frac{\partial E}{\partial t}\right)_s \quad (11.159)$$

The integrals over $d\varepsilon$ converge within a few thermal energies of the chemical potential. The density of states is assumed to be smoothly varying over this small energy range, so remove $N(\varepsilon)$ from the integral. The remaining integrals over $d\varepsilon$ have the form

$$\int d\varepsilon n_F(\varepsilon)[1 - n_F(\varepsilon - \hbar\omega)] = \hbar\omega n_{Be}(\omega) \quad (11.160)$$

$$\int d\varepsilon n_F(\varepsilon)[1 - n_F(\varepsilon + \hbar\omega)] = \hbar\omega [n_{Be}(\omega) + 1] \quad (11.161)$$

where $n_{Be}(\omega)$ is the Bose-Einstein occupation factor at the electron temperature T_e, and $n_{Bp}(\omega)$ is the Bose-Einstein occupation factor at the phonon temperature T_p. Evaluate this integral, collect terms, and get the simple relation

$$C\frac{dT_e}{dt} = -8\pi\hbar^2 N(0) \int_0^{\omega_D} d\omega \omega^2 \alpha^2 F(\omega) [n_{Be}(\omega) - n_{Bp}(\omega)] \quad (11.162)$$

This formula is evaluated in the limits of low, or high, temperature. If $T_e = T_p$ the right-hand side vanishes, as it should in equilibrium. A further simplification occurs because the heat capacity is also proportional to the density of states $C = 2\pi^2 k_B^2 T N(0)/3$:

$$\frac{dT_e}{dt} = -\frac{12\hbar^2}{\pi k_B^2 T_e} \int_0^{\omega_D} d\omega \omega^2 \alpha^2 F(\omega) [n_{Be}(\omega) - n_{Bp}(\omega)] \quad (11.163)$$

The information regarding the actual material is in $\alpha^2F(\omega)$. Now examine this formula in the limit of high and low temperatures.

1. At high temperature, use the expansion $n_B(\omega) \approx k_BT/\hbar\omega$ to write the expression as

$$\frac{dT_e}{dt} = -\frac{12\hbar}{\pi k_B T_e}(T_e - T_p)\int_0^{\omega_D} d\omega\omega\alpha^2F(\omega) \tag{11.164}$$

$$= -\frac{T_e - T_p}{\tau_{ep}} \tag{11.165}$$

$$\frac{1}{\tau_{ep}} = \frac{6\hbar}{\pi k_B T_e}\lambda\langle\omega^2\rangle \tag{11.166}$$

$$\lambda\langle\omega^2\rangle = 2\int_0^{\omega_D} d\omega\omega\alpha^2F(\omega) \tag{11.167}$$

The answer is proportional to the McMillan function $\lambda\langle\omega^2\rangle$. Equation (11.165) describes a standard process of exponential decay toward thermal equilibrium.

2. In the limit of low temperature, the integral is dominated by the low-frequency properties of $\alpha^2F(\omega) \to G\omega^2$, where G is due to the three acoustical phonons. At low temperature the upper limit of the integral can be extended to infinite frequency, and

$$\frac{dT_e}{dt} = -\Gamma(T_e^5 - T_p^5) \tag{11.168}$$

$$\Gamma = \frac{12k_B^3G}{\pi T_e\hbar^3}I_4 \tag{11.169}$$

$$I_4 = \int_0^\infty \frac{dxx^4}{e^x - 1} = 4!\zeta(5) \tag{11.170}$$

The energy exchange between the electrons and phonons obeys a T^5 law. The energy exchange is quite slow, and it is possible to cool, say, the electron system, far below the temperature of the phonon system. That is not possible at room temperature, where the energy exchange is much faster.

11.8 Ballistic Transport

Ballistic transport is when the particle, electron or phonon, traverses the sample without scattering. That happens whenever the mean free path of the particle exceeds the length of the material. It is a common occurrence at very low temperature, but has also been observed in a few systems at room temperature.

Rolf Landauer [17] formulated the first theory of ballistic transport in one dimension. The formula for the electrical current is

$$J = \frac{e}{L}\sum_{k\sigma} v_k n_F(k) = \frac{2e}{L}\sum_k v_k n_F(k) \tag{11.171}$$

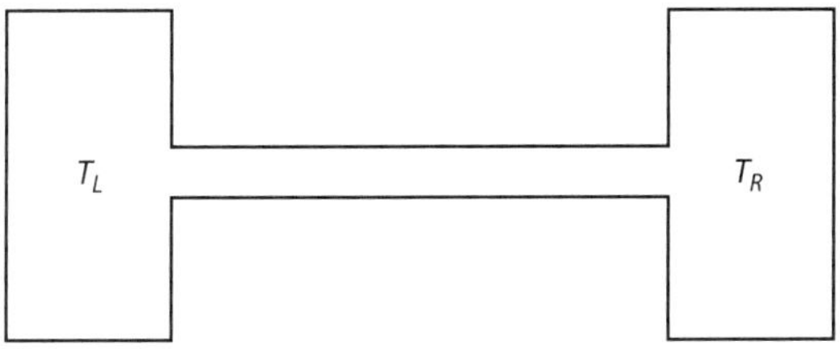

FIGURE 11.10. A one-dimensional channel connecting the left and right reservoirs.

where the factor of two is from the summation over spins. Figure 11.10 shows a one-dimensional channel for electrons that connect the left (L) and right (R) reservoirs. The two reservoirs might be at different temperatures (T_L, T_R), and might have a voltage difference between them, which causes a difference in their chemical potential ($\mu_L - \mu_R = eV$). So rewrite the above formula as

$$J = \frac{2e}{L}\sum_{k>0} v_k[n_{FL}(k) - n_{FR}(k)] \tag{11.172}$$

$$n_{FL}(k) = \frac{1}{e^{\beta_L(\varepsilon(k)-\mu_L)}+1}, \quad n_{FR}(k) = \frac{1}{e^{\beta_R(\varepsilon(k)-\mu_R)}+1} \tag{11.173}$$

The electrons going to the right come from the left reservoir, while those going to the left come from the right reservoir. If ($T_L = T_R$, $\mu_L = \mu_R$) then no net current will flow since the same number of electrons flow in both directions. A net current flows whenever $T_L \neq T_R$, or $\mu_L \neq \mu_R$. Assume that both V and $\delta T = T_L - T_R$ are small, and linearize the above formula:

$$J = -\sigma\left[\frac{dV}{dx} + S\frac{dT}{dx}\right] \tag{11.174}$$

However, in ballistic transport the current does not depend on the length L since the electrons go from one end to the other without scattering. So rewrite the above formula as

$$J = -\sigma[V + S\Delta T] \tag{11.175}$$

Here σ has the units of siemens, and S is still volts per degree. The Landauer formalism can be used to find σ and S for ballistic transport. Define the average temperature T and average chemical potential μ:

$$T = \frac{1}{2}(T_L + T_R), \quad T_L = T + \frac{\Delta T}{2}, \quad T_R = T - \frac{\Delta T}{2} \tag{11.176}$$

$$\mu = \frac{1}{2}(\mu_L + \mu_R), \quad \mu_L = \mu + \frac{eV}{2}, \quad \mu_R = \mu - \frac{eV}{2} \tag{11.177}$$

Expand the difference in occupation factors in powers of V and δT:

$$n_{FL}(k) - n_{FR}(k) = \left[\xi(k)\frac{\Delta T}{T} - eV\right]\left(-\frac{dn_F(\xi)}{d\xi}\right) \tag{11.178}$$

$$J = \frac{2e}{L}\sum_{k>0} v_k\left[\xi(k)\frac{\Delta T}{T} - eV\right]\left(-\frac{dn_F(\xi)}{d\xi}\right) \tag{11.179}$$

Since the velocity is $\hbar v_k = d\xi(k)/dk$ the integral can be converted to the energy variable:

$$\frac{1}{L}\sum_{k>0} v_k = \int_0^{\pi/a}\frac{dk}{2\pi}v_k = \frac{1}{2\pi\hbar}\int_{-\mu}^{W} d\xi \tag{11.180}$$

$$J = \frac{2e}{h}\int_{-\mu}^{W} d\xi\left[\xi(k)\frac{\Delta T}{T} - eV\right]\left(-\frac{dn_F(\xi)}{d\xi}\right) \tag{11.181}$$

The two integrals are

$$\int_{-\mu}^{W} d\xi\left(-\frac{dn_F(\xi)}{d\xi}\right) = n_F(-\mu) - n_F(W) = 1 \tag{11.182}$$

$$\int_{-\mu}^{W} \xi d\xi\left(-\frac{dn_F(\xi)}{d\xi}\right) = 0 \tag{11.183}$$

where W is the maximum band energy. Therefore,

$$J = -\sigma V,\quad \sigma = \frac{2e^2}{h} = 2\sigma_0,\quad \sigma_0 = \frac{e^2}{h} \tag{11.184}$$

The minus sign occurs on the current, since a positive value of eV was assigned to the left reservoir ($e<0$), which is a negative voltage. So the current goes to the left. The quantity σ_0 is called the quantum of conductance, and is often observed in low-temperature measurements on quantum wires. Its inverse is the quantum of resistance $\rho_0 = h/e^2 = 25{,}812.8\ \Omega$. It is interesting that this result is independent of the material that makes up the quantum wire.

The integral in (11.183) shows that the Seebeck coefficient is zero in ballistic transport. A similar calculation can be done for the energy current:

$$J_Q = \frac{2}{L}\sum_{k>0} v_k\xi(k)\left[n_{FL}(k) - n_{FR}(k)\right] \tag{11.185}$$

$$J_Q = \frac{2}{h}\int_{-\mu}^{W} d\xi\xi\left[\xi(k)\frac{\Delta T}{T} - eV\right]\left(-\frac{dn_F(\xi)}{d\xi}\right) \tag{11.186}$$

$$= -\sigma STV - K\Delta T$$

Again $S=0$ and the quantum of thermal conductance is

$$K = \frac{2}{hT}\int_{-\mu}^{W} d\xi\xi^2\left(-\frac{dn_F(\xi)}{d\xi}\right) = 2K_0 \tag{11.187}$$

$$K_0 = \frac{\pi^2}{3h}k_B^2 T \tag{11.188}$$

K_0 has the units of watts per degree.

The derivation is for the quantum of heat carried by electrons. The same quantum of thermal conductance is found for heat carried by phonons, which is a homework problem. The quantum of thermal conductance in an insulator (semiconductor) was measured in ref. [29], which confirmed the above prediction.

11.9 Carrier Drag

Most extrinsic semiconductors have a dominant type of impurity, which has been intentionally added to the material during the growth of the crystal. If the impurities are donors, then the electrons in the conduction band are the majority carrier, and the holes are called the minority carrier. If the impurities are acceptors, the holes are the majority carrier and the electrons are the minority carriers. In many semiconductor devices, such as *p–n* diodes, the minority carriers play an important role in the operation of the electronic device.

Consider an idealized system of a single hole as a minority carrier in an *n*-type semiconductor. Apply a static electric field **E** to the material. Which way does the hole move? If by itself, the hole, with its positive charge, would move in the direction of the electric field. The majority of electrons, with their negative charge, move in a direction opposite to the electric field. The electrons interact with the hole due to Coulomb interactions. When the electrons move in a negative direction, they exert a force on the hole. It is a type of microscopic wind, which tends to move the hole in the same direction as the electrons are moving. This wind has several names, such as *carrier drag, current drag,* and *Coulomb drag.* Is this wind strong enough to force the hole to move in the same direction as the electrons? Can the electron–hole interaction force the hole to move in the opposite direction that it would move if their were no electrons? This interesting question was answered in an experiment by Höpfel et al. [15]. Their result is shown in fig. 11.11.

The experiments were done in *p*-type GaAs, so electrons were the minority carrier. The crystals contained two-dimensional quantum wells grown by molecular beam epitaxy (MBE). Very large values of the electron and hole mobilities were achieved by *modulation doping.* The acceptors that create the holes were added to the barrier layers between the quantum wells. Since they were in the barriers, where the electrons and holes have a small amplitude of their eigenfunction, the scattering from these acceptors was weak. Hole and electron lifetimes are long and the mobilities are high. This increased the magnitude of the carrier drag, which enabled the experiments to observe a large effect.

The minority electrons were created by a pulsed laser narrowly focused in real space. The electrons have a short lifetime due to the recombination with the holes. This recombination radiation can be observed as a function of electric field strength. The spatial position of the observed photoluminescence shows whether the electrons are drifting to the right or to the left. The results depend on temperature and also on the strength of the electric field. At low temperatures, the carrier drag effect is large and the electrons drift in the other direction. At room temperature, the electrons drift in their natural direction.

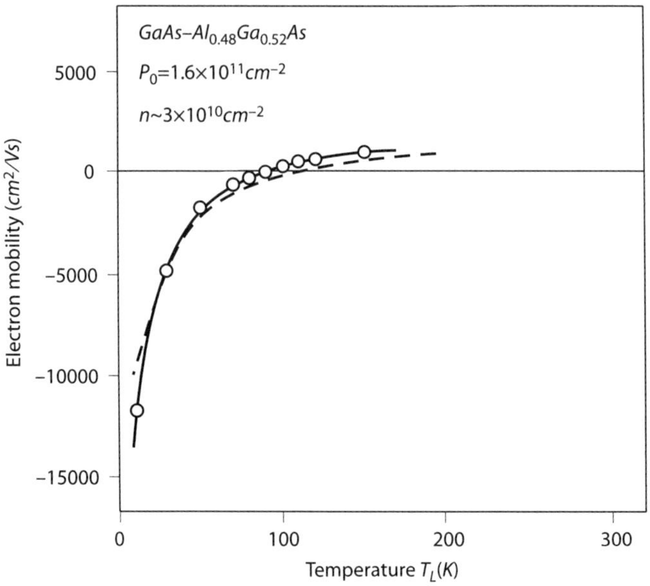

FIGURE 11.11. Electron mobility as a function of the phonon temperature in a GaAs/AlAs quantum well, for low values of the electric field. Circles are data points, and the solid line is theory. Hole density is $p = 1.6 \times 10^{11}$ cm^{-2}. Taken from Höpfel et al., *Phys. Rev. Lett.* **56**, 2736 (1986). Used with permission of the American Physical Society.

The authors present a phenomenological theory that agrees well with their data. They write down a Drude equation of motion for the electron that contains two relaxation times: the scattering between the electron and phonons τ_{ep}, and the Coulomb scattering between the electron and hole τ_{eh}. A similar equation for the relaxation of the hole contains the scattering of the holes by the phonons. The hole–hole Coulomb interaction does not greatly affect the resistivity, since such scattering does not alter the momentum of the hole system, so does not alter the current. With the convention that $e > 0$ the Drude-like equations are

$$\frac{dv_e}{dt} = -\frac{v_e}{\tau_{ep}} - \frac{(v_e - v_h)}{\tau_{eh}} - \frac{eE}{m_c} \tag{11.189}$$

$$\frac{dv_h}{dt} = -\frac{v_h}{\tau_{hp}} + \frac{eE}{m_v} \tag{11.190}$$

The first term on the right in eqn. (11.192) is the usual electron scattering by phonons. The second term is carrier drag, which is trying to make the electrons travel at the same drift velocity as the holes. There is no term in eqn. (11.190) of the electron velocity, since a single electron does not influence the average hole.

The equations are solved assuming steady motion, so all time derivatives are zero. Equation (11.190) then gives $v_h = \tau_{hp} eE/m_v$. This is inserted into eqn. (11.189), which gives

$$v_e = eE\frac{\tau_{ep}}{\tau_{eh} + \tau_{ep}}\left[\frac{\tau_{hp}}{m_v} - \frac{\tau_{eh}}{m_c}\right] \tag{11.191}$$

Define the electron mobility μ_e as $v_e = -\mu_e E$, so the electron goes in the negative E direction when the mobility is positive. Then

$$\mu_e = -e\frac{\tau_{ep}}{\tau_{eh}+\tau_{ep}}\left[\frac{\tau_{hp}}{m_v}-\frac{\tau_{eh}}{m_c}\right] \tag{11.192}$$

The last bracket is the key to the change in sign.

- If $\tau_{hp}/m_v > \tau_{eh}/m_c$, the velocity is positive and the mobility is negative. The electron travels along with the holes, and carrier drag is large. This happens at low temperature.
- If $\tau_{hp}/m_v < \tau_{eh}/m_c$, the velocity is negative and the mobility is positive. The electron drifts to the left. This is the expected behavior without carrier drag. It happens at room temperature.

The relaxation times for electron–phonon and hole–phonon were calculated above. In GaAs at room temperature, the dominant scattering is by optical phonons through the polar interaction. Then $\tau_{ep}, \tau_{hp} \propto 1/N_0$, where N_0 is the thermally activated density of optical phonons. This goes as $\tau_{ep}, \tau_{hp} \propto \exp(\hbar\omega_0/k_B T)$ and they grow exponentially at low temperatures. That is why τ_{hp} is large at low temperature. The deformation potential scattering by acoustical phonons becomes large at low temperature. All of these calculations for holes must be done for holes on the Fermi surface. The temperature dependence of the electron–hole scattering goes as $1/\tau_{eh} \propto k_B T$, which is a much slower variation.

Another form of Coulomb drag is between parallel electron gases. A current in one gas will induce a current in a parallel gas. The experiments are usually performed in quantum wells, where two parallel quantum wells are separated by a barrier region of width W. The first question is: how do electrons in the two cells communicate?

- Electrons can tunnel from one well to the other. Usually the width W is made large to avoid this process.
- Pogrebinskii [22] first suggested that the Coulomb interaction will act between electrons in parallel wells. He showed that it leads to a resistive coupling between layers.
- Another consequence of electron–electron interactions is a van der Waals interaction between the electrons in the two parallel layers. The coefficient of the van der Waals interaction changes when a current flows in either or both layers.
- Phonon-mediated interactions can couple the electrons in the two layers.

This topic is an active area of research. See Rojo [27] for a review.

11.10 Electron Tunneling

Quantum mechanics permits an electron to have a nonzero probability of going through a forbidden barrier. For a simple square barrier of height V_0 and width d, the tunneling probability is proportional to the square of the amplitude:

$$|T^2| \propto \exp[-2\kappa d], \quad \kappa = \sqrt{\frac{2m}{\hbar^2}(V_0 - E)} \tag{11.193}$$

The symbol κ has the units of wave vector, and it is the imaginary part of the electron wave vector while in the barrier region. For other tunneling barriers the WKBJ approximation usually gives a good estimate of the transmission probability.

11.10.1 *Giaever Tunneling*

Ivar Giaever [12, 13] and Leo Esaki [11] shared the Nobel Prize in physics, along with Brian Josephson, for their demonstration of electron tunneling in important physical systems. Giaever tunneling is discussed first, since it is conceptually easiest (see fig. 11.12a). The experimental system is two metals separated by an oxide, which is often the oxide of one of the metals. The samples are made by three steps: (i) a metal is evaporated on a glass slide, (ii) it is allowed to oxidize in air, and (iii) the second metal is evaporated on top of the oxide. Gold wires are soldered to the metals, and current is driven through the oxide region.

The tunneling system is described by the *tunneling Hamiltonian* first proposed by Bardeen [5]:

$$H = H_L + H_R + H_T \tag{11.194}$$

$$H_T = \sum_{\mathbf{k}\mathbf{p}\sigma} T_{\mathbf{k}\mathbf{p}} \left[C^\dagger_{\mathbf{k}\sigma} C_{\mathbf{p}\sigma} + C^\dagger_{\mathbf{p}\sigma} C_{\mathbf{k}\sigma} \right] \tag{11.195}$$

The metal on the left has conduction states described by the Hamiltonian H_L and wave vector $\mathbf{k}$, while those on the right have Hamiltonian H_R and wave vector $\mathbf{p}$. Electron spin is conserved while tunneling. Using the golden rule of quantum mechanics, the rate of tunneling is

$$w = \frac{2\pi}{\hbar} \sum_{\mathbf{k},\mathbf{p},\sigma} |T_{\mathbf{k}\mathbf{p}}|^2 \delta[E_L(\mathbf{k}) - E_R(\mathbf{p})][n_{\mathbf{k}}(1 - n_{\mathbf{p}}) - n_{\mathbf{p}}(1 - n_{\mathbf{k}})]$$

$$w = \frac{2\pi}{\hbar} \sum_{\mathbf{k},\mathbf{p},\sigma} |T_{\mathbf{k}\mathbf{p}}|^2 \delta[E_L(k) - E_R(\mathbf{p})][n_{\mathbf{k}} - n_{\mathbf{p}}] \tag{11.196}$$

The term $n_{\mathbf{k}}(1-n_{\mathbf{p}})$ has an electron tunneling from left to right, so the left side is occupied ($n_{\mathbf{k}}$) and the right side is empty ($1-n_{\mathbf{p}}$). The other term has an electron tunneling from right to left, so the right side is occupied ($n_{\mathbf{p}}$) and the left side is empty ($1-n_{\mathbf{k}}$). The quadratic terms $n_{\mathbf{k}}n_{\mathbf{p}}$ cancel.

It is convenient to measure electron energy on each side of the barrier with respect to the chemical potential on each side:

$$\xi_L = E_L(\mathbf{k}) - \mu_L, \quad n_{\mathbf{k}} = \frac{1}{e^{\beta\xi_L} + 1} \equiv n_F(\xi_L) \tag{11.197}$$

$$\xi_R = E_R(\mathbf{p}) - \mu_R, \quad n_{\mathbf{p}} = \frac{1}{e^{\beta\xi_R} + 1} \equiv n_F(\xi_R) \tag{11.198}$$

Current flows in response to an applied voltage, which makes the chemical potentials differ $eV = \mu_L - \mu_R$. Now the above formula is written as

$$w = \frac{2\pi}{\hbar} \sum_{\mathbf{k},\mathbf{p},\sigma} |T_{\mathbf{k}\mathbf{p}}|^2 \delta[\xi_L - \xi_R + eV][n_F(\xi_L) - n_F(\xi_R)] \tag{11.199}$$

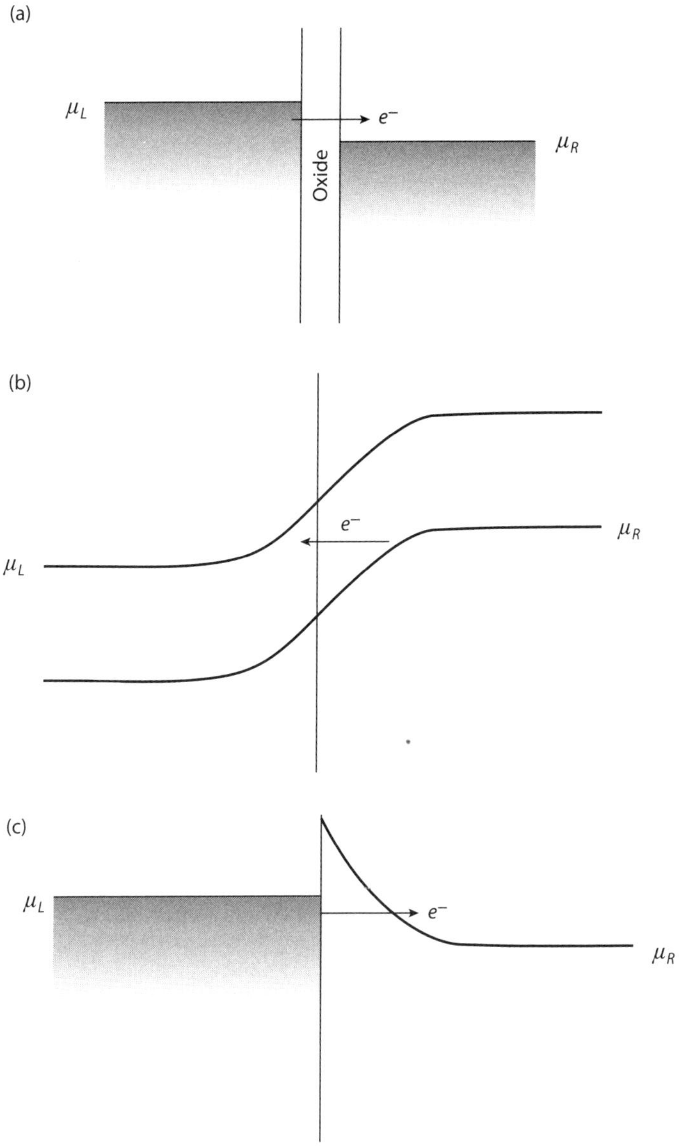

FIGURE 11.12. Three tunneling geometries: (a) Giaever tunneling between two metals, separated by a metal oxide. (b) Esaki tunneling in a $p-n$ junction in a semiconductor. (c) Schottky barrier tunneling through the depletion region of a metal–semiconductor interface.

The summation over wave vectors gives the densities of states on the two sides of the junction:

$$\sum_{\mathbf{k}} \delta(\xi_L - E_L(\mathbf{k}) + \mu_L) = N_L(\xi_L) \tag{11.200}$$

$$\sum_{\mathbf{p}} \delta(\xi_R - E_R(\mathbf{p}) + \mu_R) = N_R(\xi_R) \tag{11.201}$$

$$w = \frac{4\pi}{\hbar}\int d\xi_L N_L(\xi_L)\int d\xi_R N_R(\xi_R)|T_{\mathbf{kp}}|^2\delta[\xi_L - \xi_R + eV][n_F(\xi_L) - n_F(\xi_R)]$$

$$= \frac{4\pi}{\hbar}|T|^2\int d\xi_L N_L(\xi_L) N_R(\xi_L + eV)[n_F(\xi_L) - n_F(\xi_L + eV)] \tag{11.202}$$

The tunneling matix element $T_{\mathbf{kp}}$ is taken to be a constant near the energy of the two metal Fermi surfaces. Most experiments are done at low temperatures, where the occupation factors can be replaced by step functions, so that

$$[n_F(\xi_L) - n_F(\xi_L + eV)] = \begin{cases} 1 & -eV < \xi_L < 0 \\ 0 & \text{elsewhere} \end{cases} \tag{11.203}$$

$$w = \frac{4\pi}{\hbar}|T|^2\int_{-eV}^{0} d\xi_L N_L(\xi_L)\, N_R(\xi_L + eV) \tag{11.204}$$

If the densities of states are constants over the small voltages used in tunneling, they can be removed from the integral. Also, the current I, in amperes, is just $I = ew$, so

$$I = \sigma V, \quad \sigma = \frac{4\pi e^2}{\hbar} N_L(0)\, N_R(0)|T|^2 \tag{11.205}$$

The current is proportional to the voltage, and given by the tunneling conductance σ. A Giaever junction acts as a resistor when measured between two normal metals. The value of the resistance depends on the thickness of the oxide layer, which is in the factor of $|T|^2$.

11.10.2 Esaki Diode

Esaki reported the electron tunneling in a *p-n* diode. The geometry is shown in fig. 11.12b. One side of a semiconductor is doped *n*-type, while the other side is doped *p*-type. In the middle is a region of high electric field. An electron can tunnel through the forbidden band gap, from the valence band to the conduction band, at constant energy. The tunneling matrix element is usually evaluated using WKBJ theory:

$$T_{\mathbf{kp}} \propto \exp\left[-\int dx\, \kappa(x)\right] \tag{11.206}$$

What is the proper form for the effective wave vector $\kappa(x)$ while tunneling from the valence band to the conduction band? The answer is provided by $k \cdot p$ theory. Ignore the free particle kinetic energy, so the electron dispersion for the two bands is

$$0 = (E - E_{c0})(E - E_{v0}) - \frac{\hbar^2 p_{cv}^2}{m^2}k^2 \tag{11.207}$$

Turn this into an equation for the wave vector:

$$k^2 = \frac{m^2}{\hbar^2 p_{cv}^2}(E - E_{c0})(E - E_{v0}) \tag{11.208}$$

A graph of this formula is shown in fig. 11.13. The result for k^2 is positive in the conduction band ($E > E_{c0}$) and in the valence band ($E < E_{v0}$). In the band gap ($E_{c0} > E > E_{v0}$) the value of

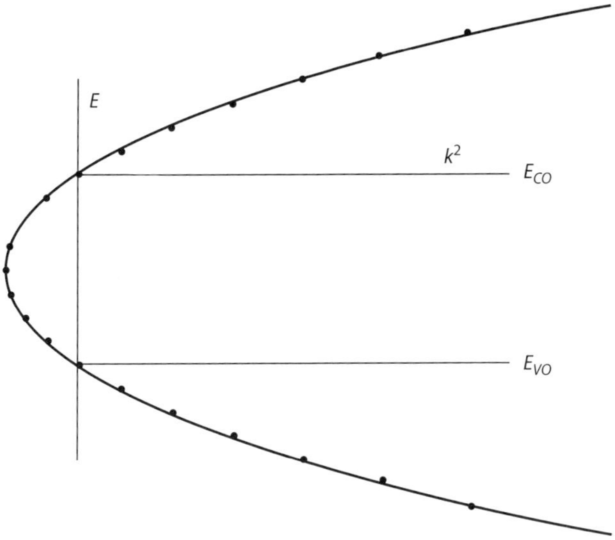

FIGURE 11.13. Electron dispersion $k^2(E)$

k^2 is negative. The negative values of k^2 make k imaginary, which is κ. The electron follows this trajectory of wave vector in tunneling from the valence to the conduction band:

$$w = 2\int dx\kappa(x) = \frac{2m}{\hbar p_{cv}}\int dx\sqrt{(E_{c0} - E)(E - E_{v0})} \tag{11.209}$$

The energy varies with position $E(x) = E_{v0} + V(x)$. The usual modeling of the Esaki diode assumes that the potential function $V(x) = Fx$ is due to a constant field F. A constant field is a reasonable aproximation of the actual potential, since it is needed over only a small distance in space:

$$E(x) = E_{v0} + xF \tag{11.210}$$

$$\int dx\sqrt{(E_{c0} - E)(E - E_{v0})} = \int_0^{E_G/F} dx\sqrt{(E_G - xF)xF} \tag{11.211}$$

$$= E_G\int_0^L dx\sqrt{\frac{x}{L}\left(1 - \frac{x}{L}\right)} = \frac{\pi L}{8}E_G$$

$$w = \frac{\pi m_e L E_G}{4\hbar p_{cv}} \tag{11.212}$$

where $L = E_G/F$ is the width of the tunneling region. Clean up the result by recalling that the matrix element p_{cv} is related to the effective mass m_c of the conduction band. In the present approximation of neglecting the kinetic energy $\varepsilon(k)$,

$$\frac{p_{cv}}{m_e} = \sqrt{\frac{E_G}{2m_c}} \tag{11.213}$$

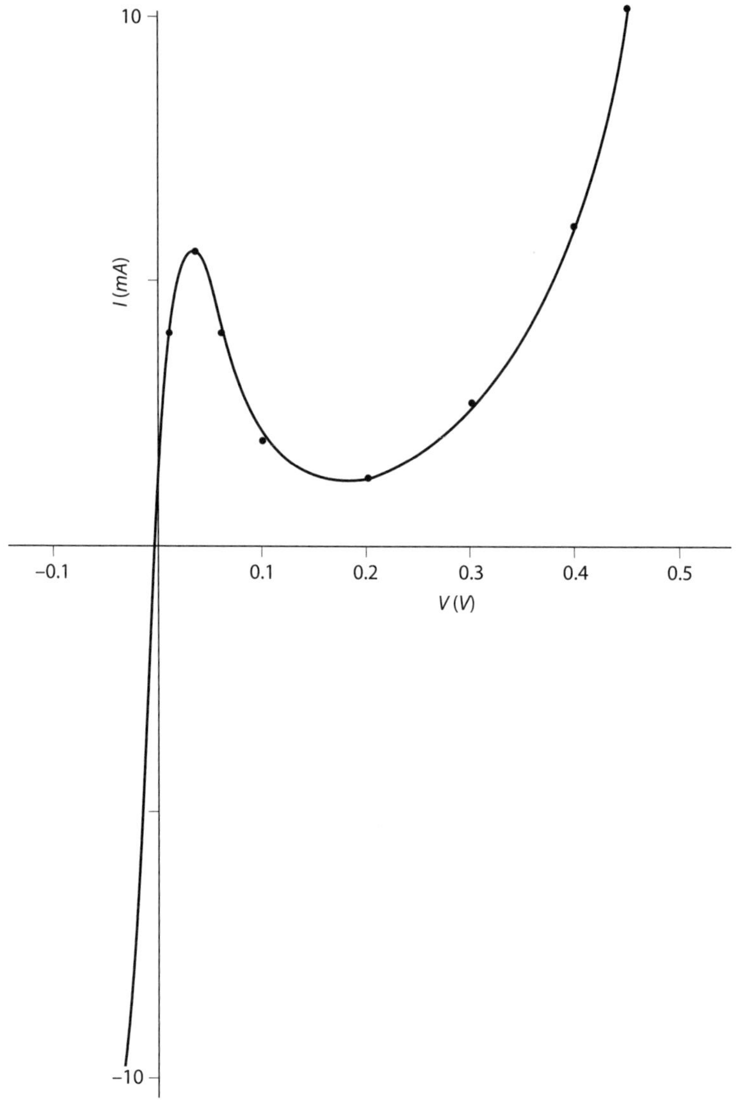

FIGURE 11.14. Esaki tunneling current as measured in GaAs. From Esaki, *Phys. Rev.* **109**, 603 (1958). Used with permission of the American Physical Society.

$$w = \frac{\pi}{4} \frac{E_G^{3/2} \sqrt{2m_c}}{\hbar F} \tag{11.214}$$

$$|T_{\mathrm{kp}}|^2 \propto e^{-w} \tag{11.215}$$

The tunneling exponent w is inversely proportional to the electric field F. The tunneling is more likely when the field is high, since then the tunneling distance L is shorter. The tunneling is also more likely for small values of the effective mass and energy gap.

An experimental measurement of the tunneling current is shown in fig. 11.14. Note the region of negative resistance, due to the lack of electron states in the band gap. Negative resistance regions are useful for making circuits that oscillate.

11.10.3 *Schottky Barrier Tunneling*

A Schottky barrier is formed when simple metals, such as gold, copper, tin, or lead, are put on the surface of most group IV or III–V semiconductors. Typical examples are Au on Si, and Pb on GaAs. The barrier is in the semiconductor region and results from trying to match the electron states of the metal onto the electron states in the semiconductor. Usually the chemical potential of the metal is pinned to the surface-state energy of the semiconductor, which is in the band gap:

$$E_{c0} > E_S > E_{v0} \tag{11.216}$$

The surface state has a high density of electron states at the surface. The electrons in the metal start to fill up this surface state, which charges up the interface. This charging retards other electrons from occupying the surface states, and a balance is achieved. The chemical potential in the bulk of the semiconductor usually has another value. The energy bands of the semiconductor must bend near the interface, which creates the Schottky barrier. This phenomenon was described in the previous chapter.

When a voltage V_a is applied between the metal and the semiconductor, it changes the width of the depletion region. The physical picture is that the energy bands in the semiconductor move rigidly:

- Upward in *forward bias*, which makes the depletion region thinner.
- Downward in *reverse bias*, which makes the depletion region wider.

The rate of electron tunneling is determined by the potential function $V(z)$ acting on an electron in the depletion region. The solution to Poisson's equation gives

$$V(z) = \frac{e^2 N_D}{2\varepsilon(0)}(z_0 - z)^2 \tag{11.217}$$

where $z = 0$ is at the interface. The value of the potential energy at the barrier height is

$$E_B = V(0) = \frac{e^2 N_D}{2\varepsilon(0)}(z_0)^2 \tag{11.218}$$

$$z_0 = \sqrt{\frac{2\varepsilon(0)\, E_B}{e^2 N_D}} \tag{11.219}$$

which gives the width of the depletion region. E_B was called V_{Bn} in an earlier chapter. With an applied voltage, this expression changes to

$$z_0(V_a) = \sqrt{\frac{2\varepsilon(0)}{e^2 N_D}(E_B - eV_a)} \tag{11.220}$$

The depletion region vanishes when $E_B = eV_a$, which is called the *flat band voltage*.

The potential $V(z)$ is a quadratic function of position, which enters into the formula for the WKBJ exponent $w(V_a)$:

$$E(z) = E_{c0} - V(z) \tag{11.221}$$

$$w = \frac{2m}{\hbar p_{cv}} \int_0^{z_0} dz \sqrt{V(z)[E_G - V(z)]} \tag{11.222}$$

The tunneling current is a function of the applied voltage V_a between the metal and the semiconductor. Figure 11.12c shows the energy bands in forward bias, where the voltage V_a raises the chemical potential of the semiconductor with respect to that of the metal. Assume that electrons tunnel from the conduction-band minimum of the semiconductor to the metal. Let $s = z_0 - z$, and for an n-type barrier

$$w = \frac{2m}{\hbar p_{cv}} \Lambda \int_0^{z_0} s ds \sqrt{z_G^2 - s^2} \tag{11.223}$$

$$\Lambda = \frac{e^2 N_D}{2\varepsilon(0)}, \quad z_G^2 = E_G/\Lambda \tag{11.224}$$

The distance z_G is the distance to tunnel from the conduction band to the valence band. An electron tunnels the distance $z_0(V_a)$ before entering the metal.

The integral gives

$$w = \frac{2m}{3p_{cv}} \Lambda \left[z_G^{3/2} - (z_G^2 - z_0^2)^{3/2} \right] \tag{11.225}$$

$$w = \frac{2}{3} \sqrt{\frac{2m_c}{E_G \Lambda}} \left[E_G^{3/2} - (E_G + eV_a - E_B)^{3/2} \right] \tag{11.226}$$

When the applied voltage V_a is positive, the value of w decreases and the tunneling becomes more likely. When $eV_a = E_B$ the depletion region vanishes, as does the tunneling barrier.

In Schottky barrier tunneling, the WKBJ exponent depends on the applied voltage. Padovani and Stratton used the voltage dependence of w to measure the function $\kappa(E)$, which maps out how the electrons tunnel from the conduction to the valence band. Their results are shown in fig. 11.15. Their initial measurements did not agree well with the predictions of $k \cdot p$ theory. Later investigators found better agreement. However, fig. 11.15 is historic.

11.10.4 *Effective Mass Matching*

When an electron tunnels from a metal to a semiconductor, the effective mass changes dramatically. In simple metals the effective mass is about one electron mass m_e. In the conduction band of GaAs the effective mass is $m_c \approx m_e/16$. This difference in the effective mass influences how one matches the wave functions at the interface between the two materials. Normally in quantum mechanics, one matches the wave function and its normal derivative on the two sides of the junction. When the effective masses are different on the two sides of the junction, this procedure gives the wrong answer. Current is not conserved: more particles approach the junction than leave, or vice versa.

When the effective masses are different on the left (L) and right (R) sides of the junction, the proper matching at $x = 0$ uses the effective masses.

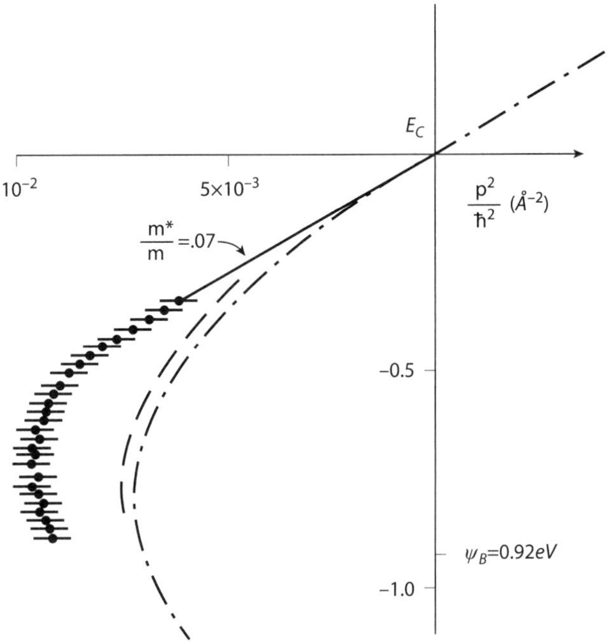

FIGURE 11.15. Tunneling dispersion through the forbidden band gap of GaAs, as measured in a Schottky barrier of GaAs. From Padovani and Stratton, *Phys. Rev. Lett.* **16**, 1202 (1966). Points are data, line is theory. Used with permission of the American Physical Society.

- Match the two wave functions: $\psi_L(0^-) = \psi_R(0^+)$.
- Match derivatives normalized by the effective mass

$$\frac{1}{m_L}\left(\frac{d\psi_L(x)}{dx}\right)_{x=0^-} = \frac{1}{m_R}\left(\frac{d\psi_R(x)}{dx}\right)_{x=0^+} \tag{11.227}$$

As a simple example, consider a junction with a flat band: planes waves come in from the left and leave on the right. The wave vector is k on the left and p on the right:

$$\psi(x) = \begin{cases} e^{ikx} + \mathrm{Re}^{-ikx} & x < 0 \\ Te^{ipx} & x > 0 \end{cases} \tag{11.228}$$

The above matching gives

$$I + R = T \tag{11.229}$$

$$v_L(I - R) = v_R T, \quad v_L = \frac{\hbar k}{m_L}, \quad v_R = \frac{\hbar p}{m_R} \tag{11.230}$$

with a solution

$$R = \frac{v_L - v_R}{v_L + v_R}, \quad T = \frac{2v_L}{v_L + v_R} \tag{11.231}$$

The particle currents on the two sides are

$$j_L = v_L(1 - R^2) = j_R = v_p T^2 \tag{11.232}$$

Using the expressions for R and T, one can verify that the currents are the same on the two sides of the interface. This identity is achieved only when using the effective mass in matching the derivatives of the wave functions.

11.11 Phonon Transport

11.11.1 Transport in Three Dimensions

All solids have a thermal conductivity K_p due to phonons. In metals the electronic term K_e is often much larger than K_p. Spin waves can carry heat in a magnet. In nonmagnetic insulators, all heat is carried by phonons. The phenomenological starting point is

$$K_p = \frac{1}{3}\bar{v}\bar{\ell}C_V \tag{11.233}$$

where C_V is the heat capacity:

- In gases, $\bar{v} = \sqrt{3k_BT/M}$ and $\bar{\ell}$ are the mean velocities and mean free path of the atoms. The heat capacity is that of the gas. For a gas of single atoms, at room temperature, it is $C_V = 3nk_B/2$, where n is the density of atoms.
- For electrons in a metal, $\bar{v} = v_F$ is the average Fermi velocity. The average mean free path of electrons at the Fermi energy is $\bar{\ell} = v_F\tau_t(k_F)$.
- In an insulator, the excitations are phonons. So $\bar{v}$ and $\bar{\ell}$ are the average velocity and average mean free path, of the phonons. The heat capacity at room temperature is that of the phonons $C_V = 3nk_B$, where n is the density of atoms. The mean free path is obtained from a solution of the phonon Boltzmann equation.

Figure 11.16 shows Glen Slack's data [14] for the thermal conductivity of silicon as a function of temperature. A log scale is used for temperature to cover a wide range. The three letters denote different scattering regimes.

- A is the low-temperature regime. The acoustic phonons can travel long distances without scattering. They mainly scatter from the walls of the crystal, which yields a $K_p \propto T^3$ dependence.
- C is the high-temperature regime. Here anharmonic effects are important: two phonons combine into one, or one decays into two. In this region the inverse lifetime is proportional to the number of phonons, which is proportional to temperature. So $\tau \propto 1/T$, which provides the basic temperature dependence.
- The peak region B is where size effects (region A) and anharmonic effects (region C) are both small. Often the main scattering of phonons in pure samples is by the isotopic mass fluctuations.

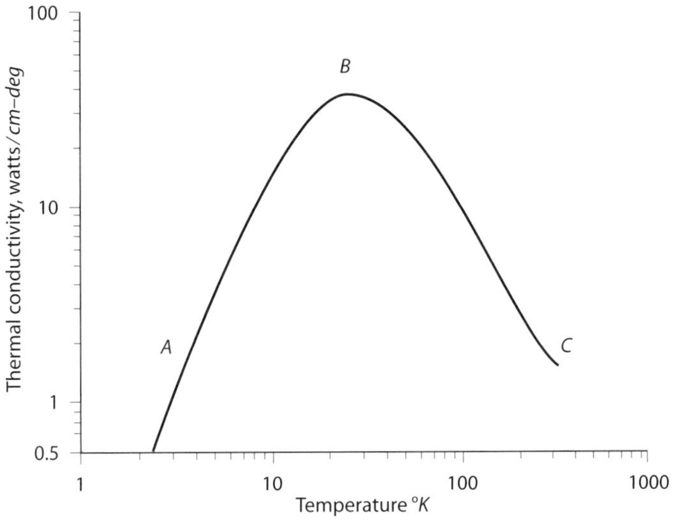

FIGURE 11.16. Typical result for the thermal conductivity of an insulator as a function of temperature. The letters A, B, C refer to different scattering regimes. Data are for silicon. From Slack et al., *Phys. Rev.* **134**, A1058 (1964). Used with permission of the American Physical Society.

Diamond has the largest thermal conductivity of any solid at room temperature: $K = 1.8$ kW/m·K. Natural diamond is 99% carbon-12 and 1% carbon-13. Tom Anthony grew an isotopically pure crystal of diamond composed of carbon-12 [2]. The room-temperature thermal conductivity rose to 3.3 kW/m·K. The small amount of carbon-13 in natural diamond provides much of the scattering mechanism at room temperature. The isotopically pure sample of diamond is the material with the largest value of thermal conductivity at room temperature. Isotopically pure samples of other crystals have also been measured. Figure 11.17 shows results for different samples of germanium. From the top down, the measurements are (i) 99.99% of ^{70}Ge; (ii) 96.3% of ^{70}Ge; (iii) 86% of ^{76}Ge; (v) natural Ge with five isotopes, and (vi) a mixture of two isotopes 70/76 (from Asen-Palmer et al. [3]). The dashed line on the left shows the T^3 law for boundary scattering. The dashed line on the right shows the $K \propto 1/T$ law for anharmonic scattering.

11.11.2 *Minimum Thermal Conductivity*

Some of the smallest thermal conductivity values are found for amorphous materials, such as glasses. The *minimum thermal conductivity* is calculated in the following fashion at room temperature.

- The internal energy is taken to be $3k_BT$ per atom, typical of a local oscillator, so the heat capacity is $C = 3k_B/\Omega_a$, where Ω_a is the volume per atom.
- The mean free path $\ell = a$, the interatomic distance. It is assumed that the phonons scatter every time they meet another atom.

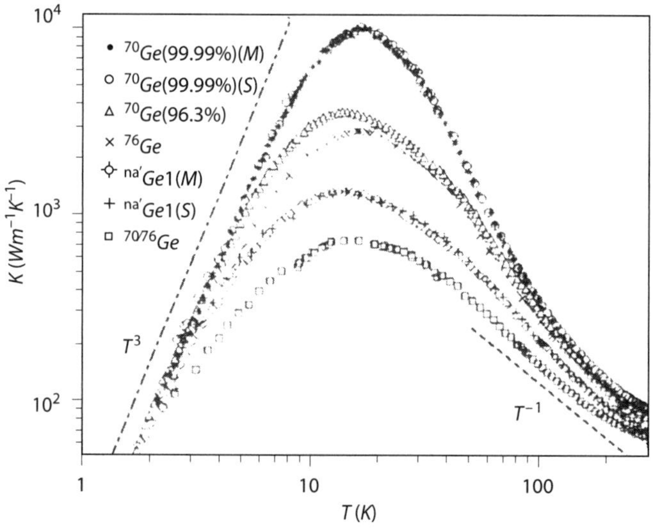

FIGURE 11.17. Thermal conductivity of germanium for different isotopic samples. From the top down, the measurements are (i) 99.99% of ^{70}Ge; (ii) 96.3% of ^{70}Ge; (iii) 86% of ^{76}Ge; (v) natural Ge with five isotopes, and (vi) a mixture of two isotopes 70/76. From Asen-Palmer et al., *Phys. Rev. B* **56**, 9431 (1997). Used with permission of the American Physical Society.

- The average velocity $\bar{v}$ is the average of the three acoustic modes. Taking values such as $\bar{v} = 3$ km/s, $\Omega_a = a^3$, $a = 3$ Å, gives

$$K_{\min} \approx 0.5 \frac{W}{m \cdot K} \tag{11.234}$$

This value is typical of amorphous materials.

The thermal conductivity in such materials is not sensitive to temperature, except at low temperatures where the heat capacity decreases.

11.11.13 *Kapitza Resistance*

Kapitza [16] discovered that a temperature discontinuity ΔT existed at the interface between superfluid liquid helium and its metal container. He thought it was a superfluid phenomenon, but a similar temperature jump occurs whenever heat flows (J_Q) between two different materials:

$$\Delta T = R J_Q \tag{11.235}$$

$$J_Q = \sigma \Delta T, \quad \sigma = \frac{1}{R} \tag{11.236}$$

R is called the Kapitza resistance, and has the units of degrees area per watt. See Pollack [23] for a review of boundary resistance between liquid He and various metals.

Table 11.1 Room temperature Kapitza boundary conductances for several systems, in units of MW/m²·K

Solid 1	Solid 2	σ
Pb	diamond	20
Al	Al_2O_3	200
Si	Si	700

Note. Data from Cahill et al. [9]

There is a Kapitza resistance at a grain boundary between the same material. In solids, heat is carried by both electrons and phonons. Here the discussion is limited to insulators, where the heat is carried by phonons. The thermal resistance is caused by the feature that phonons are partially reflected at an interface. The constant σ can be calculated from a knowledge of the phonon transmission and reflection at an interface. Table 11.1 shows a few experimental values. The first two cases have electronic heat flow as well as phonon heat flow. The Si/Si boundary is for a low-index grain boundaries.

There are several theories of the Kapitza resistance. The earliest theory used elasticity and considered only the transmission of sound waves. One matches the normal component of the elastic displacement vector $\mathbf{u}$ as well as the stress tensor. The transmission coefficient for longitudinal sound waves is

$$|\mathcal{T}|^2 = \frac{4Z_L Z_R}{(Z_L + Z_R)^2}, \quad Z_j = \rho_j C_j \tag{11.237}$$

where Z_j is the acoustic impedence on the two sides of the interface. The impedance is the product of the sound velocity and the density (kg/volume).

Consider that the boundary between two solids is a plane, with unit normal vector $\hat{n}$. The energy current of phonons approaching the interface from the right side is

$$J_{QR} = \sum_{\eta} \int \frac{d^3q}{(2\pi)^3} \hbar\omega_\eta(\mathbf{q})\, \hat{n}\cdot \mathbf{v}_\eta(\mathbf{q})\, n_B[\omega_\eta(\mathbf{q})] \tag{11.238}$$

where η is the polarization. The integral is over the part of the Brillouin zone that has $\hat{n}\cdot v_\eta(\mathbf{q}) > 0$, so only phonons going toward the interface are included. Part of the phonon energy is transmitted and part is reflected. So include the fraction $|\mathcal{T}|^2$ that is transmitted to get the heat flowing through the interface:

$$J_{QR} = \sum_{\eta} \int \frac{d^3q}{(2\pi)^3} \hbar\omega_\eta(\mathbf{q})\, \hat{n}\cdot \mathbf{v}_\eta(\mathbf{q})\, n_B[\omega_\eta(\mathbf{q})]\, |\mathcal{T}|^2 \tag{11.239}$$

The net heat flow through the interface is $J_Q = J_{QR} - J_{QL}$. At high temperature the Bose-Einstein occupation factor is approximated as $n_B[\omega_\eta(\mathbf{q})] \approx k_BT/\hbar\omega_\eta(\mathbf{q})$. The phonon energy is canceled. The difference between $T_R - T_L = \Delta T$. The formula for the Kapitza conductance is

$$J_Q = \sigma \Delta T \tag{11.240}$$

$$\sigma = \frac{k_B}{2} \sum_{\eta} \int \frac{d^3q}{(2\pi)^3} |\hat{n} \cdot \mathbf{v}_\eta(\mathbf{q})| |\mathcal{T}|^2 \tag{11.241}$$

Since half of the phonons are going toward the surface and half away from it, one can replace the requirement that $\hat{n} \cdot \mathbf{v}_\eta(\mathbf{q}) > 0$ by the expression $|\hat{n} \cdot \mathbf{v}_\eta(\mathbf{q})|/2$. The above formula gives the Kapitza conductance as a function of phonon properties. Young and Maris [32] were the first to calculate $\mathcal{T}$ by matching the displacements of atoms in the surface region. Their method included all phonon modes and all wave vectors. It is a much better theory of the Kapitza resistance than acoustic mismatch.

The above formula suggests a new quantity called the *maximum Kapitza conductance* σ_{max}, which is found by assuming that all phonons get through the interface without scattering, or $|\mathcal{T}| = 1$:

$$\sigma_{max} = \frac{k_B}{2} \sum_{\eta} \int \frac{d^3q}{(2\pi)^3} |\hat{n} \cdot \mathbf{v}_\eta(\mathbf{q})| \approx \frac{k_B \langle C_s \rangle}{2\Omega_0} \tag{11.242}$$

where Ω_0 is the volume of the Wigner-Seitz cell, and $\langle C_s \rangle$ is an average phonon velocity. Its numerical value is about 1 $\mathrm{GW/m^2 \cdot K}$ for most solids. Its inverse is the minimum value of thermal resistance $R_{min} = 1/\sigma_{max}$.

11.11.4 Measuring Thermal Conductivity

The traditional method of measuring the thermal conductivity K of a material is to have a bar of length L and area A. A temperature difference ΔT is maintained between the ends of the bar. The rate of flow of heat $\dot{Q}$, whose units are watts, is given by

$$\dot{Q} = KA \frac{\Delta T}{L} \tag{11.243}$$

This method is best, but has some problems. It is quite expensive in terms of manpower to grow a new material into a large single crystal, from which one could cut a bar. Indeed, most new materials are prepared either in powder form or as a thin film grown on a substrate. How does one measure the thermal conductivity in these cases? An additional problem is the possible Kapitza resistance between the two ends of the bar and the thermal reservoirs at each end. How much of the ΔT is along the length of the bar and how much is due to the boundary resistance? This problem is acute when measuring diamond samples.

Powder samples are measured using the *hotwire method*, as shown in fig. 11.18a. A hollow cylinder is packed with the powder material. An current pulse $I(t)$ is sent along a wire that runs down the cener of the cylinder. The Joule heating of the wire heats the powder, and the heat diffuses outward. A measurement of the temperature of the wire, as a function of time, gives information regarding the value of thermal diffusion D. The temperature of the wire is found from its resistivity $\rho(T)$.

A proper model of this device must be solved using cylindrical coordinates. That complicated solution is not given here, since it involves Bessel functions. Instead, solve the

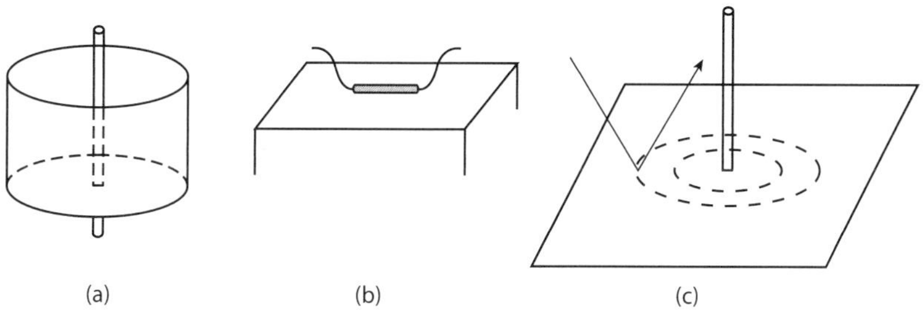

FIGURE 11.18. Methods of measuring thermal conductivity: (a) hot wire, (b) 3ω, (c) mirage.

equivalent differential equation in one dimension. The thermal diffusion equation, with a short pulse at the origin has a solution

$$\frac{\partial T}{\partial t} - D\frac{\partial^2 T}{\partial x^2} = S_0\delta(x)\delta(t) \tag{11.244}$$

$$T(x,t) = \frac{S_0}{\sqrt{\pi Dt}}\exp\left[-\frac{x^2}{4Dt}\right]\Theta(t) \tag{11.245}$$

At long times the heat flow extends over large distances. This feature is inconvenient for the measurement of thin films. A much better method uses an ac driving source at the surface. Consider the solution to

$$\frac{\partial T}{\partial t} - D\frac{\partial^2 T}{\partial x^2} = S_0\delta(x)\sin(\omega t)\Theta(t) \tag{11.246}$$

$$T(x,t) = \frac{S_0}{\sqrt{\omega D}}e^{-x/L}\sin(x/L - \omega t + \pi/4) \tag{11.247}$$

$$L = \sqrt{\frac{2D}{\omega}} \tag{11.248}$$

With an ac driving force at the surface, the heat flow penetrates only a distance L into the material. One can make L small by having a high frequency ω. This type of solution to the diffusion equation is called a *diffusion wave.*

This observation is the basis of the 3ω method of measuring the thermal diffusion constant D. One evaporates a thin wire strip on the surface of the material. The temperature dependence of the resistivity $\rho(T)$ is measured. Then one puts an alternating current $I(t) = I_0\sin(\omega t)$ through the wire strip. The Joule heating of the wire is initially

$$S(t) = \rho I_0^2\sin^2(\omega t) = \frac{\rho I_0^2}{2}[1 - \cos(2\omega t)] \tag{11.249}$$

The first term on the right will cause a general heating of the strip. The term with $\cos(2\omega t)$ causes a diffusion wave. It will cause a term in the resistance

$$\rho(t) = \rho(T_0) + \delta T\left(\frac{d\rho}{dT}\right)\cos(2\omega t) \tag{11.250}$$

Multiply this resistance by the current to get the voltage $V(t)$. The voltage has a term

$$\cos(2\omega t)\,\sin(\omega t) = \tfrac{1}{2}[\sin(3\omega t) - \sin(\omega t)] \tag{11.251}$$

The term in the voltage that oscillates with a frequency of 3ω gives the response of the wire to the diffusion wave. It is an accurate method of measuring D, and works on thin films. See Cahill [8] for the details.

The *mirage method* of measuring the thermal diffusivity makes no contacts to the sample. It has no problems with Kapitza resistance and was the first accurate measurement of the thermal conductivity of isotopically pure diamond. The geometry is shown in fig. 11.18c. A focused laser sends periodic pulses with a frequency ω to a central spot on the surface. A diffusion wave is generated in the material, which also propagates outward along the surface. A second laser probes the temperature at the surface as a function of distance from the central spot. The distance λ from the central spot to the first maximum of the diffusion wave provides a direct measurement of D. See Anthony et al. [2] for experimental details.

Finally, there are other modern methods of measuring thermal diffusion. An infrared camera can detect small temperature variations over micron distances at millisecond time scales. One can excite one spot with a laser pulse and photograph the sample as a function of time. One sees the entire heat flow, from which one can deduce the thermal diffusion.

11.12 Thermoelectric Devices

There are several types of thermoelectric devices. Several of them are solved here in one dimension, where heat is flowing down a bar. The device is operated in a steady-state fashion, so there is no time dependence. From the equation of continuity, in one dimension,

$$0 = \frac{\partial}{\partial t}\rho(x, t) + \frac{\partial J}{\partial x} = \frac{\partial J}{\partial x} \tag{11.252}$$

The current density J (amperes per area) is a constant. All of the devices require a knowledge of the temperature profile $T(x)$. One envisions a bar of length L where the cold (T_c) end is at $x = 0$ and the hot (T_h) end is at $x = L$. The conservation of heat flow is

$$-K\frac{d^2T}{dx^2} = \rho J^2 \tag{11.253}$$

where K is the thermal conductivity, ρ is the resistivity, and the term on the right is from Joule heating. Other parameters are the Seebeck coefficient S and *figure of merit* $Z = S^2/(\rho K)$. Good thermoelectric materials have a large value of Z. The quantity ZT is dimensionless, and good thermoelectric materials are those with $ZT \geq 1$.

The above equation assumes that the Thomson coefficient dS/dT is negligible. The current J is a constant. The material parameters (K, ρ, S, Z) depend upon temperature, but this dependence is ignored in the present discussion.

The solution to the above equation is

$$T(x) = T_c + \frac{x}{L}\Delta T + \frac{\rho J^2}{2K}x(L - x), \quad \Delta T = T_h - T_c \tag{11.254}$$

The heat current is

$$J_Q = T(x)JS - K\frac{dT}{dx} \tag{11.255}$$

$$= JS\left[T_c + \frac{x}{L}\Delta T + \frac{\rho J^2}{2K}x(L-x)\right] - K\frac{\Delta T}{L} - \frac{\rho J^2}{2}(L-2x)$$

These formulas are used below.

11.12.1 *Maximum Cooling*

What is the lowest temperature one can achieve with a one-stage thermoelectric device, starting at room temperature ($T_h = 300$ K)? The geometry is shown in fig. 11.19a. The hot end is connected to a reservoir, but the cold end is in vacuum. If there is no load on the cold end, then no heat flows at $x = 0$:

$$0 = J_Q(0) = JST_c - K\frac{\Delta T}{L} - \frac{\rho J^2 L}{2} \tag{11.256}$$

The above equation can be solved for T_c as a function of the applied current J:

$$T_c(J) = \frac{T_h + J^2(\rho L^2/2K)}{1 + (JSL/K)} \tag{11.257}$$

Vary J to find the minimum value of T_c. It is convenient to define a dimensionless variable $y = JSL/K$ and to write the above formula as

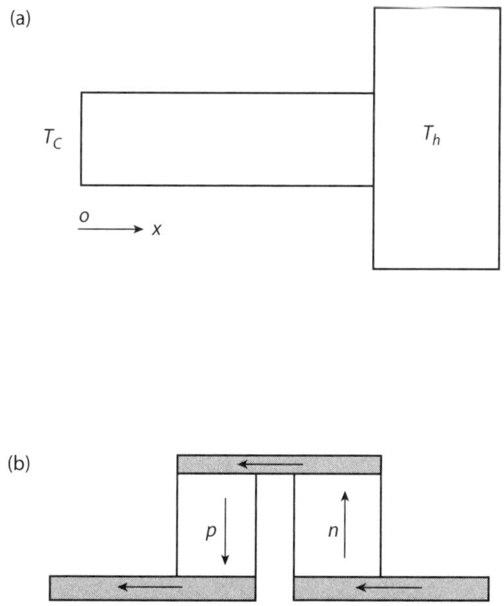

FIGURE 11.19. (a) Drawing of a bar of thermoelectric material (shaded region) connected to a hot reservoir. (b) Shows an n-type and p-type device in series.

$$ZT_c(y) = \frac{ZT_h + y^2/2}{1 + y} \tag{11.258}$$

The minimum value of y, called y_0, is found using

$$0 = \frac{d}{dy}(ZT_c) = \frac{1}{(1+y)^2}\left[y(1+y) - (ZT_h + y^2/2)\right] \tag{11.259}$$

$$y_0 = \sqrt{1 + 2ZT_h} - 1 \tag{11.260}$$

$$T_c(y_0) = \frac{1}{Z}\left[\sqrt{1 + 2ZT_h} - 1\right] \tag{11.261}$$

Note that if $ZT_h = 1$ then $T_c = 220$ K, which is correct. A single-stage thermoelectric module can cool down to about $T_c = -53$°C. The formula for the temperature difference is

$$\Delta T = \frac{1}{2Z}\left[\sqrt{1 + 2ZT_h} - 1\right]^2 \tag{11.262}$$

If $Z \to \infty$, then $T_c \to 0$. This will not happen since Z never diverges. Also, the temperature dependence of Z is pronounced at low temperature.

All thermoelectric devices have the feature that they are more efficient when ZT is as large as possible. Figure 11.20 shows experimental values of ZT vs. temperature for several materials. Near room temperature, the best material is an alloy of bismuth and antimony telluride. They have the same crystal structure, and the alloys are single crystals. At higher temperatures, other materials are good, such as PbTe and the filled skutterudite

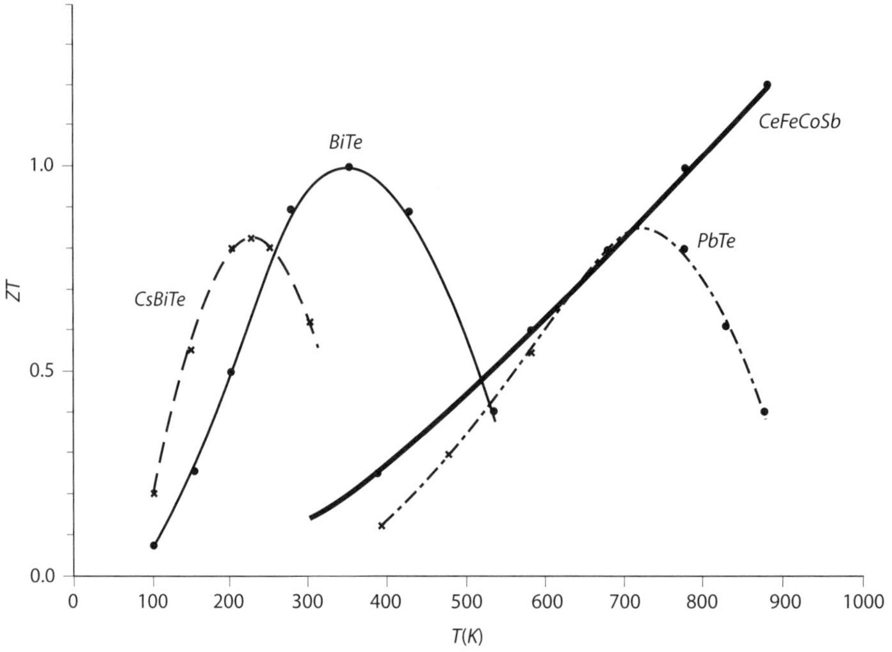

FIGURE 11.20. ZT as a function of T for several good thermoelectric materials, including Bi_2Te_3 and the filled skuderudite $CeFe_3CoSb_{12}$.

$CeFe_3CoSb_{12}$ [28]. See the review by Mahan for more details. Each material has a high value of ZT over a limited range in temperature.

11.12.2 *Refrigerator*

A thermoelectric refrigerator pumps heat from the cold reservoir to the hot reservoir. The temperature profile is again

$$T(x) = T_c + \frac{x}{L}\Delta T + \frac{\rho J^2}{2K}x(L - x), \quad \Delta T = T_h - T_c \tag{11.263}$$

The voltage variation in the bar is given by solving

$$J = -\sigma\left[\frac{dV}{dx} + S\frac{dT}{dx}\right] \tag{11.264}$$

$$\frac{dV}{dx} = -S\frac{dT}{dx} - \rho J \tag{11.265}$$

The total voltage drop is

$$\Delta V = -\int_0^L dx\frac{dV}{dx} = S\Delta T + \rho LJ \tag{11.266}$$

The Carnot definition of efficiency η is the heat pumped from the cold reservoir $J_{Qc} = J_Q(x = 0)$ divided by the input power $J\Delta V$:

$$J_{Qc} = J_Q(x = 0) = SJT_c - \frac{K\Delta T}{L} - \frac{\rho J^2}{2}L \tag{11.267}$$

$$\eta = \frac{J_{Qc}}{J\Delta V} = \frac{SJT_c - K\Delta T/L - \rho J^2 L/2}{J(S\Delta T + \rho JL)} \tag{11.268}$$

The current J is varied to find the value J_0 that gives the maximum efficiency. The maximum efficiency is called the *coefficient of performance*, and is usually abreviated COP = $\eta(J_0)$. Multiply the above equation by ρL in numerator and denominator, and define a new variable $Y = \rho JL$:

$$\eta = \frac{ST_c Y \quad \rho K\Delta T - Y^2/2}{Y(S\Delta T + Y)} \tag{11.269}$$

Set $d\eta/dY = 0$ and find $[\bar{T} = (T_c + T_h)/2]$:

$$Y_0 = \frac{\rho K\Delta T}{S\bar{T}}[1 \pm \gamma] \tag{11.270}$$

$$J_0 = \frac{Y_0}{\rho L} = \frac{K\Delta T}{S\bar{T}L}[1 + \gamma] \tag{11.271}$$

$$\gamma = \sqrt{1 + Z\bar{T}} \tag{11.272}$$

$$\mathrm{COP} = \frac{\gamma T_c - T_h}{\Delta T(1+\gamma)} \tag{11.273}$$

Getting to the last formula from $\eta(Y_0)$ takes some algebra, but the result is a simple formula. The best thermoelectric materials have $Z\bar{T} \approx 1$ so that $\gamma \approx \sqrt{2}$. The ideal Carnot efficiency $\eta_C = T_c/\Delta T$ is obtained in the limit of a perfect material ($Z \to \infty$). Clearly, practical values of $Z\bar{T} \approx 1$ are far from this ideal. The COP for compressor-based refrigerators is higher than that of a thermoelectric refrigerator, which is why the cooler in your kitchen uses compressors. However, thermoelectric devices, which have no moving parts, are very reliable. They are used in situations where reliability is more important than efficiency, such as in space stations.

The typical voltage scale of a thermoelectric device is $\Delta V \sim S\Delta T$. A good material has $S \sim 250\ \mu\mathrm{V/K}$. If the temperature variation in the refrigerator is $\Delta T \sim 40$ K, then $\Delta V \sim 0.01$ V, which is small. The ideal current J_0 is found from eqn. (11.271). A typical low value of thermal conductivity is $K \sim 1.0$ W/K m, which gives $J_0 \sim 14\ \mathrm{amps/cm^2}$, which is high. Thermoelectric devices have small voltages and large currents. So many identical modules are connected in series. Figure 11.20b shows a *pn couple* composed of a *n*-type semiconductor and a *p*-type semiconductor connected in series. The arrows show the flow of electrical current, and the shaded regions are copper. Holes carry heat in the direction of current ($S_h > 0$), while electrons carry it in the opposite direction ($S_e < 0$). A thermoelectric module is composed of 20–50 such couples in series.

11.12.3 *Power Generation*

Another thermoelectric device also has a module between a hot and a cold reservoir. Heat is allowed to flow from the hot to the cold side of the thermoelectric material. This heat flow drags current, which serves as a power generator. The Carnot definition of efficiency, in this case, is the power generated divided by the heat flow from the hot side:

$$J_{Qh} = J_Q(x = L) = SJT_h - \frac{K\Delta T}{L} + \frac{\rho J^2}{2}L \tag{11.274}$$

$$\eta = \frac{J\Delta V}{J_{Qh}} = \frac{J(S\Delta T + \rho J L)}{SJT_h - \frac{K\Delta T}{L} + \frac{\rho J^2}{2}L} \tag{11.275}$$

The current J is varied to find the value J_0 that gives the maximum efficiency. Multiply the above equation by ρL, in numerator and denominator, and define a new variable $Y = \rho J L$:

$$\eta(Y) = \frac{Y(S\Delta T + Y)}{ST_h Y - \rho K\Delta T + Y^2/2} \tag{11.276}$$

$$0 = \frac{d\eta}{dY} \to Y_0 = \frac{\rho K \Delta T}{S\bar{T}}[1 \pm \gamma] \tag{11.277}$$

In this case take the last factor to be $(1-\gamma)$. Since $\gamma > 1$, the current is negative:

$$J_0 = -\frac{K\Delta T}{S\bar{T}L}[\gamma - 1] \tag{11.278}$$

$$\eta(Y_0) = \frac{\Delta T(\gamma - 1)}{\gamma T_h + T_c} \tag{11.279}$$

Given a thermoelectric module between a hot and a cold reservoir, running the electrical current in one direction produces refrigeration, while running the electrical current in the opposite direction produces electrical power.

The maximum efficiency of the power generator is given by the last formula. Again, a perfect thermoelectric has $Z \rightarrow \infty$, which gives the ideal Carnot efficiency of $\eta_C = \Delta T/T_h$. For actual devices, with $\gamma \approx \sqrt{2}$, the power conversion has a low efficiency.

References

1. P. B. Allen, Theory of thermal relaxation of electrons in metals. *Phys. Rev. Lett.* **59**, 1460–1463 (1987)
2. T. R. Anthony, W. F. Banholzer, J. F. Fleischer, L. Wei, P. K. Kuo, R. L. Thomas, and R. W. Pryor, Thermal diffusivity of isotopically enriched ^{12}C diamond. *Phys. Rev. B* **42**, 1104–1111 (1990)
3. M. Asen-Palmer, K. Bartkowski, E. Gmelin, M. Cardona, A. P. Zhernov, A. V. Inyushkin, A. Taldenkov, V. I. Ozhogin, K. M. Itoh, and E. E. Haller, Thermal conductivity of germanium crystals with different isotope compositions. *Phys. Rev. B* **56**, 9431–9447 (1997)
4. A. Bailyn, Transport in metals: effect of the nonequilibrium phonons. *Phys. Rev.* **112**, 1587–1598 (1958)
5. J. Bardeen, Tunnelling from a many-particle point of view. *Phys. Rev. Lett.* 6, 57–59 (1961)
6. N. Cusack and P. Kendall, The absolute scale of thermoelectric power at high temperatures. *Proc. Phys. Soc.* **72**, 898–901 (1958)
7. M. Ben Dahan, E. Peik, J. Reichel, Y. Castin, and C. Salomon, Bloch oscillations of atoms in an optical potential. *Phys. Rev. Lett.* **76**, 4508–4511 (1996)
8. D. G. Cahill, Thermal conductivity measurement from 30 to 750 K: the 3ω method. *Rev. Sci. Instrum.* **61**, 802–808 (1990)
9. D. G. Cahill, W. K. Ford, K. E. Goodson, G. D. Mahan, A. Majumdar, H. J. Maris, R. Merlin, and S. R. Phillpot, Nanoscale thermal transport. *J. Appl. Phys.* **93**, 793–818 (2003).
10. J. W. Conley and G. D. Mahan, Tunneling spectroscopy in GaAs. *Phys. Rev.* **161**, 681–695 (1967)
11. L. Esaki, New phenomenon in narrow germanium *p-n* junctions. *Phys. Rev.* **109**, 603–604 (1958)
12. I. Giaever, Energy gap in superconductors measured by electron tunneling. *Phys. Rev. Lett.* **5**, 147–148 (1960)
13. I. Giaever, Electron tunneling between two superconductors. *Phys. Rev. Lett.* **5**, 464–466 (1960)
14. C. J. Glassbrenner and G. A. Slack, Thermal conductivity of silicon and germanium from 3 K to the melting point. *Phys. Rev.* **134**, A1058–A1069 (1964)
15. R. A. Höpfel, J. Shah, P. A. Wolff, and A. C. Gossard, Negative absolute mobility of minority electrons in GaAs quantum wells. *Phys. Rev. Lett.* **56**, 2736–2739 (1986)
16. P. L. Kapitza, *J. Phys. USSR* **4**, 181 (1941)
17. R. Landauer, Can a length of perfect conductor have a resistance. *Phys. Lett.* **85A**, 91–93 (1981)

18. G.D. Mahan, Good Thermoelectrics. *Solid State Physics*, Vol. 51, pp. 82–157 (Academic Press, 1997)
19. J. H. Mooij, Electrical conduction in concentrated disordered transition metal alloys. *Phys. Status Solidi A* **17**, 521–530 (1973)
20. V. U. Nazarov, J. M. Pitarke, Y. Takada, G. Vignale, and Y. C. Chang, Time-dependent current density functional theory: application to the stopping power of electron liquids. *Phys. Rev. B* **76**, 205103 (2007).
21. F. A. Padovani and R. Stratton, Experimental energy-momentum relationship determination using Schottky barriers. *Phys. Rev. Lett.* **16**, 1202–1204 (1966)
22. M. B. Pogrebinskii, Mutual drag of carriers in a semiconductor-insulator-semiconductor system. *Sov. Phys. Semicond.* **11**, 372–376 (1977)
23. G. L. Pollack, Kapitza Resistance. *Revs. Mod. Phys.* **41**, 48–81 (1969)
24. M. J. Puska and R. M. Nieminen, Atoms embedded in an electron gas: Phase shifts and cross sections. *Phys. Rev. B* **27**, 6121–6128 (1983)
25. D. L. Rode, Electron mobility in Ge, Si, and GaP. *Phys. Status Solidi B* **53**, 245–254 (1972)
26. D. L. Rode, Low Field Transport, in *Semiconductors & Semimetals*, Vol. 10, pp. 1–89 (Academic Press, New York, 1975)
27. A. G. Rojo, Electron-drag effects in coupled electron systems. *J. Phys. Condens. Matter* **11**, R31–R52 (1999)
28. B. C. Sales, D. Mandrus, and R. K. Williams, Filled skutterudite antimonides: a new class of thermoelectric materials. *Science* **272**, 1325–1328 (May 31, 1996)
29. K. Schwab, E. A. Hendricksen, J. M. Worlock, and M. L. Roukes, Measurement of the quantum of thermal conductance. *Nature*, **404**, 974 (27 April 2000)
30. P. M. Smith, M. Inoue, and J. Frey, Electron velocity saturation in Si and GaAs at very high electric fields. *Appl. Phys. Lett.* **37**, 797–799 (1980)
31. T. M. Tritt, ed., *Thermal Conductivity* (Kluwer/Plenum, New York, 2004)
32. D. A. Young and H. J. Maris, Lattice dynamical calculation of the Kapitza resistance between fcc lattices. *Phys. Rev. B* **40**, 3685–3693 (1989)
33. G. Wiedemann and R. Franz, On the thermal conductivity of metals. *Ann. Phys.* **89**, 497–531 (1853)

Homework

1. Assume the lifetime of an electron is from impurity scattering, and that the T-matrix has the feature that T_{kp} is a constant. Calculate the electrical conductivity and Seebeck coefficient of a metal.

2. The transport lifetime of an electron in a semiconductor has a power law dependence on the wave vector k, $\tau_t = wk^r$. How do the electrical conductivity and Seebeck coefficient depend on r and other parameters?

3. The electrical conductivity is $\sigma = ne^2\tau/m = \omega_p^2\tau/4\pi$. The electrical resistivity $\rho = 1/\sigma$ is the inverse of this quantity:

$$\tau = \frac{4\pi}{\rho\omega_p^2} \tag{11.280}$$

Metal	ρ ($\mu\Omega$ cm)	$\hbar\omega_p$ (eV)	r_s
Na	4.2	5.8	3.96
Mg	4.45	10.6	2.66
Al	2.65	15.3	2.06

We list some room-temperature values of ρ (*Handbook of Chemistry and Physics*) and other values (*Many-Particle Physics*) for three simple metals. Use them to calculate the numerical value of the electron lifetime τ and mean free path $\ell = v_F\tau$ for the three metals. Note that ρ is listed in laboratory units, but must be converted to cgs units of seconds to evaluate the expression.

4. Assume the distribution function in the Boltmann equation has the form of a *drifted Maxwellian*:

$$f(v) = g_0 \exp\left[-\frac{m(\mathbf{v}-\mathbf{v}_0)^2}{2k_BT}\right] \tag{11.281}$$

where the particles have an average drift velocity $\mathbf{v}_0$. Do the integrals for the density n and current $(\mathbf{J})$ to express the current in terms of n and $\mathbf{v}_0$.

5. The formula for heat current carried by phonons is

$$J_Q = \frac{1}{L}\sum_{q\lambda} v_{k\lambda}\hbar\omega_\lambda(q)\, n_B[\omega_\lambda(q)] \tag{11.282}$$

where $v_{k\lambda}$ is the phonon veloicity of a phonon of mode λ with wave vector k. Find the quantum of thermal conductance for phonon heat transport.

6. Calculate the lifetime τ_{eh} for the case of holes in a single band of mass m_v of density n_h, and a single minority electron near the bottom of the conduction band of mass m_c. Use the imaginary part of the screened exchange interaction.

7. Consider a model metal in which the electron–phonon coupling is given by

$$\alpha^2F(\omega) = \begin{cases} g\left(\dfrac{\omega}{\omega_D}\right)^2 & 0 < \omega < \omega_D \\ 0 & \omega > \omega_D \end{cases} \tag{11.283}$$

- Derive λ.
- Derive $\lambda\langle\omega^2\rangle$.
- Derive a numerical value, at room temperature, for the lifetime for temperature relaxation τ_{ep} using $g = 1$ and $\hbar\omega_D = 10$ meV:

$$\frac{1}{\tau_{ep}} = \frac{6\hbar}{\pi k_BT}\lambda\langle\omega^2\rangle \tag{11.284}$$

8. Another term in the current is due to diffusion of inhomogeneous density:

$$\mathbf{J} = -D\nabla n \tag{11.285}$$

Use the Boltzmann equation to derive an expression for D. Evaluate it for (i) semiconductors, and (ii) metals. *Hint*: Assume that $n(r)$ is caused by $\mu(r)$, so that the derivative $\nabla_r f^{(0)}(k)$ produces a term in $\nabla\mu$. Then use the relation between $\delta\mu$ and δn to get ∇n.

9. Fowler-Nordheim tunneling is when electrons tunnel out of a metal surface due to an electric field at the surface. The energy diagram is shown in fig. 11.21. Use WKBJ to derive the tunneling exponent w.

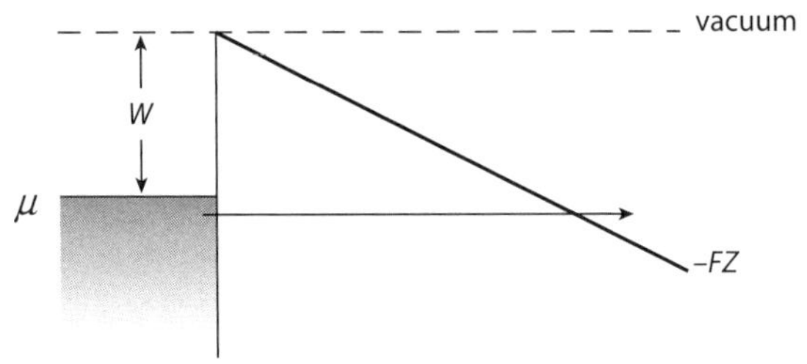

FIGURE 11.21. Fowler-Nordheim tunneling geometry.

10. For Schottky barrier tunneling in n-type semiconductors, in the case of forward bias, the electrons in the semiconductor tunnel into the unoccupied states of the metal, above the chemical potential of the metal. Draw a diagram of reverse bias $V_a < 0$, and show the electrons in the metal now tunneling into the empty states in the semiconductor band. Derive the WKBJ exponent for this case.

11. Use the formula $K = DC_V$ to estimate the thermal conductivity of air at $T = 300$ K and at a pressure of one atmosphere. *Hint*: First show that it can be written as

$$K = \frac{\bar{v}k_B}{2\sigma} \tag{11.286}$$

where σ is the cross section for the scattering of two molecules.

12. Thermoelectric devices are operated in series, with alternate n- and p-type semiconductors connecting the cold and hot reservoirs. A pair of such devices is called a couple. n-Type materials have parameters ρ_n, K_n, $S_n < 0$, and p-type materials have parameters ρ_p, K_p, $S_p > 0$. Let each side of the couple have the same cross-sectional area, so they have the same current density J. Calculate the efficiency of this couple. Show that it operates as having

- An effective resistivity of $\rho = \rho_n + \rho_p$
- An effective thermal conductivity of $K = K_n + K_p$
- An effective Seebeck coefficient of $S = S_p - S_n$.

12 Optical Properties

12.1 Introduction

Solids can be divided into conductors, semiconductors, and insulators. Their optical properties can be described by the dielectric tensor $\varepsilon_{ij}(\omega)$, where the subscripts (ij) denote the xyz-coordinates. Here we discuss only isotropic solids [$\varepsilon_{ij} = \delta_{ij}\varepsilon$], where the subscripts are unnecessary.

The dielectric function actually depends on wave vector and frequency $\varepsilon(k, \omega)$. For photons in the visible range of frequency, the wave vector k is quite small and is usually set equal to zero. The frequency $\omega \sim 10^{15}$ cycles/s, and the wave vector $k = \omega/c \sim 10^5$/cm. Although this wave vector is large, it is very small compared to those in solids, which are on the scale of the Brillouin zone $G \sim \pi/a \sim 10^8$/cm. For high-frequency photons, such as x-rays, the wave vector dependence is important.

12.1.1 *Optical Functions*

The dielectric function has real and imaginary parts, which are called ε_1 and ε_2, respectively:

$$\varepsilon(\omega) = \varepsilon_1(\omega) + i\varepsilon_2(\omega) \tag{12.1}$$

Two other optical constants are the refractive index n and extinction coefficient κ (which is not the wave vector!). They are the real and imaginary parts of the square root of the dielectric function:

$$\hat{n} = n + i\kappa = \sqrt{\frac{\varepsilon_1 + i\varepsilon_2}{\varepsilon_0}} \tag{12.2}$$

$$\varepsilon_1/\varepsilon_0 = n^2 - \kappa^2, \quad \varepsilon_2/\varepsilon_0 = 2n\kappa \tag{12.3}$$

This combination arises from Maxwell's equations. For an electromagnetic wave going through a dielectric medium, the photon wave vector is

$$k^2 = \frac{\varepsilon(\omega)}{\varepsilon_0}\frac{\omega^2}{c^2} \tag{12.4}$$

$$k = \frac{\omega}{c}[n + i\kappa] \tag{12.5}$$

Another important optical parameter is the absorption coefficient $\alpha(\omega)$. It is related to the extinction coefficient. Beer's law gives for the decay of the intensity of the electromagnetic field, when going a distance z,

$$I(z) = I_0 e^{-\alpha z} \tag{12.6}$$

Since the intensity is proportional to the square of the electric field, and the field decays according to $\exp[-z(\omega\kappa/c)]$, then

$$\alpha(\omega) = 2\kappa\frac{\omega}{c} = \frac{\omega}{nc}\frac{\varepsilon_2(\omega)}{\varepsilon_0} \tag{12.7}$$

If electromagnetic radiation is normally incident on the smooth face of a solid, the reflected intensity is

$$R = \left|\frac{\hat{n}-1}{\hat{n}+1}\right|^2 \tag{12.8}$$

This assumes that the material has infinite thickness, so there is no transmission out the back side. Another important dielectric function is

$$\Im\left(\frac{1}{\varepsilon(\omega)}\right) = \frac{-\varepsilon_2(\omega)}{\varepsilon_1^2 + \varepsilon_2^2} \tag{12.9}$$

Peaks in this function are interpreted as longitudinal oscillations of the material. Examples of such oscillations are plasmons in a metal and longitudinal optical phonons in an insulator. These definitions will be used in discussing the optical properties.

An important restriction on the form of the dielectric function in a metal is *sum rules*. The sums are integrals over frequency:

$$\int_0^\infty \omega d\omega \Im\left(\frac{\varepsilon_0}{\varepsilon(\omega)}\right) = -\frac{\pi}{2}\omega_p^2 \tag{12.10}$$

$$\int_0^\infty \frac{d\omega}{\omega}\Im\left(\frac{\varepsilon_0}{\varepsilon(\omega)}\right) = -\frac{\pi}{2} \tag{12.11}$$

where ω_p is the plasma frequency of the metal.

Maxwell's equations are

$$\nabla\cdot D(\mathbf{r},t) = \rho(\mathbf{r},t), \qquad \nabla\times\mathbf{E}(\mathbf{r},t) = -\frac{\partial\mathbf{B}(\mathbf{r},t)}{\partial t} \tag{12.12}$$

$$\nabla \cdot B(\mathbf{r}, t) = 0, \qquad \nabla \times \mathbf{H}(\mathbf{r}, t) = \frac{\partial \mathbf{D}(\mathbf{r}, t)}{\partial t} + \mathbf{j}(\mathbf{r}, t) \tag{12.13}$$

In cubic crystals set $\mathbf{D} = \varepsilon\mathbf{E}$ and $\mathbf{B} = \mu\mathbf{H}$. There are vector (**A**) and scalar (ϕ) potentials to express the fields:

$$\mathbf{E} = -\frac{\partial A}{\partial t} - \nabla\phi \tag{12.14}$$

$$\mathbf{B} = \nabla \times \mathbf{A} \tag{12.15}$$

The potential ϕ has the units of volts, while the vector potential **A** has the units of volts · s/m, so that $\mathbf{v} \cdot \mathbf{A}$ is in volts.

Photons are the quantized excitations of the electromagnetic field. They are described by raising ($a^{\dagger}_{k,m}$) and lowering ($a_{k,\lambda}$) operators, where k is the photon wave vector, and λ denotes the polarization. These operators obey boson statistics, similar to those for phonons. The vector potential for the electromagnetic field in free space is

$$\mathbf{A}(\mathbf{r}, t) = \sum_{\mathbf{k},\lambda} C(\mathbf{k}, \lambda) A(k, \lambda, t) e^{i\mathbf{k}\cdot\mathbf{r}} \hat{\xi}(\mathbf{k}, \lambda) \tag{12.16}$$

$$C(\mathbf{k}, \lambda) = \sqrt{\frac{\hbar}{2\varepsilon_0 \omega(\mathbf{k}, \lambda)\Omega}} \tag{12.17}$$

$$A(\mathbf{k}, \lambda, t) = a_{\mathbf{k},\lambda} e^{-i\omega(\mathbf{k},\lambda)t} + a^{\dagger}_{-\mathbf{k},\lambda} e^{i\omega(-k,\lambda)t} \tag{12.18}$$

In free space the two polarization vectors $\xi(\mathbf{k}, \lambda)$ for ($\lambda = 1, 2$) are both perpendicular to the wave vector **k** and perpendicular to each other.

12.1.2 *Kramers-Kronig Analysis*

Optical absorption experiments are often difficult because the absorption is too strong: the light cannot get through the sample, so there is no signal! This problem is aleviated by using thin films, but they are often not the same optical quality as bulk materials. Assume that the optical conductivity $\sigma(\omega)$ can be measured over a wide range of frequency ω. How are the results of this measurement used to obtain both parts of the dielectric function $\varepsilon_1(\omega)$, $\varepsilon_2(\omega)$? A general method uses the properties of analytic functions of frequency. These functions must have the following properties:

- $\mathcal{J}(\omega)$ must vanish when $\omega \to \infty$.
- $\mathcal{J}(\omega{=}0)$ must not diverge.
- $\mathcal{J}(\omega)$ has no poles in the upper-half plane of complex frequency. This constraint is satisfied by causal function.

Functions that obey these conditions have the *Kramers-Kronig relations* between the real $\mathcal{J}_1(\omega)$ and imaginary $\mathcal{J}_2(\omega)$ parts:

$$\mathcal{J}(\omega) = \mathcal{J}_1(\omega) + i\mathcal{J}_2(\omega) \tag{12.19}$$

$$\mathcal{J}_1(\omega) = \int_{-\infty}^{\infty} \frac{d\omega'}{\pi} \mathcal{J}_2(\omega') \mathcal{P} \frac{1}{\omega' - \omega} \tag{12.20}$$

$$\mathcal{J}_2(\omega) = -\int_{-\infty}^{\infty} \frac{d\omega'}{\pi} \mathcal{J}_1(\omega') \mathcal{P} \frac{1}{\omega' - \omega} \tag{12.21}$$

The last two equations constitute the Kramers-Kronig equations. They are derived in an appendix. If one knows either the real or imaginary part of a complex, causal, function of frequency, the other can be calculated using these identities. Although the current $\mathcal{J}(\omega)$ was used as an example, any causal, complex function of frequency obeys the same theorem. Examples are $[\varepsilon_1(\omega) - \varepsilon_0, \varepsilon_2(\omega)]$ and $[n(\omega) - 1, \kappa(\omega)]$.

Assume that an experimentalist has measured the optical absorption $\alpha(\omega)$ and refractive index $n(\omega)$. Recall that

$$\sqrt{\frac{\varepsilon_1 + i\varepsilon_2}{\varepsilon_0}} = \hat{n} = n + i\kappa \tag{12.22}$$

$$n\alpha = 2\kappa n \frac{\omega}{c} = \frac{\varepsilon_2}{\varepsilon_0} \frac{\omega}{c} \tag{12.23}$$

$$\frac{\varepsilon_2(\omega)}{\varepsilon_0} = \frac{c}{\omega} n(\omega) \alpha(\omega) \tag{12.24}$$

Then apply the Kramers-Kronig formula with $\varepsilon_2(-\omega) = -\varepsilon_2(\omega)$ to get

$$\varepsilon_1(\omega) = \varepsilon_0 + \frac{2}{\pi} \int_0^{\infty} d\omega' \varepsilon_2(\omega') \mathcal{P} \frac{\omega'}{(\omega')^2 - \omega^2} \tag{12.25}$$

If there is no absorption and $\varepsilon_2 = 0$, then $\varepsilon_1 = \varepsilon_0$, which is the dielectric function of vacuum.

Herb Philipp and Ernie Taft [17] applied the Kramers-Kronig method to obtain the first optical spectra of solids. They measured the reflected intensity $R(\omega)$ over a wide range of frequency, from infrared to far ultraviolet. Light is reflected from a surface in vacuum according to the amplitude:

$$r = \frac{\hat{n} - 1}{\hat{n} + 1} = |r| e^{i\eta(\omega)} = \exp[\ln(|r|) + i\eta] \tag{12.26}$$

An intensity measurement obtains $R = |r|^2$. Philipp and Taft made the complex variables

$$\ln(r) = \frac{1}{2} \ln(R) + i\eta \tag{12.27}$$

$$\eta(\omega) = -\int_{-\infty}^{\infty} \frac{d\omega'}{2\pi} \ln[R(\omega')] \mathcal{P} \frac{1}{\omega - \omega'} \tag{12.28}$$

Once they obtained $|r|$ and η, they could find $\hat{n}$ and $\varepsilon = \hat{n}^2$. In this way they obtained the first measurement of the optical constants of many solids over a wide range of frequency. Their results for silver are shown in fig. 12.1. The divergence at low frequency is due to the Drude behavior that $\kappa \sim \omega_p/\omega$.

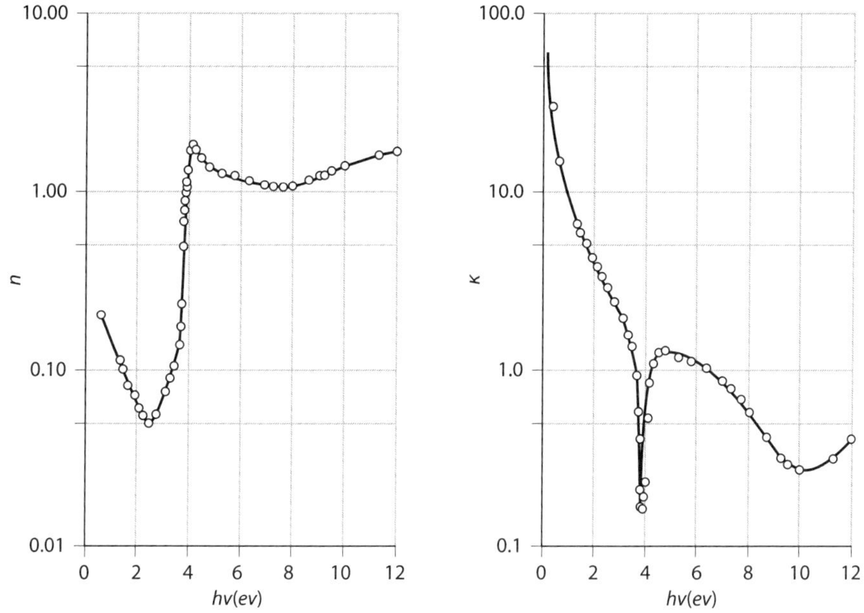

FIGURE 12.1. Optical constants of silver. Measured by Philipp and Taft, *Phys. Rev.* **121**, 1100 (1961). Used with permission of the American Physical Society. (a) Refractive index $n(\omega)$, (b) extinction coefficient $\kappa(\omega)$.

12.2 Simple Metals

Simple metals are those without *d*- or *f*-electrons. Examples are sodium, magnesium, and aluminum. Their optical properties can be divided into two features: Drude and interband.

12.2.1 *Drude*

If the electrons are undamped, the dielectric function of a metal at long wavelength is the contribution from plasma oscillations:

$$\varepsilon(\omega)/\varepsilon_0 = 1 - \frac{\omega_p^2}{\omega^2} \tag{12.29}$$

This form has the feature that

- For $\omega < \omega_p$, ε is real and negative, so that $n = 0$ and $\kappa \neq 0$. Then the reflectivity $R = 1$.
- For $\omega > \omega_p$, $\varepsilon(\omega)$ is real and positive. Then $n \neq 0$ and $\kappa = 0$ and the reflectivity is less than one.

This general behavior is observed in simple metals.

Of course, metals do have damping. According to the Drude theory, as discussed in the previous chapter, the electrical conductivity has a frequency dependence of

$$\sigma(\omega) = \frac{\sigma_0}{1 - i\omega\tau}, \quad \sigma_0 = \frac{n_e e^2 \tau}{m} = \omega_p^2 \tau \varepsilon_0 \tag{12.30}$$

The conductivity defined by the above formula has the laboratory units of siemens/m. Many optical formulas are written with the conductivity having the units of frequency: inverse seconds. That conductivity $\tilde{\sigma}$ is obtained from the one above by

$$\tilde{\sigma} = \frac{\sigma}{4\pi\varepsilon_0} = \frac{\omega_p^2 \tau}{4\pi} \tag{12.31}$$

Another way to write the dielectric function of a metal is

$$\varepsilon(\omega) = \varepsilon_0 + i\frac{\sigma}{\omega} = \varepsilon_0 + i\frac{\sigma_0}{\omega(1 - i\omega\tau)} \tag{12.32}$$

Use the above result for σ_0 to find

$$\varepsilon(\omega) = \varepsilon_0\left[1 + i\frac{\omega_p^2 \tau}{\omega(1 - i\omega\tau)}\right] \tag{12.33}$$

Compare this formula with eqn. (12.29). They become identical in the limit that $\omega\tau \gg 1$. Equation (12.33) is considered a more accurate formula and demonstrates the correct way of including damping into the nearly free electron dielectric function. One can also easily evaluate its real and imaginary parts:

$$\varepsilon_1(\omega)/\varepsilon_0 = 1 - \frac{\omega_p^2 \tau^2}{1 + \omega^2 \tau^2} \tag{12.34}$$

$$\varepsilon_2(\omega)/\varepsilon_0 = \frac{\omega_p^2 \tau}{\omega(1 + \omega^2 \tau^2)} \tag{12.35}$$

These formulas constitute the Drude theory of the dielectric function of metals. Normally they are applied in the infrared regions of the frequency spectrum, in which case $\omega_p \tau \gg \omega\tau \gg 1$. There they can be approximated as

$$\varepsilon_1(\omega)/\varepsilon_0 \approx 1 - \frac{\omega_p^2}{\omega^2} \tag{12.36}$$

$$\varepsilon_2(\omega)/\varepsilon_0 \approx \frac{\omega_p^2}{\omega^3 \tau} \tag{12.37}$$

Figure 12.2 shows $\varepsilon_2(\omega)/\varepsilon_0$ for metallic sodium. The dots are the experimental results of Neville Smith. The Drude region is to the left of the peak and is well described by the above formula. The peak starting around $\hbar\omega = 2.0$ eV was originally interpreted as due to interband transitions [2]. Al Sievers [22] made an alternate proposal that the peak was due to the absorption of surface plasmons on the metal surface. His theory is the solid line, and fits the data quite well. The surface plasmon theory also explains why the peak position depends on the dielectric properties of the surface to which the metal is evaporated.

The longitudinal excitations are found from the loss function:

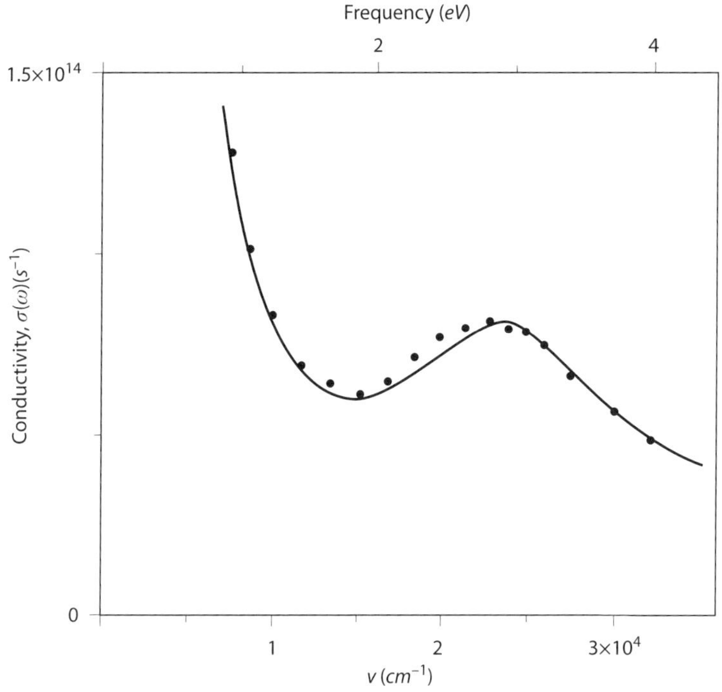

FIGURE 12.2. $\varepsilon_2(\omega)/\varepsilon_0$ for metallic sodium. The dots are data of Neville Smith, and the line is surface plasmon theory. From Sievers, *Phys. Rev. B* **22**, 1600 (1980). Used with permission of the American Physical Society.

$$\frac{\varepsilon_0}{\varepsilon} = \frac{\omega(1 - i\omega\tau)}{\omega + i\tau(\omega_p^2 - \omega^2)} \tag{12.38}$$

$$-\Im\left(\frac{\varepsilon_0}{\varepsilon(\omega)}\right) = \frac{\omega_p^2(\omega/\tau)}{(\omega^2 - \omega_p^2)^2 + (\omega/\tau)^2} \tag{12.39}$$

This function has a peak at $\omega \approx \omega_p$, with a width $\Delta\omega \sim 1/\tau$. Plasmons are important excitations of metals.

12.3 Force-Force Correlations

Several features of the optical properties of metals are discussed using the *force–force correlation function*. In a Green's function approach, the electrical conductivity is given by a *current–current correlation function*:

$$\Pi_{\mu\nu}(t - t') = -i\Theta(t - t')\langle[j_\mu(t), j_\nu(t')]\rangle \tag{12.40}$$

$$\Pi_{\mu\nu}(\omega) = \int_{-\infty}^{\infty} dt' e^{i\omega(t-t')}\Pi_{\mu\nu}(t - t') \tag{12.41}$$

$$\Re\{\sigma(\omega)\} = -\frac{\Im\{\Pi_{\mu\nu}(\omega)\}}{\omega} \tag{12.42}$$

The current operator is correlated with itself at two different times. This quantity is Fourier transformed, and its imaginary part is ω times the real part of the electrical conductivity.

The force–force correlation function is derived by remembering that the current is the summation of the momentum ($\mathbf{p}_i$) of the particles, and the derivative of momentum is the force:

$$\mathbf{j} = \frac{e}{m\Omega}\sum_i \mathbf{p}_i \tag{12.43}$$

$$\frac{\partial \mathbf{j}}{\partial t} = \frac{e}{m\Omega}\sum_i \frac{\partial \mathbf{p}_i}{\partial t} = \frac{e}{m\Omega}\sum_i \mathbf{F}_i \tag{12.44}$$

For a light wave of frequency ω, the time derivative of the current is $-i\omega\mathbf{j}$. The force–force correlation function is

$$\Pi_{\mu\nu}(\omega) = i\left(\frac{e}{m\omega\Omega}\right)^2 \int_{-\infty}^{t} dt' e^{i\omega(t-t')}\langle[F_\mu(t), F_\nu(t')]\rangle \tag{12.45}$$

This formula is an expansion in inverse powers of the frequency. The formula is accurate in the limit of high frequency $[\omega\tau \gg 1]$. It should not be used in the limit of zero frequency! It will now be derived for several contributions to the force.

12.3.1 *Impurity Scattering*

The crystal has impurities randomly located at sites $\mathbf{R}_i$. The Fourier transform of the electron–impurity interaction is $V_I(q)$. The term in the Hamiltonian is

$$V(\mathbf{r}) = \frac{1}{\Omega}\sum_{\mathbf{q}p\sigma i} V_I(q)\, e^{i\mathbf{q}\cdot(\mathbf{r}-\mathbf{R}_i)} C^\dagger_{\mathbf{p}+\mathbf{q},\sigma} C_{\mathbf{p}\sigma} \tag{12.46}$$

$$\mathbf{F}(\mathbf{r}) = -\nabla V = -\frac{i}{\Omega}\sum_{\mathbf{q}p\sigma i} \mathbf{q} V_I(q)\, e^{i\mathbf{q}\cdot(\mathbf{r}-\mathbf{R}_i)} C^\dagger_{\mathbf{p}+\mathbf{q},\sigma} C_{\mathbf{p}\sigma} \tag{12.47}$$

$$\mathbf{F}(\mathbf{q}) = -\frac{i}{\Omega}\mathbf{q} V_I(q)\rho(q)\sum_i e^{-i\mathbf{q}\cdot\mathbf{R}_i} \tag{12.48}$$

where $\rho(\mathbf{q})$ is the electron density operator. The force–force correlation function becomes a density–density correlation function, which is related to the inverse dielectric function. The correlation function contains the factor

$$\begin{aligned}&\frac{1}{\Omega^2}\sum_{ij}\sum_{\mathbf{q},\mathbf{q}'} e^{i(\mathbf{q}\cdot\mathbf{R}_i+\mathbf{q}'\cdot\mathbf{R}_j)}\langle[\rho(\mathbf{q},t), \rho(\mathbf{q}',t')]\rangle \\ &= \frac{\delta_{\mathbf{q}=-\mathbf{q}'}}{\Omega^2}\sum_{ij}\delta_{ij}\langle[\rho(\mathbf{q},t), \rho(-\mathbf{q},t')]\rangle\end{aligned} \tag{12.49}$$

The summation over impurity sites gives the number of sites N_i. One factor of volume Ω is used to define the impurity concentration $n_i = N_i/\Omega$. The final correlation function is

$$\Pi_{\mu\nu} = \frac{n_i}{\Omega}\left(\frac{e}{m\omega}\right)^2 \sum_q q_\mu q_\nu \frac{|V_I(q)|^2}{v_q}\left[\frac{\varepsilon_0}{\varepsilon(q,\omega)} - 1\right] \tag{12.50}$$

where $v_q = e^2/\varepsilon_0 q^2$ comes from the definition of the longitudinal dielectric function $\varepsilon(q,\omega)$.

The conductivity is the imaginary part of this expression. The only imaginary term is the dielectric function:

$$\Im\left[\frac{\varepsilon_0}{\varepsilon(q,\omega)} - 1\right] = \frac{-\varepsilon_0\varepsilon_2}{\varepsilon_1^2 + \varepsilon_2^2} \tag{12.51}$$

The denominator goes into the definition of the screened electron–impurity potential, and the numerator is, for small values of frequency,

$$\varepsilon_2(q,\omega) = v_q \omega \frac{m^2\varepsilon_0}{2\pi q\hbar^3} \tag{12.52}$$

$$V_{SI}(q) = \frac{V_I\varepsilon_0}{\varepsilon} \tag{12.53}$$

$$\Re\{\sigma\} = \frac{n_i}{2\pi\hbar^3}\left(\frac{e}{\omega}\right)^2 \int \frac{d^3q}{(2\pi)^3}\frac{q_\mu q_\nu}{q}|V_{SI}(q)|^2 \tag{12.54}$$

Set $\mu = \nu$ and average over angles, which is appropriate for cubic solids:

$$\Re\{\sigma\} = \frac{n_i}{6\pi\hbar^3}\left(\frac{e}{\omega}\right)^2 \int \frac{d^3q}{(2\pi)^3} q|V_{SI}(q)|^2 \tag{12.55}$$

Recall the two expressions for $\varepsilon_2(0,\omega)$ relevant to optical measurements:

$$\varepsilon_2(\omega) = \frac{\Re\{\sigma\}}{\omega} = \frac{\omega_p^2\varepsilon_0}{\omega^3\tau} \tag{12.56}$$

Combining these expressions gives a new formula for the inverse lifetime:

$$\frac{1}{\tau} = \frac{n_i m}{6\pi n_e\hbar^3}\int \frac{d^3q}{(2\pi)^3} q|V_{SI}(q)|^2 \tag{12.57}$$

This formula is the exact result if the screened impurity interaction is replaced by the *T*-matrix for scattering. It gives the correct formula for the transport lifetime, with the factor of $[1 - \cos(\theta)]$.

This term in the electron lifetime is part of the Drude contribution. Another contribution to the Drude expression is the electron lifetime from electron scattering by phonons. One can also derive an expression for $1/\tau$ using the force–force correlation function. In this case, the formula is *not* identical to that derived from the Boltzmann equation. However, the two expressions are numerically similar.

The Boltzmann equation derives a lifetime for dc transport. The force–force correlation function derives a lifetime for high-frequency scattering: usually the frequencies ω are much larger than the highest phonon frequency ω_D. Experimentally it is found that the lifetimes measured in these two experiments, dc transport and the Drude term $\varepsilon_2(\omega)$, are the same. That is actually amazing. However, evaluating these two theoretical expressions also finds they are the same within a few percent.

12.3.2 *Interband Transitions*

A free electron cannot absorb electromagnetic radiation since it cannot satisfy energy and momentum conservation. Photons provide a source of energy for the electrons. The absorption of photons can occur whenever the electron can find a source of momentum to help it satisfy momentum conservation. In the previous subsection, the momentum was provided either by scattering from impurities or by phonons. Here the momentum is supplied by the periodic potential of the lattice. These processes are considered to be interband transitions, but are actually absorption processes where the wave vector is a reciprocal lattice vector.

An electron in a crystal has a potential energy that is periodic and can be expanded in a Fourier series using reciprocal lattice vectors **G**:

$$V(\mathbf{r}) = \sum_j U(\mathbf{r}-\mathbf{R}_j) = \sum_{\mathbf{G}} e^{i\mathbf{G}\cdot\mathbf{r}} U_{\mathbf{G}} \tag{12.58}$$

$$\mathcal{V} = \int d^3 r \rho(\mathbf{r}) V(\mathbf{r}) = \sum_{\mathbf{G}} U_{\mathbf{G}} \rho(\mathbf{G}) \tag{12.59}$$

where $U(\mathbf{r})$ is the potential from a single atom, and $U_{\mathbf{G}}$ is its Fourier transform. This force also involves the electron density operator $\rho(\mathbf{G})$. The density–density correlation function is given by the inverse dielectric function at wave vector **G**. The force and force–force correlation function are

$$\mathbf{F}(\mathbf{r}) = -\nabla V(\mathbf{r}) = -i\sum_{\mathbf{G}} \mathbf{G} U_{\mathbf{G}} e^{i\mathbf{G}\cdot\mathbf{r}} \tag{12.60}$$

$$\Pi_{\mu\nu} = \frac{1}{4\pi m^2\omega^2}\sum_{\mathbf{G}} |U_{\mathbf{G}}|^2 G^2 G_\mu G_\nu \left[\frac{1}{\varepsilon(G,\omega)} - 1\right] \tag{12.61}$$

In the last equation, set $\mu = \nu$ and average over the three directions in the crystal:

$$\Re\{\sigma(\omega)\} = \frac{1}{12\pi m^2\omega^3}\sum_{\mathbf{G}} |U_{\mathbf{G}}|^2 G^4 \frac{\varepsilon_2(\mathbf{G},\omega)}{|\varepsilon(\mathbf{G},\omega)|^2} \tag{12.62}$$

This formula accurately describes interband transitions in simple metals such as the alkali atoms. Since **G** is a large wave vector and the frequencies are also high, the correct expression for $\varepsilon_2(\mathbf{G},\omega)$ is

$$\varepsilon_2(\mathbf{G},\omega) = \frac{e^2 m}{G^3}\left[k_F^2 - \left(\frac{m}{\hbar^2 G}\right)^2 (\hbar\omega - \varepsilon_G)^2\right], \varepsilon_G = \frac{\hbar^2 G^2}{2m} \tag{12.63}$$

which is valid when $G > 2k_F$.

The formula for $\Re\{\sigma\}$ was first derived by Paul Butcher [2] using the conventional method of the golden rule. The derivation by the force–force correlation function is far easier.

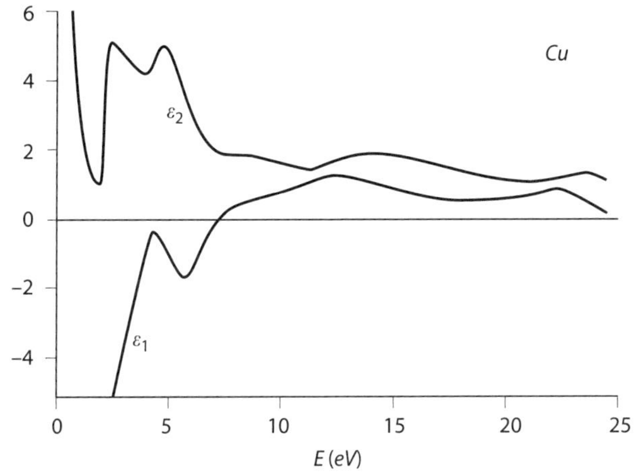

FIGURE 12.3. Dielectric functions in Cu. Peaks are due to interband transitions. From Ehrenreich and Philipp, *Phys. Rev.* **128**, 1622 (1962). Used with permission of the American Physical Society.

The energy bands associated with *d*-orbitals typically do not have much dispersion. They are almost constant in energy when graphed on an energy band diagram. The above theory does not apply to them. In metals where the *d*-bands are fully occupied, there are interband transitions from the *d*-bands to unoccupied *s–p* bands higher in energy. Figure 12.3 shows $\varepsilon_1(\omega)$, $\varepsilon_2(\omega)$ in Cu. The lowest interband transitions are from the occupied *d*-bands to the unoccupied *s–p* bands higher in energy. Cu and Au have a similar threshold energy around $\hbar\omega_T \approx 2$ eV, compared to Ag at $\hbar\omega_T \approx 4$ eV. Of course, that is why copper and gold have a yellow color, while silver does not.

12.4 Optical Absorption

In insulators and semiconductors, the most common optical measurement is interband absorption. The valence bands are all completely occupied by electrons, and the conduction bands are all completely unoccupied. A photon of the appropriate energy can be absorbed by an electron in the valence band and be raised in energy to the conduction band. At a later time, the same electron could leave the conduction band and return to the valence band by the emission of a photon.

12.4.1 *Interband Transitions in Insulators*

A one-photon absorption or emission process has a matrix element given by the $\mathbf{p}\cdot\mathbf{A}$ interaction, where $\mathbf{A}$ is the vector potential of the photon, and $\mathbf{p}$ is the momentum operator that acts on the electron. Using the golden rule for the transition rate gives

$$V = -\frac{e}{m}\mathbf{p}\cdot\mathbf{A} \tag{12.64}$$

$$w(q) = \frac{4\pi\Omega}{\hbar}\int\frac{d^3k}{(2\pi)^3}|\langle f|V|i\rangle|^2\delta[\hbar\omega + E_v(\mathbf{k}) - E_c(\mathbf{k}+\mathbf{q})] \tag{12.65}$$

The factor of $4\pi = 2\cdot 2\pi$, where the additional factor of two is for the two spin states of the electron. The factor of volume (Ω) comes from converting the summation over electron states $\mathbf{k}$ to a continuous integral.

The quantity $w(\mathbf{q})$ is the rate of absorption of a photon of wave vector $\mathbf{q}$ and frequency $\omega = cq/n$, where n is the refractive index. The functions $|i\rangle$ and $|f\rangle$ include the Bloch functions for the two bands, and the state function $|n_{\mathbf{q}}\rangle$ of the photon:

$$|i\rangle = \frac{1}{\sqrt{\Omega}}e^{i\mathbf{k}\cdot\mathbf{r}}u_v(\mathbf{r})|n_{\mathbf{q}}\rangle \tag{12.66}$$

$$|f\rangle = \frac{1}{\sqrt{\Omega}}e^{i(\mathbf{k}+\mathbf{q})\cdot\mathbf{r}}u_c(\mathbf{r})|n_{\mathbf{q}} - 1\rangle \tag{12.67}$$

$$\mathbf{A}(\mathbf{r}) = e^{i\mathbf{q}\cdot\mathbf{r}}\hat{\xi}X_q(a_{\mathbf{q}} + a^{\dagger}_{-\mathbf{q}}),\quad X_q = \sqrt{\frac{\hbar}{2\varepsilon_0\omega\Omega}} \tag{12.68}$$

The photon raising $a^{\dagger}_{-\mathbf{q}}$ and lowering $a_{\mathbf{q}}$ operators either emit or absorb one photon. The photon has a polarization $\hat{\xi}$. The matrix element for absorption is

$$\langle f|V|i\rangle = \frac{e}{m}X_q\langle n_q - 1|a_{\mathbf{q}}|n_q\rangle M_{cv} \tag{12.69}$$

$$M_{cv} = \frac{\hbar}{i}\int\frac{d^3r}{\Omega}u_c^*(\mathbf{r})e^{-i(\mathbf{k}+\mathbf{q})}\hat{\xi}\cdot\nabla\left[e^{i(\mathbf{k}+\mathbf{q})\cdot\mathbf{r}}u_v(\mathbf{r})\right] \tag{12.70}$$

$$= \frac{\hbar}{i}\int\frac{d^3r}{\Omega}u_c^*(\mathbf{r})[i\hat{\xi}\cdot(\mathbf{k}+\mathbf{q})u_v(\mathbf{r}) + \hat{\xi}\cdot\nabla u_v(\mathbf{r})] \tag{12.71}$$

The photon matrix element gives

$$\langle n_q - 1|a_{\mathbf{q}}|n_q\rangle = \sqrt{n_{\mathbf{q}}} \tag{12.72}$$

When the matrix element is squared, this factor is $n_{\mathbf{q}}$. The golden rule is that $w(\mathbf{q}) = -\dot{n}_{\mathbf{q}}$. It is the rate of absorption of photons, which must be proportional to the the number of such photons $n_{\mathbf{q}}$. Rewrite eqn. (12.65) as

$$\frac{dn_{\mathbf{q}}}{dt} = -w = -\gamma n_{\mathbf{q}} \tag{12.73}$$

whose solution is

$$n_{\mathbf{q}} = n_{0\mathbf{q}}e^{-\gamma t} = n_{0\mathbf{q}}e^{-\alpha x} \tag{12.74}$$

$$x = \frac{ct}{n},\quad \alpha = \frac{n\gamma}{c} \tag{12.75}$$

$$\alpha(\omega) = \frac{2\pi ne^2}{\varepsilon_0\omega cm^2}\int\frac{d^3k}{(2\pi)^3}|M_{cv}|^2\delta[\hbar\omega + E_v(\mathbf{k}) - E_c(\mathbf{k}+\mathbf{q})] \tag{12.76}$$

where n is the refractive index, which enters into the velocity of light c/n in the medium. The parameter α is the absorption coefficient. Equation (12.74) is *Beers law* for the absorption of light in a material.

The matrix element M_{cv} in eqn. (12.71) has two terms. The first one is zero since

$$0 = \int d^3r u_c^*(\mathbf{r}) u_v(\mathbf{r}) \tag{12.77}$$

These two functions are different eigenstates of the same Hamiltonian, and different eigenstates are orthogonal. Actually, u_v is an eigenstate with wave vector $\mathbf{k}$, while u_c is an eigenstate with wave vector $\mathbf{k} + \mathbf{q}$. However, for optical processes, the wave vector $\mathbf{q}$ is so small that the theorem is still approximately satisfied. For photons in the x-ray region, where wave vectors are of the order $O(10^8/\text{cm})$, the first term cannot be neglected. However, for optical processes, the second term dominates:

$$M_{cv} = \hat{\xi} \cdot \mathbf{p}_{cv} \tag{12.78}$$

$$\mathbf{p}_{cv} = \frac{\hbar}{i} \int_{\Omega_0} \frac{d^3r}{\Omega_0} u_c^*(\mathbf{r}) \nabla u_v(\mathbf{r}) \tag{12.79}$$

where Ω_0 is the volume of one unit cell, and the integral is over one unit cell. The matrix element $\mathbf{p}_{cv}$ was introduced in section 3.4 during the discussion of the effective mass. For symmetry points such as $\mathbf{k} = 0$, the eigenfunctions $u_{c,v}(\mathbf{r})$ often have parity, and the above integral is nonzero only if the conduction and valence band have opposite parity. However, at general points in the Brillouin zone, parity is not a valid quantum number and $\mathbf{p}_{cv}$ is not zero.

For transitions between the valence and conduction bands, the photon wave vector $\mathbf{q}$ is small and is neglected. The matrix element is a function of $\mathbf{k}$, and the integral is

$$B(\omega) = \frac{2\pi}{m} \int \frac{d^3k}{(2\pi)^3} \left| \hat{\xi} \cdot \mathbf{p}_{cv}(\mathbf{k}) \right|^2 \delta[h\omega + E_v(\mathbf{k}) - E_c(\mathbf{k})] \tag{12.80}$$

$$\alpha(\omega) = \frac{ne^2}{\varepsilon_0 \omega c m} B(\omega) \tag{12.81}$$

The function $B(\omega)$ is evaluated by an integral over the Brillouin zone, selecting only those points that satisfy the delta function for energy conservation. It is the optical equivalent of McMillan's function $\alpha^2 F(\omega)$ for the electron–phonon interaction. Here $B(\omega)$ is the similar function for the electron–photon interaction. It has the dimensional units of inverse volume, and the absorption coefficient is inverse length. $B(\omega)$ is easily calculated with existing band structure computer codes. Once this function is known, the absorption coefficient $\alpha(\omega)$ is found easily.

A simple yet important case is when the valence band maximum and the conduction band minimum occur at the same point $\mathbf{k}_0$ in the Brillouin zone. This point is often at the zone center $\mathbf{k}_0 = 0$, but sometimes is found at the edge of the Brillouin zone. In this case the matrix element $\mathbf{p}_{cv}$ can be approximated as a constant:

$$E_c(\mathbf{k}) = E_{c0} + \frac{\hbar^2}{2m_c} (\mathbf{k} - \mathbf{k}_0)^2 \tag{12.82}$$

$$E_v(\mathbf{k}) = E_{v0} - \frac{\hbar^2}{2m_v}(\mathbf{k} - \mathbf{k}_0)^2 \tag{12.83}$$

$$B(\omega) = \frac{(\hat{\xi}\cdot\mathbf{p}_{cv})^2}{4\pi^2 m}\int d^3k\delta\left[\hbar\omega - E_G - \frac{\hbar^2}{2\mu}(k-k_0)^2\right] \tag{12.84}$$

$$= \frac{(\hat{\xi}\cdot\mathbf{p}_{cv})^2}{4\pi^2 m}\left(\frac{2\mu}{\hbar^2}\right)^{3/2}\sqrt{\hbar\omega - E_G}\,\Theta(\hbar - E_G)$$

$$\mu = \frac{m_c m_v}{m_c + m_v}, \quad E_G = E_{c0} - E_{v0} \tag{12.85}$$

This theory predicts that the absorption near a energy band minimum goes as the square root of $\hbar\omega - E_G$, as long as this difference is positive. The symbol μ is the reduced mass, and E_G is the energy gap.

The theory is completely wrong, since the experimental measurements show a much different behavior. The error was the omission of exciton effects.

12.4.2 *Wannier Excitons*

The prior section took a one-electron view of the absorption process. A photon enters the crystal, is absorbed by an electron in the valence band, and that electron is excited to a state in the conduction band.

Wannier [29] took a two-particle view of the same process. If the valence band is full, then removing an electron from the valence band creates a hole in that band. View the absorption of the photon as creating two particles: the electron in the conduction band, and the hole in the valence band. These two particles both have a positive mass, but opposite charges. There is a Coulomb attraction between them. The two particles can form a hydrogenic-like bound state, which is called the *Wannier exciton*. Its Hamiltonian in effective mass theory for a direct gap semiconductor is

$$H_{\rm ex} = E_G - \frac{\hbar^2}{2m_c}\nabla_e^2 - \frac{\hbar^2}{2m_v}\nabla_h^2 - \frac{e^2}{4\pi\varepsilon(0)|\mathbf{r}_e - \mathbf{r}_h|} \tag{12.86}$$

$$H_{\rm ex}\psi(\mathbf{r}_e, \mathbf{r}_h) = E\psi(\mathbf{r}_e, \mathbf{r}_h) \tag{12.87}$$

The kinetic energy terms are better expressed in terms of center-of-mass coordinates:

$$\mathbf{R} = \frac{m_c\mathbf{r}_e + m_v\mathbf{r}_h}{M}, \quad \mathbf{r} = \mathbf{r}_e - \mathbf{r}_h, \quad M = m_c + m_v, \quad \mu = \frac{m_c m_v}{M} \tag{12.88}$$

$$H_{\rm ex} = E_G - \frac{\hbar^2}{2\mu}\nabla_r^2 - \frac{\hbar^2}{2M}\nabla_R^2 - \frac{e^2}{4\pi\varepsilon(0)r} \tag{12.89}$$

The center-of-mass motion has only the kinetic energy term. In the matrix element for optical absorption, the wave vector of the photon $\mathbf{q}$ becomes the center-of-mass wave vector of the exciton. The center-of-mass energy is $E_{\rm CM}(q) = \hbar^2 q^2/2M = (\hbar\omega)^2/2Mc^2$, which is generally a very small energy and is usually ignored.

The relative motion of the two particles has an effective Hamiltonian similar to that of positronium, which is another particle–antiparticle bound state:

$$\psi(\mathbf{r}_e, \mathbf{r}_h) = e^{i\mathbf{q}\cdot\mathbf{R}}\phi_n(\mathbf{r}) \tag{12.90}$$

$$E = E_G + E_{CM}(q) - \frac{E^*_{Ry}}{n^2} \tag{12.91}$$

$$E^*_{Ry} = \frac{\mu}{m_e} E_{Ry}\left(\frac{\varepsilon_0}{\varepsilon(0)}\right)^2 \tag{12.92}$$

The symbol E_{Ry} denotes the Rydberg energy for the hydrogen atom, which is 13.60 eV. The symbol E^*_{Ry} denotes the effective Rydberg for the Wannier exciton. It is reduced from the atomic value by the small value of the reduced mass μ, and further reduced by the large value of the static dielectric constant $\varepsilon(0)/\varepsilon_0 \sim 10$. Typically, $E^*_{Ry} \approx 10^{-3} E_{Ry} \sim 10$ meV in common semiconductors. The above eigenfunction and eigenvalue for the exciton apply to the bound state of the electron and hole. At higher photon energies, the electron and hole are no longer bound. Then their relative wave function is the scattering state of the Coulomb potential, which is a Whittaker function.

How is the optical absorption altered by the inclusion of exciton effects? The formula was given by Elliott [5], who borrowed an idea of Fermi's from nuclear physics. The spectral function $B(\omega)$ is changed to

$$B(\omega) = 2\pi \frac{(\hat{\xi}\cdot \mathbf{p}_{cv})^2}{m} \sum_n |\phi_n(0)|^2 \delta\left[\hbar\omega - E_G + \frac{E^*_{Ry}}{n^2}\right] \tag{12.93}$$

Instead of a continous summation over wave vector **k**, there is now a summation over the discrete bound states n of the Wannier exciton. Furthermore, the absorption process contains the relative wave function $\phi_n(r)$ evaluated at $r = 0$. This latter feature is initially surprising, but makes physical sense. The matrix element for emission and absorption must be identical to preserve detailed balance. For emission, the electron and hole mutually annihilate and form a photon. For them to have this event, they must both be at the same point in space $\mathbf{r}_e = \mathbf{r}_h$, which is the probability $|\phi_n(0)|^2$. Only hydrogen states with zero angular momentum, s-states, have a nonzero value at $r = 0$:

$$|\phi_{nS}(0)|^2 = \frac{1}{\pi (a^*_B)^3 n^3}, \quad a^*_B = \frac{m_e}{\mu}\frac{\varepsilon(0)}{\varepsilon_0} a_B \tag{12.94}$$

$$E_n = -\frac{E^*_{Ry}}{n^2}, \quad \frac{dE_n}{dn} = \frac{2E^*_{Ry}}{n^3} \tag{12.95}$$

$$|\phi_n(0)|^2 = \frac{1}{2\pi (a^*_B)^3 E^*_{Ry}} \frac{dE_n}{dn} \tag{12.96}$$

The effective Bohr radius a^*_B for the Wannier exciton is typically 100 times larger than the atomic value. The wave function $|\phi_n(0)|^2$ is proportional to the derivative of the hydrogenic energy. In the summation over n, since $\Delta n = 1$, write it as

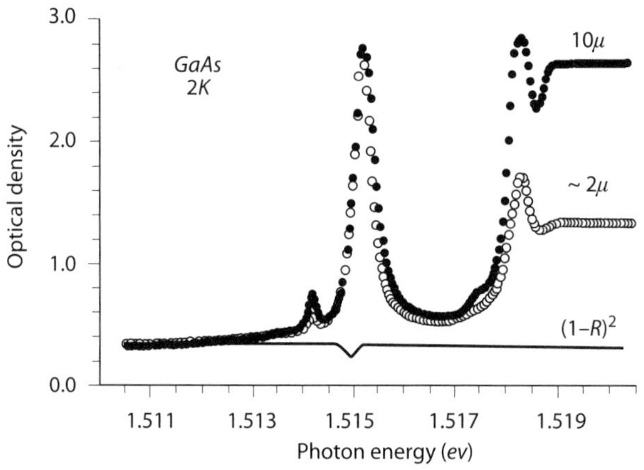

FIGURE 12.4. Optical absorption by excitons in GaAs. From D. D. Sell, *Phys. Rev. B* **6**, 3750 (1972). Exciton 1*s* state is at 1.515 eV, and 2*s* is at 1.518 eV. Weak low-frequency peak is from a donor. Data are from two samples, of thickness 2μ and 10μ. Used with permission of the American Physical Society.

$$\sum_n \Delta n |\phi_n(0)|^2 \delta[\hbar\omega - E_G - E_n] = \frac{1}{2\pi (a_B^*)^3 E_{\mathrm{Ry}}^*} \sum_n \Delta n \frac{dE_n}{dn} \delta[\hbar\omega - E_G - E_n] \quad (12.97)$$

$$= \frac{1}{2\pi (a_B^*)^3 E_{\mathrm{Ry}}^*} \int dE_n \delta[\hbar\omega - E_G - E_n] = \frac{1}{2\pi (a_B^*)^3 E_{\mathrm{Ry}}^*} \quad (12.98)$$

$$B(\omega) = \frac{(\hat{\xi} \cdot \mathbf{p}_{cv})^2}{m (a_B^*)^3 E_{\mathrm{Ry}}^*} \Theta(E_G - \hbar\omega) \quad (12.99)$$

The discrete summation is changed to a continuous integral. The Rydberg series runs over all positive values of n. For higher values of n, the separation between energy states is very small. Since the states have a nonzero lifetime, and finite width, they run together in energy. In that case, changing the summation to a continuous integral makes sense. Then one finds that the spectral function $B(\omega)$ is a constant, independent of photon frequency. This is the observed frequency spectrum. It does not have a square root dependence $\sqrt{\hbar\omega - E_G}$ near thereshold.

An experimental spectrum for GaAs is shown in fig. 12.4. The discrete absorption peak is the 1*s* exciton state. There is a small gap of low absorption, and another rise to the 2*s* peak. The energy separation between these peaks is 3.15 ± 0.15 meV. There is a third rise is in the vicinity of the 3*S* exciton. Above that rise the absorption spectrum is a constant, where the 4*s*, 5*s*, etc., states run together. The exciton binding energy is $\frac{4}{3}$ of the 1*s*–2*s* splitting and is estimated to be $E_X = 4.2 \pm 0.2$ meV.

The interband matrix element $(\hat{\xi} \cdot \mathbf{p}_{cv})^2$ is the same one that enters the formula for the effective masses m_c, m_v. From a knowledge of these masses, one can calculate the interband matrix element and get a theoretical estimate of the absorption rate that is in excellent agreement with the experimental value.

The above analysis is for the bound exciton states and covers the spectral range of $\hbar\omega < E_G$. The Elliott formula (12.93) is valid for $\hbar\omega > E_G$, where there is a summation over the relative wave vector **k** between the electron and hole. The hydrogenic eigenfunction $\phi_k(r)$ is a Whittaker function. The absorption is a continous function of frequency at $\hbar\omega = E_G$.

12.4.3 *Frenkel Excitons*

Wannier excitons are found in crystals where the electron and hole are weakly bound. The effective Bohr radius of the bound state extends over many lattice constants, which justifies the use of effective mass theory in calculating the exciton properties. Wannier excitons are primarily found in semiconductors. There are many insulating crystals in which the electron and hole are bound strongly and the effective Bohr radius is the size of the Wigner-Seitz unit cell. These excitations are called *Frenkel excitons.*

An example of Frenkel excitons is found in alkali halide crystals. Figure 12.5 shows the interband absorption spectra of RbI. The peak at $\hbar\omega = 5.7$ eV is the Frenkel exciton. It is the bound state of the electron and hole, and the exciton radius is the size of the crytal unit cell. The peak at 6.12 eV is where the excited excitons states are observed. These states are well described by the Wannier model, since the orbits extend over many unit cells.

Frenkel excitons are also found in molecular crystals. They are crystals formed from molecules, such as N_2, O_2, C_6H_6, etc. The individual molecules have optical absorption lines between occupied and empty molecular orbitals. The same absorption lines are observed in the crystal phase, and the frequencies are slightly changed from the values of the individual molecules. However, in the crystal, the Frenkel exciton can move from molecule to molecule, and is a true polarization wave of the solid. Most molecular crystals do not have cubic symmetry, and the absorption frequencies are slightly different for different directions of polarization of the light beam. Figure 12.6 shows optical absorption in crystalline anthracene ($C_{14}H_{10}$), which has a strong electronic dipole transition at

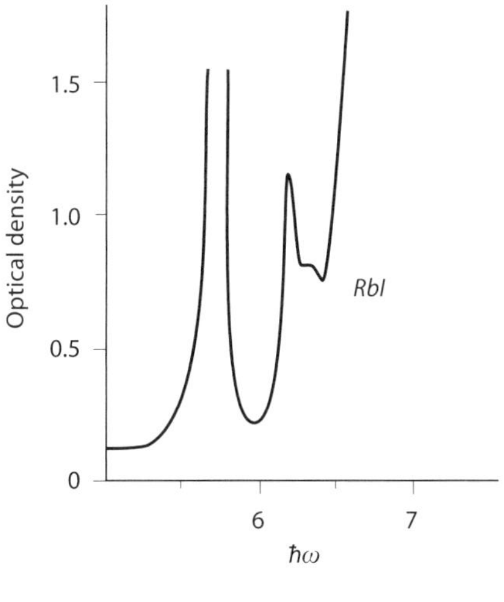

FIGURE 12.5. Interband absorption spectra of RbI at $T = 10$ K. The peak at $\hbar\omega = 5.7$ù eV is Frenkel exciton: the bound state of electron and hole. The second peak at 6.12 eV is the frequency where excited exciton states start to absorb, which is described by the Wannier model. Data from Huggett and Teegarden, *Phys. Rev.* **141**, 797 (1966). Used with permission of the American Physical Society.

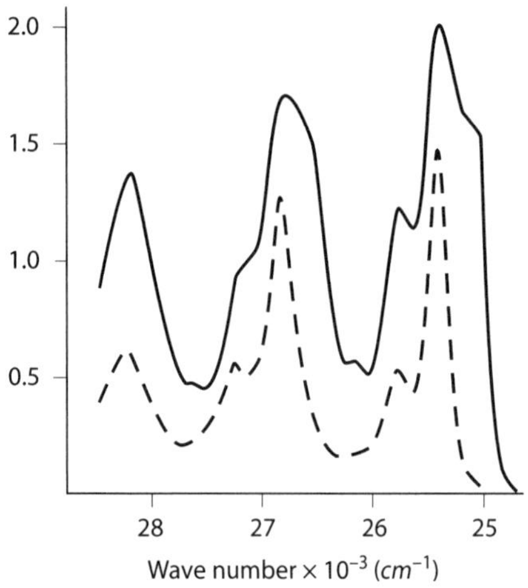

FIGURE 12.6. Optical absorption of crystalline anthracene. Solid and dashed lines are crystal spectra in two different directions of polarization. From Ferguson and Schneider, *J. Chem. Phys.* **28**, 761 (1958). Used with permission of the American Institute of Physics.

25,000 $\text{cm}^{-1} \sim 3$ eV. The two lines are for two different polarizations. Note that frequency increases to the left, and the repeating peaks are excitations of molecular vibrations.

12.5 X-ray Edge Singularity

The previous section on excitons showed that the optical absorption process can be viewed as creating two particles: an electron in the conduction band and a hole in the valence band. The Coulomb potential between these two particles of opposite charge had a major change in the absorption $\alpha(\omega)$ near the interband threshold.

Even more dramatic effects occur when the hole is in the inner core level of the atom, such as the 1*s* or 2*p* state. Since the hole states are localized to a single atomic core, they cannot move and have an infinite effective mass ($m_v = \infty$). Electrons in these states have a large binding energy, so that photons of high frequency are needed to put the electron into the conduction band. In this case, the photons are usually in the x-ray region of frequency.

- If the material is an insulator or a semiconductor, the theory is the same as for Wannier excitons.

  —If the hole band and the conduction band have different parity, the absorption matrix element is proportional to $\hat{p} \cdot \mathbf{p}_{cv} \psi_n(0)$, exactly as in the Wannier case. Since $m_v = \infty$, the center of mass exciton kinetic energy is zero, and the reduced mass equals the conduction band mass.

  —If often happens that the core–hole band and the conduction band have the same parity. In that case $\mathbf{p}_{cv} = 0$ and the absorption at the band minimum is very weak.

- The interesting new case is when the conduction band is a metal, such as sodium or aluminum. At low temperature the conduction electrons occupy the band states below the chemical potential $\varepsilon_c(\mathbf{k}) < \mu$. The conduction-band electron created during the absorption process must go into an unoccupied state $\varepsilon_c(\mathbf{k}) > \mu$. This one-electron transition is shown in fig.

12.7a. Absorption data from Callcott et al. [3] is shown for sodium in fig. 12.7b. Mahan [14] considered exciton states for this transition. He predicted an absorption resonance at threshold that has the form of a power law:

$$\alpha(\omega) \sim \left(\frac{W}{\omega - \omega_T}\right)^{\beta} \Theta(\omega - \omega_T) \tag{12.100}$$

where ω_T is the threshold frequency shown in fig. 12.7a, and W is a typical bandwidth energy. This exciton threshold effect is routinely observed in simple metals, as shown in fig. 12.7b, and is called a *Mahan singularity*. The form for the exponent β is derived below.

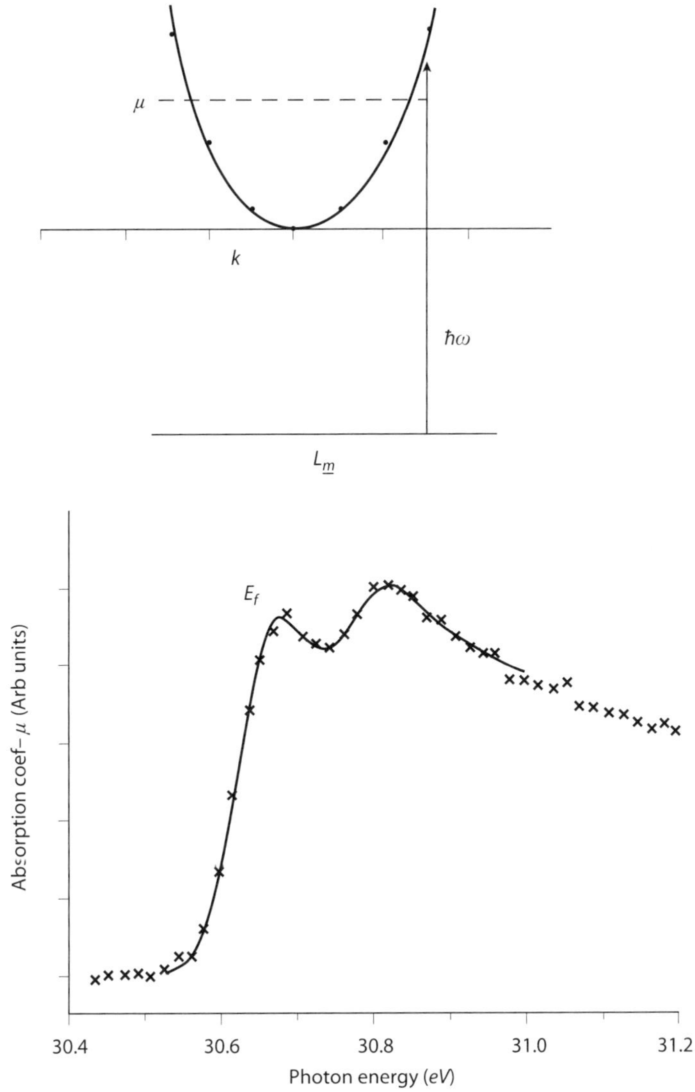

FIGURE 12.7. (a) X-ray transition from an inner core level to the conduction band of a free electron metal. (b) Experimental x-ray absorption data for L_{II}, L_{III}, edges of metallic sodium. From Callcott et al., *Phys. Rev. B* **18**, 6622 (1978). Used with permission of the American Physical Society.

The original derivation of eqn.(12.100) used perturbation theory. The Coulomb scattering between the electron and hole was derived for the first few Feynmann diagrams, and the form of the series suggested a method of summing all terms. Later derivations solved the problem exactly without using perturbation theory. The method of Pardee and Mahan (1973) gives the correct answer with a few lines of algebra. It uses ideas from scattering theory, and the analytical properties of scattering matrices.

The T-matrix for s-wave elastic scattering of a particle of wave vector k from a central potential has the phase factors for s-wave scattering ($\ell = 0$)

$$T(k, \ell = 0) \propto 2ie^{i\delta_0(k)} \sin[\delta_0(k)] = \left(e^{2i\delta_0} - 1\right) \tag{12.101}$$

These two factors are interpreted as follows: (i) the factor of "1" denotes the incoming spherical wave, and (ii) the factor of $\exp(2i\delta_0)$ is the outgoing spherical wave. The incoming wave has no phase shift, since it has not yet entered the region in space where there is a potential. When it does feel the potential, it has two phase shifts: one on the way in and the other on the way out. In x-ray absorption, the conduction electron is created in the core of the atom and becomes an outward spherical wave. There is only one factor of phase shift. The matrix element should have the factor of

$$M_{cv} \propto \xi \cdot \mathbf{p}_{cv} e^{i\delta_0(k)} \tag{12.102}$$

In x-ray absorption in metals, the phase shift should be zero for $\omega < \omega_T$ since there can be no scattering in this case. Treat the exponent as an analytic function of frequency. If its imaginary part is the phase shift, what is its real part?

$$M_{cv} \propto \xi \cdot \mathbf{p}_{cv} e^{\Gamma(\omega)} \tag{12.103}$$

$$\Im\{\Gamma(\omega)\} = i\delta_0(k)\Theta(\omega - \omega_T) \tag{12.104}$$

Write the exponent as

$$\Gamma(\omega) = \int_{\omega_T}^{W} \frac{d\omega'}{\pi} \frac{\delta_0(\omega')}{\omega' - \omega - i\delta} \tag{12.105}$$

$$= i\delta_0(k)\Theta(\omega - \omega_T) + \int_{\omega_T}^{W} \frac{d\omega'}{\pi} \delta_0(\omega') \mathcal{P} \frac{1}{\omega' - \omega}$$

The integral (12.105) has its pole in the lower half plane ($\omega = \omega' - i\delta$), as required for causal functions. Its imaginary part is eqn. (12.104). The real part can be evaluated approximately by expanding the phase shift in a Taylor series about the point ω:

$$\delta_0(\omega') = \delta_0(\omega) + (\omega' - \omega)\left(\frac{d\delta_0}{d\omega}\right)_\omega + O(\omega' - \omega)^2 \tag{12.106}$$

$$\Re\{\Gamma(\omega)\} = \frac{\delta_0(\omega)}{\pi} \ln\left[\frac{W - \omega}{\omega_T - \omega}\right] + \frac{1}{\pi}\left(\frac{d\delta_0}{d\omega}\right)_\omega (W - \omega_T) + \cdots \tag{12.107}$$

The first term in the above series provides the Mahan singularity. The absorption is the decay of the light intensity, which is proportional to $|M_{cv}|^2$:

$$\alpha(\omega) \propto e^{2\Re[\Gamma]} = \left[\frac{W-\omega}{\omega_T-\omega}\right]^{2\delta_0(\omega)/\pi} \Theta(\omega-\omega_T) \tag{12.108}$$

The Coulomb interaction between the electron and hole is used to calculate the value of the phase shift. Since they interact in a metallic environment, the interaction is screened and has a relatively short range. In this case, phase shifts are a valid concept and are easily calculated. They obey the Friedel sum rule. The results are in reasonable agreement with the observed x-ray edge absorption spectrum of simple metals. An experimental spectra for metallic sodium is shown in fig. 12.7b. The edge singularity is apparent.

12.6 Photoemission

There are many scattering experiments using photons. In Raman scattering, the incident light of frequency ω gets scattered to a final frequency of $\omega \pm \Delta\omega$ by the excitation $(-\Delta\omega)$ or destruction $(+\Delta\omega)$ of an excitation in the solid. Often the excitation is a phonon, and $\Delta\omega$ is the frequency of an optical phonon. Photoemission is another scattering experiment, where a photon is sent into the crystal and an electron leaves. The absorption of the photon creates excited electron states, and most of these electrons stay within the crystal. Only the few electrons excited near the surface and whose velocity is directed toward the surface manage to leave the surface.

In the original, historic experiments, the number of electrons leaving the surface were very few. All of them were collected by using a hemispherical metal collector. Using wire meshes for retarding voltages, one could measure the kinetic energy E_f of the electrons leaving the surface. One could also measure the work function W, the minimum photon energy required to detect emitted electrons. The final energy of the electron *inside of the crystal* is $E_f' = E_f + W$. The initial energy of the electron before absorbing the photon of frequency ω is $E_i' = E_f' - \hbar\omega$. These experiments provide a measurement of the function $B(\omega)$, the optical joint density of states between the valence bands and the unoccupied conduction bands.

The usefulness of photoemission increased with the development of synchrotron light sources. High-intensity, monochromatic, polarized sources of photons of variable frequency became available. The number of photoemitted electrons increased, and it became possible to perform angle-resolved experiments. The experimentalists could measure the energy E_f and direction (θ, ϕ) of the emitted electron in relation to the orientation of the crystal surface. *Band mapping* was born.

- The energy gives the final wave vector $k = \sqrt{2mE_f/\hbar^2}$.
- The angles give all three components of wave vector:

$$k_x = k\sin(\theta)\cos(\phi) \tag{12.109}$$

$$k_y = k\sin(\theta)\sin(\phi) \tag{12.110}$$

$$k_z = k\cos(\theta) \tag{12.111}$$

- Assuming the z-direction is normal to the surface, the transverse wave vectors $(k_x, k_y, 0)$ are unchanged when the electron goes through the surface. The z-component is changed by the work function:

$$k_z' = \sqrt{\frac{2m}{\hbar^2}\,[E_f \cos^2(\theta) + W]} \tag{12.112}$$

- Therefore, one knows the electron energy E_f' and wave vector (k_x, k_y, k_z') inside the crystal. That is a point on a conduction band.

- The initial state has energy E_i' and wave vectors (k_x, k_y, k_z'), which is a point on the valence band.

By varying the photon energy, the final kinetic energy of the electron, and the angles, one can map the energy bands of the solid. One has a map of (*i*) occupied valence bands, and (*ii*) conduction bands whose energy is above the vacuum level. Conduction-band states below the vacuum level have electrons that cannot get out of the surface. Figure 12.8 shows a mapping of the valence bands in AlAs in the (001) direction. The point Γ is the valence-band maximum.

Some angular dependence is quite simple. Consider interband optical absorption in a simple metal. Usually energy bands are drawn in a reduced zone and the optical transitions are vertical lines, since the photon wave vector is so small. However, if the interband transitions are drawn in an extended zone scheme, the optical transition actually changes the electron wave vector by a reciprocal lattice vectors **G**. Let the final wave vector be **k** and the initial wave vector be **k** − **G**. Energy conservation in a nearly free electron metal is

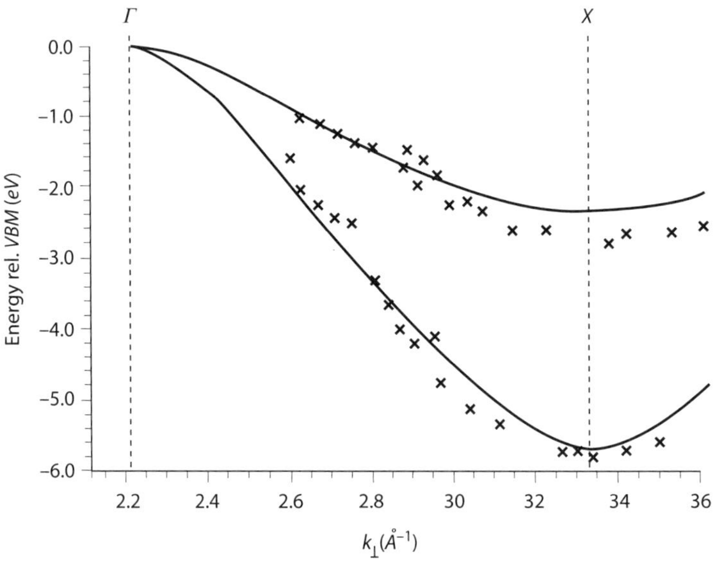

FIGURE 12.8. Band mapping of valence bands in AlAs by Kanski et al., *Solid State Commun.* 77, 617 (1991). Used with permission of Elsevier.

$$\frac{\hbar^2(\mathbf{k}-\mathbf{G})^2}{2m}+\hbar\omega=\frac{\hbar^2k^2}{2m} \tag{12.113}$$

Cancel the final energy from both sides and derive

$$\frac{\hbar^2}{m}kG\cos(\theta)=E_G+\hbar\omega,\quad E_G=\frac{\hbar^2G^2}{2m} \tag{12.114}$$

$$E_f'=\frac{\hbar^2k^2}{2m}=\frac{(E_G+\hbar\omega)^2}{4E_G\cos^2(\theta)} \tag{12.115}$$

The final energy of the electron E_f' is determined by the photon energy $\hbar\omega$ and the angle θ that $\mathbf{k}$ makes with the reciprocal lattice vector $\mathbf{G}$. All electrons of the same final energy form a conical distribution around the vector $\mathbf{G}$. These *Mahan cones* are observed in the photoemission of simple metals.

12.7 Conducting Polymers

Conducting polymers have interesting optical properties that depend on the number of monomers N in the chain. If the monomers are insulating and the chain is an insulator, the polarizability α_N of the chain of N monomers is usually $\alpha_N \sim N\alpha_1$, where α_1 is the polarizability of a single monomer.

The result is very different for a conducting polymer. To be conducting, the system has a one-dimensional electronic energy band that is partially filled. In a simple system with one conduction electron per monomer, the band will be half-full. Then the application of an electric field E along the length of the polymer causes the entire polymer to polarize, and conducting electrons can gather at one end. Then the polarizability is a nonlinear function of N. In a simple band model, as derived below, one predicts that $\alpha_N \propto N^3$. Most one-dimensional polymers with a half-full band undergo a Peierls transition to a dimerized configuration. The largest polarizabilities are found in polymers where the dimerization does not occur.

The energy of the system in the electric field is

$$E_N(F)=E_N(0)-\frac{\alpha_N}{2}E^2-\frac{\gamma_N}{4}E^4+O(E^6) \tag{12.116}$$

where $E_N(0)$ is the ground-state energy when the field is zero. The constant γ_N is the *hyperpolarizability*, and is an important parameter in nonlinear optical response. A simple band model predicts that $\gamma_N \propto N^5$. Experiments find the actual behavior as $\gamma_N \propto N^b$, where b ranges from 4 to 6 for different polymers. This result is pretty spectacular, since a molecule with ten monomers has a nonlinear response a million times larger than the single monomer. Similarly, the linear polarizability $\alpha_N \propto \alpha_1 N^3$ has a polarizability a thousand times larger than the monomer. Conducting polymers have the largest nonlinear optical response found in nature. Figure 12.9 shows the ratio α_N/α_1 for thiophene polymers, as measured by Zhao et al. [30]. The experimental behavior is nonlinear and is closer to an

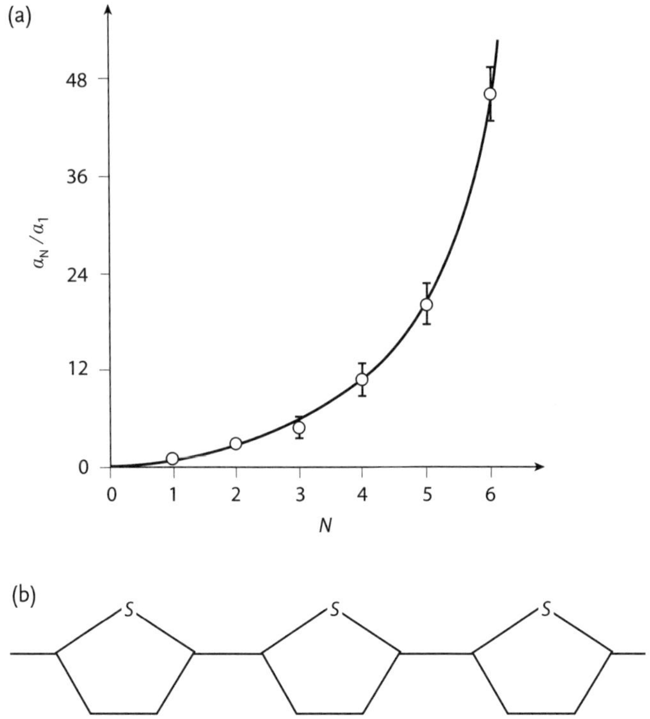

FIGURE 12.9. (a) Linear polarizability α_N as a function of N for thiophene polymers. From Zhao et al., *J. Chem. Phys.* **89**, 5535 (1988). Used with permission of the American Institute of Physics. (b) A short strand of polythiophene.

exponent of two, rather than three. Thiophene is used since its polymer does not have a Peierls transition. The thiophene molecule is a ring with one sulfur and four carbons. The sulfur atom fixes the bond lengths in the ring, and the polymer does not distort. The sulfur atom is not part of the one-dimensional conducting chain, which is composed of carbon p_z orbitals. The conducting chain is along the carbons shown in fig. 12.9b.

This phenomenon can be derived using a simple model of an electron in a one-dimensional box of length L, where $0 < x < L$. If the eigenfunction is zero at both ends of the box, the eigenfunctions and eigenvalues are

$$E_\alpha = \alpha^2 E_1, \quad E_1 = \frac{\hbar^2\pi^2}{2mL^2} \tag{12.117}$$

$$\phi_\alpha(x) = \sqrt{\frac{2}{L}}\sin(k_\alpha x), \quad k_\alpha = \frac{\pi\alpha}{L} \tag{12.118}$$

The electric field gives a term in the Hamiltonian:

$$V = eE\left(x - \frac{L}{2}\right) \tag{12.119}$$

The matrix elements are

$$M_{\alpha\beta} = \int_0^L dx V(x)\phi_\alpha(x)\phi_\beta(x) \tag{12.120}$$

$$M_{\alpha\beta} = \begin{cases} 0 & \alpha = \beta \\ 0 & \alpha \pm \beta = \text{even} \\ \frac{8eEL}{\pi^2}\frac{\alpha\beta}{(\alpha^2-\beta^2)^2} & \alpha \pm \beta = \text{odd} \end{cases} \tag{12.121}$$

The interaction $V(x)$ has odd parity as measured from the center of the box $x = L/2$. The eigenfunctions $\phi_\alpha(x)$ have even or odd parity, around the center of the box, depending on whether α is an odd or even integer. If both eigenvalues (α, β) are even or both are odd, the matrix element is zero by parity arguments. It is nonzero when one is an even integer and the other is an odd integer.

Evaluate second-order perturbation theory with this interaction, which derives the term $\alpha_N E^2/2$. This gives a formula for the polarizability:

$$\alpha_N = \frac{2}{E^2}\sum_{\alpha,\beta} n_\alpha[1 - n_\beta]\frac{M^2_{\alpha\beta}}{E_\beta - E_\alpha} \tag{12.122}$$

$$= \frac{256}{\pi^6 a_0} L^4 \sum_{\alpha,\beta} n_\alpha[1 - n_\beta]\frac{\alpha^2\beta^2}{(\beta^2 - \alpha^2)^5}$$

where a_0 is the Bohr radius, and an additional factor of two is added for spin degeneracy. The dimensions of the polarizability are volume. The factor of n_α means to sum α over occupied electron states, while the factor $[1 - n_\beta]$ means to sum β over unoccupied electron states. Since the length of the box $L \propto N$, it appears as if the polarizability goes as N^4. This guess is incorrect, since the double summation also depends on N.

The values of (α, β) run over positive integers. Let n_F be the integer of the highest occupied electron state, so the Fermi wave vector is $k_F = \pi n_F/L$. The largest value of α is n_F, and the smallest value of β is $n_F + 1$. One of these is odd and the other is even, so this pair (α, β) represents a nonzero value of $M_{\alpha\beta}$. The value of the double summation for this one pair of values is

$$\frac{\alpha^2\beta^2}{(\beta^2 - \alpha^2)^5} = \frac{n_F^2(n_F + 1)^2}{(2n_F + 1)^5} \sim \frac{1}{32n_F} \tag{12.123}$$

The latter estimate is for long chains when n_F is a large integer. The polarizability is estimated to be

$$\alpha_N \sim \frac{8L^3}{\pi^5 k_F a_0} \sim N^3 \tag{12.124}$$

As the chain of monomers becomes longer, the Fermi wave vector k_F stays the same, so that $n_F \sim L$. The polarizability scales with N^3.

12.8 Polaritons

Polaritons are electromagnetic excitations in solids. They are a mixture of photon modes and polarization modes. Two kinds of polarization modes are (*i*) optical phonons in an ionic crystal, and (*ii*) plasmons in a metal. Both cases are discussed below.

12.8.1 *Phonon Polaritons*

Polariton properties can be derived by starting with the solution to Maxwell's equations, in which the wave vector is given by

$$\varepsilon_0 k^2 = \frac{\omega^2}{c^2}\varepsilon(k,\omega) \approx \frac{\omega^2}{c^2}\varepsilon(\omega) \tag{12.125}$$

For phonon polaritons use the dielectric function of an insulator, including the contribution from optical phonons derived in chapter 7:

$$\varepsilon_0 k^2 = \frac{\omega^2}{c^2}\varepsilon(\infty)\left[1 + \frac{\Omega_p^2}{\omega_{\text{TO}}^2 - \omega^2}\right] \tag{12.126}$$

$$\Omega_p^2 = \left(\frac{\varepsilon(\infty) + 2\varepsilon_0}{3\varepsilon_0}\right)^2 \frac{\omega_{pi}^2 \varepsilon_0}{\varepsilon(\infty)} = \omega_{\text{LO}}^2 - \omega_{\text{TO}}^2 \tag{12.127}$$

Equation (12.126) is solved for the frequency. Introduce the notation

$$\omega(k) = ck\sqrt{\frac{\varepsilon_0}{\varepsilon(\infty)}} = \frac{ck}{n} \tag{12.128}$$

where n is the refractive index. The solution $\omega = \omega(k)$ describes a pure photon mode.

Equation (12.126) is a quadratic equation for ω^2:

$$0 = \omega^4 - \omega^2\left[\omega_{\text{TO}}^2 + \omega(k)^2 + \Omega_p^2\right] + \omega_{\text{TO}}^2\omega^2(k) \tag{12.129}$$

with two roots:

$$\omega_\pm^2(k) = \frac{1}{2}\left[B \pm \sqrt{B^2 - 4\omega_{\text{TO}}^2\omega^2(k)}\,\right] \tag{12.130}$$

$$B = \omega_{\text{TO}}^2 + \omega^2(k) + \Omega_p^2 = \omega_{\text{LO}}^2 + \omega^2(k) \tag{12.131}$$

These two solutions are graphed in fig. 12.10 for GaP. Two limits of this solution are as follows:

- At $k = 0$, $\omega(k) = 0$, the two solutions are

$$\omega_-^2 = 0 \tag{12.132}$$

$$\omega_+^2 = \omega_{\text{TO}}^2 + \Omega_p^2 = \omega_{\text{LO}}^2 \tag{12.133}$$

At high values of wave vector, where $\omega^2(k) \gg \omega_{\text{LO}}^2$, the roots are

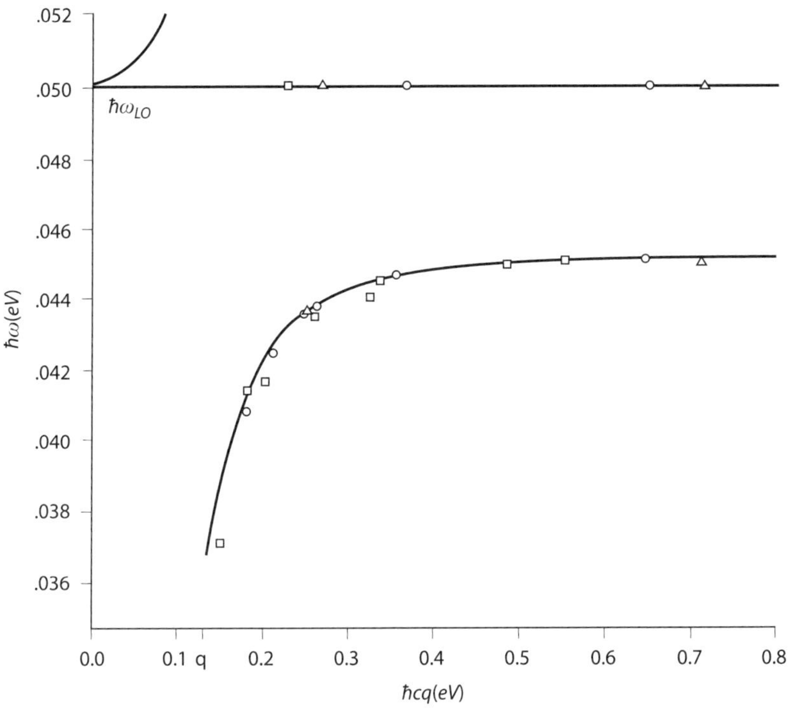

FIGURE 12.10. Phonon-polariton dispersion in GaP. Solid lines are theory, points are Raman experiments of Henry and Hopfield, *Phys. Rev. Lett.* **15**, 964 (1965) (Used with permission of the American Physical Society).

$$\omega_-^2 = \omega_{\text{TO}}^2 \tag{12.134}$$

$$\omega_+^2 = \omega^2(k) + \Omega_p^2 \tag{12.135}$$

The two solutions $\omega_{\pm}(k)$ are the excitations observed in the crystal. They result from the mode mixing between the phonons and the light waves. This mixing occurs because both modes create electromagnetic fields. Similar polaritons occur from the mixing of photons with other polarization waves in the crystal, such as exciton waves. There are no modes in the frequency range of $\omega_{\text{TO}} < \omega < \omega_{\text{LO}}$. Light completely reflects in this band of frequencies. The polariton concept was first reported in the book by Born and Huang [1].

12.8.2 Plasmon Polaritons

A similar analysis can be done for plasmons in a metal. The polariton equation is solved using the dielectric function for a metal:

$$\varepsilon_0 k^2 = \frac{\omega^2}{c^2}\varepsilon(\omega) = \frac{\omega^2}{c^2}\varepsilon_0\left(1 - \frac{\omega_p^2}{\omega^2}\right) \tag{12.136}$$

$$\omega^2 = k^2c^2 + \omega_p^2 \tag{12.137}$$

Photons can propagate only when they have a frequency $\omega \geq \omega_p$. For $\omega < \omega_p$, then, $k^2 < 0$ and k is imaginary. Photons are absorbed in the metal, and the reflectivity is unity. All of the incident photons are reflected, since there are no photon modes in the metal that can carry energy away from the surface [24].

12.9 Surface Polaritons

Solid surfaces have their own polariton modes that are called *surface polaritons.* In general, there is a surface mode for each type of bulk polariton. They are a mixture of photons and polarization waves. Electromagnetic modes on surfaces have been known since 1909, when Arnold Sommerfeld wrote about radio waves traveling along the earth's surface.

Rayleigh waves are surface acoustical waves. They are predicted by elasticity theory and are not electromagnetic in origin. They are not classified as surface polaritons. They are described in many monographs on elasticity, such as the one by Landau and Lifshitz [12].

Surface plasmons were predicted independently by Rufus Ritchie [19] and Ed Stern [25], and surface optical phonons were first predicted by Ron Fuchs and Ken Kliewer [7]. Surface polaritons have the feature that the dielectric function is negative at the surface mode frequency $\omega_{Si} : \varepsilon(\omega_{Si}) < 0$, where $i = O$ or P. The reason for this requirement is shown below.

There are several kinds of surface polaritons. The first distinction is between longitudinal and transverse modes. Longitudinal surface modes result from poles in the surface dielectric function. Recall that if a charge q is a distance d outside a metal surface, the electrostatic potential contains two terms: a direct interaction and an image interaction:

$$\phi(\mathbf{r}) = \frac{q}{4\pi\varepsilon_0}\left(\frac{1}{r} - \frac{\lambda}{r_I}\right) \tag{12.138}$$

$$r = \sqrt{\rho^2 + (d - z)^2}, \quad r_I = \sqrt{\rho^2 + (d + z)^2} \tag{12.139}$$

$$\lambda = \frac{\varepsilon_m - \varepsilon_0}{\varepsilon_m + \varepsilon_0} \tag{12.140}$$

As a charged particle approaches the surface, it can lose energy by exciting surface plasmons. They arise as longitudinal excitations of the interaction with the image charge:

$$\Im\{\lambda\} = \Im\left\{\frac{\omega_{sp}^2}{\omega^2 - \omega_{sp}^2}\right\} = -\frac{\pi}{2}\omega_{sp}\delta(\omega - \omega_{sp}) \tag{12.141}$$

$$\omega_{sp} = \frac{\omega_p}{\sqrt{2}} \tag{12.142}$$

Transverse surface polaritons are derived from Maxwell equations. Figure 12.11 shows the surface of a planar dielectric, where $z > 0$ is vacuum, and $z < 0$ is a material with a dielectric function $\varepsilon(\omega)$. Assume the mode is traveling in the x-direction with a phase factor of $\exp[i(qx - \omega t)]$. Derivatives with respect to t give $-i\omega$, and derivatives with respect to y give zero. Then Maxwell's equations

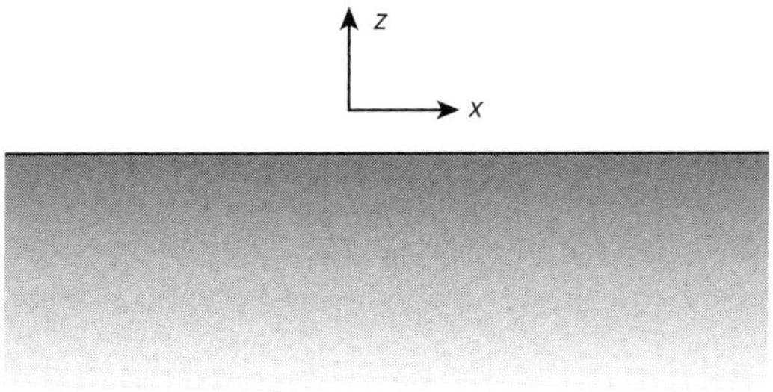

FIGURE 12.11. Geometry for the discussion of surface polaritons.

$$\nabla \times \mathbf{E} = \frac{\partial \mathbf{B}}{\partial t} \tag{12.143}$$

$$\nabla \times \mathbf{B} = -\varepsilon \mu_0 \frac{\partial \mathbf{E}}{\partial t} \tag{12.144}$$

separate into transverse electric (TE) and transverse magnetic (TM) modes.

- TE modes obey three equations for the field variables (B_x, E_y, B_z):

$$-i\omega B_x = -\frac{\partial}{\partial z} E_y \tag{12.145}$$

$$-i\omega B_z = \frac{\partial}{\partial x} E_y = iqE_y \tag{12.146}$$

$$i\omega \varepsilon \mu_0 E_y = \frac{\partial}{\partial z} B_x - \frac{\partial}{\partial x} B_z \tag{12.147}$$

TM modes obey three equations for the field variables (E_x, B_y, E_z):

$$i\varepsilon \mu_0 \omega E_x = -\frac{\partial}{\partial z} B_y \tag{12.148}$$

$$i\varepsilon \mu_0 \omega E_z = \frac{\partial}{\partial x} B_y = iqB_y \tag{12.149}$$

$$i\omega B_y = \frac{\partial}{\partial z} E_x - \frac{\partial}{\partial x} E_z \tag{12.150}$$

The surface polaritons are always TM modes. The TE modes have no solution for the above equations. Proving this statement is a homework problem.

For TM modes, $B_y(z)$ is used as the basic field function. It obeys an equation

$$0 = \left[\frac{\varepsilon}{\varepsilon_0} \frac{\omega^2}{c^2} - q^2 + \frac{d^2}{dz^2} \right] B_y(z) \tag{12.151}$$

$$B_y(z) = B_0 e^{-\gamma z}, \quad z > 0 \tag{12.152}$$

$$= B_0 e^{\bar{\gamma} z}, \quad z < 0$$

$$\gamma = \sqrt{q^2 - \omega^2/c^2}, \quad \bar{\gamma} = \sqrt{q^2 - \varepsilon\omega^2/\varepsilon_0 c^2} \tag{12.153}$$

Since $B_y(z)$ is conserved at the interface ($z = 0$), the same prefactor B_0 is used for $z > 0$ and $z < 0$. The parallel component of the electric field is also conserved at the surface. It is derived from eqn. (12.148):

$$E_x(z) = -i\frac{\gamma c^2}{\omega} B_0 e^{-\gamma z}, \quad z > 0 \tag{12.154}$$

$$E_x(z) = i\frac{\varepsilon_0 \bar{\gamma} c}{\varepsilon\omega} B_0 e^{\bar{\gamma} z}, \quad z < 0 \tag{12.155}$$

$$0 = \gamma + \frac{\varepsilon_0 \bar{\gamma}}{\varepsilon(\omega_{sp})} \tag{12.156}$$

The last equation is the eigenvalue equation that comes from matching $E_x(z = 0)$. A solution requires that $\varepsilon(\omega_{sp}) < 0$. Write the equation as $\gamma\varepsilon = -\bar{\gamma}\varepsilon_0$ and square both sides:

$$\varepsilon^2[c^2q^2 - \omega^2] = \varepsilon_0^2 c^2 q^2 - \omega^2 \varepsilon\varepsilon_0 \tag{12.157}$$

This equation has a solution $\varepsilon = \varepsilon_0$. That is not a surface mode, since they require $\varepsilon < 0$. Factor $\varepsilon - \varepsilon_0$ out of the above equation:

$$(\varepsilon - \varepsilon_0)[\varepsilon(c^2q^2 - \omega^2) + \varepsilon_0 c^2 q^2] = 0 \tag{12.158}$$

$$\varepsilon(c^2q^2 - \omega^2) + \varepsilon_0 c^2 q^2 = 0 \tag{12.159}$$

The latter equation is the eigenvalue equation for surface modes.

12.9.1 *Surface Plasmons*

The eigenvalue equation is solved using the dielectric function for simple metals $\varepsilon/\varepsilon_0 = 1 - \omega_p^2/\omega^2$:

$$\left(1 - \frac{\omega_p^2}{\omega^2}\right)[c^2q^2 - \omega^2] + c^2q^2 = 0 \tag{12.160}$$

Multiply the entire equation by ω^2 and then arrange the terms with like powers of frequency:

$$0 = \omega^4 - \omega^2(\omega_p^2 + 2c^2q^2) + c^2q^2\omega_p^2 \tag{12.161}$$

with a solution first found by Ed Stern [25]:

$$\omega_{sp}^2(q) = \frac{1}{2}\left[\omega_p^2 + 2c^2q^2 - \sqrt{\omega_p^4 + 4c^4q^4}\right] \tag{12.162}$$

The square root should have a $\pm\sqrt{}$, but the choice of a plus sign does not give $\varepsilon(\omega_{sp}) < 0$, so choose the negative sign. Figure 12.12 shows the surface plasmon dispersion relation at two different scales of wave vector. Two interesting limits are the following:

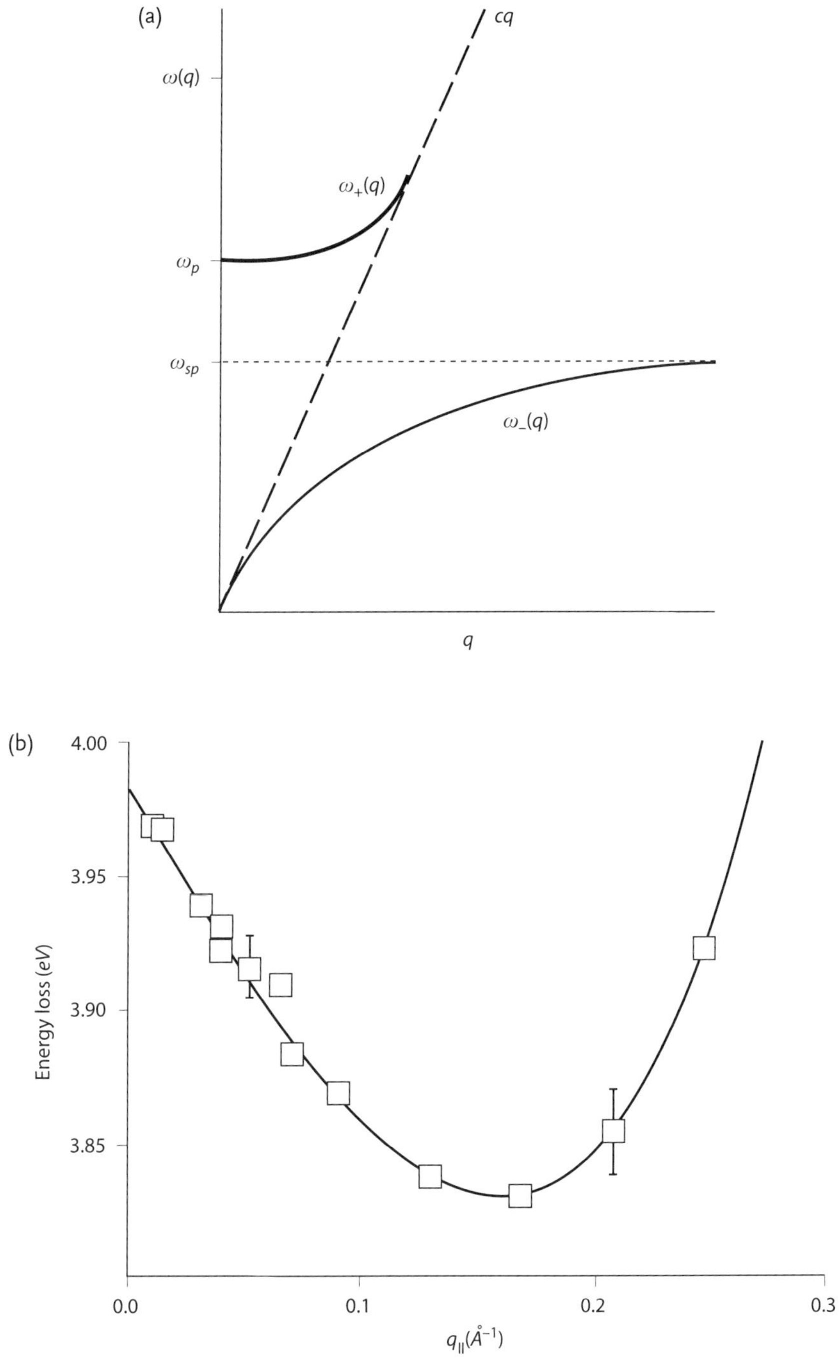

FIGURE 12.12. Surface plasmon dispersion relation. (a) Two solutions to eqn. (12.162). (b) Experimental results for metallic sodium from Plummer's group. Tsuei et al., *Phys. Rev. Lett.* **63**, 2256 (1989). Used with permission of the American Physical Society.

- For small wave vectors with $cq \ll \omega_p$ the solution is

$$\lim_{cq<\omega_p} \omega_{sp} \to cq - O\left(\frac{c^3q^3}{\omega_p^2}\right) \tag{12.163}$$

The dispersion is photon-like, with a velocity given by the speed of light.

- For large wave vectors with $cq \gg \omega_p$ the solution is

$$\lim_{cq>\omega_p} \omega_{sp} \to \frac{\omega_p}{\sqrt{2}} - O\left(\frac{\omega_p^3}{c^2q^2}\right) \tag{12.164}$$

Here the definition of "small" or "large" wave vectors is compared to $q_p = \omega_p/c \sim 10^6$ cm^{-1}. That is a small wave vector on the scale of the Brillouin zone, where $\pi/a \sim 10^8$ cm^{-1}. This curve is shown in fig. 12.12a. For large wave vectors, the surface plasmon has a dispersion $\omega_{sp}(q)$. Figure 12.12b shows the results for metallic sodium as measured by Ward Plummer's group. [27] In the limit of large wave vector $(q \gg q_p)$, the eigenvalue equation simplifies to

$$\lim_{q\gg q_p} \begin{cases} \gamma & \to & q \\ \bar{\gamma} & \to & q \\ 0 = \gamma + \bar{\gamma}/\varepsilon & \to & \varepsilon(\omega_{sp}) = -1 \end{cases} \tag{12.165}$$

Surface polaritons are easy to measure since they show up in a variety of experiments. Low-energy electron scattering from a surface, in the reflection geometry, shows these excitations quite clearly. The electrons are scattered from the surface in the reflection geometry. Typical data are shown in fig. 12.13 for the $\hbar\omega_{SO} = 69$ meV surface optical phonon in ZnO, as measured by Harold Ibach [10].

Irving Langmuir [13] discovered oscillations in gas discharges and coined the name "plasmon." He also found there was more than one plasma frequency when the density of charges was not uniform. He called these additional oscillations *multimode plasmons.* At the surface of a metal, the electron density changes from its full value inside the metal to its value of zero outside over a distance of less than a nanometer. This variation in charge density can, for some metals, induce additional longitudinal surface excitations that are multimode plasmons. They may have been observed on the surface of the alkali metals [28].

12.9.2 *Surface Optical Phonons*

Optical phonons in polar materials have a dielectric function at long wavelength of

$$\varepsilon(\omega) = \varepsilon(\infty) + \frac{\Omega_p^2 \varepsilon_0}{\omega_{\rm TO}^2 - \omega^2} \equiv \varepsilon(\infty)\frac{\omega_{\rm LO}^2 - \omega^2}{\omega_{\rm TO}^2 - \omega^2} \tag{12.166}$$

$$\Omega_p^2 = \frac{e^2}{\mu\varepsilon_0\Omega_0}\left(\frac{\varepsilon(\infty) + 2\varepsilon_0}{3\varepsilon_0}\right)^2 = \omega_{\rm TO}^2 \frac{\varepsilon(0) - \varepsilon(\infty)}{\varepsilon_0} \tag{12.167}$$

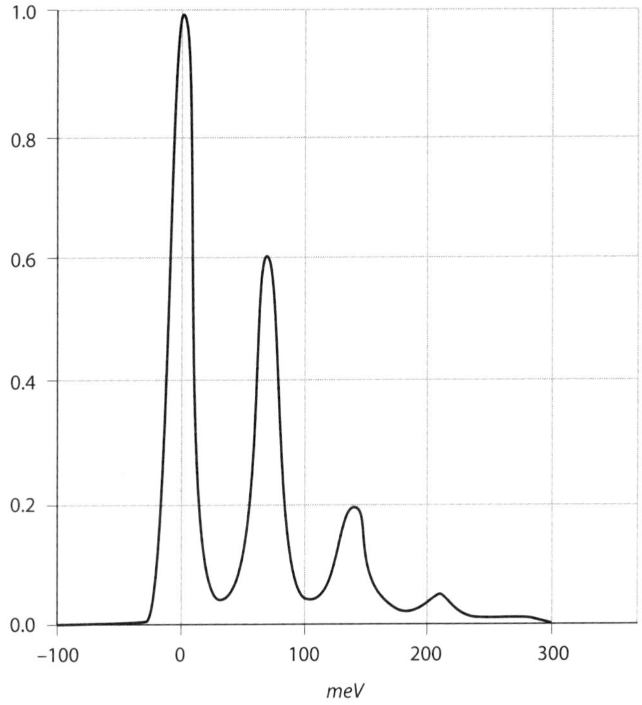

FIGURE 12.13. Low-energy electron loss spectra data for surface optical phonons in ZnO. Surface optical phonon energy is 69 meV. From Ibach, *Phys. Rev. Lett.* **24**, 1416 (1970). Used with permission of the American Physical Society.

This formula is inserted into the eigenvalue equation for surface modes (12.159) After multiplying out the various terms, one finds a quadratic equation for ω^2:

$$0 = \varepsilon(\infty)\omega^4 - \omega^2\left[c^2q^2(\varepsilon(\infty) + \varepsilon_0) + \varepsilon(\infty)\omega_{LO}^2\right] + c^2q^2\left[\varepsilon(\infty)\omega_{LO}^2 + \varepsilon_0\omega_{TO}^2\right]$$

Use the Lyddane-Sachs Teller relation to rewrite the last term as $c^2q^2\omega_{TO}^2(\varepsilon(0) + \varepsilon_0)$. The solution is

$$\omega_S^2 = \frac{1}{2}\left[\Omega(q)^2 + \omega_{LO}^2 - \sqrt{(\Omega^2(q) + \omega_{LO}^2)^2 - 4\Omega^2(q)\omega_{SO}^2}\right] \tag{12.168}$$

$$\Omega(q)^2 = \frac{\varepsilon(\infty) + \varepsilon_0}{\varepsilon(\infty)}c^2q^2 \tag{12.169}$$

$$\omega_{SO}^2 = \omega_{TO}^2\frac{\varepsilon(0) + \varepsilon_0}{\varepsilon(\infty) + \varepsilon_0} \tag{12.170}$$

Figure 12.14 shows the surface optical phonon dispersion for NaCl. The dispersion starts at a nonzero wave vector at the frequency ω_{TO}. Surface modes require $\varepsilon(\omega) < 0$, which requires that $\omega_{LO}^2 > \omega_S^2 > \omega_{TO}^2$. Put the latter inequality in the dispersion relation and rearrange terms:

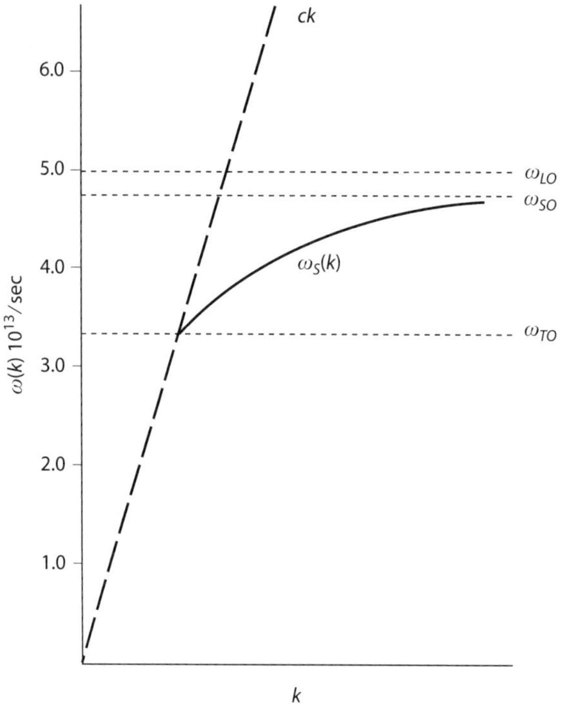

FIGURE 12.14. Surface optical phonon dispersion in NaCl.

$$\sqrt{(\Omega^2(q) + \omega_{\rm LO}^2)^2 - 4\Omega^2(q)\omega_{\rm SO}^2} < \Omega^2(q) + \omega_{\rm LO}^2 - 2\omega_{\rm TO}^2 \tag{12.171}$$

Square this equation and cancel like terms from both sides. After much canceling of various factors, one finds $q^2c^2 > \omega_{\rm TO}^2$, which is the condition to have the surface mode frequency higher than that of the TO phonon. The minimum value of wave vector is $q_c = \omega_{\rm TO}/c$.

At large wave vector the surface optical phonon has the frequency $\omega_{\rm SO}$. This is the value of frequency throughout most of the BZ. It is the frequency observed when low-energy electrons are scattered from the surface of polar crystals. It can be derived simply by setting the dielectric function equal to $-\varepsilon_0$:

$$-\varepsilon_0 = \varepsilon(\infty) + \frac{\omega_{pi}^2}{\omega_{\rm TO}^2 - \omega_{\rm SO}^2} \tag{12.172}$$

When examining a dielectric function of frequency from optical phonons,

- TO modes are where $\varepsilon \to \infty$
- LO modes are where $\varepsilon = 0$
- SO modes are where $\varepsilon = -\varepsilon_0$.

Insulators with several atoms per unit cell have many different optical phonons and many different surface optical phonons.

Here the discussion of surface modes has assumed a planar geometry. Interface modes exist in many geometries, such as along wires, at edges, or on the surface of spheres. See Sernelius [21] for many of these cases.

12.9.3 *Surface Charge Density*

The surface plasmon on a metal is the result of oscillating surface charges. The surface has regions of oscillating positive and negative charge. This charge density can be deduced from Poisson's equation:

$$\varepsilon\nabla\cdot\mathbf{E} = \rho(\mathbf{r}) \tag{12.173}$$

For the TM mode $E_y = 0$. Add the equations for the other two components of electric field:

$$i\varepsilon\mu_0\omega\left[\frac{\partial E_x}{\partial x} + \frac{\partial E_z}{\partial z}\right] = -(1-1)\frac{\partial^2 B_y}{\partial x\partial z} = 0 \tag{12.174}$$

So $\nabla\cdot\mathbf{E} = 0$, and the charge density appears to be zero. That is certainly the correct answer outside of the surface, since there are no electrons in free space. It is also true inside of the metal. The charges are right on the surface. For $q \gg \omega/c$ write E_z using $\varepsilon = -\varepsilon_0$:

$$E_z = \frac{q}{\varepsilon\mu_0\omega}B_0 e^{-q|z|}\,\mathrm{sgn}(z)e^{i(qx-\omega_{sp}t)} \tag{12.175}$$

$$\mathrm{sgn}(z) = \begin{Bmatrix} +1 & z>0 \\ -1 & z<0 \end{Bmatrix},\quad \frac{d}{dz}\mathrm{sgn}(z) = 2\delta(z) \tag{12.176}$$

$$\varepsilon\frac{\partial E_z}{\partial z} = 2\frac{q}{\mu_0\omega}B_0 e^{i(qx-\omega_{sp}t)}\delta(z) + \text{other terms} \tag{12.177}$$

The other terms are the ones that cancel in eqn. (12.174):

$$\rho(\mathbf{r}) = \frac{2q}{\mu_0\omega}B_0 e^{i(qx-\omega_{sp}t)}\delta(z) \tag{12.178}$$

A surface plasmon makes periodic oscillations of surface charge on the surface of the metal. Any physical process that induces surface charges on a metal is creating surface plamons.

Computer storage devices utilize a tape with bits of information stored on the surface, such as in a magnetic domain. If the bits are separated by a distance λ, they create a surface wave vector of $q = 2\pi/\lambda$. The resulting electric and magnetic fields must satisfy LaPlace's equation $\nabla^2\phi(\mathbf{r}) = 0$ in the air space above the tape. The fields must have the spatial dependence of

$$\phi(\mathbf{r}) = \phi_0\exp[i\mathbf{q}\cdot\boldsymbol{\rho} - zq] \tag{12.179}$$

where $\mathbf{q}$ is the two-dimensional wave vector in the surface plane. The important observation is that these fields decay away in the z-direction according to the same wave vector $q = |\mathbf{q}|$. The closer one puts the bits together on the tape, the larger the wave vector and the faster the fields fall off from the surface. The recording heads that read the tape have to be nearer to the surface of the tape.

References

1. M. Born and K. Huang, *Dynamical Theory of Crystal Lattices* (Oxford University Press, 1954); polaritons discussed on p. 91
2. P. N. Butcher, The absorption of light by alkali metals. *Proc. Phys. Soc. London Ser. A* **64**, 765–772 (1951)
3. T. A. Callcott, E. T. Arakawa, and D. L. Ederer, L_{23} soft-x-ray emission and absorption spectra of Na. *Phys. Rev. B* **18**, 6622 (1978)
4. H. Ehrenreich and H. R. Philipp, Optical properties of Ag and Cu. *Phys. Rev.* **128**, 1622–1629 (1962)
5. R. J. Elliott, Intensity of optical absorption by excitons. *Phys. Rev.* **108**, 1384–1389 (1957)
6. J. Ferguson and W. G. Schneider, Absorption spectrum of crystalline anthracene. *J. Chem. Phys.* **28**, 761–764 (1958)
7. R. Fuchs and K. L. Kliewer, Optical modes of vibration in an ionic crystal slab. *Phys. Rev.* **140**, A2076–A2088 (1965)
8. C. H. Henry and J. J. Hopfield, Raman scattering by polaritons. *Phys. Rev. Lett.* **15**, 964–966 (1965)
9. G. R. Huggett and K. Teegarden, Intrinsic photoconductivity in the alkali halides. *Phys. Rev.* **141**, 797– (1966)
10. H. Ibach, Optical surface phonon in ZnO detected by slow-electron spectroscopy. *Phys. Rev. Lett.* **24**, 1416–1418 (1970)
11. J. Kanski, P. O. Nilsson, U. O. Karlsson, and S. P. Svensson, Band mapping of MBE-grown AlAs. *Solid State Commun.* **77**, 617–618 (1991)
12. L. D. Landau and E. M. Lifshitz, *Theory of Elasticity* (Addison-Wesley, London, 1959)
13. I. Langmuir, Oscillations in ionized gases. *Proc. Natl. Acad. Sci.* **14**, 627–637 (1928)
14. G. D. Mahan, Excitons in metals: infinite hole mass. *Phys. Rev.* 163, 612–617 (1967)
15. G. D. Mahan and K. R. Subbaswamy, *Local Density Theory of Polarizability* (Plenum, New York, 1990) fig. 5.5
16. W. J. Pardee and G. D. Mahan, Simplified theory of x-ray edge in metals. *Phys. Lett.* **45A**, 117–118 (1973)
17. H. R. Philipp and E. A. Taft, Optical constants of silicon in the region 1 to 10 eV. *Phys. Rev.* **120**, 37–38 (1960)
18. D. C. Reynolds and T. C. Collins, *Excitons: Their Properties and Uses* (Academic Press, New York, 1981)
19. R. Ritchie, Plasma losses by fast electrons in thin films. *Phys. Rev.* **106**, 874–881 (1957)
20. D. D. Sell, Resolved free-exciton transitions in the optical absorption spectrum of GaAs. *Phys. Rev. B* **6**, 3750–3753 (1972)
21. B. E. Sernelius, *Surface Modes in Physics* (Wiley-VCH,New York, 2001)
22. A. J. Sievers, Infrared and optical properties of Na, K, and Rb metals. *Phys. Rev. B* **22**, 1600–1611 (1980)
23. N. V. Smith, Optical constants of sodium and potassium from 0.5 to 4.0 eV by split-beam ellipsometry. *Phys. Rev.* **183**, 634–644 (1969)
24. A. Sommerfeld, Über die Ausbreitung der Wellen in der Drahtlosen Telegraphie. *Ann. Phys.* **333**, 665–736 (1909)
25. E. A. Stern, quoted by R. A. Ferrell, Radiation of plasma oscillations in metal films. *Phys. Rev.* **111**, 1214–1222 (1958)

26. E. A. Taft and H. R. Philipp, Optical constants of silver. *Phys. Rev.* **121**, 1100–1103 (1961)
27. K.-D. Tsuei, E. W. Plummer, and P. J. Feibelman, Surface-plasmon dispersion in simple metals. *Phys. Rev. Lett.* **63**, 2256–2259 (1989)
28. K.-D. Tsuei, E. W. Plummer, A. Liebsch, E. Pehlke, K. Kempa, and P. Bakshi, Normal modes at the surface of simple metals. *Surf. Sci.* **247**, 302–326 (1991)
29. G. H. Wannier, The structure of electronic excitation levels in insulating crystals. *Phys. Rev.* **52**, 191–197 (1937)
30. M. T. Zhao, B. P. Singh, and P. N. Prasad, A systematic study of polarizability and microscopic third-order optical nonlinearity in thiophene oligomers. *J. Chem. Phys.* **89**, 5535–5541 (1988)

Homework

1. Assume the imaginary part of the dielectric function has the following frequency dependence:

$$\varepsilon_2(\omega)/\varepsilon_0 = \begin{cases} \lambda & \omega_1 < \omega < \omega_2 \\ -\lambda & -\omega_2 < \omega < -\omega_1 \\ 0 & \text{elsewhere} \end{cases} \tag{12.180}$$

Derive an expression for $\varepsilon_1(\omega)$.

2. If the dielectric function is given by a single resonance, it has the form

$$\varepsilon(\omega)/\varepsilon_0 = 1 + \frac{\omega_p^2}{\omega_0^2 - \omega^2 - i\gamma\omega} \tag{12.181}$$

(a) What are $\varepsilon_1(\omega)$ and $\varepsilon_2(\omega)$?

(b) Show they obey a Kramers-Kronig relation.

3. The dielectric function of a simple metal can be described by the Drude model:

$$\varepsilon(\omega)/\varepsilon_0 = 1 - \frac{\omega_p^2}{\omega(\omega + i/\tau)} \tag{12.182}$$

where τ is a constant. Does this form obey the Kramers-Kronig relation?

4. The angle $\theta(\omega)$ is defined as the phase of the reflectivity

$$r = \frac{n + i\kappa - 1}{n + i\kappa + 1} \equiv |r(\omega)| \exp[i\theta(\omega)] \tag{12.183}$$

Make a graph of $\theta(x)$ vs. $x = \omega\tau$ using the dielectric function of the prior problem. Assume that $\omega_p\tau = 50$.

5. The skin depth $\delta = c/(\kappa\omega)$ is the distance an electromagnetic wave penetrates into the surface of a metal. Show that in the Drude regime $(1\backslash\tau < \omega < \omega_p)$ it is nearly independent of frequency. Find its numerical value for Al and Na.

6. The *absorbance* is defined as $A = 1 - R$.

- Show that $A = 0$ using the formula $\varepsilon = 1 - \omega_p^2/\omega^2$ whenever $\omega < \omega_p$.
- Using the Drude model,

$$\varepsilon(\omega) = 1 - \frac{\omega_p^2}{\omega(\omega + i/\tau)} \tag{12.184}$$

show in the limit $1 \ll \omega\tau \ll \omega_p\tau$, that

$$A = \frac{C}{\omega_p \tau} \tag{12.185}$$

and find C. In the infrared, the absorbance is found to agree with this formula: it is a constant independent of frequency.

7. Show that the sum rule in eqn. (12.11) is obeyed by the Drude form of the dielectric function.

8. The formula for the threshold energy $\hbar\omega_T$ for interband transitions in soldium is given in the text. Calculate the numerical value in electron volts. Sodium has the bcc structure, with a lattice constant $a = 4.225$ Å at low temperature. Which reciprocal lattice vector **G** gives the lowest threshold energy?

9. Two formulas were derived for the electron transport lifetime from impurity scattering: one from the Boltzmann equation and one from the force–force correlation function. Show they are identical.

10. Use $k \cdot p$ theory to evaluate the integral:

$$M_{cv}(\mathbf{q}) = \int d^3 r u_c^*(\mathbf{k} + \mathbf{q}, \mathbf{r}) u_v(\mathbf{k}, \mathbf{r}) \tag{12.186}$$

Assume that **q** is a photon wave vector and is small. *Hint*: The $k \cdot p$ theory can be evaluated around any k-point.

11. Consider that a semiconductor such as GaAs is doped with impurities so that the conduction electrons have a finite density n_0 and a plasmon frequency $\omega_p^2 = e^2 n_0 / m\varepsilon_0$. The dielectric function including phonon and electron contributions is

$$\varepsilon(\omega) = \varepsilon(\infty) + \frac{\varepsilon(0) - \varepsilon(\infty)}{1 - (\omega/\omega_{TO})^2} - \varepsilon_0 \frac{\omega_p^2}{\omega^2} \tag{12.187}$$

Solve the equation $\varepsilon_0 k^2 = \omega^2 \varepsilon(\omega)/c^2$ to find the polariton modes of the crystal. This experiment was first done by Mooradian and Wright. Next, find the poles of $1/\varepsilon(\omega)$.

12. Derive an expression for the surface optical phonon frequency by solving $\varepsilon = -\varepsilon_0$, where

$$\varepsilon(\omega) = \varepsilon(\infty) + \frac{\varepsilon(0) - \varepsilon(\infty)}{1 - (\omega/\omega_{TO})^2} \tag{12.188}$$

13. Show that there are no transverse electric (TE) surface modes.

14. Two parallel, identical, metal plates have a narrow vacuum gap $2d$ between them. The plates have a dielectric function of $\varepsilon(\omega)/\varepsilon_0 = 1 - \omega_p^2/\omega^2$. Find the electromagnetic modes that run along the surfaces between the two plates. There are two modes. Derive the two eigenvalue equations but do not try to solve them.

15. The piezoelectric coupling contributes to the dielectric function at low frequencies ($\omega \ll \omega_{TO}$):

$$\varepsilon(q,\omega) = \varepsilon(0) + \frac{H^2 q^2 \varepsilon_0}{c s^2 q^2 - \omega^2} \tag{12.189}$$

where H is a constant with dimensions meter/second.

- Find the dispersion of the surface mode in the limit that $q^2 \gg \omega^2/c^2$.
- Find the poles of $1/\varepsilon(q,\omega)$.
- Find the dispersion of the surface polariton mode.

16. If $(a, a^\dagger)$ are photon operators and $(b, b^\dagger)$ are operators for optical phonons, show that the interaction between them can be written as

$$H = \sum_k \left[\hbar\omega_k a_k^\dagger a_k + \hbar\omega_0 b_k^\dagger b_k - \frac{i}{2}\omega_{pi}\sqrt{\frac{\omega_0}{\omega_k}}\left(a_k + a_{-k}^\dagger\right)\left(b_{-k} - b_k^\dagger\right) \right.$$

$$\left. + \frac{\omega_{pi}^2}{4\omega_k}\left(a_k + a_{-k}^\dagger\right)\left(a_{-k} + a_k^\dagger\right) \right] \tag{12.190}$$

Next, solve this equation and find the eigenvalues. You get a set of four coupled equations by taking $-i\omega C = i[H, C]$, where C signifies the four operators a, $a^\dagger$, b, $b^\dagger$. You should reproduce the classical equation that $\omega_k^2 = \omega^2 \varepsilon(\omega)$.

13 Magnetism

13.1 Introduction

Magnetism is one of the most important topics in condensed matter physics. There are many different kinds of magnetic phenomena, found in many different kinds of materials. There are some materials, such as iron or nickel, that magnetically order so they have a net magnetic moment. Other materials can be induced to have magnetic moments by the application of a magnetic field. Giant magnetoresistance is an important technology for computer memories. Magnets can be insulators, metals, or semiconductors.

13.2 Simple Magnets

13.2.1 *Atomic Magnets*

Some materials are insulators at all temperatures. Some of their atoms may have magnetic moments, and these moments may align into a magnetically ordered structure. The first step is to understand why some atoms have magnetic moments. They arise from partially filled atomic shells. Each atomic state has a set of orbital quantum numbers, such as 3*d* or 4*p*. The first number is the principal quantum number, while the second symbol, a letter, denotes the angular momentum: $\ell = 0$ is *s*, $\ell = 1$ is *p*, $\ell = 2$ is *d*, and $\ell = 3$ is *f*. The orbital degeneracy is $2\ell + 1$ different values of magnetic quantum number *m*. There is also a spin degeneracy of two for each orbital state. So each atomic shell has $N_\ell = 2(2\ell + 1)$ electron states. They fill up in a particular order, which is given by Hund's rules.

13.2.2 *Hund's Rules*

The general theme of these rules is that when electrons are added to an orbital state, the spins prefer to be parallel. If the first electron in the orbital has spin-up, then that is the preferred spin state for the second electron, and for the third electron. Add electrons until all $2\ell + 1$ orbital states will have spin-up. Finally, when the next electron, number $2\ell + 2$, is added it finds all up states occupied, and it must have spin-down. The remaining electrons also have spin-down. The filling of states must be in accord with the Pauli principle that no two electrons can be in the same quantum state. Here state means values of (ℓ, m_ℓ, m_s).

For *s*-states ($\ell = 0$), the two electrons must have opposite spins in the spin singlet arrangement. Hund's rules apply to this case but are trivial. Usually they are applied to values of $\ell = 1, 2$, or 3.

1. If S_z is the total magnetic quantum number of electron spin, choose it to be as large as possible:

$$S_z = \sum_i m_{si} \tag{13.1}$$

For one electron $S_z = \frac{1}{2}$, for two $S_z = \frac{1}{2} + \frac{1}{2} = 1$, etc. The value of $S_z = S$.

2. If L_z is the *z*-value of orbital angular momentum, choose it be be as large as possible, consistent with rule 1:

$$L_z = \sum_i m_{\ell i} \tag{13.2}$$

For example, for one electron choose $m_s = \frac{1}{2}$, $m_\ell = \ell$. For the second electron, choose $m_s = \frac{1}{2}$, $m_\ell = \ell - 1$. The value of $\ell_z = L$.

3. Once S and L are selected, the total angular momentum is J:

- $J = |L - S|$ if the orbital band is less than half-full.
- $J = S$ if the orbital band is half-full. In this case $L = 0$.
- $J = L + S$ if the orbital states are more than half-full.

Table 13.1 shows the results of these rules for the *p*-orbital. Note that the total spin S increases with the number n of electrons, until the orbital is half-full at $S = \frac{3}{2}$. Then adding further electrons smoothly reduces S. The orbital value L is unity for one and two electrons: the first has $m_\ell = 1$, while the second has $m_\ell = 0$. The third electron has $m_\ell = -1$, so that $L = 0$ for this case, when the orbital is half-full.

When Hund's rules are applied to the periodic table for the transition metal ions ($\ell = 2$) or the rare earth ions ($\ell = 3$), the predictions are very accurate. These atoms have magnetic moments because S or J are nonzero.

Hund's rules come from the Pauli principle of quantum mechanics. Two electrons of the same spin cannot be at the same point in space at the same time. So electrons with parallel spins have a correlated motion, where they try to avoid each other. This exchange process lowers the repulsive Coulomb interaction between electrons of parallel spins. Electrons with antiparallel spins have a relative motion that is much less correlated, so the Coulomb repulsion between them is larger.

Each atom has a magnetic moment $\boldsymbol{\mu}$. If they have a density n_m and are all aligned in the same direction, the magnetization per unit volume is

Table 13.1 Hund's rules for $\ell=1$

n	m_s	S	m_ℓ	L	J
0	0	0	0	0	0
1	$\frac{1}{2}$	$\frac{1}{2}$	1	1	$\frac{1}{2}$
2	$\frac{1}{2}$	1	0	1	0
3	$\frac{1}{2}$	$\frac{3}{2}$	-1	0	$\frac{3}{2}$
4	$-\frac{1}{2}$	1	1	1	2
5	$-\frac{1}{2}$	$\frac{1}{2}$	0	1	$\frac{3}{2}$
6	$-\frac{1}{2}$	0	-1	0	0

Table 13.2 Units of electromagnetic quantities

Symbol	Name	Units
e	electron charge	C
E	electric field	V/m
B	magnetic field	T
H	magnetic field	A/m
M	magnetization/volume	A/m
μ	magnetic moment	$\text{A}\cdot\text{m}^2$
χ	susceptibility	dimensionless

Note. C, Coulomb; V, volt; J, joule; A, ampere; m, meters; T, tesla; $\text{T} = \text{N}/(\text{A}\cdot\text{m})$.

$$\mathbf{M} = n_m \boldsymbol{\mu} \tag{13.3}$$

$$\mathbf{B} = \mu_0(\mathbf{H} + \mathbf{M}) \tag{13.4}$$

$$\mu_0 = 4\pi \times 10^{-7}\,\text{N/amp}^2 \tag{13.5}$$

Such parallel alignment is called ferromagnetic. Other arrangements are possible, as discussed below. The units of some electromagnetic quantities are listed in table 13.2.

13.2.3 *Curie's Law*

The easiest model to solve for a collection of N spins is when they all act independently. There is no interaction between nearby spins. Assume there is a magnetic field B acting on the atom. Each has a magnetic moment μ. For a total angular momentum J, the possible values of magnetic moment are

$$M_m = m\mu B, \quad -J \leq m \leq J \tag{13.6}$$

Assume each spin is attached to a thermal reservoir and is in thermal equilibrium. The partition function for a single spin is then

$$Z = \sum_{m=-J}^{J} e^{-\Delta m}, \quad \Delta = \frac{\mu B}{k_B T} \tag{13.7}$$

$$Z = \frac{\sinh[\Delta(J + 1/2)]}{\sinh(\Delta/2)} \tag{13.8}$$

The last expression is called a *Brillouin function.* It accurately predicts many properties of a collection of independent magnetic moments. The net magnetization per unit volume is

$$M = \frac{N\mu}{Z\Omega}\sum_m me^{-m\Delta} = -\frac{N\mu}{\Omega}\frac{d\ln(Z)}{d\Delta} \tag{13.9}$$

$$= n\mu\{(J + 1/2)\,\mathrm{cth}[\Delta(J + 1/2)] - (1/2)\,\mathrm{cth}[\Delta/2]\}$$

$$n = \frac{N}{\Omega} \tag{13.10}$$

This formula has two interesting limits:

- At low temperature, the dimensionless Δ becomes large and the hyperbolic cotangents become one. Then

  $$\lim_{T\to 0} M(T) = n\mu J \tag{13.11}$$

  The moments are all aligned in the state with $m = J$.

- At high temperature or small magnetic field, $\Delta J \ll 1$ and a simple answer is obtained by expanding the hyberbolic cotangent:

$$\lim_{z\ll 1} \mathrm{cth}(z) = \frac{1}{z} + \frac{z}{3} + O(z^3) \tag{13.12}$$

$$\lim_{\Delta\to 0} M = \frac{n\mu\Delta}{3}[(J + 1/2)^2 - (1/2)^2] = J(J + 1)\frac{n\mu^2}{3k_B T}B$$

The *susceptibility* is defined as the ratio $\chi = M/H$ in the limit of small magnetic field $B = \mu_0 H$. In the above example it is

$$\chi = J(J + 1)\frac{n\mu_0\mu^2}{3k_B T} \tag{13.13}$$

The numerator $n\mu_0\mu^2$ has the units of energy (joules), and χ is dimensionless. This formula is known as *Curie's law.* It is observed experimentally in many systems, which indicates that the magnetic moments have only weak interactions among themselves. The experimental signature is found by graphing $1/\chi$ as a function of temperature. A straight line at high temperature, if it extrapolates to the origin, is a sign of Curie's law.

13.2.4 Ferromagnetism

Many ferromagnets are metallic conductors. The understanding of these systems is hard, since some of the magnetic moment comes from electrons localized on atomic orbitals, while other moments come from conduction electrons that freely roam the solid. This section considers only insulators. The magnetic moments on different sites mutually interact, perhaps with the Heisenberg interactions. This mutual interaction causes them to spontaneously align at a temperature T_c. The phase diagram is shown in fig. 13.1 The vertical axis is the magnetization per unit volume. If the magnetic atoms each have a magnetic moment μ and a density n_0 (atoms per volume), the maximum value of $M = n_0\mu$ occurs when all moments are aligned parallel. If $n_\uparrow$, $n_\downarrow$ are the densities of up and down spins, then

$$M = \mu(n_\uparrow - n_\downarrow), \quad n_0 = n_\uparrow + n_\downarrow \tag{13.14}$$

The solid curve shows how $M(T)$ varies from no magnetization at $T = T_c$ to full magnetization at $T = 0$. This curve is valid only if there is no external magnetic field on the ferromagnetic. In the presence of an applied magnetic field, the transition from disordered to ordered state is gradual with temperature—there is no sharp transition at any temperature.

In most magnets, the definition of "up" or "down" for magnetic alignment is completely arbitrary. For $T > T_c$ there are equal numbers of magnetic moments pointing in all directions, except for the usual fluctuations. As the temperature is lowered through T_c, the moments start aligning in some direction. How they choose this direction seems quite arbitrary and is a wonderful example of symmetry breaking. For $T > T_c$ the system has cubic isotropy—the three crystal axes (x, y, z) in space are equivalent. For $T < T_c$, one direction in space is preferred.

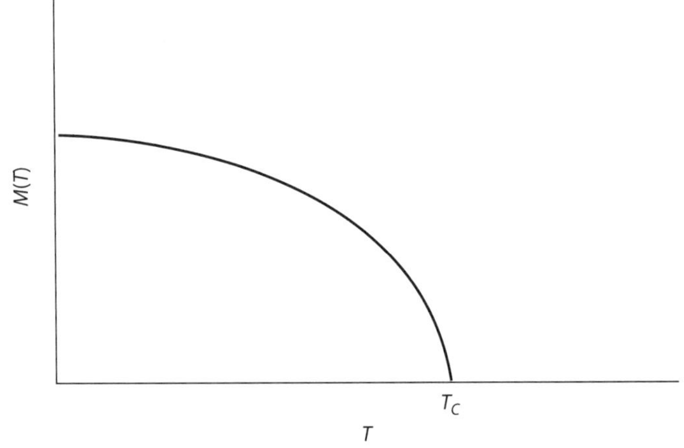

FIGURE 13.1. Magnetization as a function of temperature for a ferromagnet. The horizonal axis is temperature and the vertical axis is magnetic moment per unit volume.

Metallic iron has $T_c = 1043$ K. At room temperature it is a ferromagnet. Yet many pieces of iron appear to have no magnetic moment. The piece of iron is composed of many small crystallites. Each crystallite is a magnet, but the magnetic moment is a vector that can point in any direction. The magnetic moments of the thousands of crystallites point in all different directions and effectively cancel out. So the chunk of iron in your hand appears to be unmagnetized.

The chunk of iron can be made into a macroscopic magnet by aligning all of the magnetic moments. This is done by putting the iron in a large magnetic field, which will tend to make the moments of the individual crystallites align along the external field. This alignment is a high-energy state, while having the macroscopic field zero is a low-energy state.

13.2.5 *Antiferromagnetism*

Another common type of magnetic ordering is when the magnetic atoms align their spins alternately up and down. Each up spin has down spins as neighbors, and vice versa. This latter arrangement is called *antiferromagnetic*. Not all crystal lattices permit a simple antiferromagnetic arrangement. Those that do are called *bipartite lattices*: they have two identical sublattices. In three dimensions the sc and bcc lattices are bipartite, but the fcc is not. In two dimensions, the sq and hc lattices are bipartite, but the pt lattice is not, as illustrated in fig. 13.2. Nonbipartite lattices, such as pt and fcc, produce *frustrated* antiferromagnetic ordering.

Another feature of antiferromagnetic ordering is that often the atomic unit cell changes size as a result of the magnetic ordering. In the two-dimensional square lattice, the lattice constant is a and the area of the unit cell is $A_0 = a^2$ when all sites are identical. But if atomic sites have a crystalline ordering of alternate up and down spins, then the unit cell has two sites and the new cell area is $A_0' = 2a^2$. This arrangement is shown in fig. 13.3 for the 2D sq lattice. The magnetic ordering is again symmetry breaking—not only in choosing which sites are up and which are down, but in changing the area of the unit cell, in creating a new set of reciprocal lattice vectors, and in a smaller Brillouin zone.

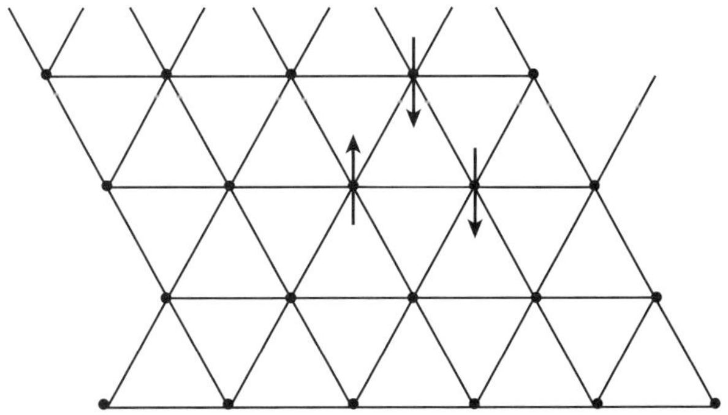

FIGURE 13.2. The pt lattice is nonbipartite. If two adjacent spins have opposite ordering, they have a common neighbor that is confused regarding its preferred arrangement.

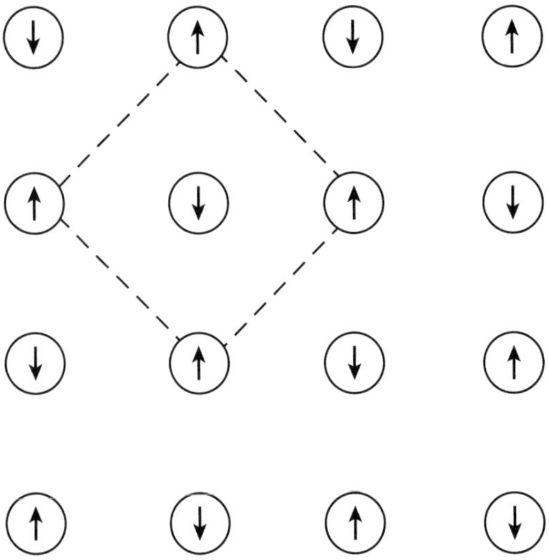

FIGURE 13.3. Dashed lines show new unit cell in antiferromagnetic alignment on a square lattice.

A lattice that has two sites per unit cell in the disordered state, such as the 2D hc lattice or the 3D diamond lattice, does not change its area when antiferromagnetically ordering. Each of the two sites in each cell has alternate spin arrangements. The temperature at which antiferromagnets start their alignment is called the *Neel temperature*.

13.3 3*d* Metals

An interesting series of magnetic compounds is formed by the 3*d*-transition metals. Table 13.3 shows these atoms and their most common valence states. The series of elements Ti through Cu form metals and, except for copper, titanium, and vanadium, they all have interesting magnetic properties.

- Titanium in its metallic form has no magnetic moment and no magnetic phases. It has the hcp structure and is a paramagnetic metal.
- Vanadium has no local moment. It has the bcc structure and is a paramagnetic metal.
- Chromium is a bcc metal. It has a Neel temperature of $T_N = 312$ K and is antiferromagnetic for $T < T_N$. The spin ordering is actually a spin density wave, with a wavelength that is incommensurate with the lattice period.
- Manganese has a complex cubic structure (α-Mn) with 58 atoms per unit cell. Its Neel temperature is $T_N = 58$ K.
- Iron (Fe) has a Curie temperature of $T_C = 1044$ K and is ferromagnetic below this temperature. It is a bcc metal for most temperatures, but has an fcc phase for $1084 < T < 1665$ K. Its

Table 13.3 Spin configuration of 3*d* atoms in various spin $(3d)^n$ and charge states, according to Hund's rules

n	S	L	J	$(3d)^{n+}$
1	$\frac{1}{2}$	2	$\frac{3}{2}$	Ti^{3+}, V^{4+}
2	1	3	2	V^{3+}
3	$\frac{3}{2}$	3	$\frac{3}{2}$	V^{2+},Cr^{3+}, Mn^{4+}
4	2	2	0	Cr^{2+}, Mn^{3+}
5	$\frac{5}{2}$	0	$\frac{5}{2}$	Mn^{2+}, Fe^{3+}
6	2	2	4	Fe^{2+}
7	$\frac{3}{2}$	3	$\frac{9}{2}$	Co^{2+}
8	1	3	4	Ni^{2+}
9	$\frac{1}{2}$	2	$\frac{5}{2}$	Cu^{2+}

magnetic susceptibility is largely independent of temperature. For temperatures very close to T_C it has the form

$$\chi = \frac{K}{(T - T_c)^{\alpha}}, \quad \alpha = \frac{4}{3} \tag{13.15}$$

The exponent of $\alpha = \frac{4}{3}$ is discussed in a later section.

- Cobalt has a Curie temperature of $T_C = 1388$ K. It is hcp for $T < 293$ K and fcc for $T > 293$ K. Its susceptibility has an exponent of $\alpha = 1.2$.
- Nickel is fcc with a Curie temperature of $T_C = 628$ K. Its susceptibility has an exponent of $\alpha = 1$.

This range of behavior provides a good introduction to the magnetic metals. A similar list could be prepared for the 4*d*, 5*d*, 4*f*, and 5*f* metals. The latter are the rare earth series.

13.4 Theories of Magnetism

There are papers published every week on aspects of magnetic theory. The subject is large and contains many sophisticated methods. Here we present some traditional theories that are relatively simple and intuitive.

13.4.1 *Ising and Heisenberg Models*

Many crystals have a magnetic ion as a constituent of the atomic unit cell. The magnetic atoms are close together and interact. Curie's law must be altered due to the interactions.

A simple model of a magnet is

- an insulator
- with one atom per unit cell,
- which is magnetic.

There are no examples in nature. Most atoms with magnetic moments, when made into a solid of one element, are conductors: e.g., iron, nickel, and cerium. There are many magnetic insulators, but all have several atoms per unit cell. Nevertheless, this ideal model has important features and has been well studied theoretically.

Since the atoms are immediate neighbors, the atomic orbitals of neighboring atoms overlap. The spin configuration on one atom will influence the spin energy of the neighboring atoms. Most of the energy comes from the exchange interaction. The most important Hamiltonian for this case is called the *Heisenberg model*:

$$H = -\frac{J}{\hbar^2}\sum_{j\delta} \mathbf{S}_j \cdot S_{j+\delta} \tag{13.16}$$

The summation over j is over all lattice sites $\mathbf{R}_j$. The summation over $\boldsymbol{\delta}$ is over all of the first neighbors $\mathbf{R}_{j+\delta}$ of $\mathbf{R}_j$. The interaction depends on the total spin $\mathbf{S}_j$, and not on the orbital angular momentum $\mathbf{L}_j$, nor on the total angular momentum $\mathbf{J}_j$. The dependence on spin comes from the exchange interaction. Another interaction between spins is due to the dipole–dipole interaction of the magnetic moments. However, the magnetic dipole–dipole interaction is very weak compared to the exchange interaction, and dipolar interactions are usually neglected.

The constant J has units of energy and is derived from the exchange energy between electrons on neighboring atoms. It is not a symbol for the total angular momentum when used in this model. Putting the negative sign in front of this term is conventional. In most cases, this results in $J > 0$. Then the lowest energy is when all spins point in the same direction, which is the *ferromagnetic* arrangement. If it is found that $J < 0$, then the lowest energy is found when the spins alternate in the up and down position.

There are many crystals in which the magnetic atoms are not first neighbors, but are second neighbors. An example is the high-temperature superconductors, which have planes of copper oxide. The copper atoms are in a sq lattice, and each pair of neighboring coppers have an oxygen atom as a bridge: the plane has the chemical formula CuO_2. This structure was shown in fig. 3.5. It is found that when two atoms are first neighbors, the exchange constant is negative (as in $-J$). When there is one bridging atom between two magnetic atoms, it is found that one has $[(-J)(-J)] = +J^2$, and the Heisenberg interaction is positive. That makes the spin arrangement antiferromagnetic. In cuprate superconductors, the copper atoms are doubly ionized (Cu^{2+}), which makes the d-shell lack an electron—there is a d-hole. They form an antiferromagnetic arrangement at low temperature. The process of having a magnetic interaction through the bonding of a nonmagnetic bridging atom is called *superexchange*. The transition metal oxides, such as NiO or CrO, are other examples of superexchange giving antiferromagnetic arrangements of magnetic moments on the transition metal atoms.

The spin $\mathbf{S}_j$ is a three-dimensional vector. The Heisenberg interaction can be written in terms of its components or in terms of spin raising and lowering operators. The (x, y, z) components are denoted by superscripts:

$$\mathbf{S}_j \cdot \mathbf{S}_{j+\delta} = S_j^{(x)} S_{j+\delta}^{(x)} + S_j^{(y)} S_{j+\delta}^{(y)} + S_j^{(z)} S_{j+\delta}^{(z)} \tag{13.17}$$

$$= \tfrac{1}{2}\left[S_j^{(+)} S_{j+\delta}^{(-)} + S_j^{(-)} S_{j+\delta}^{(+)} \right] + S_j^{(z)} S_{j+\delta}^{(z)}$$

$$S_j^{(\pm)} = S_j^{(x)} \pm i S_j^{(y)} \tag{13.18}$$

Separate terms in the interaction are models with their own names:

- The Ising model is

$$H_I = -\frac{J}{\hbar^2} \sum_{j\delta} S_j^{(z)} S_{j+\delta}^{(z)} \tag{13.19}$$

  The Ising model can be solved analytically in one or two dimensions, but not in three dimensions. Accurate computer solutions have been done in three dimensions.

- The *XY*-model is

$$H_{XY} = -\frac{J}{2\hbar^2} \sum_{j\delta} \left[S_j^{(+)} S_{j+\delta}^{(-)} + S_j^{(-)} S_{j+\delta}^{(+)} \right] \tag{13.20}$$

  It was solved in one dimension using a Jordan-Wigner transformation to fermion operators.

Some of these models are solved below. First, some general properties of magnetism are discussed.

13.4.2 Mean Field Theory

The easiest theory of magnetism is called *mean field theory*. It is applied to the Ising model. If all magnetic atoms are at identical sites with z neighbors, the energy of that moment is approximately

$$E_j = \frac{zJ}{\hbar^2} S_j^{(z)} \langle S^{(z)} \rangle \tag{13.21}$$

where $\langle S^{(z)} \rangle$ is the average magnetic moment of the neighboring atoms. We assume this average is proportional to the net magnetization in the ferromagnetic state:

$$\langle S^{(z)} \rangle = \lambda M \tag{13.22}$$

The constant λ is required to keep correct dimensional units. The maximum value of $\langle S^{(z)} \rangle$ is $\hbar S$, while the maximum value of M is $n_0 \mu S$. Since the maximum $\langle S^{(z)} \rangle$ is obtained at the maximum M, divide these two limits and get

$$\lambda = \frac{\hbar}{n_0 \mu} \tag{13.23}$$

Use the Brillouin function for the magnetization:

$$M = n_0\mu\{(S+\tfrac{1}{2})\text{cth}[(S+\tfrac{1}{2})\Delta] - (\tfrac{1}{2})\text{cth}[(\tfrac{1}{2})\Delta]\} \tag{13.24}$$

$$\Delta = \frac{zJ}{k_B T}\frac{M}{n_0\mu} \tag{13.25}$$

Equation (13.24) is a self-consistent equation for M. At low temperature it gives that $M = n_0\mu S$, while at high temperature the only solution is $M = 0$. It is simple to solve for the special case of $S = \frac{1}{2}$, where the hyperbolic functions have an alternate, simple form

$$M = \frac{n_0\mu}{2}\tanh\left(\frac{\Delta}{2}\right) \tag{13.26}$$

Rewrite the above equation as

$$\phi = \tanh(\varepsilon\phi) \tag{13.27}$$

$$\phi = \frac{2M}{n_0\mu}, \quad \varepsilon = \frac{zJ}{4k_B T} \tag{13.28}$$

Figure 13.4 shows a graphical solution of this equation. For $\varepsilon < 1$ there is no solution except $M = 0$. For $\varepsilon > 1$ there is a solution with a nonzero value of $\phi \propto M$. The transition temperature occurs at $\varepsilon = 1$, or

$$T_c = \frac{zJ}{4k_B} \tag{13.29}$$

Another feature is the behavior at temperatures near to T_c, which is called the *critical region*. Define a small parameter:

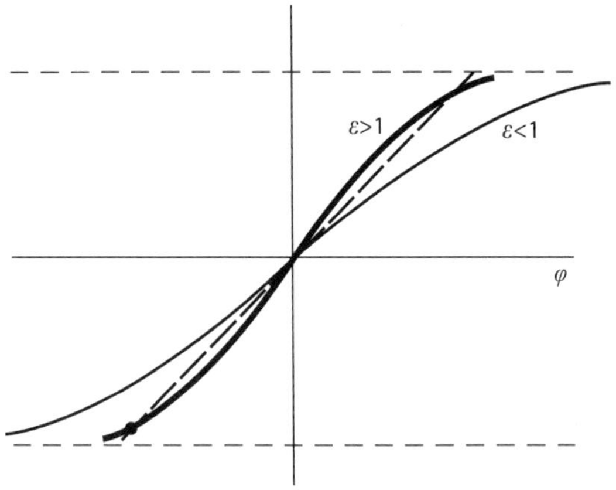

FIGURE 13.4. Graphical solution of mean field equation showing curve for $\varepsilon < 1$ and $\varepsilon > 1$.

$$t = \frac{T_c - T}{T_c} \tag{13.30}$$

$$\varepsilon = \frac{zJ}{4k_B T_c (T/T_c)} = \frac{1}{1-t} \approx 1 + t \tag{13.31}$$

Expand eqn. (13.27) in power series in t. Remember that in the critical region ϕ is also small:

$$\phi = \tanh[\phi(1+t)] \approx \tanh(\phi) + \frac{\phi t}{\cosh^2(\phi)} + O(t^2) \tag{13.32}$$

$$\phi = \phi - \frac{\phi^3}{3} + \frac{\phi t}{1 + \phi^2/2} + \cdots \tag{13.33}$$

Cancel ϕ from each term and from each side. The solution for small t is

$$\phi^2 = 3t \tag{13.34}$$

$$\phi = \sqrt{3t} \tag{13.35}$$

$$M(T) \approx \frac{n_0 \mu}{2} \sqrt{3} \sqrt{\frac{T_c - T}{T_c}} \tag{13.36}$$

For temperatures slightly below T_c, the magnetization is predicted to rise as the square root of the temperature difference ($\sim \sqrt{T_c - T}$). The full curve is shown in fig. 13.1. This critical behavior is found in some, but not all ferromagnets. Other theories of magnetism make other predictions.

The above results can also be derived starting from the free energy F. For spin $-\frac{1}{2}$,

$$Z = e^{-\beta F} = [\cosh(\Delta/2)]^N \tag{13.37}$$

$$F = -k_B T \ln[\cosh(\Delta/2)] \tag{13.38}$$

$$M = \frac{1}{\Omega} \frac{\delta(\beta F)}{\delta(\beta H)} = \frac{n\mu}{2} \tanh(\Delta/2) \tag{13.39}$$

Employing the free energy is useful for critical phenomena, which is treated in a latter section.

Another important quantity is the magnetic susceptibility for $T > T_c$. Assume that a small magnetic field B is applied to the ferromagnet. The individual spins experience a Pauli interaction $H_I = \mu \mathbf{S} \cdot \mathbf{B}/\hbar$. The energy of a single spin is

$$E_j = \frac{1}{\hbar} S_j^{(z)} \left[\frac{zJ}{\hbar} \langle S^{(z)} \rangle + \mu B \right] \tag{13.40}$$

The magnetization is given by the Brillouin function. For $S = \frac{1}{2}$ again start with

$$M = \frac{n_0 \mu}{2} \tanh\left(\frac{\Delta}{2}\right) \tag{13.41}$$

$$\Delta = \frac{1}{k_B T}\left[zJ \frac{M}{n_0 \mu} + \mu B \right] \tag{13.42}$$

For $T > T_c$, the magnetization $M = 0$ if $B = 0$. A small value of B creates a small value of M. Assume $\Delta \ll 1$; expand $\tanh(z) \approx z$ and find

$$M = \frac{n_0 \mu}{4k_B T}\left[zJ \frac{M}{n_0 \mu} + \mu B \right] \tag{13.43}$$

$$M\left[1 - \frac{zJ}{4k_B T}\right] = \frac{n_0 \mu^2}{4k_B T} B \tag{13.44}$$

$$\chi = \frac{M}{H} = \frac{n_0 \mu^2 \mu_0}{4k_B(T - T_c)}, \quad \text{for } T > T_c \tag{13.45}$$

The experimental method is to graph χ^{-1} versus temperature. If the curve goes to zero at a positive temperature T_c, that is usually near to the ferromagnetic transition temperature. The above formula is called the *Curie-Weiss law*. This behavior is illustrated in fig. 13.5. A similar derivation for an antiferromagnetic system on a bipartite lattice gives

$$\chi = \frac{M}{H} = \frac{n_0 \mu_0 \mu^2}{4k_B(T + T_c)}, \quad \text{for } T > 0 \tag{13.46}$$

Here a graph of χ^{-1} vs. T has a linear slope that intercepts the origin at negative temperatures. This behavior is also illustrated in fig. 13.5.

Mean field theory describes the ferromagnetic transition: how the magnetic moments are random for $T > T_c$, and how they start to align themselves into a magnetic state for $T < T_c$. This alignment is a second-order phase transition. A first-order phase transition has latent heat, but the magnetic ordering does not, which makes it second order.

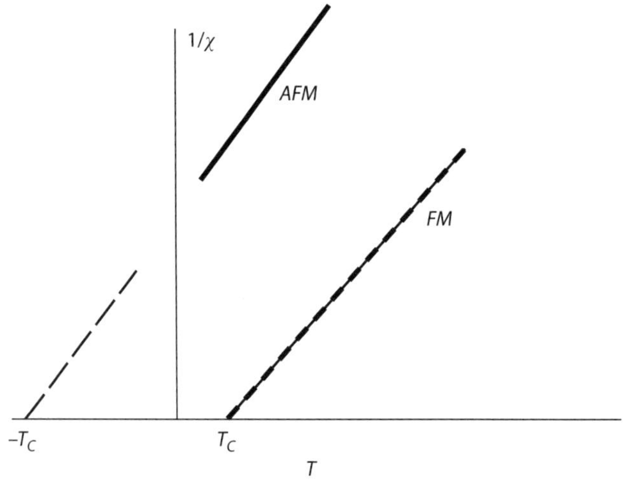

FIGURE 13.5. Solid line is χ^{-1} for ferromagnetic system, while dashed line is for an antiferromagnetic system.

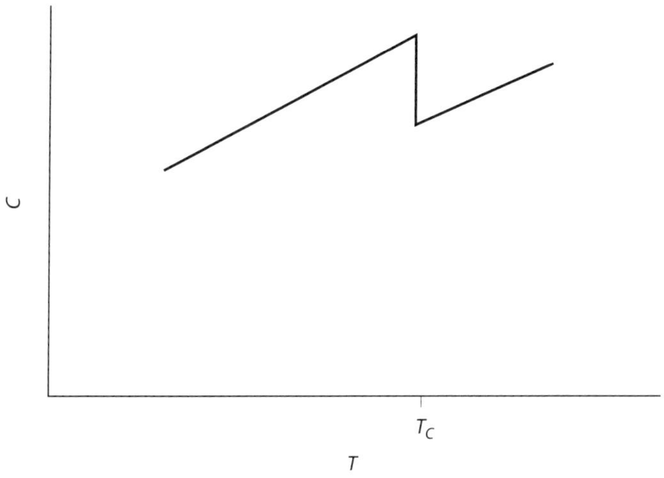

FIGURE 13.6. Mean field prediction of heat capacity of ferromagnetic ordering as a function of temperature.

An important thermodynamic quantity is the heat capacity, which is the temperature derivative of the ground-state internal energy. Write this energy for $S=\frac{1}{2}$ as

$$E = -N\frac{zJ}{2}\langle S^{(z)}\rangle^2 = -N\frac{zJ}{2}\left(\frac{\hbar}{n_0\mu}\right)^2 M^2(T) \tag{13.47}$$

$$= -N\frac{zJ}{8}\tanh^2\left(\frac{\Delta}{2}\right)$$

The factor of 1/2 is from counting each pair interaction only once. Recall that $M^2 = M_0(T_c - T)$ below the transition temperature, but is zero above. Then the heat capacity from magnetic ordering goes as

$$C = \frac{dE}{dT} = \begin{cases} 0 & T > T_c \\ C_0 & T < T_c \end{cases} \tag{13.48}$$

$$C_0 = N\frac{zJ}{2}\left(\frac{\hbar}{n_0\mu}\right)^2 M_0 \tag{13.49}$$

Mean field theory predicts that the heat capacity has a discontinuity at T_c. The above formulas apply only near to T_c. The full curve is shown in fig. 13.6. The experimental result is slightly different, as is discussed below.

13.4.3 *Landau Theory*

An important phenomenological theory of phase transitions was introduced by Landau. Here it is applied to ferromagnetism, but a similar theory was applied to ferroelectrics in chapter 4. It has been applied to other systems: liquid crystals, superconductors, etc. As shown in fig. 13.1, the magnetization $M(T)$ of a ferromagnet in zero magnetic field is a function of temperature. It is zero for $T > T_c$ and is nonzero for $T < T_c$. In Landau theory the free energy is an analytic function of the magnetization. It is treated as a scalar quantity:

$$\mathcal{F}(M) = A_0 + A_2M^2 + A_4M^4 - HM + \cdots \tag{13.50}$$

where H is the magnetic field. Only even powers of M^n are contained in the series: the energy should not change if M is changed to $-M$, while H is changed to $-H$. The key constants are A_2 and A_4. $A_4 > 0$ for the system to be stable at a nonzero value of M.

The system will choose the value of M that minimizes this free energy:

$$0 = \frac{\delta \mathcal{F}}{\delta M} = 2A_2M + 4A_4M^3 - H \tag{13.51}$$

If there is no external magnetic field ($H = 0$), a phase change occurs at T_c because $A_2(T)$ changes sign at this point. If $A_2 > 0$, $A_4 > 0$, the only solution is $M = 0$. If $A_2 < 0$, $A_4 > 0$, there is a solution

$$A_2(T) = a_2\left(\frac{T - T_c}{T_c}\right) \tag{13.52}$$

$$M(T) = \pm\sqrt{\frac{a_2}{2A_4}\left(\frac{T_c - T}{T_c}\right)}\Theta(T_c - T) \tag{13.53}$$

Since A_2 changes sign at T_c, it is reasonable to expect it to depend on the difference $T - T_c$.

Landau theory predicts that the magnetization is zero for $T > T_c$, while it increases according to $\sqrt{T_c - T}$ for $T < T_c$. This causes a discontinuous jump in the heat capacity, as was found in the ferromagnetic transition. These features agree with mean field theory. When there is a magnetic field, several results are as follows:

- At $T = T_c$, where $A_2 = 0$, then

$$M(T_c) = \left(\frac{H}{4A_4}\right)^{1/3} \tag{13.54}$$

- For $T \neq T_c$, the first term in eqn. (13.51) dominates, and

$$M = \frac{H}{a_2}\left(\frac{T_c}{T - T_c}\right) \tag{13.55}$$

The magnetic susceptibility is

$$\chi = \left(\frac{\delta M}{\delta H}\right)_{H=0} = \frac{1}{a_2}\left(\frac{T_c}{T - T_c}\right) \tag{13.56}$$

These predictions do not agree with experiments on magnetic systems, as discussed below.

13.4.4 *Critical Phenomena*

Define a symbol for the parameter:

$$t = \left(\frac{T - T_c}{T_c}\right) \tag{13.57}$$

The *critical region* of a phase change is where t is a small number. Much work has been expended by theorists and experimentalists to discern the behavior of measurable quantities in the critical region of a magnetic phase transition. Define some exponents for experimental parameters.

- The heat capacity has the exponent α:

  $$C = \frac{C_0}{|t|^\alpha} \tag{13.58}$$

  where C_0 is a constant. Actually, a separate exponent is sometimes defined for $T > T_c(\alpha)$ and $T < T_c(\alpha')$, but it is found for most cases that $\alpha = \alpha'$. Landau theory predicts that $\alpha = 0$.

- The magnetization at zero magnetic field has the exponent β:

  $$M(T) = M_0|t|^\beta \quad \text{for } T < T_c \tag{13.59}$$

  Landau theory predicts that $\beta = \frac{1}{2}$.

- At $T = T_c$ the magnetization depends on magnetic field with an exponent $1/\delta$:

  $$M \propto |H|^{1/\delta} \tag{13.60}$$

  Landau theory predicts that $\delta = 3$.

- The magnetic susceptibility has an exponent of $-\gamma$:

  $$\chi = \left(\frac{\delta M}{\delta H}\right)_{H=0} \propto \frac{1}{|t|^\gamma} \tag{13.61}$$

  Landau theory predicts that $\gamma = 1$.

These various exponents are collected in table 13.4. The two dimensional Ising model can be solved exactly, as first shown by Onsager [13]. The exponents listed as 2D Ising are from his solution and are exact. The three-dimensional Ising and Heisenberg models cannot be solved exactly. Approximate solutions have been obtained by a variety of methods, such as high-temperature series expansions or numerical renormalization group. In this case the various exponents, such as $\alpha = \frac{1}{8}$ or $\beta = \frac{5}{16}$ are guesses derived from numerical results. Some theorists think the three-dimensional Heisenberg lattice has critical exponents of $\alpha = 0$, $\beta = \frac{1}{3}$, $\delta = 5$, $\gamma = \frac{4}{3}$.

Neither of the two Ising models agrees with Landau theory. The exponent $\alpha = 0$ in Landau theory and in the 2D Ising model. These two solutions are different definitions of zero! In Landau theory, the heat capacity is discontinuous at T_c, but is a constant away from it: the constant means $\alpha = 0$. In the 2D Ising model, the dependence is

Table 13.4 Critical exponents for some models of magnetism. The 3D cases are by numerical simulation, and have typical error bars of ±0.01 for the Ising model, and ±0.1 for the Heisenberg model

exp	Landau	2D Ising	3D Ising	3D Heisenberg
α	0	0	$\frac{1}{8}$	0
β	$\frac{1}{2}$	$\frac{1}{8}$	$\frac{5}{16}$	0.36
δ	3	15	5	4.75
γ	1	$\frac{7}{4}$	$\frac{5}{4}$	1.39

$$C \propto -\ln(|t|) \tag{13.62}$$

which is very singular as $T \to T_c$. The logarithmic dependence is designated $\alpha = 0$. The only assumption in Landau theory is that one can expand the magnetization in a power series in the order parameter M, and the series has only terms of $O(M^{2n})$. Since the Landau theory fails in the critical region, this assumption is not valid.

The results in the critical region are lattice independent. In two dimensions, the Ising model can be solved for such lattices as square (sq), honeycomb (hc), or plane triangular (pt), and all yield the same critical exponents. This feature is called *universality*.

Workers such as Griffiths, Rushbrooke, and Widom have derived a number of relationships between these exponents:

$$2 = \alpha + 2\beta + \gamma \tag{13.63}$$

$$\delta = 1 + \frac{\gamma}{\beta} \tag{13.64}$$

These identities are obeyed for the models in table 13.4.

13.5 Magnetic Susceptibility

The magnetic susceptibility was discussed earlier in relation to the Curie laws. Those discussions considered only the magnetic moment due to the electron spin. There is also a contribution to the magnetic susceptibility from the orbital motion of the electrons.

Pauli diamagnetism of the electrons in the conduction band can be found from the Pauli interaction:

$$H' = -2\mu_B m_s B = \pm\Delta, \quad \Delta = \mu_B B, \quad \mu_B = \frac{e\hbar}{2m} \tag{13.65}$$

where $m_s = \pm\frac{1}{2}$ are the spin states of the electron. The Bohr magneton is μ_B. The total magnetization is the magnetic moment μ_B times the difference in the number of electrons with up spins, minus those with down spins:

$$M = \frac{\mu_B}{\Omega}[N_+ - N_-] \tag{13.66}$$

$$= \mu_B \int dEN_0(E)\left[\frac{1}{1+e^{\beta(E-\mu-\Delta)}} - \frac{1}{1+e^{\beta(E-\mu+\Delta)}}\right]$$

$$\approx 2\mu_B \Delta \int dEN_0(E)\left(-\frac{dn_F(E)}{dE}\right)$$

$$\approx 2\mu_B^2 BN_0(\mu)$$

$$\chi = 2\mu_0\mu_B^2 N_0(\mu) \tag{13.67}$$

The symbol $N_0(E)$ is the density of states for electrons of one spin and has the units of $1/(J \cdot \Omega)$. $N_0(\mu)$ is the value at the Fermi surface.

Van Vleck paramagnetism is found in atoms. Using the $p \cdot A$ term as a perturbation, and doing second-order perturbation theory, an atom has an energy term of order B^2 from

$$E_n = E_{n0} - \sum_{m \neq n} \frac{\left|\left\langle m\left|\frac{e}{m}\mathbf{p}\cdot\mathbf{A}\right|n\right\rangle\right|^2}{E_{m0} - E_{n0}} = E_{n0} - \frac{\chi}{2\mu_0}B^2 \tag{13.68}$$

If $|n\rangle$ is the electronic ground state of the atom and $|m\rangle$ are excited states, the energy denominator is positive. The entire expression is negative, so it contributes a positive magnetic susceptibility. Since $\mathbf{A} = \mathbf{B} \times \mathbf{r}/2$ and $\mathbf{r} \times \mathbf{p} = \mathbf{L}$, the above susceptibility is

$$\chi = \frac{\mu_0}{2}\left(\frac{e}{m}\right)^2 \sum_{m \neq n} \frac{|\langle m|L_z|n\rangle|^2}{E_{m0} - E_{n0}} \tag{13.69}$$

where L_z is the component of angular momentum parallel to the magnetic field. The dimensions of χ are volume. A dimensionless susceptibility is obtained by multipling the above expression by the density n of such atoms. If we include the Pauli term in the Hamiltonian, the above result is changed to

$$\chi = \frac{\mu_0}{2}\left(\frac{e}{m}\right)^2 \sum_{m \neq n} \frac{|\langle m|L_z + 2S_z|n\rangle|^2}{E_{m0} - E_{n0}} \tag{13.70}$$

The prefactor is the Bohr magneton $\mu_B = e\hbar/2m$, which gives

$$\chi = 2\mu_0 \frac{\mu_B^2}{\hbar^2} \sum_{m \neq n} \frac{|\langle m|L_z + 2S_z|n\rangle|^2}{E_{m0} - E_{n0}} \tag{13.71}$$

This term is largest in atoms in which the outer atomic shell is only partially filled with electrons.

Landau diagmagnetism is found in free electron metals. The application of a magnetic field causes orbital motion of the conduction electrons, which contribute to the net magnetization and net susceptibility:

$$\chi = -\frac{2\mu_0\mu_B^2}{3} N_0(\mu) \tag{13.72}$$

This result is $-\frac{1}{3}$ of the diamagnetism from the Pauli term, so that the total susceptibility of the conduction electrons is

$$\chi = \frac{4\mu_0\mu_B^2}{3} N_0(\mu) \tag{13.73}$$

The orbital motion makes an important contribution to the susceptibility of conduction electrons.

13.6 Ising Model

The Ising model is the simplest model of magnetism. It can be solved exactly in one and two dimensions. In three dimensions it can be solved numerically.

13.6.1 *One Dimension*

The Ising model can be solved exactly in one dimension. This case is rather dull since there is no magnetic phase transition at nonzero temperature. However, the method of solution is quite easy and should be familiar to the student.

Assume there is a chain of N spins labeled $j = 1, 2, \ldots, N$. Write the interaction energy as

$$H = -J\sum_{j=1}^{N-1} \sigma_j\sigma_{j+1}, \quad \sigma_j = \pm 1 \tag{13.74}$$

The partition function is

$$Z = \mathrm{Tr}\{e^{-\beta H}\} = \Pi_j \sum_{\sigma j=\pm 1} \exp[K\sigma_j\sigma_{j+1}] \tag{13.75}$$

where $K = J/k_BT$ is a dimensionless constant. The partition function is easily evaluated. Start at one end, say $j = 1$, and average that site:

$$\sum_{\sigma_1=\pm 1} \exp[K\sigma_1\sigma_2] = 2\cosh[K\sigma_2] = 2\cosh(K) \tag{13.76}$$

The value of $\sigma_2 = \pm 1$. The hyperbolic cosine is a symmetric function, $\cosh(-K) = \cosh(+K)$, and the same value is found for either value of σ_2. Next evaluate the $j = 2$ site:

$$\sum_{\sigma_2=\pm 1} \exp[K\sigma_2\sigma_3] = 2\cosh[K\sigma_3] = 2\cosh(K) \tag{13.77}$$

Continue down to the end of the chain, where each site average gives $2\cosh(K)$. The last average ($j = N$) gives two. The partition function is

$$Z = 2[2\cosh(K)]^{N-1} \tag{13.78}$$

which is the exact result. The partition function can be used to calculate quantities such as the average magnetization.

Another method of solving the Ising model in one dimension is using transfer matrices. It is useful for the case that there is an external magnetic field. Start with a Hamiltonian:

$$H = -J\sum_{j=1}^{N-1}\sigma_j\sigma_{j+1} - h\sum_j \sigma_j, \quad h = \mu B \tag{13.79}$$

Rewrite the Hamiltonian as

$$H = \sum_j H_j, \quad H_j = -J\sigma_j\sigma_{j+1} - \frac{h}{2}(\sigma_j + \sigma_{j+1}) \tag{13.80}$$

Write $\exp(-\beta H_j)$ as a 2×2 matrix M_j:

$$M_j = \begin{pmatrix} e^{K+\beta h} & e^{-K} \\ e^{-K} & e^{K-\beta h} \end{pmatrix} = \begin{pmatrix} a_{11} & a_{12} \\ a_{21} & a_{22} \end{pmatrix} \tag{13.81}$$

The element a_{11} is the value when both spins σ_j, σ_{j+1} are $+1$. The element a_{22} is the value when both spins σ_j, σ_{j+1} are -1. The elements a_{12}, a_{21} are when $\sigma_j = 1, \sigma_{j+1} = -1$, etc. The matrix M_j is the same for all values of j, except for $j = N$. Therefore, the partition function is found as the product of $N-1$ of these identical matrices

$$Z = 2\mathrm{Tr}\left[M_j^{N-1}\right] = 2\left[\lambda_1^{N-1} + \lambda_2^{N-1}\right] \tag{13.82}$$

The eigenvalues λ_j of the transfer matrix are

$$0 = \det\begin{vmatrix} e^{K+\beta h} - \lambda & e^{-K} \\ e^{-K} & e^{K-\beta h} - \lambda \end{vmatrix} \tag{13.83}$$

$$\lambda = e^K\cosh(\beta h) \pm \sqrt{e^{2K}\sinh^2(\beta h) + e^{-2K}} \tag{13.84}$$

For zero magnetic field ($h = 0$) the two eigenvalues are $\lambda = 2\cosh(K)$, $2\sinh(K)$. The first is the largest and dominates the answer:

$$Z = 2[2\cosh(K)]^{N-1} \tag{13.85}$$

The average magnetization due to the magnetic field is

$$\sigma = \frac{1}{N}\frac{d}{d\beta h}\ln(Z) = \frac{d}{d\beta h}\ln(\lambda) = \frac{e^K\sinh(\beta h)}{\sqrt{e^{2K}\sinh^2(\beta h) + e^{-2K}}} \tag{13.86}$$

$$\lim_{h\to\infty}\sigma = 1 \tag{13.87}$$

At large magnetic field, all of the spins become aligned and the the average magnetization is unity. At $h = 0$, there are as many up as down spins and the average is zero.

13.6.2 Two and Three Dimensions

The Ising model has been discussed earlier using mean field theory and Landau theory. Those discussions were for ferromagnetic ordering. Antiferromagnetic ordering is quite

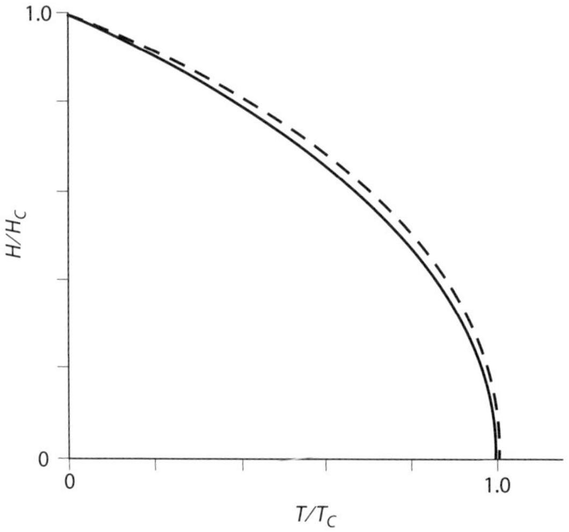

FIGURE 13.7. Antiferromagnetic phase boundary of the Ising model in two dimensions in a magnetic field. Solid line is square lattice, dashed line is honeycomb lattice. From Subbaswamy and Mahan, *Phys. Rev. Lett.* **37**, 642 (1976). Used with permission of the American Physical Society.

similar when solved without an external magnetic field. However, the ferromagnetic ordering and antiferromagnetic ordering behave quite differently in an external magnetic field.

An applied magnetic field destroys the sharp phase transition of ferromagnetic ordering. Instead, the ordering is continuous with decreasing temperature. There is no single temperature where a sharp change occurs.

Antiferromagnetic Ising systems have a sharp phase transtion in a magnetic field. The phase diagram is shown in fig. 13.7 for the two-dimensisonal Ising model. The horizontal axis is relative temperature, and the vertical axis is magnetic field. The enclosed region is the ordered state. The application of a large magnetic field destroys the ordered phase, and this boundary is quite sharp at all temperatures below T_c.

Figure 13.7 has a resemblance to fig. 13.1. The latter is the magnetization $M(T)$ vs. T in a ferromagnet in zero magnetic field. The net magnetization is zero in the antiferromagnetic state, since there are as many spins up as down.

One could define an *order parameter* $\mathcal{M}$ for the antiferromagnet. It is the fraction of A sites that have up spins as a function of temperature. It would resemble fig. 13.1 for zero magnetic field. The same curve would be the fraction of spins at B sites that point down as a function of temperature. In the AF state, alternate spins are generally pointing in the opposite direction. The magnetic field wants to align all spins in the same direction. As the magnetic field increases, there is a value where the spin energy from lining them up exceeds the local energy from the antiferromagnetic ordering. This transition is easy to find at very low temperatures, where thermal fluctuations are unimportant.

- In the antiferromagnetic ordering, N spins with z neighbors have an energy

$$E_{AF} = -zJN \tag{13.88}$$

There is no magnetic energy since half the spins are up and half are down.

- If the spins with moment μ are all aligned in the magnetic field,

$$E_B = zNJ - \mu BN \tag{13.89}$$

- These two configurations are equal at

$$B_c = \frac{2zJ}{\mu} \tag{13.90}$$

At low temperature, the spins are aligned for $B > B_c$ and they are in the antiferromagntic alignment for $B < B_c$. At higher temperatures, the value of the critical field $B_c(T)$ declines due to fluctuations.

13.6.3 *Bethe Lattice*

The Ising model can be solved exactly for the Bethe lattice. Each site in this lattice has z first neighbors. However, there is only one path between any two sites on the lattice. There is a spin one-half particle at each lattice site. The z-component of spin can point either up or down, which is called $\sigma = \pm 1$. The spins on neighboring sites $\langle ij \rangle$ interact by $-J\sigma_i\sigma_j$. The Hamiltonian is

$$H = -J\sum_{\langle ij\rangle} \sigma_i\sigma_j - h\sum_j \sigma_j \tag{13.91}$$

where $h \equiv \mu H$ is the Zeeman energy in a magnetic field.

The parameter z is the coordination number. A related parameter is $r = z - 1$ is the *branching ratio.* Figure 13.8a shows a Bethe lattice with $r = 3$. The Bethe lattice has no edges, and extends forever. A related network is the *Cayley tree,* which is drawn in fig. 13.8b. It represents the edge of the Bethe lattice, if it had one. The Cayley tree has more than half of its sites on the surface! The bottom of the figure is called the *boundary* of the lattice. The partition function is obtained by averaging the spins at the boundary and then moving inward row by row. A similar procedure was used in the 1D Ising model.

First average all of the spins in the $j = 0$ row. Each spin in the $j = 1$ row is connected to r-spins in the $j = 0$ row. Let σ denote a spin in the $j = 1$ row, and let $(\sigma_1, \sigma_2, \ldots, \sigma_r)$ denote the r-spins connected to it in the $j = 0$ row. Averaging just this small complex of spins gives a contribution to the partition function:

$$\sum_{\sigma_1 \cdots \sigma_r = \pm 1} \exp\{\beta J\sigma(\sigma_1 + \sigma_2 + \cdots + \sigma_r) + \beta h(\sigma + \sigma_1 + \sigma_2 + \cdots + \sigma_r)\}$$
$$= e^{\beta h\sigma}[2\cosh(\beta J\sigma + \beta h)]^r \tag{13.92}$$

Each spin has an effective interaction with its r-neighbors in the lower row of $\sigma[h + J(\sigma_1 + \cdots + \sigma_r)]$. In the nonmagnetic state, the sum over the neighbors will average out to

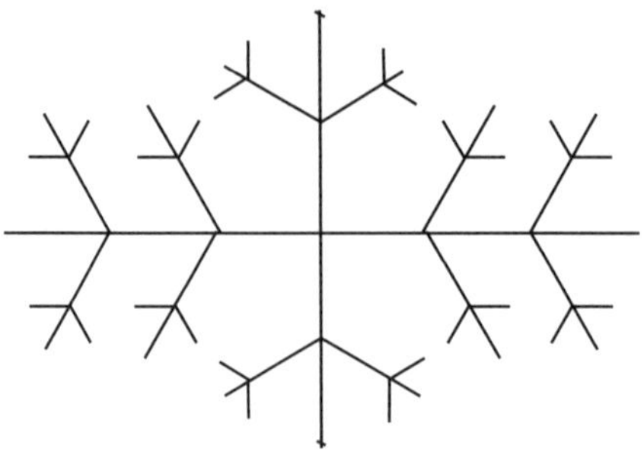

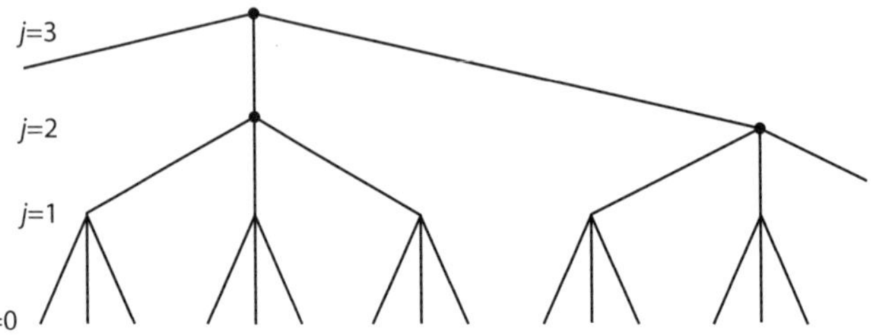

FIGURE 13.8. (a) Bethe lattice with $r = 3$. (b) A Cayley tree is a Bethe lattice with a boundary.

a small number. In the magnetic state, the sum over the neighbors could add up to $\pm r$. Define the effective magnetic energy h_j that acts on the spins in row j. A shorthand notation will be that $B_j = \beta h_j$. For the first row there are no spins below, so that $h_0 = h$, $B_0 = \beta h$. For the next row, eqn. (13.92) is used to define B_1 as

$$A_1 e^{B_1\sigma} = e^{\beta h\sigma}[2\cosh(B_0 + \beta J\sigma)]^r \tag{13.93}$$

Setting $\sigma = \pm 1$ gives two equations, which are solved for the two unknowns (A_1, B_1):

$$A_1 e^{B_1} = e^{\beta h}[2\cosh(B_0 + \beta J)]^r \tag{13.94}$$

$$A_1 e^{-B_1} = e^{-\beta h}[2\cosh(B_0 - \beta J)]^r \tag{13.95}$$

$$B_1 = \beta h + \frac{r}{2}\ln\left[\frac{\cosh(B_0 + \beta J)}{\cosh(B_0 - \beta J)}\right] \tag{13.96}$$

$$A_1 = 2^r\left[\cosh(B_0 + \beta J)\cosh(B_0 - \beta J)\right]^{r/2} \tag{13.97}$$

Equation (13.96) can be manipulated by taking

$$\frac{\cosh(B_0 + \beta J)}{\cosh(B_0 - \beta J)} = \frac{\cosh(B_0)\cosh(\beta J) + \sinh(B_0)\sinh(\beta J)}{\cosh(B_0)\cosh(\beta J) - \sinh(B_0)\sinh(\beta J)} \tag{13.98}$$

$$= \frac{1 + \tanh(B_0)\tanh(\beta J)}{1 - \tanh(B_0)\tanh(\beta J)}$$

$$\ln\left[\frac{1+x}{1-x}\right] = 2\tanh^{-1}x \tag{13.99}$$

These identities combine to give the result

$$B_1 = \beta h + r\tanh^{-1}[\tanh(B_0)\tanh(\beta J)] \tag{13.100}$$

The second step is to average the spins in the row $j=1$ to give the effective field $B_2 = \beta h_2$ in the next row. The procedure is exactly the same. The scaling equations are a modified version of eqn. (13.92):

$$\sum_{\sigma_1\cdots\sigma_r=\pm1} \exp\{\beta J\sigma(\sigma_1+\sigma_2+\cdots+\sigma_r) + \beta h\sigma + B_1(\sigma_1+\sigma_2+\cdots+\sigma_r)\}$$

$$= e^{\beta h\sigma}[2\cosh(\beta J\sigma + B_1)]^r = A_2 e^{B_2\sigma} \tag{13.101}$$

$$B_2 = \beta h + r\tanh^{-1}[\tanh(B_1)\tanh(\beta J)] \tag{13.102}$$

The general recursion relation as the rows are averaged one by one is

$$B_{j+1} = \beta h + r\tanh^{-1}[\tanh(B_j)\tanh(\beta J)] \tag{13.103}$$

Successive averaging causes the effective field $B_j = \beta h_j$ to converge to the value $B^* = \beta h^*$ in the interior of the Cayley tree. It obeys the self-consistent nonlinear equation

$$B^* = \beta h + r\tanh^{-1}[\tanh(B^*)\tanh(\beta J)] \tag{13.104}$$

$$\tanh[(B^* - \beta h)/r] = \tanh(B^*)\tanh(\beta J) \tag{13.105}$$

The solutions to this equation describe the magnetic state of the Ising model on the Bethe lattice.

Now calculate the transition temperature. Assume zero magnetic field ($h=0$) and ferromagnetic coupling ($J>0$). The transition temperature T_c ($\beta_c = 1/k_BT_c$) is where the ordering begins as one lowers the temperature. At the transition temperature the order parameter B^* is zero and it increases in value as the temperature is lowered. It has an infinitesimal value at a temperature that is infinitesimally smaller than T_c. In this case the above equation is

$$B^* = rB^*\tanh(\beta_c J) \tag{13.106}$$

$$\tanh(\beta_c J) = \frac{1}{r} \tag{13.107}$$

$$K_c = \beta_c J = \frac{1}{2}\ln\left(\frac{r+1}{r-1}\right) \tag{13.108}$$

The case for $r=1$ is a one-dimensional chain. In this case $T_c=0$ and there is no ordered state at nonzero temperature. For all other branching ratios $r>1$ there is a well-defined transition temperature.

Table 13.5 compares this prediction of the Bethe lattice with the results for the Ising model on crystalline lattices. What is actually tabulated is the quantity K_c for various values

Table 13.5 Ising model ferromagnetic transition temperatures ($K_c = J/k_BT_c$)

d	Crystal	r	K_c	$K_c(B)$
2	hc	2	0.659	0.549
2	sq	3	0.441	0.347
2	pt	5	0.274	0.203
3	sc	5	0.222	0.203
3	bcc	7	0.157	0.144
3	fcc	11	0.102	0.091

Note. The dimension is d and r is the branching ratio. The column K_c is the exact crystalline result, while $K_c(B)$ is the result from the Bethe lattice.

of r. Exact results are also shown for crystalline lattices in two and three dimensions. The lattices are honeycomb (hc), square (sq), plane-triangular (pt), simple cubic (sc), body-centered cubic (bcc), and face-centered cubic (fcc). The value of K_c for the Bethe lattice is typically below the crystalline result by 10–20%. However, the Bethe lattice has the correct trend that K_c decreases and T_c increases as the branching ratio r increases. The crystalline results are analytically known in two dimensions, since the Ising model can be solved exactly. For example, $K_c = (\frac{1}{2})\ln(2+\sqrt{3})$ for the hc lattice and $K_c = (\frac{1}{2})\ln(1+\sqrt{2})$ for the sq lattice.

Another prediction of eqn. (13.104) is for the limit that the temperature goes to zero when the magnetic field is zero:

$$\tanh(B^*/r) = \tanh(B^*)\tanh(K) \tag{13.109}$$

When T goes to zero, the argument of the hyperbolic tangent function becomes large and its value approaches one. Recall that $B^* = h^*/k_BT$, and the magnetization h^* becomes constant while the temperature goes to zero:

$$\lim_{K\to\infty} \tanh(K) = \frac{e^K - e^{-K}}{e^K + e^{-K}} = 1 - 2e^{-2K} + O(e^{-4K}) \tag{13.110}$$

Apply this limit to all three $\tanh(z)$ functions:

$$e^{-2B^*/r} = e^{-2B^*} + e^{-2K} \tag{13.111}$$

$$B^* \approx rK \tag{13.112}$$

$$h^* = rJ \tag{13.113}$$

The solution to (13.109) is that the effective local field from the r-spins in the row below is just $h^* = rJ$. At zero temperature, all of the spins are aligned in the ferromagnetic arrangement. What is interesting about this result is that the number of neighbors is $z = r + 1$ and one might expect that $h^* = zJ$, but that is not correct.

The critical properties of this model obey mean-field (Landau) theory. Several homework problems derived that the critical exponents are $\alpha = 0$, $\beta = \frac{1}{2}$, $\delta = 3$, etc.

13.6.4 Order–Disorder Transitions

There are several crystals in which the atomic arrangements have an order–disorder transition. A popular example is alloys of copper and gold. At high temperatures, the alloy Cu_xAu_{1-x} has a random arrangment of the copper and gold atoms on the fcc lattice. However, for temperatures below about 400°C the alloys form ordered arrangments. There are three, depending on relative concentrations x: AuCu, Au_3Cu, and $AuCu_3$. These alloys have ordered arrangements: in AuCu the atoms alternate lattice sites. Since fcc is not bipartite, the arrangements are complicated.

The Ising model can be used to model this order–disorder transition:

- Spin-up ($\sigma = +1$) denotes a Cu atom at the lattice site.
- Spin-down ($\sigma = -1$) denotes a Au atom at the lattice site.

When any atom sits at a lattice site, its energy is determined only by interaction with its first neighbors. Denote the two types of atoms as A or B:

- If an atom and its neighbor are both type A, the potential energy is V_{AA}.
- If an atom and its neighbor are both type B, the potential energy is V_{BB}.
- If one atom is A and its neighbor is B, the potential energy is V_{AB}.

These energies are converted into an Ising spin-$\frac{1}{2}$ system by defining

$$f(\sigma_A, \sigma_B) = C_0 + C_1(\sigma_A + \sigma_B) + C_2\sigma_A\sigma_B \tag{13.114}$$

The above options give

$$V_{AA} = f(1, 1) = C_0 + 2C_1 + C_2 \tag{13.115}$$

$$V_{BB} = f(-1, -1) = C_0 - 2C_1 + C_2 \tag{13.116}$$

$$V_{AB} = f(1, -1) = C_0 - C_2 \tag{13.117}$$

The coefficients can be solved:

$$C_0 = \tfrac{1}{4}(V_{AA} + V_{BB} + 2V_{AB}) \tag{13.118}$$

$$C_1 = \tfrac{1}{4}(V_{AA} - V_{BB}) \tag{13.119}$$

$$C_2 = \tfrac{1}{4}(V_{AA} + V_{BB} - 2V_{AB}) \tag{13.120}$$

The alloy problem can be represented by an Ising Hamiltonian in a magnetic field:

$$H = E_0 - J\sum_{j\sigma}\sigma_j\sigma_{j+\delta} - h\sum_j \sigma_j \tag{13.121}$$

where E_0 comes from the constant term C_0, $-J$ comes from C_2, and $-h$ comes from C_1.

The sign of J (i.e., C_2), determines whether the Ising interaction is ferromagnetic or antiferromagnetic. If it is ferromagnetic $C_2 < 0$, there will not be an order–disorder transition. An example is alloys of silver and gold. However, if $C_2 > 0$, then antiferromagnetic-type ordering will occur and the alloy has a order–disorder transition, as in copper and gold.

Let C_A be the fraction of sites occupied by A atoms, and C_B the fraction occupied by B atoms. Since all sites are occupied, $C_A + C_B = 1$. The average spin occupation is their difference:

$$\langle \sigma \rangle = \frac{1}{N} \sum_j \sigma_j = C_A - C_B \tag{13.122}$$

$$C_A = \tfrac{1}{2}[1 + \langle \sigma \rangle] \tag{13.123}$$

$$C_B = \tfrac{1}{2}[1 - \langle \sigma \rangle] \tag{13.124}$$

Figure 13.9 shows the order–disorder phase boundary of the alloy problem. Numerical results are obtained for the sq lattice in two dimensions. The ordered phase is found in a narrow range of concentration $0.4 < C < 0.6$.

Actual alloy phase diagrams for binary alloys show a variety of possible behaviors:

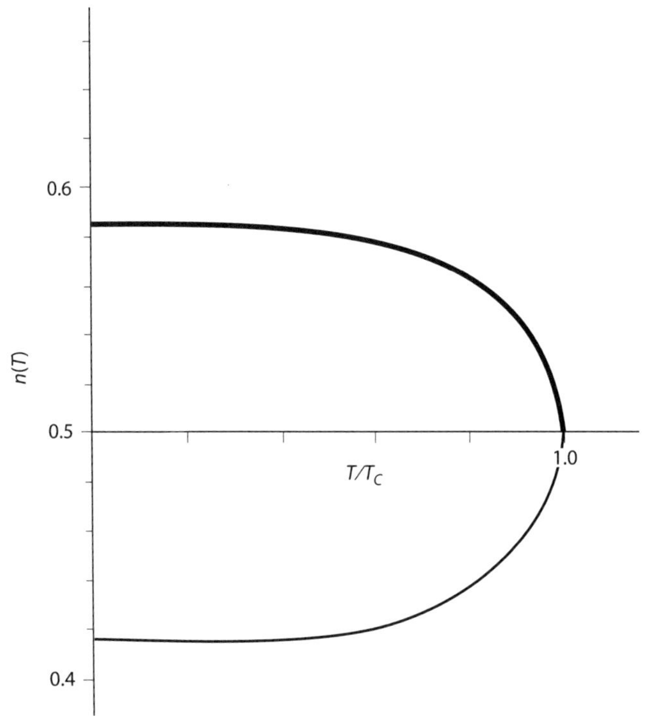

FIGURE 13.9. Order--disorder phase boundary for the two dimensional square lattice. The ordered region is $0.4 < C < 0.6$. From Subbaswamy and Mahan, *Phys. Rev. Lett.* **37**, 642 (1976). Used with permission of the American Physical Society.

- The two elements are immiscible, so one forms a pure phase at one end of the crystal and the other has a pure phase at the other end.
- The two elements alloy randomly with no long-range ordering.
- The two elements form intermetallic compounds, such as AB, A_2B, and AB_3. They are usually immiscible.

The behavior of AuCu alloys with an order–disorder transition actually seldom occurs in nature.

13.6.5 *Lattice Gas*

The lattice gas is another model based on atomic arrangements. In the alloy problem, the site has either one atom or the other. In the lattice gas, it may be empty or occupied. The crystal has N sites, and N_A atoms to fill these sites, with $N > N_A$. At low temperature, the occupied sites are ordered, while at high temperature they are disordered.

In two and three dimensions, use spin-$\frac{1}{2}$ operators to represent particle operators. In this case, the particle operators have funny commutation relations. They are, like spins, neither fermions nor bosons. This transformation associates the creation operator $C_\ell^\dagger$, which puts an atom on the site, with $S^{(+)}$. A destruction operator C, which removes an atom from the site, is represented by $S^{(-)}$:

$$C_\ell^\dagger = S_\ell^{(+)} \tag{13.125}$$

$$C_\ell = S_\ell^{(-)} \tag{13.126}$$

$$n_\ell = C_\ell^\dagger C_\ell = S_\ell^{(+)} S_\ell^{(-)} = \tfrac{1}{2} + S_\ell^{(z)} \tag{13.127}$$

Spin-$\frac{1}{2}$ operators are used since only one atom is allowed at a site.

This transformation preserves all the commutation relations for spin operators. For example, one has that

$$\{C_\ell, C_\ell^\dagger\} = 1 \tag{13.128}$$

$$[C_\ell, C_m^\dagger] = -2\left(\frac{1}{2} - n_\ell\right)\delta_{\ell m} \tag{13.129}$$

$$[C_\ell, n_\ell] = \left[S_\ell^{(-)}, S_\ell^{(z)}\right] = C_\ell \tag{13.130}$$

The particle operators C_ℓ and $C_m^\dagger$ anticommute if they are on the same site and commute if they are on different sites. They are neither fermions nor bosons. Collective operators such as

$$C_{\mathbf{k}} = \frac{1}{\sqrt{N}} \sum_j e^{i\mathbf{k}\cdot\mathbf{r}_j} C_j \tag{13.131}$$

have funny commutation relations. Nevertheless, this Hamiltonian is popular and is called the *lattice gas*. It applies to atoms diffusing on a lattice. The atoms may be considered classical particles, which commute on different sites. However, there may be no more than one atom on each site, since the atoms are large, substantial objects. "No more than one atom on each site" is an exclusion principle, which is represented by the anticommutation relations on the same site. The same physics is contained in a model that has the particle obey purely boson statistics but with the provision that there is a strong repulsive interaction U if two particles are on the same atomic site. The lattice gas results would be obtained in the limit $U \rightarrow \infty$. The statistics of anticommutation relations on the same site merely represent the strong repulsive interaction between atoms at close separation.

For electrons on a lattice, the Hamiltonian is called the Hubbard model for finite U and the Falicov-Kimble model for $U \rightarrow \infty$. When applied to highly correlated atoms it is the lattice gas model.

13.7 Topological Phase Transitions

In one dimension there is no long-range ordering of magnetic systems, while there is in three dimensions. The interesting case is two dimensions. The Ising model has a phase transition in two dimensions. Another Hamiltonian is the *XY*-model. Originally it was believed that it lacked any kind of phase transition. However, Kosterlitz and Thouless [8] showed that the *XY*-model in 2D had a new type of phase transition, which they named a topological transition. It appears in many two-dimensional systems besides magnets: liquid crystals, dislocations, etc. It is based on an excitation that resembles a vortex. A related theory was discussed earlier by Berezinskii [1].

The lack of long-range order is due to spin waves. Let the two sites $\mathbf{R}_j$ and $\mathbf{R}_\ell$ be far apart. Then their spin correlation will be a function of the separation:

$$\langle \mathbf{S}_j \cdot \mathbf{S}_\ell \rangle = f(\mathbf{R}_j - \mathbf{R}_\ell) \propto \frac{1}{|R_{j\ell}|^\alpha} \tag{13.132}$$

The last result comes from spin waves: the correlation decreases as a power law of the distance. At large distance the spin orientation is uncorrelated, so spin waves destroy long-range order. This result is an example of a theorem due to Mermin and Wagner [10] that there is no long-range order in two-dimensional magnetic systems.

The lack of long-range order in two dimensions was initially interpreted as proving there is no phase transition. However, Kosterlitz and Thouless showed there existed a topological phase transition.

Experimentalists find phase transitions in two dimensions. A film of liquid helium is a superfluid at low temperature. Magnetic films show ordinary phase transitions. These ordered systems have short-range order. Do they have long-range order? What length scale is "long"? The topic of phase transitions in two dimensions continues to be an area of research.

13.7.1 Vortices

Vortices were first observed in fluids. The discussion begins by briefly reviewing the vortex properties of a uniform, incompressible fluid in two dimensions. Let $\mathbf{u}(\mathbf{r})$ be the velocity vector of the fluid in steady-state motion. The circulation of the fluid is described by the curl of the velocity:

$$\nabla \times \mathbf{u} = \hat{z}\left[\frac{\partial u_x}{\partial y} - \frac{\partial u_y}{\partial x}\right] \tag{13.133}$$

In a vortex, the fluid rotates in a circle, so the only nonzero component of $\mathbf{u}$ is u_θ. The vorticity κ is defined as the integral over all two-dimensional space of the circulation:

$$\kappa = \int dS\hat{z}\cdot\nabla\times\mathbf{u} = \oint d\vec{\ell}\cdot\mathbf{u} \tag{13.134}$$

Using Stoke's theorem, the area integral is converted to a line integral around any closed circle. Since u_θ is a constant around the circle,

$$u_\theta(r) = \frac{\kappa}{2\pi r} \tag{13.135}$$

The divergence as $r \to 0$ is removed by cutting off the flow for a small radius r_0.

The kinetic energy of the vortex is found from

$$E_K = \frac{\rho}{2}\int d^2 r u_\theta^2 = \frac{\rho\kappa^2}{4\pi}\int_{r_0}^{R}\frac{dr}{r} = \frac{\rho\kappa^2}{4\pi}\ln\left|\frac{R}{r_0}\right| \tag{13.136}$$

where ρ is the fluid density, and R is the radius of the container holding the fluid. An important feature of this result is the logarithmic dependence on radius. Since r_0 is a microscopic radius and R is a macroscopic radius, the ratio R/r_0 is a very large number. A single vortex is an excitation that requires a large amount of energy.

The vortex can rotate in either the clockwise or the counterclockwise direction. The former is denoted with negative vorticity. Consider the fluid velocity of a pair of vortices separated by the vector $\mathbf{d}$, which have equal and opposite vorticity:

$$\mathbf{u}(r) = \frac{\kappa}{2\pi}\left[\frac{\hat{z}\times\mathbf{r}}{r^2} - \frac{\hat{z}\times(\mathbf{r}-\mathbf{d})}{|\mathbf{r}-\mathbf{d}|^2}\right] \tag{13.137}$$

$$u^2(\mathbf{r}) = \left(\frac{\kappa}{2\pi}\right)^2\left[\frac{1}{r^2} + \frac{1}{|\mathbf{r}-\mathbf{d}|^2} - \frac{2\mathbf{r}\cdot(\mathbf{r}-\mathbf{d})}{r^2|\mathbf{r}-\mathbf{d}|^2}\right] \tag{13.138}$$

where $\hat{z}\times\hat{r} = \hat{\theta}$. The kinetic energy is evaluated as above. The integrals are easy in the physical case that $r_0 \ll d \ll R$:

$$E_K = \frac{\rho\kappa^2}{2\pi}\ln\left|\frac{d}{2r_0}\right| \tag{13.139}$$

The energy no longer depends on R, since the fluid does not rotate at large distance: the motion from the two vortices cancel. The energy is now a smaller number, which depends on the vorticity κ and the separation d. The vortex–antivortex pair is a fundamental excitation of the two-dimensional fluid. In three dimensions, an excitation of finite energy is the smoke ring: the line of vortex motion forms a circle.

Another system with similar statistical mechanics is a two-dimensional Coulomb gas. The system has equal numbers of positive and negative particles of charge $q_j = \pm q$. The interaction between particles is $q_i q_j \ln(r_{ij}/a)$. The interaction is not Coulomb's law $q_i q_j / r_{ij}$ between point particles, but is appropriate if the objects are charged rods, where the rods are perpendicular to the plane. The rods are infinite in length in the z-direction and q is the charge per unit length. The statistical mechanics of this system has been studied at length: see Minnhagen [11] for a review.

13.7.2 XY-Model

Let $\mathbf{S}_j$ be the two-dimensional spin vector, which is confined to be in the plane of the two-dimensional lattice. It has length S and makes an angle θ_j with the x-axis. The classical scalar product of two such spins is

$$\mathbf{S}_j \cdot \mathbf{S}_i = S^2 \cos(\theta_i - \theta_j) \tag{13.140}$$

$$H = -J \sum_{j\delta} \cos(\theta_j - \theta_{j+\delta}) \tag{13.141}$$

Kosterlitz-Thouless took as their Hamiltonian the above equation. The spins are classical objects, not quantum operators. Assume that $J > 0$. The ferromagnetic ground state has all spins aligned. Then $\theta_j = \theta_0$ and the ground-state energy is

$$E_G = -J \sum_{j\delta} = -JzN \tag{13.142}$$

Figure 13.10 shows a spin vortex on the pt lattice. The center spin can be in any direction and is shown along the x-direction. All other spins are aligned perpendicular to the radial vector r from this central site.

At large distance from the center of the vortex, the neighbors have similar angles. Making the small-angle approximation gives

$$\cos(\theta_i - \theta_j) \approx 1 - \frac{1}{2}(\theta_i - \theta_j)^2 \tag{13.143}$$

$$H = E_G + \frac{J}{2} \sum_{j\delta} (\theta_j - \theta_{j+\delta})^2 \tag{13.144}$$

At large values of r, consider that the lattice is a polygon inscribed in a circle of radius r of $N_r = 2\pi r/a$ sides of length a. Here a is the lattice constant. One segment of the polygon subtends an angle $\Delta\theta = a/r$, which is also the angle between neighboring spins. See fig. 13.10b. Summing over all of the angles as one goes around the polygon gives

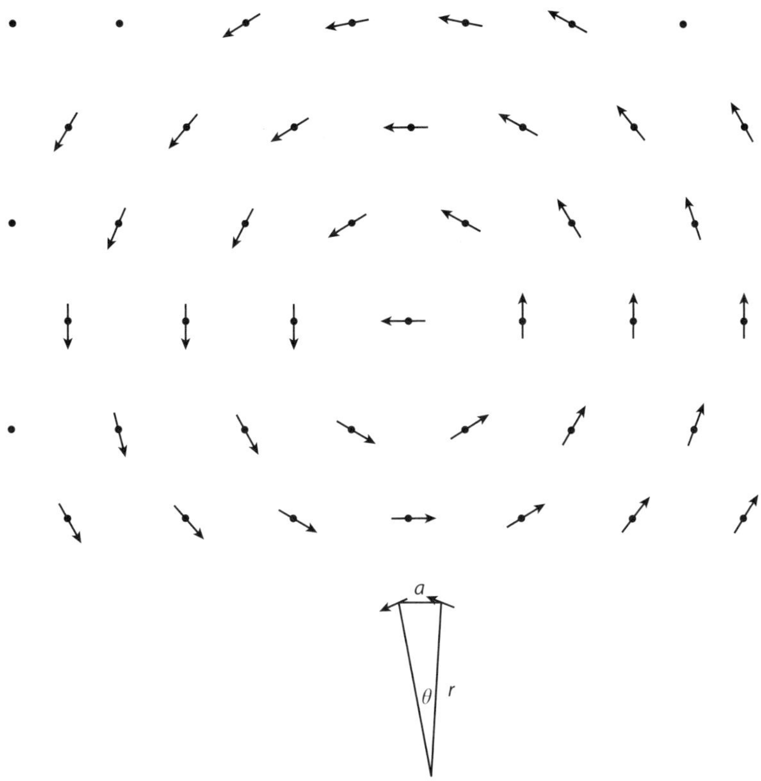

FIGURE 13.10. (a) Spin vortex on the pt lattice. (b) At large r, $\Delta\theta = a/r$.

$$E \sim \sum_{j\delta} (\theta_j - \theta_{j+\delta})^2 = \int \frac{d^2 r}{A_0} \left(\frac{a}{r}\right)^2 = \frac{2\pi a^2}{A_0} \ln(R/a) \tag{13.145}$$

where A_0 is the area of a unit cell. Another contribution is from nearby spins on parallel polygons, but they have a similar dependence. Since the area A_0 is proportional to a^2, Kosterlitz and Thouless found that the interaction energy between two spin vortices is

$$\Delta E = 2\pi J q_i q_j \ln\left|\frac{\mathbf{R}_i - \mathbf{R}_j}{a}\right| \tag{13.146}$$

where the direction of the flow is given by $q_i = \pm 1$. This energy relation is the same as found for the two-dimensional fluid vortices or the two-dimensional Coulomb gas. The statistical mechanics will be similar.

Figure 13.11 shows another topological spin arrangment. In 13.11a, all spins point toward a common central site. In 13.11b, they all point away from a common central site. These arrangements are like spin monopoles with either a plus or minus sign. Take case (b). The angle $\theta_{n,m}$ of a site $r = a(n, m)$ from the central site is

$$\theta_{n,m} = \tan^{-1}\left(\frac{m}{n}\right) \tag{13.147}$$

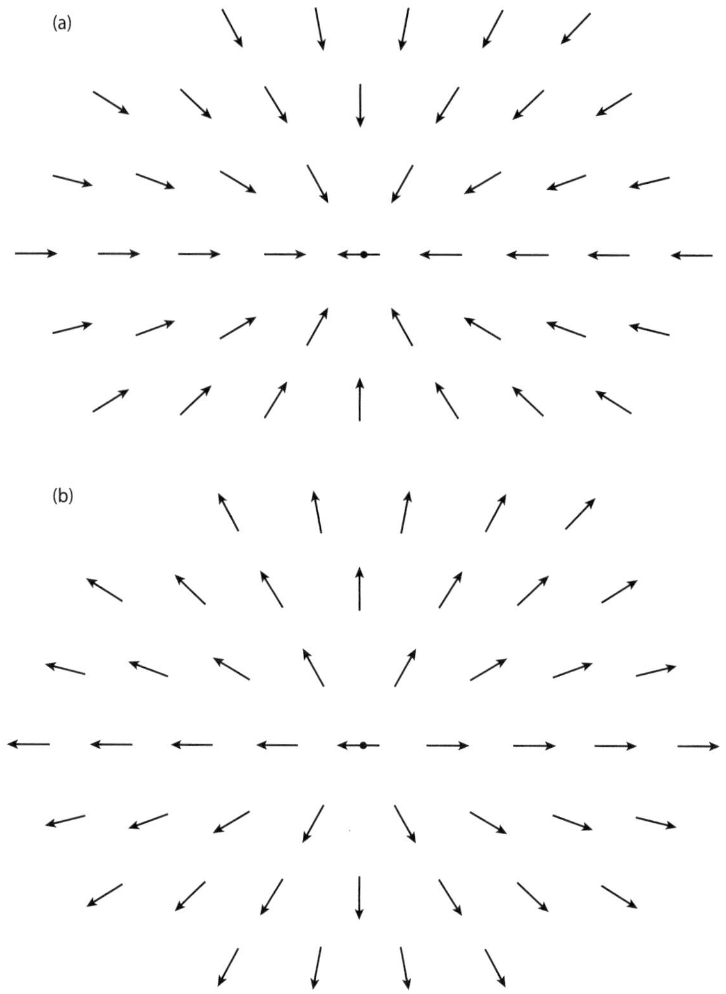

FIGURE 13.11. Another topological spin arrangement. (a) All spins point to a common center. (b) All spins point away from a common center.

Its first neighbors have the angle

$$\theta_{n,m\pm 1} = \tan^{-1}\left(\frac{m\pm 1}{n}\right) \approx \tan^{-1}\left(\frac{m}{n}\right) \pm \frac{n}{n^2+m^2} \tag{13.148}$$

$$\theta_{n\pm 1,m} = \tan^{-1}\left(\frac{m}{n\pm 1}\right) \approx \tan^{-1}\left(\frac{m}{n}\right) \mp \frac{m}{n^2+m^2} \tag{13.149}$$

The sum of the interactions with its four first neighbors is

$$\begin{aligned} &\cos(\theta_{n,m}-\theta_{n+1,m}) + \cos(\theta_{n,m}-\theta_{n-1,m}) + \cos(\theta_{n,m}-\theta_{n,m+1}) + \cos(\theta_{n,m}-\theta_{n,m+1}) \\ &= 2\cos\left(\frac{m}{n^2+m^2}\right) + 2\cos\left(\frac{n}{n^2+m^2}\right) \approx 4 - \frac{1}{n^2+m^2} \qquad (13.150) \\ &\approx 4 - \frac{a^2}{r^2} \end{aligned}$$

This energy is the same as the vortex arrangement. The lone spin monopole also has an energy that depends on $2\pi J\ln(R/r_0)$. The monopole flow pattern can be written in the continuous limit as

$$\mathbf{u} = \frac{\mathbf{r}}{r^2} \tag{13.151}$$

This flow is found when fluids flow outward from a point source. At a distance r from the source, the product of the velocity u_r and circumference $2\pi r$ must be a constant, so that $u \propto 1/r$.

The monopole–antimonopole pair has a flow pattern

$$\mathbf{u} = \frac{\mathbf{r}}{r^2} - \frac{\mathbf{r} - \vec{d}}{|\mathbf{r} - \vec{d}|^2} \tag{13.152}$$

$$u^2(\mathbf{r}) = \frac{1}{r^2} + \frac{1}{|\mathbf{r} - \vec{d}|^2} - 2\frac{\mathbf{r}\cdot(\mathbf{r} - \vec{d})}{r^2|\mathbf{r} - \vec{d}|^2} \tag{13.153}$$

This formula has the same spatial dependence as eqn. (13.138) for the vortex–antivortex pair. It will give the same energy for the pair, $\sim\ln(d/2r_0)$.

Let $\theta_{n,m}$ denote the polar angle that the site $\mathbf{R}_{n,m}=a(n,m)$ makes with the origin of the vortex. Let $\phi_{n,m}$ be the angle that the spin is canted at that site. The following cases were discussed:

- The vortex has $\phi_{n,m}=\theta_{n,m}+\pi/2$.
- The antivortex has $\phi_{n,m}=\theta_{n,m}-\pi/2$.
- The positive monopole has $\phi_{n,m}=\theta_{n,m}$
- The negative monopole has $\phi_{n,m}=\theta_{n,m}+\pi$

All of these cases have $\phi_{n,m}=\theta_{n,m}+\alpha$, where α is a constant angle. Since the energy of the arrangement depends on $\cos(\phi_{n,m}-\phi_{n\pm1,m\pm1})$, the constant angle α cancels. All of these spin arrangements have the same energy.

The statistical mechanics is not discussed in great detail, but some important features are summarized.

1. The entropy for making the vortex is quite large. The center point can be on any lattice site, so $S=k_B\ln(N)$, where N is the number of lattice sites. If $A\sim R^2$ is the total area of the crystal, then $N=A/A_0\sim R^2/a^2$ so that the entropy goes as

$$S = 2k_B\ln\left|\frac{R}{a}\right| \tag{13.154}$$

The free energy is

$$F(T) = 2[\pi J - Tk_B]\ln\left|\frac{R}{a}\right| \tag{13.155}$$

At a temperature $T\sim\pi J/k_B$ the entropy term is larger and the system starts making a large number of vortices.

2. At low temperature, the excitations of the system are vortex–anti-vortex pairs separated by a distance d

$$\Delta E = 4\pi J \ln\left|\frac{d}{2a}\right| \tag{13.156}$$

3. At high temperatures, the vortices unbind and one has a gas of unpaired vortices with roughly equal numbers of positive and negative vorticity.

The topological phase transition is between the last two states of matter: the transition between (i) vortex pairs, and (ii) a gas of unpaired vortices. This calculation of this transition is complicated. For example, in the unpaired state at high temperature, the interaction between vortices is screened by the other vortices. The analogy is the screening that conducting electrons provide Coulomb interactions in a metal. The dielectric screening of vortices is an interesting topic.

The correspondence with the two-dimensional Coulomb gas is instructive. In the low-temperature phase, the binding of the plus and minus rods is similar to a molecule of positronium. In the high-temperature phase it is a plasma of positive and negative charges. Numerical simulations find the phase transition temperature at $k_B T_c = 0.887J$. See the review by Minnhagen [11].

13.8 Kondo Effect

Chapter 11 on Transport Phenomena discussed the temperature dependence of the electrical resistivity $\rho(T)$ for metals:

$$\rho(T) = \rho_0 + \rho_p(T) \tag{13.157}$$

The first term ρ_0 comes from impurity scattering. It is proportional to the concentration of impurities and is usually independent of temperature. The second term $\rho_p(T)$ is due to phonons and increases with increasing temperature. This kind of temperature dependence is found in most metals, and is illustrated in fig. 11.9.

Experimentalists found a few cases where $\rho(T)$ had a minimum in the resistivity. A typical example is shown in fig. 13.12 for Fe impurities in Cu. The minimum occurs when transition metal atoms are impurities in nonmagnetic metals, such as Cu or Au. The transition metal atoms maintained their magnetic moment in these metals. An example is manganese in copper. Atomic Mn has an outer electronic configuration of $(3d)^5(4s)^2$. The ion Mn^{2+} ionizes its two $4s$ electrons, and the $(3d)^5$ configuration has, according to Hund's rules, $S = \frac{5}{2}$, $L = 0$, $J = S$. In other cases, as when Al is the host, no magnetic moments are detected from transition metal impurities.

Kondo [7] showed that an electron while scattering from an impurity with a magnetic moment has spin-flip events, and these cause a temperature-dependent lifetime and a temperature-dependent resistivity from impurity scattering. His theory explained the resistivity minimum. The resistivity minimum is now called the *Kondo effect*.

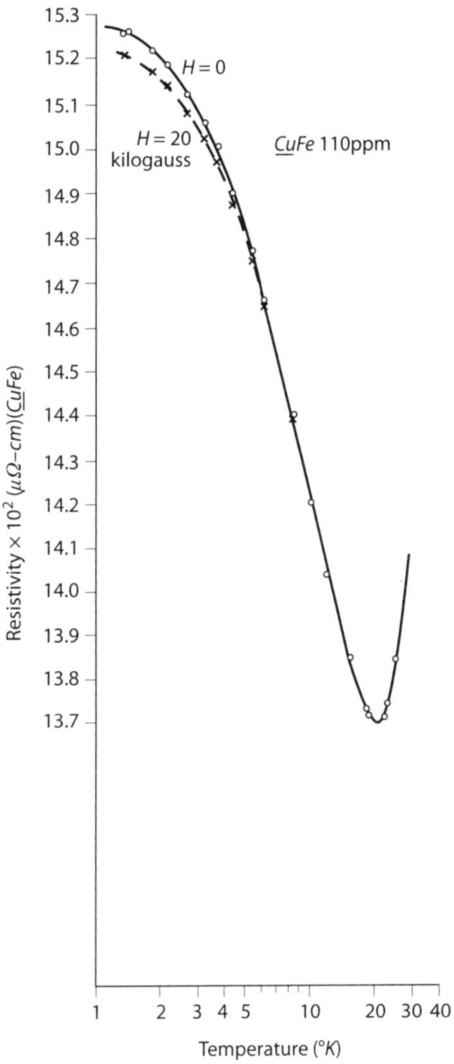

FIGURE 13.12. $\rho(T)$ for Fe impurities in Cu, showing a resistivity minimum. Data from Monod, *Phys. Rev. Lett.* **19**, 1113 (1967). Used with permission of the American Physical Society).

13.8.1 sd-Interaction

The interaction between the conduction electron and the impurity atom has two terms.

- The first is the charge interaction, since the impurity usually has a different valence than the host ions in the metal. The Coulomb interaction is screened and becomes short-ranged. Its scattering is described by phase shifts and its contribution to the resistivity is temperature independent.

- There is also an exchange interaction between the conduction electron and the electrons on the impurity. The exchange interaction can be expressed according to the product $\mathbf{S} \cdot \mathbf{s}$ of the spin $\mathbf{S}$ of the impurity ion, and the spin $\mathbf{s}$ of the conduction electron. This interaction is called V_{sd}, since the conduction electron is taken to be in an *s*-orbital and the local spin is due to a partially filled *d*-shell:

$$V_{sd}(\mathbf{r}) = -J\sum_j \mathbf{S}_j \cdot \mathbf{s}\,\delta(\mathbf{r}-\mathbf{R}_j) \tag{13.158}$$

where $\mathbf{R}_j$ are the random locations of the impurities. The conduction electron spin $\mathbf{s}$ is one-half, but the local spin $\mathbf{S}$ can be any integer or half-integer; e.g., Mn^{2+} has $S=\frac{5}{2}$. The above interaction is averaged with the electron density operator:

$$V_{sd} = \int d^3r\rho(\mathbf{r})V_{sd}(\mathbf{r}) \tag{13.159}$$

$$= -\frac{J}{N}\sum_j\sum_{\mathbf{kq}}\left[S_j^{(z)}\left(C^\dagger_{\mathbf{k+q}\uparrow}C_{\mathbf{k}\uparrow} - C^\dagger_{\mathbf{k+q},\downarrow}C_{\mathbf{k}\downarrow}\right)\right.$$

$$\left. + S_j^{(+)}C^\dagger_{\mathbf{k+q},\downarrow}C_{\mathbf{k}\uparrow} + S_j^{(-)}C^\dagger_{\mathbf{k+q},\uparrow}C_{\mathbf{k}\downarrow}\right]e^{i\mathbf{q}\cdot\mathbf{R}_j}$$

The first term on the right is $S^{(z)}s^{(z)}$, where $s^{(z)}$ is $+\frac{1}{2}$ if the conduction electron has spin up, and is $-\frac{1}{2}$ if the conduction electron has spin-down. The factor of $\frac{1}{2}$ is absorbed in J. The last two terms describe spin flip. The local spin flips up if the conduction electron spin flips down, or vice versa. The *sd* interaction forms the basis for many calculations of conduction electrons interacting with local magnetic moments. The local spin operators have an eigenstate $|S, m\rangle$ and obey the usual rules:

$$S^{(z)}|S,m\rangle = m|S,m\rangle \tag{13.160}$$

$$S^{(+)}|S,m\rangle = \sqrt{S(S+1)-m(m+1)}\,|S,m+1\rangle \tag{13.161}$$

$$S^{(-)}|S,m\rangle = \sqrt{S(S+1)-m(m-1)}\,|S,m-1\rangle \tag{13.162}$$

The factor of $\hbar$ has been absorbed in J.

13.8.2 Spin-flip Scattering

The *sd* interaction is evaluated using perturbation theory. First-order perturbation theory gives

$$\Sigma^{(1)} = \langle V_{sd}\rangle = -\frac{J}{N}\sum_j\sum_{\mathbf{k}} m_j[n_{\mathbf{k}\uparrow} - n_{\mathbf{k}\downarrow}] \tag{13.163}$$

where $\langle C^\dagger_{\mathbf{k+q}\uparrow}C_{\mathbf{k}\uparrow}\rangle = \delta_{q=0}n_{\mathbf{k}\uparrow}$. The summation over j gives N_i the number of impurities $N_i = c_iN$. The above expression averages to zero twice. In a paramagnetic system, the density of up spin electrons $n_{\mathbf{k}\uparrow}$ equals the number of down spin electrons $n_{\mathbf{k}\downarrow}$. Secondly, the magnetic quantum number m_j of the local spin varies from site to site and averages to zero. So $\Sigma^{(1)}=0$.

Second-order perturbation theory involves a summation over intermediate states:

$$\Sigma^{(2)} = -\sum_I \frac{|\langle I|V_{sd}|g\rangle|^2}{E_I - E_g} \tag{13.164}$$

Consider a conduction electron approaching the impurity with (i) spin-up and (ii) wave vector k. In the intermediate state it scatters to state $\mathbf{k}+\mathbf{q}\equiv\mathbf{k}'$.

- The term $S^{(z)}C^{\dagger}_{\mathbf{k}+\mathbf{q}\uparrow}C_{\mathbf{k}\uparrow}$ contributes to the second-order energy:

$$\Sigma_a^{(2)} = -\left(\frac{J}{N}\right)^2 \sum_j m_j^2 \sum_{\mathbf{k}'} \frac{1}{E(\mathbf{k}') - E(\mathbf{k})} \tag{13.165}$$

- The term $S^{(+)}C^{\dagger}_{\mathbf{k}+\mathbf{q}\uparrow}C_{\mathbf{k}\uparrow}$ makes the the conduction electron go to a down spin configuration in the intermediate state, while the local spin flips up one value of m_j. This term contributes to the second-order energy:

$$\Sigma_b^{(2)} = -\left(\frac{J}{N}\right)^2 \sum_j [S(S+1) - m_j(m_j+1)] \sum_{\mathbf{k}'} \frac{1}{E(\mathbf{k}') - E(\mathbf{k})} \tag{13.166}$$

These two are the only possible events for an incoming conduction electron with spin-up. Add these two terms: the factors of m_j^2 cancel and the factor of m_j averages to zero. The final result is

$$\Sigma^{(2)} = -S(S+1)\frac{c_i J^2}{N}\sum_{\mathbf{k}'}\frac{1}{E(\mathbf{k}') - E(\mathbf{k})} \tag{13.167}$$

$$= -S(S+1)c_i J^2 \int dE' \frac{N(E')}{E' - E}$$

where $N(E')$ is the density of states. The electron lifetime is a similar expression:

$$\frac{\hbar}{\tau^{(2)}} = S(S+1)c_i J^2 \int dE' N(E') 2\pi\delta(E' - E) \tag{13.168}$$

$$= 2\pi S(S+1)c_i J^2 N(E)$$

So far the result is rather dull. The electron lifetime is a constant, independent of temperature. This term contributes to the constant term ρ_0 in the electrical resistivity.

Kondo carried the calculation to the third-order of perturbation theory, and found terms that depend on temperature. This calculation is complicated, and we spare the details– they are found in section 6.1 of *Many-Particle Physics* [9]. The self-energy is

$$\Sigma^{(3)}(E) = -S(S+1)\frac{c_i J^3}{N^2}\sum_{\mathbf{k}'\mathbf{p}}\frac{1}{E - E(\mathbf{k}')}\left[\frac{1}{E - E(\mathbf{p})} - \frac{4n_F(E(\mathbf{p}))}{E(\mathbf{p}) - E(\mathbf{k}')}\right] \tag{13.169}$$

The interesting feature of this formula is the factor of the Fermi-Dirac occupation number $n_F(E')$ in the last term. When the conduction electron scatters through intermediate states $\mathbf{p}$ and $\mathbf{k}'$, it can scatter only if the final state is empty $(1 - n_F)$ and the initial state is occupied (n_F). In first- and second-order perturbation theory, all factors of n_F cancel. They do not cancel in third order, for spin-flip scattering. They cancel in all orders of perturbation theory for Coulomb scattering. The integral has terms such as

$$\frac{1}{N}\sum_{\mathbf{p}}\frac{n_F(E(\mathbf{p}))}{E(\mathbf{p}) - E(\mathbf{k}')} = \int dE' N(E') \frac{n_F(E')}{E' - E(\mathbf{k}')} \tag{13.170}$$

$$= \int_{-W}^{\mu} dE' N(E') \frac{1}{E' - E(\mathbf{k}')} \approx N(\mu) \ln\left|\frac{E(\mathbf{k}') - \mu}{W}\right|$$

where W is a bandwidth. The third-order lifetime has a constant term and a term that is not constant. The latter is

$$\frac{\hbar}{\tau^{(3)}(\xi)} = 8\pi S(S+1)c_i J^3 N(\mu)^2 \ln\left|\frac{\xi}{W}\right| \tag{13.171}$$

where $\xi = E - \mu$ is the energy of the conduction electron with respect to the chemical potential. For electrons near the Fermi surface, the ratio $|\xi|/W \ll 1$ and the logarithm is a negative number. The negative sign is compensated by the fact that $J < 0$. The Kondo effect is found only in magnetic impurities that have an antiferromagnetic coupling constant ($J < 0$).

If we average this result at nonzero temperature, where the Fermi surface has a thermal width, then

$$\frac{\hbar}{\tau^{(3)}(T)} = \int d\xi \frac{\hbar}{\tau^{(3)}(\xi)}\left(-\frac{dn_F(\xi)}{d\xi}\right) \tag{13.172}$$

$$\approx 8\pi S(S+1)c_i J^3 N(\mu)^2 \ln\left|\frac{k_B T}{W}\right|$$

This temperature dependence of the electron lifetime from spin-flip scattering explains the minimum in the electrical resistance as a function of temperature. At low temperatures, where $k_B T/W \ll 1$, the argument of the logarithm is small, so this function is a large negative number, and the inverse lifetime is a large positive number (since $J < 0$). It is interesting to combine the lifetimes from second- and third-order perturbation theory:

$$\frac{\hbar}{\tau} = 2\pi S(S+1)c_i J^2 N(\mu)\left[1 + 4N(\mu)J\ln\left|\frac{\xi}{W}\right|\right] \tag{13.173}$$

According to Matthiessen's Rule, add contributions to the resistivity. Spin-flip scattering gives a term

$$\delta\rho(T)_K = \frac{m}{ne^2\tau} = \frac{2\pi m}{ne^2\hbar}c_i S(S+1)N(\mu)^2 J^2\left[1 + 4N(\mu)J\ln\left|\frac{k_B T}{W}\right|\right] \tag{13.174}$$

This formula is used to compare with experiment.

13.8.3 Kondo Resonance

Many theorists have carried the calculation of the *sd*-interaction to higher orders of perturbation theory. They propose that the actual self-energy function can be written as

$$\Sigma(\xi) = \frac{c_i J}{1 - 2N(\mu)J\ln|\xi/W| - i\pi S(S+1)N(\mu)J} \tag{13.175}$$

The spectral function is two times the imaginary part of this expression:

$$A(\xi) = \frac{2\pi c_i J^2 N(\mu)S(S+1)}{[1 - 2N(\mu)J\ln|\xi/W|]^2 + [\pi S(S+1)N(\mu)J]^2} \tag{13.176}$$

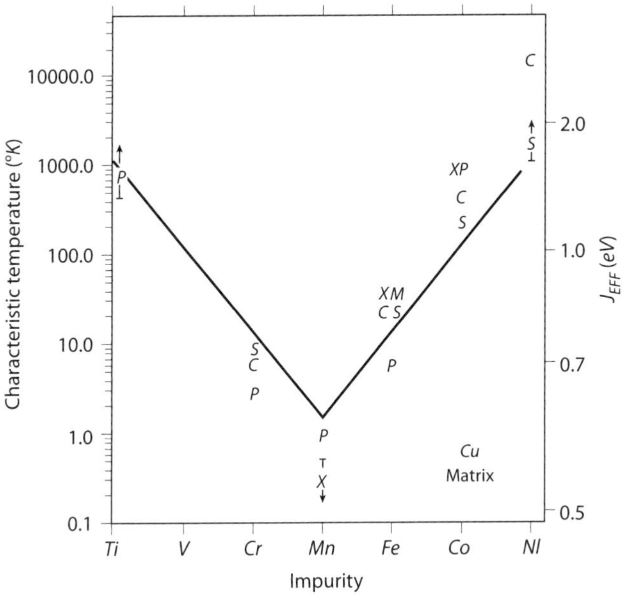

FIGURE 13.13. Kondo temperatures T_K of iron group impurities in copper. Values range from 10^{-2} to 10^4 K. From Daybell and Steyert, *Rev. Mod. Phys.* **40**, 380 (1968). Used with permission of the American Physical Society.

The bracket in the denominator can be written as

$$\left[1 - 2N(\mu)J\ln\left|\frac{\xi}{W}\right|\right] = -2N(\mu)J\ln\left|\frac{\xi}{k_BT_K}\right| \tag{13.177}$$

$$k_BT_K = W\exp\left[\frac{1}{2JN(\mu)}\right] = W\exp\left[-\frac{1}{2|J|N(\mu)}\right] \tag{13.178}$$

T_K is called the *Kondo temperature.* It sets the temperature scale of the resistivity upturn at low temperature. Experimental values for T_K for copper as the host are shown in fig. 13.13. They range in value from small to large temperatures. Similar results are found for gold as the host.

The spectral function is rewritten by factoring $2NJ$ from every term:

$$N(\mu)A(\xi) = \frac{\pi c_i S(S+1)/2}{\ln^2|\xi/k_BT_K| + [\pi S(S+1)/2]^2} \tag{13.179}$$

The expression on the right describes a resonance that occurs at $\xi = k_BT_K$, where $\ln^2|\xi/k_BT_K| = 0$. It is called the *Kondo resonance.* This resonance is found in the density of states of Kondo alloys. It has a huge effect on the heat capacity, magnetic susceptibility, and other properties. The resonance is approximately a Lorentzian-shaped peak, where the peak is near to the Fermi energy. The above formula makes it appear as if there are two peaks, at $\xi = \pm k_BT_K$, but only one is to be retained. In some cases the peak may occur right at the Fermi energy. The peak is quite sensitive to temperature, and disappears rapidly

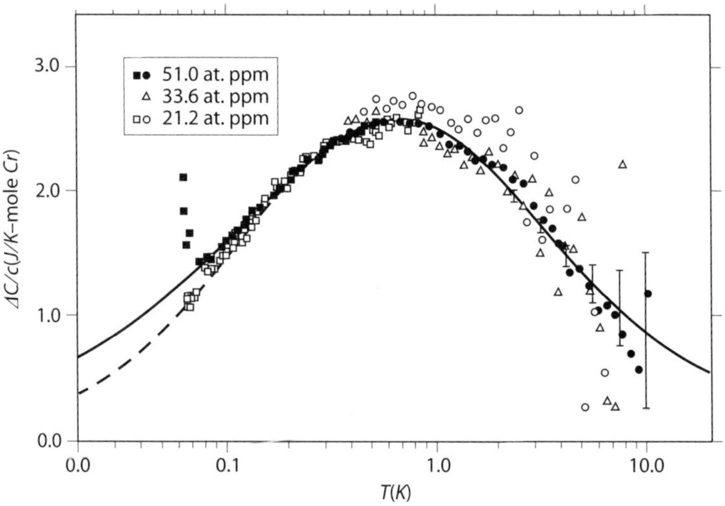

FIGURE 13.14. Excess heat capacity of Cr in Cu, per mole of Cr. Data from Triplett and Phillips, *Phys. Rev. Lett.* **27**, 1001 (1971). Used with permission of the American Physical Society.

for $T > T_K$. The peak is caused by spin fluctuations associated with the magnetic impurity. Figure 13.14 shows the excess heat capacity of Cr impurities in copper, normalized per mole of Cr. The results for different impurity concentrations follow the same curve. This system has a Kondo temperature at $T_K \sim 2$ K.

The periodic Kondo lattice is found in crystals where there are periodic, well-separated atoms with magnetic moments. Crystalline iron and nickel are *not* examples, since their atoms are not well separated, and the *d*-electrons form a tight-binding band. A good example are the cerium-115 crystals such as $CeCoIn_5$. Atomic Ce has an electron configuration $(4f)(5d)(6s)^2$. It is usually found in compounds as a Ce^{3+} ion with a single $(4f)$ electron. It is the first element in the rare earth series. The crystal has a periodic array of $(4f)$ electrons that are separated and do not have bond overlap to form a tight-binding band. However, they can coherently interact through the RKKY interaction, and at low temperature one finds that the *f*-electrons form a band with their own Fermi surface, which is different than the Fermi surface of the other electrons. At higher temperatures, the RKKY interaction gets washed out by thermal fluctuations, and the *f*-electrons cease to behave instead coherently, and behave as individual Kondo impurities.

13.9 Hubbard Model

The Hubbard model was introduced independently by John Hubbard [7] and Martin Gutzwiller [3] in 1963. It has been widely studied by theorists in one, two, and three dimensions. The Hamiltonian can be solved exactly in one dimension using the Bethe Ansatz method. Only approximate analytical solutions are available in two and three dimensions. There are accurate numerical solutions.

The Hamiltonian has the form

$$H = \sum_j \left[\sum_s E_0 n_{js} + U n_{j\uparrow} n_{j\downarrow} + W \sum_{\delta,s} C^\dagger_{j+\delta,s} C_{js} \right] + V \tag{13.180}$$

$$V = \frac{e^2}{2} \sum_{j \neq \ell} \frac{1}{R_{j\ell}} \left[\sum_s n_{js} - n \right] \left[\sum_{s'} n_{\ell s'} - n \right] \tag{13.181}$$

As usual, j denotes the site $\mathbf{R}_j$, $\boldsymbol{\delta}$ is the vector to the neighbor of j, and s is spin. The Hamiltonian describes electrons moving on a tight-binding lattice. The site energy is E_0, and the matrix element for hopping to a neighbor is W. The onsite Coulomb interaction U is found whenever two electrons are on the same atomic site. The electron orbital is assumed to be an s-wave state. The spin state is a singlet, and two electrons can both be on this site only if they have opposite spin configurations. The last term V is the long-range Coulomb interaction between electrons when they are on different atomic sites. It is normalized to the average electron occupation n, which is provided by the ion charge. The above Hamiltonian is called *the extended Hubbard model*. The extended part is V, the long-range Coulomb term.

The usual form of the Hubbard model lacks the term V. Without this term, the Hamiltonian does not properly describe charge fluctuations, such as plasma oscillations. Indeed, it predicts that the charge fluctuations have linear dispersion, which is wrong: plasmons have a constant frequency at long wavelength. However, the Hubbard model is used to describe spin fluctuations, and these they get correctly.

13.9.1 U = 0 *Solution*

Without the onsite interaction U, the Hubbard term is easy to solve. Go to a wave vector description and the resulting Hamiltonian is diagonal:

$$C_{js} = \frac{1}{\sqrt{N}} \sum_{\mathbf{k}} C_{\mathbf{k},s} e^{i\mathbf{k}\cdot\mathbf{R}_j} \tag{13.182}$$

$$H_0 = \sum_{\mathbf{k},s} E(\mathbf{k}) n_{\mathbf{k},s} \tag{13.183}$$

$$n_{\mathbf{k},s} = C^\dagger_{\mathbf{k},s} C_{\mathbf{k},s} \tag{13.184}$$

$$E(\mathbf{k}) = E_0 + W\gamma(\mathbf{k}), \quad \gamma(\mathbf{k}) = \sum_\delta e^{i\mathbf{k}\cdot\delta} \tag{13.185}$$

Define a density of states $N(\mathrm{E})$ for an electron of a single spin as

$$N(E) = \int \frac{d^3k}{(2\pi)^3} \delta[E - E(\mathbf{k})] \tag{13.186}$$

The density of states has nonzero values over the interval $-zW < E - E_0 < zW$. Evaluate the following three integrals:

$$1 = \int_{E_0 - zW}^{E_0 + zW} dE N(E) \tag{13.187}$$

$$n = 2\int_{E_0 - zW}^{\mu} dEN(E) \tag{13.188}$$

$$E_G = 2\int_{E_0 - zW}^{\mu} EdEN(E) \tag{13.189}$$

The first integral states that the energy band has one electron per site, of each spin. The second integral defines the chemical potential μ at zero temperature. The third gives the ground-state energy at zero temperature: The factor of two is spin degeneracy.

13.9.2 *Atomic Limit*

Another exactly solvable model is when $W = 0$, which is called the *atomic limit*. The Hamiltonian is solved in real space, since each atom acts independently. In a spin singlet, each site can have 0, 1, or 2 electrons. The case of one electron has two possibilities, with spin either up or down. The partition function Z_1 for a single site is

$$Z_1 = 1 + 2e^{-\beta(E_0 - \mu)} + e^{-\beta(2E_0 - 2\mu + U)} \tag{13.190}$$

The partition function for the N-atom lattice is $Z_N = Z_1^N$. Various thermodynamic quantities can be calculated using this function.

13.9.3 *U > 0*

The physics becomes more interesting when including the Hubbard interaction U. The energy of an electron of spin-up, at site $\mathbf{R}_j$ depends on the occupation $n_{j\downarrow}$ of electrons with the opposite spin. The Hartree-Fock approximation is to replace this operator by its average: $n_{j\downarrow} \rightarrow \langle n_{j\downarrow}\rangle \equiv n_\downarrow$. Define n_σ as the average number of electrons of spin σ in the lattice. The two possible values of spin configuration yield

$$n = n_\uparrow + n_\downarrow \tag{13.191}$$

Use an overbar to denote a negative sign, so $\bar{\sigma} = -\sigma$. The energy of a particle in the Hartree-Fock approximation is

$$E_\sigma(\mathbf{k}) = E_0 + W\gamma(\mathbf{k}) + Un_{\bar{\sigma}} \tag{13.192}$$

This equation has several possible solutions. One of them is magnetic.

The various states are discussed using a specific model. The density of states is a constant $N(E) = 1/W$ over the energy interval $-W/2 < E < W/2$. The integrals in eqn. (13.188) are

$$1 = \int_{-W/2}^{W/2} \frac{dE}{W} = 1 \tag{13.193}$$

$$n = \frac{2}{W}\int_{-W/2}^{\mu} dE = \frac{2}{W}\left[\mu + \frac{W}{2}\right] \tag{13.194}$$

$$\mu = \frac{W}{2}[n - 1] \tag{13.195}$$

$$E_G = \frac{2}{W}\int_{-W/2}^{\mu} EdE = \frac{1}{W}[\mu^2 - (W/2)^2] \tag{13.196}$$

$$= -\frac{W}{4}n(2-n)$$

Examine how these results are altered by the Hubbard interaction. Assume that $n < 1$, so the band is less than half-full.

1. The paramagnetic arrangement has equal numbers of up and down spins. The the energy is the same as above, plus the Hubbard term. The average ground-state energy per site is $Un_\uparrow n_\downarrow = U(n/2)^2$, since $n_\uparrow = n_\downarrow = n/2$:

$$E_P = -\frac{W}{4}n(2-n) + U\left(\frac{n}{2}\right)^2 \tag{13.197}$$

2. The ferromagnetic solution has all spins aligned: say in the up position. Then $n_\downarrow = 0$ and there is no Hubbard interaction energy. One has paid a price, in that the chemical potential is changed to a higher value $\bar{\mu}$, since only particles of the same spin occupy the band. This new chemical potential is

$$n = \frac{1}{W}\int_{-W/2}^{\bar{\mu}} dE = \frac{1}{W}\left[\bar{\mu} + \frac{W}{2}\right] \tag{13.198}$$

$$\bar{\mu} = W\left[n - \frac{1}{2}\right] \tag{13.199}$$

$$E_F = \frac{1}{W}\int_{-W/2}^{\bar{\mu}} EdE = \frac{1}{2W}[\bar{\mu}^2 - (W/2)^2] \tag{13.200}$$

$$= -\frac{W}{2}n(1-n) \tag{13.201}$$

3. It is possible that the system is only partially polarized, so that $n_\uparrow > n_\downarrow > 0$. That case is not considered here, although it is important for getting the full phase diagram.

Which of these two states has the lowest energy? Examine the case that

$$E_P > E_F \tag{13.202}$$

$$-\frac{W}{4}n(2-n) + U\left(\frac{n}{2}\right)^2 > -\frac{W}{2}n(1-n) \tag{13.203}$$

The terms linear in n cancel from both sides. Then n^2 is in all remaining terms and factors. One is left with the result that

$$\frac{1}{4}[W + U] > \frac{W}{2} \tag{13.204}$$

so that $U > W$. If $U/W > 1$, the ferromagnetic state has the lowest energy. If $U/W < 1$, the paramagnetic state has the lowest energy. Whether or not the system is magnetic depends on the ratio U/W. Since $1/W = N(\mu)$, it is magnetic if $UN(\mu) > 1$. This is called the *Stoner criterion* for magnetism.

13.9.4 *Half-filling*

An important special case is when the energy band is half-full of electrons: when $n = 1$. Review the possible ground states.

1. The ferromagnetic ground-state energy, in eqn. (13.201), is $E_F = 0$. All electrons have the same spin polarization, say spin-up. Only one electron of this spin can be on each site. Each site has one electron, and there are no empty sites. An electron cannot move to a neighboring site, since that violates the exclusion principle, which forbids two particles with the same spin and orbital quantum numbers from being in the same state. There is no kinetic energy. There is no Hubbard energy, since there are no down spins. The only energy for each electron is E_0, which is defined as the zero of energy.

2. The paramagetic energy, from eqn. (13.197) when $n = 1$, is

$$E_P(n = 1) = -\frac{1}{4}[W - U] \tag{13.205}$$

This result shows the paramagnetic state is preferred if $W > U$. This was true for other values of n. However, it also states that the ferromagnetic state is preferred if $U > W$. This conclusion is not correct.

3. Next consider the antiferromagnetic state. Reconsider the case that $U > W$. The system has equal numbers of up and down spin electrons. Assume the lattice is bipartite and arrange the electrons antiferromagnetically: alternate sites have electrons with spin-up and spin-down. Both the paramagnetic and antiferromagntic states have equal numbers of up and down spins. In paramagnetic ordering the spins have no long-range order, while they do in the antiferromagnetic state.

Consider the three energy terms. Each electron has E_0, which we are taking to be zero. There is no Hubbard energy, since each site has only one electron. Start from the atomic limit and let the hopping term be the perturbation. In second-order perturbation theory, the hopping term can take an electron from its site to one of the z-neighboring sites and then back to its orginal site. When the electron hops to its neighbor, which has the opposite spin, it goes to an energy state U. In second-order perturbation theory, the self-energy of each electron from hopping to the z neighbors and back is

$$\Sigma = -\frac{zW^2}{U} \tag{13.206}$$

So the antiferromagnetic arrangement has a negative energy for each electron. It is lower in energy than the ferromagnetic state. For $U > W$ it is lower in energy than the paramagnetic state.

The above energy Σ is from an electron, say with spin-up, hopping to its neighbors, which have spin-down. This process is illustrated in fig. 13.15. Another event is where the up

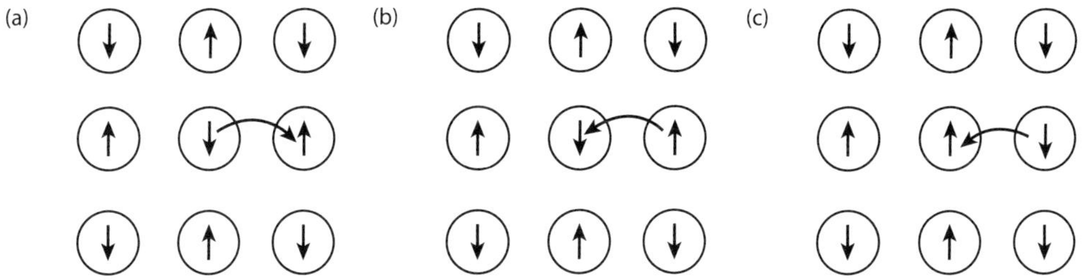

FIGURE 13.15. Electron hopping in an antiferromagnetic lattice. (a) An electron with spin-down hops to a neighbor. This intermediate state has energy $U + O(W)$. (b) The same spin-up electron hops back, which contributes to its second-order energy. (c) The spin-down electron hops to the empty site, so the two electrons exchange places.

spin hops to the neighbor, but the neighbor hops back, as in fig. 13.15c. The two spins have exchanged places. That process has the same matrix element $M = W^2/U$. From the viewpoint of an external observer, it appears as if the two electrons have mutually flipped spins. That process is described in the Heisenberg interaction $-J\mathbf{S}_j \cdot \mathbf{S}_{j+\delta}$. The Heisenberg constant $J = W^2/U$. The Hubbard model contains the Heisenberg model as a special case. The spin-flip processes also causes a Kondo resonance. It is not found in the Hartree-Fock approximation, which was evaluated in the above example. However, it is found when evaluating higher-orders of perturbation theory for the correlaton energy.

Many crystals exhibit antiferromagnetic ordering when the energy band has an average of one electron per site. An important example is the cuprate superconductors in the pure, undoped state. The crystal lattice contains layers of copper and oxygen. The conduction of electrons is in these planes of copper and oxygen. Each copper atom in the pure crystal is in the Cu^{2+} valence state, with one d-hole. These holes align antiferromagnetically. The cuprates must be doped to change the filling factor n away from $n = 1$, in order to get to the superconducting state.

There are many crystals with half-filling that exhibit neither ferromagnetic nor antiferromagnetic ordering. An example is the π-bands of graphite and graphene. In these cases, the density of conducting states is quite low, so that $UN(\mu) < 1$. They are not magnetic, in agreement with the Stoner criterion.

Next consider the case where the filling factor is over half-full: $1 < n < 2$. The Hubbard model exhibits particle–hole symmetry. Let the density of holes be $\nu = 2 - n$. The holes can have spin-up or spin-down: $\nu_\uparrow + \nu_\downarrow = \nu$. For $n > 1$ the Hubbard Hamiltonian for elecrons is transformed to a Hamiltonian for holes having a hole density $0 < \nu < 1$. The theory maps onto the problem of having the electron band less than half-full. Figure 13.16 shows the density of states of the two Hubbard bands. The one on the left assumes that the particles move such that there are no doubly occupied sites, so there are no electrons with energies $\sim U$. Once this band is filled, all other electrons are on doubly occupied sites and have an energy U, which displaces the band upward. If $W > U$ these two band overlap, and there are no magnetic phases. However, when $U > W$ the two bands are separate. In that case they are called the *lower Hubbard band* and *upper Hubbard band*, respectively. The holes are in the upper Hubbard band.

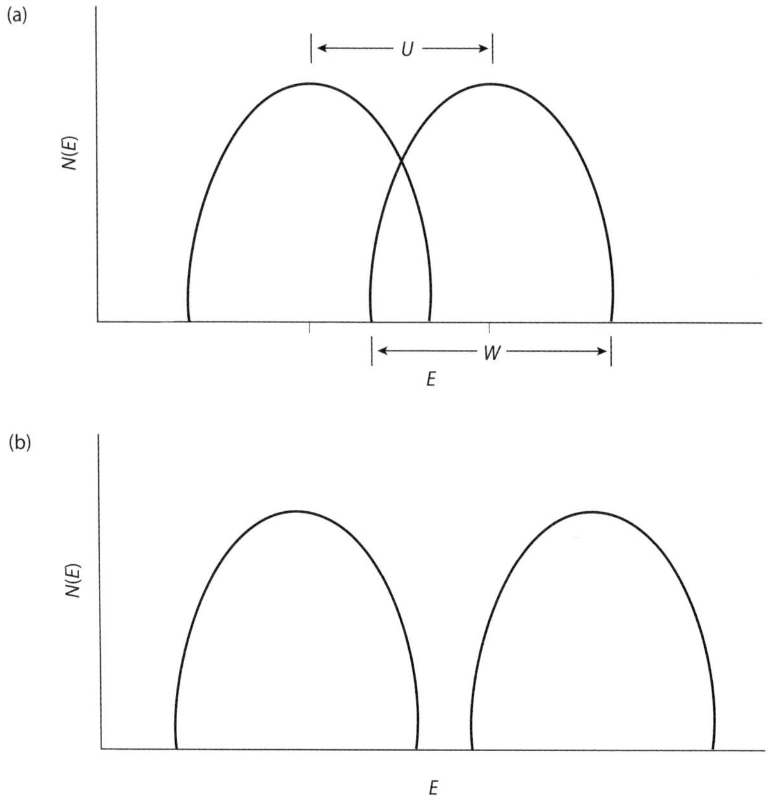

FIGURE 13.16. Density of states for two Hubbard bands. (a) $W > U$, (b) $W < U$.

It would be helpful to show a figure of the zero-temperature phase diagram of the two-dimensional Hubbard model. This plot has the horizontal axis as $0 > n > 2$ and the vertical axis as U/W. Unfortunately, no one can agree on the shape of this graph. Numerical solutions for the square lattice, such as those by Hirsch [5], find no phase boundaries at all.

References

1. V. L. Berezinskii, Destruction of long-range order in one-dimensional and two-dimensional systems having a continuous symmetry group, I: classical systems; II: quantum systems. *Sov. Phys. JETP* **32**, 493 (1971); **34**, 610 (1972)
2. M. D. Daybell and W. A. Steyert, Localized magnetic impurity states in metals. *Rev. Mod. Phys.* **40**, 380–389 (1968)
3. M. C. Gutzwiller, Effect of correlation on the ferromagnetism of transition metals. *Phys. Rev. Lett.* **10**, 159–162 (1963)
4. A. J. Heeger, Localized moments and nonmoments in metals: the Kondo effect. In *Solid State Physics*, Vol. **23**, F. Seitz, D. Turnbull, and H. Ehrenreich, eds. (Academic Press, New York, 1969), pp. 284–412
5. J. E. Hirsch, Two dimensional Hubbard model–numerical simulation study. *Phys. Rev. B* **31**, 4403–4419 (1985)

6. J. Hubbard, Electron correlations in narrow energy bands. *Proc. R. Soc. London Ser. A* **276**, 238–257 (1963)
7. J. Kondo, Theory of dilute magnetic alloys. In *Solid State Physics*, Vol. **23**, F. Seitz, D. Turnbull, and H. Ehrenreich, eds. (Academic Press, New York, 1969), pp. 184–283
8. J. M. Kosterlitz and D. J. Thouless, Ordering, metastability, and phase transitions in two-dimensional systems. *J. Phys. C* **6**, 1181–1203 (1973)
9. G. D. Mahan, *Many-Particle Physics*, 3rd ed. (Plenum/Kluwer, New York, 2000)
10. N. D. Mermin and H. Wagner, Absence of ferromagnetism or antiferromagnetism in one-or two-dimensional isotropic Heisenberg models. *Phys. Rev. Lett.* **17**, 1133–1136 (1966)
11. P. Minnhagen, Two dimensional Coulomb gas, vortex unbinding, and superfluid-superconducting films. *Rev. Mod. Phys.* **59**, 1001–1066 (1987)
12. P. Monod, The Kondo resistivity minimum in CuFe and CuMn alloys. *Phys. Rev. Lett.* **19**, 1113–1117 (1967)
13. L. Onsager, Crystal statistics: a two dimensional model with an order–disorder transition. *Phys. Rev.* **65**, 117–149 (1944)
14. K. R. Subbaswamy and G. D. Mahan, Renormalization group results for lattice-gas phase boundaries in two dimensions. *Phys. Rev. Lett.* **37**, 642–644 (1976)
15. B. B. Triplett and N. E. Phillips, Calorimetric evidence for a singlet ground state in *CuCr and CuFe*. *Phys. Rev. Lett.* **27**, 1001–1004 (1971)

Homework

1. Use Hund's rule to deduce the order of atomic filling for *d*-orbitals ($\ell = 2$).

2. Derive the mean field theory for the 3D Ising model of antiferromagnetism on a bipartite lattice. Find the transition temperature and the susceptibility. There are two sublattices: (i) up spins are on A, and (ii) down spins are on B. The neighbors of A are all B, and vice versa.

3. Onsager [13] first solved analytically the partition function of the Ising model in two dimensions on the square lattice. He showed that the specific heat went as $\ln(|t|)$, which has a critical exponent of $\alpha = 0$. The spontaneous magnetization for $T < T_c$ is given by the simple function

$$M(T) = (1 - k^2)^{1/8}, \quad k = \frac{1}{\sinh^2(2K)}, \quad K = \frac{J}{k_B T} \tag{13.207}$$

Show this implies that the critical temperature is $\sinh(2K_c) = 1$. Find the critical exponent β.

4. For the Ising model on the Bethe lattice, derive the critical exponent β. Let $t = (T_c - T)/T_c \ll 1$ and show that the internal field $h^* = D\sqrt{t}$ and find D. Assume $h = 0$.

5. For the Ising model on the Bethe lattice, derive the critical exponent δ.

6. Solve the Landau theory of phase transition when the external magnetic field H is nonzero in the vicinity of $T \approx T_c$. Does the specific heat have a discontinuity in this case?

7. Derive the partition function for the Ising model for three spins on an equilateral triangle

$$Z = \mathrm{Tr}\{e^{-\beta H}\} \tag{13.208}$$

$$\beta H = K(\sigma_1\sigma_2 + \sigma_2\sigma_3 + \sigma_3\sigma_1) + h(\sigma_1 + \sigma_2 + \sigma_3) \tag{13.209}$$

Derive an expression for the average spin

$$\langle\sigma\rangle = \frac{1}{3}\langle\sigma_1 + \sigma_2 + \sigma_3\rangle = -\frac{1}{3}\frac{\partial}{\partial h}\ln(Z) \tag{13.210}$$

and graph it as a function of h for small temperatures ($K, h \gg 1$). The coupling is antiferromagnetic.

8. Derive the partition function for an antiferromagnetic chain of spins that obey the Ising model. Do it (a) at zero magnetic field, and (b) nonzero magnetic field. (c) Find the average magnetization σ.

9. What is the partition function of the one-dimensional Ising model of spin 1? The operator $S_j^{(z)}$ has possible eigenvalues $(-1, 0, 1)$. There is no magnetic field.

10. Verify the integral in eqn. (13.139) for the energy of the spin vortex:

(a) show that the central spin has a zero interaction with the ring of first neighbors.

(b) For the sq lattice, show for spins that are n sites from the center, along either the x- or y-axis, the sum over neighbors $\Sigma\cos(\theta_{ij}) = 4 - (a/r)^2$.

11. The atom H^- exists as two electrons bound to a proton. The ionization energy of the first electron is 0.75 eV, and that of the second electron is E_{Ry}. What is the numerical value of U for the hydrogen atom?

12. Show that the Hubbard model has particle–hole symmetry. If $d_{j\sigma}^{\dagger}$ creates a hole, then $d_{j\sigma}^{\dagger} = C_{j\bar{\sigma}}$. What is the relationship between the chemical potential for the electrons and holes?

13. Solve the Hubbard model for a system of two sites. Find all of the eigenvalues when there are (a) one electron on the two sites, or (b) two electrons on the two sites.

14 Superconductivity

14.1 Discovery of Superconductivity

14.1.1 *Zero Resistance*

Superconductivity was discovered in the laboratory of Kamerlingh Onnes of Leiden in 1911 [21]. He was the first to liquefy helium, and the first person to reach the temperature of a few kelvins. His group discovered that the resistivity $\rho(T)$ vanished at a temperature T_c for common metals such as mercury, tin, and lead. T_c is different for each metal, but all have values that are a few kelvins. The electrical conductivity σ is the inverse of ρ. When ρ goes to zero, the conductivity goes to infinity. The metal becomes a perfect conductor; i.e., a "super"-conductor. In some cases, a current of electrons, once started in a superconductor, is found not to decay by resistive losses over a time period on the order of the age of the earth!

Another classical feature of the superconducting state is the measurement of the heat capacity as a function of temperature, as shown in fig. 14.1 for aluminum. In the normal state of a metal, at low temperature the specific heat has two contributions:

- The acoustic phonon term gives $C = gT^3$
- Electrons on the Fermi surface give $C = \gamma T$

The experimental data are often plotted as C/T vs. T^2:

$$\frac{C}{T} = \gamma + gT^2 \tag{14.1}$$

That is not done in fig. 14.1, since the phonon term hardly shows up over the small range of temperature. The dashed line is the heat capacity in the normal state of the metal. The

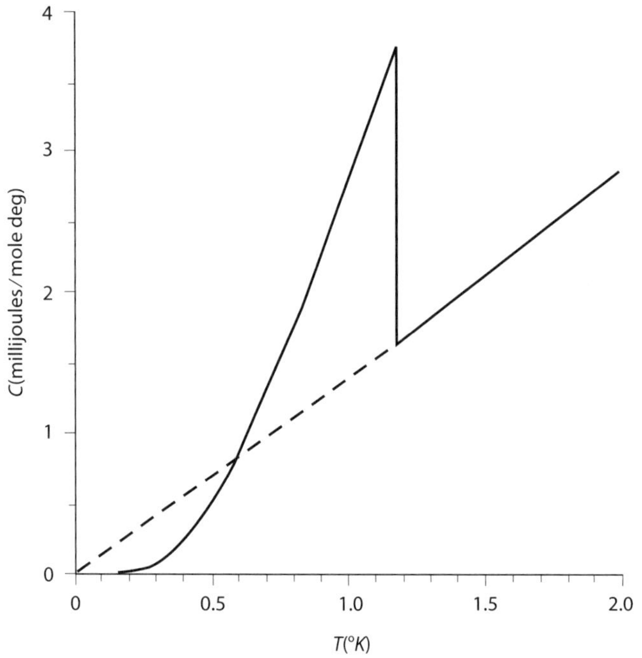

FIGURE 14.1. Heat capacity of superconducting aluminum as a function of temperature. Dashed line is for normal state. T_c is about 1 K. Adapted from Phillips, *Phys. Rev.* **114**, 676 (1959). Used with permission of the American Physical Society.

normal state is achieved by the application of a small magnetic field (300 G). In the superonducting state (no magnetic field), the heat capacity has a discontinuous jump at T_c, and then it smoothly goes to zero at zero temperature. This behavior is predicted by the Landau theory of phase transitions in chapter 13. The Landau mean field theory actually gives a good description of phase transitions in superconductors.

14.1.2 Meissner Effect

The next major discovery was by Meissner and Ochsenfeld in 1933, and is now called the *Meissner effect* [19]. They applied a static magnetic field H to a superconductor. The magnetic field inside is $B = \mu_0(H + M)$, where M is the magnetization. Experimentally they found that $B = 0$, and the magnetic flux is expelled. Another way to state the result is $M = -H$. The susceptibility is $\chi = M/H = -1$. A superconductor is a perfect diamagnet. Even today, if a new material is discovered to have zero resistivity at low temperature, a measurement of the Meissner effect is required to show that it is indeed a superconductor. There are other reasons why the resistivity may get small, but perfect diamagnetism is found only in superconductors.

The exclusion of magnetic flux is an important feature of the superconducting state. This experiment can be done several different ways, as shown in fig. 14.2.

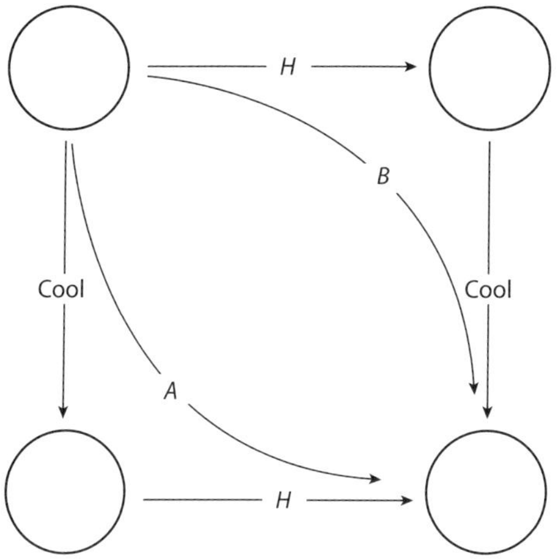

FIGURE 14.2. Two paths to go from a normal metal with no magnetic field to a superconducting metal in a magnetic field: (a) path A: one cools first, then applies the field; (b) path B: one applies the field first, then cools.

1. *Path A*: First cool the sample to the superconducting state. Second, apply the magnetic field. Imposing a magnetic field on any conductor generates induction currents. According to Lenz's law, the currents act to generate a magnetic field that opposes the one imposed from the outside. In a perfect conductor, these currents will exist for all time and explain the perfect diamagnetism. In doing the experiment this way, it is not obvious that the exclusion of magnetic flux is a feature of the state, or just a result of infinite conductivity so the eddy currents never die out.

2. *Path B*: This issue is clarified by doing the experiment a second way. First impose the external magnetic field while the sample is in the high-temperature normal state. The currents generated by the time-varying field will decay away to zero due to resistive losses. Then the magnetic field penetrates throughout the sample. Secondly, cool it into the superconducting state. It is found that the magnetic flux is excluded anyway. This way of doing the experiment makes it clear that flux exclusion is an important feature of the superconducting state.

14.1.3 Three Eras of Superconductivity

The history of experiments in superconductivity can be divided into three historical eras.

The Elemental Metals

The first known superconductors were the simple metals. They generally have a low transition temperature, as shown in table 14.1. A theory of superconductivity was proposed by Bardeen, Cooper, and Schrieffer (BCS) [5], which is described below. A refinement was proposed by Eliashberg [10]. This combined theory shows that the transition temperature T_c depends on only two material parameters: the characteristic phonon energy, and the electron–phonon coupling constant λ. Table 14.1 also lists the Debye energy and λ. The theory suggests a formula for the transition temperature of the form

Table 14.1. Second column is superconducting transition temperature; third column is Debye temperature; last column is the dimensionless electron–phonon coupling constant

Metal	T_c (K)	Θ_D (K)	λ
Mg	0.0	400	0.35
Zn	0.91	327	0.37
Cd	0.56	209	0.40
Al	1.2	428	0.43
Ga	1.1	320	0.97
In	3.37	108	0.8
Tl	2.38	78.5	0.8
Sn	3.73	200	0.72
Hg	4.16	71.9	1.6
Pb	7.22	105	1.55

$$T_c = \Theta_D f(\lambda) \tag{14.2}$$

Figure 14.3 gives a graph of the above function using experimental data from table 14.1. The vertical axis is $(100T_c/\Theta_D)$. The dashed line is a theoretical curve, which is derived using the technique of Allen and Mitrovic [3]. It can be derived from two limiting behaviors:

$$T_c = \Theta_D \begin{cases} \alpha_1 e^{-1/\lambda} & \lambda < \lambda_0 \\ \alpha_2 \sqrt{\lambda - \lambda_0} & \lambda > \lambda_0 \end{cases} \tag{14.3}$$

where α_1, α_2, and λ_0 are parameters. The top formula (small λ) was derived by McMillan [18], while the lower formula (large λ) is derived by Allen and Mitrovic [3]. The above curve was drawn with $\alpha_2 = 6.2$, $\lambda_0 = 0.54$ by fitting to the top four data points. The only outlier is gallium, which has a relatively high value of λ, but a low value of T_c. Also, the theory predicts Mg is a superconductor with a small T_c, but it is not superconducting. Since most simple metals lie close to this theoretical curve, the properties of these superconductors are understood.

A-15 Compounds

The first era of *high-temperature superconductivity* was the discovery of compounds with a crystal structure called A-15. These materials have transition temperatures over 10 K, and in some cases over 20 K, as shown in table 14.2. Niobium (Nb) has the highest value ($T_c = 9$ K) of any element. An early discovery was that NbN, with the NaCl structure, had $T_c = 15$ K. This ignited a search for other superconducting compounds of niobium.

The A-15 crystal structure is cubic. For A_3B, the B atoms are in a bcc structure. The A atoms are found in pairs on the faces of the unit cube, as shown in fig. 14.4a. On one pair of opposing faces, the pair are aligned along the x-axis. On another pair of faces the

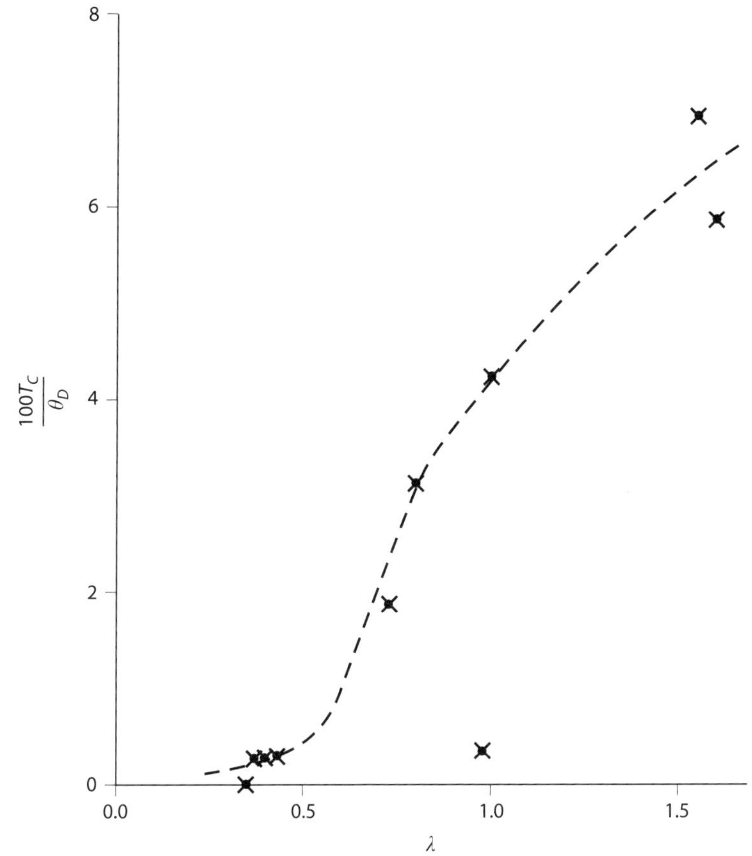

FIGURE 14.3. The transition temperature of superconductors as a function of λ. Vertical axis is $(100\ T_c/\Theta_D)$. Points are experimental. Dashed line is theory.

alignment is along the y-axis, and on the third pair of faces they are along the z-axis. The A atoms form chains in the material. The A atoms are always one of the transition metal sequence, while the B atoms can be many different elements.

In spite of this simple crystal structure, the mechanism of superconductivity in these materials is not understood. The electron–phonon coupling λ is too small to give these

Table 14.2 Superconducting transition temperatures of some A-15 compounds

	T_c		T_c
Nb_3Ga	20	V_3Al	13
Nb_3Ge	22	V_3Ga	15
Nb_3Si	17	V_3Si	17
Nb_3Sn	18	V_3Sn	5

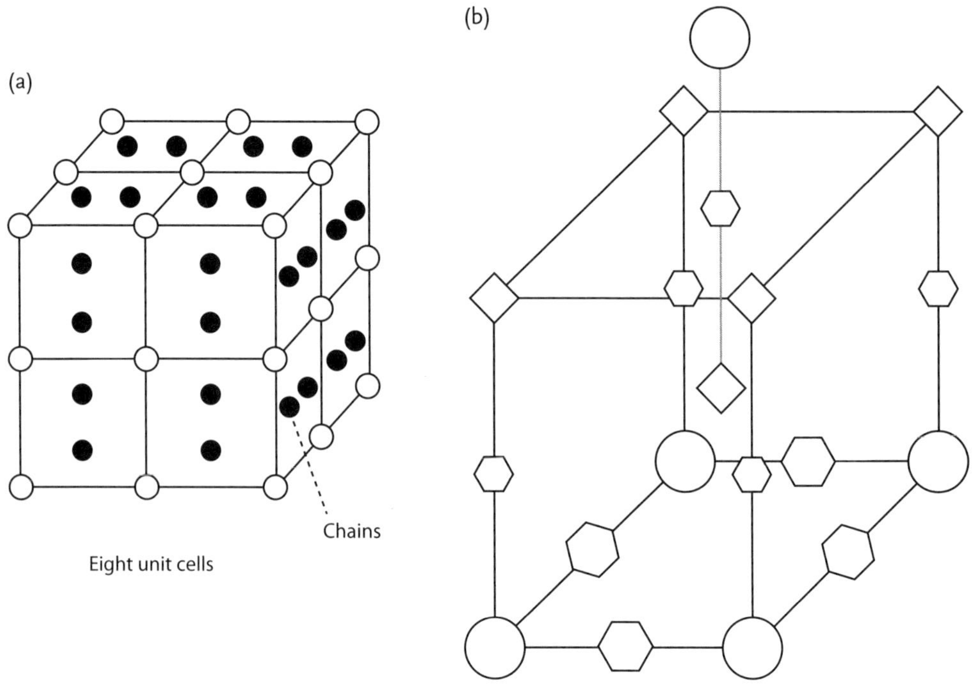

FIGURE 14.4. (a) The crystal structure of A-15 compounds, showing the chains of atoms along the faces of the cube. (b) Crystal structure of La_2CuO_4. Circles are copper, diamonds are lanthanum, large hexagons are oxygens between copper atoms in the plane, and smaller hexagons are oxygen atoms on other planes.

high temperatures. The electrons are paired by an attractive force, which must be provided by an intermediate boson. If not a phonon, other possible bosons are plasmons, spin waves, or other types of spin fluctuations. None of these other proposed bosons was found to explain the pairing. The mechanism is still unknown. There is no theory for these materials that explains the high value of T_c. Work on these materials more or less ceased in 1986, when the high-temperature cuprate superconductors were discovered.

Cuprate Superconductors

The first cuprate superconductor was discovered by Bednorz and Müller in 1986 [6]. They were investigating La_2CuO_4. The materials consist of layers of CuO_2, with La and O ions between the layers, as shown in fig. 14.4b. The layers are conducting, and become superconducting. The copper atoms are in a two-dimensional square lattice, and the oxygen atoms bridge each pair of first neighbor coppers. In the pure state, the copper atom is doubly ionized, and Cu^{2+} has a *d*-hole. The holes antiferromagnetically align since the sq lattice is bipartite. In the perfect crystal, the material is an insulator. When the material is doped with strontium to change the carrier concentration, the magnetic state is destroyed, and a superconductor is found at $T_c = 38$ K. Later, other cuprate superconductors were discovered with much higher temperatures. Table 14.3 gives a few examples of the many such compounds that have been found. These high temperatures must be due to a strong

Table 14.3 Some cuprate superconductors and their transition temperatures

Crystal	T_c (K)
$La_{2-x}Sr_xCuO_4$	38
$Y Ba_2Cu_3O_7$	92
$Tl_2Ba_2Ca_2Cu_3O_{10}$	125
$Bi_2Sr_2Ca_2Cu_3O_{10}$	110
$TlBa_2Ca_3Cu_4O_{11}$	122

attractive force between the conduction electrons. As in the A-15 compounds, this force is still unknown for the cuprate superconductors.

Recently other exotic materials have been found to be superconducting at high temperatures. The compound SmOFeAs, doped with F, has been found to be a superconductor at $T_c = 55$ K by Ren et al. [23]. Iron is usually magnetic, and magnetism and superconductivity rarely coexist. It is surprising that a material containing iron is a high-temperature superconductor.

14.2 Theories of Superconductivity

This section describes three theories of superconductivity. All are phenomenological and all are considered valid in their own way. The BCS theory is described later. It is a microscopic theory, which tends to validate these earlier phenomenological theories.

14.2.1 *London Equation*

The London brothers published a theory of the Meissner effect in 1935 [16]. They assumed that n_s is the density of "superconducting electrons," and $\mathbf{v}_s$ is their average velocity. The velocity is accelerated by an electric field, and the current is due to these electrons:

$$m\frac{d\mathbf{v}_s}{dt} = e\mathbf{E} \tag{14.4}$$

$$\mathbf{j} = en_s\mathbf{v}_s \tag{14.5}$$

The density n_s is expected to be zero at T_c and to increase as the temperature is lowered toward absolute zero. The superconducting electrons are the Cooper pairs. Today n_s is interpreted as the density of Cooper pairs, e as twice the electron charge, and m as twice the electron mass.

Combine these two equations:

$$\frac{d\mathbf{j}}{dt} = \frac{n_s e^2}{m}\mathbf{E} \tag{14.6}$$

The electric field is given by a vector and scalar potential. Selecting a gauge that has no scalar potential gives $\mathbf{E} = -\dot{\mathbf{A}}$:

$$\frac{d\mathbf{j}}{dt} = -\frac{n_s e^2}{m}\frac{d\mathbf{A}}{dt} \tag{14.7}$$

Cancel the time derivative from both sides, and derive a dc equation:

$$\mathbf{j} = -\frac{n_s e^2}{m}\mathbf{A} \tag{14.8}$$

which is London's equation. This equation is sufficient to describe the Meissner effect.

In the absence of an electric field, one of Maxwell's equations is

$$\nabla \times \mathbf{B} = \mu_0 \mathbf{j} = \nabla \times (\nabla \times \mathbf{A}) = \nabla(\nabla \cdot \mathbf{A}) - \nabla^2 \mathbf{A} \tag{14.9}$$

Using the Coulomb gauge $\nabla \cdot \mathbf{A} = 0$ derives another equation relating $\mathbf{j}$ and $\mathbf{A}$. Combining them gives

$$\frac{\mu_0 n_s e^2}{m}\mathbf{j} = \nabla^2 \mathbf{j} = \frac{n_s e^2}{mc^2 \varepsilon_0}\mathbf{j} \tag{14.10}$$

Figure 14.5 shows the geometry of the surface region. The external magnetic field is in the $\hat{z}$ direction and varies as a function of x: $\mathbf{B} = \hat{z}B(x)$. The current and vector potential are in the $\hat{y}$ direction: $\mathbf{j} = \hat{y}j(x)$. The above differential equation has the solution

$$\mathbf{j} = \hat{y}j_0 e^{-x/\Lambda}, \quad \Lambda^2 = \frac{mc^2\varepsilon_0}{n_s e^2} \approx \left(\frac{c}{\omega_{sp}}\right)^2 \tag{14.11}$$

where ω_{sp} is the plasma frequency of the superconducting electrons. At zero temperature, $n_s = n_e/2$ is half of the density of all conduction-band electrons, $e \rightarrow 2e$, $m \rightarrow 2m$, and $\omega_{sp} = \omega_p$ equals the plasma frequency of the metal in the normal state. The quantity Λ is called the penetration depth, and is an important length scale in the superconductor. It is the distance over which the magnetic field penetrates into the superconductor. At T_c, $n_s = 0$ and Λ is infinite. Near zero temperature it is $\Lambda = c/\omega_p$, which is the value quoted for a superconductor.

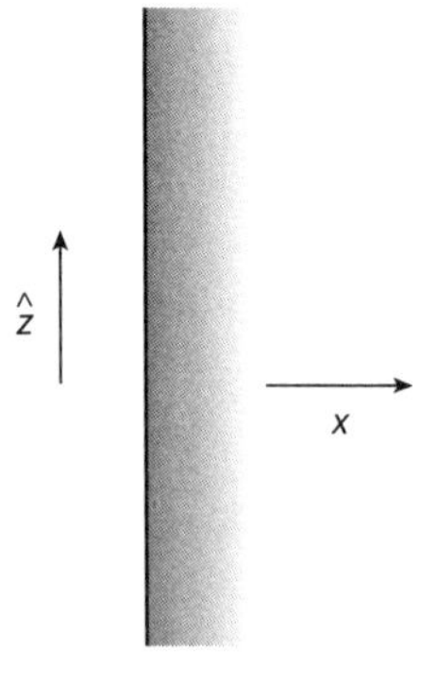

FIGURE 14.5. Surface of a superconductor. Magnetic field is in the $\hat{z}$ direction, and currents are in the $\hat{y}$ direction.

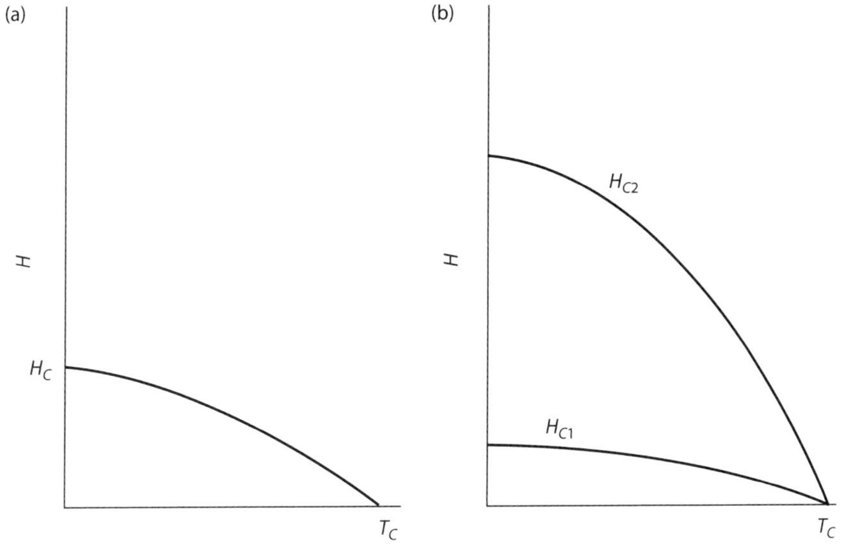

FIGURE 14.6. Magnetic field dependence of a superconductor. (a) Type I superconductor with a single phase boundary. (b) Type II superconductors have two phase boundaries, denoted H_{c1} and H_{c2}.

The picture of the Meissner effect is that actual currents swirl around the surface of the superconductor. These permanent eddy currents create a magnetic field that cancels the external magnetic field and causes the net magnetic field inside to become zero. These currents do not dissipate in time and are part of the ground state of the superconductor in a magnetic field.

Figure 14.6 shows the magnetic field dependence of the superconducting state. Figure 14.6a shows a single curve $H_c(T)$. It is the critical magnetic field as a function of temperature. For $H < H_c(T)$, all magnetic flux is excluded. For $H > H_c(T)$, the magnetic field destroys the superconducting state: the metal returns to the normal state with magnetic flux penetrating throughout the volume.

14.2.2 *Ginzburg-Landau Theory*

The Ginzburg-Landau theory [15] of superconductivity is another phenomenological theory, along the lines of London theory. The G-L theory is an extension of the Landau theory of phase transitions that was presented in the chapter on magnetism. Another term was added to Landau theory to allow the order parameter to vary in space.

Coherent states are rather unique. They are found in superfluids. An example is the condensate in a Bose-Einstein superfluid. All of the particles in the condensate have a common phase factor, which extends throughout the superfluid.

Ginzburg-Landau theory introduced a *superconducting wave function* $\psi(\mathbf{r})$. It is not the wave function of individual electrons, but rather the wave function that defines the coherent state that makes the collection of electrons into a superfluid. In G-L theory the free energy density of the superconductor is

$$\mathcal{F} = \alpha|\psi|^2 + \frac{\beta}{2}|\psi|^4 + \frac{B^2}{2\mu_0} + \frac{1}{2m}|(\mathbf{p} - q\mathbf{A})\psi|^2 \tag{14.12}$$

$\mathcal{F}$ has the units of energy per unit volume. $|\psi|^2$ has the units of inverse volume. Then α has the units of energy, and β has the units of energy times volume.

G-L took the charge of the system to be the electron charge $q = -e$, where $e > 0$. However, it is now known that in the superfluid state the electrons are paired, so that $q = -2e$, $m = 2m_e$. The constants α and β depend only on temperature. As in Landau theory, assume $\beta > 0$, and α changes sign at the superconducting transition temperature: $\alpha \propto (T - T_c)$. Again minimize the free energy. Since ψ is complex, we can separately minimize with respect to its real and imaginary parts. Instead, treat ψ and ψ^* as being independent. Then

$$0 = \frac{\delta \mathcal{F}}{\delta \psi^*} = \left[\alpha + \beta|\psi|^2 + \frac{1}{2m}(\mathbf{p} - q\mathbf{A})^2\right]\psi \tag{14.13}$$

For $T > T_c$, when α is positive, the only solution is $\psi = 0$. For $T < T_c$, when α is negative, there is a solution when $\mathbf{A} = 0$, $\nabla^2\psi < 0$ of

$$|\psi|^2 = -\frac{\alpha}{\beta} \tag{14.14}$$

This solution applies whenever $B = 0$ and the order parameter is a constant.

The factor

$$|\psi(\mathbf{r})|^2 = n_s(\mathbf{r}) = -\frac{\alpha}{\beta} \tag{14.15}$$

is the density of superconducting electrons. It is the density of superconducting electron pairs. At zero temperature it becomes half the density of all conduction electrons. However, at nonzero temperatures $n_s(\mathbf{r})$ is less than the density of all conduction electrons. Allow this density to vary with position. The most general form for this wave function has an amplitude and a phase

$$\psi(\mathbf{r}) = \sqrt{n_s(\mathbf{r})}\, e^{i\phi(\mathbf{r})} \tag{14.16}$$

Two equations of motion are obtained by minimizing the free energy with respect to several variables. Consider explicitly the case of the superconductor in a constant magnetic field $\mathbf{B} = \nabla \times \mathbf{A}$.

1. A variation with respect to the vector potential gives a Maxwell equation:

$$0 = \frac{\delta \mathcal{F}}{\delta A_\mu} \tag{14.17}$$

$$0 = \frac{1}{\mu_0}\nabla \times \mathbf{B} - \mathbf{j} \tag{14.18}$$

$$\nabla \times \mathbf{B} = \mu_0 \mathbf{j} = \mu_0 \left[\frac{\hbar e}{2mi} [\psi^\dagger \nabla \psi - \psi \nabla \psi^\dagger] + \frac{(2e)^2}{2m} \mathbf{A} |\psi|^2 \right] \tag{14.19}$$

$$\mathbf{j}(\mathbf{r}) = n_s(\mathbf{r}) \left[\frac{e\hbar}{m} \nabla \phi(\mathbf{r}) + \frac{(2e)^2}{2m} \mathbf{A}(\mathbf{r}) \right] = 2en_s \mathbf{v}_s \tag{14.20}$$

$$\mathbf{v}_s = \frac{\hbar}{2m} \nabla \phi(\mathbf{r}) + \frac{e}{m} \mathbf{A}(\mathbf{r}) \tag{14.21}$$

Equation (14.16) is used for $\psi(\mathbf{r})$. If the first term on the right in eqn. (14.20) is zero ($\nabla \phi = 0$), then the above expression is London's equation. For many years it was a mystery why the first term was zero, since London's equation clearly explained the Meissner effect. That $\nabla \phi = 0$ was shown by BCS theory. The last equation has the formula for the superconducting velocity $\mathbf{v}_s$.

2. A variation of the free energy with respect to ψ^* was derived in eqn. (14.13). Examine the solution of this equation for the case of no magnetic field ($\mathbf{A} = 0$). Divide the entire equation by $|\alpha|$ and obtain a new length scale ξ:

$$\xi^2 = \frac{\hbar^2}{2m|\alpha|} \tag{14.22}$$

$$0 = \left\{ \frac{1}{ns} |\psi(\mathbf{r})|^2 - 1 - \xi^2 \nabla^2 \right\} \psi(\mathbf{r}) \tag{14.23}$$

The quantity ξ is called the *coherence length.*

As an example of the coherence length, assume that the superconducting wave function vanishes at the surface of the superconductor. How does it rise up to its full value inside the interior? If the interior direction is x, the equation is

$$0 = \left\{ \frac{1}{n_s} |\psi(x)|^2 - 1 - \xi^2 \frac{d^2}{dx^2} \right\} \psi(x) \tag{14.24}$$

The solution to this equation, with the boundary condition $\psi(x = 0) = 0$, is

$$\psi(x) = \sqrt{n_s} \tanh\left(\frac{x}{\sqrt{2}\, \xi} \right) \tag{14.25}$$

The coherence length is the distance over which the superconducting wave function rises to its bulk value.

Ginzburg-Landau assumed that the wave function did vanish at the surface of the superconductor. Experiments show it does not vanish. However, the coherence length remains an important parameter in superconductors.

Superconductors have two length scales: Λ and ξ. Their ratio is a dimensionless constant:

$$\kappa = \frac{\Lambda}{\xi} \tag{14.26}$$

Use the definitions of Λ and ξ to find κ^2:

$$\kappa^2 = \frac{mc^2}{n_s(2e)^2}\frac{2m|\alpha|}{\hbar^2} \tag{14.27}$$

Remember that $|\alpha|/n_s = \beta$, and the definition of the Bohr magneton $\mu_B = e\hbar/2m_e$, to get

$$\kappa^2 = \frac{\beta}{2\mu_0\mu_B^2} \tag{14.28}$$

This ratio plays an important role in the next subsection.

14.2.3 Type II

The coherence length ξ was introduced in the prior section. The Bardeen-Schrieffer-Cooper (BCS) theory of superconductivity is described below [5]. An important feature is that electrons near in energy to the chemical potential pair up to form bound states. The coherence length is the average spatial separation of two electrons in a pair. A formula for ξ is derived by using the uncertainty principle and dimensional analysis:

$$\Delta t \Delta E \geq \hbar \tag{14.29}$$

For ΔE use the energy gap of the superconductor, which is called Δ. For the time, use the period of a bound-state orbit: ξ is the diameter of the orbit, so the period is $t = \pi\xi/v_F$, where v_F is the Fermi velocity. The final expression is

$$\xi = \frac{\hbar v_F}{\pi\Delta} \tag{14.30}$$

Equate this result to the formula in the above section, which derives an expression for the Ginzburg-Landau parameter α:

$$\xi^2 = \left(\frac{\hbar v_F}{\pi\Delta}\right)^2 = \frac{\hbar^2}{2m|\alpha|} \tag{14.31}$$

$$\alpha = \frac{\pi^2\Delta^2}{4E_F} \propto (T_c - T) \tag{14.32}$$

where $E_F = mv_F^2/2$ is the Fermi energy. The superconducting energy gap goes as $\Delta \propto \sqrt{T_c - T}$.

The first superconductors to be discovered were the metallic elements, such as mercury, tin, and lead. They all have a small value of gap Δ. The coherence length was much larger than the penetration depth and $\kappa < 1$.

In 1957 Abrikosov published an important paper on superconductivity [1]. He noted that the gap Δ is proportional to T_c. As new materials are discovered with higher values of T_c, their coherence length will be shorter. It might happen that the coherence length would be shorter than the penetration depth ($\kappa > 1$). In that case, he suggested that the superconductor would have a different form of the Meissner effect. He proposed the phase diagram shown in fig. 14.6b, where there are two phase lines called $H_{c1}(T)$ and

$H_{c2}(T)$. At lower magnetic fields ($H < H_{c1}(T)$), one finds the usual Meissner effect with total exclusion of magnetic flux. In the intermediate regime ($H_{c1} < H < H_{c2}$) some magnetic flux penetrates through the superconductor. For higher fields ($H > H_{c2}$) the metal is in the normal state. The flux penetration in the intermediate phase is through the creation of microscopic current loops. These loops make small magnetic flux lines that are called vortices. Each vortex has one quantum of magnetic flux. The current flow is cut off at at small distances r_0, which has atomic dimensions. In an actual material, there are many such vortices, which arrange themselves into a pt lattice.

In layered superconductdors, such as the cuprates, the curve for H_{c1} differs by a factor of three, depending on whether the magnetic field is parallel or perpendicular to the plane.

Abrikosov called his proposed new phase *type II superconductors*. The old ones, with a simple Meissner effect, he called *type I*. The distinction between these phases is

- Superconductors with $\kappa < 1/\sqrt{2}$ are type I
- Superconductors with $\kappa > 1/\sqrt{2}$ are type II

When Abrikosov made his proposal, most known superconductors were type I. However, since 1957 many more materials have been found to be superconducting, and the vast majority are type II. The high-temperature superconductors, with T_c on the order of 100 K, have a very short coherence length of a few nanometers. They are very much type II. The study of their vortex lattice is a major area of research.

14.3 BCS Theory

14.3.1 *History of Theory*

Bardeen, Cooper, and Schrieffer (BCS) published a theory of superconductivity in 1957 [5]. It successfully explains superconductivity in simple metals. They proposed that the electrons formed bound pairs, and the attractive interaction was provided by the electron–phonon interaction. The bound pairs are part of a many-electron eigenfunction that they constructed. Later it was realized that their theory applied only to metals in which the electron–phonon interaction was relative weak, such as aluminum. A subsequent theory by Eliashberg applied to other metals, and is called *Strong coupling theory* [10]. Eliashberg theory explained superconductivity in type I metals, such as lead and tin.

The BCS and Eliashberg theories are able to explain only some features of high-temperature superconductors. They have electron pairing and energy gaps. However, the force that binds the electrons into pairs is believed not to be due to phonon exchange. The actual force is unknown, but likely involves spin fluctuations.

The isotope effect was discovered in 1950. Many atoms have different isotopes. Making a metal of a pure isotope gives one value of T_c, but another isotope of the same metal gives a slightly different value of T_c. Metals of different isotopes of the same atomic number have the same lattice constant and identical electronic properties. The small change in ion mass does change the phonons. The isotope effect defines a dimensionless parameter:

$$\alpha = \left(\frac{\Delta T_c}{T_c}\right) \Big/ \left(\frac{\Delta M}{M}\right) \tag{14.33}$$

where ΔM is the change in isotope mass, and ΔT_c is the change in transition temperature. The original experiments on mercury gave $\alpha = -0.5$. Fröhlich first noted [11] that electrons could exchange phonons and this interaction could be attractive. The isotope effect is simply explained by assuming that the transition temperature is proportional to the Debye frequency:

$$k_B T_c \propto \hbar\omega_D \sim \hbar\sqrt{\frac{K}{M}} \tag{14.34}$$

Fröhlich suggested that superconductivity was caused by the electron–phonon interaction, but did not construct a suitable superconducting wave function. Before the BCS theory, it was generally believed that electrons did pair up. The pairs obey boson statistics, and could undergo Bose-Einstein condensation. The achievement of BCS was to construct the actual theory with a many-pair wave function.

14.3.2 *Effective Hamiltonian*

BCS constructed a many-particle wave function for the superconducting state. Two electrons by themselves would form a bound state similar to positronium if they had an attractive interaction. However, a system of many such pairs is a many-electron state, which has to obey Fermi statistics and be antisymmetric under the exchange of any two electrons. Obviously, such a state is far more complicated than a positronium eigenfunction.

The interaction between two electrons by exchanging a phonon has the form

$$V_{ep}(\mathbf{q}, \omega) = \sum_{\eta} |M_\eta(\mathbf{q})|^2 \frac{2\omega_\eta(\mathbf{q})}{\omega^2 - \omega_\eta(\mathbf{q})^2} \tag{14.35}$$

where η denotes the polarization of the phonon. The interaction is attractive at small frequencies. BCS simplified the problem by treating it as a constant $(-V_0)$ for frequencies less than the Debye energy:

$$V_{ep}(\mathbf{q}, \omega) = \begin{cases} -V_0 & \omega < \omega_D \\ 0 & \omega > \omega_D \end{cases} \tag{14.36}$$

Making these approximations is the high art of theoretical physics. In fact, the approximation is quite good. The BCS theory makes a number of predictions that agree with experiment. The Eliashberg strong coupling theory uses the full interaction $V_{ep}(\mathbf{q}, \omega)$. It is a more accurate theory, but is similar to BCS in many of its predictions.

BCS started with the Hamiltonian

$$H = \sum_{p\sigma} \xi(p) C^\dagger_{p\sigma} C_{p\sigma} - \frac{V_0}{2\Omega} \sum_{\mathbf{pkq}ss'} C^\dagger_{\mathbf{p+q},s} C^\dagger_{\mathbf{k-q},s'} C_{\mathbf{k},s'} C_{\mathbf{p}s} \tag{14.37}$$

It has the appearance of a standard electron–electron Hamiltonian, but with an attractive interaction.

14.3.3 Pairing States

There have been previous discussions of operators obtained by pairing a creation and a destruction operator:

- The number operator is

$$n_{\mathbf{p}\sigma} = C^{\dagger}_{\mathbf{p}\sigma} C_{\mathbf{p}\sigma} \tag{14.38}$$

$$N_e = \sum_{\mathbf{p}\sigma} n_{\mathbf{p}\sigma} \tag{14.39}$$

- A density fluctuation is

$$\rho_{\sigma}(\mathbf{q}) = \sum_{\mathbf{p}} C^{\dagger}_{\mathbf{p}+\mathbf{q},\sigma} C_{\mathbf{p}\sigma} \tag{14.40}$$

- A current fluctuation is

$$\mathbf{j}_{\sigma}(\mathbf{q}) = \frac{e\hbar}{m} \sum_{\mathbf{p}} \left(p + \frac{\mathbf{q}}{2}\right) C^{\dagger}_{\mathbf{p}+\mathbf{q},\sigma} C_{\mathbf{p}\sigma} \tag{14.41}$$

- A spin pairing operator is

$$m(\mathbf{q}) = \sum_{\mathbf{p}} C^{\dagger}_{\mathbf{p}+\mathbf{q},\uparrow} C_{\mathbf{p}\downarrow} \tag{14.42}$$

BCS proposed that the electrons were paired into a spin singlet with spin angular momentum $S=0$. Of all the many crystals that have been found to be superconducting, only one (Sr_2RuO_4) has been demonstrated as being in a spin triplet state—the rest are singlets. They also assumed that the total center-of-mass momentum of the pair would be zero. So an electron with $(\mathbf{p}\sigma)$ would pair with one on the other side of the Fermi sphere, with $(-\mathbf{p}, -\sigma)$. Therefore, they constructed the operators

$$\Delta_{\uparrow} = \frac{V_0}{\Omega} \sum_{\mathbf{p}} C_{-\mathbf{p},\downarrow} C_{\mathbf{p},\uparrow} \tag{14.43}$$

$$\Delta^{\dagger}_{\uparrow} = \frac{V_0}{\Omega} \sum_{\mathbf{p}} C^{\dagger}_{\mathbf{p},\uparrow} C^{\dagger}_{-\mathbf{p},\downarrow} \tag{14.44}$$

These operators have two creation operators, or two destruction operators. Since the electrons exist as pairs, $\Delta^{\dagger}_{\sigma}$ creates a pair state, while Δ_{σ} destroys it. Since fermion operators anticommute, note that

$$\begin{aligned}\Delta_{\uparrow} &= -\frac{V_0}{\Omega} \sum_{\mathbf{p}} C_{\mathbf{p},\uparrow} C_{-\mathbf{p},\downarrow} = -\frac{V_0}{\Omega} \sum_{\mathbf{p}'} C_{-\mathbf{p}',\uparrow} C_{\mathbf{p}',\downarrow} \\ &= -\Delta_{\downarrow}\end{aligned} \tag{14.45}$$

Rewrite the Hamiltonian while retaining only the pairing states in the interaction term, which is achieved by setting $\mathbf{k} = -\mathbf{p}$, $s' = -s$:

$$H = \sum_{\mathbf{p}\sigma} \xi(p) C^{\dagger}_{\mathbf{p}\sigma} C_{\mathbf{p}\sigma} - \frac{V_0}{2\Omega} \sum_{\mathbf{pkq}s} C^{\dagger}_{\mathbf{p}+\mathbf{q},s} C^{\dagger}_{-\mathbf{p}-\mathbf{q},-s} C_{-\mathbf{p},-s} C_{\mathbf{p}s} \tag{14.46}$$

$$\approx \sum_{p\sigma} \xi(p) C^{\dagger}_{p\sigma} C_{p\sigma} - \frac{1}{2} \sum_{ps} \left[\Delta_s C^{\dagger}_{p,s} C^{\dagger}_{-p,-s} + \Delta^{\dagger}_{-s} C_{-p,-s} C_{ps} \right]$$

We have made a Bogoliubov approximation that pairs of operators have been replaced by their average Δ_s. This replacement is standard in mean field approximations, which are used in theories of magnetism. BCS is a mean field theory of electron pairing in superconductors.

The next step is to make a Bogoliubov transformation to new operators $(\gamma^{\dagger}, \gamma)$:

$$C_{p\uparrow} = u_p \gamma_{p\uparrow} + v_p \gamma^{\dagger}_{-p,\downarrow} \tag{14.47}$$

$$C_{-p,\downarrow} = u_p \gamma_{-p,\downarrow} - v_p \gamma^{\dagger}_{p\uparrow} \tag{14.48}$$

$$C^{\dagger}_{p\uparrow} = u_p \gamma^{\dagger}_{p\uparrow} + v_p \gamma_{-p\downarrow} \tag{14.49}$$

$$C^{\dagger}_{-p,\downarrow} = u_p \gamma^{\dagger}_{-p,\downarrow} - v_p \gamma_{p\uparrow} \tag{14.50}$$

where $u_p = \cos(\theta)$, $v_p = \sin(\theta)$ are constants. Note that terms with down spin have a minus sign before v_p. This transformation maintains the anticommutation relations, such as

$$\{C_{p\sigma}, C^{\dagger}_{p'\sigma'}\} = u_p u_{p'} \{\gamma_{p\sigma}, \gamma^{\dagger}_{p'\sigma'}\} + v_p v_{p'} \{\gamma^{\dagger}_{-p,-\sigma}, \gamma_{-p',-\sigma'}\} \tag{14.51}$$

$$= \delta_{p=p'} \delta_{\sigma=\sigma'} \left[u_p^2 + v_p^2 \right] = \delta_{p=p'} \delta_{\sigma=\sigma'}$$

Apply this transformation to the Hamiltonian. For the term H_0

$$H_0 = \sum_{p\sigma} \xi(p)\, C^{\dagger}_{p\sigma} C_{p\sigma} = \sum_{p} \xi(p) \left[C^{\dagger}_{p\uparrow} C_{p\uparrow} + C^{\dagger}_{p,\downarrow} C_{p,\downarrow} \right] \tag{14.52}$$

$$= \sum_{p} \xi(p) \left[\left(u_p \gamma^{\dagger}_{p\uparrow} + v_p \gamma_{-p,\downarrow} \right) \left(u_p \gamma_{p\uparrow} + v_p \gamma^{\dagger}_{-p,\downarrow} \right) \right.$$

$$\left. + \left(u_p \gamma^{\dagger}_{p\downarrow} - v_p \gamma_{-p,\uparrow} \right) \left(u_p \gamma_{p\downarrow} - v_p \gamma^{\dagger}_{-p,\uparrow} \right) \right]$$

$$= \sum_{p} \xi(p) \left[\left(u_p^2 - v_p^2 \right) \left(\gamma^{\dagger}_{p\uparrow} \gamma_{p\uparrow} + \gamma^{\dagger}_{p\downarrow} \gamma_{p\downarrow} \right) \right.$$

$$\left. + 2 u_p v_p \left(\gamma^{\dagger}_{p\uparrow} \gamma^{\dagger}_{-p\downarrow} + \gamma_{-p\downarrow} \gamma_{p\uparrow} \right) \right]$$

Do the same for the interaction terms:

$$V = -\frac{1}{2} \sum_{p} \left[\Delta_{\uparrow} C^{\dagger}_{p\uparrow} C^{\dagger}_{-p\downarrow} + \Delta_{\downarrow} C^{\dagger}_{p,\downarrow} C^{\dagger}_{-p\uparrow} + \text{h.c.} \right] \tag{14.53}$$

Make the Bogoliubov transformation, recall that $\Delta_{\downarrow} = -\Delta_{\uparrow}$, and find

$$V = -\frac{\Delta_{\uparrow}}{2} \left[-4 u_p v_p \left(\gamma^{\dagger}_{p\uparrow} \gamma_{p\uparrow} + \gamma^{\dagger}_{p\downarrow} \gamma_{p\downarrow} \right) + 2 \left(u_p^2 - v_p^2 \right) \left(\gamma^{\dagger}_{p\uparrow} \gamma^{\dagger}_{-p\downarrow} + \gamma_{-p\downarrow} \gamma_{p\uparrow} \right) \right]$$

Combine $H = H_0 + V$. Note that $2u_p v_p = \sin(2\theta)$ and $u_p^2 - v_p^2 = \cos(2\theta)$. Choose θ to eliminate the terms with two raising or two lowering operators:

$$0 = \sum_{\mathbf{p}} (\gamma^{\dagger}_{\mathbf{p}\uparrow}\gamma^{\dagger}_{-\mathbf{p}\downarrow} + \gamma_{-\mathbf{p}\downarrow}\gamma_{\mathbf{p}\uparrow})][\xi(p)\sin(2\theta) - \Delta_{\uparrow}\cos(2\theta)] \tag{14.54}$$

$$\tan(2\theta) = \frac{\Delta_{\uparrow}}{\xi(p)} \tag{14.55}$$

$$\sin(2\theta) = \frac{\tan(2\theta)}{\sqrt{1+\tan^2(2\theta)}} = \frac{\Delta_{\uparrow}}{E(p)}, \quad E(p) = \sqrt{\xi(p)^2 + \Delta_{\uparrow}^2} \tag{14.56}$$

$$\cos(2\theta) = \frac{\xi(p)}{E(p)} \tag{14.57}$$

The remaining terms in the Hamiltonian are

$$H = (\gamma^{\dagger}_{p\uparrow}\gamma_{p\uparrow} + \gamma^{\dagger}_{p\downarrow}\gamma_{p\downarrow})[\xi(p)\cos(2\theta) + \Delta_{\uparrow}\sin(2\theta)] \tag{14.58}$$
$$= \sum_{\mathbf{p}} E(p)(\gamma^{\dagger}_{\mathbf{p}\uparrow}\gamma_{\mathbf{p}\uparrow} + \gamma^{\dagger}_{\mathbf{p}\downarrow}\gamma_{\mathbf{p}\downarrow})$$

The energy $E(p)$ is the energy of an excited electron quasiparticle in the superconductor. Most of the electrons are paired up in energy states located at the chemical potential μ (where else?). Their energy is μ per electron, or $\xi = 0$. If a pair is broken, both electrons get excited to quasiparticle states of energy $E(\mathbf{p})$, $E(-\mathbf{p})$. Each electron must overcome an energy gap $\Delta \equiv |\Delta_{\uparrow}|$. To break a pair requires an energy gap of $E_G = 2\Delta$. The operators $(\gamma^{\dagger}, \gamma)$ are the creation and destruction operators for excited quasiparticles. An operator term such as

$$\gamma^{\dagger}_{\mathbf{p}\uparrow}\gamma^{\dagger}_{-\mathbf{p}\downarrow} \tag{14.59}$$

creates two quasiparticles by breaking a pair, while

$$\gamma_{-\mathbf{p}\downarrow}\gamma_{\mathbf{p}\uparrow} \tag{14.60}$$

describes the recombination process where two quasiparticles meet, join together, and form a bound pair at the chemical potential: two quasiparticles have been removed.

The constants (u_p, v_p) are called *coherence factors,* although they have nothing at all to do with coherence. They can now be evaluated:

$$u_p^2 = \cos^2(\theta) = \frac{1}{2}[1 + \cos(2\theta)] = \frac{1}{2}\left[1 + \frac{\xi(p)}{E(p)}\right] \tag{14.61}$$

$$v_p^2 = \sin^2(\theta) = \frac{1}{2}[1 - \cos(2\theta)] = \frac{1}{2}\left[1 - \frac{\xi(p)}{E(p)}\right] \tag{14.62}$$

The coherence factors are needed to evaluate correlation functions. This process is described below.

14.3.4 Gap Equation

The energy gap $\Delta(T)$ plays an important role in the BCS theory. It is determined by deriving and solving an equation. Recall its definition:

$$\Delta_\uparrow = V_0 \sum_{p} C_{-p\downarrow} C_{p\uparrow} \tag{14.63}$$

Make the Bogoliubov transformation, and retain only terms with one raising and one lowering operator:

$$\Delta_\uparrow = V_0 \sum_{p} \left(u_p \gamma_{-p\downarrow} - v_p \gamma^\dagger_{p\uparrow}\right)\left(u_p \gamma_{p\uparrow} + v_p \gamma^\dagger_{-p\downarrow}\right) \tag{14.64}$$

$$= V_0 \sum_{p} u_p v_p \left[\gamma_{-p\downarrow}\gamma^\dagger_{-p\downarrow} - \gamma^\dagger_{p\uparrow}\gamma_{p\uparrow}\right]$$

The thermal averages are

$$\langle \gamma^\dagger_{p\uparrow}\gamma_{p\uparrow}\rangle = n_p = \frac{1}{e^{\beta E(p)} + 1} \tag{14.65}$$

$$\langle \gamma_{-p\downarrow}\gamma^\dagger_{-p\downarrow}\rangle = 1 - n_p \tag{14.66}$$

$$\Delta_\uparrow = \frac{V_0}{2} \sum_{p} [1 - 2n_p] \sin(2\theta) \tag{14.67}$$

$$= \frac{\Delta_\uparrow V_0}{2} \sum_{p} \frac{1}{E(p)} [1 - 2n_p]$$

Cancel the common factor of $\Delta_\uparrow$, which gives the BCS gap equation:

$$1 = \frac{V_0}{2} \sum_{p} \frac{1}{E(p)} [1 - 2n_p] \tag{14.68}$$

It is an implied equation for Δ, which appears in $E(p)$ and n_p.

First solve for the energy gap at $T \to 0$. In that case $n_p \to 0$ since $E(p) > \Delta$ and

$$n_p < \frac{1}{e^{\beta\Delta} + 1} \to 0 \tag{14.69}$$

The integral in eqn. (14.73) is evaluated:

$$1 = \frac{V_0}{2} \int \frac{d^3p}{(2\pi)^3} \frac{1}{E(p)} = \frac{N(0)V_0}{2} \int_{-\hbar\omega_D}^{\hbar\omega_D} \frac{d\xi}{\sqrt{\xi^2 + \Delta^2}} \tag{14.70}$$

where $N(0)$ is the density of states per spin at the Fermi energy. The energy integral extends only to the Debye energy, which is the range of the phonon-mediated interaction. The integral can be done exactly:

$$\int \frac{d\xi}{\sqrt{\xi^2 + \Delta^2}} = \ln\left[\xi + \sqrt{\xi^2 + \Delta^2}\right] \tag{14.71}$$

$$1 = \frac{N(0)V_0}{2} \ln\left[\frac{\sqrt{\hbar^2\omega_D^2 + \Delta^2} + \hbar\omega_D}{\sqrt{\hbar^2\omega_D^2 + \Delta^2} - \hbar\omega_D}\right] \tag{14.72}$$

Usually $\hbar\omega_D \gg \Delta$,

$$\sqrt{\hbar^2\omega_D^2 + \Delta^2} \approx \hbar\omega_D + \frac{\Delta^2}{2\hbar\omega_D} + \cdots \tag{14.73}$$

$$1 = N(0)V_0 \ln\left(\frac{2\hbar\omega_D}{\Delta}\right) \tag{14.74}$$

$$\Delta = 2\hbar\omega_D \exp\left[-\frac{1}{N(0)V_0}\right] \tag{14.75}$$

The latter equation is the BCS formula for the gap Δ. It could not be derived by perturbation theory, since $\Delta(V_0)$ does not have a Taylor series.

Another temperature at which the gap equation can be solved is T_c, where $\Delta = 0$. At any nonzero temperature

$$1 - 2n_p = \frac{e^{\beta E} - 1}{e^{\beta E} + 1} = \tanh[\beta E(p)/2] \tag{14.76}$$

$$1 = N(0)V_0 \int_0^{\hbar\omega_D} \frac{d\xi}{E(p)} \tanh[\beta E(p)/2] \tag{14.77}$$

At T_c where $\Delta = 0$, this integral becomes

$$1 = N(0)V_0 \int_0^{\hbar\omega_D} \frac{d\xi}{\xi} \tanh[\beta_c \xi/2] \tag{14.78}$$

where the integral now defines $\beta_c = 1/k_B T_c$. Change the integration variable to

$$x = \beta_c \xi/2, \quad x_D = \frac{\hbar\omega_D}{2k_B T_c} \tag{14.79}$$

$$1 = N(0)V_0 \int_0^{x_D} \frac{dx}{x} \tanh(x) \tag{14.80}$$

Integrate by parts: $\int u dv = uv - \int v du$, where $u = \tanh(x)$, $v = \ln(x)$, and find

$$1 = N(0)V_0 \left[\tanh(x_D)\ln(x_D) - \int_0^{x_D} dx \frac{\ln(x)}{\cosh^2(x)}\right] \tag{14.81}$$

The constant $x_D \gg 1$. In this case $\tanh(x_D) \approx 1$, and the integral goes to a constant that is written as $\ln(0.44)$:

$$\int_0^{\infty} dx \frac{\ln(x)}{\cosh^2(x)} = -0.82 = \ln(0.44) \tag{14.82}$$

$$1 = N(0)V_0 \ln\left[\frac{x_D}{0.44}\right] \tag{14.83}$$

$$x_D = 0.44 \exp\left[\frac{1}{N(0)V_0}\right] = \frac{\hbar\omega_D}{2k_B T_c} \tag{14.84}$$

$$k_B T_c = 1.14\hbar\omega_D \exp\left[-\frac{1}{N(0)V_0}\right] \tag{14.85}$$

The energy gap $E_G = 2\Delta$ can be easily measured using electron tunneling, as described below. The ratio of this gap to $k_B T_c$ is

$$\frac{E_G}{k_B T_c} = \frac{4.0}{1.14} = 3.51 \tag{14.86}$$

This ratio is found to be correct in weak superconductors such as aluminum. This agreement is another success of BCS theory. The ratio increases in value as T_c increases. This increase is predicted by Eliashberg's strong coupling theory. The term $N(0)V_0 = \lambda$, the electron–phonon coupling constant. The transition temperature depends on $\exp(-1/\lambda)$ for small λ.

14.3.5 *d*-Wave Energy Gaps

The high T_c superconductors based on cuprates have the conduction electrons confined to planes containing copper and oxygen. The electron motion is very two dimensional in the (x, y)-plane. Many of the cuprates have the copper atoms in a perfect square lattice. In this case, a wave vector point is defined by two variables: (i) the magnitude of the two-dimensional wave vectors k, and (ii) the azimuthal angle θ defined by $\tan(\theta) = k_y/k_x$.

The energy gap depends on this angle: $\Delta(\theta)$. This was realized even for metals such as aluminum and lead. Their Fermi surface is not a perfect sphere, but is intersected by Bragg planes into interesting shapes and pockets. The superconducting gap varies slightly around these Fermi surfaces. For the cuprates, several gap functions have been proposed:

- The *isotropic gap*, where Δ does not depend on angle.
- The *anisotropic s-wave* gap has the form

  $$\Delta_s(\theta) = \Delta_0 + \Delta_4 \cos(4\theta) + \Delta_8 \cos(8\theta) + \cdots \tag{14.87}$$

  or else

  $$\Delta_s(\theta) = \Delta_0 + \Delta_1[\cos(\theta_x) + \cos(\theta_y)] \tag{14.88}$$

  The gap varies around the Fermi circle, but has the same value on the $\pm\hat{x}$- and $\pm\hat{y}$-axes. This gap maintains the square symmetry of the lattice.
- The *p-wave* order parameter has spins in a triplet state with $S = 1$. This spin state has even parity, so the orbital part of the two-electron eigenfunction has to have odd parity. Then the orbital part is *p*-wave. This option was soon discarded by an experiment where a Josephson junction was measured between lead and a cuprate superconductor. Pb is an isotropic *s*-wave superconductor, with spins in the singlet state. Pair tunneling occurs only to other superconductors with the spin singlet confiuration, since tunneling does not flip spin. This experiment ruled out having a *p*-wave order parameter in the cuprates.
- The *d-wave gap* breaks square symmetry and has the possible forms

$$\Delta_{d1} = \Delta_d \cos(2\theta), \quad \Delta_d' [\cos(\theta_x) - \cos(\theta_y)] \tag{14.89}$$

$$\Delta_{d2} = \Delta_d \sin(2\theta) \tag{14.90}$$

d-Waves have even parity orbital parts and spin singlets.

Experiments have shown that the cuprate superconductors have the form Δ_{d1} with the dependence on $\cos(2\theta)$.

How can an angular dependent gap $\Delta(\theta)$ be derived from the BCS gap equation? First restore the angular integrals to the gap equation. In two dimensions, the gap equation at zero temperature is

$$\Delta(\theta) = N(0)\int_0^{\xi_c} d\xi \int_0^{2\pi} \frac{d\theta}{2\pi} \frac{V(\theta,\theta')\Delta(\theta')}{\sqrt{\xi^2 + \Delta(\theta')^2}} \tag{14.91}$$

The quasiparticle interaction $V(\theta, \theta')$ must depend on angle. If it does not have any angular dependence, the only solution is that Δ is independent of angle. To get a *d*-wave gap, this dependence must be something like

$$V(\theta, \theta') = V_2 \cos[2(\theta - \theta')] \tag{14.92}$$

which is obviously quite anistropic. An argument against pairing by phonon forces is that the electron–phonon interaction preserves square symmetry, and can never have a term that gives *d*-wave pairing. This symmetry argument is extended to every mechanism that involves charge fluctuations as the intermediate boson, such as plasmons. Only spin fluctuations can break the square symmetry of the lattice.

The precise mechanism for *d*-wave pairing is not understood. For that reason, in eqn. (14.91) the upper limit of the energy integral is given as ξ_c, which is unknown since the mechanism is unknown.

14.3.5 Density of States

The density of states of the normal metal is $N(\xi)$. It is assumed to be a smooth function of energy $\xi = \varepsilon - \mu$ near to the chemical potential, and is approximated by a constant $N(0)$. In the superconducting state, there is another function $\rho(E)$, which is superimposed on this smooth background, so the density of states is $N(E) = N(0)\rho(E)$. The superconducting contribution is found from the integral

$$\rho(E) = \int_0^{\xi_0} d\xi \delta\left(E - \sqrt{\xi^2 + \Delta^2}\right) \tag{14.93}$$

$$= \frac{E}{\sqrt{E^2 - \Delta^2}} \Theta(E - \Delta)$$

The theta function is zero if $E < \Delta$ and is one if $E > \Delta$. The function $\rho(E)$ has a square root singularity at the gap $E = \Delta^+$. This function can be measured directly in electron tunneling between a normal metal and a superconductor, as described below. This curve is shown in fig. 14.7a.

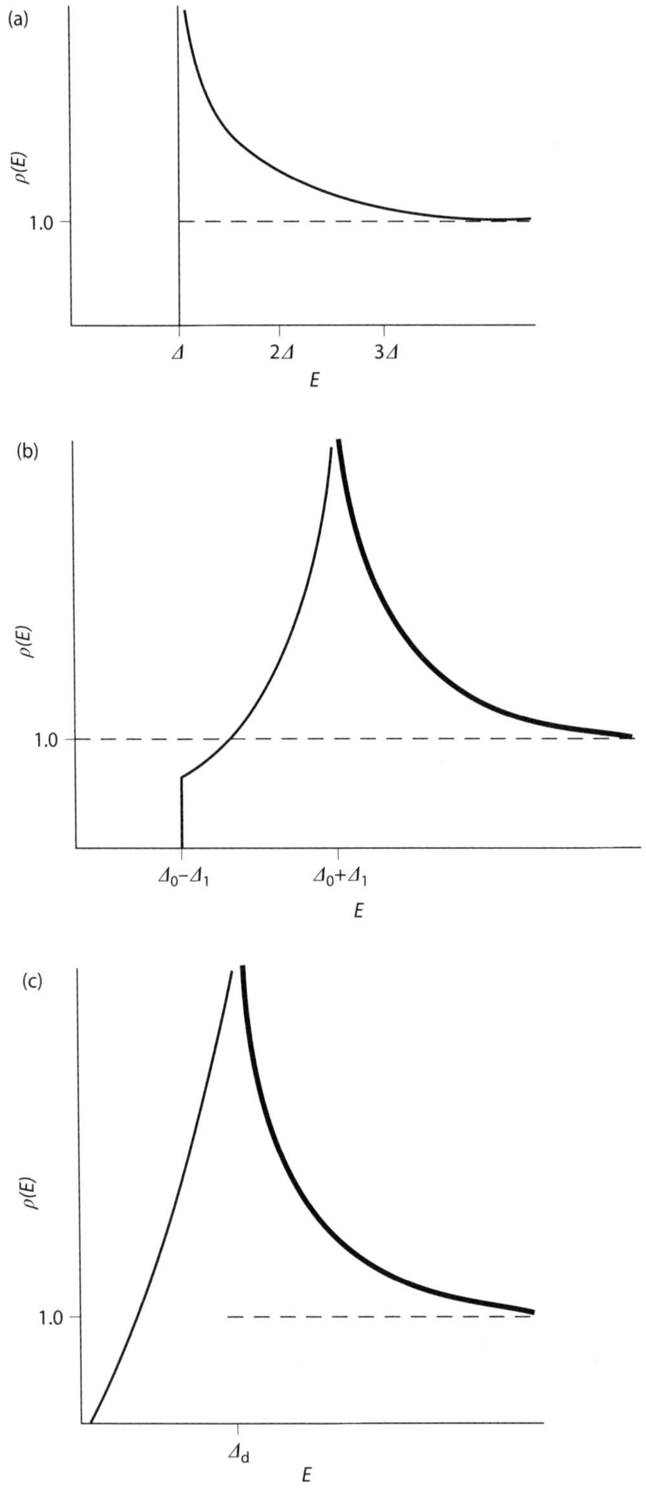

FIGURE 14.7. Superconducting density of states of (a) an isotropic *s*-wave superconductor; (b) an anisotropic *s*-wave superconductor, and (c) a *d*-wave superconductor.

The above density of states is appropriate for a superconductor with an isotropic energy gap. For high-T_c materials, the gap depends on the angle around the Fermi surface. The evaluation of the density of states requires another integral, over angle

$$\rho(E) = \frac{2E}{\pi}\int_0^{\pi/2} \frac{d\theta}{\sqrt{E^2 - \Delta^2(\theta)}}\Theta[E - \Delta(\theta)] \tag{14.94}$$

For an anisotropic *s*-wave superconductor, write $\Delta(\theta) = \Delta_0 + \Delta_1\cos(4\theta)$. There is a log-singularity at $E = \Delta_0 + \Delta_1$ and a discontinuity at $E = \Delta_0 - \Delta_1$. See fig. 14.7b.

For a *d*-wave superconductor, write $\Delta(\theta) = \Delta_0\cos(2\theta)$, and the integral depends on whether E is larger or smaller than Δ_0:

$$\rho(E) = \frac{2}{\pi}\begin{cases} K\left(\frac{\Delta_0}{E}\right) & E > \Delta_0 \\ \frac{E}{\Delta_0}K\left(\frac{E}{\Delta_0}\right) & E < \Delta_0 \end{cases} \tag{14.95}$$

where $K(y)$ is an elliptic integral that diverges as $y = 1$. The divergence is a logarithm ~ ln $|E/\Delta_0|$, which is a soft singularity. The *d*-wave density of states is shown in fig. 14.7c.

14.3.7 *Ultrasonic Attenuation*

After the publication of the BCS paper, the first experimental measurement of the gap was by ultrasonic attenuation. This historically important experiment is now described.

A typical ultrasonic wave has a frequency of $\nu = 5 - 10$ MHz. The quantum of energy $h\nu = 2 - 4 \times 10^{-8}$ eV is far too small to break a pair and excite a quasiparticle, since gaps are typically milli-elcctron volts, not nano-electron volts. The scattering of the phonons is caused by the excited quasiparticles of density n_p.

The electron–phonon interaction has the generic form of

$$H_{ep} = \frac{1}{\sqrt{\Omega}}\sum_{\mathbf{q}\eta} M_\eta(\mathbf{q})A_{\mathbf{q}\eta}\rho(\mathbf{q}) \tag{14.96}$$

$$\rho(\mathbf{q}) = \sum_{\mathbf{p}\sigma} C^\dagger_{\mathbf{p}+\mathbf{q},\sigma}C_{\mathbf{p}\sigma}, \quad A_{\mathbf{q}\eta} = a_{\mathbf{q}\eta} + a^\dagger_{-\mathbf{q},\eta} \tag{14.97}$$

Evaluate the electron density operator by making the Bogoliubov transformation

$$\begin{aligned}\rho(\mathbf{q}) &= \left(C^\dagger_{\mathbf{p}+\mathbf{q}\uparrow}C_{\mathbf{p}\uparrow} + C^\dagger_{\mathbf{p}+\mathbf{q}\downarrow}C_{\mathbf{p}\downarrow}\right) \\ &= \sum_{\mathbf{p}}\Big[\left(u_{p+q}\gamma^\dagger_{p+q\uparrow} + v_{p+q}\gamma_{-p-q\downarrow}\right)\left(u_{\mathbf{p}}\gamma_{\mathbf{p}\uparrow} + v_{\mathbf{p}}\gamma^\dagger_{-\mathbf{p}\downarrow}\right) \\ &\quad + \left(u_{p+q}\gamma^\dagger_{p+q\downarrow} - v_{p+q}\gamma_{-p-q\uparrow}\right)\left(u_{\mathbf{p}}\gamma_{\mathbf{p}\downarrow} - v_{\mathbf{p}}\gamma^\dagger_{-\mathbf{p}\uparrow}\right)\Big] \\ &= \rho_1 + \rho_2 + \rho_3\end{aligned} \tag{14.98}$$

$$\rho_1 = \sum_{\mathbf{p}}\left(u_{\mathbf{p}+\mathbf{q}}u_{\mathbf{p}} - v_{\mathbf{p}+\mathbf{q}}v_{\mathbf{p}}\right)\left(\gamma^\dagger_{\mathbf{p}+\mathbf{q}\uparrow}\gamma_{\mathbf{p}\uparrow} + \gamma^\dagger_{\mathbf{p}+\mathbf{q}\downarrow}\gamma_{\mathbf{p}\downarrow}\right) \tag{14.99}$$

$$\rho_2 = \sum_{\mathbf{p}} u_{\mathbf{p+q}} v_{\mathbf{p}} \left(\gamma^{\dagger}_{\mathbf{p+q}\uparrow} \gamma^{\dagger}_{-\mathbf{p},\downarrow} - \gamma^{\dagger}_{\mathbf{p+q}\downarrow} \gamma^{\dagger}_{-\mathbf{p},\uparrow}\right) \tag{14.100}$$

$$\rho_3 = \sum_{\mathbf{p}} u_{\mathbf{p}} v_{\mathbf{p+q}} \left(\gamma_{-\mathbf{p-q}\downarrow} \gamma_{\mathbf{p}\uparrow} - \gamma_{-\mathbf{p-q}\uparrow} \gamma_{\mathbf{p}\downarrow}\right) \tag{14.101}$$

The term ρ_1 scatters quasiparticles from state $(\mathbf{p}\sigma)$ to $(\mathbf{p}+\mathbf{q},\sigma)$ and is the important term in ultrasonic attenuation. The term ρ_2 breaks a pair and creates two quasiparticles. There is not enough energy in the phonon to do this. Similarly, ρ_3 is the inverse process of creating a pair state by recombination of two quasiparticles and emitting a phonon to take up the gap energy. This event is allowed, but does not contribute to ultrasonic absorption.

In ρ_1 the phonon wave vector $\mathbf{q}$ is very small and it is a good approximation to set it equal to zero. Then the matrix element for ultrasonic absorption is

$$u_{\mathbf{p}}^2 - v_{\mathbf{p}}^2 = \cos(2\theta_{\mathbf{p}}) = \frac{\xi(p)}{E(p)} \tag{14.102}$$

The above analysis is an example of using coherence factors. Using the golden rule of quantum mechanics, the ultrasonic attenuation is

$$\frac{\hbar}{\tau(\mathbf{q}\eta)} = 2\pi |M_\eta(\mathbf{q})|^2 \int \frac{d^3p}{(2\pi)^3} \frac{\xi(p)^2}{E(p)^2} \{n_F[E(\mathbf{p})] - n_F[E(\mathbf{p}+\mathbf{q})]\} \tag{14.103}$$
$$\times \{\delta[\hbar\omega(\mathbf{q}\eta) + E(\mathbf{p}) - E(\mathbf{p}+\mathbf{q})] - \delta[\hbar\omega(\mathbf{q}\eta) - E(\mathbf{p}) + E(\mathbf{p}+\mathbf{q})]\}$$

The first delta function is phonon absorption, and the second is phonon emission. This expression can be evaluated, and gives

$$\frac{\hbar}{\tau(\mathbf{q}\eta)} = \frac{m^2 c_s}{\pi\hbar^3} |M_\eta(\mathbf{q})|^2 n_F(\Delta) \tag{14.104}$$

where $n_F(\Delta) = [e^{\beta\Delta} + 1]^{-1}$.

The ultrasonic attenuation is proportional to $n_F(\Delta)$ and provides a direct measurement of $\Delta(T)$ in the superconducting state. To account for all of the other factors in the expression, the result is written as the ratio of the attenuation in the superconducting state, divided by that in the normal state:

$$\frac{\alpha_S}{\alpha_N} = 2n_F(\Delta) \tag{14.105}$$

At T_c, where $\Delta \to 0$, $2n_F(0) = 1$. Figure 14.8 shows the ultrasonic attenuation as a function of temperature, and the resulting curve for $\Delta(T)$. The results are in excellent agreement with the BCS theory for $\Delta(T)$.

14.3.8 Meissner Effect

The Meissner effect was described at the beginning of this chapter. The explanation depends on London's equation, which can be derived from BCS theory. Begin with a general expression for the microscopic current operator due to electrons of charge e, velocity $\mathbf{v}_i$, energy E_i, at position $\mathbf{r}_i$:

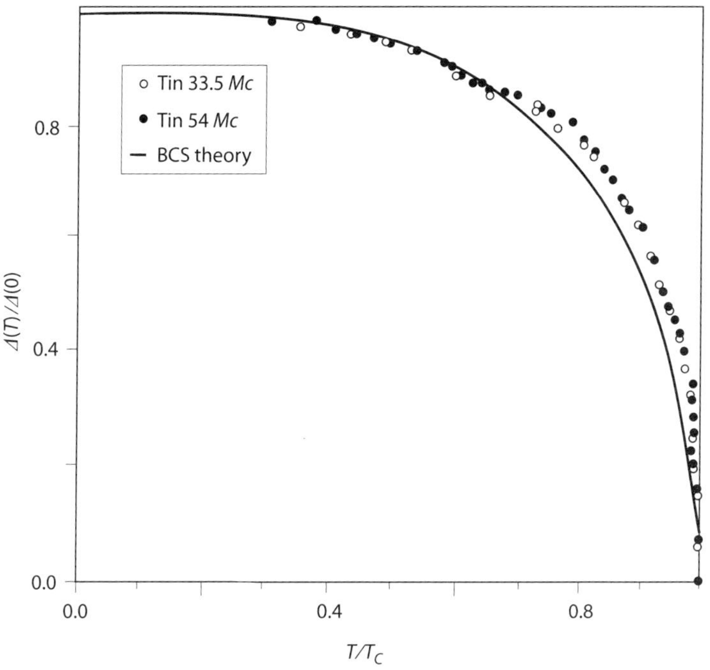

FIGURE 14.8. The gap function $\Delta(T)$ in tin as derived from ultrasonic attenuation. From Morse and Bohm, *Phys. Rev.* **108**, 1094 (1957). Used with permission of the American Physical Society.

$$\mathbf{j}(\mathbf{r}, t) = e\sum_i \mathbf{v}_i \delta(\mathbf{r} - \mathbf{r}_i) e^{-iE_i t/\hbar} \tag{14.106}$$

Take a Fourier transform

$$\mathbf{j}(\mathbf{q}, t) = \int d^3 r e^{i\mathbf{q}\cdot\mathbf{r}} \mathbf{j}(\mathbf{r}, t) = e\sum_i \mathbf{v}_i \exp[i(\mathbf{q}\cdot\mathbf{r}_i - E_i t/\hbar)] \tag{14.107}$$

In quantum mechanics, the velocity is given in terms of the momentum and the vector potential:

$$\mathbf{v}_i = \frac{1}{m}\left[\mathbf{p}_i - e\mathbf{A}(\mathbf{r}_i, t)\right] \tag{14.108}$$

$$\mathbf{j}(\mathbf{q}, t) = \frac{e}{m}\sum_i \exp[i(\mathbf{q}\cdot\mathbf{r}_i - E_i t/\hbar)]\,[\mathbf{p}_i - e\mathbf{A}(\mathbf{r}_i, t)] \tag{14.109}$$

The term with the vector potential is London's equation [15]. Write the above expression as

$$\mathbf{j}(\mathbf{q}, t) = -\frac{n_s e^2}{m}\mathbf{A}(\mathbf{r}, t) + \delta\mathbf{j} \tag{14.110}$$

$$\delta\mathbf{j} = \frac{e}{m}\sum_i \mathbf{p}_i \exp[i(\mathbf{q}\cdot\mathbf{r}_i - E_i t/\hbar)] \tag{14.111}$$

In the vector potential term, the summation (i) over electrons is replaced by the density n_s of superconducting electrons. London's equation is valid when $\delta\mathbf{j} = 0$. Usually, when the current is evaluated in a normal metal, the important term is $\delta\mathbf{j}$. Now we are trying to prove it vanishes in a superconductor. That it must vanish was realized from the time London's theory was proposed. However, before BCS no one could explain why it did vanish. In normal metals it is a large contribution.

In quantum mechanics write this term as

$$\delta\mathbf{j}(\mathbf{q}) = \frac{e}{m}\sum_{\mathbf{p}\sigma}\left(\mathbf{p}+\frac{1}{2}\mathbf{q}\right)C^{\dagger}_{\mathbf{p}+\mathbf{q}\sigma}C_{\mathbf{p}\sigma} \tag{14.112}$$

which is evaluated in the superconductor by using the Bogoliubov transformation:

$$\begin{aligned}\delta\mathbf{j}(\mathbf{q}) = \frac{e}{m}\sum_{\mathbf{p}\sigma}\left(\mathbf{p}+\frac{1}{2}\mathbf{q}\right)\Big[&u_{\mathbf{p}+\mathbf{q}}u_{\mathbf{p}}\sum_{\sigma}\gamma^{\dagger}_{\mathbf{p}+\mathbf{q}\sigma}\gamma_{\mathbf{p}\sigma} + v_{\mathbf{p}+\mathbf{q}}v_{\mathbf{p}}\sum_{\sigma}\gamma_{-\mathbf{p}-\mathbf{q}\sigma}\gamma^{\dagger}_{-\mathbf{p}\sigma}\\ &+ u_{\mathbf{p}+\mathbf{q}}v_{\mathbf{p}}\left(\gamma^{\dagger}_{\mathbf{p}+\mathbf{q}\uparrow}\gamma^{\dagger}_{-\mathbf{p}\downarrow} - \gamma^{\dagger}_{\mathbf{p}+\mathbf{q}\downarrow}\gamma^{\dagger}_{-\mathbf{p}\uparrow}\right)\\ &+ v_{\mathbf{p}+\mathbf{q}}u_{\mathbf{p}}\left(\gamma_{-\mathbf{p}-\mathbf{q}\downarrow}\gamma_{\mathbf{p}\uparrow} - \gamma_{-\mathbf{p}-\mathbf{q}\uparrow}\gamma_{\mathbf{p}\downarrow}\right)\Big]\end{aligned} \tag{14.113}$$

The current $\delta\mathbf{j}$ is provide by quasiparticles, so the first two terms provide the major result. They always are given in terms of the quasiparticle density $n_{\mathbf{p}}$. Because of the energy gap Δ, this density vanishes at low temperature. So $\delta\mathbf{j}$ vanishes because the system has an energy gap for excitations to the quasiparticle state. The same phenomenon is found in an insulator or semiconductor. However, insulators do not obey London's equation since they do not have a density n_s of paired electrons to contribute to the term with a vector potential. A superconductor is a unique system in that there are paired electrons that can contribute to London's equation, and quasiparticle electrons that contribute to $\delta\mathbf{j}$ are frozen out at low temperatures.

14.4 Electron Tunneling

The introduction to this chapter mentioned the three eras of superconductivity. Besides transition temperature, the eras were characterized by different experimental techniques. In the first era, electron tunneling was a major experimental tool to understand the behavior of the superconducting state. Electron tunneling can be divided into the Josephson effect, where pairs are tunneling, and single-particle tunneling. In the era of high T_c, single-particle tunneling was almost useless for understanding the superconducting state. But Josephson tunneling was very important in the era of high T_c. An early experiment showed that there was a Josephson current between lead and a cuprate. Since lead is an *s*-wave superconductor, with the paired electrons in a spin singlet state, Josephson tunneling from Pb is allowed only to superconductors that also have the paired electrons in a spin singlet. This ruled out the possibility that the high T_c materials could be *p*-wave superconductors. The latter have a triplet arrangement of the two spins. Josephson tunneling was able to establish very early in the era of high T_c that the order parameter had to be either *s*-wave or *d*-wave, or a mixture of these two options. Later Josephson experiments established that the gap had *d*-wave symmetry.

The theory of electron tunneling uses the *Bardeen Hamiltonian*.

$$H = H_L + H_R + H_T \tag{14.114}$$

$$H_L = \sum_{p\sigma} (E_L(p) - \mu_L) c^{\dagger}_{p\sigma} c_{p\sigma} \tag{14.115}$$

$$H_R = \sum_{k\sigma} (E_R(k) - \mu_R) d^{\dagger}_{k\sigma} d_{k\sigma} \tag{14.116}$$

$$H_T = \sum_{kp\sigma} T_{kp} \left[c^{\dagger}_{p\sigma} d_{k\sigma} + d^{\dagger}_{k\sigma} c_{p\sigma} \right] \tag{14.117}$$

H_L and H_R are the Hamiltonians of electrons on the right and left sides of the tunnel junctions. They are assumed to be independent systems, and $[H_R, H_L] = 0$. A different raising and lowering operator is used for each side to emphasize the independence. The tunneling process has a matrix element T_{kp}, which can be calculated using quantum mechanical methods, such as a particle in a square repulsive barrier, or by using WKBJ for more complicated barrier shapes.

The rate of single-electron tunneling is given by the golden rule of quantum mechanics:

$$w = \frac{2\pi}{\hbar} \sum_{kp\sigma} |T_{kp}|^2 \delta[E_L(p) - E_R(k)] [n_{FL}(p) - n_{FR}(k)] \tag{14.118}$$

This gives the number of electrons that tunnel per second. Multiply the right-hand side by the electron charge and get the current in amperes:

$$I = \frac{2\pi e}{\hbar} \sum_{kp\sigma} |T_{kp}|^2 \delta[E_L(p) - E_R(k)] [n_{FL}(p) - n_{FR}(k)] \tag{14.119}$$

The energies on the left and right are

$$E_L(p) = E_{L0} + \frac{\hbar^2 p^2}{2m_L}, \quad \xi_L(p) = E_L(p) - \mu_L \tag{14.120}$$

$$E_R(k) = E_{R0} + \frac{\hbar^2 k^2}{2m_R}, \quad \xi_R(k) = E_L(p) - \mu_R \tag{14.121}$$

$$n_{FL}(p) = \frac{1}{e^{\beta\xi_L} + 1}, \quad n_{FR}(k) = \frac{1}{e^{\beta\xi_R} + 1} \tag{14.122}$$

This expression was evaluated in chapter 11 for the tunneling between two normal metals. The tunneling current acted as a simple resistor $I = \sigma V$, where the tunneling conductance is

$$\sigma = \frac{4\pi e^2}{\hbar} N_L N_R |T|^2 \tag{14.123}$$

If the interface of the oxide is atomically smooth, then transverse wave vector may be conserved. If the interface is in the (x, y)-plane and the barrier in the z-direction, then one would have that $k_x = p_x$, $k_y = p_y$. This conservation of transverse wave vector is found in perfect junctions. Then the summation over wave vector includes only one set of transverse components:

$$\sum_{\mathbf{k}\mathbf{p}\sigma} = \sum_{\sigma} \int dk_x \int dk_y \int dk_z \int dp_z \tag{14.124}$$

$$E_L(p) = E_{L0} + \frac{\hbar^2}{2m_L}\left(k_x^2 + k_y^2 + p_z^2\right) \tag{14.125}$$

$$E_R(k) = E_{R0} + \frac{\hbar^2}{2m_R}\left(k_x^2 + k_y^2 + k_z^2\right) \tag{14.126}$$

For tunneling between two simple metals one sets $m_L = m_R = m_e$, where m_e is the mass of an electron. However, one can have tunneling between a metal and a semiconductor, and then one should use the effective mass of the semiconductor band.

The tunneling current can be calculated assuming conservation of transverse momentum. It is found that the theoretical curves do not agree with the measurements. The conclusion is that transverse momentum is rarely conserved in electron tunneling. Indeed, if the interface between the metal and the oxide is rough on an atomic scale, then all the current will flow through the areas where the oxide width is the thinnest. Even if an atomically smooth surface is missing one surface atom, all of the tunneling current flows through that one spot. The thin regions act as point sources, which do not conserve transverse momentum.

If transverse momentum is not conserved, then the tunneling can usually be expressed in terms of the density of states $N_{L,R}(E)$ on the two sides of the junction:

$$\sum_{\mathbf{p}} = \int dE_L N_L(E_L) \tag{14.127}$$

$$\sum_{\mathbf{k}} = \int dE_R N_R(E_R) \tag{14.128}$$

Furthermore, the voltages used in tunneling in superconductors is usually small: the energy gap of most superconductors is a few milli-electron volts. Over that narrow range of energy, one can treat $N_L(E)$, $N_R(E)$, and $|T_{\mathbf{kp}}|^2$ as constants. If one removes $N_L(0)$, $N_R(0)$, and $|T|^2$ from the integrals over energy, then one immediately derives the tunneling conductance of eqn. (14.123).

14.4.1 Normal–Superconductor

An interesting result is obtained for the electron tunneling between a normal metal and a superconductor. Common metals, such as copper, silver, or gold, are not superconducting, so it is easy to find a normal metal to act as one side of the tunnel junction. To treat the superconducting side of the junction, return to the tunneling Hamiltonian and apply a Bogoliubov transformation. Let the superconductor be on the right:

$$H_T = \sum_{\mathbf{k}\mathbf{p}\sigma} T_{\mathbf{kp}}\left[c^{\dagger}_{\mathbf{p}\sigma} d_{\mathbf{k}\sigma} + d^{\dagger}_{\mathbf{k}\sigma} c_{\mathbf{p}\sigma}\right] \tag{14.129}$$

$$= \sum_{\mathbf{pk}} T_{\mathbf{pk}}\left[c^{\dagger}_{\mathbf{p}\uparrow}\left(u_k \gamma_{\mathbf{k}\uparrow} + v_k \gamma^{\dagger}_{-\mathbf{k}\downarrow}\right) + \left(u_k \gamma^{\dagger}_{\mathbf{k}\uparrow} + v_k \gamma_{-\mathbf{k}\downarrow}\right) c_{\mathbf{p}\uparrow}\right.$$

$$\left. + \, c^{\dagger}_{\mathbf{p}\downarrow}\left(u_k \gamma_{\mathbf{k}\downarrow} - v_k \gamma^{\dagger}_{-\mathbf{k}\uparrow}\right) + \left(u_k \gamma^{\dagger}_{\mathbf{k}\downarrow} - v_k \gamma_{-\mathbf{k}\uparrow}\right) c_{\mathbf{p}\downarrow}\right]$$

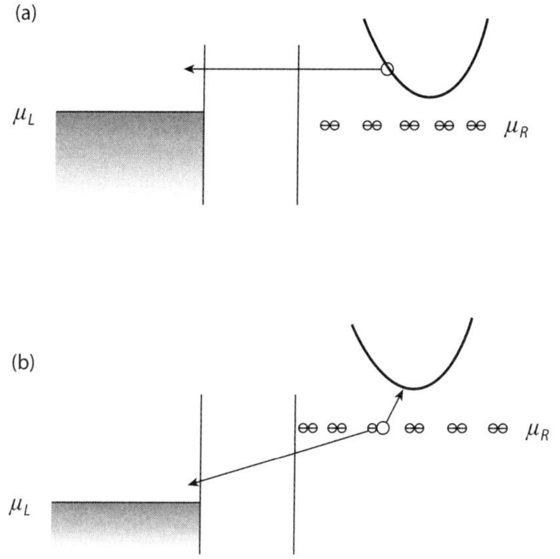

FIGURE 14.9. Tunneling processes between a normal metal on the left and a superconductor on the right.

Examine the type of terms that occur in the above expression:

- A term such as $u_k c^{\dagger}_{\mathbf{p}\sigma}\gamma_{\mathbf{k}\sigma}$ describes a process where an electron in the excitation spectrum of the superconductor tunnels and becomes a quasiparticle in the normal metal: fig. 14.9a.
- A term such as $u_k \gamma^{\dagger}_{\mathbf{k}\sigma} c_{\mathbf{p}\sigma}$ describes the reverse process where a quasiparticle in the normal metal tunnels and becomes an electron in the excitation spectrum of the superconductor. This process can happen at zero temperature.
- A term such as $v_k c^{\dagger}_{\mathbf{p}\sigma}\gamma^{\dagger}_{-\mathbf{k}-\sigma}$ describes a process where a bound pair in the superconductor is separated and one electron tunnels and the other becomes an excitation in the superconductor; see fig. 14.9b. This process can happen at zero temperature.
- A term such as $v_k \gamma_{-\mathbf{k}-\sigma} c_{\mathbf{p}\sigma}$ describes the inverse process where an electron in the normal metal tunnels and joins with an electron in the excitation spectra of the superconductor to form a bound pair at the chemical potential of the superconductor.

The next step is to derive the occupation factors.

- For NN (normal–normal) tunneling, if an electron goes from left (**p**) to right (**k**), the occupation factors are $n_p(1 - n_k)$ since there must be an occupied state on the left (n_p) and an empty state on the right $(1 - n_k)$. Electron flow in the other direction, from right to left, has the factor $n_k(1 - n_p)$. The net current is the difference of these two expressions:

$$n_p(1 - n_k) - n_k(1 - n_p) = n_p - n_k \equiv n_F(\xi_p) - n_F(\xi_k) \tag{14.130}$$

- A similar expression is found in NS tunneling for $c^{\dagger}_{\mathbf{p}\sigma}\gamma_{\mathbf{k}\sigma}$:

$$n_F(\xi_p) - n_F(E_k) \tag{14.131}$$

where $E_k \equiv E(k)$ is the excitation energy of the superconductor.

- For $c^{\dagger}_{p\sigma}\gamma^{\dagger}_{-k-\sigma}$ two quasiparticles are created in one direction of current and two destroyed in the other:

$$[1 - n_F(\xi_p)][1 - n_F(E_k)] - n_F(\xi_p)n_F(E_k) = 1 - n_F(\xi_p) - n_F(E_k) \tag{14.132}$$

Collecting all of these results derives the golden rule for NS tunneling:

$$I = \frac{2\pi e}{\hbar}\sum_{kp\sigma}|T_{pk}|^2\{u_k^2[n_F(E_k) - n_F(\xi_p)]\,\delta(eV + E_k - \xi_p) \tag{14.133}$$

$$+ v_k^2[1 - n_F(E_k) - n_F(\xi_p)]\,\delta(eV - E_k - \xi_p)\}$$

The evaluation of this expression makes the usual assumptions: (i) the summations over wave vectors can be replaced by an energy integral over the density of states, (ii) the tunneling is independent of spin, and (iii) T_{kp} can be treated as a constant.

$$I = \frac{4\pi e}{\hbar}N_L N_R|T_{RL}|^2\int d\xi_p\int d\xi_k\{u_k^2[n_F(E_k) - n_F(\xi_p)]\,\delta(eV + E_k - \xi_p)$$

$$+ v_k^2[1 - n_F(E_k) - n_F(\xi_p)]\,\delta(eV - E_k - \xi_p)\} \tag{14.134}$$

Remember the expressions for the coherence factors $(1\pm\xi_k/E_k)/2$. The factor ξ_k integrates to zero since it is an odd function of ξ_k and everything else in the integral is even. Then change the integration variable $d\xi_k = \rho(E_k)dE_k$ and find

$$I = \frac{4\pi e}{\hbar}N_L N_R|T_{RL}|^2\int_{\Delta}^{eV} dE_k\frac{E_k}{\sqrt{E_k^2 - \Delta^2}}[1 - n_F(eV - E) - n_F(eV + E)] \tag{14.135}$$

The occupation factors are such that $I(-V) = -I(V)$. For positive eV the integral can be done analytically at zero temperature:

$$I = \frac{4\pi e}{\hbar}N_L N_R|T_{RL}|^2\sqrt{(eV)^2 - \Delta^2)}\,\Theta(eV - \Delta) \tag{14.136}$$

where $\Theta(eV - \Delta)$ is a step function that is zero for $eV < \Delta$ and one for $eV > \Delta$. The conductance is

$$\frac{dI}{dV} = \frac{4\pi e^2}{\hbar}N_L N_R|T_{RL}|^2\frac{eV}{\sqrt{(eV)^2 - \Delta^2)}}\Theta(eV - \Delta) \tag{14.137}$$

$$= \sigma_0\rho(eV)\Theta(eV - \Delta) \tag{14.138}$$

where σ_0 is the conductance of the NN junction, and $\rho(eV)$ is the superconducting density of states evaluated at the energy $E = eV$. A tunneling measurement between a normal metal and a superconductor provides a direct measurement of the density of excitations in the superconductor.

It is interesting that the current is an antisymmetic function of the voltage $I(-V) = -I(V)$. Two different processes are involved for the two directions of current. As shown in fig.

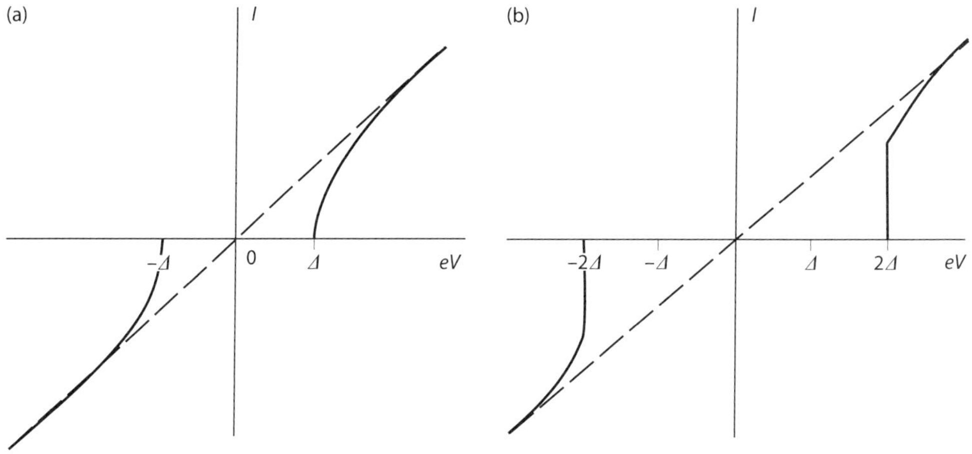

FIGURE 14.10. (a) Tunneling in an NS junction. Dashed line is the result for two normal metals $I=\sigma V$. (b) Tunneling between two superconductors.

14.9, in one direction of current an electron on the normal side tunnels directly into the excitation spectrum of the superconductor. In the other direction of current, a pair is broken in the superconductor, and one electron tunnels while the other becomes an excited quasiparticle. Figure 14.10a shows the I vs. V measurement at zero temperature between a normal metal and a superconductor. No current flows until $eV > \Delta$. The dashed line is the result for two normal metals: $I = \sigma V$.

14.4.2 Superconductor–Superconductor

The two sides of the tunnel junction could be the same superconductor or could be different metals. Either way, two basic processes are possible for the tunneling of one electron.

- An excited quasiparticle on one side can tunnel to the other and become an excited quasiparticle.
- A bound pair is broken on one side; one electron tunnels and the other becomes a quasiparticle. This process is the only one possible at low temperatures where the density of quasiparticles is very low. If the initial pair is on the left, the initial and final energies are

$$E_i = 2\mu_L \tag{14.139}$$

$$E_f = \mu_L + E_L + \mu_R + E_R \tag{14.140}$$

Since energy is conserved, $E_i = E_f$, which gives

$$\mu_L - \mu_R = eV = E_L + E_R > \Delta_L + \Delta_R \tag{14.141}$$

No tunneling occurs until the voltage is larger than the summation of the two gap functions. At the threshold voltage $eV_T = \Delta_L + \Delta_R$ the current jumps discontinuously to a nonzero value. This result is shown in fig. 14.10b.

An interesting feature of electron tunneling between two superconductors is that a dc current can flow at zero applied voltage. Bound pairs of electrons are tunneling together, which is possible when the chemical potentials are aligned on the two sides of the junction. This pair tunneling is one manifestation of the Josephson effect.

14.4.3 *Josephson Tunneling*

Josephson tunneling is due to a bound pair that tunnels together through the oxide from one superconductor to the other. If the bound pair is in a spin singlet on one side, then Josephson tunneling can occur only if the bound pair is a singlet on the other side, since tunneling does not flip any spins. In an *s*-wave or *d*-wave superconductor, the spins are in a singlet, while for a *p*-wave superconductor they are in a spin triplet. So Josephson tunneling is not permitted between a *p*-wave superconductor and one with either *s*- or *d*-wave symmetry.

A phenomenological theory of pair tunneling can be derived from Ginzburg-Landau theory. Consider two superconductors separated in space by an insulating region. The insulator is typically a metal oxide—often the oxide of one of the metals on the two sides of the junction. Each side of the junction has a superconducting wave function $\psi_{1,2}$. The two wave functions $\psi_{1,2}$ are weakly coupled by pair tunneling. This coupling can be expressed by the two coupled equations

$$\frac{\partial}{\partial t}\psi_1 = \frac{-i}{\hbar}[E_1\psi_1 + K\psi_2] \tag{14.142}$$

$$\frac{\partial}{\partial t}\psi_2 = \frac{-i}{\hbar}[E_2\psi_2 + K\psi_1] \tag{14.143}$$

The small constant $K \sim T^2$ is proportional to the tunneling probability through the insulating barrier. Write $\psi_j = \sqrt{n_j}\exp[i\phi_j]$ and take the time derivative

$$\frac{\partial}{\partial t}\psi_1 = \frac{\dot{n}_1}{2\sqrt{n_1}}e^{i\phi_1} + i\dot{\phi}_1\psi_1 = \frac{-i}{\hbar}[E_1\psi_1 + K\psi_2] \tag{14.144}$$

Multiply every term in this equation by $2\sqrt{n_1}\exp[-i\phi_1]$:

$$\dot{n}_1 + 2i\dot{\phi}_1 n_1 = \frac{-2i}{\hbar}\left[E_1 n_1 + K\sqrt{n_1 n_2}e^{i(\phi_2-\phi_1)}\right] \tag{14.145}$$

Take the real part of this equation, which eliminates most of the terms:

$$\dot{n}_1 = \frac{I_J}{2e}\sin(\phi_2 - \phi_1) \tag{14.146}$$

$$I_J = \frac{4eK}{\hbar}\sqrt{n_1 n_2} \tag{14.147}$$

The last equation describes the Josephson current, which results from the tunneling of pairs. The tunneling occurs whenever $\phi_1 \neq \phi_2$. Then pairs will try to go from one side of the junction to the other, in an effort to get the two phases to be equal. The Josephson current can be derived from the tunneling Hamiltonian (see Mahan [17]):

$$I_J = \frac{4e}{\hbar}\sum_{\mathbf{kp}} T_{\mathbf{kp}} T_{-\mathbf{k}-\mathbf{p}} \frac{\Delta_L(\mathbf{p})\,\Delta_R(\mathbf{k})}{E_L(\mathbf{p})\,E_R(\mathbf{k})\,[E_L(\mathbf{p}) + E_R(\mathbf{k})]} \tag{14.148}$$

The dc Josephson effect is found by putting a current through the tunnel junction and measuring the voltage difference. It is found that a current will flow between two superconductors with no voltage difference. This current is the tunneling of a bound pair from one superconductor to the other. The maximum value of dc current that can flow is when $\delta\phi = \phi_1 - \phi_2 = \pi/2$:

$$I < I_J \tag{14.149}$$

If the imposed current through the junction exceeds this value, the device will jump to the single-particle tunneling curve.

The ac Josephson effect occurs when a small voltage difference V is imposed between the two sides of the junction. It arises from the terms with $\dot{\phi}_i$. Derive from eqn. (14.145) the similar curve for $\dot{n}_2$ by interchanging indices one and two. Then take the complex conjugate:

$$\dot{n}_2 - 2i\dot{\phi}_2 n_2 = \frac{2i}{\hbar}\left[E_2 n_2 + K\sqrt{n_1 n_2}\, e^{i(\phi_2 - \phi_1)}\right] \tag{14.150}$$

Divide the above equation by n_2 and add it to eqn. (14.157) divided by n_1. Take the imaginary part of the resulting equation:

$$\dot{\phi}_1 - \dot{\phi}_2 = -\frac{E_1 - E_2}{\hbar} + K\cos(\phi_1 - \phi_2)\frac{n_1 - n_2}{\sqrt{n_1 n_2}} \tag{14.151}$$

Assume identical superconductors on the two sides of the junction ($n_1 = n_2$) and the last term vanishes. The phase difference between the two sides will vary in time if $E_1 \neq E_2$. The bound pairs have an energy equal to the chemical potential $E_j = 2\mu_j$, where the factor of two comes from two electrons. In equilibrium $\mu_1 = \mu_2$ and the phases do not vary in time. However, the application of a dc voltage between the two sides of the junction will cause the chemical potentials to be different:

$$E_2 - E_1 = 2(\mu_2 - \mu_1) = 2eV \tag{14.152}$$

$$\phi_1 - \phi_2 = \omega t, \quad \omega = \frac{2eV}{\hbar} \tag{14.153}$$

$$I(t) = I_J \sin(\omega t) \tag{14.154}$$

The current oscillates with a frequency $\omega = 2eV/\hbar$. The pairs tunnel through the oxide, from one chemical potential to the other. Since this process does not conserve energy, the pairs tunnel back. The back-and-forth motion creates an oscillating current, which radiates power. The prediction $\omega = 2eV/\hbar$ has been used to measure the ratio of fundamental constants e/h, since both the frequency and the voltage can be measured with great accuracy.

The Josephson current is very sensitive to magnetic fields. This dependence is large enough that the magnetic field of the earth will influence the measurement, and must be canceled by Helmholtz coils to get a null field on the junction. The magnetic field dependence can be derived by writing the phase in a gauge-invariant way. Write it as

$$\delta\phi(t) = \omega t = -\frac{2e}{\hbar}\int_0^t dt' \int_a^b dx \frac{dV}{dx} \tag{14.155}$$

The limits $a < x < b$ go from one superconductor to the other. The derivative of the voltage is an electric field, which is

$$\frac{dV}{dx} \Rightarrow -E_x = \frac{\partial A_x}{\partial t} + \frac{dV}{dx} \tag{14.156}$$

The static magnetic field is introduced through the vector potential:

$$\int_0^t dt' \frac{\partial A_x}{\partial t'} = A_x \tag{14.157}$$

The integral $\int dxA_x$ is a line integral. Using Stoke's theorem gives

$$\int dxA_x = \int d\mathbf{l} \cdot \mathbf{A} = \int d\mathbf{S} \cdot \mathbf{B} \tag{14.158}$$

$$\delta\phi(t) = \omega t - \frac{2e}{\hbar}\int d\mathbf{S} \cdot \mathbf{B} \tag{14.159}$$

The geometry is shown in fig. 14.11. The two superconductors are the shaded regions, and the strip between is the oxide tunneling barrier. The x-axis is perpendicular to the layer, while the z-axis is along the oxide barrier. The magnetic field is in the y-direction, perpendicular to the plane of the drawing. The integral is

$$\int d\mathbf{S} \cdot \mathbf{B} = \int_a^b dx' B(x') \int_0^z dz' = Bz(d + \Lambda_L + \Lambda_R) \tag{14.160}$$

$$\delta\phi(z, t) = \phi_0 + \omega t + kz \tag{14.161}$$

$$k = -\frac{2eB}{\hbar}(d + \Lambda_L + \Lambda_R) \tag{14.162}$$

$$I_J = I_J \sin(\delta\phi) \tag{14.163}$$

where d is the oxide thickness and $\Lambda_{R,L}$ are the penetration depths of the two superconductors. In a magnetic field, the phase $\delta\phi$ depends on both (z, t).

The total current must be integrated along the length of the tunnel junction, say, from $-L/2 < z < L/2$. Write the current density as

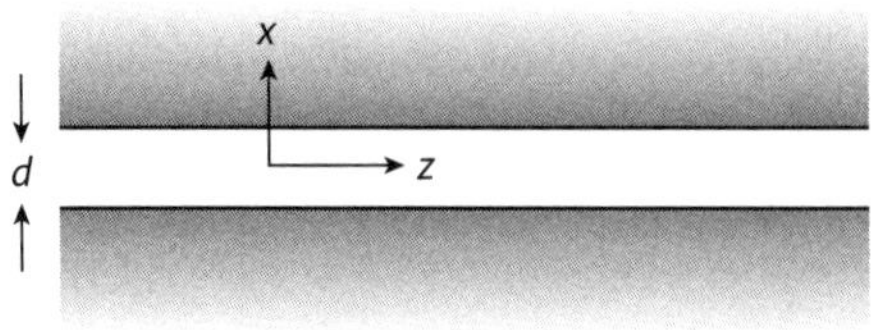

FIGURE 14.11. Coordinate system for the Josephson effect.

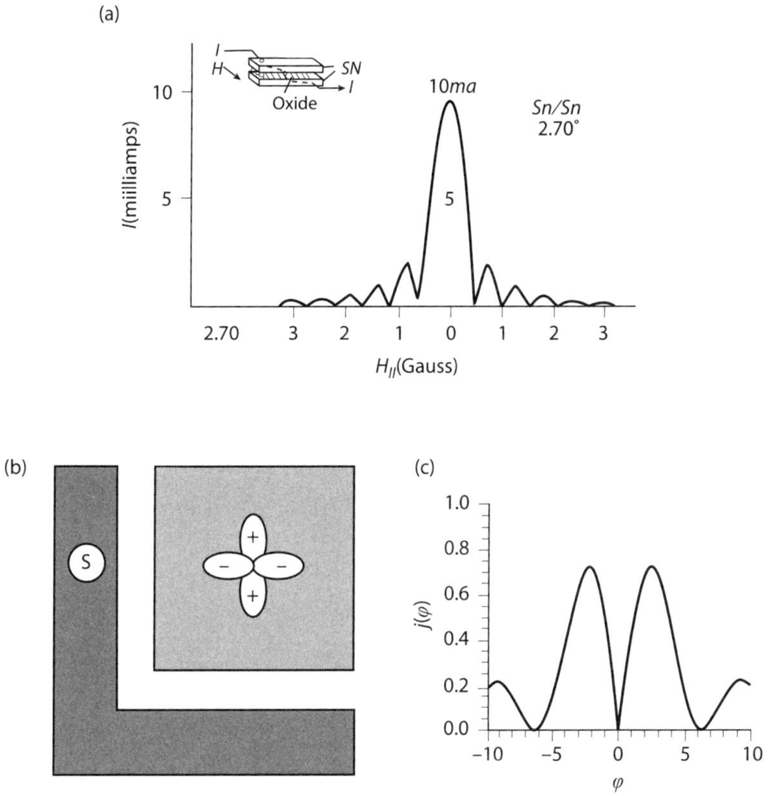

FIGURE 14.12. (a) Fraunhofer pattern between two *s*-wave superconductors. (b) Geometry of corner junction. (c) Corner junction in a magnetic field. The pattern indicates a *d*-wave superconductor.

$$J(z, t) = J_0 \sin(\phi_0 + \omega t + kz) \tag{14.164}$$

$$I = \frac{1}{L}\int_{-l/2}^{L/2} dz J(z, t) = J_0 \sin(\phi_0 + \omega t)\left[\frac{\sin(\Theta)}{\Theta}\right] \tag{14.165}$$

$$\Theta = \frac{kL}{2} \propto B \tag{14.166}$$

Usually the experiment is done at zero voltage so that $\omega = 0$. Then the maximum current has a Fraunhofer pattern $\sim \sin(\Theta)/\Theta$, which is readily observed. This measurement is a dramatic verification of Josephson's proposal of pair tunneling. Figure 14.12a shows the Fraunhofer pattern of Josephson junction in a magnetic field between two *s*-wave superconductors.

14.4.4 *Andreev Tunneling*

In 1964 Andreev proposed another tunneling process between a normal metal and a superconductor [2]. It is now called Andreev tunneling, and has been well studied. Assume the applied voltage raises the chemical potential on the normal side, but such that $eV < \Delta$. Then at zero temperature, no NS tunneling should occur. However, a small dc current is found, due to Andreev tunneling. A normal electron cannot tunnel by itself, since it lacks sufficient energy to become a quasiparticle in the superconductor. Say this electron has $\xi(p) < 0$ on the normal side, but $E(p) > \mu_R$. It teams up with another normal electron $E(p') < \mu_R$. The two electrons tunnel separately and form a bound pair in the superconductor. So a current flows. In the normal metal, there is left behind a hole where $\mathbf{p}'$ used to be and another hole where $\mathbf{p}$ used to be. For current flow in the opposite direction, a bound pair tunnels, as in the Josephson effect. However, since there is normal metal on the other side of the junction, the pair breaks up into two normal quasiparticles. Since the two electrons tunnnel separately, the matrix element for tunneling is $\sim T^2$ and the current depends on $|T|^4$. Andreev tunneling is most important in junctions with thin barriers, where T is not so large. Josephson tunneling is also done in thin barriers.

At first it seems that Josephson tunneling and Andreev tunneling are different processes. However, since pairs tunnel in both processes, they are closely related. Furusaki and Tsukada [12] calculated the rate of Andreev tunneling between two superconductors, and found

$$I = \pi\sigma(1+K)\frac{k_BT}{e}\frac{\Delta_R\Delta_L\sin(\delta\phi)}{\omega_n^2+\Delta_R\Delta_L\cos(\delta\phi)+K\Omega_{nR}\Omega_{nL}} \tag{14.167}$$

$$\omega_n = \pi k_BT(2n+1), \quad \Omega_{nL,R} = \sqrt{\omega_n^2+\Delta_{L,R}^2} \tag{14.168}$$

The quantity K controls the scattering at the interface. For $K\to\infty$ one recovers the Green's function formula (14.148) for Josephson tunneling. $\delta\phi$ is the phase difference between the two superconductors. The above formula is more accurate. The tunneling depends on the product of the two gap functions $\Delta_R\Delta_L$.

14.4.5 *Corner Junctions*

Soon after the discovery of high-temperature superconductivity in the cuprates, there began a discussion on the nature of the gap function $\Delta(\theta)$. Two main contenders were the anisotropic *s*-wave gap and the *d*-wave gap $\Delta(\theta) = \Delta_0\cos(2\theta)$, where $\tan(\theta) = k_y/k_X$ is the angle in the planes of copper and oxygen. Many different experiments, including tunneling, tried to determine which of these two gap functions were found in these materials. The definitive experiment was the corner junction of Dale Van Harlingen's group in Illinois [26].

They made a corner junction between a cuprate supeconductor and lead. Lead is a known *s*-wave superconductor. Their device is illustrated in fig. 14.12b. The square is the cuprate superconductor, and lead is wrapped around one corner of the square. Both materials had spin singlets, so Josephson tunneling is permitted. How does *d*-wave symmetry affect the Josephson tunneling? The above formula for Andreev/Josephson tunneling shows that the tunneling rate is proportional to the product of the energy gaps in the superconductors on two sides of the junction:

$$I \propto \frac{k_BT}{e}\sum_n \Delta_R\Delta_L\sin(\delta\phi) \tag{14.169}$$

The tunneling matrix element should also be proportional to $\Delta(\theta)$. In a corner junction, one arm of the corner is at $\theta = 0$ and $\Delta_d(0)$ is positive. The other arm of the corner is at $\theta = \pi/2$, $\cos(2\theta) = -1$, and $\Delta_d(\pi/2)$ is negative. When averaging the current, one gets a contribution from each arm of the corner. They have opposite signs and tend to cancel. If the two arms of the corner have the same length and the same oxide thickness, the current will be zero at zero magnetic field. The current flow at nonzero B is

$$I \propto I_0 \left[\frac{1 - \cos(\Theta)]}{\Theta} \right] \tag{14.170}$$

$$\Theta = \frac{kL}{2} \propto B \tag{14.171}$$

This behavior was observed by van Harlingen [26].

14.5 Cuprate Superconductors

The superconductors with the highest transition temperatures are the cuprates. They contain planes of copper and oxygen in the ratio CuO_2. They form a two-dimensional square lattice. These planes conduct electrons, and the superconducting currents are in these planes. There are now many different structures with these planes, with many different transition temperatures. In some cases, each CuO_2 plane is relatively isolated, while in others two, or perhaps three, planes are close together in the unit cell. In all cases, superconductivity is a three-dimensional phase. See the book by Gerry Burns [7] for a list of materials.

Another feature of these materials is that they are usually antiferromagnetic in the pure crystals. The copper ions are divalent, which creates a single hole in the atomic *d*-orbitals. The system of copper atoms forms a spin one-half antiferromagnetic whose properties are well described by the Heisenberg model. The materials are intentionally doped to change the number of electrons in the conduction band. Somewhere around $x = 10$–15% doping, the antiferromagnetic state is destroyed, and the copper oxide planes become a correlated electron gas. With a further increase in impurity levels, the planes become superconducting. A generic phase diagram is shown in fig. 14.13. The horizontal axis is labeled x to denote adding or removing electrons through intentional alloying. An example is $La_{2-x}Sr_xCuO_4$. Strontium is divalent, while lanthanum is trivalent. As x is increased, one is removing electrons, which is called *hole doping*.

In the region of doping where there is superconductivity, smaller values of x are called *underdoping*, while larger values of x are called *overdoping*. The superconducting properties are different in these two regimes.

14.5.1 *Muon Rotation*

Uemura and associates [25] measured the penetration depth using muon spin rotation. μ-Mesons are fermions with a characteristic magnetic moment. From a muon source, they are injected into the surface region of a superconductor in the presence of a magnetic field. From the precession of the magnetic moment, a measurement determines

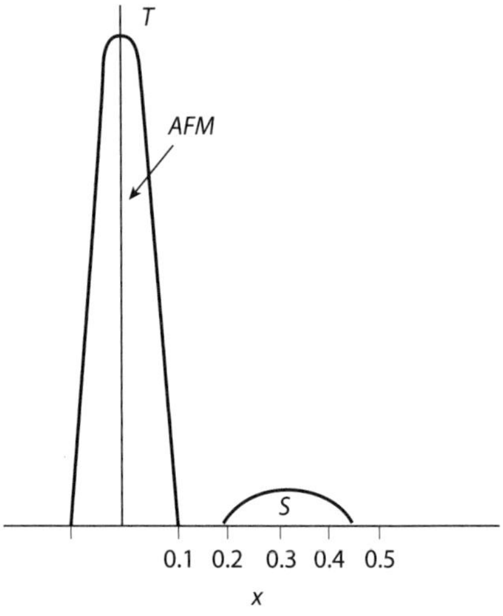

FIGURE 14.13. Phase diagram of a cuprate superconductor. The horizontal axis is the density of conducting electrons in the copper oxide plane. The vertical axis is temperature. The material is antiferromagnetic as a pure crystal. Changing the density of conduction electrons by intentional alloying destroys the AF state and then creates the superconducting state.

the penetration depth of the magnetic field. The muons penetrate much further than the field.

Recall that the inverse penetration depth is

$$\frac{1}{\Lambda^2} = \frac{n_s e^2}{\varepsilon_0 m^* c^2} \tag{14.172}$$

where n_s is the density of superconducting electrons, and m^* is the effective mass of conduction electrons in the cuprate planes. Measurements were made for many different cuprate superconductors. The different results were interpreted as caused by different values of n_s. All the cuprates have similar effective masses, since they all have identical planes of copper and oxygen. Two very important results emerged from these experiments.

1. The values of n_s are small. They are proportional to x. This result is interpreted as stating that the Hubbard model splits the energy band into an upper Hubbard band, and a lower Hubbard band. The lower Hubbard band is filled when $x = 0$, in the AF state. When hole doping, only the holes contribute to the density of supeconducting electrons—actually, to the density of superconducting holes.

2. There is a linear relation between the supercondcuting transition temperature (T_c) and the $1/\Lambda^2$. Ignoring fundamental constants, the relationship is

$$T_c \propto \frac{n_s}{m^*} \tag{14.173}$$

Figure 14.14 shows the linear relationship between T_c and the muon spin relaxation rate $\sigma \propto 1/\Lambda^2$. The series of points are different values of x for the material. In the underdoped region, T_c depends mainly on the number of conducting holes. In two dimensions, the Fermi energy is proportional to the density of electrons, so that $T_c \propto E_F$. These experimental observations argue against a phonon mechanism.

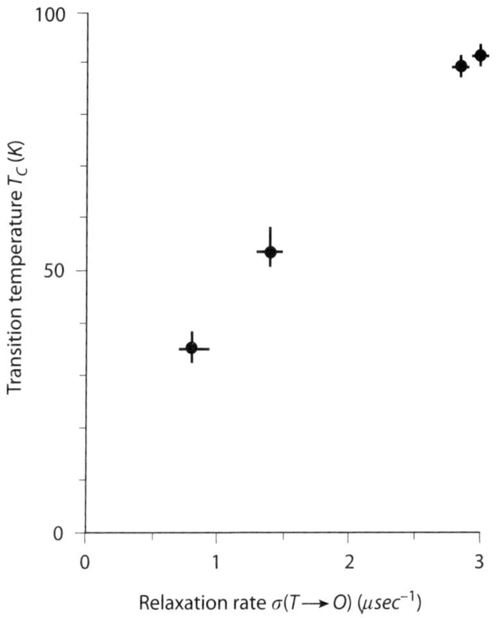

FIGURE 14.14. T_c vs. muon spin relaxation rate σ, which is proportional to n_s/m^*. Different numbers refer to different materials, and each series of points are different values of x for the same material. From Uemura et al., *Phys. Rev. B* **38**, 909 (1988). Used with permission of the American Physical Society.

As an example, the penetration depth varies as a function of temperature:

$$\frac{\Lambda^2(0)}{\Lambda^2(T)} = 1 - \left(\frac{T}{T_c}\right)^4 \propto n_s(T) \tag{14.174}$$

In $La_{2-x}Sr_xCuO_4$, it is found that $\Lambda(0) = 140$ nm. Using the formula for the penetration depth,

$$n_s(T = 0) = \frac{m^*}{m_e} 1.44 \times 10^{27} m^{-3} \tag{14.175}$$

The lattice constant in the plane is $a = 3.79$ Å, and the distance between planes is $c = 6.6$ Å. The volume of the unit cell is $\Omega_0 = a^2c$, and

$$n_s(T = 0)\,\Omega_0 = \frac{m^*}{m_e} 0.137 \approx 0.27 \sim x \tag{14.176}$$

The density of holes in the conducting plane is indeed small.

Both the AF state and the muon precession experiments emphasize the role that strong correlations play in the electronic properties of the cuprates. That $n_s \propto x$ makes sense only if the upper Hubbard band is split off from the lower Hubbard band. Antiferromagnism has the same requirement. Strong corelations are also evident in the normal state of the system when $T > T_c$.

As an example, consider La_2CuO_4. In the pure material the valence of the various ions are La^{3+}, Cu^{2+}, O^{2-}. There are 8 positive charges, and 8 negative charges, so the unit cell is charge neutral. Next, count energy bands. The lanthanum ions are out of the plane and do not enter into the energy bands of electrons or holes in the conducting plane. There are 5 *d*-orbitals from copper, and 6 *p*-orbitals from the two oxygens. Including spin degeneracy, the 11 orbitals have 22 possible electron states. There are 9 electrons in Cu^{2+} and 12 electrons in the filled shells of the two oxygens. The 21 electrons do not quite fill up the 22 available states. The highest energy band, which is antibonding, should be half-full. The earliest energy band calculations predicted this feature: a Fermi surface was generated based on a half-filled, paramagnetic, antibonding band. Since the material is antiferromagnetic and insulating, this simple energy band picture is incorrect. The material is a Mott insulator due to electron correlation. The Hubbard energy U divides the antibonding band into an upper Hubbard band and a lower Hubbard band, which do not overlap. The lower Hubbard band is completely filled, which makes it an insulator. The upper Hubbard band is the twenty-second energy band, which is completely empty.

14.5.2 *Magnetic Oscillations*

Magnetic oscillations, such as the Shubnilov-de Haas effect or the de Haas-van Alphen effect, generally require that $\omega_c\tau > 1$, where $\omega_c = eB/m^*$ is the cyclotron frequency, and τ is the hole lifetime. Since the cuprates are alloyed in order to be superconducting, the lifetime is rather small due to scattering from the alloy fluctuations. It was believed that $\omega_c\tau > 1$ was not experimentally possible. However, national magnet laboratories can create pulsed magnetic fields up to $B \sim 100$ T using large banks of capacitors. Using these facilities, an international group measured magnetic oscillations in several cuprates [4].

They reported results on two different materials. The first is 123 ($YBa_2Cu_3O_{6.5}$), and the second is 124 ($YBa_2Cu_4O_8$). They found the effective mass to be $m^* = 2.7 \pm 0.3 m_e$ in 124, where m_e is the mass of a free electron. It was $m^* = 2.0 m_e$ in 123. The most interesting result was the area of the Fermi surface:

- In 123 it was 2.0% of the area of the BZ for a hole doping of $x = 0.10$.
- In 124 it was 2.4% of the area of the BZ for a hole doping of $x = 0.14$.

These areas are obviously very small. The measurement does not provide information on where the areas are located within the BZ. In a square lattice, they could be at the corners (one pocket), in the centers of the edges (two pockets), or at a general point in the BZ (four or eight pockets). If the pockets are identical and give identical signals, such small areas do correspond to the the amount of hole doping.

The conclusion from these measurements is that the density of holes is indeed given by x, in agreement with the muon precession experiments.

14.6 Flux Quantization

An interesting phenomenon in superconductors is the quantization of magnetic flux in the center of hollow superconducting cylinders. The phenomenon was first reported simultaneously by Deaver and Fairbanks [8] and Doll and Näbauer [9].

Let $\mathcal{A}$ denote a finite area, which could be a circle or a rectangle. The scalar dS is an infinitesimal part of this area, and the total area is obtained by an integral:

$$\mathcal{A} = \int dS \tag{14.177}$$

Let Φ denote the total magnetic flux through this finite area, which is obtained by integrating the normal component of the magnetic field **B**:

$$\Phi = \int dS\hat{n} \cdot \mathbf{B} = \int dS\hat{n} \cdot \nabla \times \mathbf{A} \tag{14.178}$$

where **A** is the vector potential. Next use the Kelvin-Stokes theorem to change the surface integral into a line integral $\vec{d\ell}$ around the circumference of the area:

$$\Phi = \oint \vec{d\ell} \cdot \mathbf{A} \tag{14.179}$$

The magnetic flux in a region equals the line integral of the vector potential around the circumference.

Flux quantization starts from the Bohr-Sommerfeld quantization condition:

$$2\pi\hbar(n + 1/2) = \oint \vec{d\ell} \cdot \mathbf{p} \tag{14.180}$$

Use the fact that $\mathbf{p} = m\mathbf{v} + (q/c)\mathbf{A}$ to change the above formula to

$$\begin{aligned} h(n + 1/2) &= \oint \vec{d\ell} \cdot [m\mathbf{v} + q\mathbf{A}] \\ &= m \oint \vec{d\ell} \cdot \mathbf{v} + q\Phi \end{aligned} \tag{14.181}$$

Figure 14.15a shows a hollow cylinder of a superconductor. The magnetic field is perpendicular to the plane of the drawing, and goes along the axis of the cylinder. The above line integral is taken in a circle inside of the superconductor. It encloses the magnetic flux in the hollow part of the cylinder. The flux does not penetrate into the superconductor. Along this path, the velocity integral $\oint \vec{d\ell} \cdot \mathbf{v} = 0$. In a superconductor, the electrons are paired, so $q = 2e$, where e is the charge on an electron. The above equation is

$$\Phi = \frac{h}{2e}(n + 1/2) = \frac{\phi_0}{2}(n + 1/2) \tag{14.182}$$

This equation states that the magnetic flux Φ in the hollow cylinder is multiples of half the flux quantum $\phi_0 = h/e$. As the magnetic field is increased from zero, the flux inside the cylinder is not a linear function of field, but is a series of steps, as in a staircase. This behavior was verified by the experimental groups.

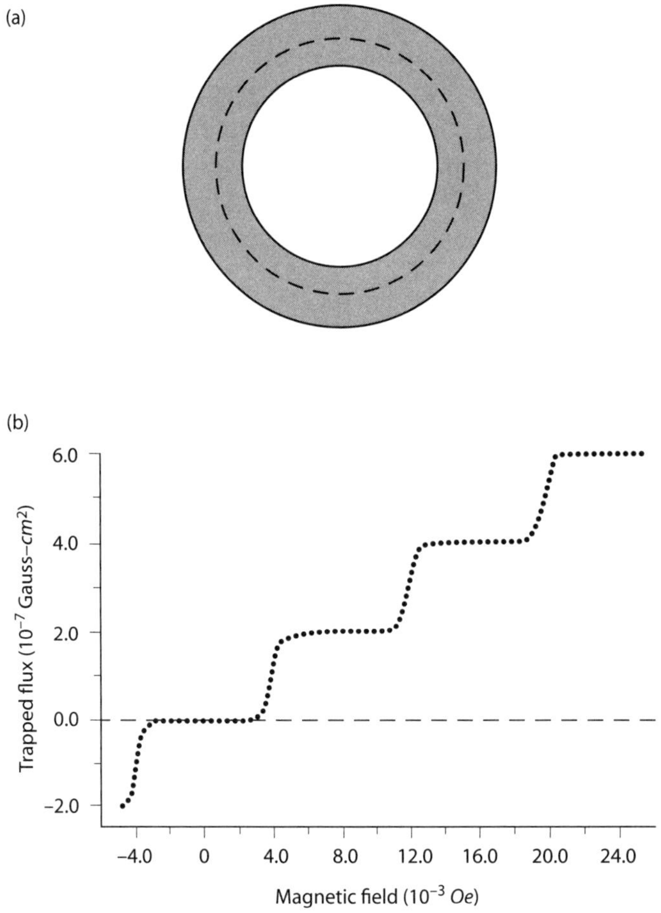

FIGURE 14.15. (a) Shaded area shows a hollow cylinder of a superconductor. Dashed line shows path of integration contour. (b) Magnetic flux quantization inside a hollow superconducting cylinder. From Goodman and Deaver, *Phys. Rev. Lett.* **24**, 870 (1970). Used with permission of the American Physical Society.

The observation is an example of a macroscopic quantum state. To see the steps, all of the electron states in the cylinder must have a coherent phase. Since the dimensions of the cylinder are millimeters, this phase is coherent over macroscopic dimensions. This is an important property of the superconducting state. Later experimental results by Deaver are shown in fig. 14.15b.

References

1. A. A. Abrikosov, On the magnetic properties of superconductors of the second group. *Sov. Phys. JETP* **5**, 1174–1183 (1957)
2. A. F. Andreev, The thermal conductivity of the intermediate state in superconductors. *Sov. Phys. JETP* **19**, 1228–1231 (1964)
3. P. B. Allen and B. Mitrovic, Theory of superconducting T_c. *Solid State Phys.* 37, 1–82 (1982)

4. A. F. Bangura et al., Small Fermi surface pockets in underdoped high-T_c superconductors: observation of Shubnikov-de Haas oscillations in $YBa_2Cu_4O_8$. *Phys. Rev. Lett.* **100**, 047004 (2008)
5. J. Bardeen, L. N. Cooper, and J. R. Schrieffer, Theory of superconductivity. *Phys. Rev.* **108**, 1175–1204 (1957)
6. G. Bednorz and K. A. Müller, Possible high-T_c superconductors in the Ba–Ca–Cu–O system. *Z. Phys. B* **64**, 189 (1986)
7. G. Burns, *High Temperature Superconductivity* (Academic Press, New York, 1992)
8. B. S. Deaver and W. M. Fairbanks, Experimental evidence for quantized flux in superconducting cylinders. *Phys. Rev. Lett.* **7**, 43–46 (1961)
9. R. Doll and M. Näbauer, Experimental proof of magnetic flux quantization in a superconducting ring. *Phys. Rev. Lett.* **7**, 51–54 (1961)
10. G. M. Eliashberg, Interactions between electrons and lattice vibrations in a superconductor. *Sov. Phys. JETP* **11**, 696 (1960)
11. H. Fröhlich, Theory of the superconducting state. *Phys. Rev.* **79**, 845–856 (1950)
12. A. Furusaki and M. Tsukada, dc Josephson effect and Andreev reflection. *Solid State Comm.* **43**, 10164–10169 (1991)
13. W. L. Goodman and B. S. Deaver, Detailed measurements of the quantized flux states of hollow superconducting cylinders. *Phys. Rev. Lett.* **24**, 870–873 (1970)
14. B. D. Josephson, Possible new effects in superconducting tunneling. *Phys. Lett.* **1**, 251 (1962)
15. L. D. Landau and V. Ginzberg, On the theory of superconductivity. *JETP-USSR* **20**, 1064–1084 (1950)
16. F. London and H. London, The electromagnetic equations of the supraconductor. *Proc. R. Soc. (London) A* **149**, 71–88 (1935)
17. G. D. Mahan, *Many-Particle Physics*, 3rd. ed. (Plenum-Kluwer, 2000)
18. W. L. McMillan, Transition temperature of strong-coupled superconductors. *Phys. Rev.* **167**, 331–344 (1968)
19. W. Meissner and R. Ochsenfeld, A new effect in penetration in superconductors. *Naturwissenschaften* **21**, 787 (1933)
20. R. W. Morse and H. V. Bohm, Superconducting energy gap from ultrasonic attenuation measurements. *Phys. Rev.* **108**, 1094–1095 (1957)
21. H. Kamerlingh Onnes, *Leiden Comm.* **120b**, **122b**, **124c** (1911)
22. N. E. Phillips, Heat capacity of aluminum. *Phys. Rev.* **114**, 676–686 (1959)
23. Z. A. Ren et al., Superconductivity at 55 K in iron-based F-doped layered quaternary compound $Sm[O_{1-x}F_x]FeAs$. *Chin. Phys. Lett.* **25**, 2215–2216 (2008)
24. M. Tinkham, *Introduction to Superconductivity*, Second Edition (McGraw-Hill, New York, 1996)
25. Y. J. Uemura et al., Universal correlations between T_c and n_s/m^* in high-T_c cuprate superconductors. *Phys. Rev. Lett.* **62**, 2317–2320 (1989)
26. D. A. Wollman, D. J. Van Harlingen, W. C. Lee, D. M. Ginzberg, and A. J. Leggett, Experimental determination of the superconducting pairing state in YBCO from the phase coherence of squids. *Phys. Rev. Lett.* **71**, 2134–2137 (1993)

Homework

1. For aluminum, (a) calculate the penetration depth in nanometers. The room-temperature plasma frequency is $\hbar\omega_p = 15.2$ eV. (b) Calculate the coherence length in nanometers. Find the Fermi velocity from $r_s = 2.07$, and the gap function using $\Delta = 1.75\ k_BT_c$.

2. At zero magnetic field, a superconducting current I (in amperes) is going through a wire of length L and cross-sectional area A. How does the phase $\Delta\phi$ vary from one end of the wire to the other? What is the superconducting velocity v_s?

3. Show that one can write the free energy density in G-L theory, in zero magnetic field, in terms of the superconducting velocity:

$$\mathcal{F} = \left[\alpha + \frac{1}{2} m v_s^2\right] |\psi|^2 + \frac{\beta}{2} |\psi|^4 \tag{14.183}$$

Vary this equation with respect to ψ^* and derive

$$j = 2en_s v_s \left[1 - \frac{m v_s^2}{2|\alpha|}\right] \tag{14.184}$$

Show that this formula predicts a maximum value for the superconducting current, which is called the critical current j_c.

4. According to BCS theory, what is the momentum distribution $n_{\mathbf{k}}$ of electrons in the superconductor at $T = 0$.

5. The entropy per unit volume of a superconductor in the BCS model is

$$S = -\frac{2k_B}{\mathcal{V}} \sum_{\mathbf{k}} \{n_F(E_k) \ln [n_F(E_k)] + [1 - n_F(E_k)] \ln [1 - n_F(E_k)]\} \tag{14.185}$$

Use this expression to derive the specific heat of the superconductor. Then use $\Delta(T) \approx \Delta_0 \sqrt{1 - T/T_c}$ to show that the specific heat is discontinuous at the critical temperature.

6. Calculate the Pauli spin susceptibility of the BCS superconducting state. Show that the formula at nonzero temperatures is

$$\chi = 2\mu_0^2 \sum_{\mathbf{k}} (d/dE_k)\, n_F(E_k) \tag{14.187}$$

7. To calculate the infrared absorption in a superconductor at zero temperature, one starts with the expression for the current operator:

$$j_\mu(\mathbf{q}) = \frac{e\hbar}{m\Omega} \left(k_\mu + \frac{1}{2} q_\mu\right) C^\dagger_{\mathbf{k}+\mathbf{q},\sigma} C_{\mathbf{k}\sigma} \tag{14.188}$$

Make the Bogoliubov transformation and identify the matrix element that creates two electrons in excitation states. Derive the matrix element in the limit that $\mathbf{q} \to 0$.

8. Derive the density of states $\rho(E)$ for a two-dimensional superconductor with an anisotropic *s*-wave energy gap: $\Delta(\theta) = \Delta_0 + \Delta_1 \cos(4\theta)$.

9. Derive the rate of single-electron tunneling between two identical superconductors. *Hint: The result is worked out in Many-particle Physics* [16].

10. Derive a formula for the Josephson current between an *s*-wave and a *d*-wave superconductor in a tunnel junction in a magnetic field when the two sides of the corner junction have unequal lengths (L_R, L_L).

15 Nanometer Physics

Nano means one-billionth. Nanoscience is concerned with phenomena that occur on the length scale of 10^{-9} meters, which is 10 Å, and denoted nm. Many different experimental systems are found under this heading.

- *Quantum dots* are small crystals whose size is several nanometers. In many respects, the electronic states and the vibrational states are similar to those of a large molecule. They are referred to as having *zero dimension*, which means they are confined in all directions.
- *Nanowires* have a diameter of a few nanometers, but their length can extend for microns or tens of microns. They are confined in two dimensions. Many of the properties resemble a system with free particles in one dimension.
- *Nanotubes* are hollow nanowires. They are made from carbon, boron nitride, or similar materials, and have many interesting properties. They are also largely one dimensional in their electronic behavior.
- *Graphene* is a sheet of carbon one atom thick with the honeycomb crystal structure. Three-dimensional graphite is composed of parallel layers of graphene. Graphene is a purely two-dimensional electronic material. It can vibrate in all three dimensions.
- *Quantum wells* are usually made in semiconductors using molecular beam epitaxy (MBE) or similar growth technology. A layer of one semiconductor is grown in a host material. The layer is typically several atoms in thickness. Often electrons are confined to these thin layers, so their free particle motion is two dimensional.
- A *superlattice* is a layered material composed of alternate layers of two materials. Each layer is several atoms thick. Electrons are usually confined to one type of layer, so that the free particle motion is again two dimensional.

Historically, quantum wells and superlattices were grown first. Experiments were done on them, and theory was developed to explain the experiments.

Quantum dots are prepared by many different methods. One way is to use silicon processing technology to isolate small regions on a silicon wafer, and these islands behave as isolated quantum systems. Another way to prepare quantum dots is to bombard a host material with ions of a different material. The impurity ions form a supersaturated system and diffuse around until they find each other. In time, they form small crystallites in the host material. The latter can be etched away, leaving the quantum dots.

Quantum wires are usually prepared using laser ablation to produce a plume of material, which attaches to metal particles. The metal particles act as a seed to grow the wire. The nanowire has the diameter of the seed particle. Repeated laser pulses provide new material for continued growth. The wires can be made very long, and often are single crystals or have only a few twin boundaries.

Graphene is produced by a simple process using Scotch tape. The tape is pressed on top of a single crystal of graphite. Then the tape is peeled off. A layer or two of carbon sticks on the tape. The tape can be dissolved, leaving a few layers of carbon. Occasionally, only one layer is left! Nanotubes are grown in furnaces.

The preparation of all of these materials involves as much art as science. The techniques are best learned from the experts.

15.1 Quantum Wells

Moleculear beam epitaxy is used to grow a material one atomic layer at a time. Molecular beams of material are directed toward the target substrate, and just enough is sent to grow a single layer. Different beam chambers are used to grow different layers, in the desired sequence. Good-quality single crystals are obtained if care is used to match the lattice constants of the different materials. Lattice matching is essential, since otherwise elastic strains develop at the interface and stretch into the layers. These strains change the electronic and vibrational properties, usually in a way that is undesirable.

15.1.1 *Lattice Matching*

Many different quantum wells and superlattices have been grown using III–V semiconductors materials. Most of them have the zincblende structure and can be alloyed with each other. If one binary material AB is alloyed with another CD, in the combination $(AB)_x(CD)_{1-x}$, the energy gap and lattice constants of the alloy are usually linear

$$E_G(x) = xE_{G,AB} + (1-x)E_{G,CD} \tag{15.1}$$

$$a(x) = xa_{AB} + (1-x)a_{CD} \tag{15.2}$$

The lattice constant of a material can be selected by alloying. Alternately, alloying can be used to choose the energy gap. Table 2.5 shows lattice constants of common III–V and II–VI semiconductors. They are often used in superlattices.

GaAs is often used for the quantum well material. The other material is called the *barrier layer*. It is selected to have two properties, compared with GaAs:

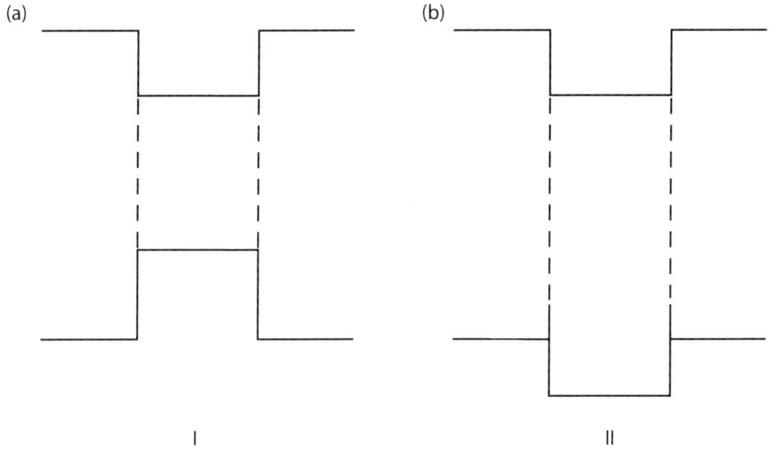

FIGURE 15.1. (a) Type I quantum wells, with both electron and holes having lower energies in the quantum well. (b) Type II wells have the electrons confined to the well, but the lowest hole bands are in the host material.

- The same lattice constant
- A larger energy gap

The obvious choice is AlAs, which has the same lattice constant. So GaAs is often embedded in AlAs as the host material. An option is to make the host material the alloy $Ga_xAl_{1-x}As$. It has the same lattice constant, but one can tune the value of the energy gap by selecting x.

By having the host material with a larger semiconductor band gap than the material of the quantum well, it is possible to confine electrons, or holes, to the quantum well. There are several physical possibilities, which are illustrated in fig. 15.1.

- In a type I QW, both electrons and holes are confined to the quantum well. The conduction band of the QW is below the value for the host, and the valence band is higher. This case is shown in fig. 15.1a. An example is a GaAs well with AlAs as the host material.
- In a type II QW, the electrons are confined to the well, but the holes are not. The lowest hole energy is in the host material. This case is shown in fig. 15.1b. An example is an InAs QW with GaSb as the host material. They are almost perfectly lattice matched.

15.1.2 *Electron States*

Denote (z) as the axis perpendicular to the QW, and (x, y) are the axes in the plane of the well. Electrons can move freely in the (x, y)-plane. Their kinetic energy is given in the effective mass approximation as

$$E(k_x, k_y) = E_{c0} + \frac{\hbar^2}{2}\left[\frac{k_x^2}{m_x} + \frac{k_y^2}{m_y}\right] \tag{15.3}$$

The effective masses (m_x, m_y) depend on those in the three-dimensional material. In GaAs, with the conduction band minimum at the center of the Brillouin zone, (m_x, m_y) equal the conduction band mass m_c of the bulk material. In silicon, with its conduction bands of six ellipsoids, the bands in the quantum well depend on the crystal orientation of the silicon layers. For example, are they grown in the (111) or the (100) direction? In general, the band structure of the quantum well is a projection onto the layer of the bulk band structure.

The electron motion in the z-direction is quantum confined. The electrons in the conduction band are confined to the QW in types I and II quantum wells. Finding the eigenstates in this direction is the simple problem of a particle in a box. This problem should be familiar to the student from a course in quantum mechanics. In the present case, there are two important features:

- The box has walls that have finite height.
- The effective mass in the box is different than outside.

The walls are of finite height, and an energetic particle can leave the QW and escape into the conduction band of the host material. Finding the height of the box requires a knowledge of *band offsets*. Although one knows the band gap of the two materials, QW and host, how they line up in energy is not obvious. If the difference in band gaps is ΔE_G, what fraction of this difference is a conduction band offset, and how much is a valence band offset? These band offsets are difficult to determine, experimentally or theoretically. They are now known for some pairs of materials.

The confinement problem of the particle in the box is solved using the effective mass approximation. The conduction band effective mass of the host material is usually different than the effective mass of the quantum well. The eigenfunction $\phi(z)$ is found inside and outside of the box. At the interface between QW and host, the wave function inside and out is matched according to the following criteria:

- The wave function $\phi(z)$ is continuous at the interface.
- The derivative of $\phi(z)$ is matched so that

$$\frac{1}{m(z)}\frac{d\phi}{dz} \tag{15.4}$$

is continuous at the interface.

The last condition uses the different conduction band masses for host and QW material in the matching of the derivative.

An example is given of this matching. Assume the bottom of the conduction band is the zero of energy, and the band offset is V_0. The energy of an electron is assumed to be in the range $0 < E < V_0$. The QW has a thickness d. Let the effective conduction band masses be m_{ci} inside, and m_{co} outside, of the QW. The eigenfunctions for symmetric states are

- Inside the QW

$$\phi(z) = A\cos(kz), \quad E = \frac{\hbar^2 k^2}{2m_{ci}} \tag{15.5}$$

- Outside the QW

$$\phi(z) = Be^{-\alpha(|z|-d/2)}, \quad V_0 - E = \frac{\hbar^2\alpha^2}{2m_{co}} \tag{15.6}$$

- Matching at the interface $z = d/2$

$$A\cos(kd/2) = B \tag{15.7}$$

$$-\frac{Ak}{m_{ci}}\sin(kd/2) = -\frac{\alpha B}{m_{co}} \tag{15.8}$$

- Divide these last two equations to cancel A and B, which gives the eigenvalue equation

$$\frac{k}{m_{ci}}\tan(kd/2) = \frac{\alpha}{m_{co}} \tag{15.9}$$

$$\sqrt{E}\tan\left[\sqrt{E/E_d}\right] = \sqrt{\frac{m_{ci}}{m_{co}}}\sqrt{V_0 - E} \tag{15.10}$$

$$E_d = \frac{\hbar^2}{2m_{ci}}\left(\frac{2}{d}\right)^2 \tag{15.11}$$

Symmetric confined states in the QW are found for every value of E that satifies this equation. A similar calculation can be done for the antisymmetric eigenstates.

Assume the QW has three bound states in the z-direction, with energies E_1, E_2, E_3. The total energy of an electron is

$$E_j(k_x, k_y) = E_j + \frac{\hbar^2}{2}\left[\frac{k_x^2}{m_x} + \frac{k_y^2}{m_y}\right] \tag{15.12}$$

The density of states for electrons confined to the quantum well, but free to move in the (x, y)-plane, is

$$N(E) = 2\sum_{j=1}^{3}\int\frac{dk_x dk_y}{(2\pi)^2}\delta[E - E_j(k_x, k_y)] \tag{15.13}$$

$$= \frac{\sqrt{m_x m_y}}{\pi\hbar^2}\sum_{j=1}^{3}\Theta(E - E_j)$$

where the $\Theta(E)$ is the step function. The density of state in two dimensions is a constant for $E > E_j$. Figure 15.2 shows the total density of states, which is a series of three steps.

15.1.3 *Excitons and Donors in Quantum Wells*

In three dimensions, the eigenstates of electrons bound to donors have an envelope function that resembles the hydrogen atom. This analogy is particularly valid for semiconductors with a single conduction-band minimum. Distances are scaled using the effective Bohr radius $a_0^* = a_0(m_e/m_c)[\varepsilon(0)/\varepsilon_0]$. It is the hydrogenic radius $a_0 = 0.0529$ nm, increased by the static dielectric constant $\varepsilon(0)/\varepsilon_0 \sim 10$, and the effective mass ratio $(m_e/m_c) \sim 10$. The effective Bohr radius is then on the order of 3–10 nm. Many QWs have a width much

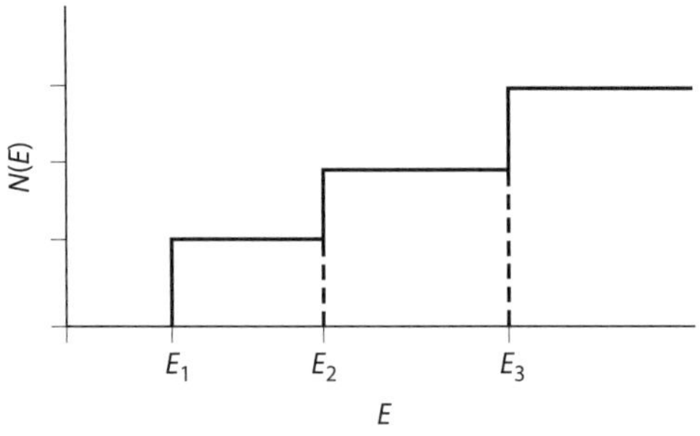

FIGURE 15.2. Density of states of an electron in a quantum well with three bands.

narrower than the donor Bohr radius. The eigenfunction will be severely affected by the quantum confinement.

The donor eigenfunction will depend on where the donor is located in the QW. A simple case is when the donor is exactly in the center, but donors are seldom that accomodating. They could be in the center if they were intentionally located there during the MBE growth process. However, for volunteer impurities, they can be located anywhere. They could be outside the QW, in the host material.

The calculation of the donor state in the QW is done by variational methods, since an exact solution has not been found. The electron must be confined to the QW in the z-direction, so the variational eigenfunction has a factor of $\phi(z)$, as derived above for free particles. There is also an exponential factor $\sim\exp(-r/a_0^*)$ from the hydrogenic eigenfunction. In cylindrical coordinates (ρ, Θ, z), some posssible variational eigenfunctions are

$$\psi_1(\mathbf{r}) = A\phi(z)\exp\left[-\lambda\rho/a_0^*\right] \tag{15.14}$$

$$\psi_2(\mathbf{r}) = A\phi(z)\exp\left[-\lambda\sqrt{\rho^2 + (z - z_i)^2}/a_0^*\right] \tag{15.15}$$

where λ is a variational parameter, and z_i is the location $(-d/2 < z\_i < d/2)$ of the impurity in the QW. The first choice uses the two-dimensional form for the hydrogen atom, while the second choice uses the three-dimensional form. Since the two-dimensional hydrogen atom has a binding energy four times larger than in three dimensions, the donor is usually more bound in the QW than in the three-dimensional material. The binding energy depends on the two parameters (d, z_i). All calculations of donor binding use an effective Hamiltonian in the effective mass approximation in the QW of

$$H = -\frac{\hbar^2\nabla^2}{2m_{ci}} - \frac{e^2}{4\pi\varepsilon(0)\sqrt{\rho^2 + (z - z_i)^2}} + V(z) \tag{15.16}$$

where $V(z)$ is the potential that confines the particle to the QW. The Coulomb potential has no image charge corrections, which is correct if the host material has the same static dielectric constant $\varepsilon(0)$ as the material in the QW. This identity is usually assumed, although it is rarely correct. Any variational calculation must be done on the computer, and the results are presented through graphs. The first careful calculations of donor binding in GaAs QWs were done by Greene and Bajaj [2]. Some of their results are shown in fig. 15.3.

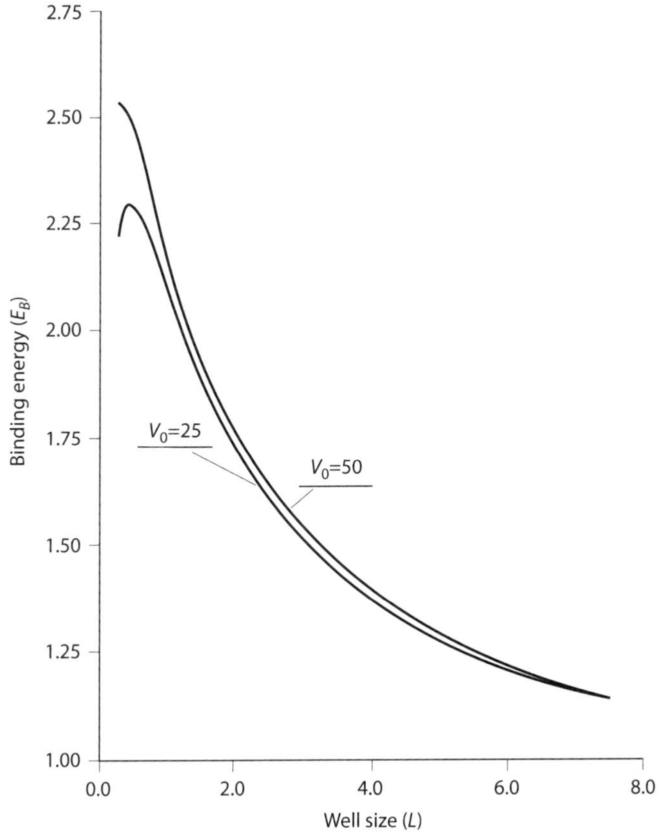

FIGURE 15.3. Donor binding in a QW. Binding energy E_B and V_0 in units of E^*_{Ry}, and well width L in units of a^*_0. From Greene and Bajaj, *Solid State Commun.* **45**, 825 (1983). Used with permission of Elsevier.

The exciton in the QW is a similar problem to the donor. In type I QWs both the electron and the hole are bound in the quantum well. Both electron and hole are confned in the z-direction, but can move freely within the plane of the QW.

In Chapter 10 on Extrinsic Semiconductors, the metal–insulator transition was described as donor impurities are added to the semiconductor. For the Mott criteria $n_D^{1/3} a_0^* > 0.25$, the donor binding energy vanishes due to screening of the Coulomb potential. It is interesting that this phenomenon does *not* happen in two dimensions. As one adds more electrons to the two-dimensional system, they provide additional screening to the Coulomb interaction. However, the donor binding energy for the screened interaction never vanishes. The Coulomb lines of force can go outside of the QW, where screening is ineffective. So one never achieves the metallic state. At all densities, every donor will bind one electron, and the system will be an insulator at low temperature. The donors will ionize at higher temperature. This gloomy prediction is avoided by a technique called modulation doping. Put the impurities in the barrier layers, so that the binding energy of the electrons is much lower.

15.1.4 *Modulation Doping*

In bulk, three-dimensional semiconductors, conduction electrons are created by intentionally adding impurities such as donors. A negative consequence of the impurities is that the electrons scatter from these same impurities, which limits the value of the mobility. In QWs, the impurities can be added to the host material during the MBE growth process, away from the QW, which is called *modulation doping*. The cross section for impurity scattering is much reduced because the impurities are in the host material, while the conducting electrons are in the QW. However, modulation doping changes the shape of the quantum well.

Figure 15.4a shows a cartoon of the energy levels before charge transfer. The circles filled with X-marks are the energy states of electrons bound to donors. These energy levels must be higher in energy than the bottom of the QW or else there will be no charge transfer. Electrons bound to donors, near to the QW, can lower their energy by hopping into the quantum well. The ionized donors create a space charge region called a *depletion region*. The depletion region raises the energy bands, as does the filling of the QW. The filling stops when the chemical potential of the QW equals the energy of the electrons bound to the donor.

This process can be described in a simple fashion for narrow quantum wells. Let μ be the chemical potential of the electrons in the QW of density n_e. For a two-dimensional electron gas with isotropic effective mass m^*, the relation between μ, n_e and the Fermi wave vector is

$$n_e = \frac{k_F^2}{2\pi} \tag{15.17}$$

$$\mu = E_j + \frac{\hbar^2 k_F^2}{2m^*} = E_j + \frac{\pi\hbar^2 n_e}{m^*} \tag{15.18}$$

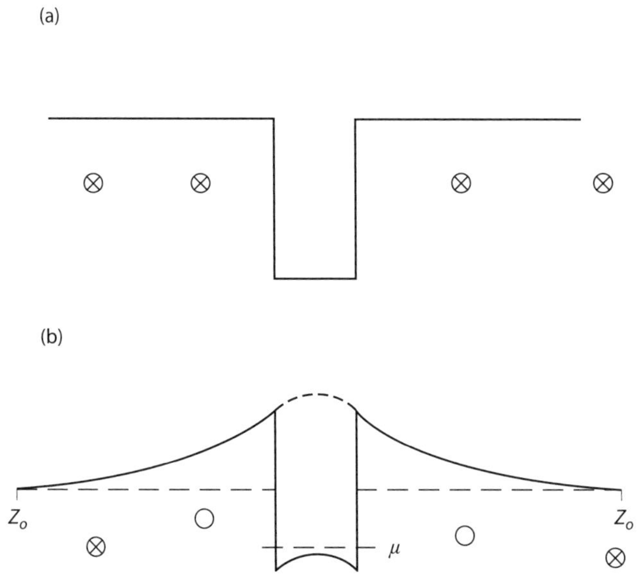

FIGURE 15.4. (a) Energy levels of bound donors in the host material. (b) Final band bending in depletion region, and in the QW.

where E_j is the minimum energy state in the QW due to confinement. The total band offset V_0 is the sum of μ, the donor binding energy E_D, and the band bending due to the depletion region V_B:

$$V_0 = \mu + E_D + V_B \tag{15.19}$$

The band bending is obtained by solving Poisson's equation for the charge density of ionized donors:

$$\frac{d^2V(z)}{dz^2} = \frac{e^2 n_D}{\varepsilon(0)} \tag{15.20}$$

$$V(z) = \frac{e^2 n_D}{2\varepsilon(0)}(z - z_0)^2 \tag{15.21}$$

where z_0 is where the band bending starts, and $z = 0$ is the edge of the QW. At the QW, the band bending is

$$V_B = \frac{e^2 n_D}{2\varepsilon(0)}(z_0)^2 \tag{15.22}$$

There is overall charge neutrality. The electron charge density n_e in the QW must equal the amount of ionized charge in the two depletion regions:

$$n_e = 2z_0 n_D \tag{15.23}$$

$$V_B = \frac{e^2}{8\varepsilon(0)n_D} n_e^2 \tag{15.24}$$

Keep in mind that n_D is a three-dimensional charge density, while n_e is two dimensional. Collecting these results in eq. (15.19) gives a self-consistent equation for the charge density in the QW:

$$V_0 = E_D + E_j + \frac{\pi\hbar^2}{m^*} n_e + \frac{e^2}{8\varepsilon(0)n_D} n_e^2 \tag{15.25}$$

A solution to the quadratic equation for n_e exists only for $V_0 > E_D + E_j$. Then it is

$$n_e = n_D a_0^* \left[\sqrt{1 + \lambda} - 1\right] \tag{15.26}$$

$$\lambda = \frac{V_0 - E_D - E_j}{\pi E_D n_D (a_0^*)^3} \tag{15.27}$$

where the formulas for the Bohr radius and donor binding energy are

$$a_0^* = \frac{4\pi\hbar^2\varepsilon(0)}{e^2 m^*}, \quad E_D = \frac{\hbar^2}{2m^*(a_0^*)^2} \tag{15.28}$$

This analysis is accurate for thin QWs. For thicker quantum wells, the potential $V(z)$ from the charge in the depletion region has to be continued into the QW, which makes calculating the electrons states in the QW much harder. Again using Poisson's equation, $V(z)$ has negative curvature in the depletion region because the donor density has positive charge. In the QW, the electrons have negative charge and the curvature in $V(z)$ is positive. Thinking of $V(z)$ as a continuous curve, the two regions of negative curvature must be connected by a region of positive curvature to get to the final result in fig. 15.4b.

In the case that the band bending term V_B is small, when $\lambda < 1$, the result is much simpler. It is given in terms of the two-dimensional density of states N_2 per spin:

$$n_e \approx \frac{1}{2} n_D a_0^* \lambda = N_2(V_0 - E_D - E_j), \quad N_2 = \frac{m^*}{2\pi\hbar^2} \tag{15.29}$$

This calculation was done assuming that all electrons in the QW were in the lowest energy-band state. If the band energies in the well have a separation $E_2 - E_1 > \mu$, the second band also starts to fill, which makes a small change in the calculation of n_e.

15.1.5 *Electron Mobility*

The mobility of conduction band electrons can be very high for transport along the plane of a quantum well. The mobility

$$\mu = \frac{e\tau}{m^*} \tag{15.30}$$

where τ is the transport lifetime. The lifetime is limited by various mechanism that scatter the electron:

- Scattering by phonons.
- Scattering by neutral and ionized impurities.
- Scattering by imperfections in the boundaries of the quantum well, which is due to the lack of ideal planar surfaces created in the MBE growth process.

The last item is quite important. An example is from GaAs/AlAs quantum wells, where the interface may have small regions, usually one atom thick, where AlAs sticks into the GaAs layer. These islands are strong scattering points for both electrons and phonons, as they each move along the quantum well.

The phonons have to be calculated for the QW or the superlattice. There are phonon modes in the QW, in the host material, and along the interface between them. Since the number of phonon modes is three times the number of atoms in the crystal, the QW has no more phonons than the same amount of bulk material. However, it has different frequencies and eigenvectors due to the presence of the QW.

15.2 Graphene

Crystalline solids composed solely of carbon were historically diamond and graphite. Both materials are important commercially, interesting scientifically, and have been extensively investigated. Then the fullerene molecule C_{60} was discovered. Related molecules such as

C_{70} were discovered later. These molecules were made into three-dimensional crystals. When compounded with alkali metals, they were found to be superconducting. The highest superconducting transition temperature of solid fullerene was higher than any material except the cuprate materials.

The nanoscience of carbon compounds took a further step with the discovery of nanotubes made solely of carbon. Carbon nanotubes are found in many different forms and have amazing transport properties. They are discussed in detail below.

A single sheet of carbon is called *graphene*. It is the ideal two-dimensional system. A single layer is one carbon atom thick. Bulk graphite is composed of many such sheets stacked in alternating layers: the unit cell of graphite has two layers of graphene.

Some properties of graphene are discussed first. An understanding of graphene is essential for the understanding of carbon nanotubes.

15.2.1 *Structure*

Carbon has four valence electrons and prefers to bond to near neighbors. The common forms of bonding are tetrahedral, as in diamond, or trigonal, as in graphene. The trigonal bond is stronger and is usually found in nature in many molecules, such as benzene. Three neighbors of each carbon are in the same plane. The three bonds in the plane, to the three first neighbors, are called *σ-bonds*. These bonds are strong and have a large binding energy. They are filled electron bands in graphite. The fourth electron for each carbon is in the p_z orbital that is perpendicular to the plane. This band is half-filled and provides the conduction and valence bands that are near to the chemical potential. They are called *π-bonds*.

In graphene the carbon atoms form the honeycomb lattice. The unit cell has two carbon atoms, as shown in fig. 15.5. The distance between neighboring atoms is a. The two lattice vectors $\mathbf{a}_{1,2}$ define the unit cell in real space:

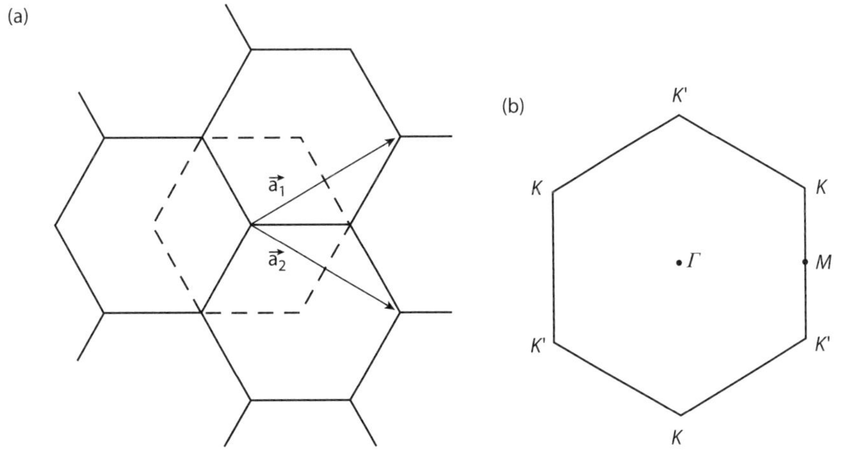

FIGURE 15.5. (a) Crystal structure of graphene. Carbon atoms at vertices. The vectors $\mathbf{a}_{1,2}$ define the unit cell in two dimensions. (b) Brillouin zone of graphene. Chemical potential is at the K- and K'-points.

$$\mathbf{a}_1 = a\left(\frac{3}{2}, \frac{\sqrt{3}}{2}\right) = b\left(\frac{\sqrt{3}}{2}, \frac{1}{2}\right) \tag{15.31}$$

$$\mathbf{a}_2 = a\left(\frac{3}{2}, -\frac{\sqrt{3}}{2}\right) = b\left(\frac{\sqrt{3}}{2}, -\frac{1}{2}\right) \tag{15.32}$$

$$A_0 = \hat{z}\cdot(\mathbf{a}_1 \times \mathbf{a}_2) = \frac{b^2\sqrt{3}}{2} \tag{15.33}$$

The lattice vectors have a length $b = |\mathbf{a}_{1,2}| = \sqrt{3}a$. The area of the unit cell in real space is A_0. The smallest, nonzero, six reciprocal lattice vectors have the form

$$\mathbf{G} = \frac{4\pi}{b\sqrt{3}}\left(\pm\frac{1}{2}, \pm\frac{\sqrt{3}}{2}\right), \quad \mathbf{G} = \frac{4\pi}{b\sqrt{3}}(\pm 1, 0) \tag{15.34}$$

Figure 15.5 shows the unit cell in real space, and the Brillouin zone in reciprocal space. Three vertices are labeled K, and three are labeled K'. The chemical potential and Fermi surface are at these points. The points labeled M are at the centers of the edges of the BZ, and the central point ($k = 0$) is Γ.

15.2.2 *Electron Energy Bands*

Because graphene is a building block of graphite, its electronic and vibrational properties have been calculated by theorists for years, well before its experimental realization. Wallace [9] first found the bands in 1947 and showed that graphene was a gapless semiconductor, with a Dirac point at the corners of the Brillouin zone. Figure 15.6 shows a

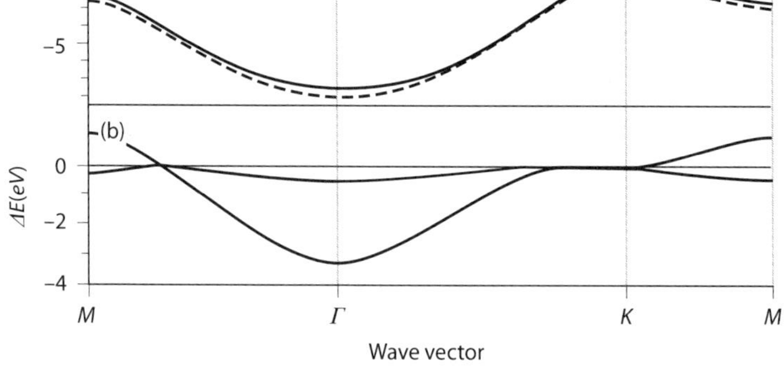

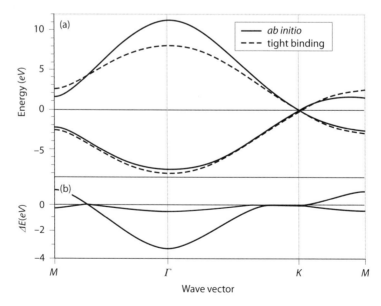

FIGURE 15.6. Calculated electronic band structure of graphene. Solid lines are *ab initio*, dashed lines are tight-binding. Sigma bands are on the bottom. From Reich et al., *Phys. Rev. B* **66**, 035412 (2002). Used with permission of the American Physical Society.

calculation by Reich et al. [7] of the electronic energy bands in graphene. The horizontal solid line shows the chemical potential. Below the chemical potential, the states are all occupied and the conduction states are above it. The solid lines are actual energy bands, and the dashed lines are the approximate π-bands found using the tight-binding model. The three bands on the bottom are the σ-bands.

The chemical potential is at the points labeled K (or K'). The bands with the lowest energy are from the σ-bands in the planes between the carbon atoms. Electrons in the σ-bands are tightly bound. The energy bands near the chemical potential are from the π-bands. They are formed from the carbon orbital p_z, which projects above and below the plane of graphene. These π-bands are well described by a nearest-neighbor tight-binding molecule.

Call the two atoms per unit cell A and B, and denote their raising and lowering operators by a and b. The tight-binding Hamiltonian is

$$H = E_0 \sum_{js} \left(a_{js}^\dagger a_{js} + b_{js}^\dagger b_{js} \right) + J_0 \sum_{js} \sum_{\delta=1}^{3} \left(a_{j+\delta,s}^\dagger b_{js} + b_{js}^\dagger a_{j+\delta,s} \right) \tag{15.35}$$

where s is the spin label. Go to collective coordinates and find

$$H = E_0 \sum_{\mathbf{k}s} \left(a_{\mathbf{k}s}^\dagger a_{\mathbf{k}s} + b_{\mathbf{k}s}^\dagger b_{\mathbf{k}s} \right) + J_0 \sum_{\mathbf{k}s} \left[\gamma(\mathbf{k})\, a_{\mathbf{k}s}^\dagger b_{\mathbf{k}s} + \gamma(\mathbf{k})^* b_{\mathbf{k}s}^\dagger a_{\mathbf{k}s} \right]$$

$$\gamma(\mathbf{k}) = \sum_{\delta=1}^{3} e^{i\mathbf{k}\cdot\delta} = e^{i\theta_x} + 2e^{-i\theta_x/2} \cos\left(\sqrt{3}\theta_y/2\right), \quad \theta_\mu = k_\mu a \tag{15.36}$$

An electron on an A atom site hops to its three neighboring B atoms, and vice versa. Since each carbon site lacks inversion symmetry, $\gamma(\mathbf{k})$ is complex. The Hamiltonian for each spin is equivalent to the matrix

$$H_\mathbf{k} = \begin{pmatrix} E_0 & J_0 \gamma(\mathbf{k}) \\ J_0 \gamma(\mathbf{k})^* & E_0 \end{pmatrix} \tag{15.37}$$

The eigenvalues are

$$E_\pm(\mathbf{k}) = E_0 \pm J_0 |\gamma(\mathbf{k})| \tag{15.38}$$

$$|\gamma(\mathbf{k})| = \sqrt{1 + 4\cos(3\theta_x/2)\cos(\sqrt{3}\theta_y/2) + 4\cos^2(\sqrt{3}\theta_y/2)} \tag{15.39}$$

The tight-binding hopping parameter is estimated to be $J_0 \approx 2.6$–3.2 eV by comparing to the band structure in fig. 15.6.

Examine the value of the energy bands at points in the BZ.

- At the Γ-point ($\mathbf{k} = 0$) then $E_\pm(0) = E_0 \pm 3J_0$. These are the extremal points of the conduction and valence band.
- There are six M-points. Two are at $\mathbf{k}_M = \pm 2\pi(1,0)/3a$. Then $|\gamma(\mathbf{k}_M)| = 1$, and the energy bands are at $E(\mathbf{k}_M) = E_0 \pm J_0$. The same energy is found at the other four values of $\mathbf{k}_M$.

- There are three K-points and three K'-points. One of them is $\mathbf{k}_K = 4\pi(1,0)/a\sqrt{27}$. Then $\Theta_x = 0$, $\sqrt{3}\,\Theta_y/2 = 2r/3$. One finds that $|\gamma(\mathbf{k}_K)| = 0$ and $E_\pm(\mathbf{k}_K) = E_0$. The conduction and valence bands are degenerate at this point. Graphene is a zero gap semiconductor.

The K- and K′-points in the Brillouin zone are where the electronic excitations are located. Thermally excited electrons and holes are near to the chemical potential. It is interesting to examine the dispersion of particles in the vicinity of this point. Let $\mathbf{p}$ be a small wave vector, and define $\mathbf{k} = \mathbf{k}_K + \mathbf{p}$. Expand $\gamma(\mathbf{k})$ in a power series in (p_x, p_y) and retain only the linear terms. Using the above value of $\mathbf{k}_K$ gives

$$\gamma(\mathbf{k}) = e^{ip_x a} + 2e^{-ip_x a/2}\cos\left(\frac{2\pi}{3} + \frac{\sqrt{3}}{2}p_y a\right) \tag{15.40}$$

$$\approx \frac{3ia}{2}(p_x + ip_y) + O(p^2) \tag{15.41}$$

For small values of $\mathbf{p}$ the Hamiltonian can be written as

$$H_{\mathbf{k}} = \begin{pmatrix} E_0 & i\hbar v_F(p_x + ip_y) \\ -i\hbar v_F(p_x - ip_y) & E_0 \end{pmatrix} \tag{15.42}$$

$$E_\pm(\mathbf{p}) = E_0 \pm \hbar v_F p \tag{15.43}$$

$$p = \sqrt{p_x^2 + p_y^2}, \quad v_F = \frac{3J_0 a}{2\hbar} \approx 840 \text{ km/s} \tag{15.44}$$

It is convenient to relabel the points K and K' as the new centers of the BZ, with $\mathbf{p}$ being the wave vector from these points. Figure 15.7 shows that the three K-points merge into a single point. The letters (a, b, c) denote regions of phase space. When linking them together by reciprocal lattice vectors, they form a continuous space, which becomes the new center of the Brillouin zone. The same happens for the three K'-points. The new BZ has two zone centers, one about K and one about K'. They are called *Dirac points*.

The Hamiltonian matrix in eqn. (15.42) is similar to the Dirac Hamiltonian for a massless, relativistic fermion. This analogy is accidental, but is often noted. The energy bands are cones around these points. The velocity v_F of these massless fermions is about 840 km/s, estimated using $J_0 = 2.6$ eV and $a = 1.42$ Å.

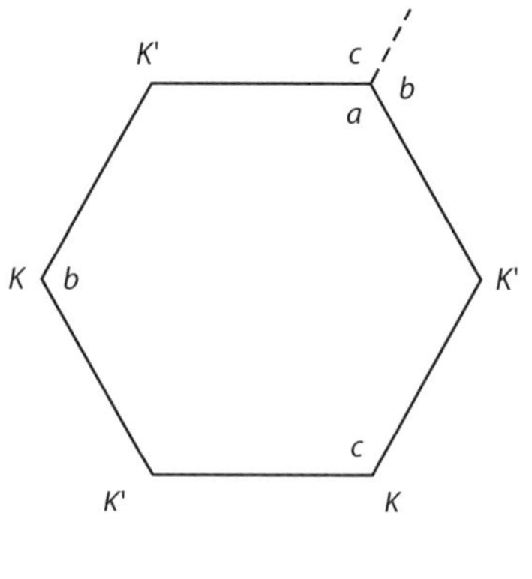

FIGURE 15.7. The three K-points merge into a single point, as do the three K'-points.

15.2.3 Eigenvectors

Return to the Hamiltonian matrix in eqn. (15.37) and find the eigenvectors. Define the phase factor $\phi(\mathbf{k})$ of the function

$$\gamma(\mathbf{k}) = |\gamma(\mathbf{k})| e^{i\phi(\mathbf{k})} \tag{15.45}$$

Let the spinor for the eigenvector have the components (u, v). The matrix equation is

$$0 = \begin{pmatrix} E_0 - E & J_0\gamma(\mathbf{k}) \\ J_0\gamma(\mathbf{k})^* & E_0 - E \end{pmatrix} \begin{pmatrix} u \\ v \end{pmatrix} \tag{15.46}$$

For the eigenvalue $E_- = E_0 - J_0|\gamma|$, cancel like factors and get

$$0 = \begin{pmatrix} 1 & e^{i\phi(\mathbf{k})} \\ e^{-i\phi(\mathbf{k})} & 1 \end{pmatrix} \begin{pmatrix} u \\ v \end{pmatrix} \tag{15.47}$$

$$\begin{pmatrix} u \\ v \end{pmatrix} = \frac{1}{\sqrt{2}} \begin{pmatrix} 1 \\ -e^{-i\phi} \end{pmatrix} \tag{15.48}$$

The last equation gives the eigenvector for E_-. A similar result is found for the eigenvectors of E_+. The phase factors $\exp[\pm i\phi(\mathbf{k})]$ appear in many formulas. To diagonalize the Hamiltonian define new operators $(\alpha_\mathbf{k}, \beta_\mathbf{k})$:

$$a_{\mathbf{k}s} = \frac{1}{\sqrt{2}}\left[\alpha_{\mathbf{k}s} + \beta_{\mathbf{k}s} e^{i\phi}\right] \tag{15.49}$$

$$b_{\mathbf{k}s} = \frac{1}{\sqrt{2}}\left[\beta_{\mathbf{k}s} - \alpha_{\mathbf{k}s} e^{-i\phi}\right] \tag{15.50}$$

Put these definitions in the Hamiltonian and find, after some algebra,

$$H = \sum_{\mathbf{k}s}\left[E_-(\mathbf{k})\alpha^\dagger_{\mathbf{k}s}\alpha_{\mathbf{k}s} + E_+(\mathbf{k})\beta^\dagger_{\mathbf{k}s}\beta_{\mathbf{k}s}\right] \tag{15.51}$$

The operators $(\alpha_\mathbf{k}, \beta_\mathbf{k})$ represent the eigenstates of the tight-binding Hamiltonian. Near points K and K', the wave vector $\mathbf{p}$ is used to denote the distance from these points. The phase factor in this case is rather simple:

$$p_x \pm ip_y = pe^{\pm i\phi(\mathbf{p})} \tag{15.52}$$

$$\tan(\phi) = \frac{p_y}{p_x} \tag{15.53}$$

15.2.4 Landau Levels

When there is a vector potential $\mathbf{A}$ from an electric or magnetic field, the off-diagonal matrix elements are

$$i\hbar v_F\left(p_x \pm ip_y + e[A_x \pm iA_y]\right) \tag{15.54}$$

This expression can be used to calculate the energy bands and Landau levels in a magnetic field. For a magnetic field perpendicular to the plane, set the vector potential

$$\mathbf{A}(\mathbf{r}) = B(y, 0, 0) \tag{15.55}$$

The Hamiltonian in eqn. (15.54) does not contain x, so p_x is a constant. Let the eigenfunction be

$$\psi(x, y) = e^{ik_x x} f(y) \tag{15.56}$$

$$H\psi = e^{ik_x x} \begin{pmatrix} E_0 & w \\ w^* & E_0 \end{pmatrix} \begin{pmatrix} f_1 \\ f_2 \end{pmatrix} \tag{15.57}$$

$$w = iv_F[\hbar k_x + ip_y + eBy] \tag{15.58}$$

$$= iv_F eB\left[y - y_0 + i\frac{p_y}{eB}\right]$$

$$y_0 = \frac{\hbar k_x}{eB} = k_x \ell^2 \tag{15.59}$$

where ℓ is the magnetic length. The above operators are the lowering operator for the harmonic oscillator:

$$a = \frac{1}{\ell\sqrt{2}}\left(y - y_0 + i\frac{p_y}{eB}\right) \tag{15.60}$$

$$w = i\lambda a, \quad \lambda = \frac{v_F \hbar\sqrt{2}}{\ell} \tag{15.61}$$

$$H\psi = e^{ik_x x} \begin{pmatrix} E_0 & i\lambda a \\ -i\lambda a^\dagger & E_0 \end{pmatrix} \begin{pmatrix} |n-1\rangle \\ \pm i|n\rangle \end{pmatrix} \tag{15.62}$$

The last column vector is the eigenvector in terms of the harmonic oscillator states $|n\rangle$. The eigenvalue of the above Hamiltonian is

$$E_n = E_0 \mp \lambda\sqrt{n}, \quad n = 0, 1, 2, \ldots \tag{15.63}$$

The energy levels are proportional to $\pm\sqrt{n}$.

15.2.5 *Electron–Phonon Interaction*

The phase factors $\phi(\mathbf{k})$ are part of the electron–phonon interaction for graphene. The most important interaction is the stretching of the σ-bands due to relative motion of the carbon atoms. The atomic positions are the equilibrium positions $\mathbf{R}_j^{(0)}$ plus the small displacements $\mathbf{Q}_j$ due to phonons. Write the hopping term as a function of the band distance $|\boldsymbol{\delta}| = a$:

$$J_0(|\boldsymbol{\delta}|) \to J_0(|\boldsymbol{\delta} + \mathbf{Q}_{B,j+\delta} - \mathbf{Q}_{Aj}|) \tag{15.64}$$

$$\approx J_0[a + \hat{\delta} \cdot (\mathbf{Q}_{B,j+\delta} - \mathbf{Q}_{Aj})]$$

$$\approx J_0(a) + \frac{dJ_0}{da} \hat{\delta} \cdot (\mathbf{Q}_{B,j+\delta} - \mathbf{Q}_{Aj})$$

$$V_{ep} = \frac{dJ_0}{da} \sum_{j\delta s} \left[b^{\dagger}_{j+\delta,s} a_{js} \hat{\delta} \cdot (\mathbf{Q}_{B,j+\delta} - \mathbf{Q}_{Aj}) + \text{h.c.} \right] \tag{15.65}$$

The electron–phonon interaction is dJ_0/da. Numerical estimates of this derivative are $dJ_0/da \approx J_0/a$.

This interaction can be written in terms of collective coordinates. Go to a wave vector representation and find for the first term

$$b^{\dagger}_{j+\delta,s} a_{js} = \frac{1}{N} \sum_{\mathbf{k}'bk} b^{\dagger}_{\mathbf{k}'s} a_{\mathbf{k}s} \exp\left[i(\mathbf{k} - \mathbf{k}') \cdot \mathbf{R}^{(0)}_j - i\mathbf{k}' \cdot \boldsymbol{\delta} \right] \tag{15.66}$$

The displacements $\mathbf{Q}_j$ must be given an A or B subscript to keep track of which atom in the sublattice is vibrating. They are also expanded in wave vectors:

$$\mathbf{Q}_{Aj} = \frac{1}{\sqrt{N}} \sum_{\mathbf{q}} e^{i\mathbf{q} \cdot \mathbf{R}^{(0)}_j} \mathbf{Q}_{A,\mathbf{q}} \tag{15.67}$$

When these definitions are collected, the first term in the electron–phonon interaction is

$$V_{ep} = \frac{dJ_0}{da} \frac{1}{N^{3/2}} \sum_{j\delta s} \sum_{\mathbf{k}\mathbf{k}'\mathbf{q}} \left\{ b^{\dagger}_{\mathbf{k}'s} a_{\mathbf{k}s} e^{i\mathbf{R}^{(0)}_j \cdot (\mathbf{k}+\mathbf{q}-\mathbf{k}') - i\delta \cdot \mathbf{k}'} \hat{\delta} \cdot \left[e^{i\mathbf{q} \cdot \delta} \mathbf{Q}_{B\mathbf{q}} - \mathbf{Q}_{A\mathbf{q}} \right] + \text{h.c.} \right\} \tag{15.68}$$

The summation over $\mathbf{R}_j$ forces $\mathbf{k}' = \mathbf{k} + \mathbf{q}$, and the summation over $\boldsymbol{\delta}$ gives $\nabla_k \gamma(\mathbf{k})$:

$$V_{ep} = \frac{dJ_0}{da} \frac{i}{aN^{1/2}} \sum_{\mathbf{k}\mathbf{q}s} \left\{ b^{\dagger}_{\mathbf{k}+\mathbf{q},s} a_{\mathbf{k}s} \left[\nabla\gamma^*(\mathbf{k}) \cdot \mathbf{Q}_{B\mathbf{q}} - \nabla\gamma^*(\mathbf{k}+\mathbf{q}) \cdot \mathbf{Q}_{A\mathbf{q}} \right] + \text{h.c.} \right\} \tag{15.69}$$

Near to the K-point, writing $\mathbf{k} = \mathbf{k}_M + \mathbf{p}$ gives

$$V_{ep} = \frac{dJ_0}{da} \frac{3}{2N^{1/2}} \sum_{\mathbf{p}\mathbf{q}s} \left\{ b^{\dagger}_{\mathbf{p}+\mathbf{q},s} a_{\mathbf{p}s} \left[\hat{x} - i\hat{y} \right] \cdot \left[\mathbf{Q}_{B\mathbf{q}} - \mathbf{Q}_{A\mathbf{q}} \right] + \text{h.c.} \right\} \quad 915.70)$$

The final result depends on the difference of the displacements of the A and B atoms. This matrix element is largest for optical phonon vibrations.

This formula is not yet in final form. The operators (a, b) are transformed into the eigenstates (α, β):

$$b^{\dagger}_{\mathbf{p}+\mathbf{q},s} a_{\mathbf{p}s} = \frac{1}{2} \left[\beta^{\dagger}_{\mathbf{p}+\mathbf{q},s} - \alpha^{\dagger}_{\mathbf{p}+\mathbf{q},s} e^{i\phi(\mathbf{p}+\mathbf{q})} \right] \left[\alpha_{\mathbf{p},s} + \beta_{\mathbf{p},s} e^{i\phi(\mathbf{p})} \right] \tag{15.71}$$

The electron–phonon interaction can scatter $\alpha \to \alpha$ and $\beta \to \beta$, but also $\alpha \to \beta$ and $\beta \to \alpha$. The latter two cases are interband transitions, across the zero gap between the two conical bands

15.2.6 *Phonons*

The phonons of graphene will be presented for a layer that floats in space, so it has no constraints on the motion. The usual experimental situation is to have the layer on a substrate. The substrate will affect the vibrations and change some frequencies. These changes depend on the nature of the substrate and how the graphene is bonded to the substrate—van der Waals, chemical, etc. Rather than deal will all possible substrates, we discuss the simple case of a free-floating single layer.

Figure 15.8 shows the phonon modes of graphene. The atoms can vibrate in all three dimensions. At long wavelength the modes are easily identified. At frequencies around $\nu = 1500\ \mathrm{cm}^{-1}$ are the optical modes, where the A and B atoms beat against each other in the plane. At $\nu = 850\ \mathrm{cm}^{-1}$ is the ZO optical phonon. It has the A and B atoms vibrating in the *z*-direction, but out of phase. Near to zero frequency, there are three modes.

- The LA mode is longitudinal acoustical vibrations in the plane. Their frequency is $\omega_{\mathrm{LA}}(q) = v_L q$.
- The TA mode is transverse acoustical vibrations in the plane. Their frequency is $\omega_{\mathrm{TA}}(q) = v_T q$.
- The flexure mode (ZA) is for acoustic oscillations of the plane in the *z*-direction. Their frequency is quadratic in wave vector:

$$\omega_F(q) = c_L q^2 a \tag{15.72}$$

where $a \sim 0.15$ nm is the thickness of the carbon layer.

The three coefficients of velocity (v_L, v_T, c_L) are defined below.

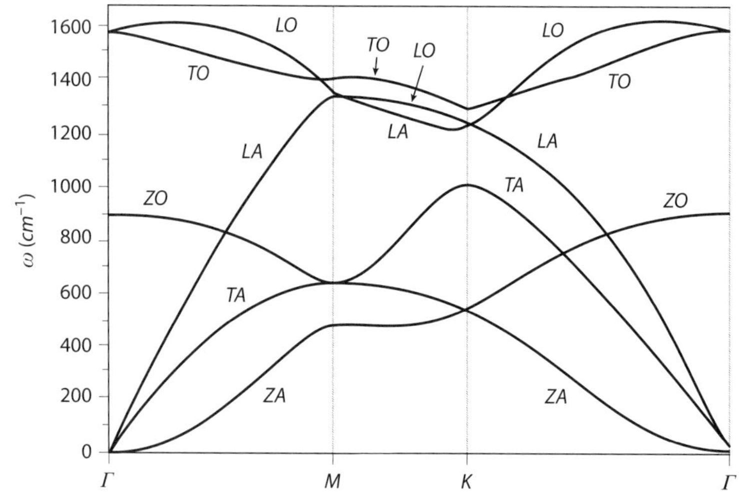

FIGURE 15.8. Phonons modes of graphene. From Wirtz and Rubio, *Solid State Commun.* **131**, 141 (2004). Used with permission of Elsevier.

Graphene has hexagonal symmetry. Layers with this symmetry have simple acoustical properties. Group theory shows that the velocity of sound is the same in all directions in the plane. It is acoustically isotropic. It is simple to derive the acoustical properties using the standard theory for isotropic materials. The vector displacement $\mathbf{u}(\mathbf{r}, t)$ of an elastic medium is given by the equation

$$\rho \frac{\partial^2}{\partial t^2} \mathbf{u} = \mu \nabla^2 \mathbf{u} + (\lambda + \mu) \nabla(\nabla \cdot \mathbf{u}) \tag{15.73}$$

The two parameters μ, λ are called Lamé constants, and ρ is the density. This equation is solved by assuming that the waves travel as $\exp[i(\mathbf{q} \cdot \mathbf{r} - \omega t)]$, so the above equation is

$$\rho \omega^2 \mathbf{u} = \mu q^2 \mathbf{u} + (\lambda + \mu) \mathbf{q} (\mathbf{q} \cdot \mathbf{u}) \tag{15.74}$$

First solve for a thick, bulk material. The longitudinal waves are found by solving for $(\mathbf{q} \cdot \mathbf{u})$. Multiplying the above equation by $\mathbf{q} \cdot$ gives

$$\omega^2 (\mathbf{q} \cdot \mathbf{u}) = V_L q^2 (\mathbf{q} \cdot \mathbf{u}), \quad V_L^2 = \frac{\lambda + 2\mu}{\rho} \tag{15.75}$$

The transverse waves are found by multiplying eqn. (15.74) by $\mathbf{q} \times$:

$$\omega^2 (\mathbf{q} \times \mathbf{u}) = V_T q^2 (\mathbf{q} \times \mathbf{u}), \quad V_T^2 = \frac{\mu}{\rho} \tag{15.76}$$

Next solve for the modes of a very thin isotropic layer. The thickness is a, and assume that $qa \ll 1$. One still solves eqn. (15.74), but now with the boundary conditions that the stresses are zero at the top and bottom of the thin layer. The derivation is given in most books on elasticity.

- The TA mode in the layer has the same velocity as the bulk material, so that $v_T = V_T$.
- The LA mode in the layer does not have the same velocity as in the bulk. The LA velocity in the thin film is

$$v_L^2 = \frac{4\mu}{\rho} \frac{\lambda + \mu}{\lambda + 2\mu} \tag{15.77}$$

  Why is the velocity in the layer different than in the bulk? LA phonons stretch or compress the layer. If one takes a thin piece of rubber and stretches it, it gets thinner. So the motion is more complicated than in the bulk. Bulk modes are solved for the case that $qa \gg 1$, where a is the thickness of the material. TA modes do not stretch or compress the layer, so their velocities are the same as in the bulk.

- The coefficient c_L in the flexure mode also has the units of velocity. It is

$$c_L = \frac{v_L}{\sqrt{12}} \tag{15.78}$$

15.3 Carbon Nanotubes

Carbon nanotubes are found in many different configurations in nature. The most simple case is the single-wall carbon tube (SWCNT). They are composed of a single strip of graphene that is rolled into a hollow tube. Other forms of carbon nanotubes (CNT) are as follows:

- Multiwall tubes have tubes inside of tubes. A double-wall tube has only two, but there are often many concentric tubes.
- Ropes are many tubes that are bunched together as strands of a rope.

Pairs of CNTs have an attractive force due to the van der Waals interaction, and will form ropes or multiwall tubes if given the opportunity. The present discussion focuses on the properties of single-wall carbon nanotubes.

15.3.1 *Chirality*

There are many different ways to cut a strip of graphene and to roll it into a tube. The different types of SWCNTs are distinguished by a *chirality index* (n, m). Recall that graphene has the honeycomb (hc) lattice. If a denotes the carbon–carbon bond length, the lattice vectors have a length $b = \sqrt{3}a$ and are

$$\mathbf{a}_1 = \frac{b}{2}(\sqrt{3}, 1), \quad \mathbf{a}_2 = \frac{b}{2}(\sqrt{3}, -1), \tag{15.79}$$

Starting from a lattice point, any other lattice point is reached by the lattice vector

$$\mathbf{R}_{n,m} = n\mathbf{a}_1 + m\mathbf{a}_2 \tag{15.80}$$

In forming the SWCNT by rolling up a strip of graphene, the tube must be periodic as you go around the circumference. One point in the lattice must be the same point after going around the circumference. This periodicity is achieved by having the lattice point $\mathbf{R}_{0,0}$ and another lattice point $\mathbf{R}_{n,m}$ be the same point in the tube. A SWCNT formed in this way is an (n, m) tube.

An easy way to understand this concept is to draw out a hc lattice on a sheet of paper. Then make a copy on a transparency sheet and roll up the transparency into a tube. It becomes immediately obvious how different values of chirality (n, m) can form a tube.

Several values of chirality have a simple structure:

- An *armchair tube* has $n = m$, so the chirality is (n, n). There are $2n$ carbon atoms in the circumference of the tube, which occur in n-pairs. It is shown in fig. 15.9a.
- A *zigzag* tube has one index zero, such as $(n, 0)$. The carbon bonds zigzag in going around the circumference of the tube. There are $2n$ carbon atoms in the zigzag path around the tube (fig. 15.9b).

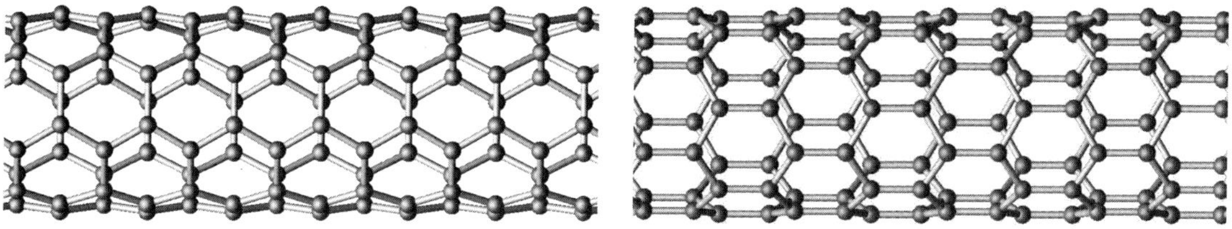

FIGURE 15.9. (a) Armchair and (b) zigzag carbon nanotubes. Courtesy of Vin Crespi.

SWCNT are grown in a furnace. This process makes a distribution of tubes with different chiralities. Many tubes seem to occur with numbers around $n \sim 10$, and these are the most studied. The smallest tubes have $n \sim 5$. Tubes with a large radius, from large values of n, are also found.

The radius of the tube depends on the chirality index. Consider a zigzag tube. Assume each carbon bond has a length a that is a straight line. Because of the zigzag of 30°, the contribution to the circumference is $\bar{a} = \sqrt{3}a/2$. If the circumference of a tube is a circle and all carbon bonds are arcs, then the radius R is given by

$$2\pi R = 2n\bar{a} \tag{15.81}$$

However, if the bonds are straight lines, the circumference is a polygon of $2n$ sides each of length $\bar{a}$. Let R be the distance from the center of the tube to a vertex of the polygon. The isosceles triangle of angle $\theta = 2\pi/2n$, with two long sides of R and the short side of $\bar{a}$, has the identity

$$\bar{a}^2 = 2R^2[1 - \cos(\theta)] = 4R^2 \sin^2(\theta/2) \tag{15.82}$$

$$R = \frac{\bar{a}}{2\sin(\pi/2n)} \tag{15.83}$$

For large values of n this expression agrees with eqn. (15.81). Similar expressions can be derived for other values of chirality.

15.3.2 *Electronic States*

In a sheet of graphene, the valence electrons are either in the p_z orbitals that are perpendicular to the plane or in the σ-bonds that lie in the plane between carbon atoms. The chemical potential is in the band of states formed from the p_z bonds, and they play the dominent role in transport phenomena. The electrons in the p_z orbitals play a major role in the electronic properties of SWCNT.

The electronic energy bands can be calculated with various degrees of accuracy. A tight-binding calculation assumes periodic boundary conditions around the circumference of the tube. This simple model gives relatively good results. A more rigorous calculation numerically solves for the energy bands of carbon atoms in the actual structure. These two methods tend to give similar results for tubes of larger radius. When the radius is small, differences occur. Here we describe the tight-binding method.

Start with graphene and the vector $\mathbf{R}_{n,m}$ that is the basis of the chirality index. Define another vector $\mathbf{T}_{n,m}$ that is perpendicular to $\mathbf{R}_{n,m}$:

$$\mathbf{R}_{n,m} = \frac{b}{2}\left[\sqrt{3}\hat{x}(n+m) + \hat{y}(n-m)\right] \tag{15.84}$$

$$\mathbf{T}_{n,m} = T_0\left[\hat{x}(n-m) - \sqrt{3}\hat{y}(n+m)\right] \tag{15.85}$$

When the strip of graphene is rolled up to form the SWCNT, the vector $\mathbf{R}_{n,m}$ becomes the circumference. The wave vector $k_\perp$ in this direction in the SWCNT is quantized using periodic boundary conditions: $k_\perp R = \alpha$, where R is the tube radius, $2\pi R = |\mathbf{R}_{n,m}|$, and α is an integer that can be positive, negative, or zero. The direction $\mathbf{T}_{n,m}$ goes along the one-dimensional axis of the SWCNT. It is assigned a continuous wave vector q. All electron states have two sets of quantum numbers. The first is the chirality index (n, m). The second set are the quantum numbers (q, α) of the states on this tube. The value of α is the angular momentum for rotations around the tube axis. Recall that graphene has two sets of energy bands with dispersion:

$$E_\pm(k_x, k_y) = E_0 \pm J_0\left[1 + 4\cos(3\theta_x/2)\cos(\phi_y) + 4\cos^2(\phi_y)\right]^{1/2} \tag{15.86}$$

$$\theta_x = k_x a, \quad \phi_y = \frac{\sqrt{3}}{2}k_y a \tag{15.87}$$

The wave vectors (k_x, k_y) are in the original (x, y) coordinates system. The same formula is used in the SWCNT, except the wave vectors must be rotated and expressed in terms of the wave vectors $(k_\perp, q)$ of the SWCNT.

These quantum numbers are simple for the two examples of armchair and zigzag tubes.

Zigzag Tubes

The case of zigzag tubes is simple. In the original (x, y)-coordinate system, motion along the $\hat{y}$-direction is zigzag. So set $k_y = k_\perp$, $k_x = q$. This choice has $m = -n$:

$$\mathbf{R}_{n,-n} = bn\hat{y}, \quad k_\perp = \frac{2\pi\alpha}{bn} = \frac{2\pi\alpha}{\sqrt{3}a} \tag{15.88}$$

$$R = \frac{n\bar{a}}{\pi} = \frac{\sqrt{3}}{2\pi}na \tag{15.89}$$

$$\phi_y = \frac{\sqrt{3}}{2}k_\perp a = \frac{\sqrt{3}}{2}\frac{\alpha a}{R} = \frac{\pi\alpha}{n} \tag{15.90}$$

$$E_\pm(q, \alpha) = E_0 \pm J_0\left[1 + 4\cos(3qa/2)\cos(\pi\alpha/n) + 4\cos^2(\pi\alpha/n)\right]^{1/2} \tag{15.91}$$

The energy bands in one dimension have a wave vector q. The Brillouin zone in one dimension extends from $-\pi/c < q < \pi/c$. For zigzag tubes, $c = 3a/2$ is the distance $\hat{x}\cdot\mathbf{a}_1$ along the tube from A atom to A atom. Write the above equation as

$$E_{\pm}(q, \alpha) = E_0 \pm J_0\left[1 + 4\cos(qc)\cos(\pi\alpha/n) + 4\cos^2(\pi\alpha/n)\right]^{1/2} \tag{15.92}$$

How many different values are allowed for α? Since there are $2n$ carbon atoms around the circumference, there must be $2n$ bands. This requires n values of α, since each value has two solutions: E_+, E_-. It can have values

$$\alpha = 0, \pm 1, \pm 2, \ldots, n/2 \tag{15.93}$$

If n is even, then $1 - n/2 \leq \alpha \leq n/2$. If n is an odd integer, $(1-n)/2 \leq \alpha \leq (n-1)/2$. For example,

- For $n = 5$ the values are $\alpha = -2, -1, 0, 1, 2$
- For $n = 6$ the values are $\alpha = -2, -1, 0, 1, 2, 3$.

Since α is in the argument of a cosine function, plus and minus integers give the same value. There are nondegenerate bands for $\alpha = 0$, $n/2$ and doubly degenerate bands for other values of α. The two cases $\alpha = 0$, $n/2$ have a simple form for the band dispersion since $\cos(\pi/2) = 0$:

$$E_{\pm}(q, 0) = E_0 \pm J_0\sqrt{5 + 4\cos(qc)} \tag{15.94}$$

$$E_{\pm}(q, n/2) = E_0 \pm J_0 \tag{15.95}$$

All zigzag tubes with an even value of n have two flat, dispersionless energy bands.

A simple band dispersion occurs in zigzag tubes whenever $\cos(\pi\alpha/n) = \frac{1}{2}$ since the argument of the square root is a perfect square. This case occurs only when n is a multiple of three, since $\cos(\pi/3) = \frac{1}{2}$. In this case

$$E_{\pm}(q, n/3) = E_0 \pm 2J_0\cos(qc/2) \tag{15.96}$$

These two bands have no energy gap and are equal at $qc = \pi$. When there is no energy gap the energy bands of the SWCNT are *metallic*. Figure 15.10 shows the energy-band structure for two zigzag tubes. One example has $n = 5$ and all bands have energy gaps. The chemical potential is in the gap. These tubes are said to be *semiconducting* since all energy bands have gaps. A graph is shown only for $q > 0$, but a symmetric curve is found for $q < 0$. In fig. 15.10b, $n = 6$ and one pair of bands cross, so this tube is metallic.

Armchair Tubes

For armchair tubes, $\mathbf{R}_{nn} = 3an\hat{x}$, where the x-direction is shown in fig. 15.5 for graphene. The nanotube wave vectors are $(k_{\perp} = k_x, q = k_y)$. The perpendicular wave vector has the values

$$k_{\perp} = \frac{2\pi\alpha}{3an} \tag{15.97}$$

$$E(q, \alpha) = E_0 \pm J_0[1 + 4\cos(\pi\alpha/n)\cos(qc) + 4\cos^2(qc)]^{1/2} \tag{15.98}$$

$$c = \frac{\sqrt{3}a}{2} \tag{15.99}$$

where α has the same set of n-values derived for zigzag tubes. The distance c is between adjacent rings of carbon atoms as one goes along the axis of the tube. The one-dimensional BZ for the tube has $-\pi/c < q < \pi/c$.

What is the value for c? Some workers take $c = b$, and there are two rows of carbon atoms for each unit cell. Other workers take $c = b/2 = \bar{a}$ and there is only one row of carbon atoms in a unit cell. These choices are not actually different.

- If $c = b = 2\bar{a}$ the BZ extends $-\pi/(2\bar{a}) < q < \pi/(2\bar{a})$. There are $4n$ energy bands.
- If $c = b/2 = \bar{a}$ the BZ extends $-\pi/(\bar{a}) < q < \pi/(\bar{a})$. There are $2n$ energy bands.

(a)

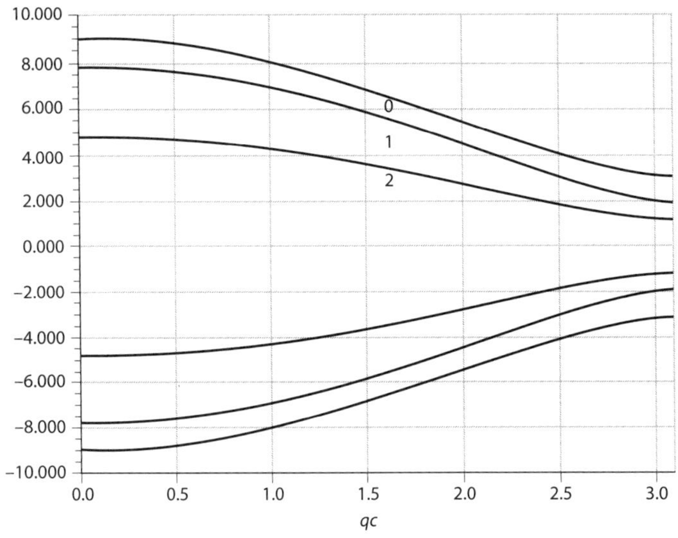

(b)

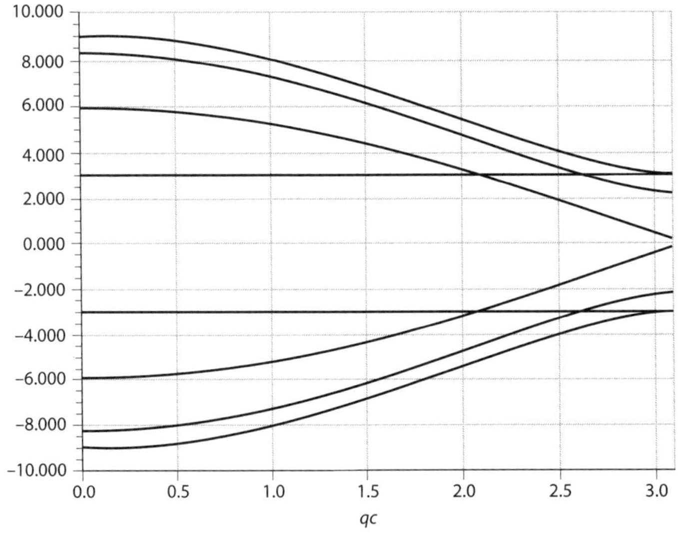

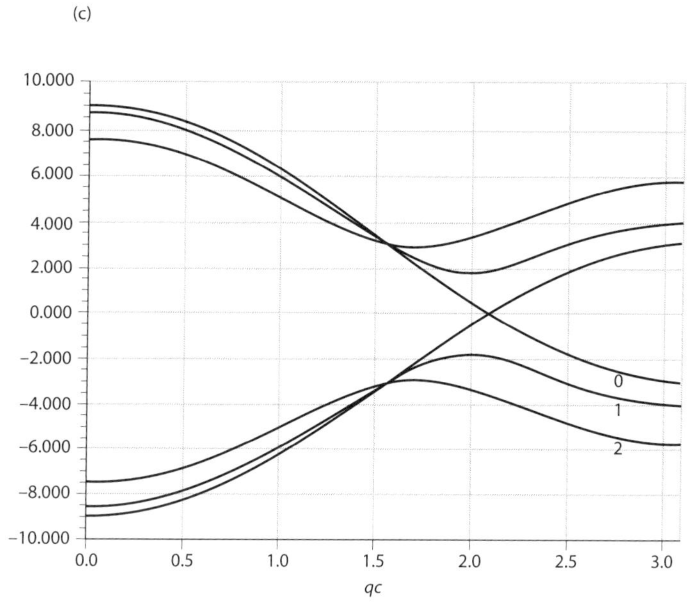

FIGURE 15.10. Energy bands of carbon nanotubes. (a) Dispersion of zigzag tube with $n = 5$, and all bands have energy gaps. (b) Dispersion of zigzag tube for $n = 6$, and one pair of bands has no gap, and the tube is metallic. (c) Dispersion of an armchair tube with $n = 5$. Numbers indicate values of α.

The number of phase space points is the same in the two cases. One method is the zone-folded version of the other. The two choices give similar bands.

The special case $\alpha = 0$ is interesting since the argument of the square root becomes a perfect square:

$$E_{\pm}(q, 0) = E_0 \pm J_0[1 + 2\cos(qc)] \tag{15.100}$$

These bands are graphed in fig. 15.10c for $n = 5$. Some special points are

$$E_{\pm}(0, 0) = E_0 \pm 3J_0 \tag{15.101}$$

$$E_{\pm}(\pm 2\pi/3c, 0) = E_0 \tag{15.102}$$

$$E_{\pm}(\pm \pi/c, 0) = E_0 \mp J_0 \tag{15.103}$$

The energy bands cross at the wave vectors $q = \pm 2\pi/3c$. These crossing points are at the location of the chemical potential. All armchair tubes are metallic tubes, since the conduction (and valence) bands intersect the chemical potential.

For other, nonzero, values of α, the dispersion of the armchair tubes can be written as

$$E(\alpha, q) = E_0 \pm J_0\{[\cos(\pi\alpha/n) + 2\cos(qc)]^2 + \sin^2(\pi\alpha/n)\}^{1/2} \tag{15.104}$$

These bands do not cross. They have an energy gap and are semiconductor bands. At the wave vector value

$$\cos(q_\alpha c) = -\tfrac{1}{2}\cos(\pi\alpha/n) \tag{15.105}$$

$$E(\alpha, q_\alpha) = E_0 \pm J_0 |\sin(\pi\alpha/n)| \tag{15.106}$$

The band gap is $E_G(\alpha) = 2J_0|\sin(\pi\alpha/n)|$. These bands are graphed in fig. 15.10c for $n=5$. This small tube is rarely found in nature—$n=10$ is very common—but it makes the graph simple. For a general chirality (n, m), the tube has metallic bands only if $(n-m)$ is a multiple of three.

15.3.3 Phonons in Carbon Nanotubes

Phonons in nanotubes have a dispersion $\omega_n(q)$, where q is the one-dimensional wave vector, and n is a band index. The phonons are are eigenstates of angular momentum. In a typical SWCNT, say an armchair (10,10) tube, there are $20 = 2n$ carbon atoms around the circumference of each layer. Each atom can move in three directions, so there are about 60 phonon bands. The number of bands is not precisely $6n$ since some collective modes have no restoring force and correspond to rigid displacements or rigid rotations of the entire tube. The number of phonon bands is quite large, and a graph of all of the phonon modes as a function of wave vector tends to have too much information! Here we shall confine our discussion to phonons that can be observed by acoustical or by optical measurements. They have small wave vectors, and most of them can be described by elasticity theory. For graphs of all phonon bands, see the paper by Jeon and Mahan [3].

Electron states in a SWCNT are described well by zone folding. The electronic energy bands of graphene are quantized in the direction of the circumference of the tube. A similar process does not work well for phonons. The difference between electron bands and phonon bands is that electron states are the solution of a scalar wave equation, while phonon states are the solution to a vector wave equation. The eigenvectors of vector equations are fundamentally changed by rolling the material into a tube!

The graphene lattice has hexagonal symmetry, which has the feature that the longitudinal velocity of sound is the same in all directions, as is the transverse velocity. Since a SWCNT is a rolled-up piece of graphene, it is similarly isotropic. Most of the long-wavelength modes are derived from elasticity theory, and are independent of chirality.

All wires have four long-wavelength, low-frequency modes that vanish at zero wave vector. This statement is true for nanowires, nanotubes, and ordinary macroscopic wires.

- The LA mode (longitudinal acoustic) has linear dispersion $\omega_{\text{LA}} = v_L q$. In a SWCNT this velocity is about $v_L = 22$ km/s. The motion corresponds to the longitudinal stretching or compression of the tube. Angular momentum $\ell = 0$.
- The TA mode is a torsional oscillation of the wire. It has linear dispersion with a frequency $\omega_{\text{TA}} = v_T q$, where $v_T = 14$ km/s is the velocity of transverse sound in graphene. Angular momentum $\ell = 0$.
- There are two flexure modes with quadratic dispersion:

$$\omega_F(q) = Rvq^2, \quad v^2 = \frac{\mu}{2\rho}\frac{3\lambda + 2\mu}{\lambda + \mu} \tag{15.107}$$

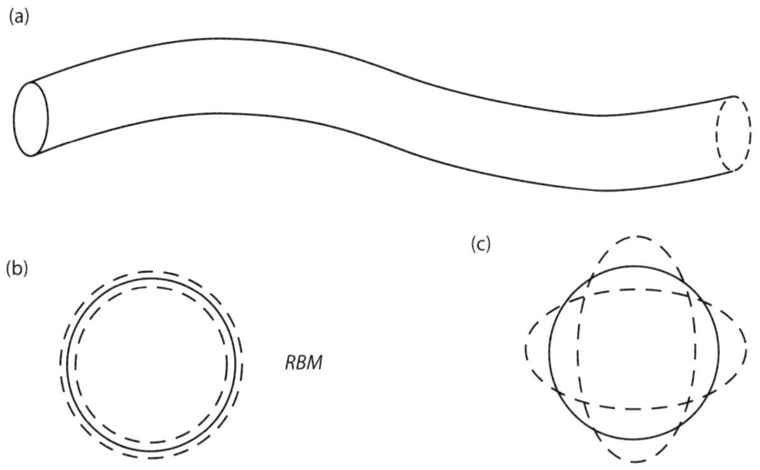

FIGURE 15.11. Solid line is a cross section of a normal tube. Dashed lines are extremes of vibrational motion. (a) Flexure mode. (b) Radial breathing mode. (c) Optical mode.

where R is the tube radius, and the velocity $v \approx 15$ km/s. They correspond to sinusodial transverse motion of the entire tube, as shown in fig. 15.11a. There are two of them, since they can oscillate in either the x- or y-direction for wave motion in the z-direction. Flexure modes have angular momentum $\ell = 1$.

- The radial breathing mode (RBM) is where the entire cross section of the tube oscillates inward or outward; see fig. 15.11b. Angular momentum $\ell = 0$. The frequency is

$$\omega_{\text{RBM}}(q=0) = \frac{v_L}{R} \tag{15.108}$$

The frequency is inversely proportional to the radius of the tube. The numerator v_L is the velocity of LA acoustic modes. The RBM mode is observed in Raman scattering. A measurement of the Raman frequency provides an easy determination of the tube radius.

- The lowest frequency optical mode is shown in fig. 15.11c. The tube oscillates without changing its circumference, and the only restoring force is bond bending. In a (10, 10) tube, $\nu = 22$ cm^{-1}. Angular momentum $\ell = 2$.

- There are high-frequency optical phonons caused by the two atoms in a unit cell oscillating out of phase. These include the in-plane vibration at 1600 cm^{-1} and the ZO out-of-plane mode at 850 cm^{-1}.

- There are also several infrared active phonon modes, whose frequencies depend on chirality.

15.3.4 Electrical Resistivity

Walter deHeer's group [1] first reported that a SWCNT had quantum conductance at room temperature. Recall that a quantum conductor has ballistic transport with a conductance

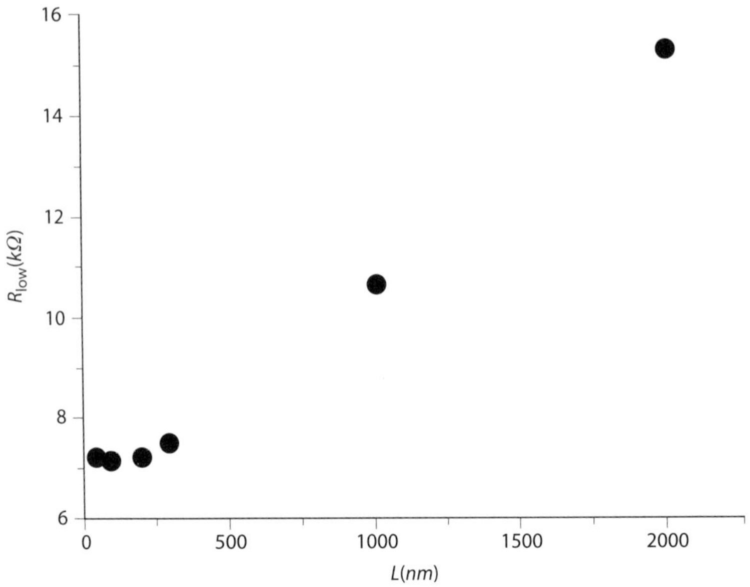

FIGURE 15.12. Resistance of a SWCNT as a function of length L from the one end. Data from Park et al., *Nanoletters* 4, 517. (2004). Used with permission of the American Chemical Society.

given by $\sigma_N = Ne^2/h$. The number of channels N is usually two (for spin) times the number of conducting bands. Armchair tubes have four energy bands that cross the Fermi surface: two carry electrons to the right, and two to the left. So they have $N = 4$ for each direction of current. Metallic zigzag tubes have only one electron energy band for each direction of current, so they have $N = 2$. Figure 15.12 shows data from Paul McEuen's group [6] on the resistivity of a SWCNT as a function of the distance (L) from one end, as measured by an AFM (atomic force microscope) tip. For small values of L, the data go to a constant value given by the choice of $N = 4$. At larger values of L, the incremental resistance becomes linear in L, as found in an ordinary resistor. The quantum of resistance is $h/e^2 = 25{,}812.8\ \Omega$, and dividing this by four gives $6453.2\ \Omega$. This value is close to the $L \to 0$ limit of fig. 15.12.

The thermoelectric power of a SWCNT was measured by Philip Kim's group [8]. They found it was usually zero at low temperature, in agreement with the predictions of ballistic transport. They found it was in quantitative agreement with a variation of the conductance $G(E)$ according to the Mott formula:

$$S = -\frac{\pi^2}{3}\frac{k_b^2 T}{|e|}\left(\frac{1}{G}\frac{dG}{dE}\right)_{E_F} \tag{15.109}$$

They varied $G(E)$ using a gate voltage, and found its variations gave nonzero values for the Seebeck. The Seebeck was zero whenever G was a constant, in agreement with ballistic transport.

References

1. S. Frank, P. Poncharal, Z. L. Wang, and W. A. de Heer, Carbon nanotube quantum resistors. *Science* **280**, 744 (12 June 1998)
2. R. L. Greene and K. K. Bajaj, Energy levels of hydrogenic impurity states in GaAs–$Ga_{1-x}Al_xAs$ quantum well structures. *Solid State Commun.* **45**, 831–835 (1983)
3. G. S. Jeon and G. D. Mahan, Flexure modes in carbon nanotubes. *Phys. Rev. B* **70**, 075405 (2004)
4. A. B. Kuzmenko, E. van Heumen, F. Carbone, and D. van der Marel, Universal optical conductance of graphite. *Phys. Rev. Lett.* **100**, 117401 (2008)
5. G.D. Mahan, Oscillations of a thin hollow cylinder: carbon nanotubes. *Phys. Rev. B* **65**, 235402 (2002)
6. J. Y. Park, S. Rosenblatt, Y. Yaish, V. Sazonova, H. Üstunel, S. Braig, T. A. Arias, P. W. Brouwer, and P. L. McEuen, Electron-phonon scattering in metallic single-walled carbon nanotubes. *Nano Letters* **4**, 517–520 (2004)
7. S. Reich, J. Maultzsch, C. Thomsen, and P. Ordejón, Tight-binding description of graphene. *Phys. Rev. B* **66**, 035412 (2002)
8. J. P. Small, K. M. Perez, and P. Kim, Modulation of thermopower of individual carbon nanotubes. *Phys. Rev. Lett.* **91**, 256801 (2003)
9. P. R. Wallace, The band theory of graphite. *Phys. Rev.* **71**, 622–634 (1947)
10. L. Wirtz and A. Rubio, The phonon dispersion of graphite revisited. *Solid State Commun.* **131**, 141–152 (2004)
11. L. M. Woods and G. D. Mahan, Electron-phonon effects in graphene and armchair (10,10) single wall carbon nanotubes. *Phys. Rev. B* **61**, 10651–10663 (2000)
12. Y. Zheng and T. Ando, Hall conductivity of a two dimensional graphite system. *Phys. Rev. B* **65**, 245420 (2002)

Homework

1. Solve for the antisymmetric eigenstates in the *z*-direction of an electron in a type I QW.

2. Calculate the two-dimensional density of conduction electrons n_e in the QW due to modulation doping, when $E_1 < E_2 < \mu < E_3$, where E_j are different band minimum in the QW.

3. For the graphene lattice, the eigenstates of the tight-binding model have phase factors $\pm\phi(\mathbf{k})$:

$$\gamma(\mathbf{k}) = |\gamma(\mathbf{k})|e^{i\phi(\mathbf{k})} \tag{15.110}$$

Derive $\phi(\mathbf{k})$.

4. For graphene, calculate the rate of optical absorption, using the golden rule, for x-polarized light. Find the transition from a cone state below the Dirac point, to one above. The result should be related to the quantum of conductance.

5. Write out the electron–phonon interaction in graphene, starting from eqn. (15.64), in terms of collective coordinates.

6. Derive the eigenvalues of a graphene ribbon, two carbon atoms in width, aligned along a zigzag direction: see fig. 15.13.

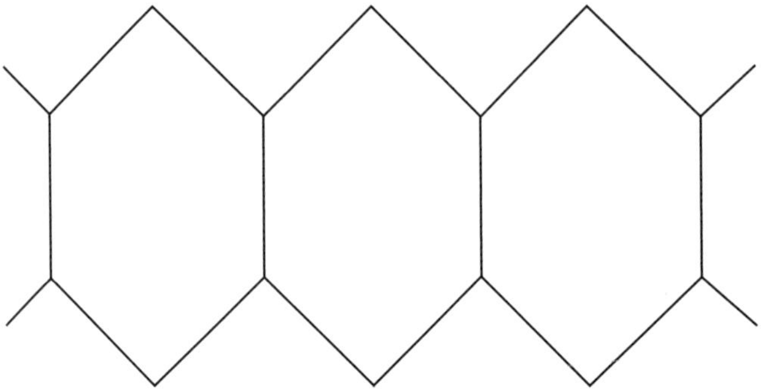

FIGURE 15.13. Graphene ribbon for problem 15.6.

7. Derive an expression for the radius of a single wall carbon nanotube in terms of the carbon–carbon bond length a and the chirality index (n, m).

8. A zigzag tube can be derived from chirality $(n, 0)$ or $(n, -n)$. Show they have the same dispersion relation. Do it for $n = 6$.

Appendix

A.1 Kramers-Kronig Relations

The Kramers-Kronig theorem is proved in two steps. If one starts a perturbation, such as an electric field at $t = 0$, then the response (such as the current) cannot happen until $t \geq 0$. Write this current as the Fourier transform of the frequency-dependent current $\mathcal{J}(\omega)$:

$$J(t) = \int_{-\infty}^{\infty} \frac{d\omega}{2\pi} e^{-i\omega t} \mathcal{J}(\omega) \tag{A.1}$$

If $J(t)$ has the property that it must vanish for $t < 0$, then $\mathcal{J}(\omega)$ must have the following property: it cannot have poles or branch cuts in the upper half plane (UHP) of complex frequency space. Treat ω as a complex variable, and the above integral as a contour integral in complex space.

- If $t < 0$, and if there is a pole at $\omega_0 + i\gamma$ in the UHP, then the contour integral can be made a closed loop by a semicircle in the UHP. In this region, the frequency factor has the exponent

  $$-i\omega t = -it(\omega_r + i\omega_i) = -it\omega_r + t\omega_i \tag{A.2}$$

  This contour is allowed only if $t < 0$ or else $\exp(t\omega_r)$ diverges at positive values of ω_r. If there is a pole or branch cut in the upper half plane, there will be a nonzero residue and $J(t) \neq 0$ for $t < 0$. Therefore, there can be no pole in the UHP. If there is no pole or cut, then closing the contour this way gives zero.

- If there is a pole or branch cut in the lower half plane (LHP), that is fine. For $t > 0$ one closes the contour in the LHP and gets a nonzero answer, which is the desired result.

The second step of the Kramers-Kronig analysis states that the function of frequency obeys the equation

$$\mathcal{J}(\omega) = \oint \frac{dz}{2\pi i} \frac{\mathcal{J}(z)}{z - \omega - i\eta} \tag{A.3}$$

The closed contour of integration goes along the real axis, from $-\infty$ to $+\infty$, and closes as a semicircle in the UHP. Since $\mathcal{J}(z)$ has no poles in the UHP, the only pole is at $z = \omega + i\eta$, where η is small and positive. The above identity is a simple result of Cauchy's theorem. During the integral along the real axis ($z = \omega'$), separate the real and imaginary parts of the denominator:

$$\frac{1}{\omega' - \omega - i\eta} = \mathcal{P}\frac{1}{\omega' - \omega} + i\pi\delta(\omega' - \omega) \tag{A.4}$$

$$\mathcal{J}(\omega) = \int_{-\infty}^{\infty} \frac{d\omega'}{2\pi i} \mathcal{J}(\omega')\mathcal{P}\frac{1}{\omega' - \omega} + \frac{1}{2}\mathcal{J}(\omega) \tag{A.5}$$

where $\mathcal{P}$ denotes the principal part of the denominator on the right. The delta function $\delta(\omega' - \omega)$ gives $\mathcal{J}(\omega)/2$ on the right, which is transferred to the left. Multiple both sides by two and get

$$\mathcal{J}(\omega) = \int_{-\infty}^{\infty} \frac{d\omega}{i\pi} \mathcal{J}(\omega')\mathcal{P}\frac{1}{\omega' - \omega} \tag{A.6}$$

The factor of i in the prefactor provides a relation between the real $\mathcal{J}_1(\omega)$ and complex $\mathcal{J}_2(\omega)$ parts of this function

$$\mathcal{J}(\omega) = \mathcal{J}_1(\omega) + i\mathcal{J}_2(\omega) \tag{A.7}$$

$$\mathcal{J}_1(\omega) = \int_{-\infty}^{\infty} \frac{d\omega}{\pi} \mathcal{J}_2(\omega')\mathcal{P}\frac{1}{\omega' - \omega} \tag{A.8}$$

$$\mathcal{J}_2(\omega) = -\int_{-\infty}^{\infty} \frac{d\omega}{\pi} \mathcal{J}_1(\omega')\mathcal{P}\frac{1}{\omega' - \omega} \tag{A.9}$$

The last two equations constitute the Kramers-Kronig equations. If one knows either the real or imaginary part of a complex, causal, function of frequency, the other can be calculated using these identities. Although the current $\mathcal{J}(\omega)$ was used as an example, any causal, complex, function of frequency obeys the same theorem. Examples are $[\varepsilon_1(\omega), \varepsilon_2(\omega)]$, and $[n(\omega), \kappa(\omega)]$.

Assume an experimentalist has just measured the optical absorption $\alpha(\omega)$ and refractive index $n(\omega)$. Recall that

$$\sqrt{\varepsilon_1 + i\varepsilon_2} = \hat{n} = n + i\kappa \tag{A.10}$$

$$n\alpha = 2\kappa n\frac{\omega}{c} = \varepsilon_2\frac{\omega}{c} \tag{A.11}$$

$$\varepsilon_2(\omega) = \frac{c}{\omega} n(\omega)\alpha(\omega) \tag{A.12}$$

One then applies the Kramers-Kronig formula to get

$$\varepsilon_1(\omega) - 1 = \int_{-\infty}^{\infty} \frac{d\omega'}{\pi} \varepsilon_2(\omega')\mathcal{P}\frac{1}{\omega' - \omega} \tag{A.13}$$

Note that the left side of the equal sign has $\varepsilon_1 - 1$, rather than just ε_1. If there is no absorption, and $\varepsilon_2 = 0$, then $\varepsilon_1 = 1$, which is the dielectric function of vacuum. In proving the Kramers-Kronig theorem, by using contours of integration in complex frequency space, we assume that functions such as $\mathcal{J}(\omega)$ vanish as $|\omega| \to \infty$. In this limit, $\varepsilon(\omega) \to 1$, so the function that converges is $\varepsilon(\omega) - 1$.

Since $\varepsilon_2(-\omega) = -\varepsilon_2(\omega)$, the integral can be converted to have only positive frequencies:

$$\varepsilon_1(\omega) = 1 + \int_0^\infty \frac{d\omega'}{\pi} \varepsilon_2(\omega') \mathcal{P} \frac{1}{\omega' - \omega} + \int_{-\infty}^0 \frac{d\omega'}{\pi} \varepsilon_2(\omega') \mathcal{P} \frac{1}{\omega' - \omega} \tag{A.14}$$

In the second integral, change $\omega' \to -\omega'$. There are four changes of sign: (1) $d\omega' = -d\omega'$, (2) $\varepsilon_2(-\omega') = -\varepsilon_2(\omega')$, (3) $\omega' - \omega = -(\omega' + \omega)$, and (4) the limit of integration is inverted:

$$\varepsilon_1(\omega) = 1 + \frac{1}{\pi} \int_0^\infty d\omega' \varepsilon_2(\omega') \mathcal{P} \left[\frac{1}{\omega' - \omega} + \frac{1}{\omega' + \omega} \right] \tag{A.15}$$

$$\varepsilon_1(\omega) = 1 + \frac{2}{\pi} \int_0^\infty d\omega' \varepsilon_2(\omega') \mathcal{P} \frac{\omega'}{(\omega')^2 - \omega^2} \tag{A.16}$$

This formula can be used to calculate the real part of the dielectric function, if one knows the imaginary part. Note that it proves that $\varepsilon_1(-\omega) = \varepsilon_1(\omega)$.

A.2 Ewald Summations

Ewald introduced a method of calculating the Coulomb interaction in an insulating crystal. A direct summation in real space converges slowly. If the summation is changed to wave vector space, it also converges slowly. The Ewald method combines the two spaces: rapid convergence is obtained by having some terms in real space, and others in reciprocal space. The technique can be applied to any dimensions. It has slightly different functions, in different dimensions, so they are discussed separately.

A.2.1 Three Dimensions

We begin by summing the wave vector transform of the Coulomb interaction over a lattice of points in three dimensions:

$$V(\mathbf{r}, \mathbf{k}) = \sum_{\mathbf{R}_j} \frac{e^{i\mathbf{k}\cdot(\mathbf{R}_j - \mathbf{r})}}{|\mathbf{r} - \mathbf{R}_j|} \tag{A.17}$$

The vector $\mathbf{r}$ ranges over any point in the crystal. Of course, if we set $\mathbf{r} \to \mathbf{R}_j$ then we must exclude that term from the summation. This important detail is handled later.

We break up the summation into two terms. One is evaluated in real space, and the other in wave vector space. This is done using the following identity

$$\frac{1}{|\mathbf{r} - \mathbf{R}_j|} = \frac{2}{\sqrt{\pi}} \int_0^\infty dt e^{-t^2(\mathbf{r} - \mathbf{R}_j)^2} \tag{A.18}$$

$$= \frac{2}{\sqrt{\pi}}\left[\int_0^{\eta} dte^{-t^2(\mathbf{r}-\mathbf{R}_j)^2} + \int_{\eta}^{\infty} dte^{-t^2(\mathbf{r}-\mathbf{R}_j)^2}\right]$$

The parameter η divides the integral into two parts. The first term is evaluated in wave vector space, while the second term is in real space. The potential $V(\mathbf{r},\mathbf{k})$ in eqn. (A.17) is a periodic function in the crystal. It can be expanded in reciprocal lattice vectors **G**. Similarly, the above integral with $0 < t < \eta$ describes a periodic function that can be expanded in reciprocal lattice vectors

$$V_1(\mathbf{r},\mathbf{k}) = \frac{2}{\sqrt{\pi}}\sum_{\mathbf{R}_j} e^{i\mathbf{k}\cdot(\mathbf{R}_j-\mathbf{r})}\int_0^{\eta} dte^{-t^2(\mathbf{r}-\mathbf{R}_j)^2} \tag{A.19}$$

$$= \sum_{\mathbf{G}} e^{i\mathbf{G}\cdot\mathbf{r}} V_{\mathbf{G}} \tag{A.20}$$

$$V_{\mathbf{G}} = \int_{\Omega_0}\frac{d^3r}{\Omega_0} e^{-i\mathbf{G}\cdot\mathbf{r}} V_1(\mathbf{r},\mathbf{k}) \tag{A.21}$$

where the integral is done over the unit cell. One can use the summation over $\mathbf{R}_j$ to extend the integral over all space

$$V_{\mathbf{G}} = \frac{2}{\sqrt{\pi}}\int_{\Omega}\frac{d^3r}{\Omega_0} e^{-i(\mathbf{k}+\mathbf{G})\cdot\mathbf{r}}\int_0^{\eta} dte^{-t^2r^2} \tag{A.22}$$

Interchange the order of the two integrals. The integral d^3r is an easy Gaussian, and the other integral over dt can also be done:

$$V_{\mathbf{G}} = \frac{4\pi}{\Omega_0(\mathbf{k}+\mathbf{G})^2}\exp\left[-\frac{(\mathbf{k}+\mathbf{G})^2}{4\eta^2}\right] \tag{A.23}$$

The choice of a Gaussian is based upon the ability to evaluate these integrals. Other functions are chosen for other dimensions. The integral over real space is a complimentary error function. We obtain $(\mathbf{g}=\mathbf{k}+\mathbf{G})$

$$V(\mathbf{r},\mathbf{k}) = \frac{4\pi}{\Omega_0}\sum_{\mathbf{G}}\frac{e^{i\mathbf{G}\cdot\mathbf{r}-g^2/4\eta^2}}{g^2} + \sum_{\mathbf{R}_j}\frac{e^{i\mathbf{k}\cdot(\mathbf{R}_j-\mathbf{r})}}{|\mathbf{r}-\mathbf{R}_j|}\operatorname{erfc}[\eta|R_j - r|] \tag{A.24}$$

The single summation in (A.17) has been replaced by two summations. However, the advantage of the above formula is that both summations converge rapidly, after a few terms. For example, if η is chosen to be an inverse lattice constant $a\eta = 1$, then both summations converge rapidly. A useful check on the answer is to vary η, since the results should not depend upon its value. It should give the same answer if $\eta = 1/a$ as it does for $\eta = 2/a$.

We evaluate the case that all lattice sites are identical. In taking the limit of $\mathbf{r}\to 0$ we must exclude the site with $\mathbf{R}_j = 0$. Also, we want to remove the factor of $\exp[-i\mathbf{k}\cdot\mathbf{r}]$. So consider the function

$$\tilde{V}(\mathbf{r},\mathbf{k}) = e^{i\mathbf{k}\cdot\mathbf{r}}\left[V(\mathbf{r},\mathbf{k}) - \frac{1}{r}\right] \tag{A.25}$$

Use the feature that $\operatorname{erfc}(z) - 1 = -\operatorname{erf}(z)$ to get

$$\tilde{V}(\mathbf{r}, \mathbf{k}) = \frac{4\pi}{\Omega_0} \sum_{\mathbf{G}} \frac{e^{i\mathbf{g}\cdot\mathbf{r} - g^2/4\eta^2}}{g^2} + \sum_{\mathbf{R}_j} \frac{e^{i\mathbf{k}\cdot\mathbf{R}_j}}{|\mathbf{r} - \mathbf{R}_j|} \operatorname{erfc}[\eta|\mathbf{R}_j - \mathbf{r}|] - \frac{1}{r} \operatorname{erf}(\eta r) \tag{A.26}$$

$$\frac{1}{r} \operatorname{erf}(\eta r) = \frac{1}{r} \frac{2}{\sqrt{\pi}} \int_0^{\eta r} dt e^{-t^2} \approx \frac{2\eta}{\sqrt{\pi}} \left[1 - \frac{(\eta r)^2}{3} + O(r^4) \right] \tag{A.27}$$

Thus, for the case $\mathbf{r} \to 0$ we find

$$\tilde{V}(\mathbf{r} = 0, \mathbf{k}) = \frac{4\pi}{\Omega_0} \sum_{\mathbf{G}} \frac{e^{-g^2/4\eta^2}}{g^2} + \sum_{\mathbf{R}_j \neq 0} \frac{e^{i\mathbf{k}\cdot\mathbf{R}_j}}{|\mathbf{R}_j|} \operatorname{erfc}[\eta|\mathbf{R}_j|] - \frac{2\eta}{\sqrt{\pi}} \tag{A.28}$$

A.2.2 Madelung Summations

The Madelung summation is done for NaCl. There are two summations, which are each evaluated using Ewald methods: one is over the fcc lattice, and the other is over the same lattice but displaced by τ:

$$\frac{\alpha_M}{d} = \sum_{\mathbf{R}_j} \frac{1}{|\mathbf{R}_j + \tau|} - \sum_{\mathbf{R}_j \neq 0} \frac{1}{|\mathbf{R}_j|}, \quad \tau = \frac{a}{2}(100) \tag{A.29}$$

$$= \frac{2\eta}{\sqrt{\pi}} - \frac{4\pi}{\Omega_0} \sum_{\mathbf{G} \neq 0} \frac{e^{-G^2/4\eta^2}}{G^2} [1 - e^{i\mathbf{G}\cdot\tau}]$$

$$- \sum_{\mathbf{R}_j \neq 0} \frac{1}{|\mathbf{R}_j|} \operatorname{erfc}[\eta|\mathbf{R}_j|] + \sum_{\mathbf{R}_j} \frac{1}{|\mathbf{R}_j + \boldsymbol{\tau}|} \operatorname{erfc}[\eta|\mathbf{R}_j + \boldsymbol{\tau}|]$$

The $\mathbf{G} = 0$ term in the summation is absent due to charge neutrality. The wave vector $\mathbf{k} = 0$. Choose $\eta = C/d$ where C is a dimensionless constant of $O(1)$. This makes the first term $(2\eta/\sqrt{\pi})$ be 90% of the answer. Then the two summations change the final answer by a few percentage points. Only a few terms are needed in each summation. The first term in each summation, when $C = 2$, $\eta = 4/a$ is as follow:

- There are eight values of the lowest nonzero $\mathbf{G} = 2\pi(111)/a$. The phase factor is $\exp[i\mathbf{G}\cdot\boldsymbol{\tau}] = \exp[i\pi] = -1$, and the term is

$$8\frac{4\pi}{(a^3/4)} 2\left(\frac{a^2}{3(2\pi)^2}\right) \exp[-(2\pi)^2 3/4^3] = \frac{0.533}{d} \tag{A.30}$$

- There are six nearest neighbors:

$$\frac{6}{d} \operatorname{erfc}(2) = \frac{0.014}{d} \tag{A.31}$$

- There are twelve second neighbors, which have the same charge as the center atom:

$$\frac{12}{\sqrt{2}d} \operatorname{erfc}(2\sqrt{2}) = \frac{0.000}{d} \tag{A.32}$$

So far the summation is

$$\alpha_M = \frac{4}{\sqrt{\pi}} - 0.533 + 0.014 = 1.738 \tag{A.33}$$

which is very close to the final answer of 1.747.... The Ewald method provides an accurate answer after a few terms in the two series.

A.2.3 Dipolar Summations

The summation over dipole–dipole interaction is derived using the above formula. We can take a double derivative of the denominator:

$$\lim_{\mathbf{r}\to 0} \nabla_\mu \nabla_\nu \frac{1}{|\mathbf{r} - \mathbf{R}_j|} = -\phi_{\mu\nu}(\mathbf{R}_j) = -\left[\frac{\delta_{\mu\nu}}{R_j^3} - \frac{3R_\mu R_\nu}{R_j^5}\right] \tag{A.34}$$

These results are used to find the wave vector transform of the dipole–dipole interaction

$$T_{\mu\nu}(\mathbf{k}) = \sum_{\mathbf{R}_j \neq 0} e^{i\mathbf{k}\cdot\mathbf{R}_j} \phi_{\mu\nu}(\mathbf{R}_j) \tag{A.35}$$

$$= \frac{4\pi}{\Omega_0} \sum_{\mathbf{G}} e^{-g^2/4\eta^2} \frac{g_\mu g_\nu}{g^2} - \frac{4\eta^3}{3\sqrt{\pi}} \delta_{\mu\nu} + \sum_{\mathbf{R}_j \neq 0} e^{i\mathbf{k}\cdot\mathbf{R}_j}$$

$$\times \left\{ \phi_{\mu\nu}(\mathbf{R}_j) \left[\mathrm{erfc}(\eta R_j) + \frac{2}{\sqrt{\pi}} (\eta R_j) e^{-(\eta R_j)^2}\right] - \frac{4\eta^3}{\sqrt{\pi}} \frac{R_\mu R_\nu}{R_j^3} e^{-(\eta R_j)^2} \right\}$$

This expression has two series, and both converge very rapidly for the choice that $\eta \sim 1/a$.

An important limit is when $\mathbf{k} \to 0$. Then the most interesting term is the first one, where the reciprocal lattice vector $\mathbf{G} = 0$ gives

$$\lim_{\mathbf{k}\to 0} T_{\mu\nu}(\mathbf{k}) = \frac{4\pi}{\Omega_0} \left[\frac{k_\mu k_\nu}{k^2} e^{-k^2/4\eta^2} + \text{other terms}\right] \tag{A.36}$$

The term on the right has the form $\hat{k}_\mu \hat{k}_\nu$, which comes from the long range of the Coulomb interaction.

A.2.4 Two Dimensions

The two-dimensional Ewald method is very similar to that for three dimensions. Write the summation as

$$V(\mathbf{r}, \mathbf{k}) = \sum_{\mathbf{R}_j} \frac{e^{i\mathbf{k}\cdot(\mathbf{R}_j - \mathbf{r})}}{|\mathbf{r} - \mathbf{R}_j|} = \sum_{\mathbf{R}_j} e^{i\mathbf{k}\cdot(\mathbf{R}_j - \mathbf{r})} \int d^3r' \frac{\rho(\mathbf{r}' - \mathbf{R}_j)}{|\mathbf{r} - \mathbf{r}'|} \tag{A.37}$$

$$\rho(\mathbf{r}' - \mathbf{R}_j) = \delta^3(\mathbf{r}' - \mathbf{R}_j) \tag{A.38}$$

Write the near-field, and far-field terms as

$$V = V_n + V_f \tag{A.39}$$

$$V_n = \sum_{\mathbf{R}_j} e^{i\mathbf{k}\cdot(\mathbf{R}_j - \mathbf{r})} \int d^3r' \frac{1}{|\mathbf{r} - \mathbf{r}'|}\left[\delta^3(\mathbf{r}' - \mathbf{R}_j) - \frac{\eta^3}{\pi^{3/2}} e^{-\eta^2(\mathbf{r}' - \mathbf{R}_j)^2}\right] \tag{A.40}$$

$$V_f = \frac{\eta^3}{\pi^{3/2}} \sum_{\mathbf{R}_j} e^{i\mathbf{k}\cdot(\mathbf{R}_j - \mathbf{r})} \int d^3r' \frac{e^{-\eta^2(\mathbf{r}' - \mathbf{R}_j)^2}}{|\mathbf{r} - \mathbf{r}'|} \tag{A.41}$$

The near-field expression has the integral

$$\begin{aligned} \frac{\eta^3}{\pi^{3/2}} \int d^3r' \frac{e^{-\eta^2(\mathbf{r}' - \mathbf{R}_j)^2}}{|\mathbf{r} - \mathbf{r}'|} &= \frac{\eta^3}{\pi^{3/2}} \int d^3y \frac{e^{-\eta^2(\mathbf{y} - \mathbf{x})^2}}{y}, \quad \mathbf{x} = \mathbf{r} - \mathbf{R}_j \\ &= 2\pi \frac{\eta^3}{\pi^{3/2}} \int_0^\infty y dy e^{-\eta^2(y^2 + x^2)} \int_{-1}^{1} d\nu e^{2xy\nu\eta^2} \\ &= \frac{\eta}{\sqrt{\pi} x} \int_0^\infty dy \left[e^{-\eta^2(x-y)^2} - e^{-\eta^2(x+y)^2}\right] \end{aligned} \tag{A.42}$$

In the first integral let $s = \eta(y - x)$, while in the second integral let $s = \eta(y + x)$. Most of the two integrals cancel. One is left with $\mathrm{erf}(\eta x)/x$. Since 1 – erf is the complimentary error function erfc, we find

$$V_n(\mathbf{r}, \mathbf{k}) = \sum_{\mathbf{R}_j} e^{i\mathbf{k}\cdot(\mathbf{R}_j - \mathbf{r})} \frac{\mathrm{erfc}(\eta|\mathbf{r} - \mathbf{R}_j|)}{|\mathbf{r} - \mathbf{R}_j|} \tag{A.43}$$

which is identical to the expression in three dimensions. This summation converges rapidly in real space.

The far-field summation is done by expanding the result in reciprocal lattice vectors:

$$V_f = \sum_{\mathbf{G}} V(\mathbf{G}) e^{i\mathbf{G}\cdot\mathbf{R}} \tag{A.44}$$

$$V(\mathbf{G}) = \int \frac{d^2r}{A_0} e^{i\mathbf{g}\cdot\mathbf{r}} \frac{\eta^3}{\pi^{3/2}} \int \frac{d^3r'}{|\mathbf{r} - \mathbf{r}'|} e^{-\eta^2(\mathbf{r}')^2} \tag{A.45}$$

where $\mathbf{g} = \mathbf{G} + \mathbf{k}$ and A_0 is the area of the unit cell in two dimensions. The reciprocal lattice vectors $\mathbf{G}$ are for the two-dimensional surface. Interchange the order of the summations, and let $\mathbf{y} = \mathbf{r} - \mathbf{r}'$:

$$V(\mathbf{G}) = \frac{\eta^3}{\pi^{3/2} A_0} \int d^3r' e^{-i\mathbf{g}\cdot\mathbf{r}' - \eta^2(r')^2} \int \frac{d^2y}{y} e^{-i\mathbf{g}\cdot\mathbf{y}} \tag{A.46}$$

$$\int \frac{d^2y}{y} e^{-i\mathbf{g}\cdot\mathbf{y}} = 2\pi \int_0^\infty dy J_0(gy) = \frac{2\pi}{g} \tag{A.47}$$

$$\frac{\eta^3}{\pi^{3/2}} \int d^3r' e^{-i\mathbf{g}\cdot\mathbf{r}' - \eta^2(r')^2} = e^{-g^2/(4\eta)^2} \tag{A.48}$$

$$V_f = \frac{2\pi}{A_0} \sum_{\mathbf{G}} \frac{e^{i\mathbf{G}\cdot\mathbf{r} - g^2/4\eta^2}}{g} \tag{A.49}$$

This expression converges rapidly in reciprocal space.

A.2.5 One Dimension

In one dimension, the dipole sums can be evaluated either in real space or in reciprocal space. They converge absolutely in real space, so it is easy to do them that way. However, there is a simple transform that allows them to be done in reciprocal space. Write the Coulomb summation as

$$V(z,\rho) = \sum_{n=-\infty}^{\infty} \frac{e^{ik(na-z)}}{\sqrt{\rho^2 + (na-z)^2}} = \sum_g V_G e^{izG} \tag{A.50}$$

where $G = 2\pi\, \alpha/a$, a is a lattice constant, and α is any integer. The Fourier coefficient is

$$V_G = \int_{-\infty}^{\infty} \frac{dz}{a} \frac{e^{igz}}{\sqrt{\rho^2 + z^2}} = \frac{2}{a} K_0(g\rho) \tag{A.51}$$

where $g = k + G$ and $K_0(z)$ is a Bessel function of imaginary argument. $K_0(z)$ declines in value rapidly for large z. Note that it diverges at zero argument, so this transform cannot be used whenever $\rho = 0$. In that case, one can just sum n directly.

A.3 Square Lattice Density of States

Here we provide the steps to derive the density of states of the square lattice in two dimensions. The band dispersion in the tight-binding model is

$$\varepsilon(\mathbf{k}) = A[\cos(\theta_x) + \cos(\theta_y)], \quad \theta_\mu = k_\mu a \tag{A.52}$$

$$N_2(E) = \int \frac{d^2k}{(2\pi)^2} \delta \tag{A.53}$$

Define $\alpha = E/A$ and change the integration range from $-\pi < \theta_\mu < \pi$ to just positive values, since negative values give the same result:

$$N_2(E) = \frac{1}{\pi^2 a^2 A} \int_0^\pi d\theta_x \int_0^\pi d\theta_y \delta[\alpha - \cos(\theta_x) - \cos(\theta_y)] \tag{A.54}$$

$$= \frac{1}{\pi^2 a^2 A} \int_0^\pi \frac{d\theta_x}{|\sin(\theta_y)|} = \frac{1}{\pi^2 a^2 A} \int_0^\pi \frac{d\theta_x}{\sqrt{1 - [\alpha - \cos(\theta_x)]^2}}$$

The integral $\int d\theta_y$ was evaluated using the delta function, which gives a $|\sin(\theta_y)|$ in the denominator. Write it as $\sin(\theta_y) = \sqrt{1 - \cos^2(\theta_y)}$ and then use the arguement of the delta function $\cos(\theta_y) = a - \cos(\theta_x)$ to turn the remaining integral into one involving only θ_x. Next change variables to $x = \cos(\theta_x)$, $dx = -di_x \sin(\theta_x)$ to get

$$N_2(E) = \frac{1}{\pi^2 a^2 A} \int_{-1}^{1} \frac{dx}{\sqrt{1 - x^2}\sqrt{(1 + \alpha - x)(1 + x - \alpha)}} \tag{A.55}$$

The stated limits to the integral are $-1 < x < +1$. This is not quite correct, since an additional constraint is that the argument of the square roots have to be positive. Although the density of states is independent of the sign of α, we must state whether it is positive or negative. The range of α is $-2 < \alpha < 2$. Say α is positive $(2 > \alpha > 0)$. Then the additional constraint is

$$1 + \alpha > x > \alpha - 1 \tag{A.56}$$

Since $1 + \alpha > 1$, then the final limits of integration are

$$N_2(E) = \frac{1}{\pi^2 a^2 A} \int_{\alpha-1}^{1} \frac{dx}{\sqrt{1 - x^2}\sqrt{(1 + \alpha - x)(1 + x - \alpha)}} \tag{A.57}$$

Change integration variables to z, defined as $x = uz + v$. We choose (u, v) by having $z = 1$ at $x = 1$, and $z = -1$ at $x = \alpha - 1$

$$1 = u + v \tag{A.58}$$

$$\alpha - 1 = - u + v \tag{A.59}$$

$$u = 1 - \alpha/2, \quad v = \alpha/2 \tag{A.60}$$

The four factors under the square root are

$$1 - x = u(1 - y) \tag{A.61}$$

$$1 + x = (1 + \alpha/2)(1 + my), \quad m = \frac{2 - \alpha}{2 + \alpha} = \frac{2A - E}{2A + E} \tag{A.62}$$

$$1 + \alpha - x = (1 + \alpha/2)(1 - my) \tag{A.63}$$

$$1 + x - \alpha = u(1 + y) \tag{A.64}$$

$$N_2(E) = \frac{1}{\pi^2 a^2 A(1 + \alpha/2)} \int_{-1}^{1} \frac{dy}{\sqrt{1 - y^2}\sqrt{1 - m^2 y^2}} \tag{A.65}$$

The last line gives the solution. The integral is twice the elliptic integral $K(m)$:

$$N_2(E) = \frac{4}{\pi^2 a^2 (2A + E)} K(m) \tag{A.66}$$

The elliptic function diverges when $m = 1$, which is at $E = 0$. The density of states is symmetric in E.

A.4 Jacobi Ellipic functions

The integral for the arcsin is

$$u = \int_0^z \frac{dt}{\sqrt{1 - t^2}} = \arcsin(z), \quad z = \sin(u) \tag{A.67}$$

In a similar way, introduce another function that depends upon the parameter k, and its inverse

$$u = \int_0^z \frac{dt}{\sqrt{(1-t^2)(1-k^2t^2)}}, \quad z = \mathrm{sn}(u) \tag{A.68}$$

The quantity sn(u) is a Jacobian Elliptic Function. Using it, one can define a family of related functions:

$$\mathrm{cn}^2(u) = 1 - \mathrm{sn}^2(u) \tag{A.69}$$

$$\mathrm{dn}^2(u) = 1 - k^2\mathrm{sn}^2(u) \tag{A.70}$$

They are periodic, with the period determined by multiples of the first elliptic integral $K(k)$:

$$K(k) = \int_0^1 \frac{dt}{\sqrt{(1-t^2)(1-k^2t^2)}} = \int_0^{\pi/2} \frac{d\theta}{\sqrt{1-k^2\sin^2\theta}} \tag{A.71}$$

$$E(k) = \int_0^1 dt\sqrt{\frac{1-k^2t^2}{1-t^2}} = \int_0^{\pi/2} d\theta\sqrt{1-k^2\sin^2\theta} \tag{A.72}$$

The period of dn(u) is $2K(k)$, while that for sn(u) and cn(u) is $4K(k)$. The second elliptic integral $E(k)$ is used below. All of these functions also depend upon k, although that dependence is not highlighted in the notation. In the limit of small values of k we get

$$\lim_{k\to 0}\begin{cases} \mathrm{sn}(u) & \to & \sin(u) \\ \mathrm{cn}(u) & \to & \cos(u) \\ \mathrm{dn}(u) & \to & 1 \end{cases} \tag{A.73}$$

The derivatives of the functions are important. Start with the definition (A.68) and take its derivative:

$$\frac{du}{dz} = \frac{1}{\sqrt{(1-z^2)(1-k^2z^2)}} \tag{A.74}$$

$$\frac{dz}{du} = \frac{d\mathrm{sn}(u)}{du} = \sqrt{(1-z^2)(1-k^2z^2)} = \mathrm{cn}(u)\,\mathrm{dn}(u) \tag{A.75}$$

The others can be derived from this result

$$\frac{d}{du}\mathrm{sn}(u) = \mathrm{cn}(u)\,\mathrm{dn}(u) \tag{A.76}$$

$$\frac{d}{du}\mathrm{cn}(u) = -\mathrm{sn}(u)\,\mathrm{dn}(u) \tag{A.77}$$

$$\frac{d}{du}\mathrm{dn}(u) = -k^2\mathrm{sn}(u)\,\mathrm{cn}(u) \tag{A.78}$$

Here we list some addition theorems for elliptic functions.

$$\mathrm{sn}(u+v) = \frac{\mathrm{sn}(u)\,\mathrm{cn}(v)\,\mathrm{dn}(v) + \mathrm{sn}(v)\,\mathrm{cn}(u)\,\mathrm{dn}(u)}{1-k^2\mathrm{sn}^2(u)\,\mathrm{sn}^2(v)} \tag{A.79}$$

$$\mathrm{cn}(u+v) = \frac{\mathrm{cn}(u)\,\mathrm{cn}(v) - \mathrm{sn}(v)\,\mathrm{dn}(v)\,\mathrm{sn}(u)\,\mathrm{dn}(u)}{1 - k^2\mathrm{sn}^2(u)\,\mathrm{sn}^2(v)} \tag{A.80}$$

$$\mathrm{dn}(u+v) = \frac{\mathrm{dn}(u)\,\mathrm{dn}(v) - k^2\mathrm{sn}(v)\,\mathrm{cn}(v)\,\mathrm{sn}(u)\,\mathrm{cn}(u)}{1 - k^2\mathrm{sn}^2(u)\,\mathrm{sn}^2(v)} \tag{A.81}$$

Other relations are

$$\mathrm{sn}(u+v)\,\mathrm{sn}(u-v) = \frac{\mathrm{sn}^2(u) - \mathrm{sn}^2(v)}{1 - k^2\mathrm{sn}^2(u)\,\mathrm{sn}^2(v)} \tag{A.82}$$

$$\mathrm{cn}(u+v)\,\mathrm{cn}(u-v) = \frac{\mathrm{cn}^2(v) - \mathrm{sn}^2(u)\,\mathrm{dn}^2(v)}{1 - k^2\mathrm{sn}^2(u)\,\mathrm{sn}^2(v)} \tag{A.83}$$

$$\mathrm{dn}(u+v)\,\mathrm{dn}(u-v) = \frac{\mathrm{dn}^2(v) - k^2\mathrm{sn}^2(u)\,\mathrm{cn}^2(v)}{1 - k^2\mathrm{sn}^2(u)\,\mathrm{sn}^2(v)} \tag{A.84}$$

The first and second derivatives are shown in Table A1. Second derivatives are needed for the equations of motion. Note that they all have, on the right, a term with a single power of a cnoidal function, and a term with a cubic power. Those are exactly the type of terms needed for equations of motion of quadratic/quartic springs.

The last three functions in Table A1 are related to the first three. The period of sn(u), cn(u) is $4K$, so that K is a quarter wave. An exact identity is

$$\mathrm{sn}(u \pm K) = \pm\mathrm{cd}(u) \tag{A.85}$$

$$\mathrm{cn}(u \pm K) = \mp k_1\,\mathrm{sd}(u) \tag{A.86}$$

$$\mathrm{dn}(u \pm K) = k_1\,\mathrm{nd}(u) \tag{A.87}$$

Any solution that contains sn(u) is also solved by the function cd(u): the same is true for cn(u) and $k_1$sd(u), and dn(u) and k_1nd(u). Note, in Table A1, that cd(u) has exactly the same second derivative as does sn(u).

Table A.1. First and second derivatives of some cnoidal functions ($k_1^2 = 1 - k^2$)

fn(u)	$\frac{d}{du}$ fn(u)	$\frac{d^2}{du^2}$ fn(u)
sn(u)	cn(u)dn(u)	$-\mathrm{sn}(u)[1 + k^2 - 2k^2\backslash \mathrm{sn}^2(u)]$
cn(u)	$-$sn(u)dn(u)	$-\mathrm{cn}(u)[1 - 2k^2 + 2k^2\backslash \mathrm{cn}^2(u)]$
dn(u)	$-k^2$, sn(u)cn(u)	$\mathrm{dn}(u)[2 - k^2 - 2\mathrm{dn}^2(u)]$
sd(u)	cd(u)nd(u)	$-\mathrm{sd}(u)[1\text{-}2k^2 + 2k^2k_1^2\mathrm{sd}^2(u)]$
cd(u)	$-k_1^2$, sd(u)nd(u)	$-\mathrm{cd}(u)[1 + k^2 - 2k^2\mathrm{cd}^2(u)]$
nd(u)	k^2, sd(u)cd(u)	$\mathrm{nd}(u)[2 - k^2 - 2k_1^2,\ \mathrm{nd}^2(u)]$

Index